THE BUILDING REGULATIONS

EXPLAINED AND ILLUSTRATED

The authors

M. J. Billington, *BSc, MRICS*, who has worked on various editions of the book since 1986, is a chartered building surveyor. He was formerly Senior Lecturer in building control and construction at De Montfort University, Leicester, and is now a private building control consultant to the Government, industry and education. He is co-author of *Means of Escape from Fire*, also published by Blackwell.

M. W. Simons, *BSc, MSc, PhD, MIOA, MCIOB*, is a Senior Lecturer at Coventry University where he is Programme Manager for degree courses in architectural design technology, building surveying and construction management. His academic specialism and research relate to the internal environment of buildings. He is co-author of the book *Sound Control in Buildings: A Guide to Part E of the Building Regulations*.

J. R. Waters, *BSc, MPhil, PhD, MCIBSE, CEng*, is Honorary Research Fellow at Coventry University, where he was formerly Head of Building Services Engineering. He is a consultant on environmental aspects of building design, has written *Energy Conservation in Buildings* and co-authored *Sound Control in Buildings*, both from Blackwell Publishing.

THE BUILDING REGULATIONS

EXPLAINED AND ILLUSTRATED

TWELFTH EDITION

M. J. Billington

M. W. Simons

J. R. Waters

Blackwell
Science

Blackwell Science Ltd, a Blackwell Publishing Company
Editorial offices:
Blackwell Science Ltd, 9600 Garsington Road, Oxford OX4 2DQ, UK
Tel: +44 (0)1865 776868
Blackwell Publishing Inc., 350 Main Street, Malden, MA 02148-5020, USA
Tel: +1 781 388 8250
Blackwell Science Asia Pty Ltd, 550 Swanston Street, Carlton, Victoria 3053, Australia
Tel: +61 (0)3 8359 1011

First edition published by Crosby Lockwood & Son Ltd 1967; Second edition 1968; Third edition with metric supplement 1970; reprinted 1972; Fourth edition 1973; Second impression 1974; Third impression (with supplement) 1976; Fifth edition published by Granada Publishing Ltd in Crosby Lockwood Staples 1978; reprinted 1979; Fifth edition (amended) 1981; Sixth edition published by Granada Publishing Ltd, Technical Books Division 1982; reprinted with amendments 1983; reprinted with minor revisions by Collins Professional Books 1984; reprinted with amendments 1985; Seventh edition 1986; reprinted 1986, 1987; reprinted by BSP Professional Books 1987; reprinted with amendments 1988; 1989 (three times); Eighth edition 1990; reprinted with updates 1991; reprinted 1992; Ninth edition published by Blackwell Scientific Publications 1992; reprinted 1993; Tenth edition 1995; Eleventh edition published by Blackwell Science Ltd 1999; reprinted 2000, 2001, 2002, 2003; Twelfth edition published by Blackwell Science, a Blackwell Publishing company 2004; reprinted 2004, 2005

Library of Congress Cataloging-in-Publication Data
Billington, M.J. (Michael J.)
The building regulations: explained and illustrated / M.J. Billington, M.W. Simons, J.R. Waters. —12th ed.
p. cm.
Rev. ed. of: The building regulations / Vincent Powell-Smith and M.J. Billington. 11th ed. 1999.
Includes bibliographical references and index.
ISBN 0-632-05837-4
1. Building laws—England. I. Simons, M.W. II. Waters, J.R. III. Powell-Smith, Vincent
Building regulations. IV. Title.

KD1140.W48 2003
343.41′07869—dc22

2003063029

ISBN-10: 1-632-05837-4
ISBN-13: 978-0632-05837-2

A catalogue record for this title is available from the British Library

Set in 10/12.5 Times by DP Photosetting, Aylesbury, Bucks
Printed and bound by Replika Press Pvt. Ltd, India

The publisher's policy is to use permanent paper from mills that operate a sustainable forestry policy, and which has been manufactured from pulp processed using acid-free and elementary chlorine-free practices. Furthermore, the publisher ensures that the text paper and cover board used have met acceptable environmental accreditation standards.

For further information visit our website:
www.thatconstructionsite.com

Contents

Preface to Twelfth Edition

The previous edition of this book was published in 1999. The intervening period has probably resulted in some of the most radical changes to the technical requirements of the Regulations and their supporting guidance since the change from prescriptive to functional regulations, which occurred in 1985.

This has resulted in the redrafting of substantial parts of the previous edition and has meant that the current (12th) edition of this book now exceeds 900 pages. As one of the principal objectives of the 1985 changes was to simplify the control system, the reader may be entitled to ask why this situation has arisen.

In part this may be put down to the effects of global warming and the Government's support for the Kyoto agreement, which has resulted in a commitment to reduce CO_2 emissions. To facilitate this, Chapter 16 dealing with conservation of fuel and power has been rewritten to reflect new regulations in Part L and guidance in Approved Documents L1 and L2. Indeed, the changes were so extensive and complex that it was felt necessary to bring in expert help in the drafting of this Chapter. For this we are extremely grateful to Dr Bob Waters for the knowledge and attention to detail that he has brought to this difficult area of control.

In a similar vein, changes to Part E of the Regulations controlling resistance to the passage of sound (Chapter 10) have meant that pre-completion sound testing of many controlled buildings is now a mandatory requirement. The range of buildings to which sound insulation requirements apply has been extended beyond normal dwellings and now includes hotels, hostels, care homes and similar premises where people live. There has also been a general tightening up of the guidance so that more stringent standards are now required to be met. Again, it was felt that the high standards that our readers have come to expect from this book could only be maintained if specialist advice could be utilised in the preparation of this chapter. Accordingly, we were delighted when Dr Martin Simons agreed to take responsibility for this.

Other significant changes have resulted in a new Chapter 7 Fire, covering Part B of the Regulations (including the changes that were introduced in March 2003), and complete redrafting of Chapter 13 Drainage and waste disposal (dealing with the 2002 revision to Part H of the regulations) and Chapter 14 Heat producing appliances (which covers the 2002 revision to Part J). In fact, Part J is now entitled 'Combustion Appliances and Fuel Storage Systems'. However, we decided to keep our original chapter title in order to preserve continuity with previous editions.

The early chapters, which set out the legal and administrative provisions of the regulations, have been revised to take account of further expansion of the Approved Inspector control system and a number of important amendments to the Building Act 1984. They also cover the consolidation of both the 1991 Building Regulations and the 1985 Building (Approved Inspectors etc.) Regulations and their subsequent amendments into their 2000 equivalents (and the inevitable amendments, yet again, of both sets of regulations).

As always, the aim is to provide a convenient and straightforward guide and reference to a complex and constantly evolving subject. It must be stressed that this book is a guide to the regulations and approved and other documents, and is not a substitute for them. We hope that it may shed light on some of the more obscure and difficult to understand parts of the source documents. It should be stressed that the guidance in the Approved Documents is not mandatory and differences of opinion can quite legitimately exist between controllers and developers or designers as to whether a particular detail in a building design does actually satisfy the mandatory functional requirements of the Building Regulations.

At the time of writing, revisions to Part A Structure and Part M Access and facilities for disabled people (see Chapters 6 and 17, respectively) are being prepared by the Office of the Deputy Prime Minister. Part A is expected to be published in February 2004 and to come into force in the late summer or early autumn of 2004; Part M is expected to be published in late 2003 and to come into force in the summer of 2004. The anticipated main changes are included at the end of Chapter 6 (Part A) and Chapter 17 (Part M).

The intended readers of this book are all those concerned with building work – architects and other designers, building control officers, approved inspectors, building surveyors, clerks of works, services engineers and contractors, etc. – as well as their potential successors, the current generation of students. This book is designed to be of use to both students and teachers and it is gratifying that successive editions are widely adopted by the various academic institutions and professional bodies.

As always, we are especially grateful to Julia Burden, Publisher at Blackwell Publishing, and her editorial team, for their help, patience, interest and sense of humour during the production of this edition.

The law is stated on the basis of cases reported and other material available to us on 1 October 2003.

M.J. Billington
M.W. Simons
J.R. Waters

Preface to Eleventh Edition

The last edition of this book was published in 1995 and has since gone through a number of reprints without the need for minor amendments. However a stage has now been reached, following the publication of the Building Regulations (Amendment) Regulations 1997 and 1998, which necessitates a revision of several chapters.

Chapter 15 (formerly entitled *Stairways, ramps and guards*) has been rewritten and retitled *Protection from falling, collision and impact* to reflect new regulations dealing with reducing the risk of injury from collisions with open windows, skylights and ventilators, or when using various types of sliding or powered doors and gates. Similarly, Chapter 18 introduces new provisions governing the safe use and cleaning of glazed elements. These new provisions apply only to workplaces and have been introduced mainly in order to ensure compliance with the Workplace (Health, Safety and Welfare) Regulations 1992.

Chapter 17 (*Access and facilities for disabled people*) has been rewritten to take account of new regulations which come into force on 25 October 1999 whereby accessibility controls are extended to dwellings, and Chapter 8 includes details of the latest amendment to the Approved Document to support regulation 7, covering materials and workmanship.

The early chapters which set out the legal and administrative provisions of the regulations have been completely rewritten to take account of the rapid expansion of the Approved Inspector control system and a number of important amendments to the Building Act 1984. Additionally, a new Chapter 5 has been included which sets out details of 18 Acts of Parliament and statutory instruments which may apply to a building project as well as, or in addition to, the Building Regulations 1991. This includes important changes to the way in which local authorities can charge for dealing with building regulation submissions. These are designed to improve competition with approved inspectors on the setting of charges.

As always, the aim is to provide a convenient and straightforward guide and reference to a complex and constantly evolving subject. It is a guide to the regulations and approved and other documents and is not a substitute for them. The intended readers are all those concerned with building work – architects and other designers, building control officers, approved inspectors, building surveyors and contractors – as well as their potential successors, the current generation of students. This book is designed to be of use to both students and teachers and it is gratifying that successive editions are widely adopted by the various academic institutions and professional bodies.

Finally, it is with sadness that I must record the death in 1997, of my co-author Vincent Powell-Smith. With Walter Whyte, Vincent had the original idea for this book in 1966 and it was already recognised as the authoritative text on the Building Regulations when I took over the technical content from Walter Whyte in 1985.

Since then I have collaborated with Vincent on four editions and this will be the first time that he had no direct involvement in its production. However, past readers will no doubt still recognise his unique style. Accordingly, I am especially grateful to Julia Burden, Deputy Publisher at Blackwell Science and Sue Moore, Senior Editor, for their help, patience, interest and sense of humour during the production of this edition.

The law is stated on the basis of cases reported and other material available to me on 1 March 1999.

Michael J. Billington
March 1999

Acknowledgements

All material reproduced from the Approved Documents is Crown copyright and is reproduced with the permission of the Controller of HMSO.

The authors are also grateful to the following for permission to reproduce copyright material:

- BRE/CRC for permission to use table 3 in IP17/01 and data from tables 2, 3 and 4 in BRE Digest 457.
- BRECSU for permission to use data from tables 3.1, 3.2, 3.3 and 3.4 in SAP 2001.
- The British Standards Institution for permission to reproduce tables from BS EN ISO 6946 under licence number 2002SK/0365; extracts from BS EN ISO 11654:1997 under licence number 2003/SK036; extracts from BS EN ISO 717-1 & 717-2: 1997 under licence number 2003SK/0121.

BRE/CRC and BRECSU publications, plus most others, may be obtained from BRE Bookshop, 151 Rosebery Avenue, London EC1R 4GB.

BSI publications can be obtained from BSI customer services, 389 Chiswick High Road, London W4 4AL (tel: +44 (0)20 8996 9001, email: cservices@bsi-global.com).

Legal and Administrative

1 Building control: an overview

1.1 Introduction

The building control system in England and Wales was radically revised in 1985. After a long period of gestation, building regulations were laid before Parliament and came into general operation on 11 November 1985. They applied to Inner London from 6 January 1986. Subject to specified exemptions, all building work (as defined in the regulations) in England and Wales is governed by building regulations.

The current regulations are the Building Regulations 2000 which came into force on 1 January 2001. The 2000 regulations have been amended twice since then, the latest being the Building (Amendment) Regulations 2002, and the provisions of all these amendments are reflected in this book.

A separate system of building control applies in Scotland and in Northern Ireland.

The power to make building regulations is vested in the Secretary of State for the Environment by section 1 of the Building Act 1984 which sets out the basic framework. Building regulations may be made for the following broad purposes:

- Securing the health, safety, welfare and convenience of people in or about buildings and of others who may be affected by buildings or matters connected with buildings.
- Furthering the conservation of fuel and power.
- Preventing waste, undue consumption, misuse or contamination of water.

The 2000 Regulations are very short and contain no technical detail. That is found in a series of Approved Documents and certain other non-statutory guidance, all of which refer to other non-statutory documents such as National Standards or Technical Specifications (e.g. British Standards or Agrément Certificates), with the objective of making the system more flexible and easier to use. The 2000 Regulations implement the final conclusions of a major review of both the technical and procedural requirements.

A significant feature of the system is that there are alternative systems of building control – one by local authorities, and the other a private system of certification which relies on 'approved inspectors' operating under a separate set of regulations called The Building (Approved Inspectors, etc.) Regulations 2000. These set out the detailed procedures for operating the system of private certification and came into effect at the same time as the main regulations.

1.2 The Building Act 1984

The Building Act 1984 received the Royal Assent on 31 October 1984 and the majority of its provisions came into force on 1 December 1984. It consolidated most, but not all, of the primary legislation relating to building which was formerly scattered in numerous other Acts of Parliament.

Part I of the Building Act 1984 is concerned with building regulations and related matters, while Part II deals with the system of private certification discussed in Chapter 4. Other provisions about buildings are contained in Part III which, amongst other things, covers drainage, and the local authority's powers in relation to dangerous buildings, defective premises, etc.

The provisions of the 1984 Act are of the greatest importance in practice, and many of them are referred to in this and subsequent chapters.

'Building' is defined in the 1984 Act in very wide terms. A building is 'any permanent or temporary building and, unless the context otherwise requires, it includes any other structure or erection of whatever kind or nature (whether permanent or temporary)'. 'Structure or erection' includes a vehicle, vessel, hovercraft, aircraft or other movable object of any kind in such circumstances as may be prescribed by the Secretary of State. The Secretary of State's opinion is, however, qualified. The circumstances must be those which 'in [his] opinion ... justify treating it ... as a building'.

The result of this definition is that many things which would not otherwise be thought of as a building may fall under the Act – fences, radio towers, silos, air-supported structures and the like. Happily, as will be seen, there is a more restrictive definition of 'building' for the purposes of the 2000 Regulations, but a comprehensive definition is essential for general purposes, e.g. in connection with the local authority's powers to deal with dangerous structures. Hence the statutory definition is necessarily couched in the widest possible terms. In general usage (and at common law) the word 'building' ordinarily means 'a structure of considerable size intended to be permanent or at least to last for a considerable time' (*Stevens* v. *Gourely* (1859) 7 CBNS 99) and considerable practical difficulties arose as to the scope of earlier building regulations which the 1984 definition has removed.

In *Seabrink Residents Association* v. *Robert Walpole Campion and Partners* (1988) (6-CLD-08-13; 6-CLD-08-10; 6-CLD-06-32) for example, the High Court held that walls and bridges on a residential development were not subject to the then Building Regulations 1972 because they were not part of 'a building'. The development was not to be considered as a homogenous whole. The then regulations, said Judge Esyr Lewis QC, were 'concerned with structures which have walls and roofs into which people can go and in which goods can be stored'. Each structure in the development must be looked at separately to see whether the regulations applied. 'Obviously a wall may be part of a building and so, in my view, may be a bridge'.

1.3 The linked powers

Local authorities exercise a number of statutory public health functions in conjunction with the process of building control, although these have been reduced in recent years; for example, controls on construction of drains and sewers. These provisions are commonly called 'the linked powers' because their operation is linked with the local authority's building control functions, both in checking deposited plans or considering a building notice, and under the approved inspector system of control. Many of the former linked powers have been brought under the building regulations, but local authorities are responsible for certain functions now found in the 1984 Act. In those cases, the local authority must reject the plans (or building notice) or the approved inspector's initial notice if relevant compliance is not achieved or else must impose suitable safeguards. The relevant provisions are:

(a) Provision of drainage. The local authority can insist that the drainage connects to a nearby public sewer. Disputes under section 21 are dealt with by a magistrates' court.
   A related provision is section 98 of the Water Industry Act 1991 under which owners or occupiers of premises can require the water authority to provide a public sewer for domestic purposes in their area, subject to various conditions which can include in an appropriate case the making of a financial contribution.
(b) Provision of water supply. Section 25 requires the local authority to reject plans of a house submitted under the building regulations unless they are satisfied with the proposals for providing the occupants with a sufficient supply of wholesome water for domestic purposes, by pipes or otherwise. Usually, of course, this will be by means of a mains supply provided by the water authority under the Water Industry Act 1991. Disputes are determined by the magistrates' court.
   A related provision is section 37 of the Water Industry Act 1991 which enables a landowner who proposes to erect buildings to require the water authority to lay necessary mains for the supply of water for domestic purposes to a point which will enable the buildings to be connected to the mains at a reasonable cost, a provision which is of considerable use to developers.

1.4 Building Regulations

The Secretary of State is given power to make comprehensive regulations about the provision of services, fittings and equipment in or in connection with buildings as well as about the design and construction of buildings. A very comprehensive list of the subject matter of building regulations is contained in Schedule 1. The Regulations are supported by approved documents, giving 'practical guidance' (see section 2.3).

Building regulations may include provision as to the deposit of plans of executed, as well as proposed work; for example where work has been done without the deposit of plans or there has been a departure from the approved plans. Broad powers are given to make building regulations about the inspection and testing of work, and the taking of samples.

Prescribed classes of buildings, services, etc. may be wholly or partially exempted from regulation requirements. Similarly, the Secretary of State may, by direction, exempt any particular building or buildings at a particular location.

Schedule 1 of the 1984 Act is a flexible provision and covers the application of the regulations to existing buildings. It enables regulations to be made regarding not only alterations and extensions, but also the provision, alteration or extension of services, fittings and equipment in or in connection with existing buildings. It also enables the regulations to be applied on a *material change of use* as defined in the regulations and, very importantly, makes it possible for the regulations to apply where re-construction is taking place, so that the regulations can deal with the whole of the building concerned and not merely with the new work.

The 1984 Act contains enabling powers for the making of regulations on a number of procedural matters.

The regulations made and currently in force are:

- The Building Regulations 2000
- The Building (Approved Inspectors, etc.) Regulations 2000 (as amended)
- The Building (Local Authority Charges) Regulations 1998
- The Building (Inner London) Regulations 1985

Most of these regulations have been amended, in some cases several times, and care should be taken to ensure that the most recent amendments are being used.

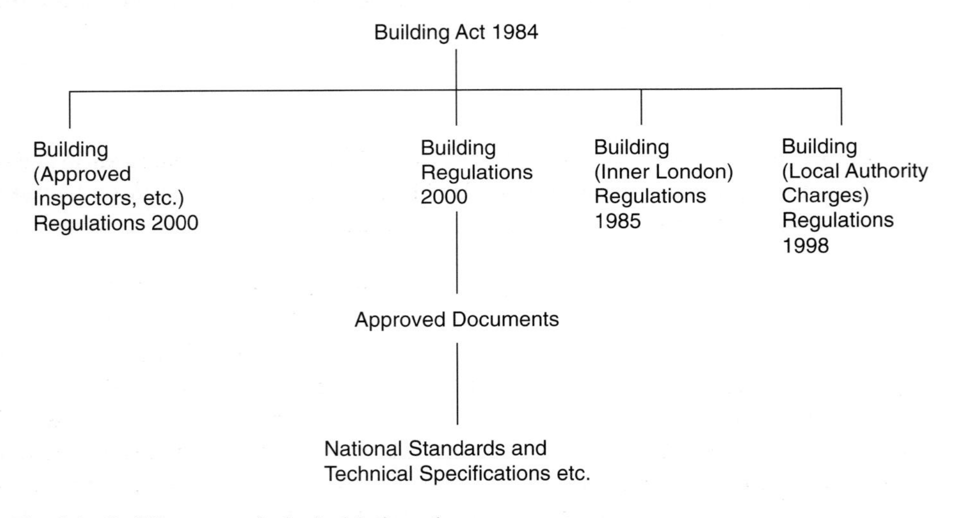

Fig. 1.1 Building control: the legislative scheme.

1.5 Building Regulations – exemptions

1.5.1 Crown immunity

The Building Regulations do not apply to premises which are occupied by the Crown. The general position regarding Crown exemption is that a statute does not bind the Crown unless it so provides either expressly or by necessary implication. In fact, there is provision in section 44 of the Building Act 1984 to apply the substantive requirements of the Regulations to Crown buildings but this has never been activated.

In recent years a number of premises have lost their Crown immunity, often in response to public opinion. These include National Health Service buildings (see section 60 of the National Health and Community Care Act 1990) and the Metropolitan Police. Additionally, the reorganisation of the Post Office has meant that whilst the Royal Mail is still regarded as Crown property, Post Office Counters is not.

In practice, it is normal for government department building work to be designed and constructed in accordance with the Building Regulations. In some areas the plans and particulars may even be submitted to the local authority for comment, although it is more usual for these to be scrutinised by specialist companies (replacing the service which was formally given by the Property Services Agency) who will also carry out on-site inspections of the works in progress. Even so, such companies have no legal control over the work and cannot take enforcement action in the event of a breach of the Regulations.

Interestingly, Crown premises are not exempt from certification under the Fire Precautions Act 1971. However, they are inspected *not* by the fire authority, but by the Crown Premises Inspection Group within the Home Office Fire Service Inspectorate, a bureaucratic anomaly which has attracted much criticism. Unfortunately, the powers of entry to premises contained in the 1971 Act do not apply to premises occupied by the Crown.

1.5.2 Building Act exemptions

Taken together, the Building Act 1984 and the Building Regulations 2000 (as amended), exempt certain uses of buildings and many categories of work from control as is illustrated by the following examples.

- *Educational buildings:* as a result of the repeal of Regulation 8 of the Education (Schools and Further and Higher Education) Regulations 1989, schools in England are not now exempt from the Building Regulations. The Department for Education and Skills has issued guidance on constructional standards for schools, indicating where it is necessary to supplement the Building Regulation Approved Documents (see sections 2.4 and 2.5 below) to take account of the specific requirements of schools. It is proposed that in due course the Approved Documents will be amended to take account of these.
- *Statutory undertakers and other public bodies:* under section 4 of the Building Act 1984, a building belonging to a statutory undertaker (as defined in section 126 of the Building Act 1984), the United Kingdom Atomic Energy Authority, or the Civil Aviation Authority is exempt from the application of building regulations. The building must be one which is held for the purposes of the undertaking (such as an electricity substation), therefore buildings such as houses, offices or showrooms are not exempt unless they:
  (a) form part of a railway station, or

(b) in the case of the Civil Aviation Authority, are on an aerodrome owned by the Authority (although the exemption does not extend to airport hotels).

The Metropolitan Police Authority has been exempted from having to comply with the procedural requirements of building control, although it is still required to comply with the substantive or technical requirements of the Regulations. As an exempt body, the Metropolitan Policy Authority is also exempt from the enforcement procedures by local authorities.

Additionally, the fact that many former public bodies are now in private hands has meant that both gas and electricity suppliers are not now regarded as statutory undertakers under section 4 of the Building Act 1984. Furthermore, it should be noted that the Post Office lost its statutory undertaker status in relation to building regulations, in March 2001.

It should be noted that local authority buildings are *not* exempt from either the procedural or substantive requirements of the Building Regulations.

1.5.3 Miscellaneous

Under section 16 of the Building Act 1984 there is power to approve the plans of a proposed building by stages. Usually, the initiative will rest with the applicant as to whether to seek approval by stages – subject to the local authority's agreement. However, local authorities may – of their own initiative – give approval by stages; they might, for example, await further information. In giving stage approval, local authorities will be able to impose a condition that certain work will not start until the relevant information has been produced.

Plans may also be approved subject to agreed modifications, e.g. where there is a minor defect in the plans.

Section 19 of the Building Act 1984 deals with the use of short-lived materials. The provision applies where plans, although conforming with the regulations, include the use of items listed in the regulations for the purpose of section 19. In such circumstances the local authority has a discretion:

- to pass the plans;
- to reject the plans; or
- to pass them subject to the imposition of a time limit, whether conditionally or otherwise.

Interestingly, the Building Regulations 2000 (as amended) contain no specific references to any particular materials; however as will be seen, regulation 7 of the 2000 Regulations requires that building work which must comply with the Schedule 1 requirements must be carried out 'with proper materials which are appropriate for the circumstances in which they are used...', and the supporting approved document deals with the use of short-lived materials.

The local authority may impose a time limit either on the whole of a building or on particular work. Additionally, they may impose conditions as to the use of a building or the particular items concerned. Appeal against the local authority's decision lies to the Secretary of State.

Eventually, section 19 will cease to have effect when section 20, which is wider in scope, is brought into force by the Secretary of State.

Building regulations may impose continuing requirements on the owners and occupiers of buildings, including buildings which were not, at the time of their erection, subject to building regulations. These requirements are of two kinds. Continuing requirements may be imposed *first*, in respect of designated provisions of the regulations to ensure that their purpose is not frustrated, e.g. the keeping clear of fire escapes; and *second*, in respect of services, fittings and equipment, e.g. a requirement for the periodical maintenance and inspection of lifts in flats.

Type relaxations may be granted by the Secretary of State; he may dispense with or relax some regulation requirement generally. A type relaxation can be made subject to conditions or for a limited period only. It should be noted that before granting a type relaxation the Secretary of State must consult such bodies as appear to him to be representative of the interests concerned and he has to publish notice of any relaxations issued.

The Building Act 1984, sections 39 to 43, contains the appeal provisions. The principal appeals to the Secretary of State are:

- appeals against rejection of plans by a local authority; and
- appeals against a local authority's refusal to give a direction dispensing with or relaxing a requirement of the regulations or against a condition attached by them to such a direction.

Interestingly, section 38 of the Building Act 1984 is concerned with civil liability but has yet to be activated. Under this section, breach of duty imposed by the regulations will be actionable at civil law, where damage is caused, except where the regulations otherwise provide. 'Damage' is defined as including the death of, or injury to, any person (including any disease or any impairment of a person's physical or mental condition). The regulations themselves may provide for defences to such a civil action and section 38 will not, when operative, prejudice any right which exists at common law.

1.6 Dangerous structures, etc.

Local authorities have power to deal with a building or structure which is in a dangerous condition or is overloaded. The procedure is for the local authority to apply to the magistrates' court for an order requiring the owner to carry out remedial works or, at his option, to demolish the building or structure and remove the resultant rubbish. The court may restrict the use of the building if the danger arises from overloading. If the owner fails to comply with the order within the time limit specified by the court, the local authority may execute the works themselves and recover the expenses incurred from the owner, who is also liable to a fine (Building Act 1984, section 77).

Under section 78 of the Building Act 1984 the local authority may take immediate action in an emergency so as to remove the danger, e.g. if a wall is in danger of imminent collapse. Where it is practicable to do so, they must give notice of the proposed action to the owner and occupier. The local authority may recover expenses which they have reasonably incurred in taking emergency action, unless the magistrates' court considers that they might reasonably have proceeded under section 77. An owner or occupier who suffers damage as a result of action taken under section 78 may in some circumstances be entitled to recover compensation from the local authority.

Section 79 of the 1984 Act empowers local authorities to deal with ruinous and dilapidated buildings or structures and neglected sites 'in the interests of amenity', which is a term of wider significance than 'health and safety': *Re Ellis and Ruislip and Northwood UDC* [1920] 1 KB 343. (Section 76 of the Act enables them to deal with defective premises which are 'prejudicial to health or a nuisance'.)

Under section 79, where a building or structure is in such a ruinous or dilapidated condition as to be seriously detrimental to the amenities of the neighbourhood, the local authority may serve notice on the owner requiring him to repair or restore it or, at his option, demolish the building or structure and clear the site.

Demolition is itself subject to control. Section 80 requires a person who intends to demolish the whole or part of a building to notify the local authority, the occupier of any adjacent building and the gas and electricity authorities of his intention to demolish. He must also comply with any requirements which the local authority may impose by notice under section 82.

The demolition notice procedure does not apply to the demolition of:

- An internal part of an occupied building where it is intended that the building should continue to be occupied.
- A building with a cubic content (ascertained by external measurement) of not more than 1750 cubic feet ($50m^3$) or a greenhouse, conservatory, shed or prefabricated garage which forms part of a larger building.
- An agricultural building unless it is contiguous to a non-agricultural building or falls within the preceding paragraph.

The local authority may by notice require a person undertaking demolition to carry out certain works,

- To shore up any adjacent building.
- To weatherproof any surfaces of an adjacent building exposed by the demolition.
- To repair and make good any damage to any adjacent building caused by the demolition.
- To remove material and rubbish resulting from the demolition and clearance of the site.
- To disconnect and seal and/or remove any sewers or drains in or under the building.
- To make good the ground surface.
- To make arrangements with the gas, electricity and water authorities for the disconnection of supplies.

- To make suitable arrangements with the fire authority (and Health and Safety Executive, if appropriate) with regard to burning of structures or materials on site.
- To take such steps in connection with the demolition as are necessary for the protection of the public and the preservation of public amenity.

1.7 Other legislation

Although the Building Act 1984 attempted to rationalise the main controls over buildings, there are in fact a great many pieces of legislation, in addition to the Building Act and the Building Regulations, which affect the building, its site and environment and the safety of working practices on and within the building. A summary of the legislation most commonly encountered is given in Chapter 5.

2 The Building Regulations and Approved Documents

2.1 Introduction

Although the statutory framework of building control is found in the Building Act 1984, the 2000 Regulations, as amended, contain the detailed rules and procedures. The regulations are comparatively short because the technical requirements have mostly been cast in a functional form.

Each technical requirement is supported by a document approved by the Secretary of State intended to give practical guidance on how to comply with the requirements. The Approved Documents refer to British Standards and other guidance material such as BRE publications and thus give designers and builders a great degree of flexibility.

The 2000 Regulations became effective on 1 January 2000, and have since been amended twice.

2.2 Division of the Regulations

There are 27 regulations, arranged logically in six parts. The division is as follows:

PART I: GENERAL

Reg. 1. Citation and commencement.
Reg. 2. Interpretation.

PART II: CONTROL OF BUILDING WORK

Reg. 3. Meaning of building work.
Reg. 4. Requirements relating to building work.
Reg. 5. Meaning of material change of use.
Reg. 6. Requirements relating to material change of use.
Reg. 7. Materials and workmanship.
Reg. 8. Limitation on requirements.
Reg. 9. Exempt buildings and work.

Part III: EXEMPTION OF PUBLIC BODIES FROM PROCEDURAL REQUIREMENTS

Reg 10. The Metropolitan Police Authority.

PART IV: RELAXATION OF REQUIREMENTS

Reg. 11. Power to dispense with or relax requirements.

PART V: NOTICES AND PLANS

Reg. 12. Giving of a building notice or deposit of plans.
Reg. 13. Particulars and plans where a building notice is given.
Reg. 14. Full plans.
Reg. 14A. Consultation with sewerage undertaker.
Reg. 15. Notice of commencement and completion of certain stages of work.
Reg. 16. Energy rating.
Reg. 16A. Provisions applicable to replacement windows, rooflights, roof windows and doors.
Reg. 17. Completion certificates.

PART VI: MISCELLANEOUS

Reg. 18. Testing of building work.
Reg. 19. Sampling of material.
Reg. 20. Supervision of building work otherwise than by local authorities.
Reg. 20A. Sound insulation testing.
Reg. 21. Unauthorised building work.
Reg. 22. Contravention of certain regulations not to be an offence.
Reg. 23 Transitional provisions.
Reg. 24. Revocations.

There are also four schedules:

SCHEDULE 1 – REQUIREMENTS

This contains technical requirements which are almost all expressed in functional terms and grouped in 13 parts set out in tabular form:

PART A: STRUCTURE – Covers loading, ground movement and disproportionate collapse.

PART B: FIRE SAFETY – Covers means of warning and escape, internal and external fire spread, and access and facilities for the fire service.

PART C: SITE PREPARATION AND RESISTANCE TO MOISTURE – Covers preparation of site, dangerous and offensive substances, subsoil drainage, and resistance to weather and ground moisture.

PART D: TOXIC SUBSTANCES – Deals with cavity insulation.

PART E: RESISTANCE TO THE PASSAGE OF SOUND – Protection against sound from other parts of a building and adjoining buildings, protection against sound emanating within relevant buildings, reverberation in the common internal parts of relevant buildings and acoustic conditions in schools.

PART F: VENTILATION – Covers means of ventilation and condensation in roofs.

PART G: HYGIENE – Deals with bathrooms, hot water storage and sanitary conveniences and washing facilities.

PART H: DRAINAGE AND WASTE DISPOSAL – Deals with foul water drainage, wastewater treatment systems and cesspools, rainwater drainage, building over sewers, separate systems of drainage and solid waste storage.

PART J: COMBUSTION APPLIANCES AND FUEL STORAGE SYSTEMS – Covers air supply, discharge of products of combustion, protection of the building, provision of information, protection of liquid fuel storage systems and protection against pollution.

PART K: PROTECTION FROM FALLING, COLLISION AND IMPACT – Covers stairs, ladders and ramps, protection from falling, vehicle barriers and loading bays, protection from collision with open windows, etc., and protection against impact from and trapping by doors.

PART L: CONSERVATION OF FUEL AND POWER – Provides that reasonable provision must be made for the conservation of fuel and power.

PART M: ACCESS AND FACILITIES FOR DISABLED PEOPLE – Deals with the provision of facilities for the disabled: access and use, sanitary conveniences, and audience or spectator seating.

PART N: GLAZING – SAFETY IN RELATION TO IMPACT, OPENING AND CLEANING – Deals with reducing the risks associated with glazing in critical locations in buildings and covers the safe operation and cleaning of windows, skylights and ventilators, etc.

SCHEDULE 2 – EXEMPT BUILDINGS AND WORK

This lists exempt buildings and work in seven classes, and one of its effects is significantly to reduce the extent of control by giving complete exemptions for certain buildings and extensions.

SCHEDULE 2A – EXEMPTIONS FROM REQUIREMENT TO GIVE BUILDING NOTICE OR DEPOSIT FULL PLANS

This schedule lists certain types of work (for example, the installation of various kinds of combustion appliances or the installation of replacement windows, doors and rooflights) and gives details of certain classes of people who can carry out the work without giving a building notice or depositing full plans with the local authority. Such individuals will need to be registered under various industry schemes appropriate to the work in order to benefit from the exemption.

SCHEDULE 3 – REVOCATION OF REGULATIONS

This lists the former regulations which are revoked, i.e. the Building Regulations 1991, and parts of other relevant regulations.

2.3 Approved Documents

There are 15 Approved Documents issued by the Office of the Deputy Prime Minister (ODPM) intended to give practical guidance on how the technical requirements of Schedule 1 may be complied with. They are written in straightforward technical terms with accompanying diagrams and the intention is that they will be quickly updated as necessary. Additionally, there are two other Approved Documents covering respectively: *Timber intermediate floors for dwellings* published by the Timber Research and Development Association, and *Basements for dwellings* published by the British Cement Association.

The status and use of Approved Documents is prescribed in sections 6 and 7 of the Building Act 1984. Section 6 provides for documents giving 'practical guidance with respect to the requirements of any provision of building regulations' to be approved by the Secretary of State or some body designated by him. The documents so far issued have been approved by the Secretary of State, although they refer to other non-statutory material.

The legal effect of 'Approved Documents' is specified in section 7. Their use is not mandatory, and failure to comply with their recommendations does not involve any civil or criminal liability, but they can be relied upon by either party in any proceedings about an alleged contravention of the requirements of the regulations. If the designer or contractor proves that he has complied with the requirements of an Approved Document, in any proceedings which are brought against him he can rely upon this 'as tending to negative liability'. Conversely, failure to comply with an Approved Document may be relied on by the local authority 'as tending to establish liability'. In other words, the onus will be upon the designer or contractor to establish that he has met the functional requirements in some other way.

The position is illustrated by *Rickards* v. *Kerrier District Council* (1987) CILL 345, 4-CLD-04-26 where it was held that if the local authority proved that the works did not comply with the Approved Document, it was then for the appellant to show compliance with the regulations. If the designer fails to follow an Approved Document, it is for him to prove (if prosecuted) that he used an equally effective method or practice.

All the Approved Documents are in a common format, and their provisions are considered in subsequent chapters. They may be summarised as follows:

A: STRUCTURE – This supports Schedule 1, A1, A2 and A3. Section 1, which deals with houses and other small buildings, contains tables for timber sizes, wall thicknesses, etc., and a lot of technical guidance. Section 2 deals with disproportionate collapse and is relevant to all types of building. It lists Codes and Standards for structural design and construction for all building types and emphasises certain basic principles which must be taken into account if other approaches are adopted.

B: FIRE SAFETY – This supports Schedule 1, B1, B2, B3, B4 and B5 and is probably the most complex part of the Regulations. B1 deals with means of warning and escape (fire alarm systems and the design of buildings to permit rapid evacuation in the event of fire). B2 and B3 covers the ability of the building to resist fire spread over the surfaces of internal walls and ceilings, the ability of a building to stand up to the effects of a fire so that it will not collapse before people have had a chance to escape and the way a building can be designed so that fire is prevented from spreading through its internal structure (in floor, ceiling and wall voids and past party walls etc.). B4 deals with the prevention of external fire spread across an open space where it might affect a neighbouring building and B5 covers ways of making buildings accessible for fire fighters when they need to save lives.

C: SITE PREPARATION AND RESISTANCE TO MOISTURE – Read in conjunction with Schedule 1, Part C, it deals with the necessary basic requirements. Section 1 covers site preparation and site drainage and Section 2 deals with contaminants, including the erection of buildings on sites affected by radon gas or the landfill gases, methane and carbon dioxide. Additionally, it covers any substances in the ground which might cause a danger to health, and its provisions effectively replace those of the repealed section 29 of the Building Act 1984. C 4 describes the measures necessary in order to prevent the passage of moisture to the inside of the building.

D: TOXIC SUBSTANCES – This supports Schedule 1, Part D and it is very short. It gives advice on guarding against fumes from urea formaldehyde foam.

E: RESISTANCE TO THE PASSAGE OF SOUND – This supports Schedule 1,

Part E and deals with the ability of a building to prevent the passage of unwanted sound from internal sources (sound penetration through external walls is covered by planning legislation, not Building Regulations). The details apply to dwellings and to rooms in other buildings which are used for residential purposes (like hotel bedrooms and similar rooms in hostels and residential homes for the elderly). This means, for example, that it is now necessary to insulate walls between hotel bedrooms and also to apply lining materials to wall surfaces of common access stairs and corridors in such buildings. E4 gives guidance on how to improve acoustic conditions in schools.

F: VENTILATION – Supporting Part F of Schedule 1 it covers means of ventilation including the precautions to be taken to prevent excessive condensation in the roof voids of dwellings.

G: HYGIENE – Supporting Part G of Schedule 1 it includes the requirements of certain repealed sections (sections 26 to 28) of the Building Act 1984 dealing with water-closets and bathrooms as well as covering unvented hot water systems.

H: DRAINAGE AND WASTE DISPOSAL – This supports Part H of Schedule 1 and covers above and below ground drainage, wastewater treatment systems and cesspools. Certain new sections which deal with building over sewers and separate systems of drainage, thereby replacing the repealed section 18 and some sub-sections of section 21 of the Building Act 1984; and solid waste storage.

J: HEAT PRODUCING APPLIANCES – Supporting Part J of Schedule 1, it deals with gas appliances up to 60 kW and solid and oil fuel appliances up to 45 kW, as well as protection of liquid fuel storage systems and protection against pollution caused by heating oil leakage.

K: PROTECTION FROM FALLING, COLLISION AND IMPACT – This supports Part K of Schedule 1 and covers the design and construction of stairs, ramps and guarding. It has been extended in the 1998 edition to cover vehicle loading bays, protection from collision with open windows, skylights and ventilators, and protection against impact from and trapping by doors.

L: CONSERVATION OF FUEL AND POWER – Supporting Part L. A new edition was issued in 2002. It is mainly concerned with making sure that buildings are reasonably efficient in their use of energy and that carbon dioxide emissions are kept to a minimum. Dwellings are covered by L1 whereas other buildings are dealt with in L2.

M: ACCESS AND FACILITIES FOR DISABLED PEOPLE – Supporting Part M of Schedule 1 this gives practical guidance on means of access, use of buildings, sanitary conveniences and audience or spectator seating.

N: GLAZING – SAFETY IN RELATION TO IMPACT, OPENING AND

CLEANING – Supporting Part N. This covers safe operation and access for cleaning windows, etc., in addition to measures designed to reduce the risks of accidents caused by contact with glazing.

There is a further Approved Document – MATERIALS AND WORKMANSHIP – to support Regulation 7 – and it is phrased in very general terms.

Relaxations of the mandatory requirements may be given only by local authorities in appropriate cases, with the possibility of an appeal against refusal to the Secretary of State. An approved inspector cannot grant a relaxation.

2.4 Definitions in the Regulations

Regulation 2 provides a number of general definitions, but not all of them are equally important or helpful. In this section full definitions are given for purposes of ease of reference, although the various special definitions will be referred to again in later chapters. The definitions are:

THE ACT – This means the Building Act 1984.

AMENDMENT NOTICE – This is a notice given by an approved inspector under section 51A of the Building Act 1984 where the scope of the work has changed to such an extent that the original Initial Notice no longer truly reflects the work actually being carried out.

BUILDING – The regulations apply only to buildings as defined. There is a narrow definition of 'building' for the purposes of the regulations:

A building is 'any permanent or temporary building but not any other kind of structure or erection'. When 'a building' is referred to in the regulations this includes a part of a building.

The effect of this definition is to exclude from control under the regulations such things as garden walls, fences, silos, air-supported structures and so forth.

BUILDING NOTICE – A notice in prescribed form given to the local authority under regulations 12(2)(a) and 13 informing the authority of proposed works.

BUILDING WORK – The regulations apply only to building work as defined in regulation 3(1); any work not coming within the definition is not controlled. Building work means:

- The erection or extension of a building;
- The material alteration of a building;
- The provision, extension or material alteration of services or fittings required by Schedule 1, Parts G, H, J or L (and called 'controlled services or fittings');

- Work required by Regulation 6 – which sets out the requirements relating to 'material change of use' (see below);
- The insertion of insulating material into the cavity wall of a building;
- Work involving the underpinning of a building; or
- Work involving replacement windows, doors and rooflights, space heating and hot water boilers, and hot water vessels in dwellings.

CONTROLLED SERVICE OR FITTING – This means services or fittings required by Parts G, H, J or L, i.e. bathrooms, hot water storage systems, sanitary conveniences, drainage and waste disposal, heat producing appliances; and replacement doors, windows and rooflights, space heating and hot water boilers, and hot water vessels.

DAY – Any period of 24 hours commencing at midnight. It does not include weekends, Bank Holidays or public holidays.

DWELLING – This includes a dwellinghouse and a flat.

DWELLINGHOUSE excludes a flat or building containing a flat.

ENERGY RATING – A numerical indication of the energy efficiency of a dwelling calculated in accordance with a procedure approved by the Secretary of State.

EUROPEAN TECHNICAL APPROVAL ISSUING BODY – This means the issue of a favourable technical assessment of the fitness for use of a construction product for the purposes of the Construction Products Directive by an authorised body.

FINAL CERTIFICATE – A certificate given by an approved inspector to a local authority under section 51 of the Building Act 1984 to indicate that a project has been successfully completed.

FLAT – Separate and self-contained premises (including a maisonette) constructed or adapted for residential purposes and forming part of a building divided horizontally from some other part (see Fig. 2.1).

FLOOR AREA – This means the aggregate area of every floor in a building or extension. The area is to be calculated by reference to the finished internal faces of the enclosed walls or, where there is no enclosing wall, to the outermost edge of the floor (see Fig. 2.2).

FRONTING – As section 203(3) of the Highways Act 1980 (includes being adjacent to).

FULL PLANS – Plans deposited with a local authority in accordance with regulations 12(2)(b) and 14. The Building Act 1984, section 126, gives a definition of

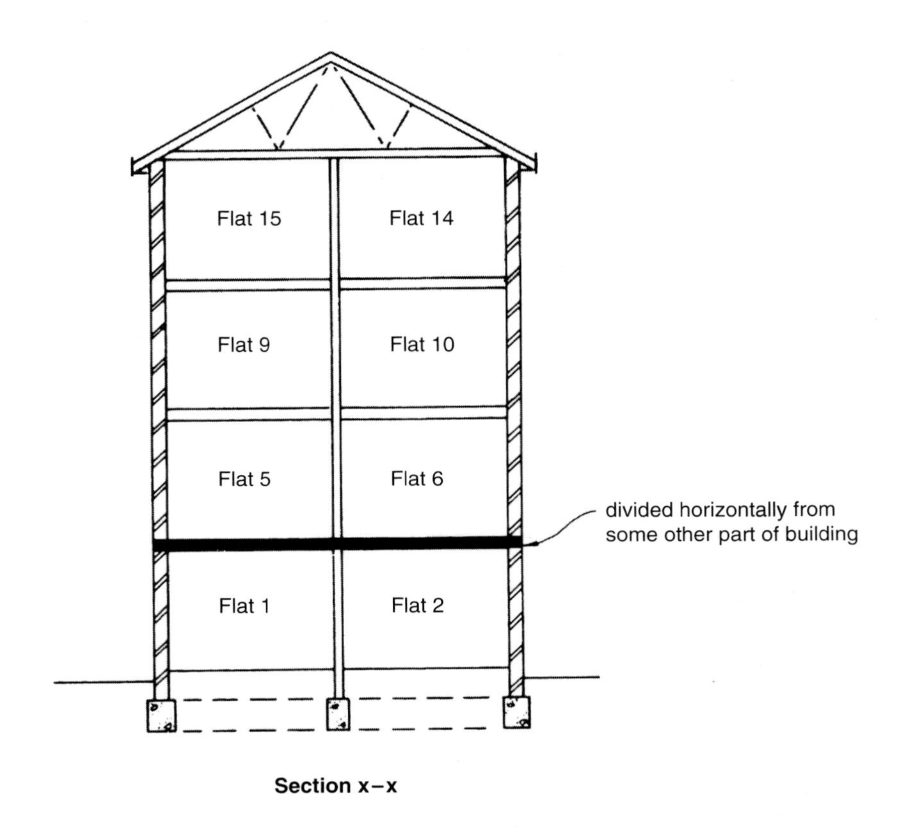

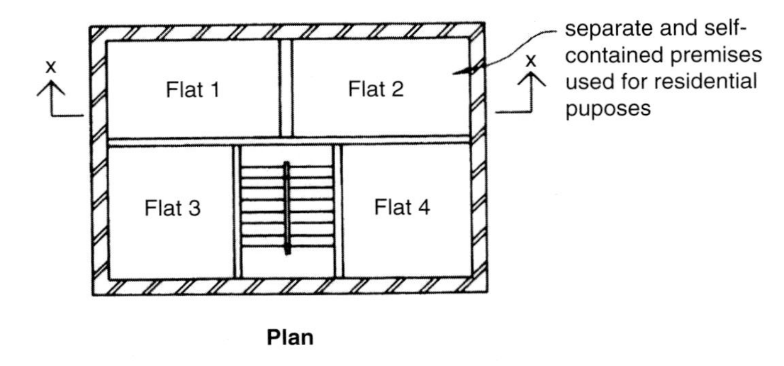

Fig. 2.1 Flat – Regulation 2.

'plans' as including drawings of any description and specifications or other information in any form.

HEIGHT – This means the height of a building measured from the mean level of the ground adjoining the outside external walls to a level of half the vertical

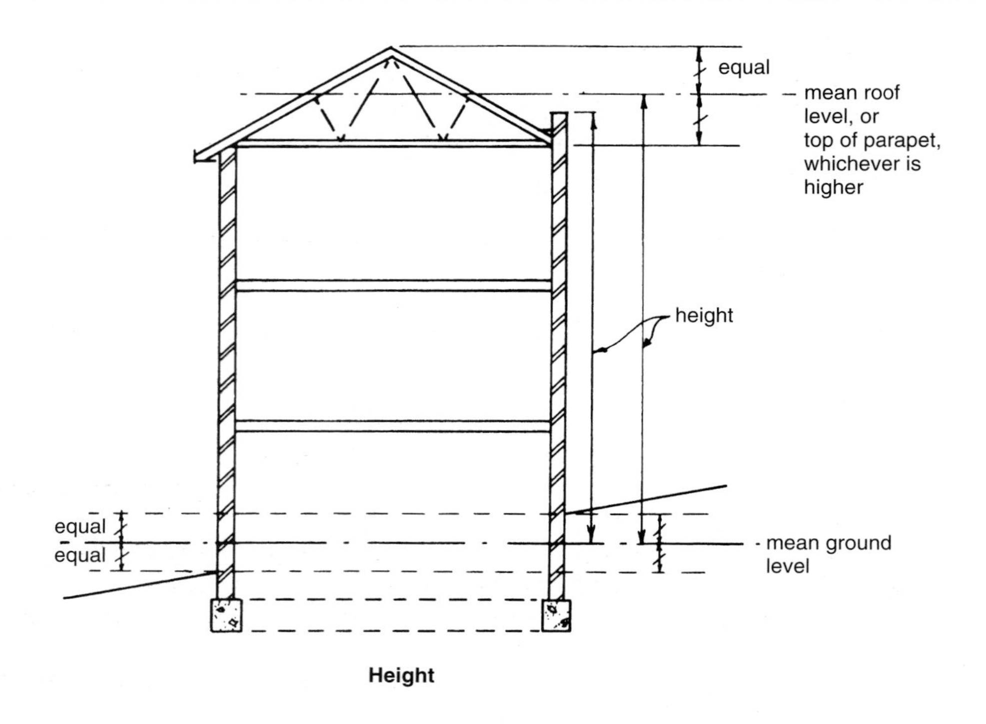

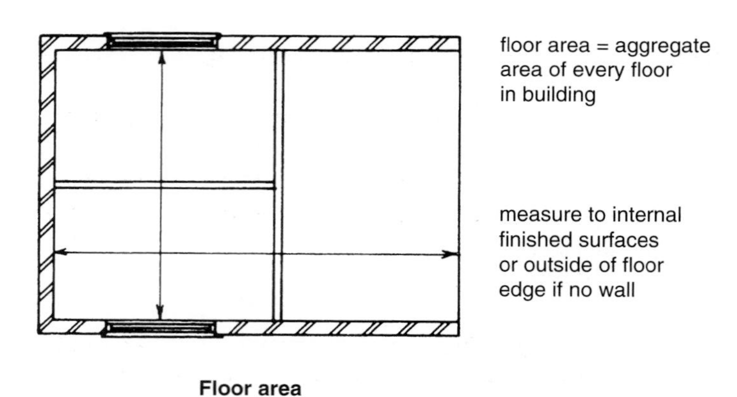

Fig. 2.2 Floor area and height – Regulation 2.

height of the roof, or to the top of any walls or parapet, whichever is the higher (see Fig. 2.2).

INITIAL NOTICE – A notice given by an approved inspector to a local authority under section 47 of the Building Act 1984.

INSTITUTION – This means a hospital, home, school, etc., used as living accommodation for, or for the treatment, care, etc., of people suffering from disabilities due to illness or old age or other physical or mental disability or who are under five years old. Those concerned must sleep on the premises and so day care centres, etc., are not included.

MATERIAL ALTERATION – This is defined in Regulation 3(2) and is described fully in section 2.6 below.

MATERIAL CHANGE OF USE – This is defined by reference to Regulation 5 and there are nine cases:

- Where a building becomes a dwelling when it was not one before.
- Where a building will contain a flat for the first time.
- Where a building becomes a hotel or boarding house, where it previously was not.
- Where a building becomes an institution, where it previously was not.
- Where a building becomes a public building and it was not before.
- Where a building was previously exempt from control (see Schedule 2, below), but is no longer so exempt.
- Where a building containing at least one dwelling is altered so that it provides more or less dwellings than before.
- Where the building contains a room for residential purposes, where previously it did not.
- Where a building containing at least one room for residential purposes is altered so that it contains more or less of such rooms than it did before.

PRIVATE STREET – As section 203(2) of the Highways Act 1980.

PUBLIC BUILDING – This means a building which consists of or contains:

- A theatre, public library, hall or other place of public resort.
- A school or other educational establishment which is not exempt under the 1984 Act, section 4(1)(a).
- A place of public worship.

The definition is restrictive because occasional visits by the public to shops, stores, warehouses or private houses do not make the building a public building.

ROOM FOR RESIDENTIAL PURPOSES – This means a room (or suite of rooms) which is not in a dwellinghouse or flat and which is used by people to live and sleep in. It includes rooms in hotels, hostels, boarding houses, halls of residence and residential homes. It does not include rooms in hospitals or other similar establishments, used for patient accommodation.

SHOP – This includes premises used by members of the public:

- For sales of food or drink for consumption on or off the premises.
- For retail sales by auction.
- As a barber's or hairdresser's business.
- For the hiring of any item.
- For the treatment or repair of goods.

2.5 Exempt buildings and work

Certain buildings and extensions are granted complete exemption from control. The exempt buildings and work fall into seven classes listed in Schedule 2:

CLASS I – BUILDINGS CONTROLLED UNDER OTHER LEGISLATION

- Buildings subject to the Explosives Acts 1875 and 1923.
- Buildings (other than dwellings, offices or canteens) on a site licensed under the Nuclear Installations Act 1965.
- Buildings scheduled under section 1 of the Ancient Monuments and Archaeological Areas Act 1979.

CLASS II – BUILDINGS NOT FREQUENTED BY PEOPLE

- Detached buildings into which people do not normally go.
- Detached buildings housing fixed plant or machinery, normally visited only intermittently for the purpose of inspecting or maintaining the plant, etc. Such buildings are only exempt where they are at least one-and-a-half times their own height from the boundary of the site or any other building frequented by people.

CLASS III – GREENHOUSES AND AGRICULTURAL BUILDINGS

- A building used as a greenhouse.

A greenhouse is not exempted if the main purpose for which it is used is retailing, packing or exhibiting, e.g. one at a garden centre.

- A building used for agriculture which is:
  Sited at a distance not less than one-and-a-half times its own height from any building containing sleeping accommodation; *and*,
  is provided with a fire exit not more than 30 metres from any point within the building.

The definition of 'agriculture' includes horticulture, fruit growing, seed growing and fish farming. Agricultural buildings are not exempted if the main purpose for which they are used is retailing, packing or exhibiting.

CLASS IV – TEMPORARY BUILDINGS

- A building intended to remain where it is erected for 28 days or less, e.g. exhibition stands.

CLASS V – ANCILLARY BUILDINGS

- Buildings on a site intended to be used only in connection with the letting or sale of buildings or building plots on that estate.
- Site buildings on all construction and civil engineering sites, provided they contain no sleeping accommodation.
- Buildings, except those containing a dwelling or used as an office or showroom, erected in connection with a mine or quarry.

CLASS VI – SMALL DETACHED BUILDINGS

- Detached single storey buildings of up to 30 m^2 floor area, with no sleeping accommodation.

For the exemption to apply, such buildings must either be:
Situated more than one metre from the boundary of their curtilage; *or*;
Constructed substantially of non-combustible material.

- Detached buildings of up to 30 m^2 intended to shelter people from the effects of nuclear, chemical or conventional weapons and not used for any other purpose. The excavation for the building must be no closer to any exposed part of another building or structure than a distance equal to the depth of the excavation plus one metre.
- Detached buildings with a floor area not exceeding 15 m^2 and which do not contain sleeping accommodation, e.g. garden sheds.

CLASS VII – EXTENSIONS

- Ground level extensions of up to 30 m^2 floor area which are conservatories, porches, covered yards or ways or a carport open on at least two sides.

A conservatory or porch which is wholly or partly glazed must satisfy the requirements of Part N.

The regulations do not apply to the erection of any building set out in Classes I to VI or to extension work in Class VII. Furthermore, they have no application at all to *any* work done to or in connection with buildings in Classes I to VII provided, of course, that the work does not involve a change of use which takes the building out of exemption, e.g. a barn conversion.

2.6 Application of the regulations

The 2000 Regulations apply only to 'building work' or to a 'material change of use', i.e. use for a different purpose. Work or a change of use not coming under these headings is not controlled.

Meaning of 'building work'
The definition of 'building work' means that the regulations apply in six cases:

ERECTION OR EXTENSION OF A BUILDING

Subject to the exemptions set out in the preceding section, the regulations apply to the erection or extension of all buildings. No attempt is made to define what is meant by 'erection of a building', nor is any definition really necessary. There is a good deal of obscure case law under other legislation as to what amounts to 'erection of a building', but none of it is particularly helpful in the light of section 123 of the Building Act 1984.

This gives a relevant statutory definition. For the purposes of Part II of the Act and for building regulation purposes, erection will include related operations 'whether for the reconstruction of a building, [and] the roofing over of an open space between walls or buildings'.

For the purposes of Part III of the 1984 Act (other provisions about buildings) which is also relevant to building control, *certain* building operations are 'deemed to be the erection of a building'. These are:

(1) Re-erection of any building or part of a building when an outer wall has been pulled or burnt down to within ten feet (3 metres) of the surface of the ground adjoining the lowest story of the building.

It follows that the outer wall must have been demolished throughout its length to within ten feet (3 metres) of ground level to constitute re-erection.

(2) The re-erection of any frame building when it has been so far pulled or burnt down that only the framework of the lowest storey remains.

(3) Roofing over any space between walls or buildings. Clearly other operations could be 'the erection of a building'.

PROVISION OR EXTENSION OF CONTROLLED SERVICES AND FITTINGS

Controlled services and fittings are those required by specified parts of Schedule 1:

- G1 – Sanitary conveniences and washing facilities.
- G2 – Bathrooms in dwellings.
- G3 – Hot water storage systems, except space heating systems, industrial systems, or those with a storage capacity of 15 litres or less.
- H – Drainage and waste disposal systems.
- J – Fixed heat producing appliances burning solid or oil fuel or gas or incinerators.
- L – In non-domestic buildings, heating and hot water systems, lighting, air conditioning and mechanical ventilation systems.
  – In dwellings, replacement windows, doors and rooflights, space heating or hot water service boilers and hot water vessels.

MATERIAL ALTERATION OF A BUILDING OR OF A CONTROLLED SERVICE OR FITTING

The material alteration of an existing building falls within the definition of building work, and is subject to the regulation requirements. Other alterations are not controlled. There are two cases where an alteration is material, namely an alteration to a building or controlled service or fitting, or part of the work involved, which would at any stage result *either*:

- In the building or controlled service or fitting not complying with the relevant requirements of Schedule 1 where it previously did comply: *or*,
- In the building, which did not comply with such requirements before work started, being made worse in relation to the requirement after the alteration.

The specified requirements (called 'relevant requirements' in the regulations) are: Part A (structure); B1 (means of warning and escape); B3 (internal fire spread – structure); B4 (external fire spread); B5 (access and facilities for the fire service); and Part M (access and facilities for disabled people).

The work done must, of course, comply with all the requirements of Schedule 1. In general, it is not necessary to bring the existing building up to regulation standards, however, it should not be made worse when measured against the standards of the relevant requirements in Schedule 1.

WORK IN CONSEQUENCE OF A MATERIAL CHANGE OF USE

When there is a material change of use, as defined in Regulation 5 (see section 2.4), work must be done to make the building comply with some of the regulations, as explained below. Such work is, of course, then subject to control, just as the material change of use is itself controlled. In practical terms, change of use is only subject to control if the change involves the provision of sleeping accommodation or use as a public building or where the building was previously exempt.

'Material change of use' requirements

Material change of use has already been defined (see section 2.4), and in the nine cases falling within that definition, specific technical requirements from Schedule 1 are made to apply in the interests of health and safety, which is the philosophy behind building control. Interestingly, there is no requirement applicable in respect of surface water drainage or stairs, nor is there any definition of 'part' of a building.

The parts of the regulations applicable are set out in Table 2.1.

INSERTION OF INSULATING MATERIAL INTO A CAVITY WALL

When there is the insertion of cavity fill in an existing wall in a building, the work done must comply with certain specific regulation requirements, namely C4 (resistance of walls to the passage of moisture) and D1 (toxic substances).

UNDERPINNING OF A BUILDING

Work involving the underpinning of an existing building is 'building work' for the purposes of the regulations and so comes under control.

2.7 Regulation requirements

The regulations impose broad general requirements on the builder. Breach of these requirements does not, of itself, involve the builder in any civil liability although such liability may arise, quite independently, at common law.

Compliance with Schedule 1 is mandatory. All building work must be carried out

Table 2.1 Requirements applicable according to material change of use.

Case	*Schedule 1 requirements*
[A] All cases (dwellings, flats, hotels, boarding houses, and institutions and public buildings, no longer exempt) where there is a change of use to the whole of the building	B1 (means of warning and escape) B2 and B3 (internal fire spread) B4(2) (external fire spread – roofs) B5 (access etc. for fire services) F1 and F2 (ventilation) G1 (sanitary conveniences & washing facilities) G2 (bathrooms) H1 (foul water drainage) H6 (solid waste storage) J1 to J3 (combustion appliances) L1 (conservation of fuel and power – dwellings) L2 (conservation of fuel and power – buildings other than dwellings)
[B] Exempt building to non-exempt, hotel, boarding house, institution, public building	As in [A] plus A1 to A3 (structure)
[C] Building more than 15 metres in height	As in [A] plus B4(1) (external fire spread – walls)
[D] Building used as a dwelling, where previously it was not	As in [A] plus C4 (resistance to weather and ground moisture)
[E] Building used as a dwelling, hotel, boarding house or containing a flat where it did not before; where more or less dwellings are provided than was originally the case; where the building contains a room for residential purposes, where previously it did not; where a building containing at least one room for residential purposes is altered so that it contains more or less of such rooms than it did before	As in [A] plus E1 to E3 (resistance to passage of sound)
[F] Change of use to public building consisting of or containing a school	As in [A] plus E4 (acoustic conditions in schools)
[G] Change of use of part only of a building	The part itself must comply with the relevant requirements as [A], [B], [D] and [E] and [F]. In [C] the whole building must comply with B4(1)

so that it complies with the requirements set out in that Schedule. The method adopted for compliance must not result in the contravention of another requirement.

The work must also be carried out so that, after completion, an existing building or controlled service or fitting to which work has been done continues to comply with the specified requirements if it previously did so comply or, if it did not so comply before in any respect, it must not be more unsatisfactory afterwards.

2.8 Schedule 1 – Technical requirements

Schedule 1 contains the technical requirements, which are discussed in Chapters 6 to 18 and which are almost all expressed functionally, e.g. C1 dealing with site preparation states that 'the ground to be covered by the building shall be reasonably free from vegetable matter'. These requirements cannot be subject to relaxation.

Which requirements apply depends on the type of building being constructed, but the majority of them is of universal application.

Materials and workmanship

Regulation 7(1) provides that any building work which is required to comply with any relevant requirement of Schedule 1

> 'shall be carried out (a) with adequate and proper materials which – (i) are appropriate for the circumstances in which they are used, (ii) are adequately mixed and prepared, and (iii) are applied, used or fixed so as adequately to perform the functions for which they are designed; and (b) in a workmanlike manner.'

This is a general statutory obligation imposed on the builder. Guidance on how the obligation may be met is contained in Approved Document to support Regulation 7 'Materials and workmanship', although that guidance is of a very general nature.

This statutory obligation is akin to a building contractor's obligation at common law when, in the absence of a contrary term in the contract, the builder's duty is to do the work in a good and workmanlike manner, to supply good and proper materials and to provide a building reasonably fit for its intended purpose: *Hancock* v. *B.W. Brazier (Anerley) Ltd* (1966) [1966] 1 WLR 1317; [1966] 2 All ER 901, CA. This threefold obligation would normally be implied in any case where a contractor was employed to both design and build, but the third limb of the duty would not arise, for example, where the client employs his own architect (*Lynch* v. *Thorne* (1956) [1956] 1 WLR 303; [1956] 1 All ER 744, CA) although the other two limbs remain.

The principal object of the regulations is to ensure that buildings meet reasonable standards of health and safety, and this is spelled out in regulation 8:

> 'Parts A to D, F to K and N (except for paragraphs H2 and J6) of Schedule 1 shall not require anything to be done except for the purpose of securing *reasonable*

standards of health and safety for persons in or about buildings (and any others who may be affected by buildings, or matters connected with buildings).'

The obligations imposed by the regulations are not therefore absolute obligations, but rather a duty to use reasonable skill and care to secure reasonable standards of health and safety of people using the building and others who may be affected by failure to comply with the requirements of the regulations.

2.9 Relaxation of regulation requirements

Section 8 of the Building Act 1984 enables the Secretary of State to dispense with or relax any requirement of the regulations 'if he considers that the operation of [that] requirement would be unreasonable in relation to the particular case'. This power has been delegated to the local authority which may grant a relaxation if, because of special circumstances, the terms of a requirement cannot be fully met.

However, the majority of regulation requirements cannot be relaxed because they require something to be provided at an 'adequate' or 'reasonable' level, and to grant a relaxation would mean acceptance of something that was 'inadequate' or 'unreasonable'.

The application procedure is laid down in sections 9 and 10 of the 1984 Act. There is no prescribed form. Only the local authority (or the Secretary of State on appeal) can grant a relaxation; approved inspectors have no power to do so.

At least 21 days before giving a decision on an application for dispensation or relaxation of any requirement, the local authority must advertise the application in a local newspaper unless the application relates only to internal work. The notice must indicate the situation and nature of the work, and the requirement which it is sought to relax or dispense with. Objections may then be made on grounds of public health or safety. No notice need be published if the effect of the proposal is confined to adjoining premises only, but notice must then be given to the owner and occupier of those premises.

Where a local authority refuse an application they must notify the applicant of his right of appeal to the Secretary of State. This must be exercised within one month of the date of refusal. The grounds of the appeal must be set out in writing, and a copy must be sent to the local authority, who must send it to the Secretary of State with a copy of all relevant documents, and any representations they wish to make. The applicant must be informed of the local authority's representations. There is no time limit prescribed for the Secretary of State's decision on the appeal.

Where a local authority fail to give a decision on an application within two months, it is deemed to be refused and the applicant may appeal forthwith.

Neither the Secretary of State nor the local authority may give a direction for any relaxation of the regulations where, before the application is made, the local authority has become statutorily entitled to demolish, remove or alter any work to which the application relates, i.e. as a result of service of a notice under section 36 of the 1984 Act. The same prohibition applies where a court has issued an injunction requiring the work to be demolished, altered or removed.

The procedure may be summarised in tabular form:

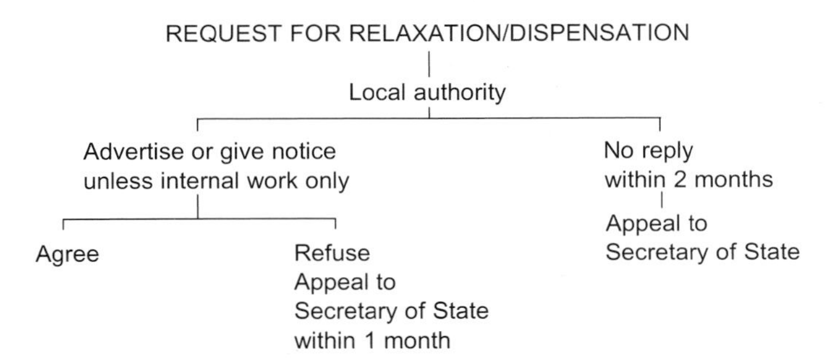

2.10 Type relaxations

The local authority's power of dispensation and relaxation must be distinguished from that of the Secretary of State to grant a type relaxation, i.e. to dispense with a requirement of the regulations generally. A type relaxation can be made subject to conditions and can be for a limited period only. It can be issued on application to the Secretary of State, e.g. from a manufacturer, in which case a fee may be charged. The Secretary of State may also make a type relaxation of his own accord. Before granting a relaxation the Secretary of State must consult such bodies as appear to him to be representative of the interests concerned and must publish notice of any relaxation issued. No such type relaxations have been granted, under the current legislation.

2.11 Continuing requirements

Under section 2 of the Building Act 1984 building regulations can impose continuing requirements on owners and occupiers of buildings. These requirements are of two kinds:

- Continuing requirements in respect of designated provisions of the building regulations, to ensure that the purpose of the provision is not frustrated.

For example, where an item is required to be provided, there could be a requirement that it should continue to be provided or kept in working order. Examples of the possible use of the power are the operation of mechanical ventilation which is necessary for health reasons or the operation of any lifts required to be provided in blocks of flats.

- Requirements with regard to services, fittings and equipment. This enables requirements to be imposed on buildings whenever they were erected and independently of the normal application of building regulations to a building.

A possible use of this power would be to require the maintenance and periodic inspection of lifts in flats if they are to be kept in use. This power of continuing requirements has, as yet, not been used.

2.12 Testing and sampling

Regulations 18 and 19 empower the local authority to test building work to ensure compliance with the requirements of regulation 7 and any applicable parts of Schedule 1 and to take samples of materials *to be used* in the carrying out of building work. The wording does not appear to cover materials which are already incorporated in the building, but this may prove to be of little importance if the provisions of section 33 of the Building Act 1984, are ever activated.

Under that section the local authority may test for compliance with the regulations. They will also be permitted to require a builder or developer to carry out reasonable tests or may carry out such tests themselves and also take samples for the purpose. Section 33(3) sets out the following matters with respect to which tests may be made:

- Test of the soil or subsoil of the site of any building.
- Tests of any material or component or combination of components.
- Tests of any service, fitting or equipment.

This is not an exhaustive description of the matters which may be subjected to tests.

The cost of testing is to be borne by the builder or developer, and there will be a right to apply to a magistrates' court regarding the reasonableness of any test required or of any decision of the local authority on meeting the cost of the test. It should be noted that the local authority will have a discretionary power to bear the whole or part of the costs themselves.

In fact the power of testing is given to 'a duly authorised officer of the local authority'. 'Authorised officer' is defined in section 126 of the Building Act 1984 as:

> '... an officer of the local authority authorised by them in writing, either generally or specially, to act in matters of any special kind, or in any specified matter; or ... by virtue of his appointment and for the purpose of matters within his province, a proper officer of the local authority...'

Section 95 of the 1984 Act confers upon an authorised officer appropriate powers of entry, and penalties for obstructing any person acting in the execution of the regulations are provided by section 112.

A duly authorised officer of the local authority must also be permitted to take samples of the materials used in works or fittings, to see whether they comply with the requirements of the regulations. In practice the authorised officer may ask the builder to have the tests carried out and to submit a report to the local authority. In any event, the builder should be notified of the result of the tests.

It should be noted, however, that regulations 18 and 19 do not apply where the work is supervised by an approved inspector (see Chapter 4) or is done under a public body's notice; however, similar powers exist in the Building (Approved Inspectors etc.) Regulations 2000.

2.13 Unauthorised building work

Regulation 21 allows local authorities retrospectively to certify unauthorised building work carried out on or after 11 November 1985. The regulation became effective on 1 October 1994 and there are prescribed fees payable. It applies to building work which should have been subject to control, but the person who carried out the work failed to deposit plans with the authority, to give a building notice or to give an initial notice jointly with an approved inspector. The regulation enables the owner of the building (the applicant) to make a written application to the authority for a regularisation certificate.

The applicant's notice should describe the unauthorised work and, if reasonably practicable include a plan of it as well as a plan showing any additional work needed to ensure compliance with the regulations. On receipt of the notice and the accompanying plans, the council may require the applicant to take reasonable steps to enable them to inspect the work, e.g., opening up, testing and sampling. The local authority will then notify the applicant of any work required to ensure compliance, with or without relaxation and when this has been carried out to their satisfaction they may issue a regularisation certificate. This is stated to be evidence (but not conclusive evidence) that the relevant specified requirements have been complied with.

2.14 Contravening works

Under section 36 of the Building Act 1984, where a building is erected, or work is done contrary to the regulations, the local authority may require its removal or alteration by serving notice on the owner of the building. Where work is required to be removed or altered, and the owner fails to comply with the local authority's notice within a period of 28 days, the local authority may remove the contravening work or execute the necessary work themselves so as to ensure compliance with the regulations, recovering their expenses in so doing from the defaulter.

A section 36 notice may not be given after the expiration of twelve months from the date on which the work was completed. A notice cannot be served where the local authority have passed the plans and the work has been carried out in accordance with the deposited plans.

The recipient of a section 36 notice has a right of appeal to the magistrates' court. The burden of proving non-compliance with the regulations lies on the authority, but if they show that the works do not comply with an approved document (under section 7) then the burden shifts. The appellant against the notice must then prove compliance with the regulations: *Rickards* v. *Kerrier District Council* (1987) CILL 345; 4-CLD-04-26.

Section 37 provides an alternative to the ordinary appeal procedure. Under that section, the owner may notify the local authority of his intention to obtain from 'a suitably qualified person' a written report about the matter to which the section 36 notice relates. Such notices are served where the local authority considers that the technical requirements of the regulations have been infringed.

The expert's report is then submitted to the local authority. In light of it the local authority may withdraw the section 36 notice and *may* pay the owner the expenses which he has reasonably incurred in consequence of the service of the notice, including his expenses in obtaining the report. Adopting this procedure has the effect of extending the time for compliance with the notice or appeal against it from 28 to 70 days.

If the local authority rejects the report, it can then be used as evidence in any appeal under section 40 and section 40(6) provides that

> 'if, on appeal . . . there is produced to the court a report that has been submitted to the local authority . . . the court, in making an order as to costs, may treat the expenses incurred in obtaining the report as expenses for the purposes of the appeal.'

Thus, in the normal course of events, if the appeal was successful, the owner would recover the cost of obtaining the report as well as his other costs.

The local authority – or anyone else – may also apply to the civil courts for an injunction requiring the removal or alteration of any contravening works. This power is exercisable even in respect of work which has been carried out in accordance with deposited plans, e.g. oversight or mistake on the part of the local authority. In such a case the court might well order the local authority to pay compensation to the owner. The twelve months' time limit does not apply to this procedure which is, however, unusual and rarely invoked in practice. The Attorney-General, as guardian of public rights, may seek an injunction in similar circumstances, and in practice proceedings for an injunction must be taken in his name and with his consent.

Where a person contravenes any provision in the building regulations, he renders himself liable to prosecution by the local authority. The case is dealt with in the magistrates' court. The maximum fine on conviction is £2000, with a continuing penalty of £50 a day (Building Act 1984, section 35).

In *Torridge District Council* v. *Turner* (1991) 9-CLD-07-21, it was held, for reasons which are not entirely clear, that breach of 'do' provisions such as requirement A1 which requires that a building 'shall be so constructed' as to meet the specified standards does not constitute a continuing offence which means that the proceedings must be commenced within six months of the commission of the alleged offence. This six month limitation period is specified by section 127(1) of the Magistrates' Courts Act 1980. The Divisional Court held that the person constructing a building commits an offence when the building works are completed in a way not complying with the regulations. He does not commit a continuing offence.

If this decision is correct it makes the enforcement of the regulations a well-nigh impossible task.

2.15 Determinations

It is sometimes the case that a local authority rejects a full plans application (or an approved inspector refuses to give a plans certificate – see Chapter 4) on the grounds that the plans show a contravention of the Building Regulations, but the applicant believes that the plans do, in fact, comply. In this case the applicant can apply to the Secretary of State for a determination as to whether or not the work complies with the regulations.

It is possible to apply for a determination at any time after the plans have been submitted to the local authority or an approved inspector has been asked for a plans certificate, but it is usually better to wait until a decision has been made and the plans have been declared unacceptable. Applications for a determination must usually be made before the commencement of work (or before commencement of that part of the work which is the subject of the determination).

The application, in the form of a letter, is made direct to the Office of the Deputy Prime Minister if the proposal is in England, or to the National Assembly for Wales if it is in Wales, and it should include the following information:

- The names and addresses of the parties involved, including any agents.
- Details of the local authority or approved inspector providing the building control service.
- The full address of where the proposed building work will be carried out.
- A statement setting out details of the building, the proposed work, and the matter in dispute.
- A statement setting out the case for compliance with the particular Regulation requirement in question.
- A copy of the plans of the proposed work and any other documents which have been submitted to the local authority with the full plans application, or those submitted to the approved inspector on which he was unable to give a plans certificate.
- A copy of all relevant correspondence with the local authority/approved inspector involved, including the notice of rejection of plans if one has been issued.
- A copy of any listed building consent if required for the proposed work and any associated planning permission relevant to the listed building.
- A copy of any other documents supporting the case for compliance, including calculations.
- Where appropriate, a location or block plan and photographs of the proposed work to illustrate particular points.
- The appropriate fee (i.e. half the local authority's plan charge, with a minimum of £50 and a maximum of £500 payable).

It is not necessary to obtain the permission of the local authority or approved inspector before making such an application.

3 Local authority control

3.1 Introduction

Local authorities have exercised control over buildings in England and Wales since 1189, but it was not until 1965 that uniform national building regulations were made applicable throughout the country generally. Inner London retained its own system based on the London Building Acts 1930 to 1978 and byelaws made thereunder until 6 January 1986. Building regulations now apply to Inner London, although many provisions of the London Building Acts continue to apply in modified form. The Building Regulations 1985 introduced a number of substantive changes to the system of local authority control, and there have been several modifications to the system since then, culminating in the Building Regulations 2000, which came into force on 1 January 2001.

Part V of the Building Regulations 2000, as amended, contains the procedural requirements which must be observed where a person proposes to undertake building work covered by the regulations and opts for local authority control. Although a great deal of building work continues to be under local authority control and supervision, an increasing volume of work is now dealt with under an alternative system – private control and supervision by an approved inspector, as explained in Chapter 4. Until January 1997 the private system was confined to house-builders under the National House Building Council (NHBC) scheme. It is now possible to use an approved inspector for any class of building work and at the date of this edition there were 11 corporate and in excess of 30 non-corporate approved inspectors operating in England and Wales.

Two main procedural options are available under the local authority system of control:

- Control based on service of a building notice.
- Control based on the deposit of full plans.

There are also a number of cases – where the work relates to the installation of gas, solid fuel or oil-fired combustion appliances, drainage and plumbing works and the installation of replacement windows, doors and rooflights – where neither notice nor deposit of plans is required. It is also possible to have an intermediate situation where plans may be passed in stages.

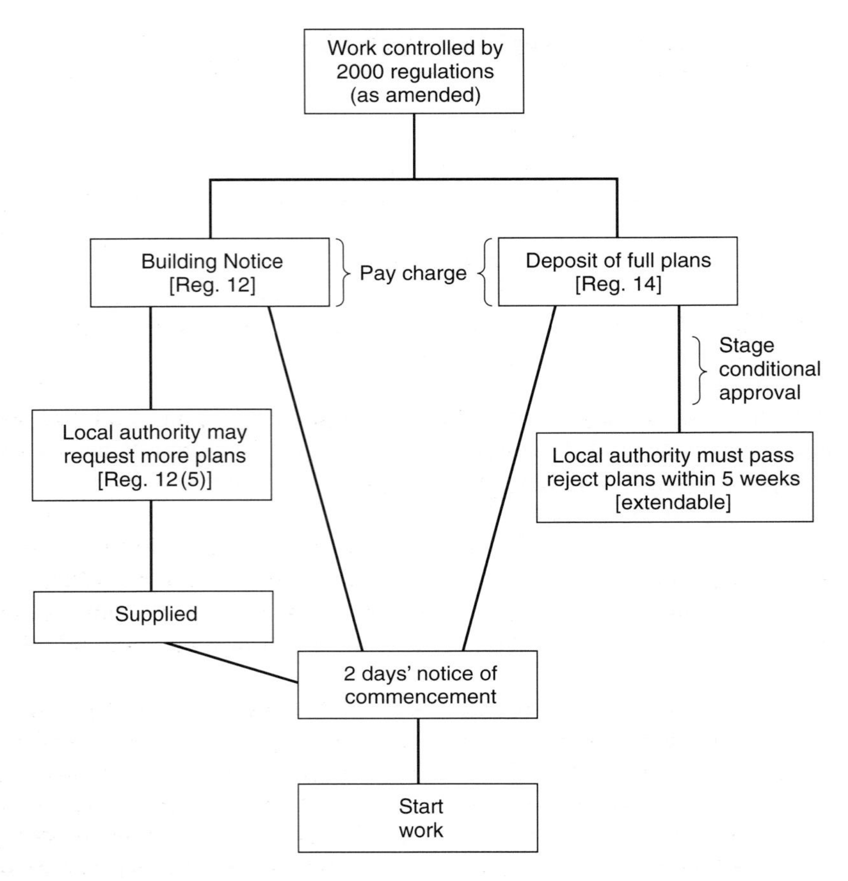

Fig. 3.1 Local authority supervision.

3.2 The local authority

The local authority for the purposes of the regulations is the district council, a London borough council, the Common Council of the City of London, the Sub-Treasurer of the Inner Temple, the Under Treasurer of the Inner Temple, and the Council of the Isles of Scilly.

3.3 Building notice procedure

The major procedural innovation introduced in 1985, and now to be found in the 2000 Regulations, is based on service of a building notice. There is no approval of plans. Interestingly, this is copied from the former Inner London system.

A person intending to carry out building work or make material change in the use of a building may give a building notice to the local authority unless the work is to a building which is put to a 'relevant use'. A 'relevant use' means a use as a workplace of a kind to which Part II of the Fire Precautions (Workplace) Regulations 1997 applies (see Chapter 5), or is one which is designated under the Fire Precautions Act 1971. The following buildings are designated under the Act:

- *Hotels and boarding houses*

Where there is sleeping accommodation for six or more guests or staff or any number of guests or staff above the first floor or below the ground floor.

- *Factories, offices, shops and railway premises*

Where more than twenty people are employed or more than ten people are employed other than on the ground floor or in factories only where explosive or highly flammable materials are stored or used.

There is no prescribed form of building notice. The notice must be signed by the person intending to carry out the work or on his behalf, and must contain or be accompanied by the following information:

- The name and address of the person intending to carry out the work.
- A statement that it is given in accordance with regulation 12(2)(a).
- A description of the proposed building work or material change of use.
- A description of the location of the building to which the proposal relates and the use or intended use of that building.
- If it relates to the erection or extension of a building it must be supported by a plan to a scale of not less than 1:1250, showing size and position of the building, its own boundaries and its relationship with adjoining boundaries, the size, position and use of every other building within its boundaries, the width and position of any streets on or within its boundaries, the number of storeys and the provisions to be made for its drainage. Where any local legislation applies, the notice must state how it will be complied with.
- Where the building notice involves cavity wall insulation, information must be given about the insulating material to be used and whether or not it has been approved by any European Technical Approval issuing body or conforms to any national standard of a member state of the European Economic Area. The name of the body which has approved the installer must also be stated on the building notice.
- If the work includes the provision of a hot water storage system covered by Schedule 1, G3 (e.g. an unventilated system with a storage capacity of 16 litres of more) details of the system and whether or not the system and its installer are approved.

The local authority is not required to approve or reject the building notice and, indeed, has no power to do so. However, it is entitled to ask for any plans it thinks are necessary to enable it to discharge its building control functions and may specify a time limit for their provision.

The regulations make plain that the building notice and plans shall not be 'treated as having been *deposited* in accordance with the building regulations'. In some ways this is an odd provision because the relevant building control sections of many of the local Acts of Parliament mentioned in Chapter 5 – and which provide for special local requirements – are triggered off by the 'deposit' of plans. At first sight, therefore, this would render such requirements inoperative, but presumably it is thought that compliance will be ensured through the requirement that the building notice must contain a statement of the steps to be taken to comply with any local enactment.

Once a building notice has been given, work can be commenced, although there is a requirement (see below) that the local authority be notified at least two days before work commences.

A building notice remains in effect for a period of three years from the date on which it was given to the local authority. If the work has not been commenced within that period or the material change of use has not been made, the building notice lapses automatically. This three-year restriction on the validity of building notices was introduced in 1991 and is in line with a full plans submission (see section 3.5).

It has already been mentioned that a building notice cannot be given for works to buildings put to a relevant use (see section 3.3 above). There are also two other instances where a building notice cannot be given and therefore full plans must be deposited as follows:

- Where the work involves the erection of a building fronting onto a private street.
- Where the work involves the building over of any sewers, disposal mains or drains shown on any map of sewers kept by a sewerage undertaker under section 199 of the Water Industry Act 1991.

3.4 Exemptions from the requirement to give a building notice or deposit full plans

A person who intends to carry out building work consisting only of the work described in the first column of Table 3.1 below is not required to give a building notice or deposit full plans if the work is to be carried out by a person described in the corresponding entry in the second column of the Table.

3.5 Deposit of plans

This is the traditional system of building control by which full plans are deposited with the appropriate local authority in accordance with section 16 of the Building Act 1984, as supplemented by regulation 14. Section 16 imposes a duty on the building control authority to either pass or reject plans deposited for the proposed work.

In *Murphy* v. *Brentwood District Council* (1990) 20 ConLR 1, CA the Court of Appeal held that the duty is imposed on the local authority itself either to pass or reject the deposited plans, and it cannot discharge its duty by delegating performance to outside consultants. If the local authority leaves it to outside consultants

Table 3.1 Exemption from the requirements to give a building notice or deposit full plans.

Type of work	Person carrying out work
Installation of a heat-producing gas appliance[1]	A person, or an employee of a person approved in accordance with regulation 3 of the Gas Safety (Installation and Use) Regulations 1998
Installation of: • an oil-fired combustion appliance which has a rated heat output of 45 kilowatts or less and which is installed in a building of three storeys or less (not counting basements); or • oil storage tanks and the pipes connecting them to combustion appliances	A person registered under the Oil Firing Registration Scheme by the Oil Firing Technical Association for the Petroleum Industry Ltd in respect of that type of work
Installation of a solid fuel burning combustion appliance which has a rated heat output of 50 kilowatts or less and which is installed in a building of three storeys or less (not counting basements)	A person registered under the Registration Scheme for Companies and Engineers involved in the Installation and Maintanance of Domestic Solid Fuel Fired Equipment by HETAS Ltd in respect of that type of work
Installation of: • a service or fitting in relation to which Part G of Schedule 1 imposes a requirement; • a foul water drainage system in relation to which paragraph H1 of Schedule 1 imposes a requirement; • a rainwater drainage system in relation to which paragraph H3 of Schedule 1 imposes a requirement; or • a hot water vessel in relation to which paragraphs L1 and L2 of Schedule 1 impose requirements, which is installed in or in connection with a building of three storeys or less (not counting basements) and which does not involve connection to a drainage system at a depth greater than 750 mm below ground level	A person registered under the Approved Contractor Person Scheme (Building Regulations) by the Institute of Plumbing in respect of that type of work
Installation, as a replacement, of a window, rooflight, roof window or door in an existing building	A person registered under the Fenestration Self-Assessment Scheme by Fensa Ltd in respect of that type of work
Any building work[2] which is necessary to ensure that any appliance, service or fitting which is installed and which is described above, complies with the applicable requirements contained in Schedule 1 of the Building Regulations 2000	The person who installs the appliance, service or fitting to which the building work relates and who is described above

Notes:
1. 'appliance' includes any fittings or services, other than a hot water storage vessel not incorporating a vent pipe to the atmosphere, which form part of the space heating or hot water system served by the combustion appliance.
2. 'building work' does not include the provision of a masonry chimney. The building work referred to in this row, which is necessary to ensure that a gas combustion appliance complies with Schedule 1 of the Building Regulations, can only be carried out by the person referred to in the second column if the appliance has a net rated heat input of 70 kilowatts or less and is installed in a building of three storeys or less (not counting basements).

to decide whether plans are passed or rejected, the local authority is vicariously responsible if the consultants are negligent, subject to proof of recoverable damage.

However, the Court of Appeal proceeded on the basis that *Anns* v. *London Borough of Merton* [1978] AC 728; [1977] 2 All ER 492, HL; 5 BLR 1 was rightly decided, and in light of the fact that *Anns* was subsequently overruled it is thought that the local authority could only be vicariously liable in these circumstances (if at all) where personal injury was suffered by the occupier or there was damage to other property. Indeed, it is probable that in the current climate of judicial opinion the local authority would be held able to discharge its section 16 duty by reliance on competent outside expertise.

If the plans submitted are not defective the authority has no alternative but to approve them unless, of course, they contravene the linked powers discussed in Chapter 1.

Where the proposed works are subject to the regulations, and it is proposed to deposit full plans, the provisions of section 16 and Regulation 14 must be observed. The local authority must give notice of approval or rejection of plans within five weeks unless the period is extended by written agreement. The extended period cannot be later than two months from the deposit of plans, and any extension must be agreed before the five-week period expires. However, the five-week period does not begin to run unless the applicant submits a 'reasonable estimate' of the cost of the works (where applicable) and pays the plan charge at the same time as the plans are deposited.

The approval lapses if the work is not commenced within a period of three years from the date of the deposit of the plans, provided the local authority gives formal notice to this effect. The local authority must pass the plans of any proposed work deposited with them in accordance with the regulations unless the plans are defective, or show that the proposed work would contravene the regulations. The notice of rejection must specify the defects or non-conformity, and the applicant may then ask the Secretary of State to determine the issue. His decision is then final. The Secretary of State may refer questions of law to the High Court and must do so if the High Court so directs.

The local authority may pass plans by stages and, where it does, it must impose conditions as to the deposit of further plans. It may also impose conditions to ensure that the work does not proceed beyond the authorised stage. It has power to approve plans subject to agreed modifications, e.g. where the plans are defective in a minor respect or show a minor contravention. However, it should be noted that local authorities are not obliged to pass plans conditionally or in stages, and the applicant must agree in writing to these procedures.

The 'full plans' required under the deposit method are the same as those required under the building notice procedure, together with such other plans as are necessary to show that the work will comply with the building regulations.

Regulation 14 specifies that the plans must be deposited in duplicate; the local authority retains one set of plans and returns the other set to the applicant. They must be accompanied by a statement that they are deposited in accordance with regulation 12(2)(b) of the 2000 Regulations and if the building is put or is intended to be put to a relevant use (see section 3.3), by a statement to that effect. Two additional copies of the plans must be submitted where Part B (Fire Safety) imposes a requirement in relation to the work and both additional plans may be retained by the local authority, although, it is not necessary to provide additional copies of the plans where the proposed work relates to the erection, extension or material alteration of a dwelling-house or flat. Where it is proposed to build over a sewer and regulation H4 applies, particulars of the precautions to be taken must be provided.

Work may be commenced as soon as plans have been deposited – although the local authority must be given notice of commencement at least 2 days before work commences – but it is an unwise practice to commence work before notice of approval is received.

If the applicant wants the authority to issue a completion certificate (see section 3.7) in due course, a request to that effect should accompany the plans.

The advantage of the full deposit of plans method of control is that if the work is carried out exactly in conformity with the plans as passed by the local authority, they cannot take any action in respect of an alleged contravention under section 36 of the Building Act 1984.

The deposit of full plans procedure and the possible alternative solutions are shown diagrammatically in Fig. 3.2.

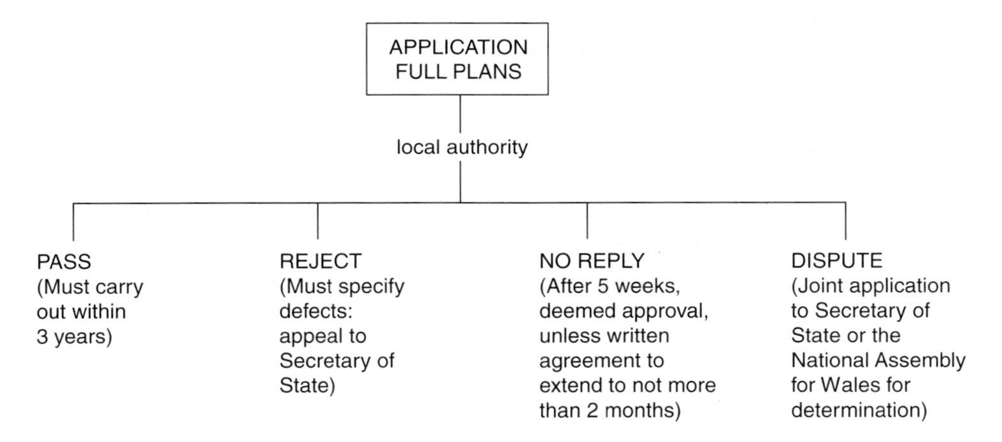

Fig. 3.2 The full plans procedure.

3.5.1 Consultation with the sewerage undertaken

Regulation 14A requires that where full plans have been deposited with the local authority and there are proposals to build over a sewer (as covered by paragraph H4 of Schedule 1 to the Building Regulations 2000), the local authority must consult the sewerage undertaker:

- as soon as practicable after the plans have been deposited; and
- before issuing any completion certificate in relation to the building work.

The local authority must give the sewerage undertaker sufficient plans to show whether the work would comply with the requirements of paragraph H4 and have regard to any views it expresses. The local authority must not pass plans or issue a completion certificate until 15 days have elapsed from the date on which the sewerage undertaker was consulted (unless, of course, the sewerage undertaker has expressed its views to the local authority before the 15 days has expired).

3.6 Notice requirements

Wherever the work is to be supervised by the local authority, in addition to the building notice or deposit of plans, the person undertaking the work must pay an inspection charge and give certain notices to the local authority. The notices must be in writing 'or by such other means as [the local authority] may agree', e.g. by telephone, but most building control authorities provide applicants who deposit plans with pre-printed postcards.

Failure to give the required notices is a criminal offence, punishable on summary conviction by a substantial fine. Under the previous regulations, the Court of Appeal has ruled that failure to deposit plans and give notices under the building regulations is not a continuing offence, with the result that magistrates have no power to try informations laid more than six months after the relevant period for compliance. The regulations require the notices to be given by a specified deadline, and once that deadline has passed, the offence has been committed: *Hertsmere Borough Council* v. *Alan Dunn Building Contractors Ltd* (1986) 84 LGR 214; 9-CLD-07-16.

In practice the majority of local authorities do not seek to enforce the penalty but rely on their powers to serve written notice requiring the person concerned within a reasonable time to cut into, lay open or pull down so much of the work as is necessary to enable them to check whether it complies with the regulations.

Where the person carrying out the work is advised in writing by the local authority of contravening works, and has rectified these as required by the local authority, he must give the authority written notice within a reasonable time after the completion of the further work.

'Reasonable time' is not defined in either situation; it is a question of fact in each case. The phrase has been judicially defined as being 'reasonable under ordinary circumstances': *Wright* v. *New Zealand Shipping Co.* [1878] AC 23.

Regulation 15 requires the giving of the following notices to the local authority:

- At least two days' notice of commencement before commencing the work.
- At least one day's notice of:
  the covering up of any foundation excavation, foundation, damp-proof course, concrete or other material laid over a site; *or*,
  covering up of any drain or private sewer subject to the regulations.

These periods of notice commence after the day on which the notice was served. The local authority must be given notice by 'the person carrying out building work' not more than five days after the completion of:

- The laying of any drain or private-sewer, including any haunching, surrounding or trench backfilling.
- Any building work covered by the regulations.
- The completion of any other work.

There is no definition of the term 'person carrying out building work' in the regulations, but in *Blaenau Gwent Borough Council* v. *Khan* (1993) 35 ConLR 65 the High Court held that the owner of a building who authorises a contractor to carry out building works on his behalf fell within the term. The court took the view that those words should not be confined so as to restrict the meaning of the phrase to the person who physically performs the work, 'but includes the owner of the premises on which the works are being performed and who had authorised the work.'

Where a building or part of a building is occupied before completion, the local authority must be given notice at least five days before occupation. This notice is additional to the required notice after completion.

Additionally, a further notice is required to be given to the local authority where a new dwelling is created either by new building work or by a material change of use. The person carrying out the building work must calculate the energy rating of the dwelling by the government approved Standard Assessment Procedure (SAP) and must give notice of this within five days of completion of the dwelling or at least five days before occupation if this occurs before completion. A notice stating the energy rating for the dwelling must be affixed in a conspicuous place in the dwelling by the person carrying out the building work.

If a person fails to comply with the notice requirements of Regulation 15(1)(2)(3) the local authority may require him by notice to cut into, lay open or pull down any work so that they may find out whether the regulations have been complied with. The person concerned must comply with the notice within a *reasonable time* which, it is suggested, will normally be short.

Where the local authority has served notice specifying that the work contravenes the regulations, on completion of the remedial work, notice must be given to the authority within a reasonable time.

There is now a definition of *day*. It means a period of 24 hours commencing at midnight, i.e. a calendar day, but Saturdays, Sundays and Bank or public holidays are excluded.

3.7 Completion Certificate

The local authority must issue a completion certificate when they are satisfied, after having taken all reasonable steps, that the Schedule 1 requirements are met. Its issue is mandatory in respect of fire safety requirements, i.e. where the building is to be put to a relevant use (see section 3.3 above), but the completion certificate need only relate to the Part B (Fire Safety) requirements. In other cases the authority need issue the certificate only where they have been requested to do so, but the completion certificate will need to relate to all applicable requirements of the regulations.

The completion certificate is evidence – but not *conclusive* evidence – that the requirements specified in the certificate have been complied with.

It should be noted that the local authority cannot be held liable for a fine if it contravenes this regulation by failing to give a completion certificate (see regulation 22).

4 Private certification

4.1 Introduction

One of the Government's aims in reforming the previous system of building control was to provide an opportunity for self-regulation by the construction industry through a scheme of private certification. This is not a complete substitute for local authority control, because local authorities will always remain responsible for taking any enforcement action which may be necessary. Indeed, in certain closely-defined circumstances they may resume their control functions.

The developer is given the option of having the work supervised privately, rather than relying on the local authority control system described in the previous chapter. Essentially, the private certification scheme is based on the proposals set out in a Government White Paper *The Future of Building Control in England and Wales* published by HMSO in February 1981.

The statutory framework of the alternative system is contained in Part II of the Building Act 1984. In broad terms, this provides that the responsibility for ensuring compliance with building regulations may, at the option of the person intending to carry out the work, be given to an approved inspector instead of to the local authority. It also enables approved public bodies to supervise their own work. Various supplementary provisions deal with appeals, offences, and the registration of certain information.

The detailed rules and procedures relating to private certification are to be found in the Building (Approved Inspectors, etc.) Regulations 2000 as amended, which also contain prescribed forms that must be used.

It has taken some considerable time for the private certification system to become fully operational even though the first approved inspector, the National House-Building Council (NHBC), was approved on 11 November 1985. Their original approval related only to dwellings of not more than four storeys but this was later extended to include residential buildings up to eight storeys and this was further extended in 1998 to include any buildings.

The approval of further corporate bodies as approved inspectors was held up by a number of factors, but was due mainly to the difficulty posed in obtaining the level of insurance cover which was required by the Department of the Environment. After a period of consultation new proposals for insurance requirements were agreed and these were implemented on 8 July 1996. At the same time the Construction Industry Council (CIC) was designated as the body for approving non-corporate inspectors, although the Secretary of State reserved the right to approve corporate bodies.

Three further corporate bodies were approved by the Secretary of State on 13

January 1997. Others (including in excess of 30 non-corporate approved inspectors) have continued to be approved since that date, but at present, NHBC BCS Ltd remain the only body insured to deal with speculative domestic construction (i.e. self-contained houses, flats and maisonettes built for sale to private individuals).

In this context, the DTLR issued insurance guidelines on 23 October 2001 that allowed approved inspectors to carry out their building control function on a range of different dwelling types, except so-called 'non-exempt' dwellings (see below). This definition excludes speculative dwellings constructed by housebuilding companies for sale to the public. Furthermore, from 1 March 1999 the CIC became responsible also, for the approval of corporate approved inspectors.

Further information on corporate and non-corporate approved inspectors may be obtained from The Association of Consultant Approved Inspectors, Lutyens House, Billing Brook Road, Weston Favell, Northampton, NN3 8NW or from their website: http//www.acai.org.uk.

4.2 Insurance requirements

All approved inspectors are required to carry insurance cover in accordance with a scheme approved by the Secretary of State.

Since the NHBC deals mainly with dwellings the insurance cover required is more extensive than that needed for other types of buildings. In fact the NHBC has to provide two different types of insurance policy:

- *Ten year no-fault insurance* against breaches of the Building Regulations relating to site preparation and resistance to moisture, structure, fire, drainage and heat producing appliances. The limit on cover is related to the original cost of the work allowing for inflation during the ten year period up to a maximum of 12% per annum compound.
- *Insurance against the approved inspector's liabilities in negligence* for fifteen years from the issue of the Final Certificate for each dwelling. The limit of cover is twice the cost of the building work (unless there is a simultaneous claim made under the no-fault policy), together with cover against claims made for personal injury (which is normally £100,000 a dwelling). This is also proof against inflation up to 12% compound per annum and is subject to a minimum of £1 million a site.

For corporate approved inspectors dealing with work other than dwellings it is necessary to provide professional indemnity insurance renewable on an annual basis. Additionally, a 10 year run-off period is required where the approved inspector fails to renew his policy. Indemnity has to extend to any claim reported in writing within 10 years from the acceptance of the final certificate, and claims from the owner of the work, his successors in title, or third parties must be met. Cover must be provided for claims against damage (including injury) resulting from the negligent performance by the approved inspector who issued the Initial Notice (see sections 4.7.1 and 4.7.2 below).

Unfortunately, approved inspectors covered under the above insurance scheme cannot undertake building control on dwellings. Thus, although all approved inspectors are currently approved without any direct limitation on their approvals, NHBC Building Control Services Ltd is the only approved inspector in a position to undertake building control on all types of buildings, including dwellings.

In an attempt to remedy this situation and to create wider consumer choice regarding building control services for dwellings a consultation document was issued by the Department in July 1999. This resulted in new insurance schemes which were approved during 2002, allowing approved inspectors to undertake building control work on a limited range of dwellings. This situation came about because the Department, in its covering letter to the guidelines listed a range of 'exempt' dwellings (i.e. those which are covered by the guidelines and could be dealt with by approved inspectors).

The complete list of exempt dwellings is as follows:

- dwellings in purpose groups 2(a) or (b) (Residential (Institutional) and Residential (Other)) as defined in Appendix D to Approved Document B (2000 Edition); also see Chapter 7, section 7.3;
- dwellings in purpose groups 1(a), (b) or (c) (flats, masonettes and dwelling houses)
  (1) which are being developed, for renting tenants, by a local authority, a registered social landlord, a housing association not registered with the Housing Corporation or a local housing company; or
  (2) which are being built, or created by conversion work, by or for a person on their own land and for their own occupation; or
  (3) which are flats, serving purposes that are functionally connected to one or more non-residential uses of the buildings in which they are situated, whether or not access to the flats involves passing through non-residential accommodation; or
  (4) which belong to schools, universities, hospitals or similar establishments and which are used as living accommodation for their staff, pupils or students; or
  (5) which are specifically designed for use as living accommodation for the staff, pupils or students of such establishments and which are subject to planning conditions or legal agreements restricting their use to such living accommodation.

'Non-exempt' dwellings are not covered by the guidelines and consequently approved inspectors are not able at present to provide a building control service for such dwellings. Non-exempt dwellings include primarily, speculative private development schemes for houses and flats etc.

Nothing in the new guidelines affects the existing status of approved inspectors, therefore the NHBC remains the only approved inspector permitted to carry out building control services on all types of housing developments. It is anticipated that further guidelines will need to be issued to correct this anomaly.

4.3 Approval of inspectors

Section 49 of the Building Act 1984 defines an 'approved inspector' as being a person approved by the Secretary of State or a body designated by him for that purpose. Part II of the Building (Approved Inspectors, etc.) Regulations 2000 (as amended) sets out the detailed arrangements and procedures for the grant and withdrawal of approval.

There are two types of approved inspector:

- Corporate bodies, such as the NHBC or Carillion Specialist Services Ltd.
- Individuals, not firms, who must be approved by a designated body.

Approval may limit the description of work in relation to which the person or company concerned is an approved inspector.

Approval of an inspector is not automatic. Any individual or corporate body wishing to operate as an approved inspector must satisfy several criteria. They must hold suitable professional qualifications, have adequate practical experience and carry suitable indemnity insurance. They must also be registered with a body designated for that purpose by the Secretary of State. At present this is the Construction Industry Council (CIC). The CIC established the Construction Industry Council Approved Inspectors Register (CICAIR) to maintain and operate the Approved Inspector Register in accordance with the responsibilities entailed by CIC's appointment as a designated body on 8 July 1996.

The CICAIR route to qualification for approved inspectors involves the following four stages of assessment.

4.3.1 Application

Completion of an application form and a detailed knowledge base. The knowledge base addresses six key areas as follows:

- Building Regulations & Statutory Control
- Law
- Construction Technology & Materials
- Fire Studies
- Foundation & Structural Engineering
- Building Service & Environmental Engineering.

4.3.2 Pre-qualification verification

On receipt of an application, the CICAIR Registrar will check for gaps in experience or qualification which may disqualify the application or cause delays at further stages.

4.3.3 Admissions panel

On successful completion of pre-qualification verification, the applicant becomes a *Candidate* for approved inspector. The papers are then considered by professional assessors who decide whether the candidate has demonstrated the necessary experience and knowledge to merit a *Professional Interview*. Assessors include experts nominated from across the range of disciplines by CIC members together with qualified approved inspectors.

4.3.4 Professional interview

Candidates granted a professional interview will be seen by an interview panel consisting of three assessors assisted by the CICAIR Registrar. The professional interview is the final stage of assessment and is an opportunity for candidates to demonstrate their knowledge and experience, and expand upon the information provided in their application.

Successful completion of the above assessment stages will result in the candidate being invited to register as an approved inspector. The approval will be for a period of five years. Further terms of approval may be sought.

If an applicant/candidate is unsuccessful at any stage in the assessment he/she will be given reasons and, on application to the Registrar, any advice CICAIR is able to give. Opportunities for appeals against decisions are provided.

The CIC can withdraw its approval – for example, if the inspector has contravened any relevant rules of conduct or shown that he or she is unfitted for the work.

More seriously, where an approved inspector is convicted of an offence under section 57 of the 1984 Act (which deals with false or misleading notices and certificates, etc.) the CIC may withdraw their approval. In this case the convicted person's name would be removed from the list for a period of five years. There is no provision for appeals or reinstatement.

The Secretary of State may himself withdraw his approval of any designated body, thus ensuring that the designated bodies act responsibly in giving approvals. Such action would not necessarily prejudice any approvals given by the designated body but the Secretary of State can, if he so desires, withdraw any approvals given by the designated body.

Provision is made for the Secretary of State to keep lists of designated bodies and inspectors approved by him, and for their supply to local authorities. He must also keep the lists up-to-date (if there are withdrawals or additions to the list) and must notify local authorities of these changes.

In a similar manner, designated bodies are required to maintain a list of inspectors whom they have approved. There is no express provision for these lists to be open to public inspection, although the designated body is bound to inform the appropriate local authority if it withdraws its approval from any inspector.

In approving any inspector, either the Secretary of State or a designated body may limit the description of work in relation to which the person concerned is approved. Any limitations will be noted in the official lists, as will any date of expiry of approval.

4.4 Approved persons and self-certification by competent persons

The following bodies, together with the Chartered Institution of Building Services Engineers, have been designated to approve private individuals who wish to become approved persons who can certify plans to be deposited with the local authority as complying with the energy conservation requirements:

- The Chartered Institute of Constructors
- The Faculty of Architects and Surveyors
- The Association of Building Engineers
- The Institution of Building Control Officers
- The Institution of Civil Engineers
- The Institution of Structural Engineers
- The Royal Institute of British Architects
- The Royal Institution of Chartered Surveyors

Additionally, the Institution of Civil Engineers and the Institution of Structural Engineers have been designated to approve persons to certify plans as complying with the structural requirements. Approved *persons* under section 16(9) of the Building Act should not be confused with approved *inspectors* under sections 47 to 54. As yet however, no approved persons have been designated in England and Wales although a pilot scheme for structural approvals has been operating in Scotland for some time with limited success.

4.5 Independence of approved inspectors

An approved inspector cannot supervise work in which he or she has a professional or financial interest, unless it is 'minor work'. In this context, 'minor work' means:

(a) The material alteration or extension of a dwelling-house (not including a flat or a building containing a flat) which has two storeys or less before the work is carried out and which afterwards has no more than three storeys. A basement is not regarded as a storey.
(b) The provision, extension or material alteration of controlled services or fittings (see section 2.4 above for definition of controlled services or fittings).
(c) Work involving the underpinning of a building.

Independence is not required of an inspector supervising minor work but the limitation on the number of storeys should be noted.

There is a broad definition of what is meant by having a professional or financial interest in the work, the effect of which is to debar the following:

- Anyone who is or has been responsible for the design or construction of the work in any capacity, e.g. the architect.
- Anyone who or whose nominee is a member, officer or employee of a company or other body which has a professional or financial interest in the work, e.g. a shareholder in a building company.
- Anyone who is a partner or employee of someone who has a professional or financial interest in the work.

However, involvement in the work as an approved inspector on a fee basis is not a debarring interest!

4.6 Approval of public bodies

Public bodies, such as nationalised industries, are able to supervise their own building work by following a special procedure, which is detailed in the regulations.

Regulation 21 empowers the Secretary of State (or the National Assembly for Wales) to approve public bodies for this purpose although, curiously, no criteria have been laid down as to the qualification and experience of the personnel involved. The regulation confers wide discretionary powers on the Secretary of State, but clearly approval will be limited to those bodies which may reasonably be expected to operate responsibly without detailed supervision.

4.7 Private certification procedure

The procedures which operate when using an approved inspector are illustrated in Fig. 4.1. The left column shows the steps applicable to the person carrying out the building work (i.e. the client) and the right column indicates the duties and responsibilities of the approved inspector. Joint actions are indicated in the centre column.

4.7.1 Initial notice

If the developer decides to employ an approved inspector, whether an individual or a corporate body, the first formal step in the process is for the applicant and the approved inspector jointly to give to the local authority in whose area the work is to be carried out an initial notice in the prescribed form. The purpose of the initial notice is to make the local authority aware that building work in their area is being properly controlled under the regulations, and to notify them of certain linked powers that they have under the Building Act 1984 and any local Acts of Parliament.

The initial notice must be signed by the approved inspector and the 'person intending to carry out the work'. This is the not usually the builder, but the person on whose behalf the work is being carried out (i.e. the client).

The initial notice must be in a prescribed form, and it is a contravention of the regulation to start work before the notice has been accepted. Fig. 4.2 shows a typical initial notice. It must contain:

- A description of the work.
- A declaration that an approved scheme of insurance applies to the work, which must be signed by the insurer.
- For approved inspectors approved under the procedures operated by the CIC, a copy of the notice of approval (this does not apply to those approved inspectors who were approved previously by the Secretary of State).

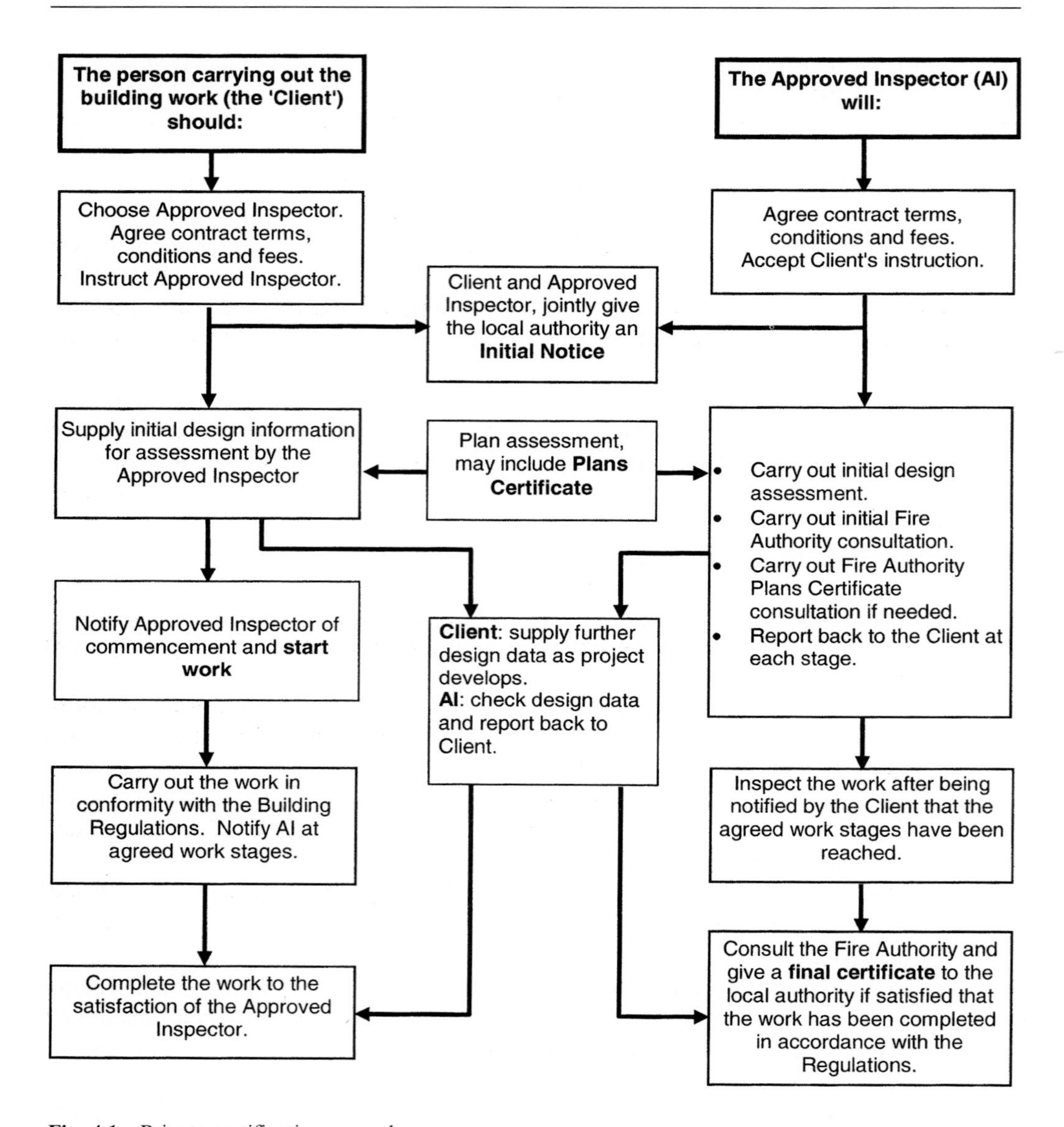

Fig. 4.1 Private certification procedure.

- In the case of a new building or extension, a site plan to a scale of not less than 1:1250 showing the boundaries and location of the site.
- Where the work includes the construction of a new drain or private sewer, a statement:
  (1) as to the approximate location of the proposed connection that is to be made to the sewer, or
  (2) where no connection is to be made to a sewer, as to the method of discharge of the proposed drain or private sewer. This could include, for example, the

location of any septic tank and secondary treatment system, or any wastewater treatment system or cesspool.

- A statement of any local legislation relevant to the work and the steps to be taken to comply with it (see section 5.8 below).
- A declaration that the approved inspector has no financial or professional interest in the work (other than for minor work).
- An undertaking that the approved inspector will consult the Fire Authority where obliged to do so.
- An undertaking that the approved inspector will consult the sewerage undertaker where obliged to do so (i.e. where the works involve building over or near a sewer).

It is essential that the initial notice is fully completed, because the local authority must reject it unless they are satisfied that the notice contains sufficient information. The local authority has five working days in which to consider the notice and may only reject it on prescribed grounds. These are:

- The notice is not in the prescribed form.
- The notice has been served on the wrong local authority.
- The person who signed the notice as an approved inspector is not an approved inspector.
- The information supplied is deficient because neither the notice nor plans show the location or contain a description of the work (including the use of any building to which the work relates).
- The approved inspector is obliged to consult the sewerage undertaker before giving a plans certificate or final certificate and the initial notice does not contain an undertaking to do so.
- Where it is intended to erect or extend a building, the local authority consider that a proposed drain must connect to an existing sewer, but no such arrangement is indicated in the initial notice.
- The initial notice is not accompanied by a copy of the approved inspector's notice of approval (this does not apply to those approved inspectors who were approved previously by the Secretary of State).
- Evidence of approved insurance is not supplied.
- The notice does not contain an undertaking to consult the fire authority (where this is appropriate).
- The notice does not contain a declaration that the approved inspector has no financial or professional interest in the work (this does not apply to minor work).
- Local legislative requirements will not be complied with.
- An earlier initial notice has been given for the work, which is still effective. This ground for rejection does not apply if:
  (1) the earlier notice has ceased to be in force and the local authority have taken no positive steps to supervise the work described in it; or
  (2) the initial notice is accompanied by an undertaking from the approved inspector who gave the earlier notice such that the earlier notice will be cancelled when the new notice is accepted.

A N APPROVED INSPECTORS LTD
INITIAL NOTICE

This notice is issued pursuant to section 47 of the Building Act 1984 ('the Act') and the Building (Approved Inspectors etc) Regulations 2000 ('the 2000 Regulations') as amended
FORM 1

To:(Note 1)
. . . .
. . . .
. . . .
. . . .

This notice relates to the following work:(Note 2)
Location: . . . .
. . . .
Description of work: . . . .
. . . .
Use of any building: . . . .

The person intending to carry out the work is:(Note 3)
Name: . . . .
Address: . . . .
. . . .
. . . . Postcode: . . . .

The Approved Inspector in relation to the work is: AN Approved Inspectors Ltd, 1 Any Street, Anytown, AB12 3CD

A declaration signed by the insurer for AN Approved Inspectors Ltd accompanies this notice and states that a named scheme of insurance approved by the Secretary of State applies in relation to the work described. AN Approved Inspectors Ltd was approved under section 3(1) of the 2000 regulations and a copy of the notice of approval accompanies this notice.

With this notice are the following documents, which are those relevant to the work described in this notice:(Note 4)

(a) in the case of the erection or extension of a building, a plan to a scale of not less than 1:1250 showing the boundaries and location of the site, and where the work includes the construction of a new drain or private sewer, an indication within that plan and/or an accompanying statement:
  (i) as to the approximate location of any proposed connection to be made to a sewer, or
  (ii) if no connection is to be made to a sewer, as to the proposals for the discharge of the proposed drain or private sewer including the location of any septic tank and associated secondary treatment system, or any wastewater treatment system or any cesspool.(Note 5)

(b) the following **Local Enactment** is relevant to the work (*delete if none applies*):
. . . .
The following steps will be taken to comply with it:
. . . .
. . . .

Fig. 4.2 Typical Initial Notice.

A N Approved Inspectors Ltd declares that:

(i) The work is/is not* minor work(Note 6)

(ii) It does not and will not while this notice is in force have any financial or professional interest in the work described.(Note 7)

(iii) It will/will not* be obliged to consult the fire authority by regulation 13 of the 2000 Regulations.(Note 8)

(iv) It undertakes that before giving a plans certificate in accordance with section 50 of the Act, or a final certificate in accordance with section 51 of the Act, it will consult the fire authority in respect of any work described above.(Note 9)

(v) It will/will not* be obliged, by regulation 13A of the AI regulations, to consult the sewerage undertaker. (*If the sewerage undertaker has to be consulted the next following declaration must be made.*)

(vi) It undertakes that before giving a plans certificate in accordance with section 50 of the Act, or a final certificate in accordance with section 51 of the Act, it will consult the sewerage undertaker in respect of any of the work described above.(Note 9)

(vii) It is aware of the obligations laid upon it by Part II of the Act and by regulation 11 of the 2000 Regulations.

(* delete whichever statement is inapplicable)

Signed .... Name ....

For and on behalf of A N Approved Inspectors Ltd Position ....

Approved Inspector Date ....

Signed .... Name ....

For and on behalf of the person intending to carry out the work(Note 10) Position ....

Date ....

Notes

(1) Insert the name and address of the local authority in whose area the work will be carried out. This should be addressed to the office of the building control department, not the general local authority address. Failure to insert the correct address may lead to delays in acceptance of this notice.
(2) Location and description of the work and the use of any building to which the work relates.
(3) Insert the name and address of the person intending to carry out the work (this will usually be the client who commissions the work, not his agents or contractors etc.).
(4) The local authority may reject this notice only on grounds prescribed by the Secretary of State. These are set out in Schedule 3 to the 2000 Regulations. They include failure to provide the relevant documents set out in this section. The documents listed in this section of the notice relevant to the work described above should therefore be sent with this notice. Any sub-paragraph which does not apply should be deleted.
(5) The design of any drainage system to which the requirements of Part H of Schedule 1 to the Building Regulations 2000 apply, and which therefore falls to be considered by the A N Approved Inspectors Ltd, will not necessarily be shown in full on the plans accompanying this notice. The plans will indicate the *location* of any connection to be made to a sewer, or the proposals for the *discharge* of any proposed drain or private sewer including the *location* of any septic tank and associated secondary treatment system, or any wastewater system or any cesspool.
(6 'Minor work' has the meaning given in regulation 10(1) of the 2000 Regulations. If the work is **not** minor work, the next following declaration **must** be made.
(7) 'Professional or financial interest' has the meaning given in regulation 10 of the 2000 Regulations.
(8) If the inspector is obliged to consult the fire authority, the next following declaration **must** be made.
(9) Delete this statement if it does not apply.
(10) The person intending to carry out the work will usually be the client who commissions the work, not his agents or contractors etc.

Fig. 4.2 (*Contd*).

If the local authority does not reject the initial notice within five working days (beginning on the day the notice is given to the local authority) it is presumed to have accepted it without imposing requirements. Therefore, an initial notice comes into force when it has been accepted by a local authority (or is deemed to have been accepted by the passing of five days). So long as the initial notice remains in force, the function of enforcing the building regulations, which is conferred on a local authority under the Building Act 1984, is not exercisable in relation to the work described in the initial notice.

Generally, the initial notice remains in force during the currency of the works. However, in certain circumstances, it may be cancelled, or cease to have effect after the lapse of certain defined periods of time where there has been a failure to give a final certificate to the local authority. The time periods depend on the circumstances, but the position may be summarised as follows:

- If a final certificate is rejected – four weeks from the date of rejection.
- Where there is a failure to give a final certificate:
  (1) Eight weeks from the date of occupation for the erection, extension or material alteration of a building. This period of time is reduced to four weeks where the building is to be put to a designated use under the Fire Precautions Act 1971, section 1, or will be a workplace subject to Part II of the Fire Precautions (Workplace) Regulations 1997.
  (2) Eight weeks after the change of use takes place where the work relates to a material change of use.

The local authority is given power to extend these time periods.

It is possible to give a final certificate for part of a building or extension if it is needed to be occupied before overall completion of the project. In these circumstances the initial notice is not cancelled but remains in force until final completion of the work.

Sometimes it may be necessary to vary work which is the subject of an initial notice (e.g. it may be decided to change the number of units being erected). In such circumstances the person who is carrying out the work and the approved inspector should give an amendment notice to the local authority.

There is a prescribed form for an amendment notice and it must contain the information which is required for an initial notice (see above) plus either:

- a statement to the effect that all plans submitted with the original notice remain unchanged; or
- amended plans are submitted with the amendment notice plus a statement that any plans not included remain unchanged.

The local authority has five working days in which to accept or reject the notice and it may only reject it on prescribed grounds. The procedure is identical to that for acceptance or rejection of an initial notice.

4.7.2 Cancellation of initial notice

In the following cases, the approved inspector must cancel the initial notice by issuing to the local authority a cancellation notice in a prescribed form. The grounds on which the initial notice must be cancelled are:

- The approved inspector has become or expects to become unable to carry out (or continue to carry out) his functions.
- The approved inspector believes that because of the way in which the work is being carried out he cannot adequately perform his functions.
- The approved inspector is of the opinion that the requirements of the Regulations are being contravened and despite giving notice of contravention to the person carrying out the work that person has not complied with the notice within the three-month period allowed (Approved Inspector Regulations, regulation 19).

It is also possible for the person carrying out the work to cancel the initial notice. This arises if it becomes apparent that the approved inspector is no longer willing or able to carry out his functions (through bankruptcy, death, illness etc.). This must be done in the prescribed form and must be served on the local authority and (where practicable), on the approved inspector.

Alternatively, it is possible for the person carrying out the work to give a new initial notice jointly with a new approved inspector, provided that the new notice is accompanied by an undertaking by the original approved inspector that he will cancel the earlier notice as soon as the new notice is accepted. Once the initial notice has ceased to have effect, the approved inspector will be unable to give a final certificate and the local authority's powers to enforce the Building Regulations can revive. In this case the local authority becomes responsible for enforcing the regulations and it must be provided on request with plans of the building work so far carried out. Additionally, it may require the person carrying out the work to cut into, lay open or pull down work so that it may ascertain whether any work not covered by a final certificate contravenes the regulations.

If it is intended to continue with partially completed work, the local authority must be given sufficient plans to show that the work can be completed without contravention of the Building Regulations. A fee, which is appropriate to that work, will be payable to it.

Where the work covered by the initial notice has not commenced within three years from the date on which the initial notice was accepted, the local authority may (not must) cancel the initial notice.

4.7.3 Functions of approved inspectors

The fees payable to an approved inspector are a matter for negotiation; there is no prescribed scale. The functions which an approved inspector must carry out are specified and detailed in regulation 11 of The Building (Approved Inspectors, etc.) Regulations 2000, and his obligation is to 'take such steps (which may include the making of tests of building work and the taking of samples of material) as are reasonable to enable him to be satisfied within the limits of professional care and skill that' specified requirements are complied with. An approved inspector is liable for negligence and it is suggested that he *must* inspect the work to ensure compliance, in contrast to local authorities who have a discretion as to whether or not to inspect.

In *NHBC Building Control Services Ltd* v. *Sandwell Borough Council* (1990) 50 BLR 101 the Divisional Court emphasised that regulation 11 does not require a system of individual inspection of every detail covered by the substantive requirements of the regulations. In principle, random sampling is sufficient, although in case of dispute it is for the approved inspector to show that adopting a system of random or selective sampling is a satisfactory way of discharging his duties.

The approach of the court to this important matter was indicated by Lord Justice Leggatt:

> 'Any system of inspection that is selective involves consideration not only of the importance of a risk against which the inspection is designed to guard, but of the likelihood of its occurrence. In my judgement the justices' conclusion that the [approved inspector's] system is an inadequate precaution is not one that can properly be based solely upon the fact that the risk was obvious and potentially fatal. That amounts to saying that failure in relation to an individual house to detect the absence of rockwool in the gap between the ceiling and wall of its garage could not have occurred unless the system was inadequate or the inspector had shown want of professional skill and care in operating the system. But the liability imposed is not absolute. The system has been impliedly approved by the Secretary of State. In the light of its experience the [inspector] determines the extent and closeness of the inspections to be conducted in respect of the work of any particular builder. Inherent in any selected system is the risk that some defects may escape detection. Except [for] the fact that the defect ... was not spotted, there is no criticism to be made of the system. It follows that the mere fact that an important defect escaped detection in a particular instance cannot ... constitute a proper basis for concluding beyond reasonable doubt that there was any failure to undertake the functions of supervision so as to render false the statement that the [inspector] had performed those functions.'

The approved inspector may arrange for plans or work to be inspected on his behalf by someone else (although only the approved inspector can give plans or final certificates), but delegation does not affect any civil or criminal liability. In particular, the 1984 Act states that:

> 'an approved inspector is liable for negligence on the part of a person carrying out an inspection on his behalf in like manner as if it were negligence by a servant of his acting in the course of his employment'.

The approved inspector must be satisfied that:

- The requirements relating to building work, material change of use, and materials and workmanship specified in Regulations 4, 6 and 7 of the 2000 Regulations are complied with.

- Regulation 12 relating to energy ratings for dwellings is complied with. This means that the person carrying out the building work must supply a SAP energy rating to the approved inspector not more than 5 days after completion of the dwelling (see also section 3.6 above).
- Regulation 12A relating to sound insulation testing is complied with. This means that the person carrying out the building work must supply a copy of the sound insulation testing results (see Chapter 10 Sound insulation) to the approved inspector not more than five days after completion of the work to which the initial notice relates.

Where cavity wall insulation is inserted, the approved inspector need not supervise the insulation work, but is required to state in his final certificate whether or not the work has been carried out.

Consultation with the fire authority

Where an initial notice or an amendment notice is to be given (or has been given) in relation to the erection, extension, material alteration or change of use of a building which:

(1) is to be put to a designated use under the Fire Precautions Act 1971, section 1, or
(2) will be a workplace subject to Part II of the Fire Precautions (Workplace) Regulations 1997,

and the Building Regulations 2000 (SI 2000/2531), Schedule 1, Part B (Fire safety) also applies, the approved inspector is required, before or as soon as practicable after giving the notice, to consult the fire authority. He must give them sufficient plans, and/or other information to show that the work described in the notice will comply with the applicable parts of the Building Regulations 2000, Schedule 1, Part B and must have regard to any views they express.

Additionally, before giving a plans certificate or final certificate to the local authority the approved inspector must allow the fire authority 15 working days to comment, and have regard to the views they express. Some local Acts of Parliament also impose extensive fire authority consultation requirements. The approved inspectors must undertake any consultation required by local legislation.

Consultations with the sewerage undertaker

Where an initial notice or amendment notice is to be given (or has been given) and it is intended to erect, extend or carry out underpinning works to a building within 3 m of the centreline of a drain, sewer or disposal main to which the Building (Amendment) Regulations 2001, Schedule 1, paragraph H4 applies, the approved inspector must consult the sewerage undertaker. The procedures and time periods involved parallel those described for fire authority consultations above.

4.8 Plans certificates

A plans certificate is a certificate issued by an approved inspector certifying that the design has been checked and that the plans comply with the 2000 Regulations. Its issue is entirely at the option of the person carrying out the work, and is issued by the approved inspector to the local authority and the building owner.

If the approved inspector is asked to issue a plans certificate and declines to do so on the grounds that the plans do not comply with the building regulations, the building owner can refer the dispute to the Secretary of State for a determination. A plans certificate can be issued at the same time as the initial notice or at a later stage, provided the work has not been carried out. There are two prescribed forms of plans certificate. There are three preconditions to its issue:

- The approved inspector must have inspected the plans specified in the initial notice.
- He must be satisfied that the plans are neither defective nor show any contravention of the regulation requirements.
- He must have complied with any requirements about consultation, etc.

If a plans certificate is issued and accepted and, at a later stage, the initial notice ceases to be effective, the local authority cannot take enforcement action in respect of any work described in the plans certificate if it has been done in accordance with those plans.

The local authority has five working days in which to reject the plans certificate, but may only do so on certain specified grounds:

- The plans certificate is not in the prescribed form.
- It does not describe the work to which it relates.
- It does not specify the plans to which it relates.
- Unless it is combined with an initial notice, that no initial notice is in force.
- The certificate is not signed by the approved inspector who gave the initial notice or that he is no longer an approved inspector.
- The required declaration of insurance is not given.
- There is no declaration that the fire authority has been consulted (if appropriate).
- The approved inspector was obliged to consult the sewerage undertaker before giving the certificate, but the certificate does not contain a declaration that he has done so.
- There is no declaration of independence (except for minor work).

When combined with an initial notice, the grounds for rejecting an initial notice specified in Schedule 3 (see section 4.7.1) also apply.

Plans certificates may be rescinded by a local authority if the work has not been commenced within three years from the date on which the certificate was accepted.

4.9 Final certificates

The final certificate should be issued by the approved inspector when the work is completed, but curiously there are no sanctions against an approved inspector who fails to issue a final certificate. The final certificate need not relate to all the work covered by the initial notice; it can, for example, be given in respect of part of a building which complies with the 2000 Regulations, or one or more of the houses on a development covered by an initial notice. Once given and accepted the initial notice ceases to apply.

It is to be issued, in a prescribed form, where an approved inspector is satisfied that any work specified in an initial notice given by him has been completed and certifies that 'the work described ... has been completed' and that the inspector has performed the functions assigned to him by the regulations. If the local authority do not reject the final certificate within ten working days they are deemed to have accepted it. A final certificate can only be rejected on limited grounds. These are:

- The certificate is not in the prescribed form.
- It does not describe the work to which it relates.
- No initial notice relating to the work is in force.
- The certificate is not signed by the approved inspector who gave the notice or he is no longer an approved inspector.
- The required declaration of insurance is not provided.
- There is no declaration of independence (except for minor works).

Once the final certificate is accepted by a local authority its powers to take proceedings against a person for contravention of building regulations in relation to the work referred to in the final certificate, are cancelled.

4.10 Public body's notices and certificates

Part VII of the Building (Approved Inspectors, etc.) Regulations 2000 is concerned with public bodies and, read in conjunction with section 54 of the Building Act 1984, its effect is to enable designated public bodies to self-certify their own work.

Public bodies are approved by the Secretary of State, and the regulations, relating to notices, consultation with the fire authority, plans certificates and final certificates mirror those of Part III dealing with approved inspectors. The grounds on which the local authority may reject a public body's notice, etc., mirror those applicable to private certification, except that:

- There is no provision for cancellation of a public body's notice.
- There is no requirement that there should be an approved insurance scheme in force.

4.11 Prescribed forms

Twelve prescribed forms are set out in Schedule 2 of the Building (Approved Inspectors, etc.) Regulations 2000. Regulation 2(2) provides that where the regulations require the use of one of the numbered forms set out in Schedule 2, 'a form substantially to the like effect may be used'. Approved inspectors, public bodies, and local authorities, etc., may therefore have their own forms printed, provided they follow the precedents laid down in Schedule 2.

5 Other legislation affecting health and safety

5.1 Introduction

Chapters 1 and 2 of this book contain information which should enable a person intending to carry out building work to assess whether or not the Building Regulations apply to that work. It will be seen that there are circumstances where the building or the work itself may be exempt from control. Even if this is the case, there may be legislation, other than the Building Regulations, which does apply.

This chapter provides a brief guide to a range of Acts of Parliament and regulations which might affect a building project, whether or not the Building Regulations also apply to that project. In practice there is a great deal of legislation which affects building development. The Acts and regulations covered in this chapter are those most commonly encountered concerning public health and safety, therefore town planning and conservation area issues, and legislation specific to a particular type of development (such as caravan sites, etc.) are not covered.

5.2 Conflicting statutory requirements

Some legislation (such as the Building Regulations) applies to a building when it is being designed and constructed whereas other different legislation will apply to it when it is being used (e.g. the Fire Precautions Act 1971). Therefore, it is possible that two (or more) pieces of legislation, (perhaps enforced by different authorities), might apply to the same building or work. In order to avoid conflicting requirements applying to a building it is often the case that a 'statutory bar' applies whereby one piece of legislation takes precedence over another. Often, conflicts are avoided by the provision in Acts of Parliament or regulations for consultation to take place between the different enforcing authorities (e.g. see section 4.7.3 above). Therefore, although a project may be exempt from the need to comply with the Building Regulations, it is possible that this may nullify the effect of a statutory bar and hence result in the need to comply with a different piece of legislation (such as the Fire Precautions Act 1971).

The Acts of Parliament and regulations referred to below apply at different stages in the normal life-cycle of the building and are often associated with the age and condition of the building. This is shown in Fig. 5.1.

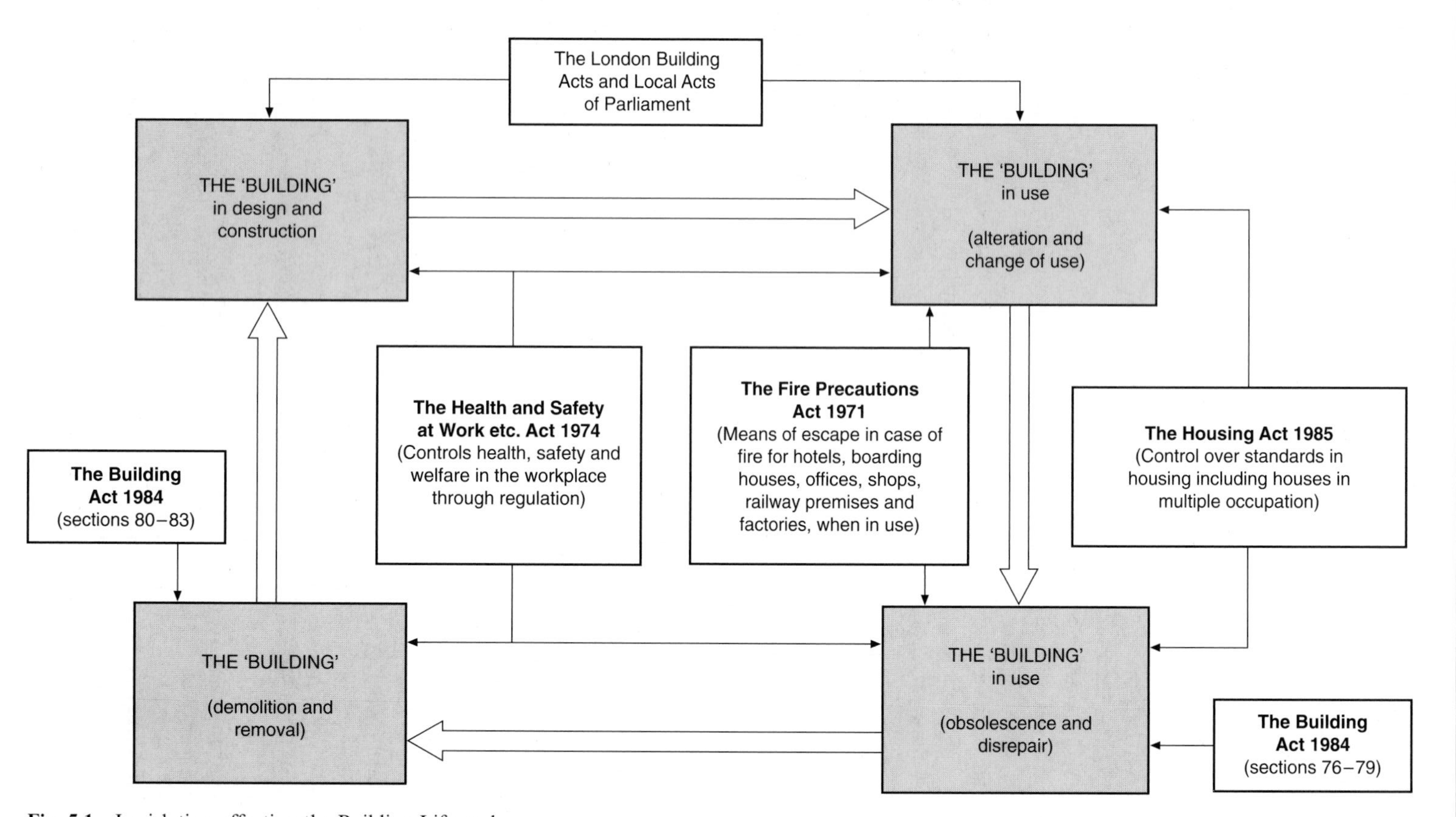

Fig. 5.1 Legislation affecting the Building Life-cycle.

5.3 The Fire Precautions Act 1971

This Act is mainly concerned with providing and maintaining safe means of escape from fire in existing buildings when they are being used. Under the Act a fire certificate is required for certain designated premises. These are:

- Hotels or boarding houses where sleeping accommodation is provided for more than six staff or guests (or some sleeping accommodation is provided above the first floor or below the ground floor); and
- Factories, offices, shops and railway premises where more than 20 people are employed or more than ten people work other than on the ground floor or in factories only where explosive or highly flammable materials are stored or used.

Since it is possible that an existing building may be put to a designated use where alterations are unnecessary, it may be the case that a fire certificate is needed. Accordingly, application must be made to the fire authority for the area in which the building is located (and which is the body which enforces the Act) in the prescribed form in accordance with the requirements of section 5 of the Act. (Of course, where work is being carried out which must comply with the Building Regulations, the local authority or approved inspector will deal with the application and carry out the necessary consultations with the fire authority. There is no need for a separate application to be made to the fire authority).

5.4 The Fire Precautions (Workplace) Regulations 1997

The Fire Precautions (Workplace) Regulations were made to fill a perceived deficiency in current legislation in implementing the general fire safety provisions of the European Framework and Workplace Directives. The Regulations require the provision of minimum fire safety standards in workplaces and impose duties on employers and on others in control of places of work. They apply to all workplaces including those covered by other fire-specific fire safety legislation (such as those for which a Fire Certificate is in force, or has been applied for under the Fire Precautions Act 1971).

Usually, employers must undertake a fire risk assessment of the workplace to establish what precautions are necessary to ensure the safety of employees in the event of fire. This could include, for example:

- The provision of fire fighting equipment, fire detectors and alarms.
- Making sure that fire fighting equipment is readily accessible, easy to use and appropriately signed.
- Ensuring that properly trained employees are nominated to put the necessary fire safety measures into practice.
- Making sure that adequate arrangements are made for contacting the emergency services.

- Providing suitable emergency routes and exits.
- Organising a suitable maintenance system so that all equipment is kept in good working order.

Fire authorities are required to enforce the Regulations in their area and they are empowered to inspect premises at any time to ensure compliance with the Regulations.

5.5 The Housing Act 1985

This Act allows housing authorities to carry out specified work in order to make houses fit for occupation. For example, houses in multiple occupation (i.e. those in which the occupants do not form part of a single household) are covered by sections 352 to 394 of the Act and among other things, the housing authority can insist that works are carried out which result in the provision of:

- Adequate storage accommodation.
- Food preparation and cooking facilities.
- Adequate toilets, baths, showers and wash hand basins with hot and cold water supplies.
- Suitable means of escape in case of fire and other fire precautions.

5.6 The Party Wall Act 1996

The Party Wall Act applies where work is being carried out to a party wall (which may be part of a building), or a boundary wall.

The Act applies in three main areas:

- It provides a method whereby negotiations may take place over the construction of new party walls.
- It deals with the situation where there is a need to excavate below the level of a neighbouring building or structure which is within 6 m of the boundary.
- It allows owners to carry out work on party walls if appropriate notices have been served.

Any disputes which arise between the parties may be dealt with by means of mechanisms in the Act which allow for the appointment of surveyors to act as adjudicators.

5.7 The London Building Acts 1930 to 1982

Originally, the Building Regulations did not apply to Inner London which continued to be dealt with by the Greater London Council under the London Building

Acts 1930 to 1982 and the building byelaws made under them. This was altered on 6 January 1986 when the Building (Inner London) Regulations came into operation. Following the abolition of the Greater London Council on 1 April 1986, its building control functions and those of district surveyors under the London Building Acts were transferred to the Common Council of the City of London and the 12 Inner London borough councils.

Certain transitional provisions were made, but as a result of the new regulations, Inner London building control procedures became essentially the same as elsewhere in England and Wales, since all the London building byelaws and many sections (but not all) of the London Building Acts were repealed, and other sections were amended.

The most important provisions of the London Building Acts which were retained are listed below:

- *Buildings in excess height and cubical content*
  Section 20 of the London Building Acts (Amendment) Act 1939 applies special fire precautions in high buildings, i.e. a building which has a storey at a greater height than 30 m (or 25 m if the area of the building exceeds 930 m^2), or is a large building, or warehouses (over 7100 m^3), by requiring that they be divided up by division walls as defined in the 1939 Act. A wide range of fire protection measures can be required, and there are extra requirements for areas of 'special fire risks' such as boiler rooms.

  Plans must be deposited before any alterations are made to buildings of excess height or cubical content, and the borough council must consult with the London Fire Authority before issuing consent.

- *Uniting of buildings: 1939 Act section 21*
  Local authority consent is required if two buildings are united by making an opening in a party wall, or external wall if access is obtained between the buildings without passing into the external air. This does not apply if the buildings are in one ownership and if united would comply with the London Building Acts.

- *Special and temporary structures: 1939 Act sections 29 to 31*
  These sections deal with the erection and retention of certain temporary buildings which need the consent of the local authority and are not covered by other legislation.

- *Dangerous and neglected structures: 1939 Act sections 60 to 70*
  The procedure includes the service of a dangerous structure notice and is far more efficacious than the procedures in sections 77 and 78 of the Building Act 1984 (which do not apply to Inner London).

5.8 Local Acts of Parliament

Although the Building Act 1984 attempted to rationalise the main controls over buildings, there are in fact a great many pieces of local legislation with the result that many local authorities have special powers relevant to building control.

Where a local Act is in force, its provisions must also be complied with, since many of these pieces of legislation were enacted to meet local needs and perceived deficiencies in national legislation. The Building Regulations make it clear that local enactments must be taken into account.

With the growth and development of Building Regulation control over fire precautions in particular, it is likely that most of the current local legislation is now outdated or has been superceded by the Building Regulations. In fact, some local enactments already contain a statutory bar which gives precedence to building regulations.

Local authorities are obliged by section 90 of the Building Act 1984 to keep a copy of any local Act provisions and these must be available for public inspection free of charge at all reasonable times.

A full list of local Acts of Parliament may be found in an Appendix to this book where it will be seen that the most common local provisions relating to building control are:

- *Special fire precautions for basement garages or for large garages*
  The usual provision is that if a basement garage for more than three vehicles or a garage for more than twenty vehicles is to be erected, the local authority can impose access, ventilation and safety requirements.

- *Fire precautions in high buildings or for large storage buildings*
  There must be adequate access for the fire brigade in certain high buildings. A high building is one in excess of 18.3 m and the local authority must be satisfied with the fire precautions and may impose conditions, e.g. fire alarm systems, fire brigade access etc. (In many cases these requirements have been superseded by Part B of Schedule 1 to the Building Regulations 2000.) Large storage buildings in excess of 14,000 m^3 are required to be fitted with sprinkler systems by some local Acts.

- *Extension of means of escape provisions*
  The Building Act 1984, section 72, is a provision under which the local authority can insist on the provision of means of escape where there is a storey which is more than 20 ft above ground level in certain types of buildings, e.g. hotels, boarding houses, hospitals, etc. Local enactments replace the 20 ft by 4.5 m and make certain other amendments to the national provisions.

- *Drainage systems*
  In some cases, local legislation requires that every building must have separate foul and surface water drainage systems. In the light of the new requirement H5 (separate systems of drainage) in Part H of Schedule 1 to the Building Regulations 2000, the Office of the Deputy Prime Minister (ODPM) is currently carrying out a consultation with a view to repealing certain Local Acts which duplicate this provision. They are:
  - the East Ham Corporation Act 1957: s 38
  - the Croydon Corporation Act 1960: s 79
  - the South Yorkshire Act 1980: s 39

- the Staffordshire Act 1983: s 18
- the Leicestershire Act 1985: s 30
- the West Yorkshire Act 1980: s 50.

- *Safety of stands at sports grounds*
  In many areas, local Acts impose controls over the safety of stands at sports grounds. Again much of this local legislation has been largely superseded by the provisions of the Fire Safety and Safety of Places of Sport Act 1987.

5.9 The Highways Act 1980

The Highways Act deals with the creation and control of highways, and the rights and duties of people who use them. Apart from trunk roads which are the responsibility of the Minister of Transport, the controlling authority will be the county council (or borough council in Greater London).

Building work on or immediately adjacent to a highway may present risks to the health and safety of people using it, therefore certain sections control such building work to the extent indicated below:

- *Builder's skips:* under sections 139 and 140 local authority permission is required to deposit these on the highway and it can impose certain conditions as to lighting and signing.
- *Dangerous land* adjoining a highway and building operations affecting public safety are dealt with under sections 165 and 168 respectively.
- *Building materials* that come into contact with the highway and might damage it, or be a danger to the public, are dealt with under section 170 (mixing of mortar, etc., on the highway), and section 171 (deposit of building materials).
- *Temporary structures* such as scaffolding (section 169) and hoardings (sections 172 and 173) are controlled by a system of licensing administered by the local authority.
- Control is also exercised by the local authority over the construction of buildings:
  (1) Over a highway, sections 176 to 178, and
  (2) Under a highway, sections 179 and 180.
- The provision of vehicular or pedestrian access to premises necessitating the construction of a carriage crossing, section 184.

5.10 The Water Industry Act 1991

Under the Water Industry Act 1991, water undertakers have responsibilities with regard to the provision of sewers, drains, discharges and drinking water. The following sections illustrate the duties of the water undertakers and the rights of building owners and occupiers under the Act and are of importance to building development:

- Duty to provide a system of public sewers, section 94.
- Requisition of a sewer from the sewerage undertaker, section 98.
- Adoption by the sewerage undertaker at the request of the owner, section 102.
- Agreement to adopt sewers at a future date, section 104.
- Owners and occupiers rights to connect to a public sewer, sections 106 to 108.
- Powers of a sewerage undertaker to require a drain or sewer to be constructed so as to form part of a general system of drainage, section 112.
- Duty to supply water for domestic purposes, section 52.
- Duty to make connections with a water main, section 45.
- The supply of water for non-domestic premises, section 55.

The old Water Byelaws were replaced by the Water Regulations made under section 74 of the Water Industry Act, on 1 July 1999.

5.11 The Clean Air Acts 1956 and 1993

The Clean Air Act deals, among other things, with the control of atmospheric pollution from furnaces and heating plant. With specific reference to building work, local authorities are empowered to control the height of chimneys for furnaces, and the choice of fuels which may be burnt.

5.12 The Environmental Protection Act 1990

The Environmental Protection Act (as amended by the Environment Act 1995) is the principal statute dealing with the collection and disposal of waste. A duty is placed on waste collection authorities, (primarily District Councils and London Boroughs) by virtue of section 45 of the 1990 Act, to collect all household waste in their areas. (Isolated areas, where other adequate disposal arrangements have been made, can be excepted). Additionally, they must collect commercial waste if requested to do so by the occupier of the premises and they may make a charge. They do not have to collect industrial waste but may do so with the consent of the relevant waste disposal authority, and they may charge for this service also.

Under sections 46 and 47, the waste disposal authorities may require the provision of suitable waste receptacles for household, commercial or industrial waste.

5.13 The Health and Safety at Work etc. Act 1974

This is the principal Act which deals with securing the health, safety and welfare of people at work and of others whose health and safety may be affected by work activities. The Act sets out certain general principles for health and safety at work, the actual details being contained in regulations which deal with particular situations and practices at work. Those of most relevance to building operations are set out below. Overall, the Act is enforced by the Health and Safety Executive, however many of its duties are carried out by local authorities (i.e. those which apply to work involving non hazardous operations).

5.14 The Workplace (Health, Safety and Welfare) Regulations 1992

Made under the Health and Safety at Work etc. Act 1974 the regulations implement provisions of European Workplace Directive 89/654/EEC. A general duty is placed on employers to ensure that workplaces comply with the requirements of the regulations including provisions for:

- Environmental measures such as ventilation, temperature control, lighting levels and adequacy of room dimensions.
- General welfare including cleanliness, sanitary conveniences and washing facilities.
- Safety measures such as the use of windows, doors, stairs, ladders and ramps, etc.

There is a connection with the Building Regulations in that new buildings which follow the guidance in the Approved Documents will, in most cases satisfy the requirements of the Workplace Regulations which, of course, apply to workplaces when they are being used.

5.15 The Construction (Design and Management) Regulations 1994

These regulations were made under the Health and Safety at Work etc. Act 1974 and implement provisions of European Directive No. 89/654/EEC. They apply to construction work, in a broad sense, and require the preparation of, and adherence to, a health and safety plan for the project by imposing duties on the client, the planning supervisor, the designer, the main contractor and on other contractors involved in the work.

The regulations do not normally apply to householders having work carried out to their own residences, or to specified small construction works.

5.16 The Construction (Health, Safety and Welfare) Regulations 1996

Made under the Health and Safety at Work etc. Act 1974, the regulations implement European Directive No. 92/57/EEC on minimum health and safety requirements for temporary or mobile construction sites for building, civil engineering or engineering construction. General duties are imposed on employers, the self-employed and others who control the way in which work is carried out. Employees are also seen to be responsible for their own actions. Enforcement is by the Health and Safety Executive.

The requirements deal with:

- The provision of safe work places (e.g. safe access and egress, sufficient and suitable working space).
- Measures to prevent falling (including through fragile materials).
- Means to ensure the stability of structures.
- Means of ensuring safe methods of demolition or dismantling.
- The safety of excavations.
- Safe traffic routes about the site.
- The provision of safety devices on doors and gates.
- Measures to reduce the risk of fire.
- The provision of fire detection and fire fighting equipment, and emergency escape routes.
- The provision of welfare facilities such as sanitary conveniences, washing facilities, drinking water, rest areas, etc.

5.17 The Building (Local Authority Charges) Regulations 1998

These regulations (which replace the Building (Prescribed Fees) Regulations 1994) authorise local authorities to fix and recover charges for the performance of their Building Regulations control functions according to a Scheme governed by principles prescribed in the regulations.

The Charges Regulations (which do not apply to Approved Inspectors) therefore have the effect of making each local authority responsible for fixing the amount of its own Building Regulations control charges for the following functions:

- A plan charge for the passing or rejection of plans of proposed work deposited with them.
- An inspection charge for the inspection of works in progress.
- A building notice charge where the building notice procedure applies. (This is the sum of the plan and inspection charges and usually becomes payable when the notice is given to the local authority, however this may be varied with the consent of the local authorities, see below.)
- A reversion charge where the approved inspector system is used and the initial notice is cancelled so that control reverts to the local authority (see section 4.7.2 above).
- A regularisation charge for unauthorised building work (see Regulation 21, section 2.13 above).

Local authorities have complete freedom to set the levels of the charges they impose, however the income derived from the charges over a continuous period of three years (commencing on the date on which the LA fixes its charges) must be not less than the cost directly or indirectly (the 'proper costs') incurred in accordance with proper accounting principles. In fact, local authorities are required to estimate the aggregate of their proper costs over the three year period before fixing their charges and they must issue a statement at the end of each financial year which sets out the details of the charging scheme and shows the amount of the income and proper costs. The following points should also be noted:

- Local authorities are not permitted to include any costs in relation to the control of work which is solely for the benefit of disabled people (they are not permitted to levy a charge for such work).
- Before bringing its charges into effect a local authority must give at least seven days' notice, although the charges can be amended whenever necessary provided the due notice is given on each occasion.
- Where the proper costs do not exceed £450,000 over the three year period or where at least 65% of the charges received are connected with work to small domestic buildings such as extensions, garages and car ports, the income derived must cover at least 90% of costs.
- Charges for work may, with the agreement of the local authority, be paid in instalments.
- Reduced charges may be levied for repetitive building work, where either it is part of the same submission (such as the construction of multiple housing units on a single estate), or it is for similar work on different sites but submitted by the same person (such as work to terrace dwellings under a refurbishment scheme).

Charges for work comprising

- The erection of small domestic buildings (up to 300 m^2 in floor area) and up to three storeys high, or
- Small detached garages and car ports with floor area up to 40 m^2, *or*
- Domestic extensions (including associated access work) with floor areas up to 60 m^2,

must be applied by reference to the floor area of the building or extension concerned. Where it is intended to erect a number of extensions to a building their floor areas must be aggregated.

All other types of work are subject to charges which must be related to an estimate of the cost of the work and local authorities are permitted to forego their inspection charge if the value of the work is less than £5000; however, a local authority may not charge for building work which comprises:

- The installation of cavity fill material, or
- The installation of unvented hot water systems,

where this work forms part of other building work and is being carried out in accordance with Parts D and G respectively of Schedule 1 to the Building Regulations 2000.

Charges are also payable where application is made to the Secretary of State for determination of questions under sections 16 and 50 of the Building Act 1984 (see section 2.15).

The Charges Regulations came into force on 1 April 1999 and prior to that date all submissions were subject to the Building (Prescribed Fees) Regulations 1994. In order to achieve some form of national consistency over levels of charges, the Local Government Association has published a 'model scheme' which local authorities are urged to follow.

5.18 The Construction Products Regulations 1991

These regulations (as amended) apply to any construction products which are produced for incorporation in a permanent manner in construction works, and which were supplied after 27 December 1991.

The regulations are designed to ensure that when products are used in construction work, the work itself will satisfy any relevant 'essential requirements' of the Construction Products Directive (see section 8.3 below). Products bearing the CE marking are presumed to satisfy the requirements of the regulations (see section 8.4.2 below).

5.19 The Gas Safety (Installation and Use) Regulations 1998

The Gas Safety Regulations (as amended) are concerned with controlling the risks which arise when using gas from either mains pipes or gas storage vessels. The work must be carried out by a competent person who is a member (or is employed by a member) of the Council for Registered Gas Installers (CORGI).

Additionally:

- Gas appliances and the pipes which supply them, must be maintained in a safe condition if they are situated in workplaces covered by the regulations.
- There are restrictions placed on certain kinds of gas appliances used in bathrooms and sleeping accommodation, or in cupboards or compartments in such rooms.
- Gas appliances and pipework installed in rented accommodation must be maintained in a safe condition (and maintenance records kept).
- Gas appliances, fittings, pipework and flues must be installed safely and checks carried out to ensure compliance with the regulations.
- Gas fires must be installed in a safe position.
- No alterations must be carried out to any premises which would adversely affect safety or cause a contravention of the regulations.
- LPG storage vessels and LPG fired appliances fitted with automatic ignition devices or pilot lights must not be installed in cellars or basements.

5.20 The Disability Discrimination Act 1995

The purpose of the Act is to end the discrimination faced by disabled people in the areas of employment, the provision of goods, facilities and services and in the renting or buying of property or land.

From the viewpoint of construction work, after 2004 service providers will have to ensure that there are no physical features which make it unreasonably difficult for disabled people to use their services. For example, this may involve installing accessible entrances, lifts, etc., in existing buildings.

II Technical

6 Structural stability

6.1 Introduction

Part A of Schedule 1 to the 2000 Regulations is concerned with the strength, stability and resistance to deformation of the building and its parts. The loads to be allowed for in the design calculations are specified, and recommendations as to construction are given in Approved Document A.

In line with the Government's intention to remove from the regulations those matters which are not directly concerned with public health and safety or the conservation of fuel and power, the previous requirements regarding the ability of a building structure or foundation to resist *damage* due to settlement, etc., have been omitted.

Additionally, control of deflection or deformation of the building structure under normal loading conditions is only relevant if it would impair the stability of another building.

It is conceivable, therefore, that a building constructed under the regulations could be safe and stable but could settle and deflect to such an extent that it would be unusable. In that event, of course, the owner would probably have redress against the designer and/or builder under the general law by way of an action for damages. Insurance cover might be somewhat hard to obtain for such a building.

The section of the regulations dealing with disproportionate collapse was simplified by the revocation (in the 1994 amendment) of paragraph A4 of Schedule 1. This was concerned with maintaining structural stability in public buildings and shops in the event of roof failures, where roof spans exceeded nine metres.

The original requirements concerning the failure of long span roof structures were introduced in response to a number of roof collapses which occurred in the 1970s and which led to the banning of high alumina cement in structural work. As this ban still exists the problem of such failures seems largely to have been solved without additional regulatory safeguard.

A further change introduced by the 1991 Regulations concerns the application of the structural requirements to certain buildings subject to a material change of use (see section 2.4 regulations 5 and 6). In these cases it is necessary to carry out structural appraisals of the existing buildings to see if they are capable of coping with the changed loading conditions necessitated by the change of use. Guidance concerning this may be found in the following documents:

- BRE Digest 366: *Structural appraisal of existing buildings for change of use.*
- The Institution of Structural Engineers Report, *Appraisal of existing structures*, 1980.

The Institution of Structural Engineers' report contains an item on design checks where a choice of various partial factors should be made to suit the individual circumstances of each case. Since the report was published in 1980 many of the BS Codes and Standards quoted will have been revised. ADA1/2 advises that the latest versions of these documents should be used.

6.2 Loading

Buildings must be constructed so that all dead, imposed and wind loads are sustained and transmitted to the ground:

- safely; and,
- without causing such settlement of the ground, or such deflection or deformation of the building, as will impair the stability of any other building.

The imposed and wind loads referred to above are those to which the building is likely to be subjected in the normal course of its use and for the purpose for which it is intended.

6.3 Ground movement

In addition to the provisions of paragraph A1 above regarding loading, there are requirements in paragraph A2 of Schedule 1 that the building shall be so constructed that movement of the ground caused by:

- swelling, shrinking or freezing of the subsoil; or
- landslip or subsidence (other than subsidence arising from shrinkage),

will not impair the stability of any part of the building.

It should be noted that the requirement as to landslip and subsidence applies to the extent that the risk can be reasonably foreseen.

Structural safety depends on the successful interrelationship between design and construction, particularly with regard to:

- degree of loading – dead and imposed loads should be assessed in accordance with BS 6399: Parts 1 and 3, and wind loads to CP3: Chapter V: Part 2;
- the properties of the materials chosen;
- the design analysis used;
- constructional details;
- safety factors; and
- standards of workmanship.

It is essential that the numeric values of the safety factors which are used are derived from a consideration of the above factors, since a change in any one of these could disturb the safety of the structure as a whole.

Additionally, loads used in calculations should take account of possible dynamic, concentrated and peak loads which may arise.

Approved Document A1/2 is arranged in four sections and gives guidance that may be adopted, if relevant, at the discretion of the designer. Where precise guidance is not given, due regard should be paid to the factors listed above.

- Section 1 allows the sizes of certain structural members to be assessed in small buildings of traditional masonry construction.
- Section 2 was added by the 1991 amendment and gives guidance on the fixing and support of external wall cladding.
- Section 3, also added by the 1991 amendment, makes it clear that certain roof re-covering operations may constitute a material alteration to the building and gives guidance to that effect.
- Section 4 lists various codes and standards for structural design and construction and is relevant to all types of buildings. Further information sources are included for the first time regarding landslip and structural appraisal of existing buildings subject to a change of use.

6.4 Design of structural members in houses and other small buildings

6.4.1 Definitions

The following definitions apply throughout section 1 of AD A1/2.

BUTTRESSING WALL – A wall which provides lateral support, from base to top, to another wall perpendicular to it.

CAVITY WIDTH – The horizontal distance between the leaves in a cavity wall.

COMPARTMENT WALL – See Chapter 7, Fire.

DEAD LOAD – The load due to the weight of all roofs, floors, walls, services, finishes and partitions i.e. all the permanent construction.

IMPOSED LOAD – The load assumed to be produced by the intended occupancy or use, including moveable partitions, distributed, concentrated, impact, inertia and snow loads, but *excluding* wind loads.

PIER – An integral part of a wall which consists of a thickened section occurring at intervals along a wall to which it is bonded or securely tied so as to afford lateral support.

SEPARATING WALL – A wall which is common to two adjoining buildings (see Chapter 7, Fire).

SPACING – The centre to centre distance between two adjacent timbers measured in a plane parallel to the plane of the structure of which they form part.

SPAN – The distance measured along the centreline of a member between centres of adjacent bearings. (However, it should be noted that the spans given in the tables for floor joists, rafters, purlins, ceiling joists, binders and roof joists are *clear spans* i.e. measured between the faces of supports.)

SUPPORTED WALL – A wall which is supported by buttressing walls, piers or chimneys, or floor or roof lateral support arrangements.

WIND LOADS – All loads due to the effect of wind pressure or suction.

6.5 Structural stability

The basic stability of a small house of traditional masonry construction is largely dependent on the provision of a braced roof structure which is adequately anchored to walls restrained laterally by buttressing walls, piers or chimneys. If this can be achieved then it should not be necessary to take additional precautions against wind loading.

A traditional fully boarded or hipped roof provides in-built resistance to instability. However, where this is not provided then extra wind bracing may be required.

Trussed rafter roofs have, in the past, been susceptible to collapse during high winds. If this form of construction is used it should be braced in accordance with BS 5268 *Structural use of timber*, Part 3: 1985 *Code of practice for trussed rafter roofs*. The recommendations of this code may also be used for traditional roofs where bracing is inadequate.

Small buildings of masonry construction having walls designed in accordance with section 1C of ADA1/2 and roofs and floors designed in accordance with section 1B of ADA1/2 will be satisfactory with regard to structural stability if the roof is braced as mentioned above.

6.5.1 Structural work of timber in single family houses

Section 1B of ADA1/2 provides that if the work concerned is in a floor, ceiling or roof of a single occupancy house of not more than three storeys, that work will be satisfactory if the grades and dimensions of the timbers used are at least equal to those given in Tables A1 to A24 of Appendix ADA1/2 and if the work complies in other respects with BS 5268 *Structural use of timber* Part 2: 1991 *Code of practice for permissible stress design, materials and workmanship*.

In effect this means that for a house of this type it is not necessary to calculate the size of joists, rafters, purlins, etc.; one merely selects the appropriate sizes from the tables in ADA1/2. Unusual load or support conditions might necessitate a check calculation by the recommendations in BS 5268: Part 2: 1991.

Tables A1 to A24 apply to all floor, ceiling and roof timbers in a single occupancy house of three storeys or less.

The timber used for any binder, beam, joist, purlin or rafter must be of a species, origin and grade specified in Table 1 to ADA1/2 (see below) or as given in the more comprehensive Tables of BS 5268: Part 2: 1991.

When using Tables A1 to A24 the following points should also be taken into account:

- The imposed load to be sustained by the floor, ceiling or roof of which the member forms part should not exceed:
  - (a) In the case of a floor: 1.5 kN/m^2 (Tables A1 and A2).
  - (b) In the case of a ceiling: 0.25k N/m^2 and a concentrated load of 0.9 kN acting with the imposed load (Tables A3 and A4).
  - (c) In the case of a flat roof with access not limited to the purposes of maintenance or repair: 1.5 kN/m^2 or a concentrated load of 0.9 kN (Tables A21 and A22).
  - (d) In the case of a roof (flat or pitched up to 45°) with access only for maintenance: 0.75 kN/m^2 or 1.00 kN/m^2, measured on plan (depending on the location, see below), or a concentrated load of 0.9 kN (Tables A5 to A20 inclusive).
  - (e) In the case of a roof supporting sheeting or decking pitched at between 10° and 35°: 0.75 kN/m^2 or 1.00 kN/m^2 measured on plan (depending on the location, see below), or a concentrated load of 0.9 kN (Tables A23 and A24).

The loading variations on the roofs mentioned in (d) and (e) above are due to the different imposed snow loadings which vary with altitude and location in England and Wales.

Diagram 2 from ADA1/2 is reproduced below and shows how the values of 0.75 kN/m^2 and 1.00 kN/m^2 are chosen. It would seem that designers will now be required to establish the level of their site above ordnance datum when making a building regulation submission.

- Floorboarding is assumed to comply with BS 1297: 1987 *Specification for tongued and grooved softwood flooring* and Table 6.1 below.
- As stated in the footnotes to Tables A1 to A24 in Approved Document A1/2, the cross-sectional dimensions given are applicable to either basic sawn or regularised sizes from BS 4471: 1987 Specification for sizes of sawn and processed softwood. For North American timber (CLS/ALS) the tables apply to surface sizes only unless the timber has been resawn to BS 4471 requirements.
- Notches and holes in floor and roof joists should comply with Fig. 6.1. However, no notches or holes should be cut in rafters except for birdsmouths at supports. The rafter may be birdsmouthed to a depth of up to one third the rafter depth. Notches and holes should not be cut in purlins or binders unless checked by a competent person.
- Bearing areas and workmanship should be in accordance with BS 5268: Part 2: 1991 and the following minimum bearing lengths should be provided:

AD A1/2

Table 1 Common species/grade combinations which satisfy the requirements for the strength classes to which tables A1–A24 in Appendix A relate.

Species	Origin	Grading Rules	Grades to satisfy strength class				
			SC3			SC4	
Redwood or whitewood	Imported	BS 4978	GS	MGS	M50	SS	MSS
Douglas Fir	UK	BS 4978	M50	SS	MSS	—	—
Larch	UK	BS 4978	GS	MGS	M50	SS	MSS
Scotch Pine	UK	BS 4978	GS	MGS	M50	SS	MSS
Corsican Pine	UK	BS 4978		M50		SS	MSS
European Spruce	UK	BS 4978		M75			
Sitka Spruce	UK	BS 4978		M75			
Douglas Fir-Larch Hem-Fir Spruce-Pine-Fir	CANADA	BS 4978	GS	MGS	M50	SS	MSS
Douglas Fir-Larch Hem-Fir Spruce-Pine-Fir	CANADA	NLGA	Joist & Plank Struct. L.F.	No.1 & No.2 No.1 & No.2		Joist & Plank Struct. L.F.	Select Select
Douglas Fir-Larch Hem-Fir Spruce-Pine-Fir	CANADA	MSR		Machine Stress-Rated 1450f-1.3E			Machine Stress-Rated 1650f-1.5E
Douglas Fir-Larch	USA	BS 4978	GS	MGS		SS	MSS
Hem-Fir	USA	BS 4978	GS	MGS	M50	SS	MSS
Western Whitewoods	USA	BS 4978	SS	MSS		—	—
Southern Pine	USA	BS 4978	GS	MGS		SS	MSS
Douglas Fir-Larch	USA	NGRDL	Joist & Plank Struct. L.F.	No.1 & No.2 No.1 & No.2		Joist & Plank Struct. L.F.	Select Select
Hem-Fir	USA	NGRDL	Joist & Plank Struct. L.F.	No.1 & No.2 No.1 & No.2		Joist & Plank Struct. L.F.	Select Select
Western Whitewoods	USA	NGRDL	Joist & Plank Struct. L.F.	Select Select		— —	
Southern Pine	USA	NGRDL	Joist & Plank	No.3 Stud grade		Joist & Plank	Select
Douglas Fir-Larch Hem-Fir Southern Pine	USA	MSR		Machine Stress-Rated 1450f-1.3E			Machine Stress-Rated 1650f-1.5E

Notes: The common species/grade combinations given in this table are for particular use with the other tables in Appendix A and for cross section sizes given in those tables.

Definitive and more comprehensive tables for assigning species/grade combinations to strength classes are given in BS 5268: Part 2: 1991.

The grading rules for American and Canadian Lumber are those approved by the American Lumber Standards (ALS) Board of Review and the Canadian Lumber Standards (CLS) Accreditation Board respectively (see BS 5268: Part 2: 1991).

NGLA denotes the National Lumber Grading Association

NGRDL denotes the National Grading Rules for Dimension Lumber

MSR denotes the North American Export Standard for Machine Stress-Rated Lumber.

AD A1/2

Diagram 2 Imposed snow roof loading

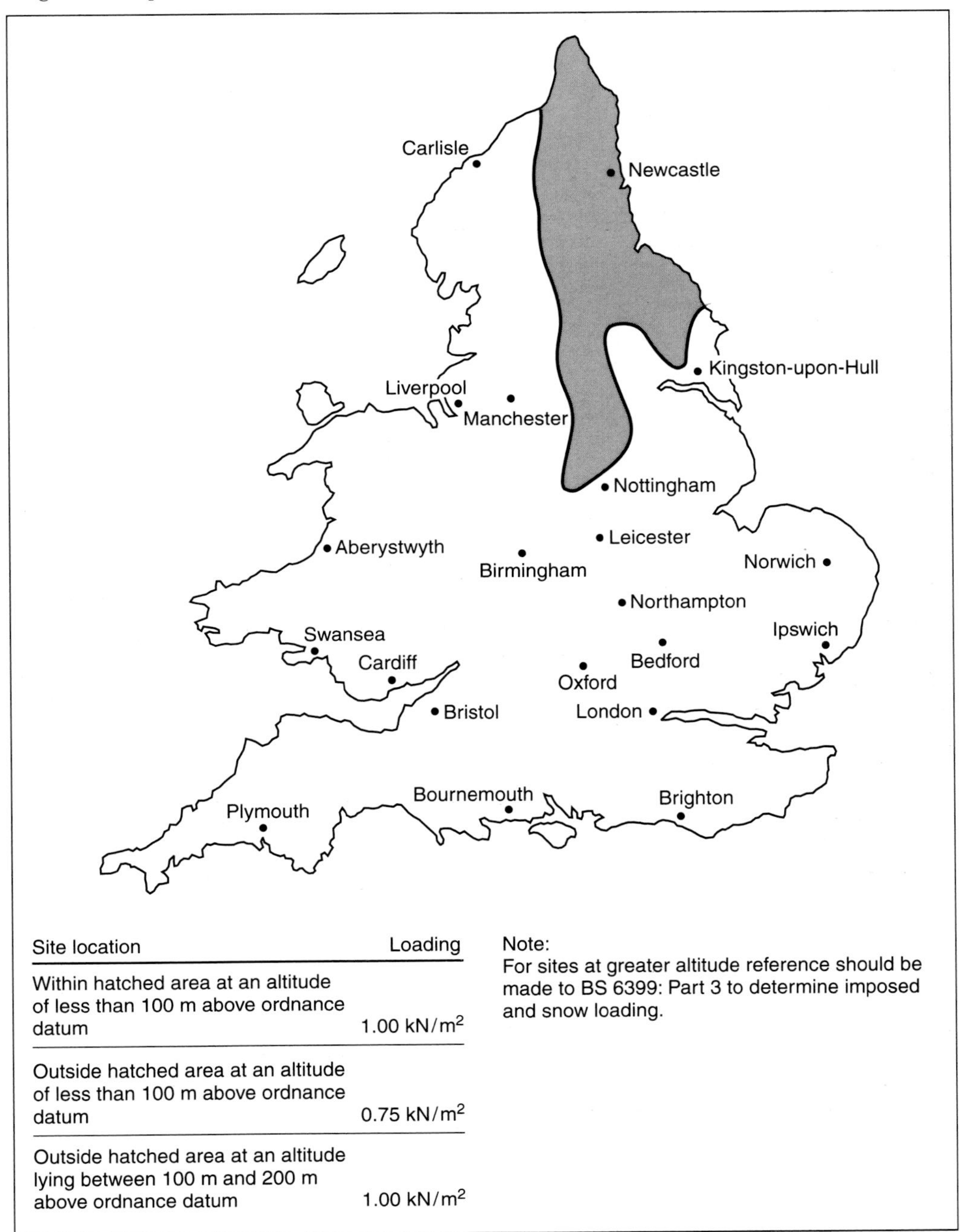

Site location	Loading
Within hatched area at an altitude of less than 100 m above ordnance datum	1.00 kN/m^2
Outside hatched area at an altitude of less than 100 m above ordnance datum	0.75 kN/m^2
Outside hatched area at an altitude lying between 100 m and 200 m above ordnance datum	1.00 kN/m^2

Note:
For sites at greater altitude reference should be made to BS 6399: Part 3 to determine imposed and snow loading.

Table 6.1 Softwood floorboards (tongued and grooved).

Finished thickness of board (mm)	Maximum span of board (centre to centre of joists) (mm) up to
16	500
19	600

(a) Floor, ceiling and roof joists – 35 mm
(b) Rafters and binders – 35 mm
(c) Purlins – 50 mm.

- If the spans of purlins or rafters are unequal it is permissible to choose the section sizes for each span separately or to use the longer span. However, for ceiling joists and binders the longer span only should be used.
- On floor joists no allowances have been made for additional loadings due to baths or partitions. It is recommended that the joists should be duplicated for bath support but no advice is given regarding partition loads. Similarly, when choosing ceiling joist sizes no account has been taken of trimming or of other additional loads such as water tanks.
- Purlins are assumed to be placed perpendicular to the roof slope and adequate connections between the various roof members should be provided as appropriate.
- Tables A1 to A24 are not applicable to trussed rafter roofs.

Example applications of Tables A1 to A24 are given in Fig. 6.2. It should be remembered that all spans, except for floorboards, are measured as the clear dimension between supports, and all spacings are the dimensions between longitudinal centres of members.

6.5.2 Strutting of joists

Where floor joists span more than 2.4m they should be strutted with one or more rows of:

- solid timber at least 38 mm wide and 0.75 times the joist depth; or
- herringbone strutting in 38 mm × 38 mm timber except where the distance between the joists is greater than three times the joist depth.

In this latter case the alternatives to timber herringbone strutting are not specified and it is not clear if solid timber strutting would be recommended or whether proprietary steel herringbone strutting, for example, could be used.

One row of strutting at mid-span is recommended for joist spans between 2.5 m and 4.5 m. Above 4.5 m, two rows of strutting at the one third positions would be required.

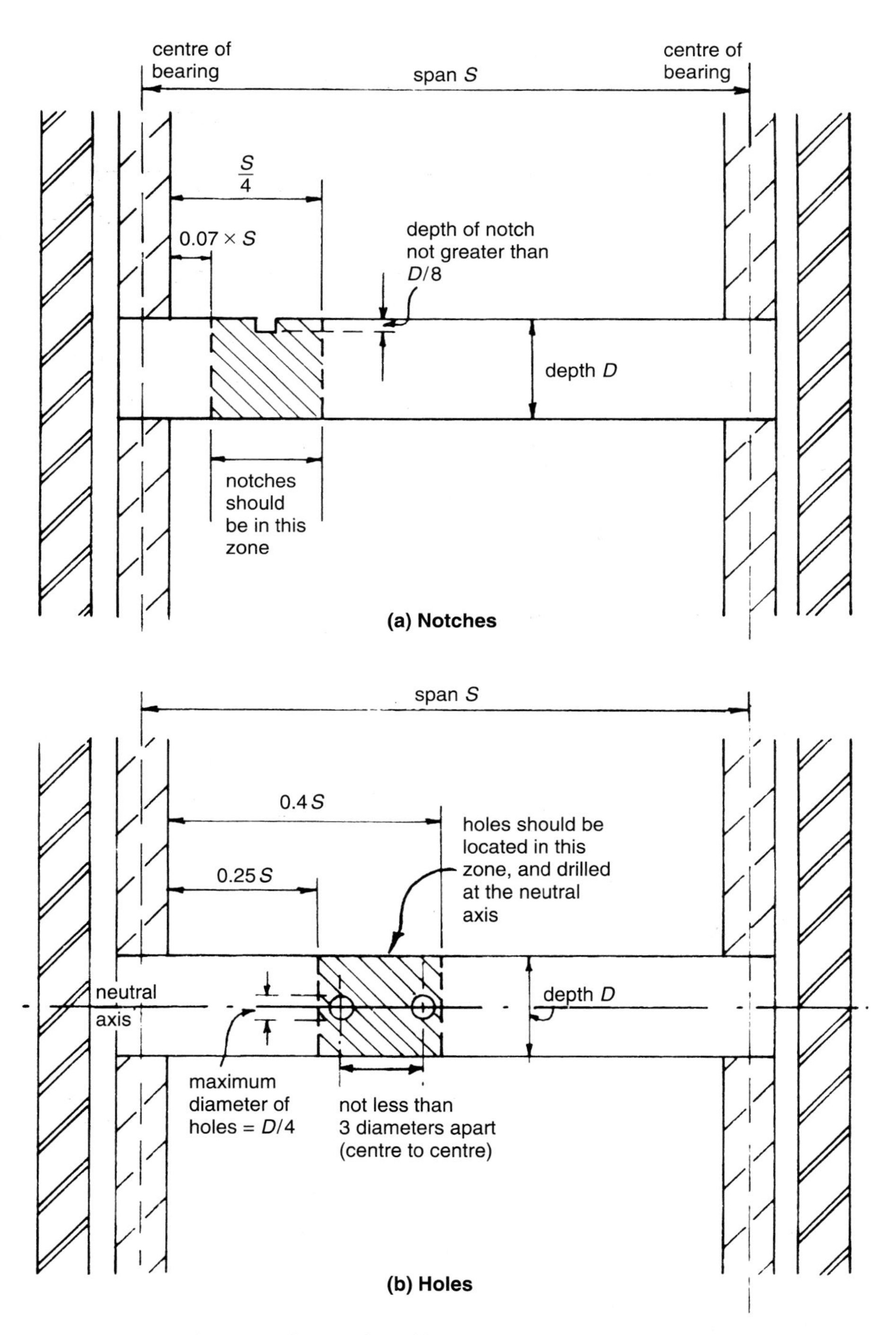

Fig. 6.1 Notches and holes in floor and roof joists.

(a) Floor joists, small house

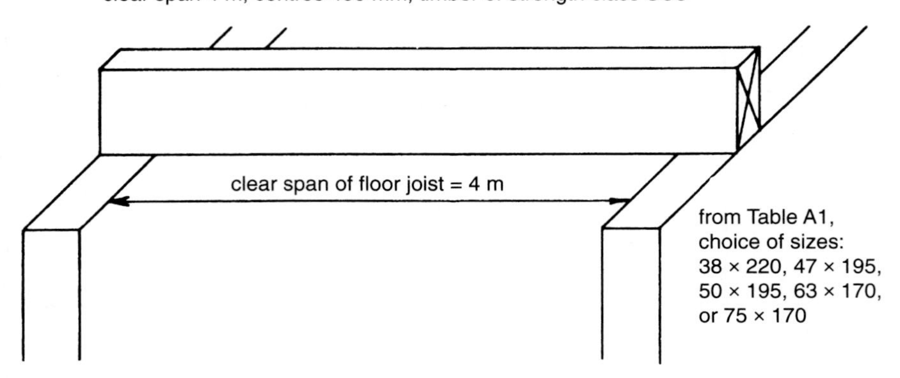

(b) Rafter, small house

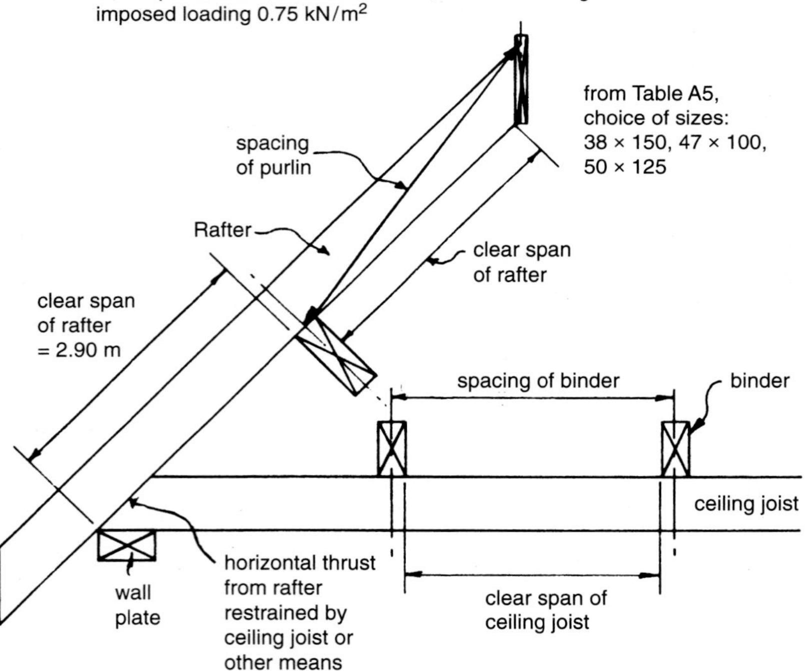

Fig. 6.2 Example of application of Tables A1 to A24.

AD A1/2, Appendix A

Table A1 Floor joists

Maximum clear span of joist (m) Timber of strength class SC3 (see Table 1)

Size of joist (mm × mm)	Dead Load [kN/m²] excluding the self weight of the joist								
	Not more than 0.25			More than 0.25 but not more than 0.50			More than 0.50 but not more than 1.25		
	Spacing of joists (mm)								
	400	450	600	400	450	600	400	450	600
38 × 97	1.83	1.69	1.30	1.72	1.56	1.21	1.42	1.30	1.04
38 × 122	2.48	2.39	1.93	2.37	2.22	1.76	1.95	1.79	1.45
38 × 147	2.98	2.87	2.51	2.85	2.71	2.33	2.45	2.29	1.87
38 × 170	3.44	3.31	2.87	3.28	3.10	2.69	2.81	2.65	2.27
38 × 195	3.94	3.75	3.26	3.72	3.52	3.06	3.19	3.01	2.61
38 × 220	4.43	4.19	3.65	4.16	3.93	3.42	3.57	3.37	2.92
47 × 97	2.02	1.91	1.58	1.92	1.82	1.46	1.67	1.53	1.23
47 × 122	2.66	2.56	2.30	2.55	2.45	2.09	2.26	2.08	1.70
47 × 147	3.20	3.08	2.79	3.06	2.95	2.61	2.72	2.57	2.17
47 × 170	3.69	3.55	3.19	3.53	3.40	2.99	3.12	2.94	2.55
47 × 195	4.22	4.06	3.62	4.04	3.89	3.39	3.54	3.34	2.90
47 × 220	4.72	4.57	4.04	4.55	4.35	3.79	3.95	3.74	3.24
50 × 97	2.08	1.97	1.67	1.98	1.87	1.54	1.74	1.60	1.29
50 × 122	2.72	2.62	2.37	2.60	2.50	2.19	2.33	2.17	1.77
50 × 147	3.27	3.14	2.86	3.13	3.01	2.69	2.81	2.65	2.27
50 × 170	3.77	3.62	3.29	3.61	3.47	3.08	3.21	3.03	2.63
50 × 195	4.31	4.15	3.73	4.13	3.97	3.50	3.65	3.44	2.99
50 × 220	4.79	4.66	4.17	4.64	4.47	3.91	4.07	3.85	3.35
63 × 97	2.32	2.20	1.92	2.19	2.08	1.82	1.93	1.84	1.53
63 × 122	2.93	2.82	2.57	2.81	2.70	2.45	2.53	2.43	2.09
63 × 147	3.52	3.39	3.08	3.37	3.24	2.95	3.04	2.92	2.58
63 × 170	4.06	3.91	3.56	3.89	3.74	3.40	3.50	3.37	2.95
63 × 195	4.63	4.47	4.07	4.44	4.28	3.90	4.01	3.85	3.35
63 × 220	5.06	4.92	4.58	4.91	4.77	4.37	4.51	4.30	3.75
75 × 122	3.10	2.99	2.72	2.97	2.86	2.60	2.68	2.58	2.33
75 × 147	3.72	3.58	3.27	3.56	3.43	3.13	3.22	3.09	2.81
75 × 170	4.28	4.13	3.77	4.11	3.96	3.61	3.71	3.57	3.21
75 × 195	4.83	4.70	4.31	4.68	4.52	4.13	4.24	4.08	3.65
75 × 220	5.27	5.13	4.79	5.11	4.97	4.64	4.74	4.60	4.07
38 × 140	2.84	2.73	2.40	2.72	2.59	2.17	2.33	2.15	1.75
38 × 184	3.72	3.56	3.09	3.53	3.33	2.90	3.02	2.85	2.47
38 × 235	4.71	4.46	3.89	4.43	4.18	3.64	3.80	3.59	3.11

AD A1/2, Appendix A

Table A2 Floor joists

Maximum clear span of joist (m) Timber of strength class SC4 (see Table 1)

Size of joist (mm × mm)	Dead Load [kN/m²] excluding the self weight of the joist								
	Not more than 0.25			More than 0.25 but not more than 0.50			More than 0.50 but not more than 1.25		
	Spacing of joists (mm)								
	400	450	600	400	450	600	400	450	600
38 × 97	1.94	1.83	1.59	1.84	1.74	1.51	1.64	1.55	1.36
38 × 122	2.58	2.48	2.20	2.47	2.37	2.08	2.18	2.07	1.83
38 × 147	3.10	2.98	2.71	2.97	2.85	2.59	2.67	2.56	2.31
38 × 170	3.58	3.44	3.13	3.43	3.29	2.99	3.08	2.96	2.68
38 × 195	4.10	3.94	3.58	3.92	3.77	3.42	3.53	3.39	3.07
38 × 220	4.61	4.44	4.03	4.41	4.25	3.86	3.97	3.82	3.46
47 × 97	2.14	2.03	1.76	2.03	1.92	1.68	1.80	1.71	1.50
47 × 122	2.77	2.66	2.42	2.65	2.55	2.29	2.38	2.27	2.01
47 × 147	3.33	3.20	2.91	3.19	3.06	2.78	2.87	2.75	2.50
47 × 170	3.84	3.69	3.36	3.67	3.54	3.21	3.31	3.18	2.88
47 × 195	4.39	4.22	3.85	4.20	4.05	3.68	3.79	3.64	3.30
47 × 220	4.86	4.73	4.33	4.71	4.55	4.14	4.26	4.10	3.72
50 × 97	2.20	2.09	1.82	2.08	1.98	1.73	1.84	1.75	1.54
50 × 122	2.83	2.72	2.47	2.71	2.60	2.36	2.43	2.33	2.06
50 × 147	3.39	3.27	2.97	3.25	3.13	2.84	2.93	2.81	2.55
50 × 170	3.91	3.77	3.43	3.75	3.61	3.28	3.38	3.25	2.94
50 × 195	4.47	4.31	3.92	4.29	4.13	3.75	3.86	3.72	3.37
50 × 220	4.93	4.80	4.42	4.78	4.64	4.23	4.35	4.18	3.80
63 × 97	2.43	2.32	2.03	2.31	2.19	1.93	2.03	1.93	1.71
63 × 122	3.05	2.93	2.67	2.92	2.81	2.55	2.63	2.53	2.27
63 × 147	3.67	3.52	3.21	3.50	3.37	3.07	3.16	3.04	2.76
63 × 170	4.21	4.06	3.70	4.04	3.89	3.54	3.64	3.51	3.19
63 × 195	4.77	4.64	4.23	4.61	4.45	4.05	4.17	4.01	3.65
63 × 220	5.20	5.06	4.73	5.05	4.91	4.56	4.68	4.51	4.11
75 × 122	3.22	3.10	2.83	3.09	2.97	2.71	2.78	2.68	2.43
75 × 147	3.86	3.72	3.39	3.70	3.57	3.25	3.34	3.22	2.93
75 × 170	4.45	4.29	3.91	4.27	4.11	3.75	3.86	3.71	3.38
75 × 195	4.97	4.83	4.47	4.82	4.69	4.29	4.41	4.25	3.86
75 × 220	5.42	5.27	4.93	5.25	5.11	4.78	4.88	4.74	4.35
38 × 140	2.96	2.84	2.58	2.83	2.72	2.47	2.54	2.44	2.17
38 × 184	3.87	3.72	3.38	3.70	3.56	3.23	3.33	3.20	2.90
38 × 235	4.85	4.71	4.31	4.70	4.54	4.12	4.24	4.08	3.70

AD A1/2, Appendix A

Table A3 Ceiling joists

Maximum clear span of joist (m) Timber of strength class SC3 and SC4 (see Table 1)

Size of joist (mm × mm)	Dead Load [kN/m²] excluding the self weight of the joist											
	Not more than 0.25			More than 0.25 but not more than 0.50			Not more than 0.25			More than 0.25 but not more than 0.50		
	Spacing of joists (mm)											
	400	450	600	400	450	600	400	450	600	400	450	600
38 × 72	1.15	1.14	1.11	1.11	1.10	1.06	1.21	1.20	1.17	1.17	1.16	1.12
38 × 97	1.74	1.72	1.67	1.67	1.64	1.58	1.84	1.82	1.76	1.76	1.73	1.66
38 × 122	2.37	2.34	2.25	2.25	2.21	2.11	2.50	2.46	2.37	2.37	2.33	2.22
38 × 147	3.02	2.97	2.85	2.85	2.80	2.66	3.18	3.13	3.00	3.00	2.94	2.79
38 × 170	3.63	3.57	3.41	3.41	3.34	3.16	3.81	3.75	3.58	3.58	3.51	3.32
38 × 195	4.30	4.23	4.02	4.02	3.94	3.72	4.51	4.43	4.22	4.22	4.13	3.89
38 × 220	4.98	4.88	4.64	4.64	4.54	4.27	5.21	5.11	4.86	4.86	4.75	4.47
47 × 72	1.27	1.26	1.23	1.23	1.21	1.17	1.35	1.33	1.30	1.30	1.28	1.24
47 × 97	1.92	1.90	1.84	1.84	1.81	1.73	2.03	2.00	1.93	1.93	1.90	1.83
47 × 122	2.60	2.57	2.47	2.47	2.42	2.31	2.74	2.70	2.60	2.60	2.55	2.43
47 × 147	3.30	3.25	3.11	3.11	3.05	2.90	3.47	3.42	3.27	3.27	3.21	3.04
47 × 170	3.96	3.89	3.72	3.72	3.64	3.44	4.15	4.08	3.89	3.89	3.81	3.61
47 × 195	4.68	4.59	4.37	4.37	4.28	4.04	4.90	4.81	4.57	4.57	4.47	4.22
47 × 220	5.39	5.29	5.03	5.03	4.91	4.63	5.64	5.53	5.25	5.25	5.14	4.84
50 × 72	1.31	1.30	1.27	1.27	1.25	1.21	1.39	1.37	1.34	1.34	1.32	1.28
50 × 97	1.97	1.95	1.89	1.89	1.86	1.78	2.08	2.06	1.99	1.99	1.96	1.88
50 × 122	2.67	2.63	2.53	2.53	2.49	2.37	2.81	2.77	2.66	2.66	2.62	2.49
50 × 147	3.39	3.34	3.19	3.19	3.13	2.97	3.56	3.50	3.35	3.35	3.29	3.12
50 × 170	4.06	3.99	3.81	3.81	3.73	3.53	4.25	4.18	3.99	3.99	3.91	3.69
50 × 195	4.79	4.70	4.48	4.48	4.38	4.13	5.01	4.92	4.68	4.68	4.58	4.32
50 × 220	5.52	5.41	5.14	5.14	5.03	4.73	5.77	5.66	5.37	5.37	5.25	4.95
38 × 89	1.54	1.53	1.48	1.48	1.46	1.41	1.63	1.62	1.57	1.57	1.55	1.49
38 × 140	2.84	2.79	2.68	2.68	2.63	2.50	2.99	2.94	2.82	2.82	2.77	2.63
38 × 184	4.01	3.94	3.75	3.75	3.68	3.47	4.20	4.13	3.94	3.94	3.85	3.64
	SC3						SC4					

AD A1/2, Appendix A

Table A4 Binders supporting ceiling joists

Maximum clear span of binder (m) Timber of strength class SC3 and SC4 (see Table 1)

		Dead Load [kN/m²] excluding the self weight of the binder											
		Not more than 0.25						More than 0.25 but not more than 0.50					
		Spacing of binders (mm)											
	Size of binder (mm × mm)	1200	1500	1800	2100	2400	2700	1200	1500	1800	2100	2400	2700
SC3	47 × 150	2.17	2.05	1.96	1.88	1.81		1.99	1.87				
	47 × 175	2.59	2.45	2.33	2.24	2.15	2.08	2.37	2.23	2.11	2.02	1.94	1.87
	50 × 150	2.22	2.11	2.01	1.93	1.86		2.04	1.92	1.83			
	50 × 175	2.65	2.51	2.39	2.29	2.21	2.13	2.42	2.28	2.16	2.07	1.99	1.91
	50 × 200	3.08	2.91	2.77	2.65	2.55	2.47	2.81	2.64	2.50	2.39	2.29	2.21
	63 × 125	1.97	1.87					1.82					
	63 × 150	2.44	2.31	2.20	2.12	2.04	1.97	2.23	2.11	2.00	1.91	1.84	
	63 × 175	2.90	2.74	2.61	2.51	2.41	2.33	2.65	2.49	2.37	2.26	2.17	2.10
	63 × 200	3.37	3.18	3.03	2.90	2.79	2.69	3.07	2.88	2.74	2.61	2.51	2.42
	63 × 225	3.83	3.61	3.44	3.29	3.16	3.05	3.49	3.27	3.10	2.96	2.84	2.74
	75 × 125	2.12	2.01	1.92	1.85			1.95	1.84				
	75 × 150	2.61	2.47	2.36	2.26	2.18	2.11	2.39	2.25	2.14	2.05	1.97	1.90
	75 × 175	3.10	2.93	2.79	2.68	2.58	2.49	2.83	2.66	2.53	2.42	2.32	2.24
	75 × 200	3.59	3.39	3.23	3.09	2.98	2.88	3.27	3.08	2.92	2.79	2.68	2.58
	75 × 225	4.08	3.85	3.66	3.51	3.37	3.26	3.71	3.50	3.31	3.16	3.03	2.92
SC4	47 × 150	2.28	2.16	2.06	1.98	1.90	1.84	2.09	1.97	1.87			
	47 × 175	2.72	2.57	2.45	2.34	2.26	2.18	2.48	2.34	2.22	2.12	2.03	1.96
	50 × 150	2.33	2.21	2.11	2.02	1.95	1.89	2.14	2.02	1.92	1.83		
	50 × 175	2.78	2.63	2.51	2.40	2.31	2.23	2.54	2.39	2.27	2.17	2.08	2.01
	50 × 200	3.23	3.05	2.90	2.78	2.67	2.58	2.95	2.77	2.62	2.51	2.40	2.32
	63 × 125	2.07	1.97	1.88	1.81			1.91	1.80				
	63 × 150	2.56	2.42	2.31	2.22	2.14	2.07	2.34	2.21	2.10	2.01	1.93	1.86
	63 × 175	3.04	2.87	2.74	2.62	2.53	2.44	2.78	2.61	2.48	2.37	2.28	2.20
	63 × 200	3.52	3.32	3.16	3.03	2.92	2.82	3.21	3.02	2.86	2.73	2.63	2.53
	63 × 225	4.00	3.77	3.59	3.44	3.31	3.19	3.65	3.42	3.24	3.10	2.97	2.86
	75 × 125	2.22	2.11	2.01	1.94	1.87	1.81	2.04	1.93	1.84			
	75 × 150	2.73	2.59	2.47	2.37	2.28	2.21	2.50	2.36	2.24	2.15	2.06	1.99
	75 × 175	3.24	3.07	2.92	2.80	2.70	2.61	2.96	2.79	2.65	2.53	2.43	2.35
	75 × 200	3.75	3.54	3.37	3.23	3.11	3.00	3.42	3.22	3.05	2.92	2.80	2.70
	75 × 225	4.26	4.02	3.82	3.66	3.52	3.40	3.88	3.65	3.46	3.30	3.17	3.06

AD A1/2, Appendix A

Table A5 Common or jack rafters for roofs having a pitch more than 15° but not more than 22.5° with access only for purpose of maintenance or repair. Imposed loading 0.75 kN/m² (see Diagram 2).

Maximum clear span of rafter (m) Timber of strength class SC3 and SC4 (see Table 1)

	Size of rafter (mm × mm)	Dead Load [kN/m²] excluding the self weight of the rafter								
		Not more than 0.50			More than 0.50 but not more than 0.75			More than 0.75 but not more than 1.00		
		Spacing of rafters (mm)								
		400	450	600	400	450	600	400	450	600
SC3	38 × 100	2.10	2.05	1.93	1.93	1.88	1.75	1.80	1.75	1.61
	38 × 125	2.89	2.79	2.53	2.63	2.55	2.34	2.44	2.35	2.15
	38 × 150	3.47	3.34	3.03	3.26	3.14	2.78	3.08	2.96	2.57
	47 × 100	2.46	2.40	2.18	2.25	2.19	2.03	2.10	2.03	1.87
	47 × 125	3.10	2.99	2.72	2.92	2.81	2.56	2.78	2.67	2.41
	47 × 150	3.71	3.57	3.25	3.50	3.36	3.06	3.32	3.20	2.86
	50 × 100	2.54	2.45	2.23	2.35	2.29	2.09	2.19	2.12	1.95
	50 × 125	3.17	3.05	2.78	2.98	2.87	2.61	2.83	2.73	2.48
	50 × 150	3.78	3.64	3.32	3.57	3.43	3.12	3.39	3.26	2.94
	38 × 89	1.76	1.72	1.63	1.63	1.59	1.49	1.53	1.49	1.38
	38 × 140	3.24	3.12	2.83	3.05	2.93	2.61	2.82	2.72	2.41
SC4	38 × 100	2.42	2.33	2.11	2.28	2.19	1.99	2.16	2.08	1.88
	38 × 125	3.01	2.90	2.64	2.83	2.73	2.48	2.69	2.59	2.35
	38 × 150	3.60	3.47	3.16	3.39	3.26	2.97	3.22	3.10	2.82
	47 × 100	2.59	2.49	2.27	2.44	2.35	2.13	2.32	2.23	2.02
	47 × 125	3.22	3.11	2.83	3.04	2.92	2.66	2.89	2.78	2.53
	47 × 150	3.85	3.71	3.38	3.63	3.50	3.18	3.45	3.32	3.02
	50 × 100	2.64	2.54	2.32	2.49	2.40	2.18	2.37	2.28	2.07
	50 × 125	3.29	3.17	2.89	3.10	2.98	2.72	2.95	2.83	2.58
	50 × 150	3.93	3.78	3.45	3.70	3.57	3.25	3.52	3.39	3.09
	38 × 89	2.16	2.07	1.88	2.03	1.95	1.77	1.92	1.85	1.68
	38 × 140	3.37	3.24	2.95	3.17	3.05	2.77	3.01	2.90	2.63

AD A1/2, Appendix A

Table A6 Purlins supporting rafters to which Table A5 refers (Imposed loading 0.75 kN/m^2)

Maximum clear span of purlin (m) Timber of strength class SC3 and SC4 (see Table 1)

		Dead Load [kN/m^2] excluding the self weight of the purlin																	
		Not more than 0.50						More than 0.50 but not more than 0.75						More than 0.75 but not more than 1.00					
		Spacing of purlins (mm)																	
	Size of purlin (mm × mm)	1500	1800	2100	2400	2700	3000	1500	1800	2100	2400	2700	3000	1500	1800	2100	2400	2700	3000
SC3	50 × 150	1.90																	
	50 × 175	2.22	2.08	1.96	1.87			2.08	1.95	1.84				1.97	1.84				
	50 × 200	2.53	2.37	2.24	2.13	2.02	1.92	2.38	2.22	2.10	1.97	1.85		2.25	2.10	1.95	1.82		
	50 × 225	2.84	2.66	2.52	2.40	2.26	2.14	2.67	2.50	2.35	2.20	2.07	1.96	2.53	2.36	2.18	2.03	1.91	1.81
	63 × 150	2.06	1.94	1.83				1.94	1.82					1.84					
	63 × 175	2.41	2.26	2.13	2.03	1.95	1.87	2.26	2.12	2.00	1.91	1.82		2.14	2.01	1.90	1.80		
	63 × 200	2.75	2.58	2.44	2.32	2.22	2.14	2.58	2.42	2.29	2.18	2.08	1.97	2.45	2.29	2.16	2.05	1.93	1.83
	63 × 225	3.09	2.89	2.74	2.61	2.50	2.40	2.90	2.72	2.57	2.45	2.33	2.20	2.75	2.58	2.43	2.29	2.16	2.04
	75 × 125	1.83																	
	75 × 150	2.19	2.06	1.95	1.86			2.06	1.94	1.83				1.96	1.83				
	75 × 175	2.56	2.40	2.27	2.17	2.08	2.00	2.41	2.26	2.13	2.03	1.95	1.87	2.28	2.14	2.02	1.92	1.84	
	75 × 200	2.92	2.74	2.59	2.47	2.37	2.28	2.75	2.58	2.44	2.32	2.22	2.14	2.61	2.44	2.31	2.20	2.10	2.00
	75 × 225	3.28	3.08	2.91	2.78	2.66	2.56	3.09	2.89	2.74	2.61	2.50	2.40	2.93	2.74	2.60	2.47	2.36	2.23
SC4	50 × 150	1.99	1.86					1.87											
	50 × 175	2.32	2.17	2.05	1.95	1.87		2.18	2.04	1.92	1.83			2.06	1.93	1.82			
	50 × 200	2.64	2.48	2.34	2.23	2.14	2.05	2.49	2.33	2.20	2.09	2.00	1.92	2.36	2.20	2.08	1.98	1.89	
	50 × 225	2.97	2.78	2.63	2.51	2.40	2.31	2.79	2.62	2.47	2.35	2.25	2.16	2.65	2.48	2.34	2.22	2.12	1.94
	63 × 150	2.16	2.02	1.91	1.82			2.03	1.90					1.92	1.80				
	63 × 175	2.51	2.36	2.23	2.13	2.04	1.96	2.36	2.22	2.10	2.00	1.91	1.84	2.24	2.10	1.99	1.89	1.81	
	63 × 200	2.87	2.69	2.55	2.43	2.33	2.24	2.70	2.53	2.39	2.28	2.18	2.10	2.56	2.40	2.27	2.16	2.06	1.98
	63 × 225	3.22	3.02	2.86	2.73	2.61	2.52	3.03	2.84	2.69	2.56	2.45	2.36	2.88	2.70	2.55	2.43	2.32	2.23
	75 × 125	1.91																	
	75 × 150	2.29	2.15	1.04	1.94	1.86		2.16	2.02	1.91	1.82			2.05	1.92	1.82			
	75 × 175	2.67	2.51	2.37	2.26	2.17	2.09	2.51	2.36	2.23	2.13	2.04	1.96	2.39	2.24	2.12	2.02	1.93	1.85
	75 × 200	3.05	2.86	2.71	2.58	2.48	2.39	2.87	2.69	2.55	2.43	2.33	2.24	2.72	2.55	2.42	2.30	2.20	2.12
	75 × 225	3.42	3.21	3.04	2.90	2.78	2.68	3.22	3.02	2.86	2.73	2.62	2.52	3.06	2.87	2.72	2.59	2.48	2.38

AD A1/2, Appendix A

Table A7 Common or jack rafters for roofs having a pitch more than 15° but not more than 22.5° with access only for purposes of maintenance or repair. Imposed loading 1.00 kN/m^2 (see Diagram 2).

Maximum clear span of rafter (m) Timber of strength class SC3 and SC4 (see Table 1)

	Size of rafter (mm × mm)	Dead Load [kN/m^2] excluding the self weight of the rafter								
		Not more than 0.50			More than 0.50 but not more than 0.75			More than 0.75 but not more than 1.00		
		Spacing of rafters (mm)								
		400	450	600	400	450	600	400	450	600
SC3	38 × 100	2.10	2.05	1.90	1.93	1.88	1.75	1.80	1.75	1.61
	38 × 125	2.73	2.63	2.35	2.59	2.49	2.17	2.44	2.34	2.03
	38 × 150	3.27	3.14	2.79	3.10	2.97	2.58	2.94	2.78	2.41
	47 × 100	2.35	2.26	2.05	2.23	2.15	1.95	2.10	2.03	1.83
	47 × 125	2.93	2.82	2.56	2.78	2.68	2.41	2.66	2.56	2.26
	47 × 150	3.50	3.37	3.07	3.33	3.20	2.86	3.18	3.06	2.68
	50 × 100	2.40	2.31	2.10	2.28	2.19	1.99	2.18	2.09	1.88
	50 × 125	2.99	2.88	2.62	2.84	2.73	2.48	2.71	2.61	2.33
	50 × 150	3.57	3.44	3.13	3.40	3.27	2.95	3.25	3.12	2.76
	38 × 89	1.76	1.72	1.63	1.63	1.59	1.49	1.53	1.49	1.38
	38 × 140	3.05	2.94	2.61	2.90	2.78	2.42	2.76	2.61	2.26
SC4	38 × 100	2.28	2.19	1.99	2.16	2.08	1.89	2.07	1.99	1.80
	38 × 125	2.84	2.73	2.48	2.70	2.59	2.35	2.58	2.48	2.25
	38 × 150	3.40	3.27	2.97	3.23	3.10	2.82	3.09	2.97	2.69
	47 × 100	2.44	2.35	2.14	2.32	2.23	2.03	2.22	2.13	1.94
	47 × 125	3.04	2.93	2.67	2.89	2.78	2.53	2.77	2.66	2.42
	47 × 150	3.64	3.50	3.19	3.46	3.33	3.03	3.31	3.18	2.89
	50 × 100	2.49	2.40	2.18	2.37	2.28	2.07	2.27	2.18	1.98
	50 × 125	3.10	2.99	2.72	2.95	2.84	2.58	2.82	2.72	2.47
	50 × 150	3.71	3.57	3.26	3.46	3.40	3.09	3.38	3.25	2.95
	38 × 89	2.03	1.95	1.77	1.93	1.85	1.68	1.84	1.77	1.60
	38 × 140	3.18	3.06	2.78	3.02	2.90	2.63	2.88	2.77	2.52

AD A1/2, Appendix A

Table A8 Purlins supporting rafters to which Table A7 refers (Imposed loading 1.00 kN/m^2)

Maximum clear span of purlin (m) Timber of strength class SC3 and SC4 (see Table 1)

		Dead Load [kN/m^2] excluding the self weight of the purlin																	
		Not more than 0.50						More than 0.50 but not more than 0.75						More than 0.75 but not more than 1.00					
		Spacing of purlins (mm)																	
	Size of purlin (mm × mm)	1500	1800	2100	2400	2700	3000	1500	1800	2100	2400	2700	3000	1500	1800	2100	2400	2700	3000
SC3	50 × 175	2.09	1.95	1.84				1.97	1.85					1.88					
SC3	50 × 200	2.38	2.23	2.10	1.97	1.85		2.26	2.11	1.96	1.82			2.15	1.98	1.83			
SC3	50 × 225	2.68	2.50	2.36	2.20	2.07	1.96	2.54	2.36	2.18	2.04	1.92	1.81	2.42	2.21	2.04	1.90		
SC3	63 × 150	1.94	1.82					1.84											
SC3	63 × 175	2.27	2.12	2.01	1.91	1.83		2.15	2.01	1.90	1.81			2.05	1.92	1.81			
SC3	63 × 200	2.59	2.42	2.29	2.18	2.09	1.98	2.45	2.30	2.17	2.06	1.94	1.83	2.30	2.19	2.06	1.92	1.81	
SC3	63 × 225	2.91	2.72	2.58	2.45	2.33	2.21	2.76	2.58	2.44	2.30	2.16	2.05	2.63	2.46	2.30	2.15	2.02	1.91
SC3	75 × 150	2.07	1.94	1.83				1.96	1.84					1.87					
SC3	75 × 175	2.41	2.26	2.14	2.04	1.95	1.88	2.29	2.14	2.03	1.93	1.85		2.18	2.04	1.93	1.84		
SC3	75 × 200	2.75	2.58	2.44	2.33	2.23	2.14	2.61	2.45	2.31	2.20	2.11	2.01	2.49	2.33	2.20	2.10	1.98	1.88
SC3	75 × 225	3.09	2.90	2.74	2.61	2.50	2.41	2.93	2.75	2.60	2.48	2.36	2.24	2.80	2.62	2.48	2.35	2.21	2.09
SC4	50 × 150	1.87																	
SC4	50 × 175	2.18	2.04	1.93	1.83			2.07	1.93	1.82				1.97	1.84				
SC4	50 × 200	2.49	2.33	2.20	2.10	2.00	1.92	2.36	2.21	2.08	1.98	1.89		2.25	2.10	1.98	1.88		
SC4	50 × 225	2.80	2.62	2.48	2.36	2.25	2.16	2.65	2.48	2.34	2.23	2.13	1.95	2.53	2.36	2.23	2.12	1.91	
SC4	63 × 150	2.03	1.90	1.80				1.93	1.81					1.84					
SC4	63 × 175	2.37	2.22	2.10	2.00	1.91	1.84	2.25	2.10	1.99	1.89	1.81		2.14	2.01	1.90	1.80		
SC4	63 × 200	2.70	2.53	2.40	2.28	2.19	2.10	2.57	2.40	2.27	2.16	2.07	1.99	2.45	2.29	2.16	2.06	1.97	1.89
SC4	63 × 225	3.04	2.85	2.70	2.57	2.46	2.36	2.88	2.70	2.55	2.43	2.32	2.23	2.75	2.58	2.43	2.31	2.21	2.12
SC4	75 × 125	1.80																	
SC4	75 × 150	2.16	2.03	1.92	1.83			2.05	1.92	1.82				1.96	1.83				
SC4	75 × 175	2.52	2.36	2.24	2.13	2.04	1.96	2.39	2.24	2.12	2.02	1.93	1.86	2.28	2.14	2.02	1.92	1.84	
SC4	75 × 200	2.87	2.70	2.55	2.43	2.33	2.24	2.73	2.56	2.42	2.31	2.21	2.12	2.61	2.44	2.31	2.20	2.10	2.02
SC4	75 × 225	3.23	3.03	2.87	2.74	2.62	2.52	3.07	2.88	2.72	2.59	2.48	2.39	2.93	2.75	2.60	2.47	2.36	2.27

AD A1/2, Appendix A

Table A9 Common or jack rafters for roofs having a pitch more than 22.5° but not more than 30° with access only for purposes of maintenance or repair. Imposed loading 0.75 kN/m^2 (see Diagram 2).

Maximum clear span of rafter (m) Timber of strength class SC3 and SC4 (see Table 1)

	Size of rafter (mm × mm)	Dead Load [kN/m^2] excluding the self weight of the rafter								
		Not more than 0.50			More than 0.50 but not more than 0.75			More than 0.75 but not more than 1.00		
		Spacing of rafters (mm)								
		400	450	600	400	450	600	400	450	600
SC3	38 × 100	2.18	2.13	2.01	2.01	1.96	1.82	1.88	1.82	1.68
	38 × 125	2.97	2.86	2.60	2.74	2.66	2.44	2.54	2.46	2.25
	38 × 150	3.55	3.42	3.11	3.34	3.21	2.92	3.17	3.04	2.72
	47 × 100	2.55	2.46	2.23	2.35	2.28	2.10	2.18	2.12	1.95
	47 × 125	3.18	3.06	2.79	2.99	2.88	2.62	2.84	2.73	2.48
	47 × 150	3.80	3.66	3.33	3.57	3.44	3.13	3.39	3.27	2.97
	50 × 100	2.60	2.51	2.28	2.45	2.36	2.14	2.28	2.21	2.03
	50 × 125	3.24	3.12	2.84	3.05	2.93	2.67	2.89	2.79	2.53
	50 × 150	3.87	3.73	3.40	3.65	3.51	3.20	3.46	3.33	3.03
	38 × 89	1.82	1.79	1.69	1.69	1.65	1.55	1.59	1.55	1.44
	38 × 140	3.32	3.19	2.90	3.12	3.00	2.72	2.94	2.84	2.55
SC4	38 × 100	2.48	2.38	2.17	2.33	2.24	2.03	2.21	2.12	1.93
	38 × 125	3.08	2.97	2.70	2.90	2.79	2.53	2.75	2.65	2.40
	38 × 150	3.69	3.55	3.23	3.47	3.34	3.04	3.29	3.17	2.88
	47 × 100	2.65	2.55	2.32	2.49	2.40	2.18	2.37	2.28	2.07
	47 × 125	3.30	3.18	2.90	3.11	2.99	2.72	2.95	2.84	2.58
	47 × 150	3.94	3.80	3.46	3.71	3.58	3.26	3.53	3.40	3.09
	50 × 100	2.71	2.61	2.37	2.55	2.45	2.23	2.42	2.32	2.11
	50 × 125	3.37	3.24	2.96	3.17	3.05	2.78	3.01	2.90	2.63
	50 × 150	4.02	3.87	3.53	3.79	3.65	3.32	3.60	3.46	3.15
	38 × 89	2.21	2.12	1.93	2.07	1.99	1.81	1.97	1.89	1.72
	38 × 140	3.45	3.32	3.02	3.24	3.12	2.84	3.08	2.96	2.69

AD A1/2, Appendix A

Table A10 Purlins supporting rafters to which Table A9 refers (Imposed loading 0.75 kN/m^2)

Maximum clear span of purlin (m) Timber of strength class SC3 and SC4 (see Table 1)

		Dead Load [kN/m^2] excluding the self weight of the purlin																	
		Not more than 0.50						More than 0.50 but not more than 0.75						More than 0.75 but not more than 1.00					
		Spacing of purlins (mm)																	
	Size of purlin (mm × mm)	1500	1800	2100	2400	2700	3000	1500	1800	2100	2400	2700	3000	1500	1800	2100	2400	2700	3000
SC3	50 × 150	1.95	1.83					1.83											
SC3	50 × 175	2.27	2.12	2.01	1.92	1.83		2.13	1.99	1.88				2.02	1.89				
SC3	50 × 200	2.59	2.43	2.30	2.19	2.09	1.99	2.43	2.28	2.15	2.03	1.91	1.81	2.30	2.15	2.01	1.88		
SC3	50 × 225	2.92	2.73	2.58	2.46	2.34	2.22	2.74	2.56	2.42	2.27	2.14	2.02	2.59	2.42	2.25	2.10	1.98	1.87
SC3	63 × 150	2.12	1.98	1.88				1.99	1.86					1.88					
SC3	63 × 175	2.47	2.31	2.19	2.09	2.00	1.92	2.32	2.17	2.05	1.95	1.87		2.19	2.05	1.94	1.85		
SC3	63 × 200	2.81	2.64	2.50	2.38	2.28	2.19	2.64	2.48	2.34	2.23	2.13	2.04	2.50	2.35	2.22	2.11	1.99	1.89
SC3	63 × 225	3.16	2.97	2.81	2.68	2.56	2.47	2.97	2.78	2.63	2.51	2.40	2.28	2.82	2.64	2.49	2.37	2.23	2.11
SC3	75 × 125	1.88																	
SC3	75 × 150	2.25	2.11	2.00	1.91	1.83		2.11	1.98	1.87				2.00	1.88				
SC3	75 × 175	2.62	2.46	2.33	2.22	2.13	2.05	2.46	2.31	2.19	2.08	1.99	1.92	2.33	2.19	2.07	1.97	1.89	1.81
SC3	75 × 200	2.99	2.81	2.66	2.54	2.43	2.34	2.81	2.64	2.50	2.38	2.28	2.19	2.67	2.50	2.36	2.25	2.15	2.07
SC3	75 × 225	3.36	3.15	2.99	2.85	2.73	2.63	3.16	2.96	2.80	2.67	2.56	2.46	3.00	2.81	2.66	2.53	2.42	2.31
SC4	50 × 150	2.04	1.91	1.81				1.91						1.31					
SC4	50 × 175	2.37	2.22	2.10	2.00	1.92	1.84	2.23	2.09	1.97	1.88			2.11	1.97	1.86			
SC4	50 × 200	2.71	2.54	2.40	2.29	2.19	2.11	2.54	2.38	2.25	2.14	2.05	2.97	2.41	2.26	2.13	2.02	1.94	1.84
SC4	50 × 225	3.05	2.86	2.70	2.57	2.46	2.37	2.86	2.68	2.53	2.41	2.30	2.21	2.71	2.54	2.39	2.28	2.18	2.07
SC4	63 × 125	1.84																	
SC4	63 × 150	2.21	2.07	1.96	1.87			2.08	1.95	1.84				1.97	1.84				
SC4	63 × 175	2.57	2.42	2.29	2.18	2.09	2.01	2.42	2.27	2.15	2.04	1.96	1.88	2.29	2.15	2.03	1.93	1.85	
SC4	63 × 200	2.94	2.76	2.61	2.49	2.39	2.30	2.76	2.59	2.45	2.33	2.24	2.15	2.62	2.45	2.32	2.21	2.11	2.03
SC4	63 × 225	3.30	3.10	2.93	2.80	2.68	2.58	3.10	2.91	2.75	2.62	2.51	2.42	2.94	2.76	2.61	2.48	2.38	2.28
SC4	75 × 125	1.96	1.84					1.84											
SC4	75 × 150	2.35	2.20	2.09	1.99	1.91	1.84	2.21	2.07	1.96	1.87			2.09	1.96	1.86			
SC4	75 × 175	2.73	2.57	2.43	2.32	2.22	2.14	2.57	2.41	2.28	2.18	2.09	2.01	2.44	2.29	2.16	2.06	1.97	1.90
SC4	75 × 200	3.12	2.93	2.78	2.65	2.54	2.45	2.93	2.75	2.61	2.49	2.38	2.29	2.79	2.61	2.47	2.35	2.26	2.17
SC4	75 × 225	3.50	3.29	3.12	2.98	2.86	2.75	3.30	3.10	2.93	2.80	2.68	2.58	3.13	2.94	2.78	2.65	2.54	2.44

AD A1/2, Appendix A

Table A11 Common or jack rafters for roofs having a pitch more than 22.5° but not more than 30° with access only for purposes of maintenance or repair. Imposed loading 1.00 kN/m^2 (see Diagram 2).

Maximum clear span of rafter (m) Timber of strength class SC3 and SC4 (see Table 1)

		Dead Load [kN/m^2] excluding the self weight of the rafter								
		Not more than 0.50			More than 0.50 but not more than 0.75			More than 0.75 but not more than 1.00		
		Spacing of rafters (mm)								
	Size of rafter (mm × mm)	400	450	600	400	450	600	400	450	600
SC3	38 × 100	2.18	2.13	1.96	2.01	1.96	1.82	1.88	1.82	1.68
SC3	38 × 125	2.80	2.69	2.45	2.65	2.55	2.30	2.53	2.44	2.15
SC3	38 × 150	3.35	3.22	2.93	3.18	3.06	2.73	3.03	2.92	2.55
SC3	47 × 100	2.41	2.32	2.11	2.28	2.20	2.00	2.18	2.10	1.90
SC3	47 × 125	3.00	2.89	2.63	2.85	2.74	2.49	2.72	2.62	2.37
SC3	47 × 150	3.59	2.46	3.14	3.41	3.28	2.98	3.25	3.13	2.83
SC3	50 × 100	2.46	2.37	2.15	2.33	2.24	2.04	2.23	2.14	1.94
SC3	50 × 125	3.06	2.95	2.68	2.91	2.80	2.54	2.78	2.67	2.43
SC3	50 × 150	3.66	3.52	3.21	3.48	3.34	3.04	3.32	3.20	2.90
SC3	38 × 89	1.82	1.79	1.69	1.69	1.65	1.55	1.59	1.55	1.44
SC3	38 × 140	3.13	3.01	2.74	2.97	2.85	2.56	2.83	2.72	2.29
SC4	38 × 100	2.34	2.25	2.04	2.21	2.13	1.93	2.11	2.03	1.84
SC4	38 × 125	2.91	2.80	2.55	2.76	2.66	2.41	2.64	2.53	2.30
SC4	38 × 150	3.48	3.35	3.05	3.30	3.18	2.89	3.16	3.04	2.76
SC4	47 × 100	2.51	2.41	2.19	2.38	2.29	2.08	2.27	2.18	1.98
SC4	47 × 125	3.12	3.00	2.73	2.96	2.85	2.59	2.83	2.72	2.47
SC4	47 × 150	3.73	3.59	3.27	3.54	3.41	3.10	3.38	3.26	2.96
SC4	50 × 100	2.56	2.46	2.24	2.42	2.33	2.12	2.32	2.23	2.02
SC4	50 × 125	3.18	3.06	2.79	3.02	2.91	2.64	2.89	2.78	2.52
SC4	50 × 150	3.80	3.66	3.34	3.61	3.48	3.16	3.45	3.32	3.02
SC4	38 × 89	2.08	2.00	1.82	1.97	1.90	1.72	1.88	1.81	1.64
SC4	38 × 140	3.25	3.13	2.85	3.09	2.97	2.70	2.95	2.84	2.57

AD A1/2, Appendix A

Table A12 Purlins supporting rafters to which Table A11 refers (Imposed loading 1.00 kN/m²)

Maximum clear span of purlin (m) Timber of strength class SC3 and SC4 (see Table 1)

	Size of purlin (mm × mm)	Dead Load [kN/m²] excluding the self weight of the purlin																	
		Not more than 0.50						More than 0.50 but not more than 0.75						More than 0.75 but not more than 1.00					
		Spacing of purlins (mm)																	
		1500	1800	2100	2400	2700	3000	1500	1800	2100	2400	2700	3000	1500	1800	2100	2400	2700	3000
SC3	50 × 150	1.84																	
	50 × 175	21.4	2.00	1.89				2.03	1.89					1.93	1.80				
	50 × 200	2.45	2.29	2.16	2.05	1.93	1.82	2.31	2.16	2.03	1.89			2.20	2.05	1.89			
	50 × 225	2.75	2.57	2.43	2.29	2.15	2.04	2.60	2.43	2.26	2.11	1.99	1.88	2.48	2.29	2.11	1.97	1.85	
	63 × 150	2.00	1.87					1.89						1.88					
	63 × 175	2.33	2.18	2.06	1.96	1.88	1.80	2.20	2.06	1.95	1.85			2.10	1.96	1.85			
	63 × 200	2.66	2.49	2.35	2.24	2.14	2.05	2.51	2.35	2.22	2.12	2.00	1.90	2.40	2.24	2.12	1.99	1.87	
	63 × 225	2.98	2.80	2.65	2.52	2.41	2.29	2.83	2.65	2.50	2.38	2.24	2.12	2.69	2.52	2.38	2.22	2.09	1.98
	75 × 150	2.12	1.99	1.88				2.01	1.88					1.92					
	75 × 175	2.47	2.32	2.20	2.09	2.00	1.93	2.34	2.20	2.08	1.98	1.89	1.82	2.24	2.09	1.98	1.88	1.80	
	75 × 200	2.82	2.65	2.51	2.39	2.29	2.20	2.68	2.51	2.37	2.26	2.16	2.08	2.55	2.39	2.26	2.15	2.05	1.94
	75 × 225	3.17	2.98	2.82	2.68	2.57	2.47	3.01	2.82	2.67	2.54	2.43	2.32	2.87	2.69	2.54	2.42	2.29	2.17
SC4	50 × 150	1.92						1.82											
	50 × 175	2.24	2.10	1.98	1.89	1.80		2.12	1.98	1.87				2.02	1.89				
	50 × 200	2.56	2.39	2.26	2.15	2.06	1.98	2.42	2.26	2.14	2.03	1.94	1.86	2.31	2.16	2.03	1.93	1.81	
	50 × 225	2.87	2.69	2.54	2.42	2.32	2.22	2.72	2.55	2.40	2.29	2.18	2.09	2.59	2.42	2.29	2.17	2.04	1.83
	63 × 150	2.09	1.95	1.85				1.98	1.85					1.88					
	63 × 175	2.43	2.28	2.16	2.05	1.97	1.89	2.30	2.16	2.04	1.94	1.86		2.20	2.06	1.94	1.85		
	63 × 200	2.77	2.60	2.46	2.35	2.25	2.16	2.63	2.46	2.33	2.22	2.12	2.04	2.51	2.35	2.22	2.11	2.02	1.94
	63 × 225	3.12	2.92	2.77	2.64	2.52	2.43	2.95	2.77	2.62	2.49	2.39	2.29	2.82	2.64	2.49	2.37	2.27	2.18
	75 × 125	1.85																	
	75 × 150	2.22	2.08	1.97	1.88			2.10	1.97	1.86				2.01	1.88				
	75 × 175	2.58	2.42	2.29	2.19	2.10	2.02	2.45	2.30	2.17	2.07	1.98	1.91	2.34	2.19	2.07	1.97	1.89	1.81
	75 × 200	2.95	2.77	2.62	2.50	2.39	2.30	2.80	2.62	2.48	2.36	2.26	2.18	2.67	2.50	2.37	2.25	2.16	2.07
	75 × 225	3.31	3.11	2.94	2.81	2.70	2.59	3.14	2.95	2.79	2.66	2.55	2.45	3.00	2.81	2.66	2.53	2.42	2.33

AD A1/2, Appendix A

Table A13 Common or jack rafters for roofs having a pitch more than 30° but not more than 45° with access only for purposes of maintenance or repair. Imposed loading 0.75 kN/m^2 (see Diagram 2).

Maximum clear span of rafter (m) Timber of strength class SC3 and SC4 (see Table 1)

		Dead Load [kN/m^2] excluding the self weight of the rafter								
		Not more than 0.50			More than 0.50 but not more than 0.75			More than 0.75 but not more than 1.25		
					Spacing of rafters (mm)					
	Size of rafter (mm × mm)	400	450	600	400	450	600	400	450	600
SC3	38 × 100	2.28	2.23	2.10	2.10	2.05	1.91	1.96	1.91	1.76
SC3	38 × 125	3.07	2.95	2.69	2.87	2.77	2.52	2.65	2.56	2.35
SC3	38 × 150	3.67	3.53	3.22	3.44	3.31	3.01	3.26	3.14	2.85
SC3	47 × 100	2.64	2.54	2.31	2.45	2.38	2.17	2.28	2.21	2.04
SC3	47 × 125	3.29	3.17	2.88	3.09	2.97	2.70	2.92	2.81	2.56
SC3	47 × 150	3.93	3.78	3.45	3.69	3.55	3.23	3.50	3.37	3.06
SC3	50 × 100	2.69	2.59	2.36	2.53	2.43	2.21	2.38	2.30	2.09
SC3	50 × 125	3.35	3.23	2.94	3.15	3.03	2.76	2.98	2.87	2.61
SC3	50 × 150	4.00	3.86	3.52	3.76	3.62	3.30	3.57	3.44	3.13
SC3	38 × 89	1.91	1.87	1.77	1.77	1.73	1.62	1.67	1.62	1.50
SC3	38 × 140	3.43	3.30	3.01	3.22	3.10	2.82	3.05	2.93	2.66
SC4	38 × 100	2.56	2.47	2.24	2.40	2.31	2.10	2.28	2.19	1.99
SC4	38 × 125	3.19	3.07	2.80	2.99	2.88	2.62	2.84	2.73	2.48
SC4	38 × 150	3.81	3.67	3.35	3.58	3.45	3.14	3.39	3.27	2.97
SC4	47 × 100	2.74	2.64	2.41	2.58	2.48	2.25	2.44	2.35	2.13
SC4	47 × 125	3.41	3.29	3.00	3.21	3.09	2.81	3.04	2.93	2.66
SC4	47 × 150	4.08	3.93	3.59	3.83	3.69	3.36	3.64	3.50	3.19
SC4	50 × 100	2.80	2.70	2.45	2.63	2.53	2.30	2.49	2.40	2.18
SC4	50 × 125	3.48	3.35	3.06	3.27	3.15	2.87	3.10	2.99	2.72
SC4	50 × 150	4.16	4.01	3.66	3.91	3.77	3.43	3.71	3.57	3.25
SC4	38 × 89	2.28	2.20	2.00	2.14	2.06	1.87	2.03	1.95	1.77
SC4	38 × 140	3.56	3.43	3.13	3.35	3.22	2.93	3.17	3.05	2.77

AD A1/2, Appendix A

Table A14 Purlins supporting rafters to which Table A13 refers. (Imposed loading 0.75 kN/m^2)

Maximum clear span of purlin (m) Timber of strength class SC3 and SC4 (see Table 1)

		Dead Load [kN/m^2] excluding the self weight of the purlin																	
		Not more than 0.50						More than 0.50 but not more than 0.75						More than 0.75 but not more than 1.00					
		Spacing of purlins (mm)																	
	Size of purlin (mm × mm)	1500	1800	2100	2400	2700	3000	1500	1800	2100	2400	2700	3000	1500	1800	2100	2400	2700	3000
SC3	50 × 150	2.02	1.89					1.89											
SC3	50 × 175	2.36	2.21	2.09	1.99	1.90	1.83	2.21	2.06	1.95	1.86			2.08	1.95	1.84			
SC3	50 × 200	2.69	2.52	2.38	2.27	2.17	2.09	2.52	2.36	2.23	2.12	2.01	1.90	2.38	2.23	2.10	1.97	1.85	
SC3	50 × 225	3.02	2.83	2.68	2.55	2.44	2.34	2.83	2.65	2.50	2.38	2.24	2.12	2.68	2.50	2.36	2.20	2.07	1.96
SC3	63 × 125	1.83																	
SC3	63 × 150	2.19	2.06	1.95	1.85			2.05	1.93	1.82				1.94	1.82				
SC3	63 × 175	2.55	2.40	2.27	2.16	2.07	1.99	2.39	2.24	2.12	2.02	1.94	1.86	2.27	2.12	2.01	1.91	1.83	
SC3	63 × 200	2.91	2.74	2.59	2.47	2.37	2.28	2.73	2.56	2.42	2.31	2.21	2.13	2.59	2.42	2.29	2.18	2.09	1.98
SC3	63 × 225	3.28	3.07	2.91	2.78	2.66	2.56	3.07	2.88	2.73	2.60	2.49	2.39	2.91	2.72	2.58	2.45	2.33	2.21
SC3	75 × 125	1.94	1.82					1.82											
SC3	75 × 150	2.33	2.19	2.07	1.97	1.89	1.82	2.18	2.05	1.94	1.85			2.07	1.94	1.83			
SC3	75 × 175	2.71	2.55	2.41	2.30	2.21	2.12	2.55	2.39	2.26	2.15	2.06	1.99	2.41	2.26	2.14	2.04	1.95	1.87
SC3	75 × 200	3.10	2.91	2.75	2.63	2.52	2.43	2.91	2.73	2.58	2.46	2.36	2.27	2.75	2.58	2.44	2.33	2.23	2.14
SC3	75 × 225	3.48	3.27	3.10	2.95	2.83	2.73	3.26	3.06	2.90	2.77	2.65	2.55	3.09	2.90	2.74	2.61	2.50	2.41
SC4	50 × 150	2.11	1.98	1.87				1.98	1.85					1.87					
SC4	50 × 175	2.46	2.31	2.18	2.08	1.99	1.91	2.31	2.16	2.04	1.94	1.86		2.18	2.04	1.93	1.83		
SC4	50 × 200	2.81	2.63	2.49	2.37	2.27	2.19	2.63	2.47	2.33	2.22	2.12	2.04	2.49	2.33	2.20	2.09	2.00	1.92
SC4	50 × 225	3.16	2.96	2.80	2.67	2.56	2.46	2.96	2.77	2.62	2.50	2.39	2.30	2.80	2.62	2.48	2.36	2.25	2.16
SC4	63 × 125	1.91																	
SC4	63 × 150	2.29	2.15	2.03	1.94	1.86		2.15	2.01	1.90	1.81			2.03	1.90	1.80			
SC4	63 × 175	2.67	2.50	2.37	2.26	2.17	2.08	2.50	2.35	2.22	2.12	2.03	1.95	2.37	2.22	2.10	2.00	1.91	1.84
SC4	63 × 200	3.04	2.86	2.71	2.58	2.47	2.38	2.86	2.68	2.54	2.42	2.31	2.23	2.70	2.53	2.40	2.28	2.19	2.10
SC4	63 × 225	3.42	3.21	3.04	2.90	2.78	2.68	3.21	3.01	2.85	2.72	2.60	2.50	3.04	2.85	2.70	2.57	2.55	2.36
SC4	75 × 125	2.03	1.90	1.80				1.90						1.80					
SC4	75 × 150	2.43	2.28	2.16	2.06	1.98	1.91	2.28	2.14	2.03	1.93	1.85		2.16	2.03	1.92	1.83		
SC4	75 × 175	2.83	2.66	2.52	2.40	2.31	2.22	2.66	2.49	2.36	2.25	2.16	2.08	2.52	2.36	2.24	2.13	2.04	1.96
SC4	75 × 200	3.23	3.03	2.88	2.74	2.63	2.54	3.03	2.85	2.70	2.57	2.47	2.37	2.88	2.70	2.55	2.43	2.33	2.24
SC4	75 × 225	3.63	3.41	3.23	3.08	2.96	2.85	3.41	3.20	3.03	2.89	2.77	2.67	3.23	3.03	2.87	2.74	2.62	2.52

AD A1/2, Appendix A

Table A15 Common or jack rafters for roofs having a pitch more than 30° but not more than 45° with access only for purposes of maintenance or repair. Imposed loading 1.00 kN/m² (see Diagram 2).

Maximum clear span of rafter (m) Timber of strength class SC3 and SC4 (see Table 1)

	Size of rafter	Dead Load [kN/m²] excluding the self weight of the rafter								
		Not more than 0.50			More than 0.50 but not more than 0.75			More than 0.75 but not more than 1.00		
		Spacing of rafters (mm)								
	(mm × mm)	400	450	600	400	450	600	400	450	600
SC3	38 × 100	2.28	2.23	2.03	2.10	2.05	1.91	1.96	1.91	1.76
	38 × 125	2.90	2.79	2.54	2.75	2.64	2.40	2.62	2.52	2.26
	38 × 150	3.47	3.34	3.04	3.29	3.16	2.87	3.13	3.01	2.69
	47 × 100	2.50	2.40	2.18	2.36	2.27	2.06	2.25	2.17	1.97
	47 × 125	3.11	2.99	2.72	2.94	2.83	2.58	2.81	2.70	2.45
	47 × 150	3.72	3.58	3.26	3.52	3.39	3.08	3.36	3.23	2.94
	50 × 100	2.55	2.45	2.23	2.41	2.32	2.11	2.30	2.21	2.01
	50 × 125	3.17	3.05	2.78	3.00	2.89	2.63	2.87	2.76	2.51
	50 × 150	3.79	3.65	3.33	3.59	3.46	3.15	3.43	3.30	3.00
	38 × 89	1.91	1.87	1.77	1.77	1.73	1.62	1.67	1.62	1.50
	38 × 140	3.24	3.12	2.84	3.07	2.95	2.68	2.93	2.82	2.52
SC4	38 × 100	2.42	2.33	2.12	2.29	2.20	2.00	2.18	2.10	1.90
	38 × 125	3.02	2.90	2.64	2.86	2.75	2.50	2.72	2.62	2.38
	38 × 150	3.61	3.47	3.16	3.42	3.29	2.99	3.26	3.14	2.85
	47 × 100	2.60	2.50	2.27	2.46	2.36	2.15	2.34	2.25	2.05
	47 × 125	3.23	3.11	2.83	3.06	2.95	2.68	2.92	2.81	2.55
	47 × 150	3.86	3.72	3.39	3.66	3.52	3.21	3.49	3.36	3.06
	50 × 100	2.65	2.55	2.32	2.51	2.41	2.19	2.39	2.30	2.09
	50 × 125	3.30	3.17	2.89	3.12	3.01	2.73	2.98	2.87	2.61
	50 × 150	3.94	3.79	3.46	3.73	3.60	3.27	3.57	3.43	3.12
	38 × 89	2.16	2.08	1.89	2.04	1.96	1.78	1.95	1.87	1.70
	38 × 140	3.37	3.25	2.95	3.19	3.07	2.79	3.05	2.93	2.66

AD A1/2, Appendix A

Table A16 Purlins supporting rafters to which Table A15 refers. (Imposed loading 1.00 kN/m^2)

Maximum clear span of purlin (m) Timber of strength class SC3 and SC4 (see Table 1)

		Dead Load [kN/m^2] excluding the self weight of the purlin																	
		Not more than 0.50						More than 0.50 but not more than 0.75						More than 0.75 but not more than 1.00					
		Spacing of purlins (mm)																	
	Size of purlin (mm × mm)	1500	1800	2100	2400	2700	3000	1500	1800	2100	2400	2700	3000	1500	1800	2100	2400	2700	3000
SC3	50 × 150	1.91						1.80											
	50 × 175	2.22	2.08	1.97	1.87			2.10	1.96	1.85				2.00	1.87				
	50 × 200	2.54	2.38	2.25	2.14	2.03	1.92	2.40	2.24	2.12	1.99	1.87		2.28	2.13	1.99	1.85		
	50 × 225	2.85	2.67	2.53	2.40	2.27	2.15	2.70	2.52	2.38	2.22	2.09	1.98	2.56	2.40	2.22	2.07	1.95	1.84
	63 × 150	2.07	1.94	1.84				1.96	1.83					1.86					
	63 × 175	2.41	2.26	2.14	2.04	1.95	1.88	2.28	2.14	2.02	1.92	1.84		2.17	2.03	1.92	1.83		
	63 × 200	2.76	2.58	2.44	2.33	2.23	2.14	2.61	2.44	2.31	2.20	2.10	2.00	2.48	2.32	2.19	2.09	1.97	1.86
	63 × 225	3.10	2.90	2.75	2.62	2.51	2.41	2.93	2.74	2.59	2.47	2.36	2.23	2.79	2.61	2.47	2.33	2.20	2.08
	75 × 125	1.84																	
	75 × 150	2.20	2.07	1.96	1.86			2.08	1.95	1.85				1.98	1.86				
	75 × 175	2.57	2.41	2.28	2.17	2.08	2.00	2.43	2.28	2.15	2.05	1.96	1.89	2.31	2.17	2.05	1.95	1.87	
	75 × 200	2.93	2.75	2.60	2.48	2.38	2.29	2.77	2.60	2.46	2.34	2.24	2.16	2.64	2.47	2.34	2.23	2.13	2.04
	75 × 225	3.29	3.09	2.92	2.79	2.67	2.57	3.12	2.92	2.76	2.63	2.52	2.43	2.97	2.78	2.63	2.50	2.40	2.28
SC4	50 × 150	1.99	1.87					1.88											
	50 × 175	2.32	2.18	2.06	1.96	1.88	1.80	2.20	2.06	1.94	1.85			2.09	1.96	1.85			
	50 × 200	2.65	2.49	2.35	2.24	2.14	2.06	2.51	2.35	2.22	2.11	2.02	1.94	2.39	2.23	2.11	2.00	1.91	
	50 × 225	2.98	2.79	2.64	2.52	2.41	2.31	2.82	2.64	2.49	2.37	2.27	2.18	2.68	2.51	2.37	2.25	2.15	2.01
	63 × 125	1.81																	
	63 × 150	2.16	2.03	1.92	1.83			2.05	1.92	1.81				1.95	1.83				
	63 × 175	2.52	2.36	2.24	2.13	2.04	1.97	2.39	2.24	2.11	2.01	1.93	1.85	2.27	2.13	2.01	1.91	1.83	
	63 × 200	2.88	2.70	2.56	2.44	2.33	2.24	2.72	2.55	2.41	2.30	2.20	2.12	2.59	2.43	2.30	2.19	2.09	2.01
	63 × 225	3.23	3.03	2.87	2.74	2.62	2.52	3.06	2.87	2.71	2.59	2.48	2.38	2.92	2.73	2.58	2.46	2.35	2.26
	75 × 125	1.92						1.82											
	75 × 150	2.30	2.16	2.04	1.95	1.87		2.18	2.04	1.93	1.84			2.07	1.94	1.84			
	75 × 175	2.68	2.51	2.38	2.27	2.18	2.10	2.54	2.38	2.25	2.15	2.06	1.98	2.42	2.27	2.14	2.04	1.96	1.88
	75 × 200	3.06	2.87	2.72	2.59	2.49	2.39	2.89	2.72	2.57	2.45	2.35	2.26	2.76	2.59	2.45	2.33	2.23	2.15

AD A1/2, Appendix A

Table A17 Joists for flat roofs with access only for purposes of maintenance or repair. Imposed loading 0.75 kN/m^2 (see Diagram 2).

Maximum clear span of joist (m) Timber of strength class SC3 (see Table 1)

Size of joist (mm × mm)	Dead Load [kN/m^2] excluding the self weight of the joist								
	Not more than 0.50			More than 0.50 but not more than 0.75			More than 0.75 but not more than 1.00		
	Spacing of joists (mm)								
	400	450	600	400	450	600	400	450	600
38 × 97	1.74	1.72	1.67	1.67	1.64	1.58	1.61	1.58	1.51
38 × 122	2.37	2.34	2.25	2.25	2.21	2.11	2.16	2.11	2.01
38 × 147	3.02	2.97	2.85	2.85	2.80	2.66	2.72	2.66	2.51
38 × 170	3.63	3.57	3.37	3.41	3.34	3.17	3.24	3.17	2.98
38 × 195	4.30	4.23	3.86	4.03	3.94	3.63	3.81	3.72	3.45
38 × 220	4.94	4.76	4.34	4.64	4.49	4.09	4.38	4.27	3.88
47 × 97	1.92	1.90	1.84	1.84	1.81	1.74	1.77	1.74	1.65
47 × 122	2.60	2.57	2.47	2.47	2.43	2.31	2.36	2.31	2.19
47 × 147	3.30	3.25	3.12	3.12	3.06	2.90	2.96	2.90	2.74
47 × 170	3.96	3.89	3.61	3.72	3.64	3.40	3.53	3.44	3.23
47 × 195	4.68	4.53	4.13	4.37	4.28	3.89	4.14	4.04	3.70
47 × 220	5.28	5.09	4.65	4.99	4.81	4.38	4.75	4.58	4.17
50 × 97	1.97	1.95	1.89	1.89	1.86	1.78	1.81	1.78	1.70
50 × 122	2.67	2.64	2.53	2.53	2.49	2.37	2.42	2.37	2.25
50 × 147	3.39	3.34	3.19	3.19	3.13	2.97	3.04	2.97	2.80
50 × 170	4.06	3.99	3.69	3.81	3.73	3.47	3.61	3.53	3.30
50 × 195	4.79	4.62	4.22	4.48	4.36	3.97	4.23	4.13	3.78
50 × 220	5.38	5.19	4.74	5.09	4.90	4.47	4.85	4.67	4.25
63 × 97	2.19	2.16	2.09	2.09	2.06	1.97	2.01	1.97	1.87
63 × 122	2.95	2.91	2.79	2.79	2.74	2.61	2.66	2.61	2.47
63 × 147	3.72	3.66	3.44	3.50	3.43	3.25	3.33	3.26	3.07
63 × 170	4.44	4.35	3.97	4.16	4.07	3.74	3.95	3.85	3.56
63 × 195	5.14	4.96	4.54	4.86	4.69	4.28	4.61	4.47	4.07
63 × 220	5.77	5.57	5.10	5.46	5.27	4.82	5.21	5.02	4.59
75 × 122	3.17	3.12	3.00	3.00	2.94	2.80	2.86	2.80	2.65
75 × 147	3.98	3.92	3.64	3.75	3.67	3.44	3.56	3.48	3.27
75 × 170	4.74	4.58	4.19	4.44	4.33	3.96	4.21	4.11	3.77
75 × 195	5.42	5.23	4.79	5.13	4.95	4.53	4.89	4.72	4.31
75 × 220	6.07	5.87	5.38	5.76	5.56	5.09	5.50	5.30	4.85
38 × 140	2.84	2.79	2.68	2.68	2.63	2.51	2.56	2.51	2.37
38 × 184	4.01	3.94	3.64	3.76	3.68	3.43	3.56	3.48	3.25

AD A1/2, Appendix A

Table A18 Joists for flat roofs with access only for purposes of maintenance or repair. Imposed loading 0.75 kN/m^2 (see Diagram 2).

Maximum clear span of joist (m) Timber of strength class SC4 (see Table 1)

Size of joist (mm × mm)	Dead Load [kN/m^2] excluding the self weight of the joist								
	Not more than 0.50			More than 0.50 but not more than 0.75			More than 0.75 but not more than 1.00		
	Spacing of joists (mm)								
	400	450	600	400	450	600	400	450	600
38 × 97	1.84	1.82	1.76	1.76	1.73	1.66	1.69	1.66	1.59
38 × 122	2.50	2.46	2.37	2.37	2.33	2.22	2.27	2.22	2.11
38 × 147	3.18	3.13	3.00	3.00	2.94	2.79	2.85	2.79	2.64
38 × 170	3.81	3.75	3.50	3.58	3.51	3.30	3.40	3.32	3.12
38 × 195	4.51	4.40	4.01	4.22	4.13	3.78	3.99	3.90	3.59
38 × 220	5.13	4.95	4.51	4.85	4.67	4.25	4.59	4.44	4.04
47 × 97	2.03	2.00	1.94	1.94	1.91	1.83	1.86	1.83	1.74
47 × 122	2.74	2.70	2.60	2.60	2.55	2.43	2.48	2.43	2.30
47 × 147	3.47	3.42	3.26	3.27	3.21	3.04	3.11	3.04	2.87
47 × 170	4.15	4.08	3.76	3.89	3.81	3.54	3.69	3.61	3.36
47 × 195	4.88	4.70	4.29	4.58	4.44	4.05	4.33	4.22	3.85
47 × 220	5.48	5.29	4.83	5.18	5.00	4.56	4.94	4.76	4.33
50 × 97	2.08	2.06	1.99	1.99	1.96	1.88	1.91	1.88	1.79
50 × 122	2.81	2.77	2.66	2.66	2.62	2.49	2.54	2.49	2.36
50 × 147	3.56	3.50	3.32	3.35	3.29	3.12	3.19	3.12	2.94
50 × 170	4.26	4.18	3.83	3.99	3.91	3.61	3.78	3.69	3.43
50 × 195	4.97	4.80	4.38	4.68	4.53	4.13	4.43	4.31	3.93
50 × 220	5.59	5.39	4.93	5.28	5.09	4.65	5.04	4.85	4.42
63 × 97	2.31	2.28	2.20	2.20	2.16	2.07	2.11	2.07	1.97
63 × 122	3.10	3.05	2.93	2.93	2.88	2.74	2.80	2.74	2.59
63 × 147	3.90	3.84	3.58	3.67	3.60	3.38	3.49	3.41	3.21
63 × 170	4.65	4.51	4.12	4.35	4.26	3.89	4.13	4.03	3.70
63 × 195	5.33	5.15	4.71	5.05	4.87	4.45	4.82	4.64	4.24
63 × 220	5.98	5.78	5.30	5.67	5.47	5.00	5.41	5.22	4.76
75 × 122	3.33	3.27	3.14	3.14	3.08	2.93	2.99	2.93	2.77
75 × 147	4.17	4.10	3.78	3.92	3.84	3.57	3.73	3.64	3.40
75 × 170	4.92	4.75	4.35	4.64	4.50	4.11	4.40	4.29	3.92
75 × 195	5.61	5.42	4.97	5.32	5.14	4.70	5.08	4.90	4.48
75 × 220	6.29	6.08	5.59	5.97	5.77	5.28	5.70	5.50	5.04
38 × 140	2.99	2.94	2.82	2.82	2.75	2.63	2.69	2.63	2.49
38 × 184	4.21	4.13	3.79	3.94	3.85	3.57	3.73	3.64	3.39

AD A1/2, Appendix A

Table A19 Joists for flat roofs with access only for purposes of maintenance or repair. Imposed loading 1.0 kN/m^2 (see Diagram 2).

Maximum clear span of joist (m) Timber of strength class SC3 (see Table 1)

Size of joist (mm × mm)	Dead Load [kN/m^2] excluding the self weight of the joist								
	Not more than 0.50			More than 0.50 but not more than 0.75			More than 0.75 but not more than 1.00		
	Spacing of joists (mm)								
	400	450	600	400	450	600	400	450	600
38 × 97	1.74	1.72	1.67	1.67	1.64	1.58	1.61	1.58	1.51
38 × 122	2.37	2.34	2.25	2.25	2.21	2.11	2.16	2.11	2.01
38 × 147	3.02	2.97	2.75	2.85	2.80	2.61	2.72	2.66	2.49
38 × 170	3.62	3.49	3.17	3.41	3.31	3.01	3.24	3.17	2.88
38 × 195	4.15	3.99	3.63	3.94	3.79	3.45	3.77	3.63	3.29
38 × 220	4.67	4.49	4.09	4.44	4.27	3.88	4.25	4.09	3.71
47 × 97	1.92	1.90	1.84	1.84	1.81	1.74	1.77	1.74	1.65
47 × 122	2.60	2.57	2.45	2.47	2.43	2.31	2.36	2.31	2.19
47 × 147	3.30	3.24	2.95	3.12	3.06	2.80	2.96	2.90	2.68
47 × 170	3.88	3.74	3.40	3.69	3.56	3.23	3.53	3.40	3.09
47 × 195	4.44	4.27	3.89	4.23	4.07	3.70	4.05	3.89	3.54
47 × 220	4.99	4.81	4.38	4.75	4.58	4.17	4.55	4.38	3.99
50 × 97	1.97	1.95	1.89	1.89	1.86	1.78	1.81	1.78	1.70
50 × 122	2.67	2.64	2.50	2.53	2.49	2.37	2.42	2.37	2.25
50 × 147	3.39	3.31	3.01	3.19	3.13	2.86	3.04	2.97	2.73
50 × 170	3.96	3.81	3.47	3.77	3.63	3.30	3.61	3.47	3.16
50 × 195	4.53	4.36	3.97	4.31	4.15	3.78	4.13	3.97	3.61
50 × 220	5.09	4.90	4.47	4.85	4.67	4.25	4.65	4.47	4.07
63 × 97	2.19	2.16	2.09	2.09	2.06	1.97	2.01	1.97	1.87
63 × 122	2.95	2.91	2.70	2.79	2.74	2.57	2.66	2.61	2.46
63 × 147	3.70	3.56	3.25	3.50	3.39	3.09	3.33	3.25	2.95
63 × 170	4.26	4.10	3.74	4.06	3.91	3.56	3.89	3.74	3.41
63 × 195	4.86	4.69	4.28	4.64	4.47	4.07	4.45	4.28	3.90
63 × 220	5.46	5.27	4.82	5.21	5.02	4.59	5.00	4.82	4.39
75 × 122	3.17	3.12	2.86	3.00	2.94	2.72	2.86	2.80	2.60
75 × 147	3.90	3.76	3.44	3.72	3.59	3.27	3.56	3.44	3.13
75 × 170	4.49	4.33	3.96	4.29	4.13	3.77	4.11	3.96	3.61
75 × 195	5.13	4.95	4.53	4.89	4.72	4.31	4.70	4.53	4.13
75 × 220	5.76	5.56	5.09	5.50	5.30	4.85	5.28	5.09	4.65
38 × 140	2.84	2.79	2.62	2.68	2.63	2.48	2.56	2.51	2.37
38 × 184	3.92	3.77	3.43	3.73	3.58	3.25	3.56	3.43	3.11

AD A1/2, Appendix A

Table A20 Joists for flat roofs with access only for purposes of maintenance or repair. Imposed loading 1.0 kN/m^2 (see Diagram 2).

Maximum clear span of joist (m) Timber of strength class SC4 (see Table 1)

Size of joist (mm × mm)	Dead Load [kN/m^2] excluding the self weight of the joist								
	Not more than 0.50			More than 0.50 but not more than 0.75			More than 0.75 but not more than 1.00		
	Spacing of joists (mm)								
	400	450	600	400	450	600	400	450	600
38 × 97	1.84	1.82	1.76	1.76	1.73	1.66	1.69	1.66	1.59
38 × 122	2.50	2.46	2.37	2.37	2.33	2.22	2.27	2.22	2.11
38 × 147	3.18	3.13	2.86	3.00	2.94	2.71	2.85	2.79	2.59
38 × 170	3.77	3.63	3.30	3.58	3.45	3.13	3.40	3.30	2.99
38 × 195	4.31	4.15	3.78	4.10	3.95	3.59	3.93	3.78	3.43
38 × 220	4.85	4.67	4.25	4.61	4.44	4.04	4.42	4.25	3.86
47 × 97	2.03	2.00	1.94	1.94	1.91	1.83	1.86	1.83	1.74
47 × 122	2.74	2.70	2.55	2.60	2.55	2.42	2.48	2.43	2.30
47 × 147	3.47	3.37	3.07	3.27	3.21	2.91	3.11	3.04	2.79
47 × 170	4.03	3.88	3.54	3.84	3.70	3.36	3.68	3.54	3.22
47 × 195	4.61	4.44	4.05	4.39	4.23	3.85	4.21	4.05	3.68
47 × 220	5.18	5.00	4.56	4.94	4.76	4.33	4.73	4.56	4.15
50 × 97	2.08	2.06	1.99	1.99	1.96	1.88	1.91	1.88	1.79
50 × 122	2.81	2.77	2.60	2.66	2.62	2.47	2.54	2.49	2.36
50 × 147	3.56	3.44	3.13	3.35	3.27	2.97	3.19	3.12	2.85
50 × 170	4.11	3.96	3.61	3.92	3.77	3.43	3.75	3.61	3.28
50 × 195	4.70	4.53	4.13	4.48	4.31	3.93	4.29	4.13	3.76
50 × 220	5.28	5.09	4.65	5.04	4.85	4.42	4.83	4.65	4.23
63 × 97	2.31	2.28	2.20	2.20	2.16	2.07	2.11	2.07	1.97
63 × 122	3.10	3.05	2.81	2.93	2.88	2.67	2.80	2.74	2.56
63 × 147	3.84	3.70	3.38	3.66	3.52	3.21	3.49	3.38	3.07
63 × 170	4.42	4.26	3.89	4.21	4.06	3.70	4.04	3.89	3.54
63 × 195	5.05	4.87	4.45	4.81	4.64	4.24	4.62	4.45	4.06
63 × 220	5.67	5.47	5.00	5.41	5.22	4.76	5.19	5.00	4.56
75 × 122	3.33	3.26	2.97	3.14	3.08	2.83	2.99	2.93	2.71
75 × 147	4.05	3.91	3.57	3.86	3.72	3.40	3.71	3.57	3.25
75 × 170	4.66	4.50	4.11	4.45	4.29	3.92	4.27	4.11	3.75
75 × 195	5.32	5.14	4.70	5.08	4.90	4.48	4.88	4.70	4.29
75 × 220	5.97	5.77	5.28	5.70	5.50	5.04	5.48	5.28	4.83
38 × 140	2.99	2.94	2.72	2.82	2.77	2.59	2.69	2.63	2.47
38 × 184	4.07	3.92	3.57	3.87	3.73	3.39	3.71	3.57	3.24

AD A1/2, Appendix A

Table A21 Joists for flat roofs with access not limited to the purposes of maintenance or repair. Imposed loading 1.50 kN/m^2.

Maximum clear span of joist (m) Timber of strength class SC3 (see Table 1)

Size of joist (mm × mm)	Dead Load [kN/m^2] excluding the self weight of the joist								
	Not more than 0.50			More than 0.50 but not more than 0.75			More than 0.75 but not more than 1.00		
	Spacing of joists (mm)								
	400	450	600	400	450	600	400	450	600
38 × 122	1.80	1.79	1.74	1.74	1.71	1.65	1.68	1.65	1.57
38 × 147	2.35	2.33	2.27	2.27	2.25	2.18	2.21	2.18	2.09
38 × 170	2.88	2.85	2.77	2.77	2.74	2.64	2.68	2.64	2.53
38 × 195	3.47	3.43	3.29	3.33	3.28	3.16	3.21	3.16	3.02
38 × 220	4.08	4.03	3.71	3.90	3.84	3.56	3.75	3.68	3.43
47 × 122	2.00	1.99	1.94	1.94	1.93	1.87	1.89	1.87	1.81
47 × 147	2.60	2.58	2.51	2.51	2.48	2.40	2.44	2.40	2.31
47 × 170	3.18	3.14	3.06	3.06	3.02	2.91	2.95	2.91	2.78
47 × 195	3.82	3.78	3.54	3.66	3.61	3.40	3.52	3.46	3.28
47 × 220	4.48	4.38	3.99	4.27	4.20	3.83	4.10	4.03	3.70
50 × 122	2.06	2.05	2.00	2.00	1.98	1.93	1.95	1.93	1.86
50 × 147	2.68	2.65	2.59	2.59	2.56	2.47	2.51	2.47	2.38
50 × 170	3.27	3.23	3.14	3.14	3.10	2.99	3.04	2.99	2.86
50 × 195	3.93	3.88	3.61	3.76	3.70	3.47	3.62	3.56	3.35
50 × 220	4.60	4.47	4.07	4.38	4.30	3.91	4.21	4.13	3.78
63 × 97	1.67	1.66	1.63	1.63	1.61	1.57	1.59	1.57	1.53
63 × 122	2.31	2.29	2.24	2.24	2.21	2.15	2.17	2.15	2.07
63 × 147	2.98	2.95	2.87	2.87	2.84	2.74	2.78	2.74	2.63
63 × 170	3.62	3.59	3.41	3.48	3.43	3.28	3.36	3.30	3.16
63 × 195	4.34	4.29	3.90	4.15	4.08	3.75	3.99	3.92	3.62
63 × 220	5.00	4.82	4.39	4.82	4.64	4.22	4.62	4.48	4.08
75 × 122	2.50	2.48	2.42	2.42	2.40	2.32	2.35	2.32	2.24
75 × 147	3.23	3.19	3.11	3.11	3.07	2.96	3.00	2.96	2.84
75 × 170	3.91	3.87	3.61	3.75	3.69	3.47	3.61	3.55	3.35
75 × 195	4.66	4.53	4.13	4.45	4.36	3.97	4.28	4.20	3.84
75 × 220	5.28	5.09	4.65	5.09	4.90	4.47	4.92	4.74	4.32
38 × 140	2.19	2.17	2.12	2.12	2.10	2.04	2.07	2.04	1.94
38 × 184	3.21	3.17	3.08	3.08	3.04	2.93	2.98	2.93	2.80

AD A1/2, Appendix A

Table A22 Joists for flat roofs with access not limited to the purposes of maintenance or repair. Imposed loading 1.50 kN/m^2.

Maximum clear span of joist (m) Timber of strength class SC4 (see Table 1)

Size of joist (mm × mm)	Dead Load [kN/m^2] excluding the self weight of the joist								
	Not more than 0.50			More than 0.50 but not more than 0.75			More than 0.75 but not more than 1.00		
	Spacing of joists (mm)								
	400	450	600	400	450	600	400	450	600
38 × 122	1.91	1.90	1.86	1.86	1.84	1.79	1.81	1.79	1.73
38 × 147	2.49	2.46	2.40	2.40	2.38	2.30	2.33	2.30	2.21
38 × 170	3.04	3.01	2.93	2.93	2.89	2.79	2.83	2.79	2.67
38 × 195	3.66	3.62	3.43	3.51	3.46	3.29	3.38	3.33	3.18
38 × 220	4.30	4.25	3.86	4.10	4.04	3.71	3.94	3.87	3.58
47 × 122	2.12	2.10	2.06	2.06	2.04	1.98	2.00	1.98	1.91
47 × 147	2.75	2.73	2.66	2.66	2.62	2.54	2.57	2.54	2.44
47 × 170	3.35	3.32	3.22	3.22	3.18	3.06	3.11	3.06	2.93
47 × 195	4.03	3.98	3.68	3.85	3.80	3.54	3.71	3.64	3.42
47 × 220	4.71	4.56	4.15	4.49	4.39	3.99	4.31	4.23	3.85
50 × 122	2.19	2.17	2.12	2.12	2.10	2.04	2.06	2.04	1.97
50 × 147	2.83	2.81	2.73	2.73	2.70	2.61	2.65	2.61	2.51
50 × 170	3.45	3.41	3.28	3.31	3.27	3.15	3.20	3.15	3.01
50 × 195	4.14	4.09	3.76	3.96	3.90	3.61	3.81	3.74	3.49
50 × 220	4.83	4.65	4.23	4.61	4.47	4.07	4.42	4.32	3.93
63 × 97	1.77	1.75	1.72	1.72	1.71	1.66	1.68	1.66	1.61
63 × 122	2.44	2.42	2.36	2.36	2.34	2.27	2.30	2.27	2.18
63 × 147	3.15	3.12	3.03	3.03	2.99	2.89	2.93	2.89	2.77
63 × 170	3.82	3.78	3.54	3.66	3.61	3.41	3.53	3.47	3.29
63 × 195	4.56	4.45	4.06	4.36	4.29	3.90	4.19	4.11	3.77
63 × 220	5.19	5.00	4.56	5.00	4.82	4.39	4.84	4.66	4.24
75 × 122	2.64	2.62	2.56	2.56	2.53	2.45	2.48	2.45	2.36
75 × 147	3.40	3.36	3.25	3.27	3.23	3.11	3.16	3.11	2.98
75 × 170	4.11	4.07	3.75	3.94	3.88	3.61	3.79	3.73	3.49
75 × 195	4.79	4.70	4.29	4.67	4.53	4.13	4.49	4.38	3.99
75 × 220	5.48	5.28	4.83	5.28	5.09	4.65	5.11	4.93	4.49
38 × 140	2.32	2.30	2.25	2.25	2.22	2.16	2.19	2.16	2.08
38 × 184	3.39	3.35	3.24	3.25	3.21	3.09	3.14	3.09	2.95

AD A1/2, Appendix A

Table A23 Purlins supporting sheeting or decking for roofs having a pitch more than 10° but not more than 35°. Imposed loading 0.75 kN/m^2.

Maximum clear span of purlin (m) Timber of strength class SC3 and SC4 (see Table 1)

	Size of purlin (mm × mm)	Dead Load [kN/m^2] excluding the self weight of the purlin																	
		Not more than 0.25						More than 0.25 but not more than 0.50						More than 0.50 but not more than 0.75					
		Spacing of purlins (mm)																	
		900	1200	1500	1800	2100	2400	900	1200	1500	1800	2100	2400	900	1200	1500	1800	2100	2400
SC3	50 × 100	1.68	1.63	1.51	1.42	1.34	1.28	1.55	1.48	1.40	1.31	1.24	1.18	1.45	1.37	1.31	1.22	1.16	1.10
	50 × 125	2.24	2.03	1.88	1.77	1.67	1.60	2.06	1.88	1.74	1.63	1.54	1.47	1.91	1.77	1.63	1.53	1.44	1.37
	50 × 150	2.68	2.44	2.26	2.12	2.01	1.91	2.49	2.26	2.09	1.96	1.85	1.76	2.34	2.12	1.96	1.83	1.73	1.65
	50 × 175	3.12	2.84	2.63	2.47	2.34	2.23	2.90	2.63	2.43	2.28	2.16	2.06	2.72	2.47	2.28	2.13	2.02	1.92
	50 × 200	3.56	3.24	3.00	2.82	2.67	2.55	3.31	3.00	2.78	2.60	2.46	2.35	3.11	2.81	2.60	2.44	2.30	2.19
	50 × 225	4.00	3.63	3.37	3.17	3.00	2.86	3.71	3.37	3.12	2.93	2.77	2.64	3.49	3.16	2.92	2.74	2.59	2.47
	63 × 100	1.87	1.77	1.64	1.54	1.46	1.39	1.72	1.64	1.51	1.42	1.34	1.28	1.60	1.52	1.42	1.33	1.26	1.20
	63 × 125	2.42	2.20	2.04	1.92	1.82	1.73	2.25	2.04	1.89	1.77	1.68	1.60	2.10	1.91	1.77	1.66	1.57	1.50
	63 × 150	2.90	2.63	2.44	2.30	2.18	2.08	2.69	2.44	2.26	2.12	2.01	1.92	2.53	2.29	2.12	2.00	1.88	1.79
	63 × 175	3.37	3.07	2.85	2.67	2.54	2.42	3.13	2.84	2.63	2.47	2.34	2.23	2.94	2.67	2.47	2.32	2.19	2.09
	63 × 200	3.84	3.50	3.25	3.05	2.89	2.76	3.57	3.24	3.01	2.82	2.67	2.55	3.36	3.05	2.82	2.65	2.51	2.39
	63 × 225	4.31	3.92	3.64	3.43	3.25	3.10	4.01	3.64	3.38	3.17	3.01	2.87	3.77	3.42	3.17	2.97	2.82	2.68
SC4	50 × 100	1.79	1.71	1.58	1.48	1.40	1.34	1.64	1.57	1.46	1.37	1.30	1.23	1.53	1.45	1.37	1.28	1.21	1.15
	50 × 125	2.34	2.13	1.97	1.85	1.75	1.67	2.17	1.97	1.82	1.71	1.62	1.54	2.02	1.85	1.71	1.60	1.51	1.44
	50 × 150	2.80	2.55	2.36	2.22	2.10	2.00	2.60	2.36	2.18	2.05	1.94	1.85	2.44	2.21	2.05	1.92	1.81	1.73
	50 × 175	3.26	2.97	2.75	2.58	2.45	2.34	3.03	2.75	2.54	2.39	2.26	2.15	2.85	2.58	2.39	2.24	2.12	2.01
	50 × 200	3.72	3.38	3.14	2.95	2.79	2.67	3.45	3.13	2.90	2.73	2.58	2.46	3.25	2.94	2.72	2.55	2.42	2.30
	50 × 225	4.17	3.80	3.52	3.31	3.14	3.00	3.88	3.52	3.26	3.06	2.90	2.77	3.65	3.31	3.06	2.87	2.72	2.59
	63 × 100	1.99	1.84	1.71	1.61	1.52	1.45	1.81	1.71	1.58	1.49	1.41	1.34	1.69	1.60	1.48	1.39	1.32	1.26
	63 × 125	2.53	2.30	2.13	2.00	1.90	1.81	2.35	2.13	1.97	1.85	1.76	1.68	2.21	2.00	1.85	1.74	1.65	1.57
	63 × 150	3.02	2.75	2.55	2.40	2.28	2.17	2.81	2.55	2.37	2.22	2.10	2.01	2.64	2.40	2.22	2.08	1.97	1.88
	63 × 175	3.52	3.20	2.97	2.80	2.65	2.53	3.27	2.97	2.76	2.59	2.45	2.34	3.08	2.79	2.59	2.43	2.30	2.19
	63 × 200	4.01	3.65	3.39	3.19	3.03	2.89	3.73	3.39	3.14	2.95	2.80	2.67	3.51	3.19	2.95	2.77	2.62	2.50
	63 × 225	4.49	4.10	3.81	3.58	3.40	3.25	4.18	3.80	3.53	3.32	3.15	3.00	3.94	3.58	3.32	3.11	2.95	2.81

AD A1/2, Appendix A

Table A24 Purlins supporting sheeting or decking for roofs having a pitch more than 10° but not more than 35°. Imposed loading 1.0 kN/m².

Maximum clear span of purlin (m) Timber of strength class SC3 and SC4 (see Table 1)

		Dead Load [kN/m²] excluding the self weight of the purlin																	
		Not more than 0.25						More than 0.25 but not more than 0.50						More than 0.50 but not more than 0.75					
		Spacing of purlins (mm)																	
	Size of purlin (mm × mm)	900	1200	1500	1800	2100	2400	900	1200	1500	1800	2100	2400	900	1200	1500	1800	2100	2400
SC3	50 × 100	1.67	1.51	1.40	1.31	1.24	1.18	1.55	1.42	1.31	1.22	1.16	1.10	1.45	1.34	1.24	1.16	1.09	1.04
	50 × 125	2.08	1.88	1.74	1.64	1.55	1.47	1.95	1.77	1.63	1.53	1.45	1.38	1.85	1.67	1.54	1.44	1.36	1.30
	50 × 150	2.49	2.26	2.09	1.96	1.85	1.77	2.34	2.12	1.96	1.83	1.73	1.65	2.22	2.00	1.85	1.73	1.64	1.56
	50 × 175	2.90	2.63	2.43	2.28	2.16	2.06	2.73	2.47	2.28	2.14	2.02	1.92	2.58	2.34	2.16	2.02	1.91	1.81
	50 × 200	3.31	3.00	2.78	2.61	2.47	2.35	3.11	2.82	2.60	2.44	2.31	2.20	2.95	2.67	2.46	2.31	2.18	2.07
	50 × 225	3.72	3.37	3.12	2.93	2.77	2.64	3.49	3.16	2.93	2.74	2.59	2.47	3.31	3.00	2.77	2.59	2.45	2.31
	63 × 100	1.80	1.64	1.51	1.42	1.35	1.28	1.69	1.54	1.42	1.33	1.26	1.20	1.60	1.45	1.34	1.26	1.19	1.13
	63 × 125	2.25	2.04	1.89	1.77	1.68	1.60	2.11	1.92	1.77	1.66	1.57	1.50	2.00	1.81	1.68	1.57	1.49	1.41
	63 × 150	2.69	2.44	2.26	2.13	2.01	1.92	2.53	2.29	2.12	1.99	1.88	1.80	2.40	2.17	2.01	1.88	1.78	1.70
	63 × 175	3.13	2.85	2.64	2.48	2.35	2.24	2.95	2.67	2.47	2.32	2.20	2.09	2.80	2.53	2.34	2.20	2.08	1.98
	63 × 200	3.57	3.25	3.01	2.83	2.68	2.55	3.36	3.05	2.82	2.65	2.51	2.39	3.19	2.89	2.67	2.51	2.37	2.26
	63 × 225	4.01	3.65	3.38	3.18	3.01	2.87	3.77	3.43	3.17	2.98	2.82	2.69	3.58	3.25	3.01	2.82	2.67	2.54
SC4	50 × 100	1.74	1.58	1.46	1.37	1.30	1.24	1.64	1.48	1.37	1.28	1.21	1.16	1.53	1.40	1.30	1.21	1.15	1.09
	50 × 125	2.17	1.97	1.82	1.71	1.62	1.54	2.04	1.85	1.71	1.60	1.52	1.44	1.94	1.75	1.62	1.51	1.43	1.36
	50 × 150	2.60	2.36	2.19	2.05	1.94	1.85	2.45	2.22	2.05	1.92	1.82	1.73	2.32	2.10	1.94	1.82	1.72	1.63
	50 × 175	3.03	2.75	2.55	2.39	2.26	2.16	2.85	2.58	2.39	2.24	2.12	2.02	2.70	2.45	2.26	2.12	2.00	1.90
	50 × 200	3.46	3.14	2.91	2.73	2.58	2.46	3.25	2.95	2.73	2.56	2.42	2.30	3.08	2.79	2.58	2.42	2.28	2.17
	50 × 225	3.88	3.52	3.27	3.07	2.90	2.77	3.65	3.31	3.06	2.87	2.72	2.59	3.46	3.14	2.90	2.72	2.57	2.44
	63 × 100	1.89	1.71	1.58	1.49	1.41	1.34	1.77	1.61	1.49	1.39	1.32	1.26	1.68	1.52	1.41	1.32	1.25	1.19
	63 × 125	2.35	2.13	1.98	1.86	1.76	1.68	2.21	2.00	1.85	1.74	1.65	1.57	2.10	1.90	1.76	1.65	1.56	1.48
	63 × 150	2.81	2.55	2.37	2.22	2.11	2.01	2.65	2.40	2.22	2.08	1.97	1.88	2.51	2.27	2.10	1.97	1.87	1.78
	63 × 175	3.27	2.97	2.76	2.59	2.46	2.34	3.08	2.79	2.59	2.43	2.30	2.19	2.92	2.65	2.45	2.30	2.18	2.07
	63 × 200	3.73	3.39	3.15	2.96	2.80	2.67	3.51	3.19	2.95	2.77	2.63	2.50	3.33	3.02	2.80	2.63	2.48	2.37
	63 × 225	4.18	3.81	3.53	3.32	3.15	3.01	3.94	3.58	3.32	3.12	2.95	2.81	3.74	3.30	3.15	2.95	2.79	2.66

6.5.3 Structural work of bricks, blocks and plain concrete

If a wall of these materials comes within the scope of Section 1C of AD A1/2, it is not necessary to calculate loads or wall thicknesses, provided the wall is built with the thicknesses required by section 1C and complies with the rules therein and in all other respects complies with BS 5628 *Code of practice for use of masonry*, Part 1: 1978 *Structural use of unreinforced masonry* and Part 3: 1985 *Materials and components, design and workmanship*.

Section 1C may be applied to any wall which is:

- an external wall, compartment wall, internal load-bearing wall or separating wall of a residential building of not more than three storeys, *and*,
- an external wall or internal load-bearing wall of a small single-storey non-residential building or small annexe to a residential building (such as a garage or outbuilding) provided that:
  - (a) the building design complies with the requirements of paragraphs 1C14 to 1C17 of section 1C, *and*,
  - (b) the wall construction details comply with the requirements of paragraphs 1C18 to 1C39 of section 1C.

6.5.4 Building design requirements (section 1C, paragraphs 1C14 to 1C17)

These are concerned with the design wind speed, the imposed load, the building proportions and the plan area of each storey or sub-division.

DESIGN WIND SPEED (V_s). When determined in accordance with CP3: Chapter V *Loading*, Part 2: 1972 *Wind loads*, this should not exceed 44 metres/second.

In order to determine the value of V_s it is necessary to carry out a fairly complicated calculation from CP3 which relates the design wind speed to the basic wind speed adjusted to take account of site conditions and the design of the building. This has been simplifed in AD A1/2, the design wind speed being obtained using the following procedure:

- Use Diagram 10 (Map showing basic wind speeds in m/s) to determine the basic wind speed for the building by reference to its location.
- Look up the maximum building height permitted in either Table 8 (Maximum height of buildings on normal or slightly sloping sites) or Table 9 (Maximum height of buildings on steeply sloping sites, including hill, cliff and escarpment sites) on the line which corresponds to the basic wind speed.

The height obtained should not be exceeded in the building design.

Tables 8 and 9, and Diagram 10 are reproduced from AD A1/2.

IMPOSED LOADS. These should not exceed:

- on any floor, 2.0 kN/m^2 distributed;

AD A1/2

Table 8 Maximum height of buildings on normal or slightly sloping sites.

	Maximum building height in metres			
	Location			
Basic wind speed m/s	Unprotected sites, open countryside with no obstructions	Open countryside with scattered windbreaks	Country with many windbreaks, small towns, outskirts of large cities	Protected sites, city centres
36	15	15	15	15
38	15	15	15	15
40	15	15	15	15
42	15	15	15	15
44	15	15	15	15
46	11	15	15	15
48	9	13	15	15

AD A1/2

Table 9 Maximum height of buildings on steeply sloping sites, including hill, cliff and escarpment sites.

	Maximum building height in metres			
	Location			
Basic wind speed m/s	Unprotected sites, open countryside with no obstructions	Open countryside with scattered windbreaks	Country with many windbreaks, small towns, outskirts of large cities	Protected sites, city centres
36	8	11	15	15
38	6	9	15	15
40	4	7.5	14	15
42	3	6	12	15
44	0*	5	10	15
46	0*	4	8	15
48	0*	3	6.5	14

* Section 1C guidance is not applicable.

- on any ceiling, 0.25 kN/m^2 distributed and 0.9 kN concentrated; *and*,
- on any roof, 1.00 kN/m^2 for spans not exceeding 12 m, or 1.5 kN/m^2 for spans not exceeding 6 m.

BUILDING PROPORTIONS. For residential buildings of not more than three storeys:

- the height of any part of a wall or roof of the building should not exceed 15 m, as measured from the lowest finished surface of the ground adjoining the building;
- the width of the building should not be less than at least half the height of the building;

AD A1/2

Diagram 10 Map showing basic wind speeds in m/s.

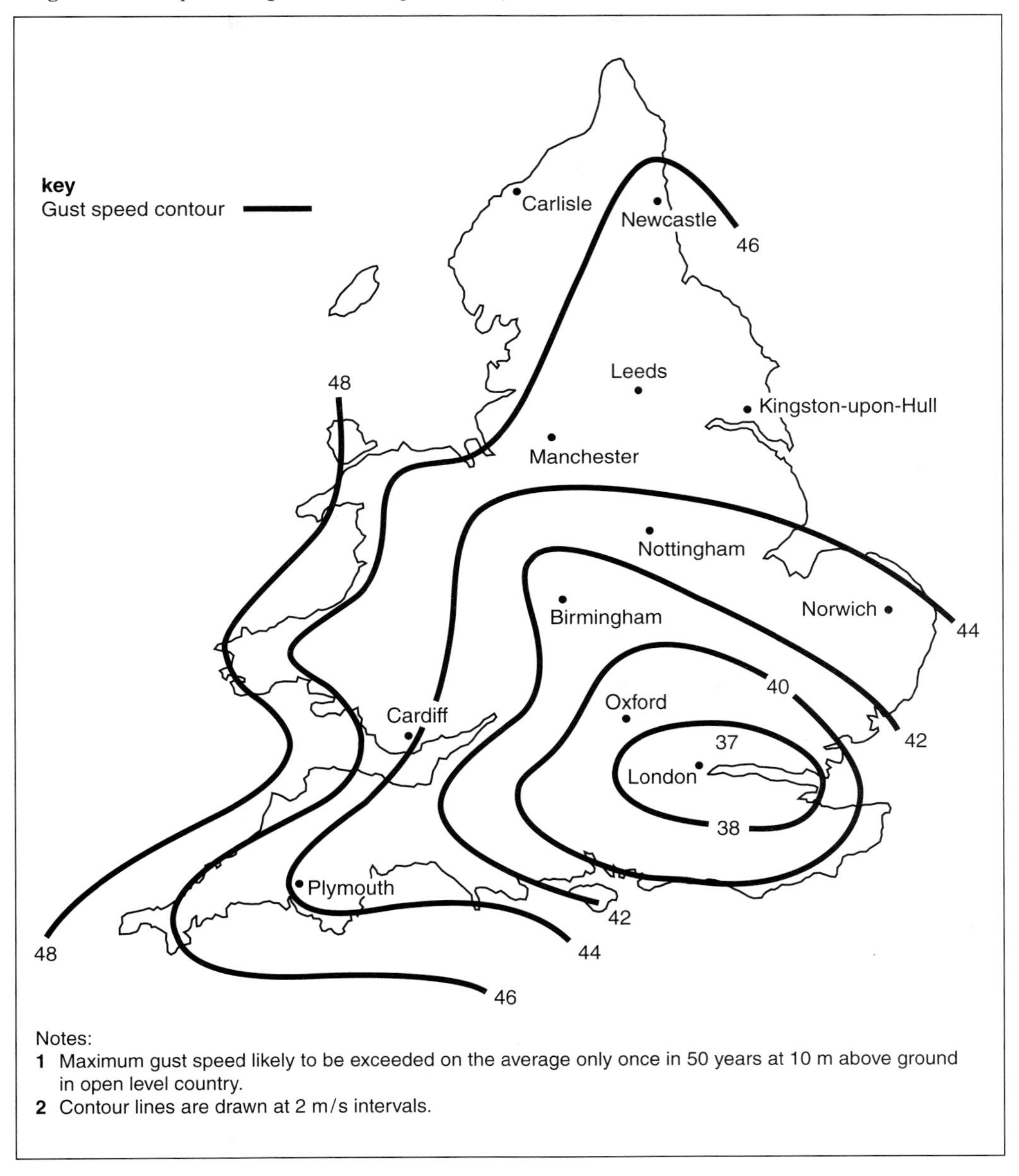

- any wing of the building which projects more than twice its own width from the remainder of the building should have a width at least equal to half its height. ('Height' is measured to the highest part of any roof or wall of the building or wing.)

For small single-storey non-residential buildings:

- the height of the building should not exceed 3 m;
- the width of the building measured in the direction of the roof span should not exceed 9 m.

For annexes attached to residential buildings the height of any part should not exceed 3 m.

The heights mentioned above may need to be reduced in line with the design wind speed requirement of 44 metres/second maximum. (See above, Tables 8 and 9 and Diagram 10.)

PLAN AREA OF STOREY. The plan area of each storey which is completely bounded by structural walls on all sides should not be more than 70 m^2. However, if the storey is bounded in this way on all sides but one, the limiting area is 30 m^2.

These requirements are summarised in Figs. 6.3 and 6.4.

6.55 Wall construction requirements (Section 1C, Paragraphs 1C18 to 1C39)

These are concerned with height and length, materials, buttressing, loading conditions, openings and recesses, and lateral support.

Height and length

The height or length of a wall should not be more than 12m and together with storey heights should be measured in accordance with the following rules:

- The height of the ground storey of a building is measured from the base of the wall to the underside of the next floor above.
- The height of an upper storey is measured from the level of the underside of the floor of that storey, in each case to the level of the underside of the next floor above.
- For a top storey which comprises a gable wall, measure to a level midway between the gable base and the top of the roof lateral support along the line of the roof slope, but if there is also lateral support at about ceiling level, to the level of that lateral support.
- Where a compartment or separating wall (as defined) comprises a gable, measure the height from its base to the base of the gable.
- Any other gable wall (except a compartment or separating wall) should be measured from its base to half the height of the gable.
- Any wall which is not a gable wall should be measured from its base to its highest part, excluding any parapet not exceeding 1.2 m in height.
- Walls are regarded as being divided into separate lengths by securely tied buttressing walls, piers or chimneys for the purposes of measuring their length.

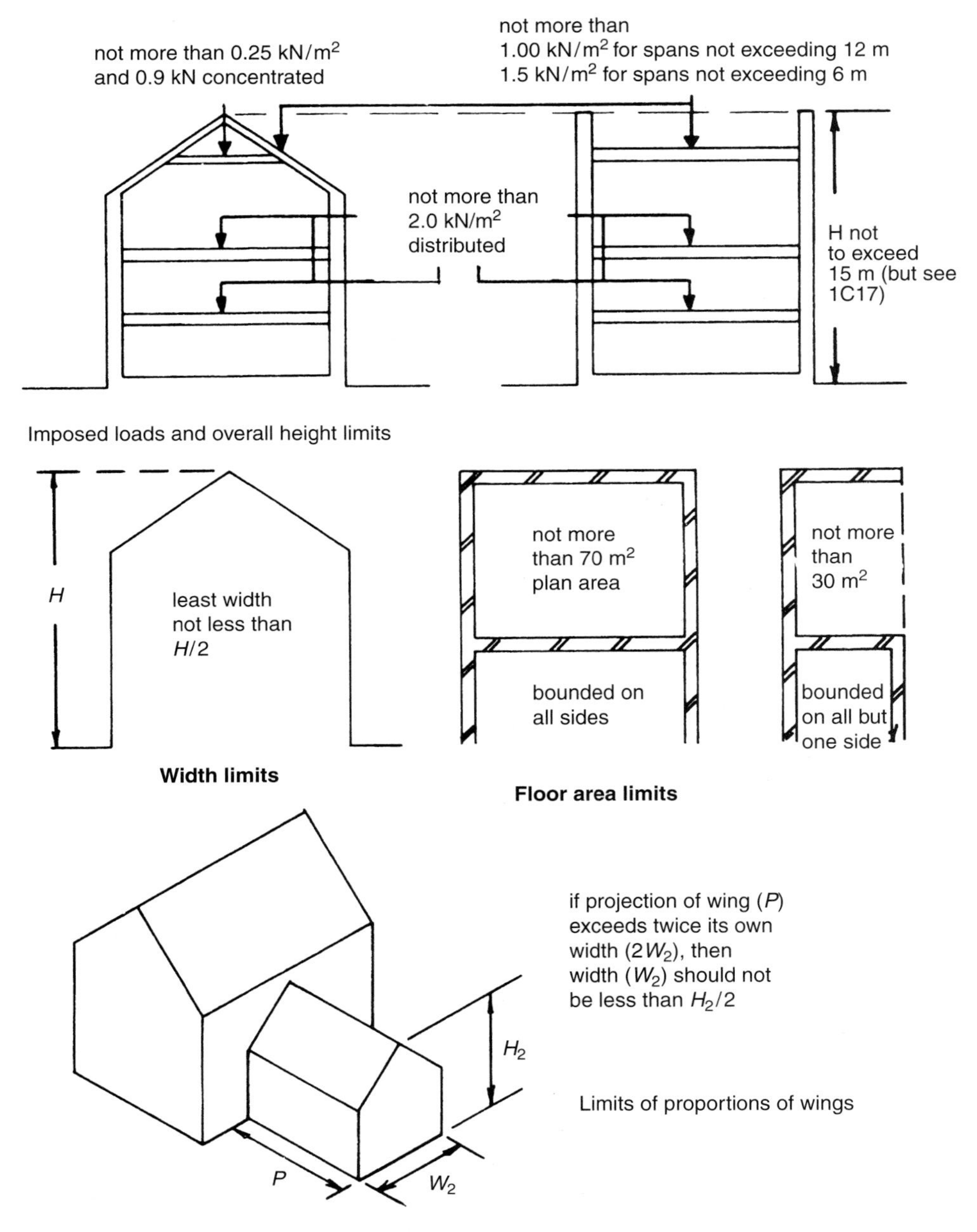

Fig. 6.3 Building design requirements for residential buildings not exceeding three storeys in height.

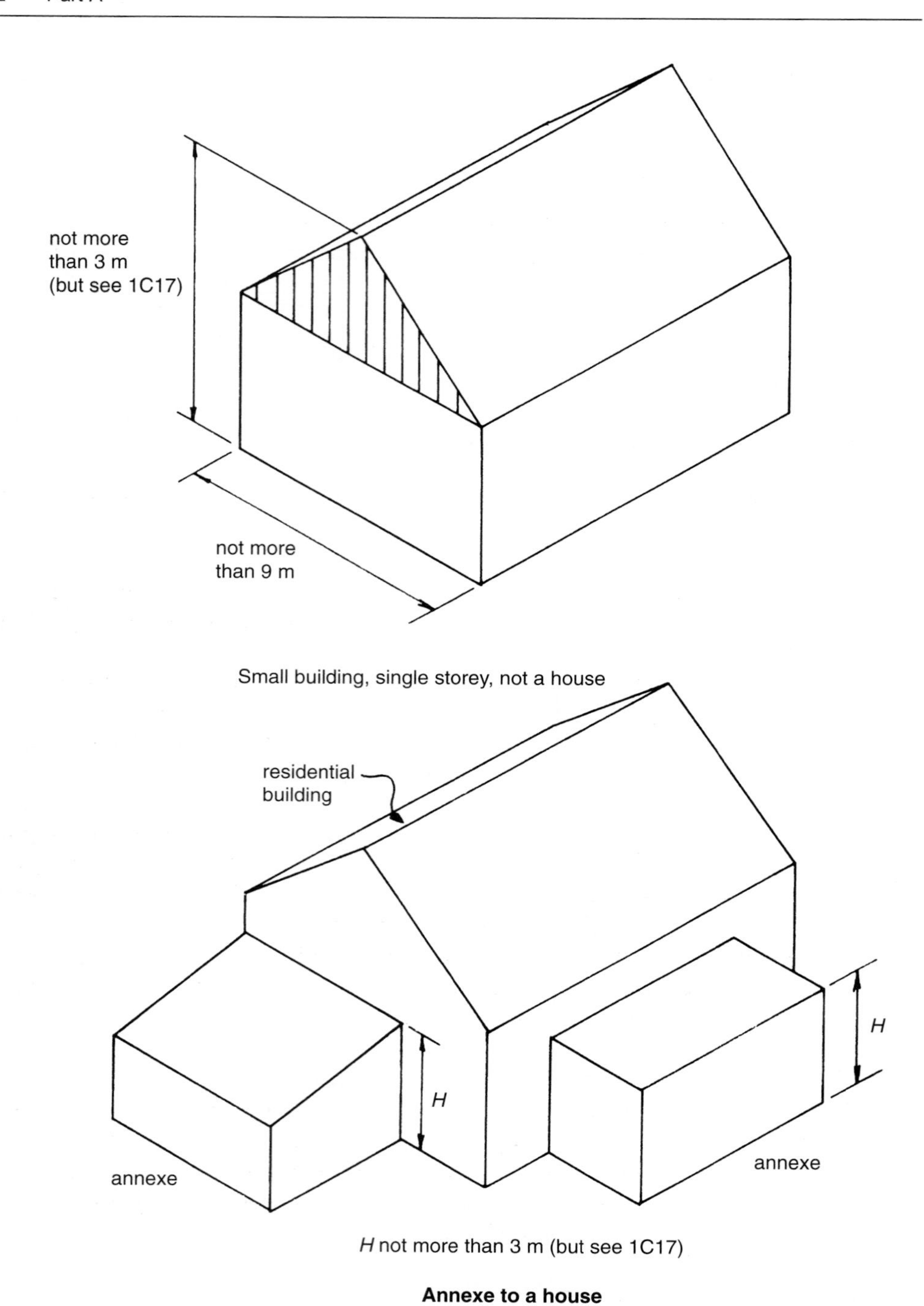

Fig. 6.4 Building design requirements for small non-residential buildings and annexes.

These separate lengths are measured centre to centre of the piers, etc. These special requirements are noted in Fig. 6.5.

Materials and workmanship

BRICKS AND BLOCKS. The wall should be constructed of bricks or blocks, properly bonded and solidly put together with mortar. The materials should comply with the following standards:

- Clay bricks or blocks to BS 3921: 1985 or BS 6649: 1985
- Calcium silicate bricks to BS 187: 1978 or BS 6649: 1985
- Concrete bricks or blocks to BS 6073: Part 1: 1981
- Square dressed natural stone to BS 5390: 1976 (1984).

Additionally, bricks or blocks should have a compressive strength of not less than the following:

(a) when used in any part of a wall with storey heights not exceeding 2.7 m (except in the outer leaf of an external cavity wall of a three-storey building), 5 N/mm^2 for bricks and 2.8 N/mm^2 for blocks;
(b) when used in the inner leaf of a ground storey external cavity wall of a three-storey building, 15 N/mm^2 for bricks and 7 N/mm^2 for blocks;
(c) when used in any circumstances other than those described in (a) or (b) above, 7 N/mm^2.

The above guidance to compressive strengths of blocks or bricks is only applicable where the roof structure is of timber construction. Additionally, the ground storey part of an internal wall in a three-storey building should be at least 140 mm thick if in blockwork or 215 mm thick if in brickwork.

It should be noted that in determining the ground storey height for the purposes of (a) above, the measurement is made from the upper surface of the ground floor and not from the base of the wall.

MORTAR. The mortar used in any wall to which section 1C of AD A1/2 applies should be at least equal in strength to a 1:1:6 Portland cement/lime/fine aggregate mortar measured by volume of dry materials, or to the proportions given in BS 5628 *Code of practice for use of masonry*, Part 1: 1978 (1985) *Structural use of unreinforced masonry* for mortar designation (iii). The mortar should be compatible with the masonry units and the position of use.

WALL TIES. These should comply with BS 1243: 1978 *Specification for metal ties for cavity wall construction* unless conditions of severe exposure occur. In that case austenitic stainless steel or suitable non-ferrous ties should be used. (Severe exposure is defined in BS 5628: Part 3: 1985.)

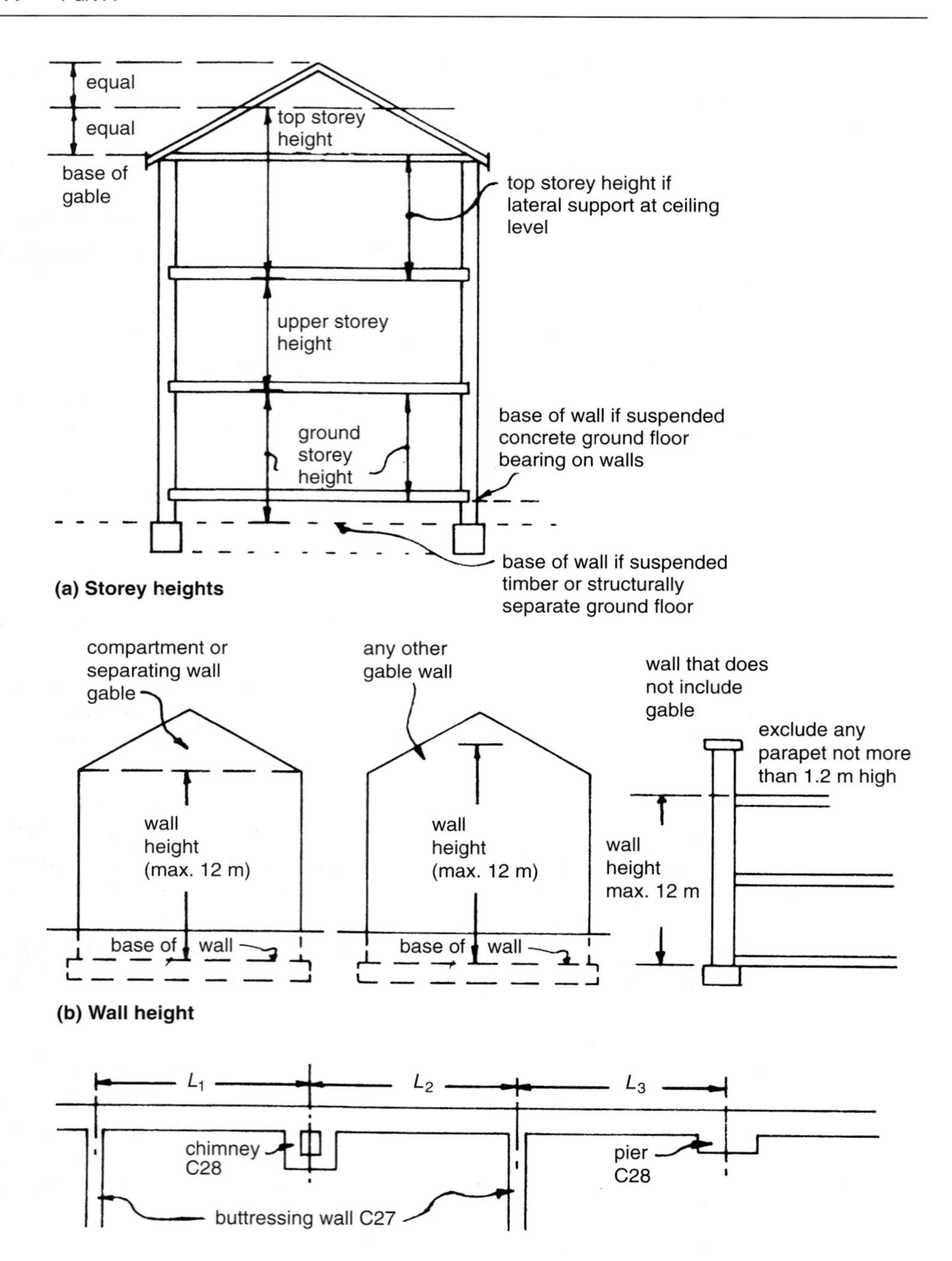

Fig. 6.5 Rules for measurement, 1C18, 1C19.

6.6 Buttressing walls, piers and chimneys

Any load-bearing wall should be bonded or securely tied at each end to a buttressing wall, pier or chimney. (This does not apply to single leaf walls less than 2.5 m in height and length which form part of a small single-storey non-residential building or annexe.) These supporting elements should be of such dimensions as to provide effective lateral support over the full wall height from its base to its top.

If, additionally, such supporting elements are bonded or securely tied to the supporting wall at intermediate points in the length of the wall, then the wall may be regarded as being divided into separate distinct lengths by these buttressing walls, piers or chimneys. Each of the distinct lengths may then be regarded as a supported wall, and the length of any wall is the distance between adjacent supporting elements.

BUTTRESSING WALLS should have:

- one end bonded or securely tied to the supported wall;
- the other end bonded or securely tied to another buttressing wall, pier or chimney;
- no opening or recess greater than 0.1 m in area within a horizontal distance of 550 mm from the junction with the supported wall, and openings and recesses generally disposed so as not to impair the supporting effect of the buttressing wall;
- a length of not less than one sixth of the height of the supported wall;
- the minimum thickness required by the appropriate rule, according to whether the buttressing wall is actually an external compartment, separating or internal load-bearing wall or a wall of a small building or annexe; but if the wall is none of these, then a thickness, t (see Fig. 6.6), of not less than the greater of:
  - (i) half the thickness required of a solid external, compartment or separating wall of the same height and length as the buttressing wall, less 5 mm; or
  - (ii) if the buttressing wall is part of a dwellinghouse and the supported wall as a whole is not more than 6 m high and 10 m in length, 75 mm; or
  - (iii) in any other case, 90 mm (see Fig. 6.6).

PIERS may project on either or both sides of the supported wall and should:

- run from the base of the supported wall to the level of the roof lateral support, or to the top of the wall if there is no roof lateral support;
- have a thickness, measured at right angles to the length of the supported wall and including the thickness of that wall, of at least three times the thickness required of the supported wall; and
- measure at least 190 mm in width (the measurement being parallel to the length of the supported wall).

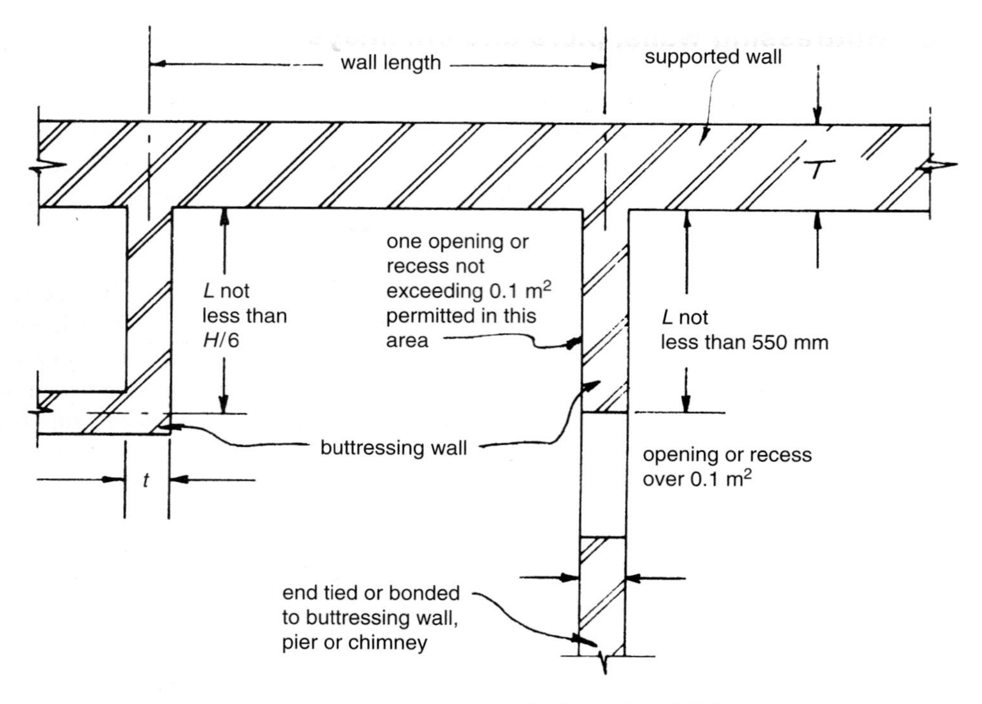

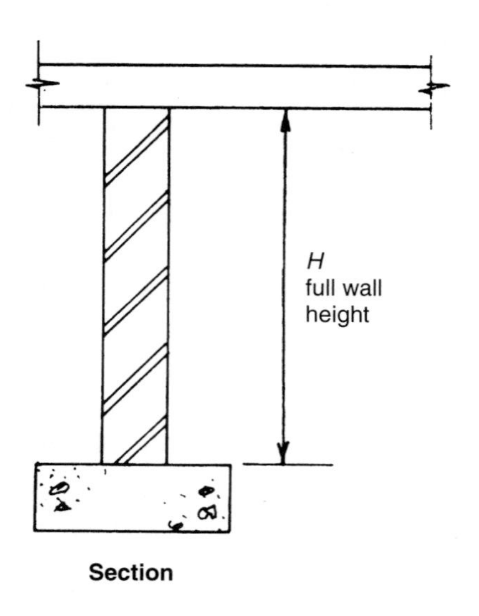

position and shape of openings or recesses should not impair lateral support provided by buttressing wall

Fig. 6.6 Buttressing walls.

CHIMNEYS should have:

(a) a horizontal cross-section area, excluding any fireplace opening or flue, of not less than the area required of a pier in the same wall; and
(b) a thickness overall of at least twice the thickness required of the supported wall.

The requirements for piers and chimneys are shown in Fig. 6.7.

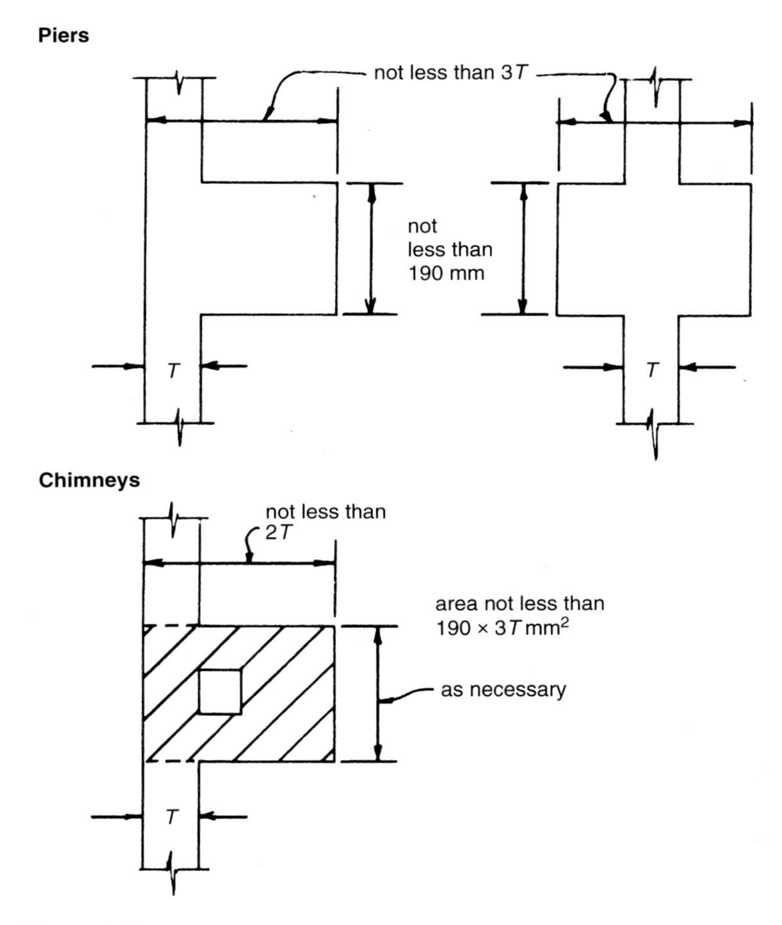

Fig. 6.7 Piers and chimneys.

It should be noted that requirements in respect of plan dimensions of piers do not apply to piers in walls of small buildings and annexes, for which there are special rules (see Fig. 6.18).

6.6.1 Loading conditions

FLOOR SPANS. The wall should not support any floor members with a span of more than 6 m. (Span is measured centre to centre of bearings.)

LATERAL THRUST. Where the levels of the ground or oversite concrete on either side of a wall differ, the thickness of the wall as measured at the higher level should not be less than one quarter of the difference in level.

In the case of a cavity wall, the thickness is taken as the sum of the leaf thicknesses. However, if the cavity is filled with fine concrete, the overall thickness may be taken.

The lateral thrust occasioned in these circumstances is the only one which a wall must be expected to sustain, apart from that due to direct wind load and the transmission of wind load.

VERTICAL LOADING. The total dead and imposed load transmitted by a wall at its base should not exceed 70 kN/m. All vertical loads carried by a wall should be properly distributed. This may be assumed for pre-cast concrete floors, concrete floor slabs and timber floors complying with Section 18 of ADA1/2. Distributed loading may also be assumed for lintels with a bearing length of 150 mm or more. Where the clear span of the lintel is 1200 mm or less the bearing length may be reduced to 100 mm.

These requirements are summarised in Fig. 6.8.

6.6.2 Openings and recesses

Openings or recesses in a wall should not be placed in such a manner as to impair the stability of any part of it. Adequate support for the superstructure should be provided over every opening and recess.

As a general rule, any opening or recess in a wall should be flanked on each side by a length of wall equal to at least one sixth of the width of the opening or recess, in order to provide the required stability. Accordingly, the minimum length of wall between two openings or recesses should not be less than one sixth of the *combined* width of the two openings or recesses.

However, where long span roofs or floors bear onto a wall containing openings or recesses it may be necessary to increase the width of the flanking portions of wall. Table 10 of Section 1C of AD A1/2 (see below) contains factors that enable this to be done.

Where several openings and/or recesses are formed in a wall, their total width should, at any level, be not more than two thirds of the length of the wall at that level.

No opening or recess should exceed 3 m in length. These requirements are illustrated in Fig. 6.9.

6.6.3 Chases

The depth of vertical chases should not be more than one third the thickness of the wall, or in a cavity wall, one third the thickness of the leaf concerned. Depth of horizontal chases should be not more than one sixth the thickness of the wall or leaf. Chases should not be placed in such a manner as to impair the stability of the wall, particularly where hollow blocks are used (see Fig. 6.10).

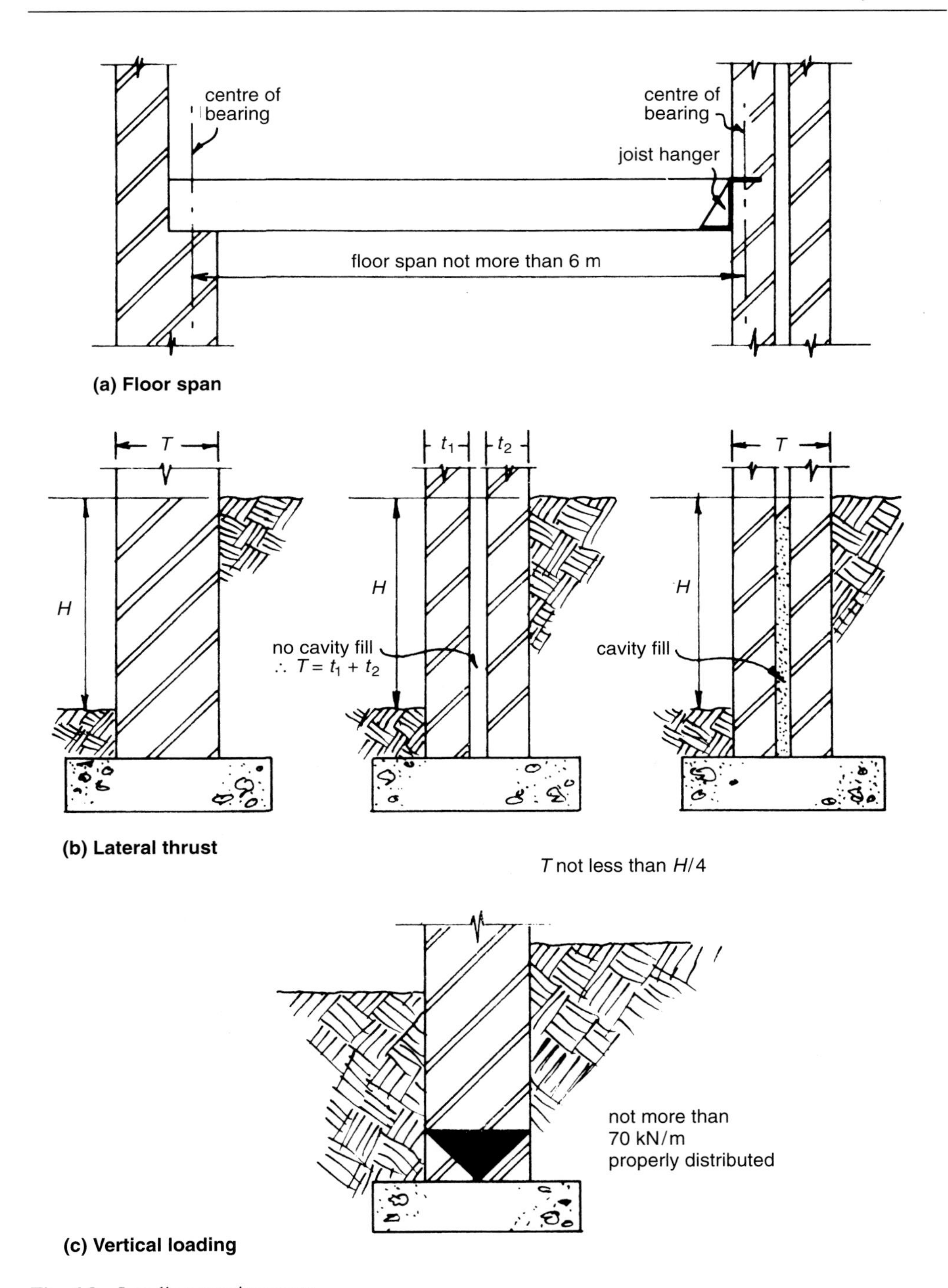

Fig. 6.8 Loading requirements.

AD A1/2, section 1C

Table 10 Value of factor 'X'.

Nature of roof span	Maximum roof span [m]	Minimum thickness of wall inner leaf [mm]	Span of floor is parallel to wall	Span of timber floor into wall		Span of concrete floor into wall	
				max 4.5 m	max 6.0 m	max 4.5 m	max 6.0 m
			Value of factor 'X'				
roof span parallel to wall	not applicable	100	6	6	6	6	6
		90	6	6	6	6	5
timber roof spans into wall	9	100	6	6	5	4	3
		90	6	4	4	3	3

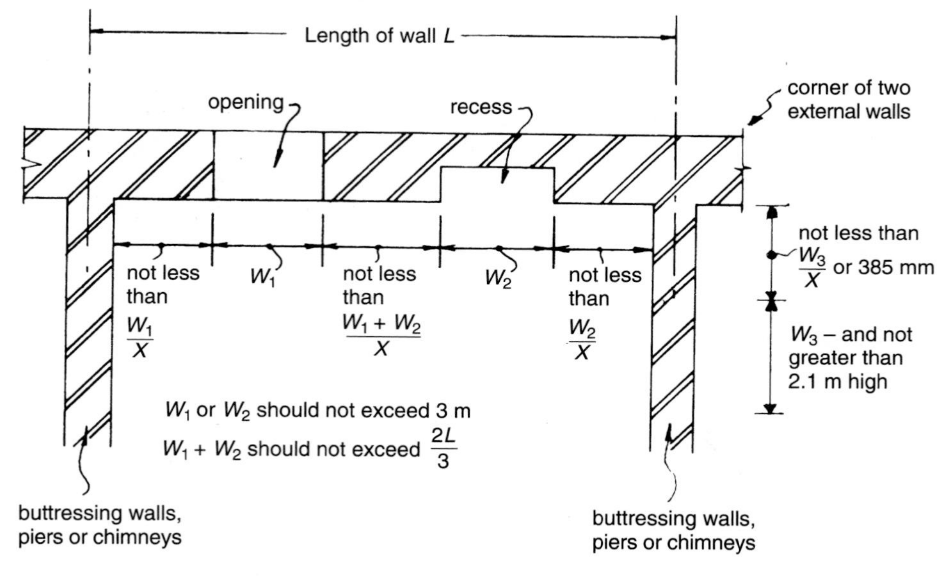

Note: value of X comes from Table 10 of Section 1C of AD A1/2 which is reproduced above OR it may be given the value 6 provided the compressive strength of the blocks or bricks (or cavity wall loaded leaf) is not less than 7 N/mm^2

Fig. 6.9 Openings and recesses.

6.6.4 Overhanging

Where a wall overhangs a supporting structure beneath it, the amount of the overhang should not be such as to impair the stability of the wall. No limits are specified, but this would generally be interpreted as allowing an overhang of one third the thickness of the wall (see Fig. 6.11).

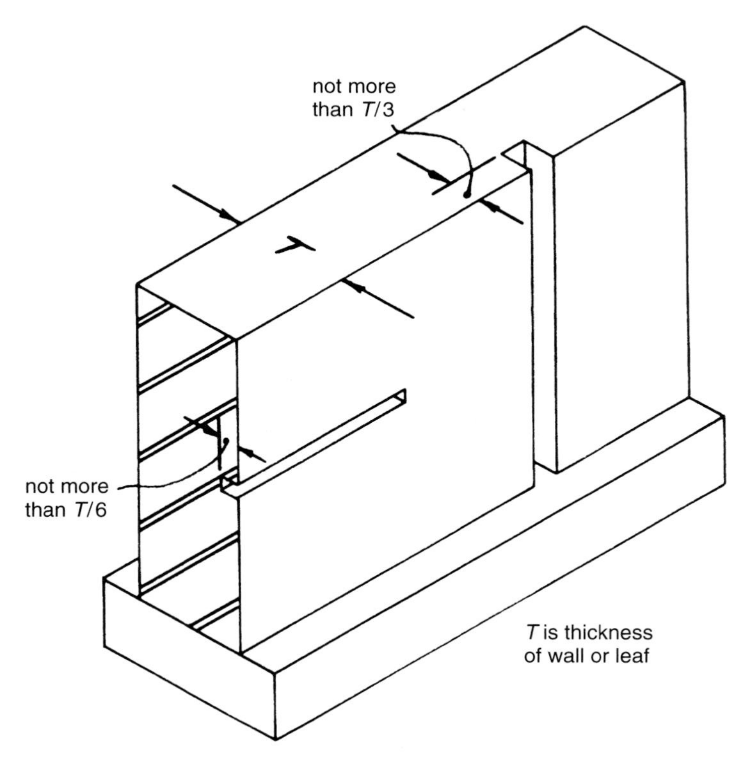

Fig. 6.10 Chases.

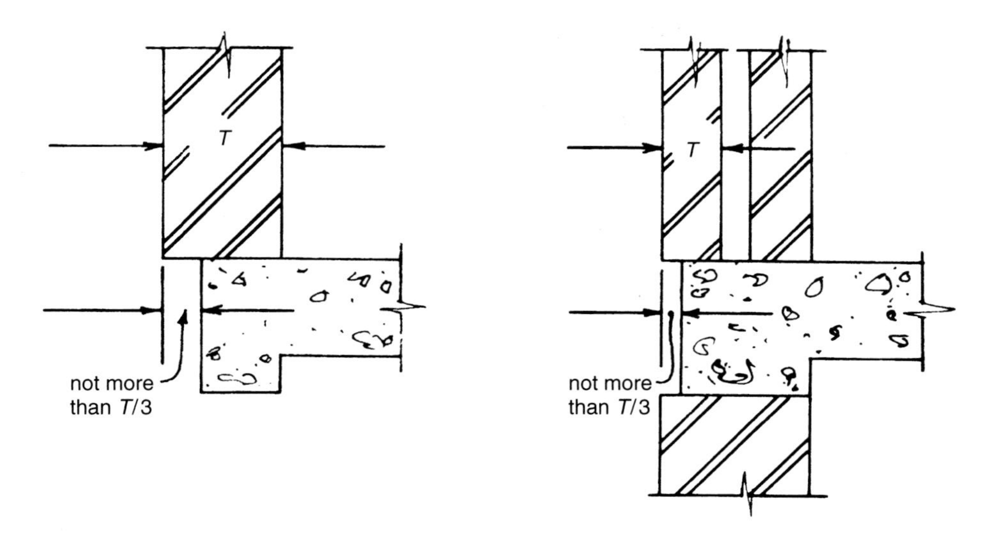

Fig. 6.11 Overhanging.

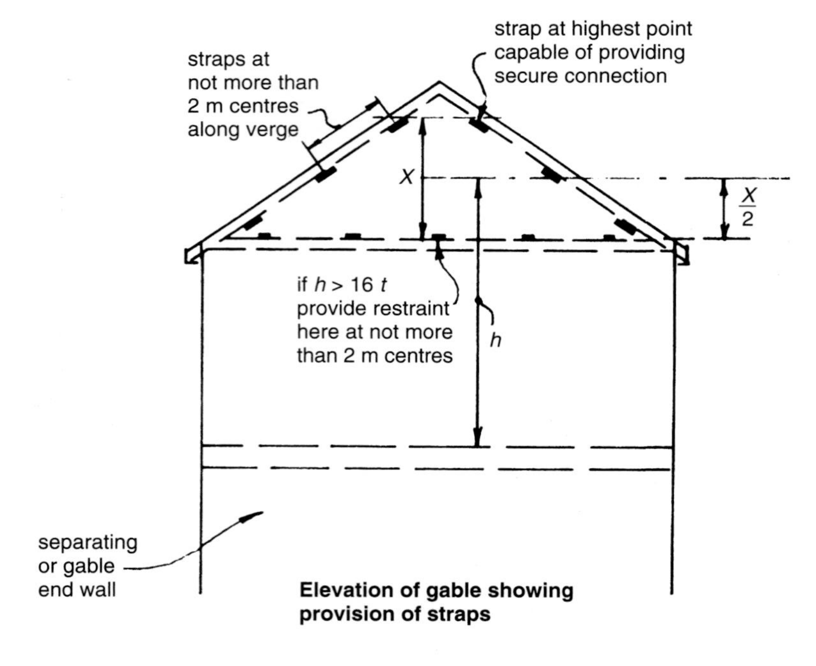

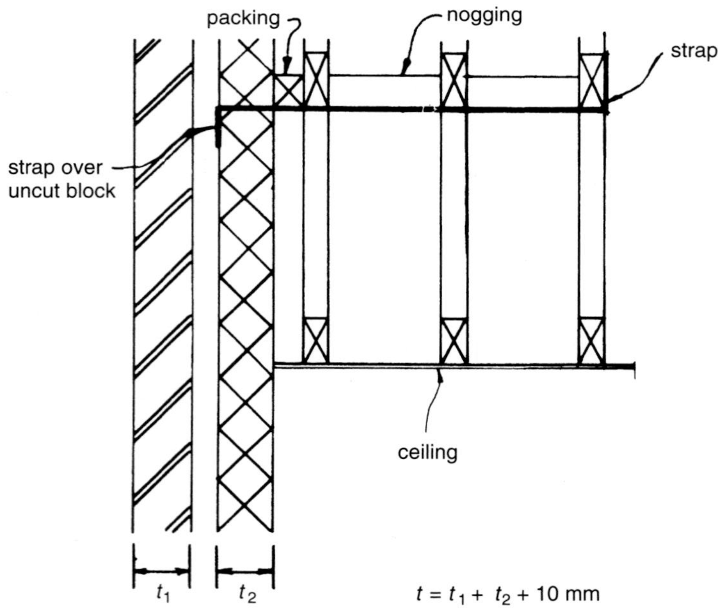

Fig. 6.12 Lateral support for roof.

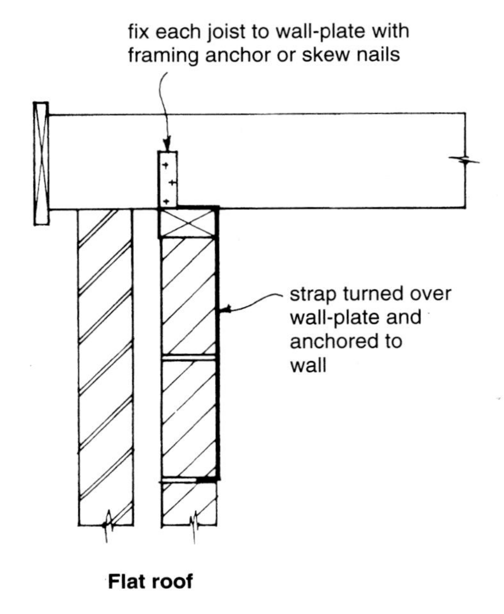

Flat roof

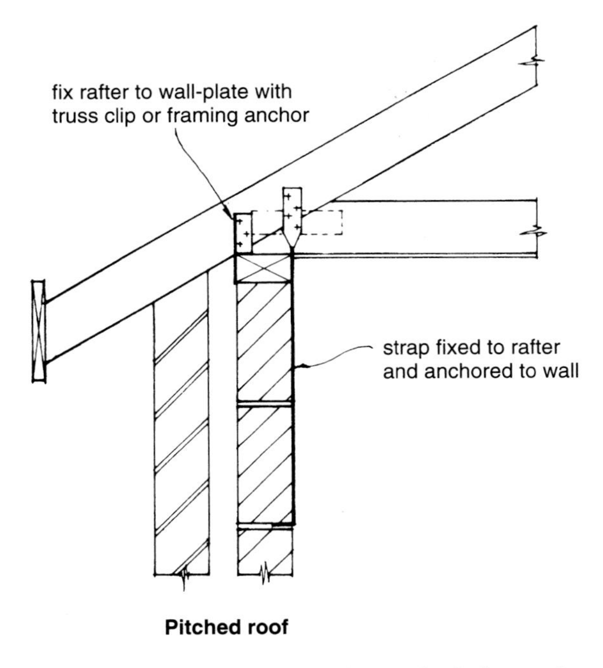

Pitched roof

(b) Section at eaves level showing method of strapping

Fig. 6.12 (*Contd*).

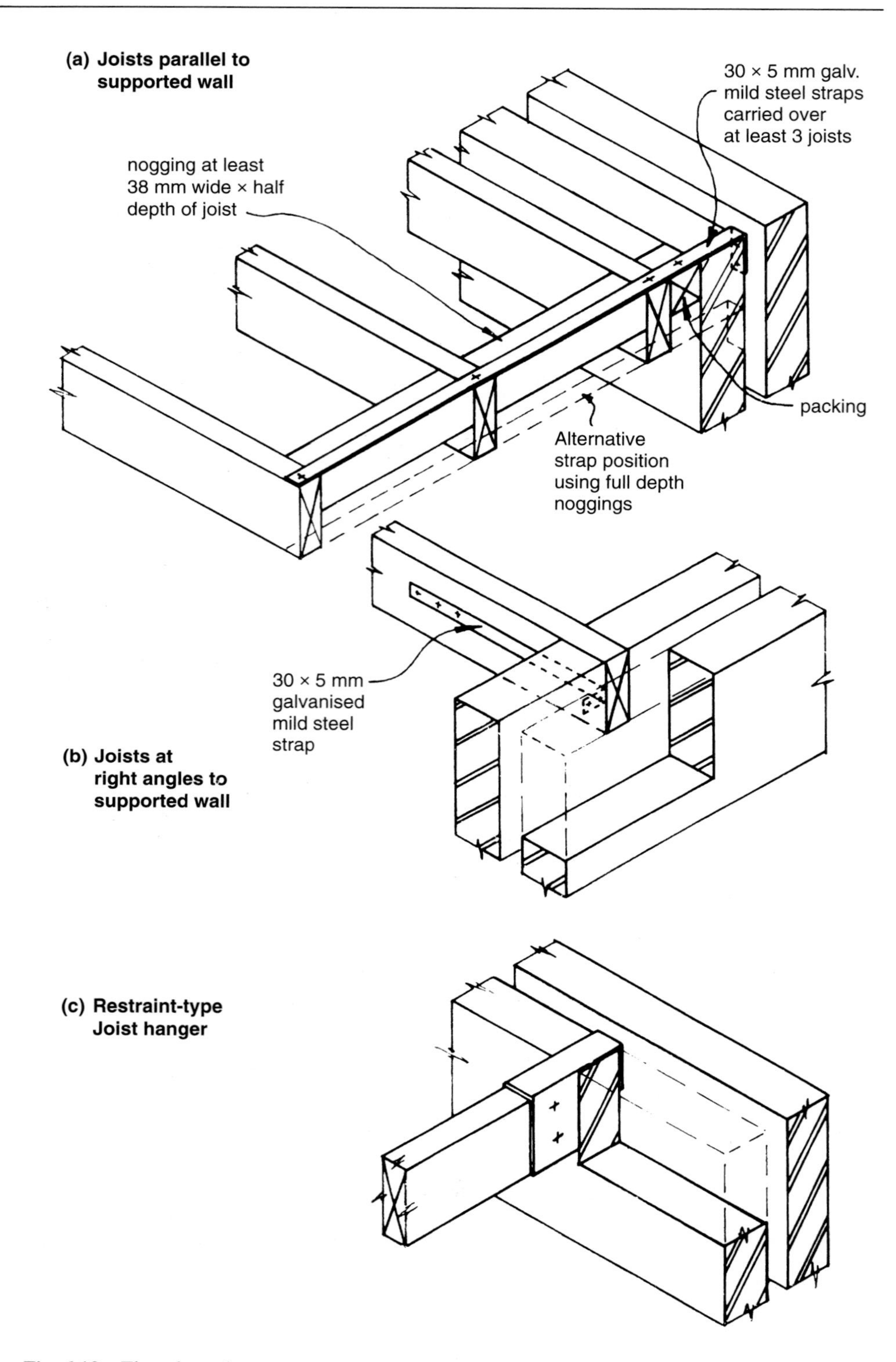

Fig. 6.13 Floor lateral support.

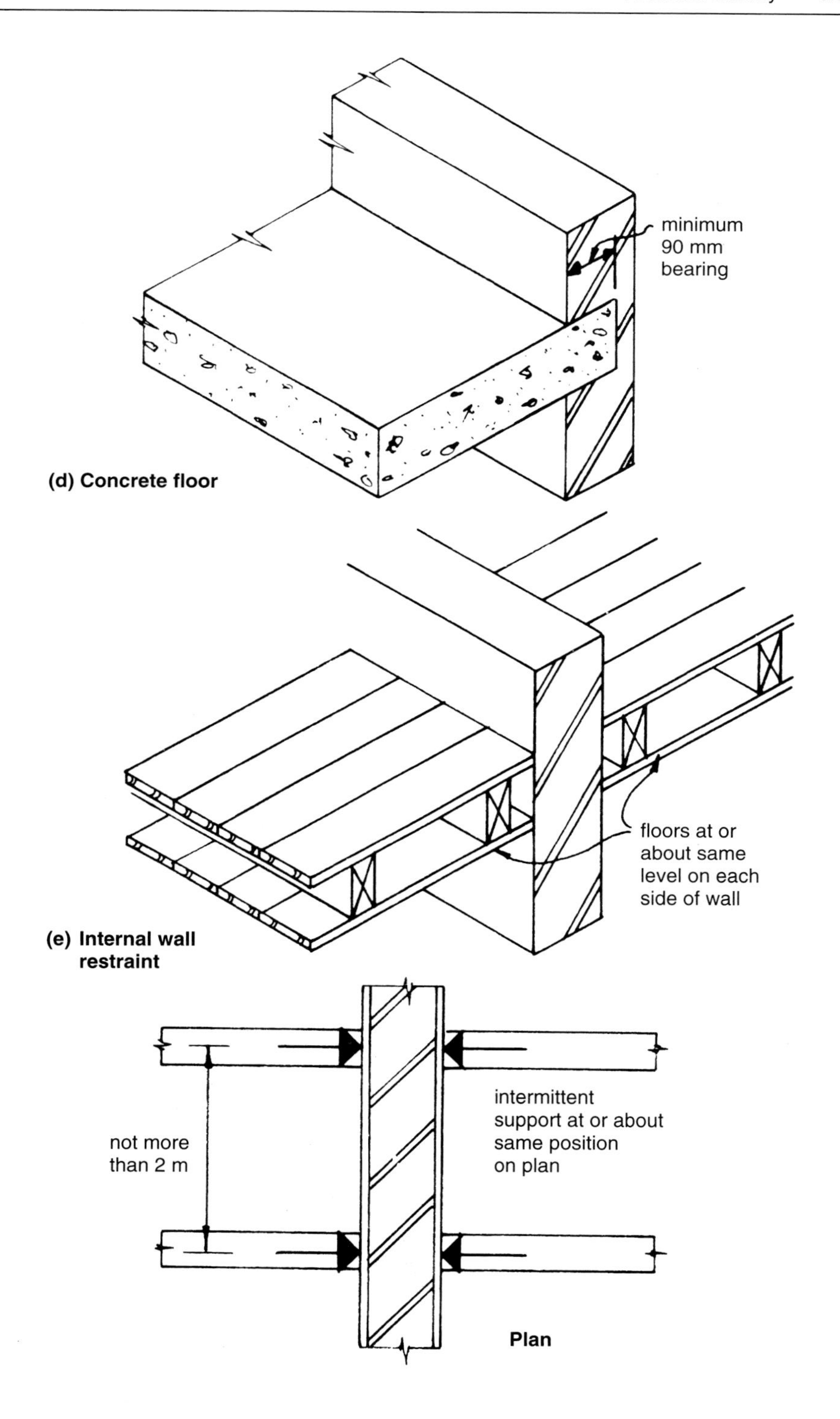

Fig. 6.13 (*Contd*).

6.6.5 Lateral support

Floor or roof lateral support is horizontal support or stiffening, intended to stabilise or stiffen a wall by restraining its movement in a direction at right angles to the wall length. The restraint or support is provided by connecting a floor or roof to the wall in such a way that the floor or roof acts as a stiffening frame or diaphragm, transferring the lateral forces to walls, buttressing walls, piers or chimneys.

ROOF LATERAL SUPPORT. This should be provided for all external, compartment, separating and internal load-bearing walls irrespective of their length, at the point of junction between the roof and supported wall (i.e. at eaves level and along the verges).

Walls should be strapped to roofs at not exceeding 2 m centres using galvanised mild steel or other durable metal straps, with a minimum cross section of 30 mm × 5 mm, and a minimum length of 1 m for eaves strapping.

Eaves strapping need not be provided for a roof which:

- has a pitch of 15° or more;
- is tiled or slated;
- is of a type known by local experience as being resistant to damage by wind gusts;
- has main timber members spanning onto the supported wall at intervals of not more than 1.2 m.

Figure 6.12 shows methods of providing satisfactory lateral support at separating or gabled end wall positions.

FLOOR LATERAL SUPPORT. This should be provided for any external, compartment or separating wall which exceeds 3 m in length.

It should also be provided for any internal load-bearing wall (which is not a compartment or separating wall) at the top of each storey, irrespective of its length.

Walls should be strapped to floors above ground level at not exceeding 2 m centres using galvanised mild steel or other durable metal straps, with a minimum cross section of 30 mm × 5 mm.

There are certain cases where, because of the nature of the floor construction, it is not necessary to provide restraint straps.

- Where a floor forms part of a house having not more than two storeys and:
  - (a) has timber members spanning so as to penetrate into the supported wall at intervals of not more than 1.2 m with at least 90 mm bearing directly on the walls or 75 mm bearing onto a timber wall plate; or
  - (b) the joists are carried on the supported wall by *restraint* type joist hangers, described in BS 5628: Part 1, at not more than 2 m centres.
- Where a concrete floor has a bearing onto the supported wall of at least 90 mm.
- Where two floors are at or about the same level on either side of a supported wall, contact between floors and wall may be continuous or intermittent. If inter-

mittent, the points of contact should be at or about the same positions on plan at intervals not exceeding 2 m. Figure 6.13 summarises these provisions.

6.6.6 Interruption of lateral support

It is clear that in certain circumstances it may be necessary to interrupt the continuity of lateral support for a wall. This occurs chiefly where a stairway or similar structure adjoins a supported wall and necessitates the formation of an opening in a floor or roof.

This is permitted provided certain precautions are taken:

- The opening extends for a distance not exceeding 3 m measured parallel to the supported wall.
- If the connection between wall and floor or roof is provided by means of anchors, these should be spaced closer than 2 m on either side of the opening so as to result in the same number of anchors being used as if there were no opening.
- Other forms of connection should be provided throughout the length of each part of the wall on either side of the opening.
- There should be no other interruption of lateral support (see Fig. 6.14).

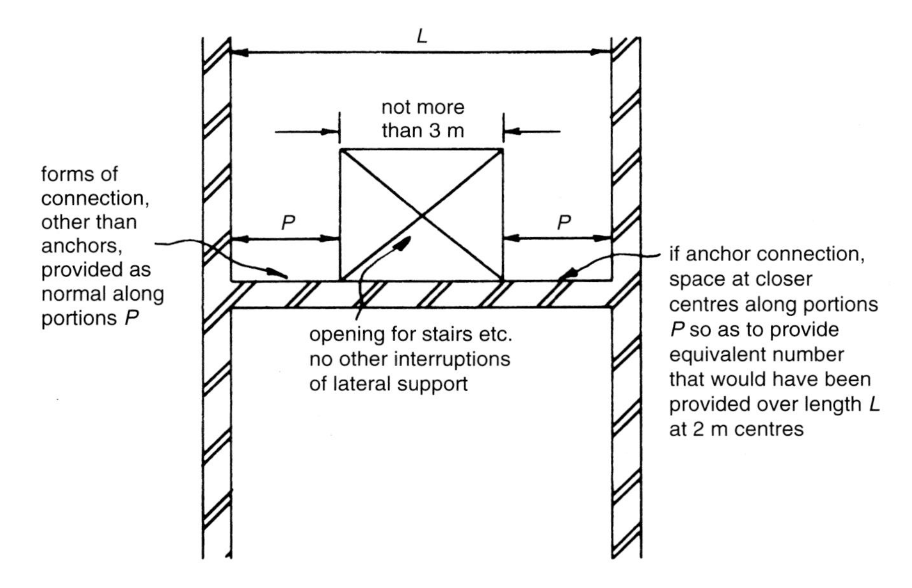

Fig. 6.14 Interruption of lateral support.

6.6.7 Thickness of walls

Provided the building design and wall construction requirements discussed above are satisfied, it is permissible to determine the thickness of a wall without calculation.

The minimum thicknesses required depend upon the wall height and length, and the rules applying to walls of bricks or blocks are set out in Table 5 of section 1C of ADA1/2 (see below) and illustrated in Fig. 6.15.

AD A1/2, section 1C

Table 5 Minimum thickness of certain external walls, compartment walls and separating walls.

(1) Height of wall	(2) Length of wall	(3) Minimum thickness of wall
Not exceeding 3.5 m ...	Not exceeding 12 m ...	190 mm for the whole of its height
Exceeding 3.5 m but not exceeding 9 m	Not exceeding 9 m ...	190 mm for the whole of its height
	Exceeding 9 m but not exceeding 12 m	290 mm from the base for the height of one storey, and 190 mm for the rest of its height
Exceeding 9 m but not exceeding 12 m	Not exceeding 9 m ...	290 mm from the base for the height of one storey, and 190 mm for the rest of its height
	Exceeding 9 m but not exceeding 12 m	290 mm from the base for the height of two storeys, and 190 mm for the rest of its height

These thicknesses do not apply to parapet walls, for which there are special rules (see below) or to bays, and gables over bay windows above the level of the lowest window sill.

As a general rule, the thickness of any storey of a brick or block wall should not be less than one sixteenth of the height of that storey. However, walls of stone, flints, clunches of bricks or other burnt or vitrified material should have a thickness of at least $1\frac{1}{3}$) times the thickness required of brick or block walls.

Irrespective of the materials used in construction, no part of a wall should be thinner than any other part of the wall that it supports.

6.6.8 Solid internal load-bearing walls which are not compartment or separating walls

For these walls the sum of the wall thickness, plus 5 mm, should be equal to at least half the thickness that would be required by Table 5 for an external wall, compartment wall or separating wall of the same height and length.

Where a wall forms the lowest storey of a three-storey building, and it carries loading from both upper storeys, its thickness should not be less than the thickness calculated above or 140 mm, *whichever is greater*. Thus there is an absolute minimum thickness of 140 mm for such walls.

6.6.9 Cavity walls

Any external, compartment or separating wall which is built as a cavity wall should consist of two leaves, each leaf built of bricks or blocks.

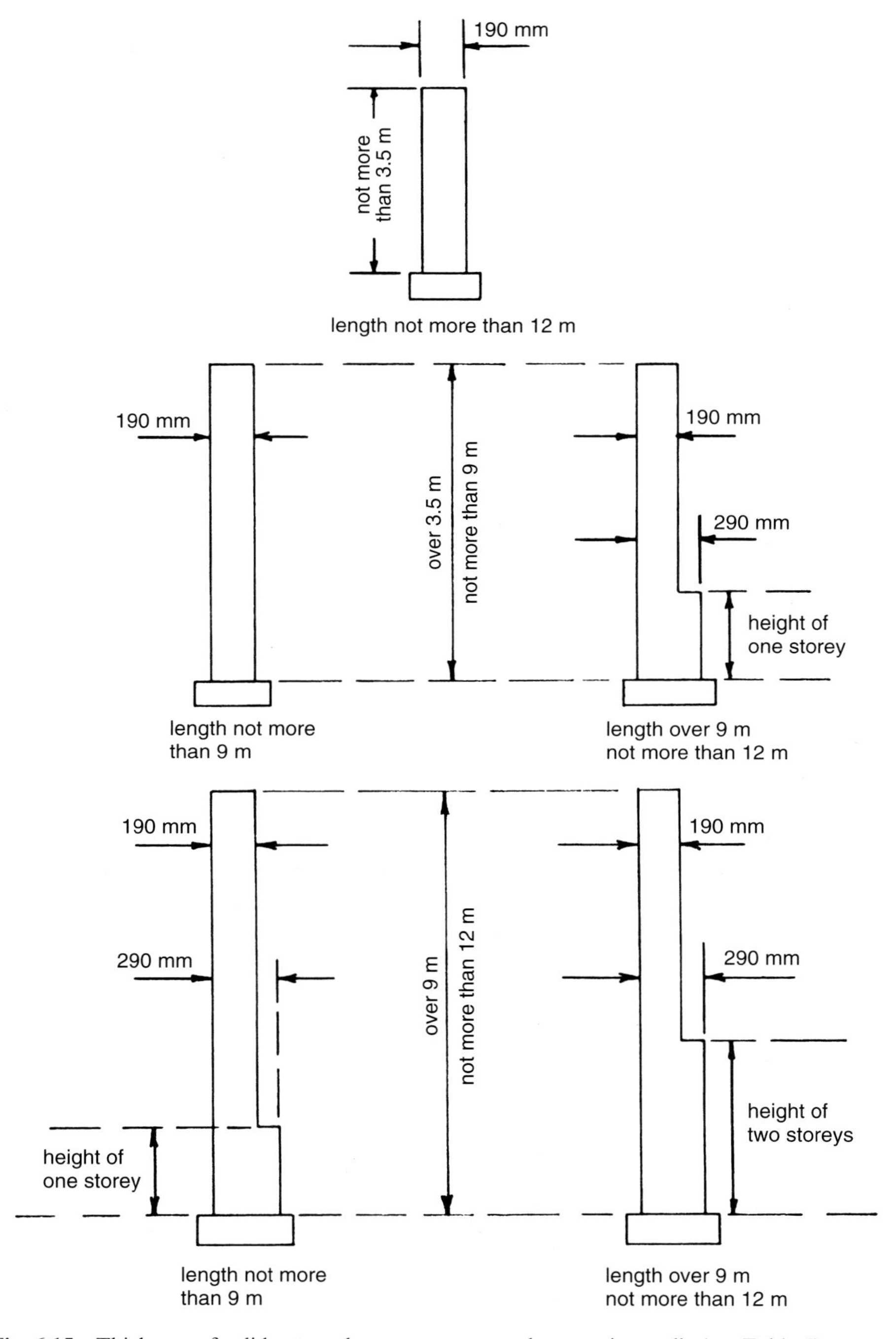

Fig. 6.15 Thickness of solid external, compartment and separating walls (see Table 5).

The leaves of these walls should be properly tied together with wall ties to BS 1243: 1978, or other not less suitable ties. Ties should be placed at centres 900 mm horizontally and 450 mm vertically, and at any opening at least one tie should be provided for each 300 mm of height within 225 mm of the opening unless the leaves are connected by a bonded jamb.

The cavity should be at least 50 mm, and not more than 75 mm, in width at any level. However, if vertical twist type ties are used, with horizontal spacing reduced to 750 mm, the cavity width may be up to 100 mm. Each leaf should be at least 90 mm thick at any level.

The sum of the thicknesses of the two leaves, plus 10mm, should not be less than the thickness required for a solid wall by Table 5. (See also Fig. 6.16.)

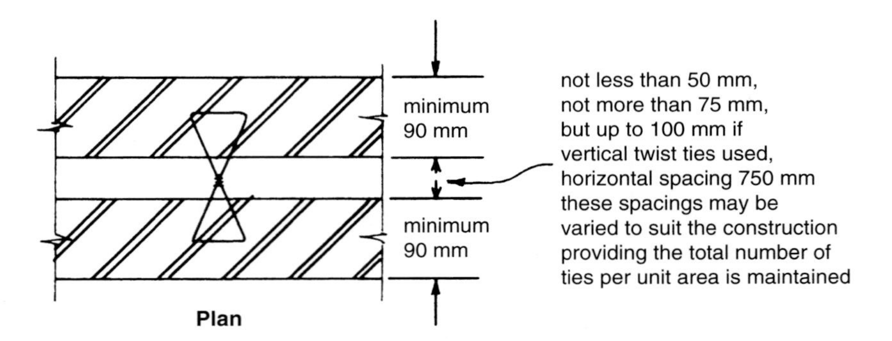

Fig. 6.16 Cavity walls.

6.6.10 Parapets

When referring to Fig. 6.17(a), the minimum thickness, t, for a solid parapet wall should not be less than the greater of:

- $H/4$; or
- 150 mm.

For a cavity parapet the minimum thickness, t, is related to the maximum parapet height as shown in Fig. 6.17(b).

6.6.11 Block and brick dimensions

The wall thicknesses specified in Section 1C relate to the *work size* of the materials used. This means the size specified in the relevant British Standard as the size to which the brick or block must conform, account being taken of any permissible deviations or tolerances specified in the British Standard.

Some walls may be constructed of bricks or blocks having modular dimensions derived from BS 6750: 1986 *Specification for modular coordination in building* without the bricks or blocks themselves being covered by a British Standard.

In these cases, the thicknesses prescribed in Section 1C may be reduced by an amount not exceeding that allowed in a British Standard for the same material.

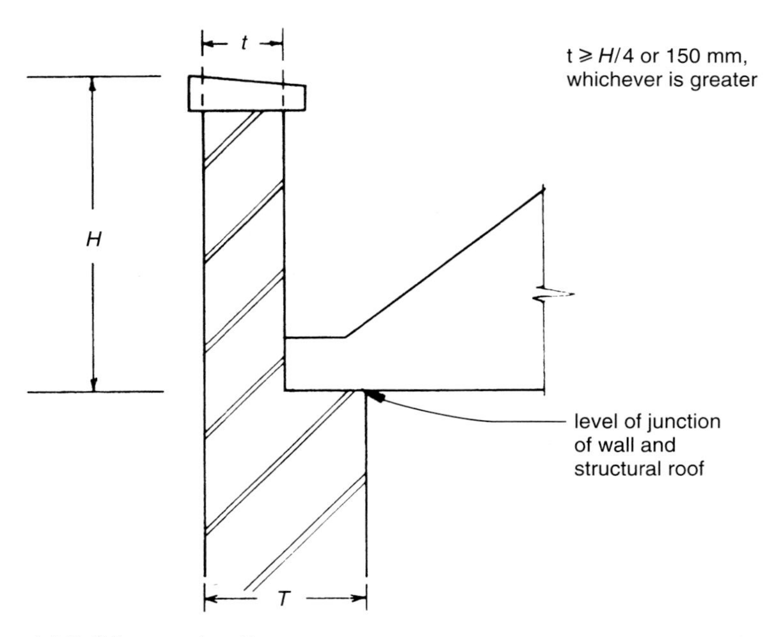

(a) Solid parapet walls

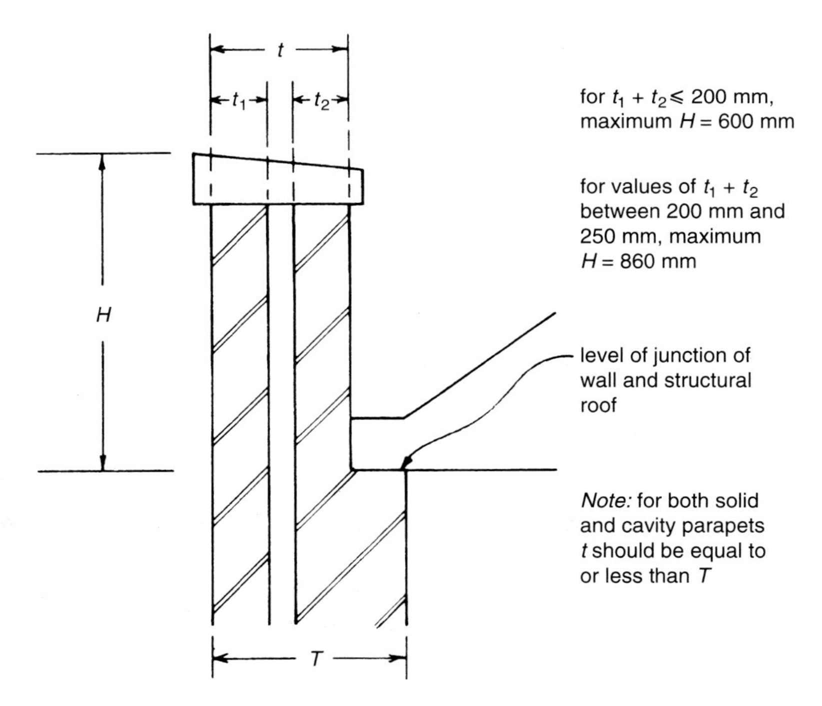

(b) Cavity parapet walls

Fig. 6.17 Height of parapet walls.

6.6.12 External walls of small buildings and annexes

The external walls of small buildings and annexes have to comply with special rules.
The external walls of such buildings may be not less than 90 mm thick if:

- the walls are bonded at each end and intermediately to piers or buttressing walls of not less than 190 mm square in horizontal section (including wall thickness); and
- the piers, etc., are positioned so that they divide the wall into distinct lengths and each length is not more than 3 m (this does not apply if the wall is less than 2.5 m high and long); and
- the enclosed floor area does not exceed 36 m^2.

The wall should be built as a solid wall of bricks and blocks and should carry only distributed loading from the roof of the building or annexe, and not be subjected to any lateral thrust from the roof (see Fig. 6.18).

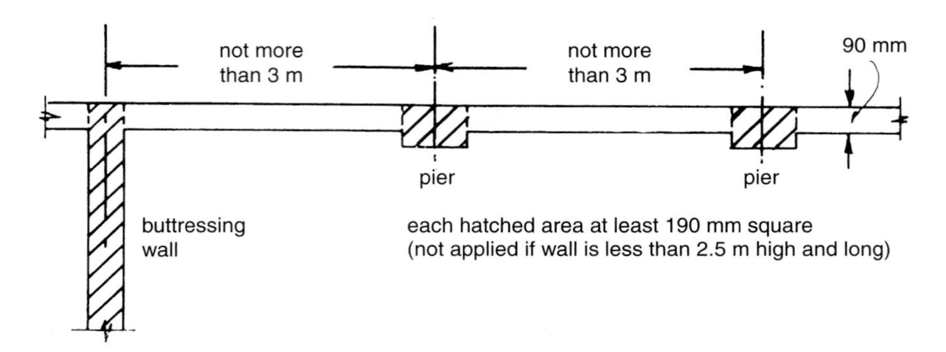

Fig. 6.18 Small buildings and annexes.

6.6.13 Dimensions of chimneys

The wholly external part of a chimney, constructed of masonry and not supported by adequate ties or otherwise stabilised will be deemed satisfactory if the width of the chimney, at the level of the highest point in the line of junction with the roof and at any higher level, is such that its height as measured from that level to the top of the external part of the chimney is not more than $4\frac{1}{2}$ times that width. That height includes any pot or flue terminal on a chimney. Additionally, the masonry should have a density greater than 1500 kg/m^3.

The width of chimney at any level is taken as the smallest width which can be shown on an elevation of the chimney from any direction. This is illustrated in Fig. 6.19.

6.6.14 Foundation recommendations

Section 1E of AD A1/2 provides rules for the construction of strip foundations of plain concrete. It should be remembered that section 1 applies to certain residential buildings of not more than three storeys, small single-storey non-residential build-

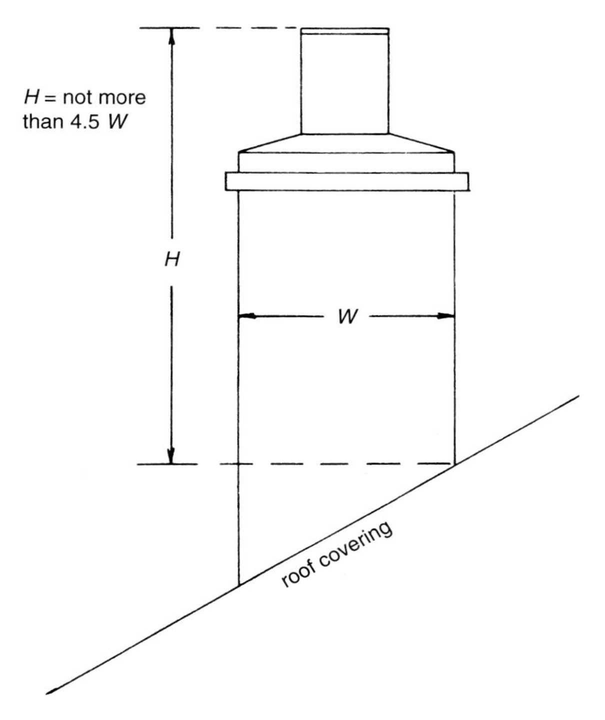

Fig. 6.19 External part of chimneys, 1D1.

ings and annexes. Strictly speaking, the guidance given in Sections 1A to 1E should not be used for any other building types.

Strip foundations of plain concrete placed centrally under the walls will be satisfactory if they comply with the following rules:

- There is no made ground or wide soil strength variation in the loaded area or weak soil patches likely to cause foundation failure.
- The width of foundation strip is in accordance with Table 12 to Section 1E of ADA1/2 which is reproduced below.
- In chemically non-aggressive soils the concrete should be composed of:
  (i) cement to BS 12: 1989 *Specification for Portland cements*; *and*,
  (ii) coarse and fine aggregate to BS 882: 1983 *Specification for aggregates from natural sources for concrete*.
- For foundations in chemically aggressive soils, the guidance in BS 5328: Part 1: 1990 *Guide to specifying concrete* should be followed.
- The concrete mix is:
  (i) in the proportion 50 kg of cement:0.1 m^3 fine aggregate: 0.2 m^3 coarse aggregate i.e. 1:3:6 or better; *or*,
  (ii) Grade ST1 concrete to BS 5328: Part 2: 1990 *Methods for specifying concrete mixes*.

- The concrete strip thickness is equal to or greater than the projection from the wall face, and never less than 150 mm.
- The upper level of a stepped foundation overlaps the lower level by twice the height of the step, by the thickness of the foundation or 300 mm, whichever is the greater.
- The height of a step is not greater than the thickness of the foundation.
- The foundation strip projects beyond the faces of any pier, buttress or chimney forming part of a wall by at least as much as it projects beyond the face of the wall proper.

Table 12 to section E1 of AD A1/2 specifies seven subsoil types, and the minimum strip widths to use vary according to the calculated load per metre run of the wall at foundation level. The Table is reproduced below.

Where a wall load exceeds 70 kN per metre run the foundation will be outside the scope of section E1 and must be properly designed on structural principles.

These requirements are illustrated in Fig. 6.20.

6.7 Design of structural members in buildings of all types

Section 1 of AD A1/2, which is discussed above, deals with a fairly restricted range of building types of traditional masonry construction. If the various parts of Section 1 are complied with it is not necessary to provide design calculations.

Building types falling outside the scope of section 1 will need full structural calculations and design. Therefore, sections 2, 3 and 4 of AD A1/2 contain references to British Standards, Codes of Practice and other sources of design information which, if used appropriately, will satisfy the requirements of Paragraphs A1 and A2 of Schedule 1 to the 2000 Regulations.

As mentioned above in section 6.3, sections 2 and 3 were added to AD A1/2 by the 1991 amendment and deal with external wall cladding and re-covering of roofs respectively.

6.7.1 External wall cladding

In recent years a number of accidents have occurred involving heavy concrete cladding panels. Failure of the fixings and deterioration of the concrete has resulted in parts and, in some cases, whole panels becoming detached with the resultant danger to people in the street below.

Guidance is provided in section 2 of AD A1/2 which relates specifically to heavier forms of cladding. However, some of the guidance is also applicable to curtain walling.

Weather-resistance of wall cladding is not covered by the guidance in AD A1/2; Approved Document C (Site preparation and resistance to moisture) should be consulted for this.

Wall cladding should be:

Table 12 Minimum width of strip foundations.

(1) Type of subsoil	(2) Condition of subsoil	(3) Field test applicable	(4) Minimum width in millimetres for total load in kilonewtons per lineal metre of load-bearing walling of not more than					
			20 kN/m	30 kN/m	40 kN/m	50 kN/m	60 kN/m	70 kN/m
I Rock	Not inferior to sandstone, limestone or firm chalk	Requires at least a pneumatic or other mechanically operated pick for excavation	In each case equal to the width of the wall					
II Gravel Sand	Compact Compact	Requires pick for excavation. Wooden peg 50 mm square in cross-section hard to drive beyond 150 mm	250	300	400	500	600	650
III Clay Sandy clay	Stiff Stiff	Cannot be moulded with the fingers and requires a pick or pneumatic or other mechanically operated spade for its removal	250	300	400	500	600	650
IV Clay Sandy clay	Firm Firm	Can be moulded by substantial pressure with the fingers and can be excavated with graft or spade	300	350	450	600	750	850
V Sand Silty sand Clayey sand	Loose Loose Loose	Can be excavated with a spade. Wooden peg 50 mm square in cross-section can be easily driven	400	600	Note: In relation to types V, VI and VII, foundations do not fall within the provisions of this Section if the total load exceeds 30 kN/m			
VI Silt Clay Sandy clay Silty clay	Soft Soft Soft Soft	Fairly easily moulded in the fingers and readily excavated	450	650				
VII Silt Clay Sandy clay Silty clay	Very soft Very soft Very soft Very soft	Natural sample in winter conditions exudes between fingers when squeezed in fist	600	850				

Examples: cavity wall 60 kN/m run in different soil types, to rules of Table 12

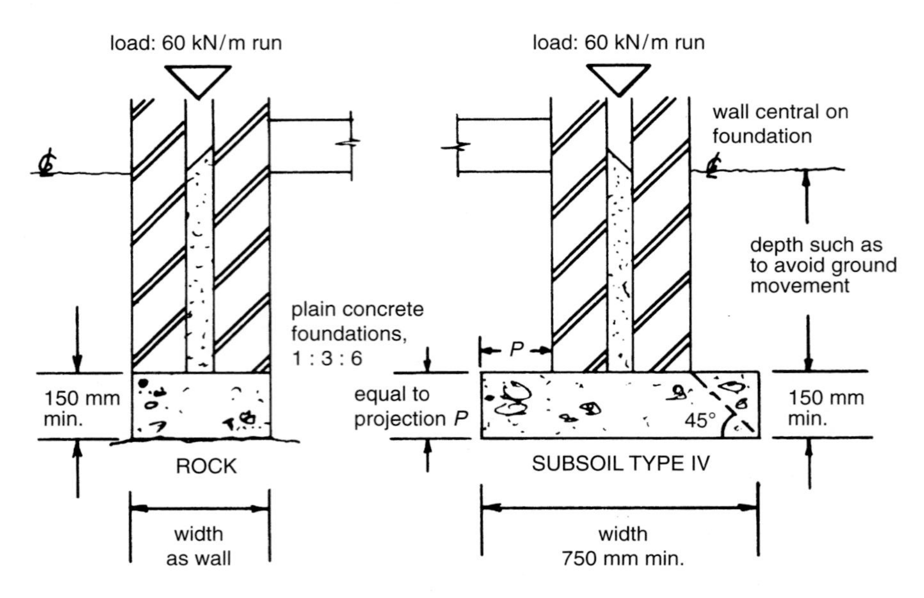

no made ground, no weak patches, no strength variation

(a) Plain strip foundation

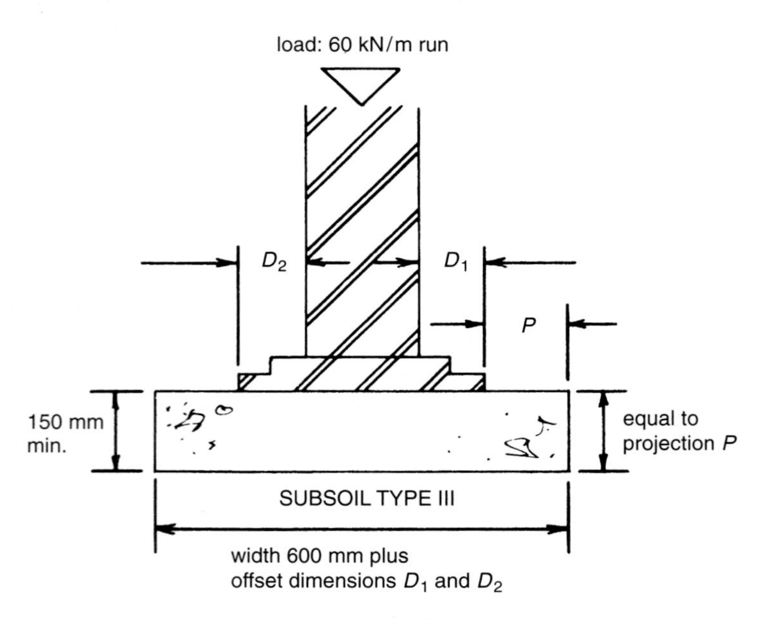

no made ground, no weak patches, no strength variation

(b) Strip foundation with footing

Fig. 6.20 Strip foundations of plain concrete.

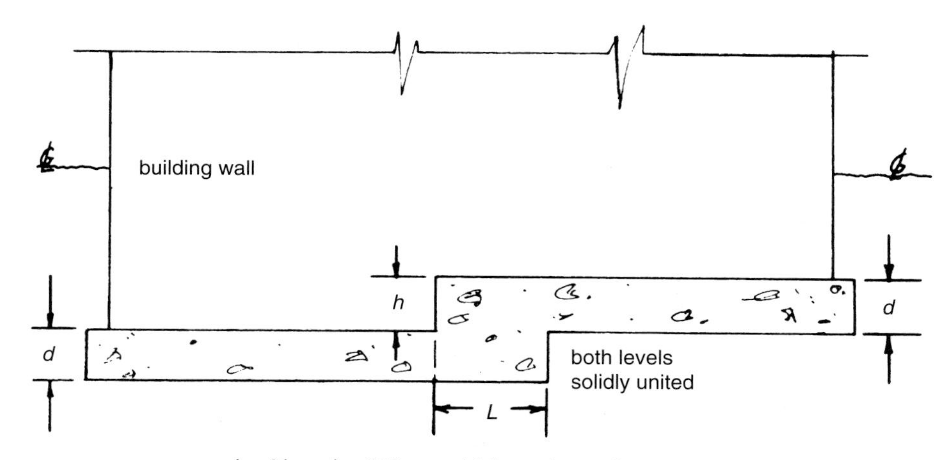

(c) Steps in foundations

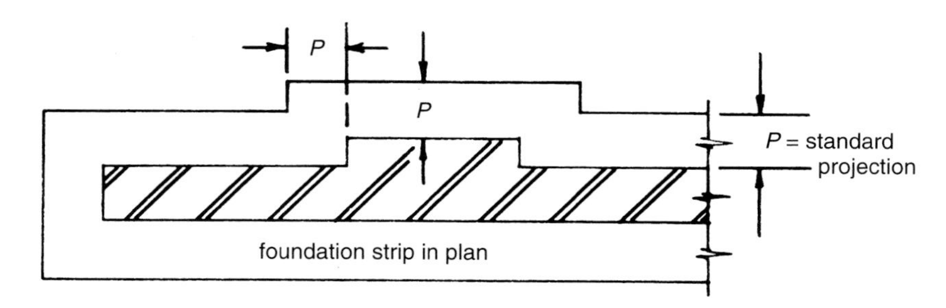

(d) Projections

Fig. 6.20 (*Contd*).

- capable of safely carrying and transmitting to the structure of the building the combined dead, imposed and wind loads.
- securely fixed to and supported by the structure of the building, the fixing comprising both vertical support and lateral restraint.
- capable of accommodating differential movement between the cladding and the building support structure.
- manufactured of durable materials (including any fixings and associated support components which should also have an anticipated life at least equal to that of the cladding).

In many cases fixings will not be easily accessible for inspection and maintenance. In these cases care is needed in the selection of materials and in the quality of workmanship. Reference should be made to the Approved Document for Regulation 7 (Chapter 8).

Assessment of loading

Apart from the dead load of the cladding itself the following loads should also be taken into account:

- wind loading – see CP3: Chapter V: *Loading*, Part 2: 1972 *Wind loads* (as amended). Use Class A building size for assessing the ground roughness factor *S*. Factor *S* must not be less than one.
- an assessment of the imposed forces from maintenance equipment such as ladders or access cradles which should be based on the actual equipment likely to be used.
- loading from fixtures such as antennae or signboards supported by the cladding.
- lateral loads where the cladding is required to act as pedestrian guarding to stairs, ramps and open wells, or as a vehicle barrier. Refer to Approved Document K (Stairs, ramps and guards) for loading requirements (see Chapter 15).
- lateral pressures from crowds where the wall cladding is required to act as spectator barriers at sports stadia requiring a safety certificate. Recommended design loadings are given in the publication entitled *Guide to Safety at Sports Grounds* (1990) published by the Home Office/Scottish Office.

Design and testing of fixings

The strength of a fixing is a function of the fixing itself and the material into which it is fixed. Therefore, its strength should be derived from tests using materials which are representative of the true in-situ condition. In this way inherent weaknesses in the support structure, such as shrinkage or flexure cracks in concrete and voids in masonry, will be highlighted and may be taken into account in the final design of the fixing.

A number of standards and references exist for assessing the strength of fixings.

- BS 5080: Part 1: 1974 (1982) *Tensile loading* describes a method for testing fixings, such as expanding anchors installed in solid materials either on site or for comparative purposes in a standard material.
- BS 5080: Part 2: 1986 *Method for determination of resistance to loading in shear* gives details of a method for conducting tests under shear force on structural fixings installed in concrete or masonry.
- British Board of Agrément MOAT No. 19: 1981 *The assessment of torque expanded anchor bolts when used in dense aggregate concrete.*

When a load is applied to a fixing from a cladding panel it is possible that the fixing will be subjected to a certain amount of eccentric loading. This can lead to local spalling in the material in which the fixing is anchored and therefore, the fixing should be designed to allow for this by assuming an increase in eccentricity equal to half the diameter of the fixing.

A further problem with fixings and their support components is that there is a possibility of unintended slippage between the parts when in use. Suitable lockable fixings should be used or other means should be sought to mechanically fix the various components together.

Commonly, panel fixings are of the expanding bolt type or are resin bonded into the substrate. When tested in shear and tension the assumed safe working strength of either should not exceed the following:

- 1/3 × (mean shear or tensile failure test load) minus the standard deviation derived from the tests;

unless, for expanding bolt fixings, the mean of the loads which causes a displacement of 0.1 mm under direct tension and 1.0 mm under direct shear is lower, in which case the lower value should be taken.

Care should be taken in the design of resin bonded fixings due to their rapid loss of strength at temperatures above 50°C.

Further guidance on fixings may be obtained from BS 8200: 1985, clause 38 and BS 8298 1989 *Code of practice for design and installation of natural stone cladding and lining*, clauses 6 and 20, and on the provision of movement joints in BS 5628: Part 3: 1985.

6.7.2 Replacement of roof coverings

It is possible that the re-roofing of a building may result in the existing roof structure having to carry substantially more or less load than it did before the works were carried out. In this case the replacement works would constitute a material alteration under the provisions of Regulation 3(2).

Section 3 of AD A1/2 contains guidance on how the existing roof structure may be assessed to see if it is capable of coping with the changed loading conditions. There are three stages to the assessment procedure:

Stage 1
Compare the proposed and original roof loadings.

Allowance should be made for the increase in loading due to water absorption which may be only 0.3% for oven dry slates but up to 10.5% for plain clay or concrete tiles. These figures are based on the dry mass per unit area of roof coverings.

Stage 2
Carry out a structural inspection on the original roof.

The roof structure must be checked to see if:

- it is capable of sustaining the increased load; *or*,
- it contains sufficient vertical restraints to cope with the wind uplift forces as a result of the lighter roof covering or addition of underlay.

Stage 3
Carry out appropriate strengthening measures.

These may include:

- replacement of defective parts of the roof, such as, structural members, nails or other fixings and vertical restraints.
- provision of additional structural members as necessary to take the increased loads, such as, rafters, purlins, binders or trusses, etc.
- provision of additional restraint straps, ties or fixings to walls as necessary to resist wind uplift forces.

6.8 Guidance on structural design in buildings of all types

6.8.1 Loading

Dead and imposed loads may be assessed by reference to BS 6399 *Loading for buildings*, Part 1: 1984 *Code of practice for dead and imposed loads*. Similarly, imposed roof loads are covered in the same code, Part 3: 1988 *Code of practice for imposed roof loads*.

Wind loads may be assessed by reference to CP 3: Chapter V: Part 2: 1972 *Wind loads*. However, the *S* factor should never be taken as less than one.

If the actual load is greater than the design load from BS 6399: Part 1: 1984, the actual load should be used. (See also section 3 of AD A1/2, roof re-covering, section 6.7.2 above.)

6.8.2 Foundations and ground movement

Foundations should be designed in accordance with BS 8004: 1986 *Code of practice for foundations*.

Paragraph A2 of Schedule 1 to the 2000 Regulations requires that ground movement caused by landslip or subsidence must not impair the stability of the building.

Guidance is given in Section 4 of AD A1/2 in the form of reference sources where information may be found regarding the more common forms of ground instability. This includes geological faults, landslides, disused mines or similar unstable strata which may affect the building site or its environs.

The following reviews of various geotechnical conditions have been carried out under the sponsorship of the Minerals and Land Reclamation Division of the Directorate of Planning Services of the DOE and information regarding their availability may be obtained from DPS/2, DOE, Room C15/19, 2 Marsham Street, London SW1P 3EB.

- *Review of research into landsliding in Great Britain.*
- *Review of mining instability in Great Britain.*
- *Review of natural underground cavities in Great Britain.*
- *Review of foundation conditions in Great Britain.*

The reviews are concerned with assessing the general state of knowledge concerning various forms of land instability in order to obtain a general picture of the scale and nature of the problems and how they might be overcome.

The results are presented as reports on a regional basis including 1:250000 scale maps and databases for use by anyone concerned with planning, development or engineering. They cover the nature and causes of instability and the consequent implications for planning and development, investigation methods and remedial preventative measures.

6.8.3 Structure above foundations

- Structural work of reinforced, prestressed or plain concrete should comply with BS 8110 *Structural use of concrete*, Part 1: 1985 *Code of practice for design and construction*, Part 2: 1985 *Code of practice for special circumstances* and Part 3: 1985 *Design charts for singly reinforced beams, doubly reinforced beams and rectangular columns*.
- Structural work of aluminium should comply with CP 118: 1969 *The structural use of aluminium*, using one of the alloys listed in section 1.1 of the code. (Under section 5.3 of the code, the structure should be classified as a safe-life structure.)
- Structural work of masonry should comply with BS 5628 *Code of practice for use of masonry*, Part 1: 1978 (1985) *Structural use of unreinforced masonry*, Part 3: 1985 *Materials and components, design and workmanship*.
- Structural work of timber should comply with BS 5268 *Structural use of timber*, Part 2: 1991 *Code of practice for permissible stress design, materials and workmanship*, Part 3: 1985 *Code of practice for trussed rafter roofs*.
- Structural work of steel should comply with BS 5950 *Structural use of steelwork in building*:
  (a) Part 1: 1990 *Code of practice for design in simple and continuous construction: hot rolled sections*.
  (b) Part 2: 1992 *Specification for materials, fabrication and erection: hot rolled sections*.
  (c) Part 3 *Design in composite construction*, section 3.1: 1990 *Code of practice for design of simple and continuous composite beams*.
  (d) Part 4: 1982 *Code of practice for design of floors with profiled steel sheeting*.
  (e) Part 5: 1987 *Code of practice for design of cold formed sections*; or,
  (f) BS 449 *Specification for the use of structural steel in building*, Part 2: 1969 *Metric units*.

6.9 Disproportionate collapse

6.9.1 Buildings of five or more storeys

In May 1968 a gas explosion on the eighteenth floor of a block of flats in London, known as Ronan Point, caused a large portion of the corner of the block to collapse.

Following on the subsequent tribunal and public inquiry into the disaster, new building regulations were formulated and introduced in 1970 with the express purpose of preventing further similar occurrences of this kind where the extent of the collapse is disproportionate to its cause.

These regulations have been updated and revised in line with current experience and knowledge, the main requirement being stated in paragraph A3 of Schedule 1 to the 2000 Regulations.

Buildings of five or more storeys (including basement storeys) are required to be constructed so that in the event of an accident the building will not suffer collapse to an extent disproportionate to the cause of that collapse.

Buildings that come into the five-storey category merely because they have a fifth floor in the roof space are excluded from the provisions if the roof has a pitch of 70° or less to the horizontal.

Approved Document A3 contains guidance on measures designed to reduce the sensitivity of a building to disproportionate collapse in the event of an accident and which may also avoid or reduce the hazards to which the building may be exposed.

Three approaches may be adopted depending on the extent to which it is possible to tie the structural members together.

(a) Provide effective horizontal and vertical ties complying with:
- Clause 2.2.2.2 of BS 8110 *Structural use of concrete*, Part 1: 1985 *Code of practice for design and construction* and clause 2.6 of Part 2: 1985 *Code of practice for special circumstances*, for structural work of reinforced, pre-stressed or plain concrete.
- Clause 2.4.5.3 of BS 5950 *Structural use of steelwork in building*, Part 1: 1990 *Code of practice for design in simple and continuous construction: hot rolled sections*, for structural work of steel.
- Clause 37 of BS 5628 *Code of practice for use of masonry*, Part 1: 1978 (1985) *Structural use of unreinforced masonry*, for structural use of masonry.

Compliance with these measures will require no further action to be taken in the structural design.

(b) Provide effective horizontal tying but vertical tying of vertical load-bearing members is not feasible:
- Assume each untied member is removed one at a time, in each storey in turn.
- Check that remainder of structure can bridge over the missing member even if it is in a substantially deformed state.
- In this deformed state certain members, such as cantilevers or simply supported floor panels, will be vulnerable to collapse.
- Check that collapse within the storey and the immediately adjacent storeys would be limited to (i) 15% of the storey area; or, (ii) 70 m^2, whichever is the less.
- If bridging, as detailed above, is not possible, design the removed member as a protected member (see below).

(c) Effective horizontal and vertical tying of load-bearing members is not feasible:

- Assume each member is removed one at a time, in each storey in turn.
- Check that area at risk of collapse within the storey and the immediately adjacent storeys is limited as in (b) above.
- If the area put at risk cannot be limited as described above when a member is notionally removed, then design the member as a protected member (see below).

6.9.2 Design of protected members

Protected members (or key elements) should be designed in accordance with the codes and standards listed under (a) above. These documents contain minimum loadings which protected members must be designed to withstand. For example, in BS 5950: Part 1: 1990, accidental loadings are referred to in clause 2.4.5.5. These should be chosen to reflect the importance of the key element and the consequences of failure, and the key element should be capable of withstanding a load of at least 34 kN/m^2 applied in any direction.

6.9.3 Long span roof structures

After the Ronan Point disaster referred to above, a number of long span buildings suffered roof collapses, this time not caused by accidents. The first of these occurred in June 1973 at the Camden School for Girls when the roof of the assembly hall collapsed.

The subsequent investigation by the Building Research Establishment revealed a number of causes, including loss of strength in the high alumina cement prestressed concrete roof beams due to conversion of the cement. Lack of adequate tying at the supports was also indicated as a contributory cause amongst others.

This failure led to the banning of high alumina cement in structural concrete work and also to the inclusion of long span roof structures in the disproportionate collapse regulations with the introduction of paragraph A4 of Schedule 1 in the 1991 Regulations.

The 1994 amendment revoked paragraph A4 and therefore the recommendations of section 6 of AD A4 no longer apply.

6.10 Forthcoming amendments to Part A and Approved Document A

At the time of going to press, Part A and Approved Document A were nearing the end of the consultation process and it is anticipated that a new edition of Approved Document A will be published in February or March 2004, possibly coming into force in late summer 2004. Amendments have become necessary as the current guidance is outdated in respect of references to British Standards and other documents, and the opportunity has been taken to carry out a general updating of the guidance in line with comments received from users and other interested parties.

The only change to the requirements of Part A is the removal of the 'Limits on application' associated with Requirement A3 *Disproportionate collapse*.

The main changes proposed to Approved Document A include the following:

- A general reordering of the material to bring it into a more logical order
- Updating of the lists of Codes, Standards and other references
- Deletion of the timber structural tables in Appendix A as a consequence of the proposed new Approved Document being produced by TRADA. This will contain similar tables and substantial additional information
- Deletion of the advice on bay window construction
- Clarification of the advice on cavity walls
- A general revision of the guidance on the design of masonry walls
- Revision of the guidance on the design of concrete strip foundations to align with current practice and the warranty scheme requirements of the NHBC and Zurich
- Revision of the advice on external wall cladding to bring it into line with advances in materials technology
- Clarification of the circumstances whereby the recovering of roofs is regarded as building work under the Building Regulations
- The introduction of advice on protection against attack from House Longhorn Beetle previously contained in Approved Document to support Regulation 7 (see section 8.4.9 below)
- Revision of the advice on the means of meeting compliance with Regulation A3 relating to disproportionate collapse as a result of the deletion of the 'Limits on application' referred to above.

7 Fire

7.1 Introduction

Part B of Schedule 1 to the Building Regulations 2000 is concerned with means of escape from buildings, fire spread within and between buildings, and access for the fire services to fight fires. Since the regulations are made in the interests of public health and safety, they do not attempt to achieve non-combustible buildings. The aims are to ensure the safety of the occupants and others who may be affected by the building, and to provide assistance for fire fighters. The protection of property, including the building itself, is the province of insurers who may require additional measures to provide higher standards before accepting the insurance risk. Guidance on the protection of property in the event of fire is given in the Loss Prevention Council's *Design guide for the fire protection of buildings*. For the protection of assets in the Civil and Defence Estates reference may be made to the *Crown Fire Standards* published by the Property Advisers to the Civil Estate (PACE).

Buildings must therefore be constructed so that, in the event of a fire:

- the occupants are given sufficient warning and are able to reach a place of safety
- they will resist collapse for a sufficient period of time to allow evacuation of the occupants and prevent further rapid fire spread
- the spread of fire within and between buildings is kept to a minimum
- there is satisfactory access for fire appliances and facilities are provided to assist firefighters in the saving of lives.

The first requirement is met by providing a means of warning and an adequate number of exits and protected escape routes. The second is met by setting reasonable standards of fire-resistance for the structural elements – the floors, roofs, load-bearing walls and frames. The third is met by:

- dividing large buildings into *compartments* and requiring higher standards of fire-resistance of the walls and floors bounding a compartment
- setting standards of fire-resistance for external walls
- controlling the surface linings of walls and ceilings to inhibit flame spread
- sealing and sub-dividing concealed spaces in the structure or fabric of a building to prevent the spread of unseen fire and smoke and
- setting standards of resistance to fire penetration and flame spread for roof coverings.

The fourth is met by:

- providing access for fire appliances and firefighting personnel;
- providing fire mains within the building; and
- making sure that heat and smoke may be vented from basement areas.

In some large and complex buildings (and in those containing different uses, such as airport terminals) the provisions contained in Approved Document (AD) B may prove inadequate or difficult to apply. In such buildings, the only viable way to achieve a satisfactory standard of fire safety may be to adopt a fire safety engineering approach which takes into account the total fire safety package. This approach may also be appropriate for solving a building design problem which otherwise follows the provisions of AD B. For a framework and guidance on the design and assessment of fire safety measures in buildings it may be useful to follow the discipline of British Standard Draft for Development (DD) 240 *Fire safety engineering in buildings*. This should enable designers and building control bodies to:

- be aware of the relevant issues
- be aware of the need to consider complete fire-safety systems; and
- follow a disciplined analytical framework.

Some difficulty may also be encountered when trying to apply the provisions of AD B to existing buildings, particularly when they are of special historic or architectural importance. In buildings of this type it may be appropriate to carry out an assessment of the potential fire hazard or risk to life, and then incorporate in the design a sufficient number of fire safety features to alleviate the danger. The risk assessment should take account of:

- the likelihood of a fire occurring
- the anticipated severity of the fire
- how well the structure of the building is able to resist the spread of smoke and flames
- the consequential danger to persons in or near the building.

Fire safety measures which can be incorporated in the design include:

- an assessment of the adequacy of the means to prevent fire
- the installation of automatic fire detection and warning systems
- the provision of adequate means of escape
- the provision of smoke control
- design features aimed at controlling the rate of growth of a fire if one does occur
- an assessment of the ability of a structure to resist the effects of fire
- the extent of fire containment offered by the building
- the fire separation from other buildings or parts of the same building
- the standard of firefighting equipment in the building
- the ease with which the fire service may gain access to fight a potential fire

- the existence of legislative controls to require staff training in fire safety and fire routines (e.g. the Fire Precautions Act 1971, the Fire Precautions (Workplace) Regulations 1997 or licensing and registration controls)
- the existence of continuing control so that fire safety systems can be seen to be maintained (as in the certification procedure under the Fire Precautions Act)
- the installation of suitable fire management procedures.

Some factors in the measures listed above may be assessed using quantitative techniques and given numerical values in some circumstances. Quantitative techniques may also be used to evaluate risk and hazard although the assumptions made need to be carefully assessed.

An example of an overall approach to fire safety may be found in BS 5588 *Fire precautions in the design, construction and use of buildings*, Part 10: 1991 *Code of practice for shopping complexes*. (See section 7.29 below for further reference to enclosed shopping complexes.)

Other examples where the guidance in AD B may be inappropriate include:

- Where a building contains an atrium which passes through compartment floors. Special fire safety measures may need to be incorporated in the design. BS 5588 *Fire precautions in the design, construction and use of buildings*, Part 7: 1997 *Code of practice for the incorporation of atria in buildings* contains guidance on suitable fire safety measures.
- Hospitals. The design of fire safety in hospitals is dealt with in Health Technical Memorandum (HTM) 81 *Fire precautions in new hospitals* (1996 revision). Part B of the Building Regulations 2000 will be satisfied where HTM 81 is used.

Fire safety in buildings is a complex matter and in recent years a number of reviews and enquiries have been set up to consider different aspects of fire safety law. Additionally, matters have been further complicated by the need to comply with various European Community directives.

Following the 2001 General Election, there was a reorganisation of Government Departmental responsibilities for fire safety. It brought the Home Office's Fire Service Inspectorate and the Building Regulations Division together, under the Department for Transport, Local Government and the Regions (DTLR) for the first time. The DTLR has since been broken up and responsibility for Building Regulations now rests with the Office of the Deputy Prime Minister (ODPM). There may therefore be a unified approach to fire safety in the future.

7.2 Terminology

Certain terms which apply generally throughout this chapter are defined here. Other terms are defined in the specific section to which they apply.

BASEMENT STOREY – A storey which has some part of the perimeter of its floor more than 1200 mm below the highest level of the ground adjoining that part of the

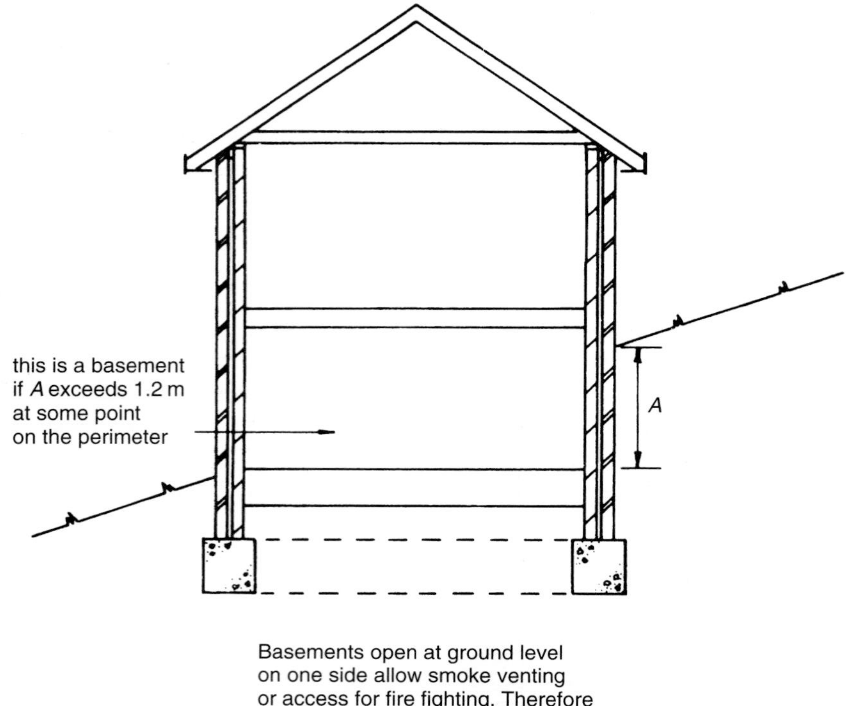

Fig. 7.1 Basement storey.

floor. A basement storey may be treated as the above ground structure for fire-resistance purposes, if one side is open at ground level for smoke venting or fire-fighting. (See Fig. 7.1.)

BOUNDARY – When referring to any side of a building or compartment (including any external wall or part), means the usual legal boundary adjacent to that side, being taken up to the centre of any abutting railway, street, canal or river.

CEILING – Includes any soffit, rooflight or other part of a building which encloses and is exposed overhead in a room, circulation space or protected shaft (but not including the surface of the frame of any rooflight, the upstand of which is considered as part of the wall).

CIRCULATION SPACE – A space (including a protected stairway) used mainly as a means of access between a room and an exit from the compartment or building.

COMPARTMENT – Any part of a building (including rooms, spaces or storeys) which is constructed to stop fire spreading to or from another part of the same building, or an adjacent building. If any part of the top storey of a building comes

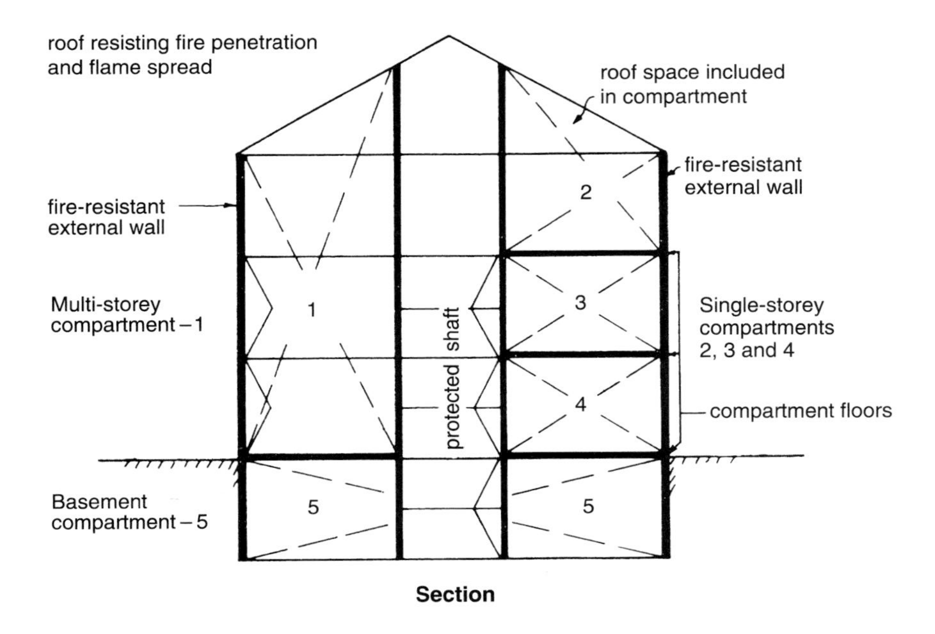

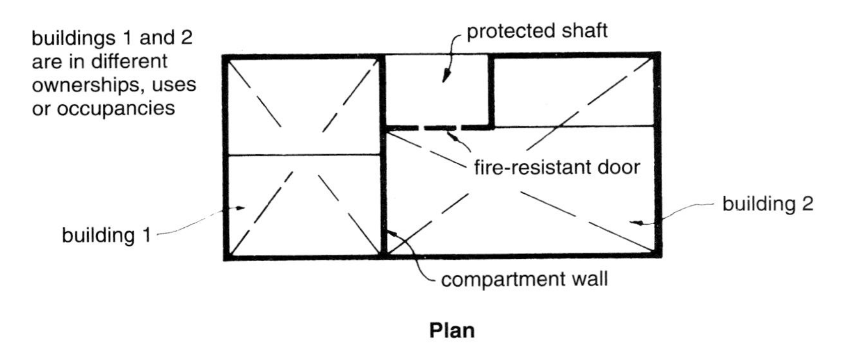

Fig. 7.2 Division of buildings into compartments.

within a compartment, that compartment is taken to include any roof space above that part of the top storey (see Fig. 7.2). See also 'Separated part' below.

COMPARTMENT WALL/FLOOR – Fire-resisting construction provided to separate one fire compartment from another for the purpose of preventing fire spread.

CONCEALED SPACE (CAVITY) – A space which is concealed by the elements of a building (such as a roof space or the space above a suspended ceiling) or contained within an element (such as the cavity in a wall). This definition does *not* include a room, cupboard, circulation space, protected shaft or space within a flue, chute, duct, pipe or conduit.

ELEMENTS OF STRUCTURE –

- Any member forming part of the structural frame of a building or any other beam or column. (This does not normally include members which form part of a roof structure only unless the roof performs the function of a floor, or the roof structure provides stability for fire-resisting walls.)
- A floor (but not the lowest floor in a building, or a platform floor).
- An external wall.
- A compartment wall (including a wall which is common to two or more buildings).
- A load-bearing wall or the load-bearing part of a wall.
- A gallery (but not a loading gallery, fly gallery, lighting bridge, stage grid or any gallery provided for similar purposes or for maintenance and repair).

These elements are illustrated in Fig. 7.3.

EXTERNAL WALL – Includes a portion of a roof sloping at 70° or more to the horizontal if it adjoins a space within the building to which persons have access, other than for occasional maintenance and repair (see Fig. 7.4).

FIRE DOOR – Includes any shutter, cover or other form of protection to an opening in any fire-resisting wall or floor of a building or in the structure surrounding a protected shaft. The fire door should be able to resist the passage of fire and/or gaseous products of combustion in accordance with specified criteria (see Fire doors in section 7.22.3 below). A fire door may have one or more leaves.

FIRE SEPARATING ELEMENT – Includes a compartment wall or floor, a cavity barrier and construction enclosing a protected escape route and/or place of special fire hazard.

GALLERY – A floor (or raised storage area) which projects into another space but has less than half the floor area of that space.

HABITABLE ROOM – A room used for dwelling purposes, including a kitchen (in Part B) but not a bathroom.

NOTIONAL BOUNDARY – A boundary which is assumed to exist between two buildings in the residential, and assembly and recreation purpose groups, on the same site where there is no actual boundary. The notional boundary line should be so placed that neither building contravenes any of the requirements of AD B relevant to the external walls facing each other (see Fig. 7.5).

OCCUPANCY TYPE – Part of a purpose group (see section 7.3 below).

PLACES OF SPECIAL FIRE HAZARD – Oil-filled switchgear and transformer rooms, boiler rooms, stores for fuel or other highly flammable substances, and

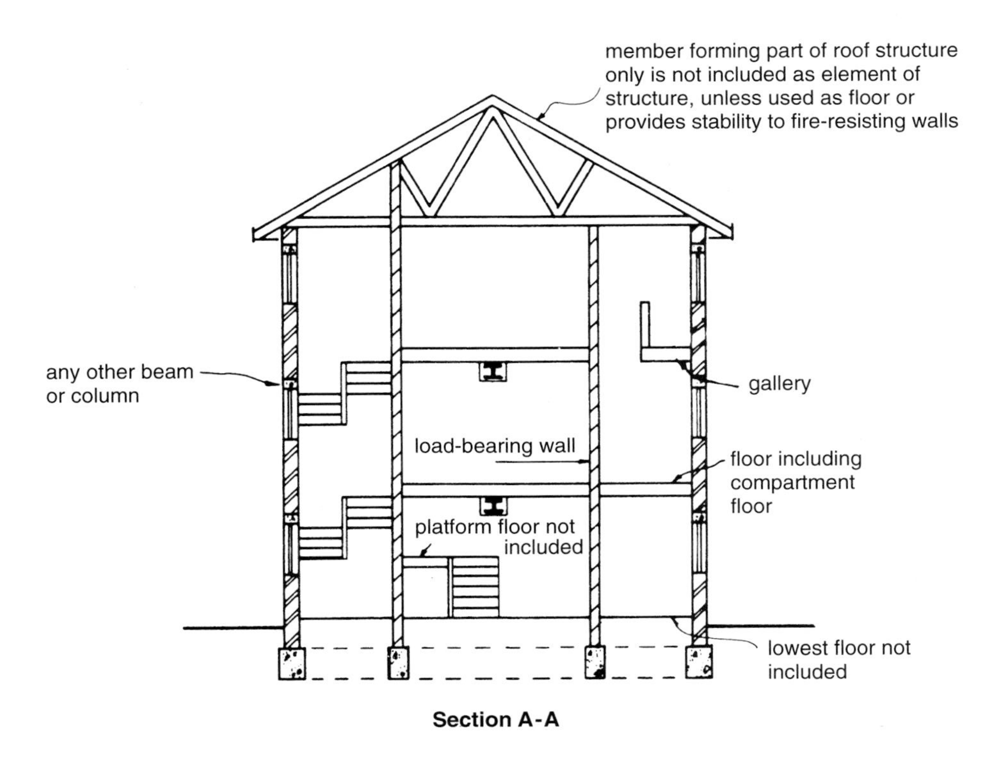

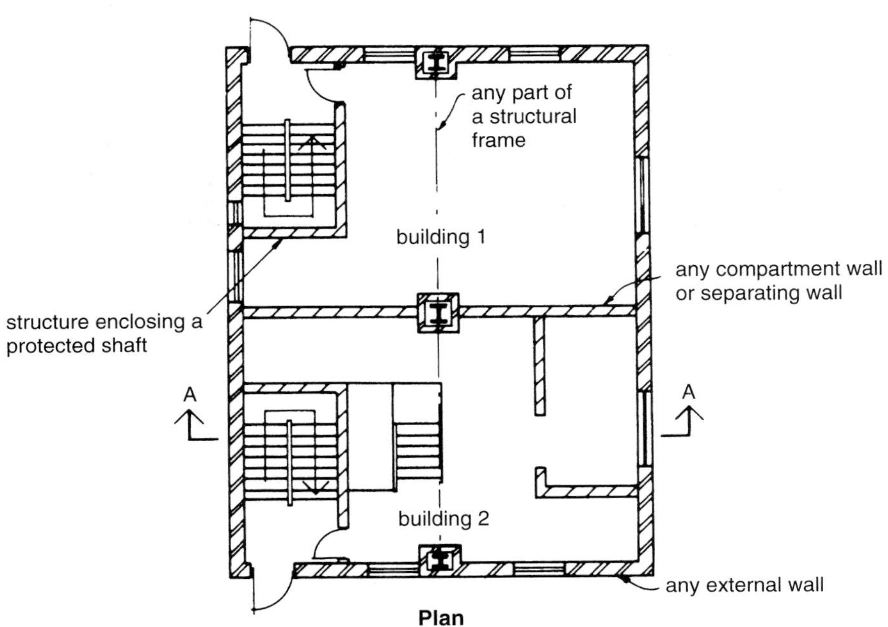

Fig. 7.3 Elements of structure.

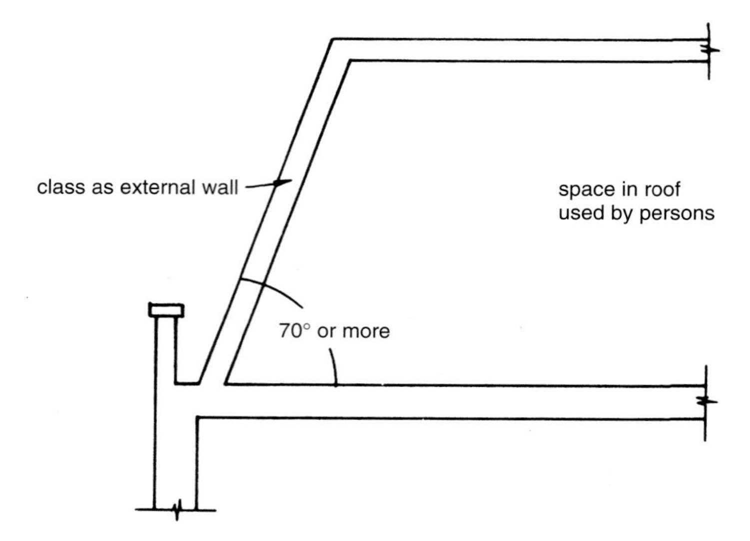

Fig. 7.4 Steeply pitched roofs.

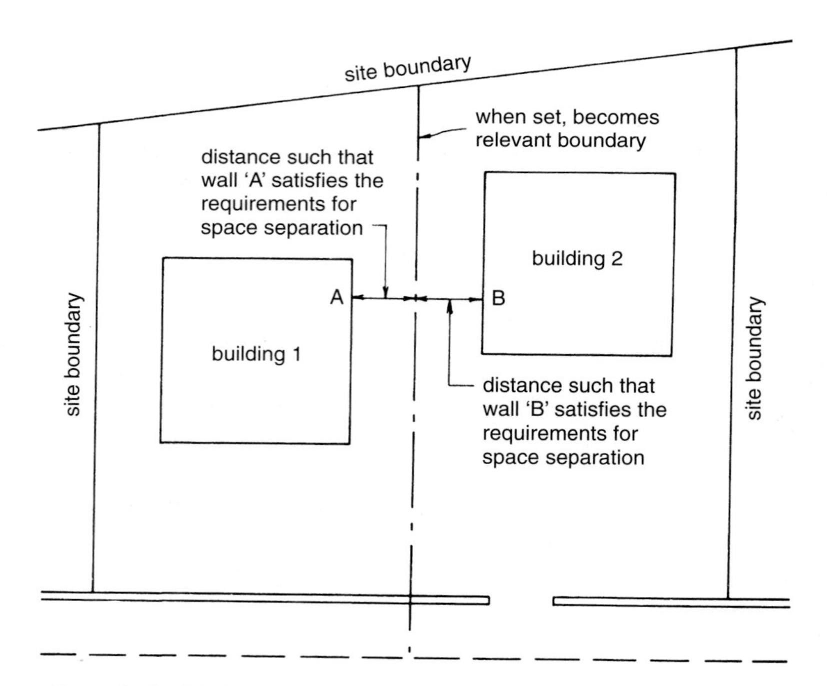

One or both of the buildings new.
One or other of the buildings of residential or assembly and recreation use.
Existing building treated as identical new building but with existing unprotected area and fire resistance in external wall.

Fig. 7.5 Notional boundary.

rooms containing fixed internal combustion engines. Additionally, in schools – laboratories, technology rooms with open heat sources, kitchens and stores for PE mats or chemicals.

PLATFORM FLOOR (sometimes called an access or raised floor) – A floor over a concealed space intended to house services, which is supported by a structural floor.

PROTECTED CORRIDOR/LOBBY – A corridor or lobby protected from fire in adjacent accommodation by fire-resisting construction.

PROTECTED SHAFT – A shaft enclosed with fire-resisting construction which enables persons, things or air to pass between different compartments.

PROTECTED STAIRWAY – A stair adequately protected with fire-resisting construction discharging to place of safety through final exit (includes passage from foot of stair to final exit).

RELEVANT BOUNDARY – For a boundary to be considered relevant it should:

- be coincident with, or
- be parallel to, or
- not make an angle of more than 80° with the external wall (see Fig. 7.6).

In certain circumstances a 'notional' boundary, as defined above, will be the relevant boundary. A wall may have more than one relevant boundary.

ROOFLIGHT – Includes any domelight, lanternlight, skylight, ridge light, glazed barrel vault or other element which is intended to admit daylight.

ROOM – An enclosed space in a building, but not one used solely as a circulation space. This term would also include cupboards that were not fittings and large rooms such as warehouses and auditoria. Excluded are voids such as ducts, roof spaces and ceiling voids.

SEPARATED PART (OF A BUILDING) – Where a compartment wall completely divides a building from top to bottom and is in one plane, the divided sections of the building are referred to as separated parts. The height of each separated part may then be treated individually (see Fig. 7.7 and section 7.4 Rules for measurement below).

SINGLE-STOREY BUILDING – A building which consists of a ground storey only. (A separated part consisting of a ground storey only and with a roof which is accessible only for the purposes of maintenance and repair may be treated as part of a single-storey building.) Basements are not included when counting the number of storeys in a building.

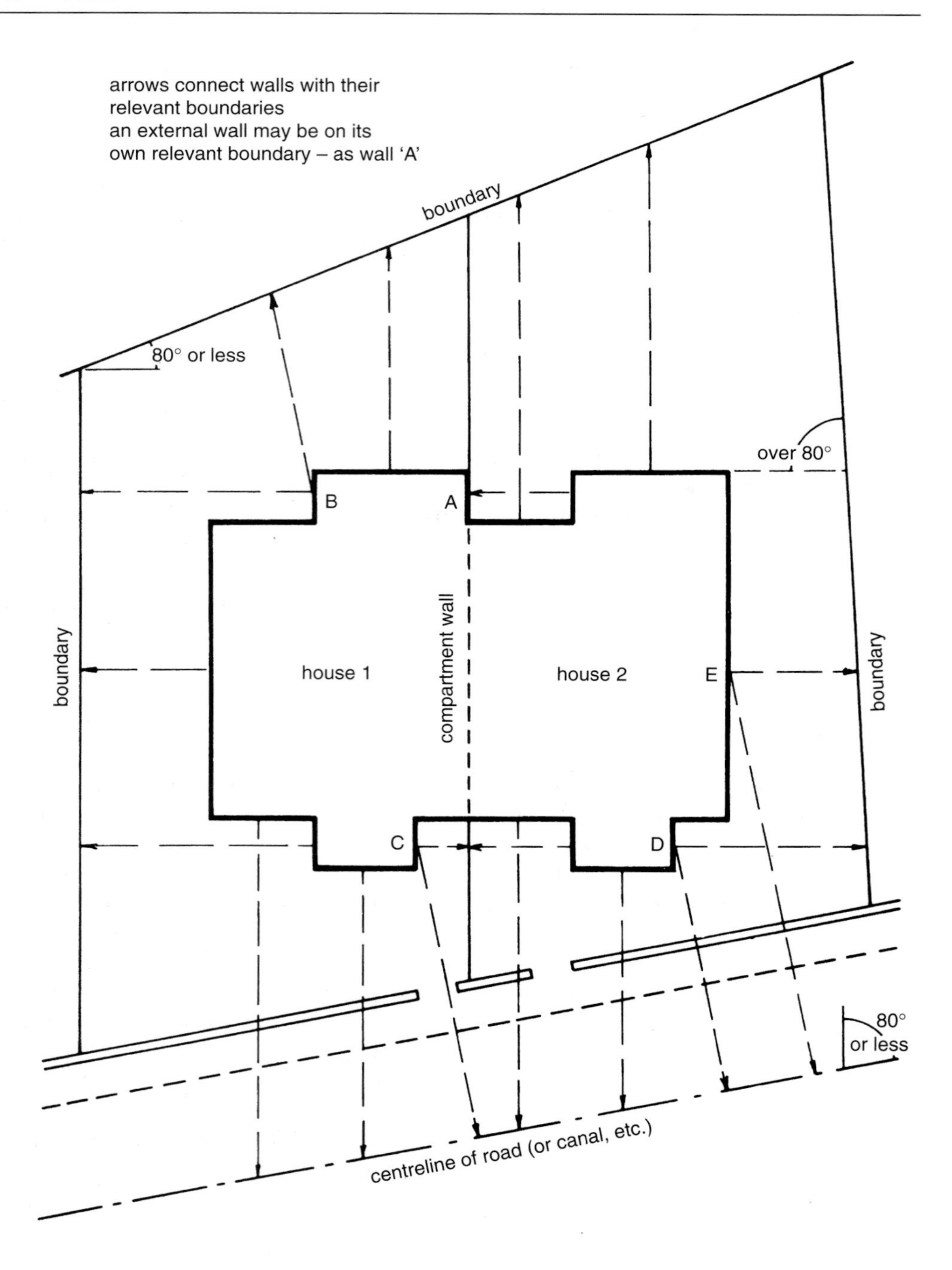

Fig. 7.6 Relevant boundaries.

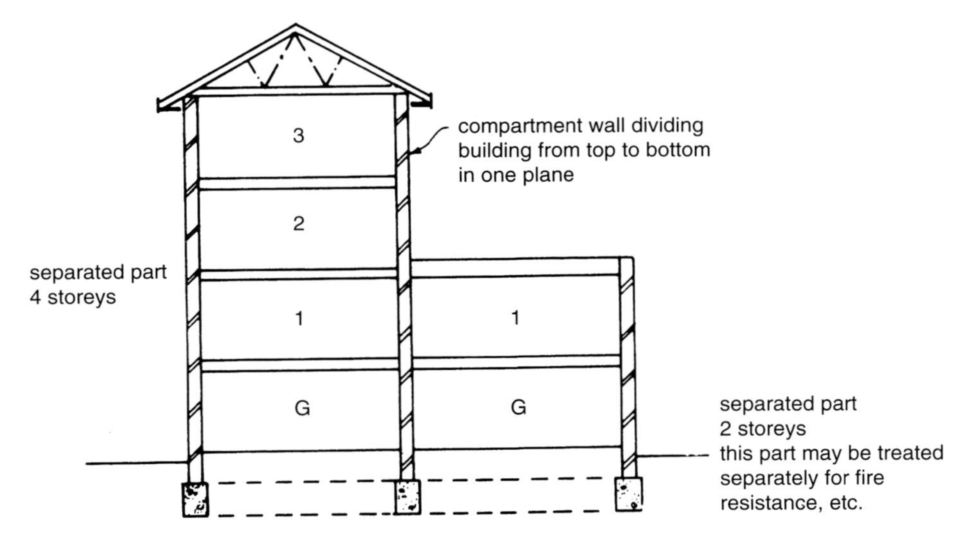

Fig. 7.7 Separated part.

SITE (OF A BUILDING) – The land occupied by the building up to the boundaries with land in other ownership.

STOREY – Included in this definition are the following:

- any gallery in an assembly building (PG5); and
- a gallery in any other building if its area exceeds half that of the space into which it projects; and
- a roof, unless used only for maintenance or repair.

UNPROTECTED AREA – In relation to an external wall or side of a building, means:

- a window, door or other opening (although windows designed and glazed to provide the necessary standard of fire resistance and which are not openable, and recessed car parking areas are not regarded as unprotected areas)
- any part of an external wall of fire-resistance less than that required by Section 13 of AD B4 (see section 7.27)
- any part of an external wall with external facing attached or applied, whether as cladding or not, the facing being of combustible material more than 1mm thick (combustible in this context means any material which does not have a Class 0 rating) (see Fig. 7.8).

7.3 Purpose Groups

The fire hazard presented by a building will, to a large extent, depend on the use to which the building is put. Many of the provisions concerning means of warning and

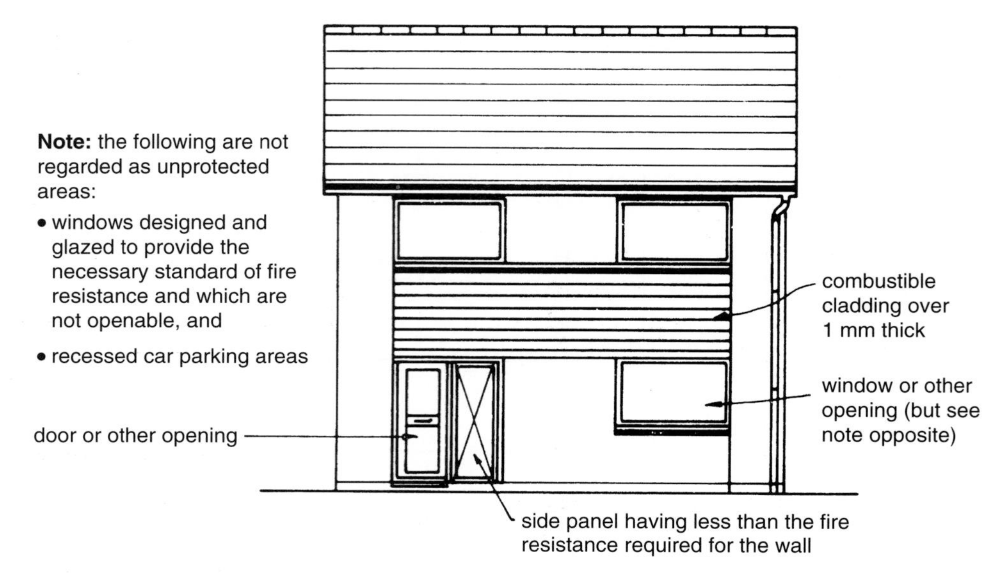

Fig. 7.8 Unprotected areas.

escape, fire-resistance, compartmentation, etc. are directly related to these use classifications, which in AD B are termed purpose groups (PG).

The seven PGs are set out in Table D1 (Classification of purpose groups) of Appendix D of AD B. They are as follows:

1 **Residential (dwellings)** – This includes parts of a dwelling used by the occupant in a professional or business capacity (such as a surgery, consulting room, office or other accommodation), not exceeding 50 m^2 in total. This group is further sub-divided into:

- **1(a)** flat or maisonette
- **1(b)** dwellinghouse (with a habitable storey more than 4.5 m above ground level)
- **1(c)** dwellinghouse (no habitable storey above 4.5 m from ground level). 1(c) also includes any detached garage or open carport not exceeding 40 m^2 in area, or a detached building consisting of a garage and open carport, neither of which exceeds 40 m^2 in area, irrespective of whether or not they are associated with a dwelling.

2(a) **Residential (institutional)** – Includes a hospital, home, school or other similar establishment. The premises will be used as living (and sleeping) accommodation for, or for the treatment, care or maintenance of:

- persons suffering from disabilities due to illness or old age or other physical or mental incapacity
- children under the age of five years
- persons in a place of lawful detention.

2(b) Residential (other) – Includes a hotel, boarding house, residential college, hall of residence, hostel, and any other residential purpose not described above.

3 Office – Includes offices or premises used for the purpose of:
- administration
- clerical work (including writing, book keeping, sorting papers, filing, typing, duplicating, machine calculating, drawing and the editorial preparation of matter for publication, police and fire service work)
- handling money (including banking and building society work)
- communications (including postal, telegraph and radio communications) radio, television, film, audio or video recording, or performance (not open to the public) and their control.

4 Shop and Commercial – Shops or premises used for a retail trade or business including:
- the sale to members of the public of food or drink for immediate consumption
- retail by auction, self-selection and over-the-counter wholesale trading
- the business of lending books or periodicals for gain
- the business of a barber or hairdresser
- premises to which the public is invited to deliver or collect goods in connection with their hire, repair or other treatment, or (except in the case of repair of motor vehicles) where they themselves may carry out such repairs or other treatments.

5 Assembly and Recreation – Places of assembly, entertainment or recreation including:
- broadcasting, recording and film studios open to the public
- bingo halls, casinos, dance halls
- entertainment, conference, exhibition and leisure centres
- funfairs and amusement arcades
- museums and art galleries
- non-residential clubs
- theatres, cinemas and concert halls
- educational establishments
- dancing schools, gymnasia, swimming pool buildings, riding schools, skating rinks, sports pavilions, sports stadia
- law courts
- churches and other buildings of worship
- crematoria
- libraries open to the public
- non-residential day centres, clinics, health centres and surgeries
- passenger stations and termini for air, rail, road or sea travel
- public toilets
- zoos and menageries.

6 Industrial – Factories and other premises used for:

- manufacturing, altering, repairing, cleaning, washing, breaking-up, adapting or processing any article
- generating power
- slaughtering livestock.

7 **Storage and other non-residential** – This group is further sub-divided into:

7(a) Place for the storage or deposit of goods or materials [other than described under 7(b)] and any building not within any of the purpose groups 1 to 6, and

7(b) Car parks designed to admit and accommodate only cars, motorcycles and passenger or light goods vehicles weighing no more than 2500 kg gross.

Normally the PG is applied to the whole building, or (where a building is compartmented) to a compartment in the building, by reference to the main use of the building or compartment. Parts of the building put to different uses can be treated as ancillary to the main use, and therefore as though they are in the same use. However, a different use in the same building is not regarded as ancillary, and is therefore treated as belonging to a PG in its own right:

- where the ancillary use is a flat or maisonette
- where the building or compartment exceeds 280 m^2 in area and the ancillary use exceeds one-fifth of the total floor area of the building or compartment
- where the building is a shop or commercial building or compartment of Purpose Group (4) and contains a storage area which exceeds one third of the total floor area of the building or compartment and the building or compartment is more than 280m^2 in area.

Where a building contains different main uses which are not ancillary to one another, each use should be considered as belonging to a purpose group in its own right.

Some large buildings, such as shopping complexes, may involve complicated mixes of purpose groups. In these cases special precautions may need to be taken to reduce any additional risks caused by the interaction of the different purpose groups.

7.4 Rules for measurement

Many of the requirements concerning compartmentation and fire resistance, etc. in AD B are based on the height, area and cubic capacity of the building, compartment or separated part. For consistency, it is necessary to have a standard way of measuring these proportions. Appendix C of AD B indicates diagrammatically how the various forms of measurement should be made. These rules can be summarised as follows:

- HEIGHT – The height of the building or part is measured from the mean level of the ground adjoining the outside of the building's external walls to the level of half the vertical height of the roof, or to the top of the walls or parapet, whichever is the higher. This rule applies to double pitch, mono-pitch, flat and mansard type roofs (see Fig. 7.9(a)).

- AREA – The area of any storey of a building, compartment or separated part should be calculated as the total area in that storey within the finished inner surfaces of the enclosing walls. If there is no enclosing wall, the area is measured to the outermost edge of the floor on that side. The area should include any internal walls or partitions.

  The area of a room, garage, conservatory or outbuilding is calculated by measuring to the inner surface of the enclosing walls.

  The area of any part of a roof should be calculated as the actual visible area of that part, as measured on a plane parallel to the roof slope. For a lean-to roof, the measurement should be taken from the wall face to the outer edge of the roof slope. For a hipped roof, the measurement should be to the outer point of the roof as a base area (see Fig. 7.9(a)).

- CUBIC CAPACITY – The cubic capacity of a building, compartment or separated part should be calculated as the volume of the space between the finished surfaces of the enclosing walls, the upper surface of its lowest floor and the under surface of the roof or ceiling surface as appropriate. If there is no enclosing wall, the measurement should be taken to the outermost edge of the floor on that side. The cubic capacity should again include space occupied by other walls, shafts, ducts or structures within the measured space (see Fig. 7.9(b)).

- NUMBER OF STOREYS – The number of storeys in a building or separated part should be calculated at the position which gives the maximum number. Basement storeys are not counted. In most purpose group buildings galleries are also not counted as storeys. However, in assembly buildings, a gallery is included as a storey unless it is a fly gallery, loading gallery, stage grid, lighting bridge or other similar gallery, or is for maintenance and repair purposes. (The common factor here is that these excluded galleries are not generally accessible to the public.) (See Fig. 7.9(b).)

- HEIGHT OF TOP STOREY – This is measured from ground level on the lowest side of the building to the upper surface of the top floor. Roof-top plant areas are excluded from this measurement as are any top storeys consisting exclusively of plant rooms.

7.5 Means of warning and escape in case of fire

7.5.1 The mandatory requirement

Buildings must be designed and constructed so that there are appropriate provisions for the early warning of fire, and appropriate means of escape in case of fire capable

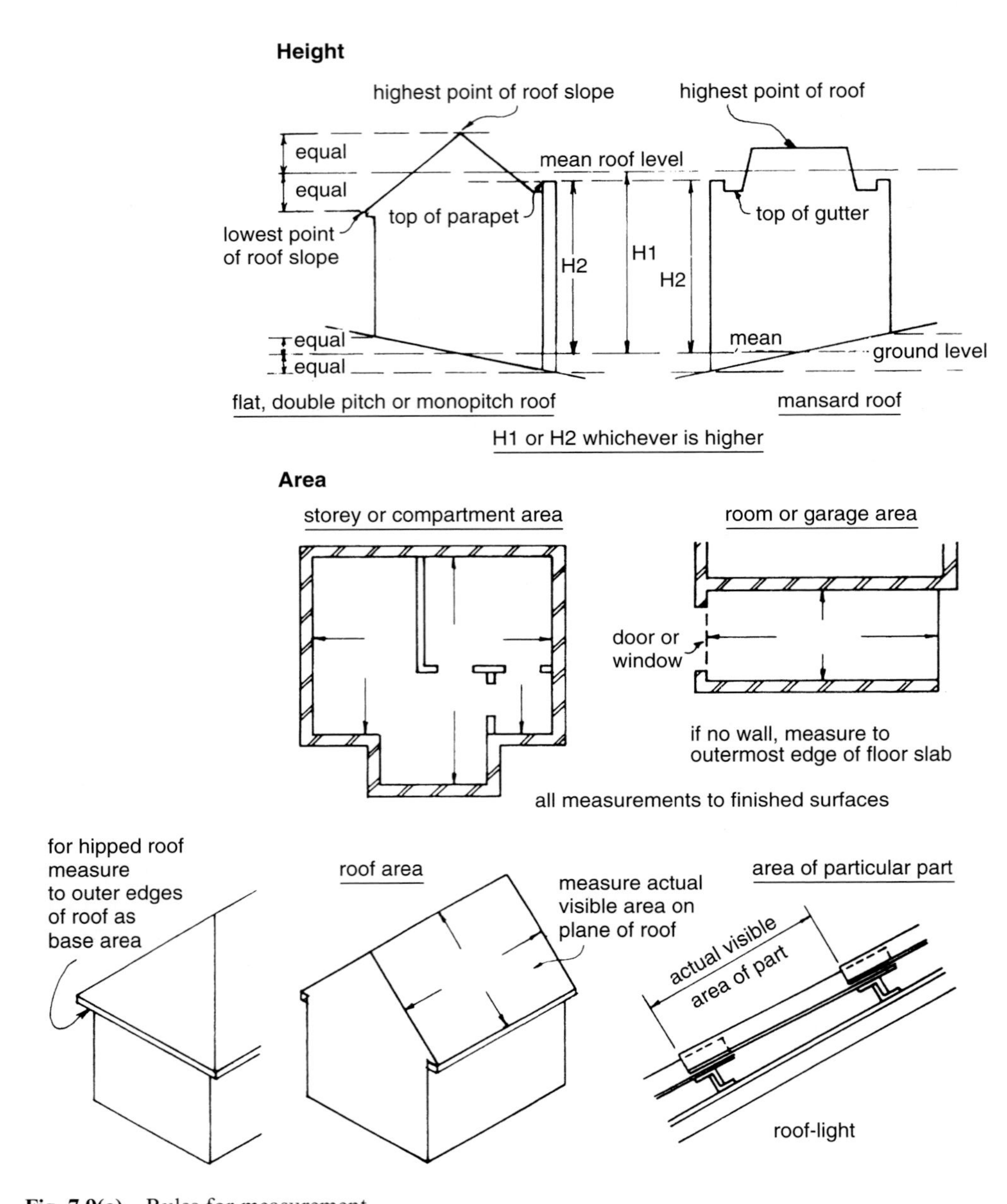

Fig. 7.9(a) Rules for measurement.

of being used safely and effectively at all material times. (No definition is given of material times.) The mandatory requirement goes on to state that the means of escape must be to a place of safety outside the building; however, it will be seen below that for certain classes of buildings the place of safety may be within the building itself. This requirement applies to all buildings except prisons provided under section 33 of the Prisons Act 1952.

Cubic capacity

cubic capacity of compartment or building

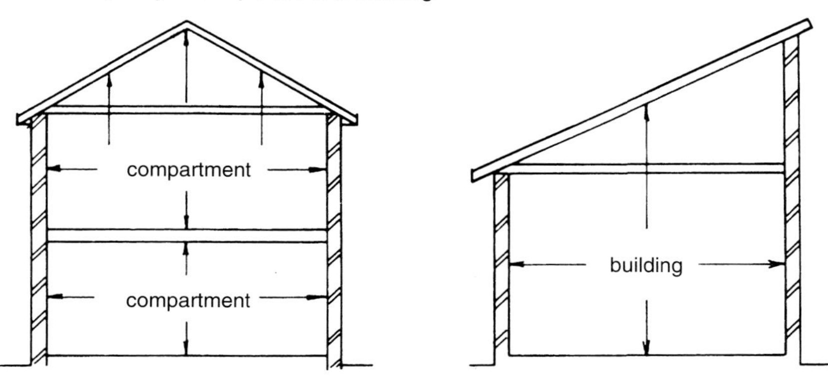

Number of storeys

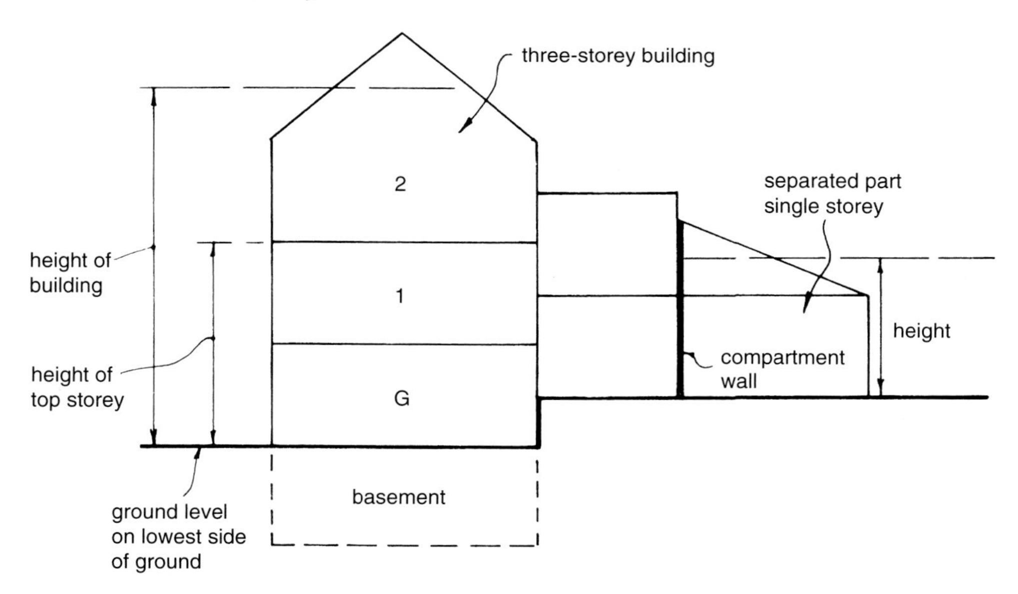

Fig. 7.9(b) Rules for measurement.

7.5.2 Standard of performance to meet the mandatory requirement

The mandatory requirement may be met by:

- providing escape routes which are suitably located and of sufficient number and size to enable the occupants to escape to a place of safety in the event of fire
- making sure that escape routes are adequately lit and suitably signed

- providing appropriate facilities to limit the ingress of smoke to the escape route or to restrict the fire and remove smoke
- enclosing escape routes where necessary, so that they are sufficiently protected from the effects of fire

to an extent necessary depending on the size, height and use of the building.

Additionally, there must be sufficient means for giving early warning of fire for the building's occupants.

Approved Document B1, Means of warning and escape, contains details of practical ways to satisfy the mandatory requirement and the performance standard referred to above. It does this in two ways:

- by giving actual recommendations in the text (described in detail below); and
- by reference to other sources of guidance.

Generally, BS 5588: *Fire precautions in the design, construction and use of buildings*, Part 0: 1996 *Guide to fire safety codes of practice for particular premises/applications* contains references to a great many codes and guides dealing with means of escape provision. If it is decided to use one of these codes or guides, the parts which are relevant to means of escape should be followed rather than a mixture of the specific publication and the relevant sections of AD B1. Having said this, it may still be necessary to supplement the Approved Document information with guidance from another publication where insufficient information is provided, or specific reference is made. For example, section 18 of AD B (which deals with access to buildings for fire-fighting personnel) recommends the use of BS 5588: Part 5 *Code of practice for firefighting stairs and lifts* since AD B contains no other guidance on the subject. Additionally, some buildings which house particular industrial or commercial activities may present special fire hazards (filling stations, etc.) and may need additional fire precautions to those detailed in AD B. BS5588: Part 0 should be consulted in these cases.

7.6 Means of escape – links with other legislation

In addition to building regulations there is a large amount of other legislation covering means of escape in case of fire. Reference to some other provisions will be found in Chapter 1 (Building control: an overview). However, the purpose of this book is not to review all the legislation on this subject, but to show how it may interact with and affect building regulations. For a comprehensive guide to means of escape legislation see *Means of escape from fire* by M.J. Billington, Anthony Ferguson and A.G. Copping (Blackwell Science 2002). Some of the most important provisions are mentioned below.

7.6.1 Fire Precautions Act 1971 and the Fire Precautions (Workplace) Regulations 1997

Together, these probably represent the most significant pieces of legislation which are linked with the Building Regulations. They are enforced by fire authorities. The Fire Precautions Act designates certain uses of premises (called relevant premises in the Building Regulations) for which a fire certificate is required (see section 3.3 and Chapter 5) and imposes a statutory duty on the occupiers of certain smaller premises to provide a reasonable means of escape. Additionally, the Fire Precautions (Workplace) Regulations 1997 (as amended in 1999) impose minimum fire safety standards on all workplaces where more than one person is employed (see Chapter 5). A duty is placed on the employer to carry out a fire risk assessment of the workplace to establish what precautions are necessary to ensure the safety of employees in the event of fire.

In the case of a workplace to which the Fire Precautions (Workplace) Regulations do not apply (e.g. a small family-run designated hotel which does not employ any staff) fire authorities cannot, as a condition of issuing a fire certificate, make requirements for structural or other alterations to means of escape if the building plans comply with the Building Regulations. This does not stop the fire authority from requiring compliance with provisions which were not required to be supplied to the local authority or approved inspector under the building regulations if they consider the means of escape to be inadequate, and in practice, this gives fire authorities considerable leeway to demand additional measures which are not covered by Building Regulations. Additionally, some things may be required by the Fire Precautions Act which are outside the scope of Building Regulations, e.g. the provision of first aid fire fighting equipment for use by the building occupants.

Therefore it is possible for a building to be subject to the Fire Precautions Act, The Fire Precautions (Workplace) Regulations and other legislation which imposes fire safety requirements at the same time. Some buildings (theatres, nursing homes etc.) may need to be registered or licensed with the local authority. It is essential in these circumstances, that adequate consultations take place during the design process with all the relevant enforcement bodies. Guidance on the procedures that should be adopted for these consultations is contained in *Building Regulations and Fire Safety – Procedural Guidance*, published jointly by the DETR (now the Office of the Deputy Prime Minister) and the National Assembly for Wales.

7.6.2 Health and Safety at Work etc. Act 1974

In the case of some highly specialised industrial or storage premises where hazardous processes are carried out or materials are kept the Health and Safety Executive may have certification responsibilities similar to those of the fire authority.

7.6.3 Housing Act 1985

Local authorities are obliged to require means of escape from houses in multiple occupation, i.e. houses which are occupied by persons who do not form part of a single household. (For guidance on the interpretation of this definition see DOE/Home Office/Welsh Office Circular *Memorandum on overcrowding and houses in multiple occupation*.) Many local authorities use the services of the fire authority in deciding on the adequacy of an existing means of escape in such buildings. If new dwellinghouses, flats and maisonettes are designed in accordance with AD B1, they should be acceptable for use as 'houses' in multiple occupation.

7.7 Interpretation of AD B1

A large number of terms are used in Approved Document Bl which relate specifically to means of escape:

ACCOMMODATION STAIR – A stair which is provided for the convenience of the occupants of a building and is additional to those required for means of escape. (When calculating the number of stairs required in a building, accommodation stairs are ignored.)

ALTERNATIVE ESCAPE ROUTES – Routes which are sufficiently separated from one another by fire-resisting construction or space and direction so that one route will still be available even if the other is affected by fire. In most buildings this will mean alternative protected corridors and stairs; however, in dwellings an alternative escape route could be via a second stair, balcony or flat roof if it enabled a person to reach a place free from danger of fire.

ALTERNATIVE EXIT – One of two or more exits, each of which is separate from the other. (There are rules in AD Bl for determining whether exits are sufficiently far enough apart to be considered as alternatives and these are illustrated in section 7.15.2 below (Fig. 7.24).)

ATRIUM – A space in a building (not necessarily vertically aligned) which passes through one or more structural floors. (Enclosed escalator wells, enclosed lift wells, building services ducts or stairways are not classified as atria.)

COMMON BALCONY – An escape route from more than one flat or maisonette which is formed by a walkway open to the air on one or more sides.

COMMON STAIR – An escape stair which serves more than one maisonette or flat.

CORRIDOR ACCESS – A common horizontal internal access or circulation space which serves each dwelling in a building containing flats. It may include a common entrance hall.

DEAD END – An area from which it is only possible to escape in one direction.

DIRECT DISTANCE – The shortest distance which can be measured from within the floor area of the inside of the building, to the storey exit. All internal walls, partitions and fittings are ignored when measuring the direct distance, except the walls enclosing the protected stairway.

DWELLING – A unit of residential accommodation which is occupied by:

- a single person or family; or
- not more than six residents living together as a single household, including a household where the residents receive care.

The dwelling need not be the sole or main residence of the occupants. (This is an important definition since it shows the difference, for the purposes of means of escape, between a dwelling and a house in multiple occupation. It also appears to indicate that certain buildings, where people receive care, may be regarded as dwellings and not institutional buildings, again, for the purposes of means of escape.)

EMERGENCY LIGHTING – Lighting which is provided to be used when the normal lighting fails.

ESCAPE LIGHTING – Part of the emergency lighting which is provided specifically to light escape routes.

ESCAPE ROUTE – That part of the means of escape from any point in the building to the final exit.

EVACUATION LIFT – A lift used to evacuate disabled people in the event of fire.

FINAL EXIT – The termination of an escape route from a building sited so that people may be able rapidly to get clear of any danger from smoke or fire in the vicinity of the building. It should give direct access to a street, passageway, walkway or open space. (It should be noted that windows are not acceptable as final exits.)

INNER ROOM – A room contained within another room (termed the access room). Escape is only possible by passing through the access room.

MAISONETTE – A 'flat' on more than one level.

MEANS OF ESCAPE – Structural means whereby a safe route or routes is or are provided for persons to travel to a place of safety from any point in the building, in the event of fire.

OPEN SPATIAL PLANNING – The internal arrangements of a building whereby a number of floors are contained within one undivided space, e.g. split-level floors. AD B1 makes a distinction between an atrium space and open spatial planning.

PROTECTED CIRCUIT – An electrical circuit which is protected against fire.

PROTECTED ENTRANCE HALL/LANDING – A hall or circulation space in a dwelling which is protected by fire-resisting construction (other than any part of the wall which is external).

SMOKE ALARM – A device for detecting smoke which gives an audible alarm. All the components will be contained within one housing (except possibly the energy source).

STOREY EXIT – A doorway giving direct access to:

- a protected stairway (defined in section 7.2)
- a firefighting lobby (defined in section 7.30.2)
- an external escape route.

Also the following:

- a final exit (see above for definition)
- a door in a compartment wall in an institutional building if the building is planned for progressive horizontal evacuation (see section 7.17.1).

TRAVEL DISTANCE – The actual distance travelled by a person from any point in the floor area to the nearest storey exit. In this case the layout of the floor in terms of walls, partitions and fittings *is* taken into account. (Cf. Direct distance above.)

7.8 General requirements for means of warning and escape

There are certain basic principles which govern the design of means of warning and escape in buildings and which apply to all building types. In general the design should be based on an assessment of the risk to the occupants should a fire occur and should take account of:

- the use of the building (and the activities of the users)
- the nature of the building structure
- the processes undertaken and/or the materials stored in the building
- the potential fire sources
- the potential for fire spread throughout the building
- the standard of fire safety management to be installed.

In assessing the above, judgements regarding the likely level of provision may have to be made when the exact details are unknown.

The following assumptions must be made in order that a safe and economical design may be achieved:

- In general, when a fire occurs, the occupants should be able to escape safely, without external assistance or rescue from the fire service or anyone else. Obviously, there are some institutional buildings where it is not practical to expect the occupants to escape unaided and special arrangements are necessary in these cases. Similar considerations apply to disabled people. Aided escape is also permitted in certain low-rise dwellings.
- Fires do not normally break out in two different parts of a building at the same time.
- Fires are most likely to occur in the furnishings and fittings of a building or in other items which are not controlled by the Building Regulations.
- Fires are less likely to originate in the building structure and accidental fires in circulation spaces, corridors and stairways are unlikely due to the restriction on the use of combustible materials in these areas.
- When a fire breaks out the initial hazard is to the immediate area in which it occurs and it is unlikely that a large area will be affected at this stage. When fire spread does occur it is usually along circulation routes.
- The primary danger in the early stages of a fire is the production of smoke and noxious gases. These obscure the way to escape routes and exits and are responsible for the most casualties. Therefore, limiting the spread of smoke and fumes is vital in the design of a safe means of escape.
- Buildings covered are assumed to be properly managed. Where there is a failure of management responsibility, the building owner or occupier may be prosecuted under the Fire Precautions Act or the Health and Safety at Work etc. Act, which may result in prohibition of the use of the building.

7.8.1 Alternative escape routes

When a fire occurs it should be possible for people to turn their backs on it and travel away from it to either a final exit or a protected escape route leading to a place of safety. This means that alternative escape routes should be provided in most situations.

The basic criteria governing the design of means of escape are as follows.

- The first part of the escape route will be within the accommodation or circulation areas and will usually be unprotected. It should be of limited length so that people are not exposed to fire and smoke for any length of time. Where the horizontal escape route is protected it should still be of limited length since there is always the risk of premature failure.
- The second part of the escape route will usually be in a protected stairway designed to be virtually ‘fire sterile’. Once inside it should be possible to proceed direct to a place of safety without rushing. Therefore, flames, smoke and gases must be excluded from these routes by fire-resisting construction or adequate smoke control measures or by both these methods. This does not preclude the use of unprotected stairs for normal everyday use; however their relative vulnerability to fire situations means that they can only be of limited use for escape purposes.

- The protected stairway should lead direct to a place of safety or it may do this via a protected corridor. The ultimate place of safety is open air clear of the effects of fire; however in certain large and complex buildings reasonable safety may be provided within the building if suitable planning and protection measures can be included in the design.

7.8.2 Dead ends

Ideally, alternative escape routes should be provided from all points in a building, since there is always the possibility that the path of a single escape route may become impassable due to the presence of fire, smoke or fumes. Escape in one direction only (a dead end) is acceptable under certain conditions depending on:

- the use of the building
- its associated fire risk
- its size and height
- the length of the dead end
- the number of people accommodated in the dead end.

7.8.3 Unacceptable means of escape

Certain paths of travel are not acceptable as means of escape, including the following.

- Lifts, unless designed and installed as evacuation lifts for disabled people in the event of fire.
- Portable or throw-out ladders.
- Manipulative apparatus and appliances such as fold-down ladders and chutes.
- Escalators. These should not be counted as additional escape routes due to the uneven nature of the top and bottom steps; however it is likely that people would use them in the event of a fire. Mechanised walkways could be acceptable if they were properly assessed for capacity as a walking route in the static mode.

7.8.4 Security

It is possible that security measures intended to prevent unauthorised access to a building may hinder the entry of the fire services when they need access to fight a fire or rescue trapped occupants. Advice may be sought from architectural liaison officers attached to most police forces so that possible conflicts between security and access may be solved at the design stage.

AD B1 does not deem it appropriate to control, under the Building Regulations, the type of lock used on the front doors to dwellings. Guidance on door security in buildings other than dwellings is given in section 7.16.2 below.

7.9 Rules for measurement

In addition to the rules for measurement described in section 7.4 above, which apply to all parts of AD B, there are certain rules which relate only to AD Bl. These are concerned with assessing the length and capacity of escape routes which may, in turn, have a bearing on the quantity of routes and the number of stairways that it is necessary to provide.

7.9.1 Occupant capacity

In order to design a safe means of escape, it is necessary to assess the number of people who are likely to be present in the different parts of the building (i.e. the occupant capacity). Occupant capacity is the total number of people that a building, storey or room is designed to contain and it depends partly on the use (or purpose group) of the building and partly on the use of individual rooms in the building. It has a direct effect on the numbers and widths of:

- the exits from any room, storey or tier; and
- escape stairs and final exits from the building.

For some building uses (such as theatres or restaurants) occupant capacity can be calculated by totalling the number of seats and then adding an allowance for staff.

In other buildings (such as speculative office developments, shops and supermarkets), the designer will not be able to do this since he will never be sure how intensively the building will be used. In these circumstances, Table 1 of Approved Document B1 provides floor space factors (expressed in m^2 per person) which will give a value for the occupant capacity when divided into the relevant floor area.

The table is based on the type of accommodation contained in the building (i.e. the use of individual rooms) and therefore the following procedure could be adopted to calculate the occupant capacity for the building:

- choose the top floor and inspect each room to determine its use
- for each room calculate its floor area and divide it by the relevant floor space factor to determine the occupant capacity for the room
- add together the individual room capacities to obtain the occupant capacity for the floor
- repeat this process for each floor
- total the occupants of all floors to obtain the occupant capacity for the building.

Floor space factors based on Table 1 of AD B1 are given in Table 7.1 below. The following should also be noted when using the floor space factors:

- It is not necessary to calculate occupancy capacities for stair enclosures, lifts or sanitary accommodation and fixed parts of the building structure may also be excluded from the calculation (although the area taken up by counters and display units etc. should not be excluded).

Table 7.1 Floor space factors.

Floor space factor (m^2 per person)	Type of accommodation/use of room	Notes
0.3	Standing spectator areas. Bars and other similar refreshment areas, without seating	
0.5	Amusement arcade, assembly hall (including a general purpose place of assembly), bingo hall, crush hall, dance floor or hall, venue for pop concert or similar events	
0.7	Concourse, queuing area or shopping mall	See also section 4 of BS 5588: Part 10: 1991 Code of practice for shopping complexes for detailed guidance on the calculation of occupancy in common public areas in shopping complexes.
1.0	Committee room, common room, conference room, dining room, licensed betting office (public area), lounge or bar (other than above), meeting room, reading room, restaurant, staff room or waiting room	In many of these uses the occupants will normally be seated. In such cases occupant capacity may be taken as the number of *fixed* seats provided
1.5	Exhibition hall or studio (film, radio, television, recording)	
2.0	Skating rink, shop sales area (1)	(1) Shops, such as supermarkets and the main sales areas of department stores; shops for personal services (e.g. hairdressers); shops for the delivery or collection of goods for cleaning, repair or other treatment either by the company or by the public themselves. For other shops see (2) below
5.0	Art gallery, dormitory, factory production area, museum or workshop	
6.0	Office	
7.0	Kitchen or library, shop sales area (2)	(2) Shops trading in large items such as furniture, floor coverings, cycles, prams, large domestic appliances or other bulky goods, or cash and carry wholesalers. For other shops see (1) above
8.0	Bedroom or study-bedroom	
10.0	Bed-sitting room, billiards or snooker room or hall	
30.0	Storage and warehousing	
	Car park	Occupant capacity based on two persons per parking space

- Where the descriptions do not completely cover the accommodation it is acceptable to use a value based on a similar use.
- Where any part of the building is likely to have multiple uses (for example, a multi-purpose hall might be used for a low density activity like gymnastics or a high density use such as a disco) the most onerous floor space factor should be used.

As an example, take an exhibition hall with a floor area (excluding stair enclosures, lifts and sanitary accommodation) of 2250 m^2. From Table 7.1, the floor space factor is 1.5 m^2 per person. Therefore, the occupant capacity is 2250 divided by 1.5 i.e. 1500 people.

Alternatively, some designers or developers may be able to obtain actual data relating to occupancy numbers, taken from existing premises which are similar to those being designed. Where this data is available it should reflect the average occupant density at a peak trading time of year.

7.9.2 Travel distance

This is measured along the shortest, most direct route. Where there is fixed seating or other fixed obstructions, the travel distance is measured along the centreline of the seating or gangways. If a stair (such as an accommodation stair) is included in the escape route, the travel distance is measured along the pitch line on the centre line of travel.

7.9.3 Width

Usually, the narrowest part of an escape route will be at the door openings which form the room or storey exits. These are measured as shown in Fig. 7.10 below.

Escape route widths are measured at a height of 1500 mm above floor level where the route is defined by walls. Elsewhere, the width will be measured between any fixed obstructions.

Stair widths are measured clear between walls and balustrades. Strings may be

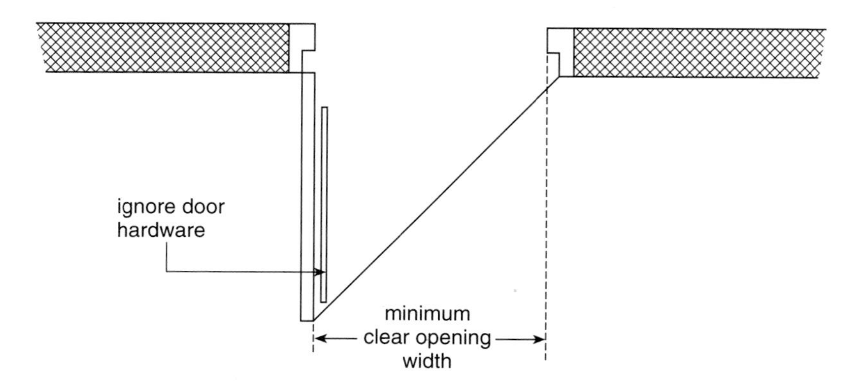

Fig. 7.10 Door width measurement.

ignored as may handrails which project less than 100 mm. Additionally, where a stairlift is installed, the guide rail may be ignored; however, the chair or carriage must be capable of being parked where it does not obstruct either the stair or the landing.

7.10 Fire alarm and fire detection systems

7.10.1 Fire detection and alarm in dwellings

Dwellinghouses

In all dwellinghouses (including bungalows), safety levels can be significantly increased by installing systems which automatically give early warning of fire. Approved Document B1 gives the following range of options for dwellinghouses:

- an automatic fire detection and alarm system in accordance with the relevant recommendations of BS 5839 *Fire detection and alarm systems for buildings*, Part 1:1988 *Code of practice for system design, installation and servicing* to at least the L3 standard specified in the code; or
- an automatic fire detection and alarm system in accordance with the relevant recommendations of BS 5839 *Fire detection and alarm systems for buildings*, Part 6:1995 *Code of practice for the design and installation of fire detection and alarm systems in dwellings* to at least a Grade E type LD3 standard; or
- a suitable number of smoke alarms provided as indicated below.

Large dwellinghouses

A large dwellinghouse is defined in AD B1 as having one or more floors greater than 200 m^2 in area. Where the house has more than three storeys (including basements) the following recommendations are given in AD B1.

- It may be fitted with an L2 system as described in BS 5839 Part 1 but the provisions in clause 16.5 regarding duration of the standby supply may be disregarded.
- Where the system is unsupervised, the standby supply should be able to automatically maintain the system in normal operation for 72 hours (but with audible and visible indication of mains failure). At the end of the 72 hours, sufficient capacity should remain to supply the maximum alarm for a minimum of 15 minutes.

Where a large house has no more than three storeys (including basements) it may be fitted with an automatic fire detection and alarm system of Grade B type LD3 in accordance with BS 5839: Part 6 instead of the L2 system referred to above.

Loft conversions

Where it is proposed to convert a loft space to habitable accommodation in a one or two storey dwellinghouse, an automatic smoke detection and alarm system based on linked smoke alarms should be installed throughout the dwellinghouse (i.e. not just in the extended part). The installation should follow the guidance described above and below for the size of dwelling which will be created by the extension (i.e. either a two storey or three storey house). Further details of the means of escape provisions for loft conversions are given in section 7.12.10 below.

Sheltered housing

Sheltered housing usually consists of a block or group or dwellings designed specifically for persons who might require assistance (such as elderly people) where some form of assistance will be available at all times (although not necessarily on the premises). Each unit of accommodation will have its own cooking and sanitary facilities and amenities common to all occupiers may also be provided such as communal lounges.

BS 5839 Part 6 recommends the installation of a Grade C, Type LD3 system. The detection equipment should have a connection to a central monitoring point (or central relay station) so that the warden or supervisor is able to identify the dwelling in which the fire has occurred. It is not intended that the provisions in AD B1 or BS 5839 Part 6 be applied to the common parts of sheltered accommodation and they do not apply to sheltered accommodation in Purpose Groups 2(a) Residential (Institutional) or 2(b) Other Residential (see section 7.3 above).

Flats and maisonettes

The principles for the provision of fire alarm and detection systems in flats and maisonettes are the same as those which apply to dwellinghouses described above, with the following additions.

- There is no need to apply the provisions to the common parts of blocks of flats.
- The systems in individual flats do not need to be interconnected.
- A maisonette (i.e. a 'flat' in which the accommodation is contained on more than one level and which may be entered from either the higher or lower level) should be treated in the same way as a house with more than one storey.

Student residential accommodation

Traditional halls of residence consisting of individual study bedrooms and shared dining and washing facilities should be designed for general evacuation in the same way as non-residential buildings (see section 7.15 below).

Where the accommodation is arranged as in a block of flats, with a group of students sharing an individual flat with its own entrance door, it is appropriate for an automatic detection system to be provided in each flat. In such cases the flats will be compartmented from each other and the automatic detection system need only give warning in the flat of fire origin.

Smoke alarms in dwellings

Smoke alarms should fulfil these criteria.

- Mains-operated, and designed and manufactured in accordance with BS 5446 *Components of automatic fire alarm systems for residential premises*, Part 1: 1990 *Specification for self-contained smoke alarms and point-type detectors*. They may have a secondary power supply (such as a capacitor or rechargeable or replaceable battery, see clause 13 of BS 5839 Part 6). BS 5446: Part 1 deals with smoke alarms based on ionization chamber smoke detectors and optical (photo-electric) smoke detectors. Each type of detector responds differently to smouldering and fast flaming fires. Therefore, optical smoke alarms should be installed in circulation spaces (e.g. hallways and landings) and ionization chamber based smoke alarms may be more appropriate if placed where a fast burning fire presents the greater danger to occupants (such as in living or dining rooms). Additionally, optical detectors are less affected by low levels of 'invisible' smoke that often cause false alarms.
- Located in circulation areas between sleeping places and places where fires are likely to start (kitchens and living rooms) and within 7.5 m of the door to every habitable room.
- Fixed to the ceiling and at least 300 mm from any walls and light fittings (unless there is test evidence to prove that the detector will not be adversely affected by the proximity of a light fitting). Units specially designed for wall mounting are acceptable provided that they are mounted above the level of doorways into the space and are fixed in accordance with manufacturers' instructions.
- Sited so that the sensor, for ceiling mounted devices, is between 25 mm and 600 mm below the ceiling (between 25 mm and 150 mm for heat detectors), assuming that the ceiling is predominantly flat and level.
- Fixed in positions that allow for routine maintenance, testing and cleaning (i.e. not over a stairwell or other floor opening).
- Sited away from areas where steam, condensation or fumes could give false alarms (this would include heaters, air-conditioning outlets, bathrooms, showers, cooking areas or garages, etc.).
- Sited away from areas that get very hot (e.g. boiler rooms) or very cold (e.g. an unheated porch). They should not be fitted to surfaces which are either much hotter or much colder than the rest of the room since air currents might be created which would carry smoke away from the unit.

The number of alarms which are installed will depend on the size and complexity of the layout of the dwelling and should be based on an analysis of the risk to life from fire. The following minimum provisions should be observed.

- At least one alarm should be installed in each storey of the dwelling.
- Where more than one smoke alarm is installed they should be interconnected so that the detection of smoke in any unit will activate the alarm signal in all of them. Provided that the lifetime or duration of any standby power supply is not

reduced, smoke alarms may be interconnected using radio-links. Manufacturers' instructions should be adhered to regarding the maximum number of units that can be interconnected.

- In open plan designs where the kitchen is not separated from the stairway or circulation space by a door, the kitchen should contain a compatible interlinked heat detector. This should be in addition to the normal provision of smoke detector(s) in the circulation space(s).

It should be noted that maintenance of the system in use is of utmost importance. Since this cannot be made a condition of the passing of plans for Building Regulation purposes it is important to ensure that developers and builders provide occupiers with full details of the use of the equipment and its maintenance (or guidance on suitable maintenance contractors). BS 5839: Parts 1 and 6 also recommend that occupiers receive manufacturers' operating and maintenance instructions.

Power supplies to smoke alarms

Power supplies for smoke alarm systems should be derived from the dwelling's mains electricity supply and connected to the smoke alarms through a separately fused circuit at the distribution board (consumer unit).

The power supply options described here are all based on using the mains supply at the normal 240 volts. Other effective (though possibly more expensive) options exist, such as, reducing the mains supply to extra low voltage in a control unit incorporating a standby trickle-charged battery, before distributing the power to the smoke alarms at that voltage.

The smoke alarm system can include a stand-by power supply which will operate during mains failure. This can allow the system to obtain its power by connection to a regularly used local lighting circuit with the advantage that the circuit will be unlikely to be disconnected for any prolonged period. Where the system does not include a stand-by power supply, no other electrical equipment should be connected to the smoke alarm circuit (except for a mains failure monitoring device, see below).

The mains supply to the smoke alarm system can be monitored by a device which will give warning in the event of failure of the supply. The warning of failure may be visible or audible (in which case it should be possible to silence it) and should be sited so that it is readily apparent to occupants. The circuit for the mains failure monitor should be designed so that any significant reduction in the reliability of the mains supply is avoided.

Ideally, the smoke alarm circuit should not be protected by any residual current device (rcd) such as a miniature circuit breaker or earth leakage trip. However, sometimes it is necessary for reasons of electrical safety that such devices be used and in these cases, either:

- the rcd should serve only the circuit supplying the smoke alarms, or
- the rcd protection of a fire alarm circuit should operate independently of any rcd protection for circuits supplying socket outlets or portable equipment.

Since it does not need any fire survival properties, the mains supply to smoke alarms, and any interconnecting wiring, may comprise any cable which is suitable for ordinary domestic mains wiring. Cables used for interconnections should be readily identifiable from those supplying power (e.g. by colour coding).

7.11 Fire detection and alarm in buildings other than dwellings

Introduction

It is extremely important to realise that there is a causal connection between the design of the means of warning and escape in a building and the eventual management of that means of warning and escape. Therefore, the escape strategy to be adopted in a particular building will be based on one of the following.

- Simultaneous evacuation; all the occupants are expected to leave the building at the same time.
- Phased evacuation; only the storeys most affected by the fire (e.g. the floor of origin and the floor above it) are evacuated immediately. Subsequently, two floors at a time are evacuated if the need arises.
- Progressive horizontal evacuation; the concept which is usually adopted in the in-patient parts of hospitals and similar healthcare premises where total evacuation of the building is inappropriate. In-patients are evacuated, in the event of fire, to adjoining compartments or sub-divisions of compartments, the object being to provide a place of relative safety within a short distance. If necessary, further evacuation can be made from these safe places but under less pressure of time.

Therefore, in buildings other than dwellings, selection of the appropriate fire detection and alarm system will depend on the means of escape strategy adopted. For example, in residential accommodation (where the occupants sleep on the premises) the threat posed by a fire will be much greater than in premises where the occupants are fully alert. In these circumstances an escape strategy based on simultaneous evacuation will mean that all fire sounders will operate almost instantaneously once a manual call point or fire detector has been activated. If however, the escape strategy is based on phased evacuation, a staged alarm system may be more appropriate. Two or more stages of alarm may be given within a particular area corresponding to 'alert' and 'evacuate' signals.

Fire alarm systems

All buildings should have arrangements for detecting fire and in most buildings this will be done directly by people through observation or smell. In many small buildings where there is no sleeping risk this may be all that is needed. Similarly, the means of raising the alarm may be simple in such buildings and where the occupants are in sight and hearing of each other a shouted warning may be sufficient. Clearly, it is necessary to assess the risk in each set of circumstances and decide standards on a case by case basis.

The risk analysis will consider the likelihood of a fire occurring and the degree to which the alarm can be heard by all the occupants. Therefore any of the following may need to be incorporated in the building.

- Manually operated sounders (e.g. rotary gongs or handbells).
- Simple manual callpoints combined with bell, battery and charger.
- Electrically operated fire warning system with manual callpoints sited adjacent to exit doors, combined with sufficient sounders to ensure that the alarm can be heard throughout the building. The system should comply with BS 5839: Part 1 (see below) and the call points with BS 5839: *Fire detection and alarm systems for buildings:* Part 2: 1983 *Specification for manual call points* or Type A of BS EN 54 *Fire detection and fire alarm systems – Part 11: Manual call points.* BS EN 54-11 covers two types of call points:
- Type A (direct operation) – the change to the alarm condition is automatic (i.e. without the need for further manual action) when the frangible element is broken or displaced, and
- Type B (indirect operation) – the change to the alarm condition requires a separate manual operation of the operating element by the user after the frangible element is broken or displaced. (Type B call points should only be used with the approval of the Building Control Body).

Four types of fire alarm and detection system are specified in BS 5839: Part 1 as follows:

- Type L – for the protection of life
- Type M – manual alarm systems
- Type P – for the protection of property
- Type X – for multi-occupancy buildings.

Type L systems are further subdivided into:

- L1 – systems installed throughout the protected building
- L2 – systems installed only in defined parts of the protected building (but always providing the coverage required of a Type L3 system)
- L3 – systems installed only for the protection of escape routes.

Type P systems are further subdivided into:

- P1 – systems installed throughout the protected building
- P2 – systems installed only in defined parts of the building.

In certain premises where large numbers of the public are present (e.g. large shops and places of assembly) it may be undesirable for an initial general alarm to be sounded since this may cause unnecessary confusion. Therefore it is essential in these circumstances that staff are trained to effect pre-planned procedures for safe evacuation. Usually, actuation of the fire alarm system will alert staff by means of discreet sounders or personal pagers first. This will enable them to be prepared and in position should it be necessary for a general evacuation to be initiated by means of sounders or an announcement over the public address system. In all other respects any staff system should conform to BS 5839: Part 1.

Voice alarm systems are useful in circumstances where it is considered that people might not respond quickly to a fire warning, or where they are unfamiliar with fire warning arrangements. An audible fire warning signal can be given via a public address system, provided that it is distinct from other signals which are in general use and is accompanied by clear verbal instructions. Voice alarms should comply with BS 5839: *Fire detection and alarm systems for buildings:* Part 8 1998 *Code of practice for the design, installation and servicing of voice alarm systems.*

Automatic fire detection and alarm systems

Automatic fire detection systems involve a sensor network plus associated control and indicating equipment. Sensors may detect heat, smoke or radiation and it is usual for the control and indicating equipment to operate a fire alarm system. It may also perform other signalling or control functions, such as, the operation of an automatic sprinkler system.

Automatic fire detection and alarm systems should be installed in Institutional (Purpose Group 2(a)) buildings and in Other Residential (Purpose Group 2(b)) occupancies.

Automatic fire detection systems are not normally needed in non-residential occupancies; however it may be desirable to install a fire detection system in the following circumstances:

- to compensate for the fact that it has not been possible to follow all the guidance in Approved Document B
- where it is necessary as part of a fire protection operating system, such as a pressurised staircase or automatic door release mechanism
- where a fire could occur unseen in an unoccupied or rarely visited part of a building and prejudice the means of escape from the occupied parts.

Where a building is designed for phased evacuation (see above) it should be fitted with an appropriate fire warning system conforming to at least the L3 standard given in BS 5839: Part 1. Additionally, an internal speech communication system (telephone, intercom etc.) should be provided so that conversation is possible between a fire warden at every floor level and a control point at the fire service access level.

Where a building contains an atrium and is designed in accordance with BS 5588: *Fire precautions in the design, construction and use of buildings,* Part 7: *Code of practice for the incorporation of atria in buildings*, then the relevant recommendations of that code should be followed for the installation of fire alarm and/or fire detection systems.

Further guidance on the standard of automatic fire detection that may need to be provided in a building can be found in:

- Home office guides that support the Fire Precautions Act 1971 and the Fire Precautions (Workplace) Regulations 1997.
- The NHS Estates 'Firecode' documents for buildings in the Institutional Purpose Group 2(a).

Design, maintenance and installation of systems

Fire warning and detection systems must be properly designed, installed and maintained. Installation and commissioning certificates should be obtained wherever a fire alarm system is installed. Additionally, third party certification schemes for fire protection products and related services offer an effective means for providing the fullest possible assurances that the required level of performance will be achieved, and that the products actually supplied are provided to the same specification or design as those that have been tested or assessed.

7.12 Means of escape in dwellinghouses

7.12.1 Introduction

Approved Document B1 deals with means of escape from dwellinghouses according to the height of the top storey above ground level (i.e. the ground level on the lowest side of the building).

This is probably a sensible approach since storey heights can vary and means of escape through upper windows become more hazardous with increasing height. Thus, the divisions chosen are:

- houses with all floors not more than 4.5 m above ground (i.e. ground and first floor only)
- houses with one floor more than 4.5 m above ground (i.e. ground floor, first floor and second floor)
- houses with two or more floors more than 4.5 m above ground (i.e. ground floor and three or more upper floors).

Therefore, as the height of the top floor increases above ground level, the means of escape provisions become more complex and these are dealt with under separate sections below. Certain recommendations however, are common to all dwellings and these include:

- the provision of an automatic fire detection and alarm system (see above)
- special provisions to deal with basements and inner rooms
- windows and external doors used for escape purposes
- balconies and flat roofs.

The guidance contained in this section is also applicable to houses in multiple occupation (HMOs) provided that there are no more than six residents. An HMO is defined in section 345 of the Housing Act 1985 as 'a house which is occupied by persons who do not form part of a single household'. For HMOs containing greater numbers of occupants, technical guidance may be sought in DOE Circular 12/92 (see below) or Welsh Office Circular 25/92.

It may also be possible to treat an unsupervised group home for mentally ill or mentally impaired people with up to six residents as an ordinary dwelling but this will depend on the nature of the occupants and their management. Such premises have to be registered, therefore the registration authority should be consulted to establish if there are any additional fire safety measures needed by the authority.

7.12.2 Basements and inner rooms

With certain dwelling designs (such as open-plan layouts and the provision of sleeping galleries) it is possible that that a situation will be created whereby the innermost room (termed the inner room) will be put at risk by a fire occurring in the room that gives access to it (termed the access room), since escape is only possible by passing through that access room. Therefore, an inner room should only be used as:

- a kitchen, laundry or utility room
- a dressing room
- a bathroom, shower room or WC
- any other room which has a suitable escape window or door (see next section for details), provided the room is in the basement, or is on the ground or first floor
- a sleeping gallery (see section 7.12.5 below for details).

Escape from a basement fire may be particularly hazardous if an internal stair has to be used, since it will be necessary to pass through a layer of smoke and hot gases. Therefore, any habitable room in a basement should have either:

- an alternative escape route via a suitable door or window, or
- a protected stairway leading from the basement to a final exit.

7.12.3 Windows and external doors used for escape purposes

To be suitable for escape purposes, windows and external doors should conform to the dimensions given in Fig. 7.11. Dormer windows and roof windows situated above the ground storey should comply with Fig. 7.16. Escape should be to a place of safety free from the effects of fire. Where this is to an enclosed back garden or yard from which escape may be made only by passing through other buildings, its length should be at least equivalent to the height of the dwelling (see Fig. 7.12).

7.12.4 Balconies and flat roofs

If used as an escape route, a flat roof should:

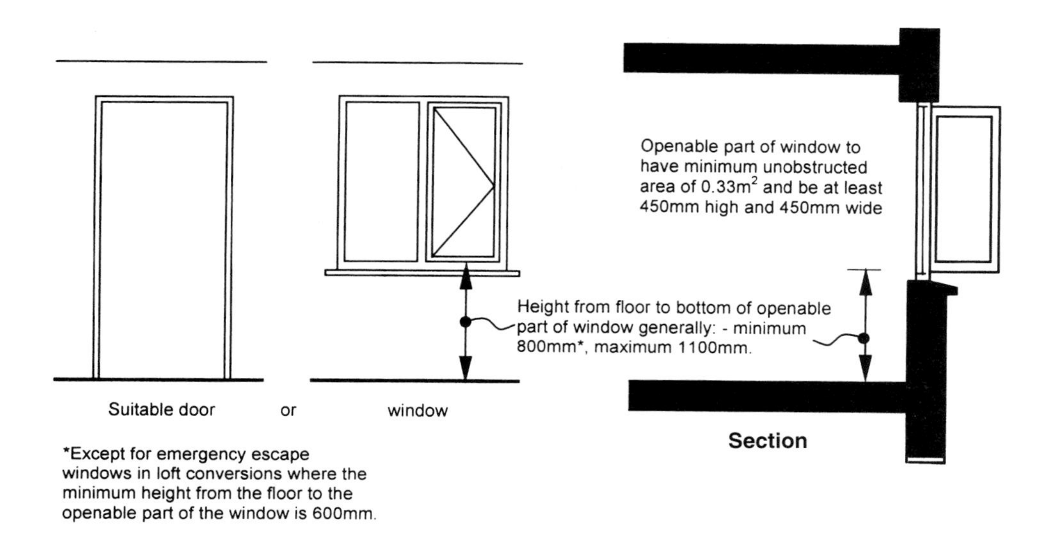

Fig. 7.11 Windows and doors for escape purposes.

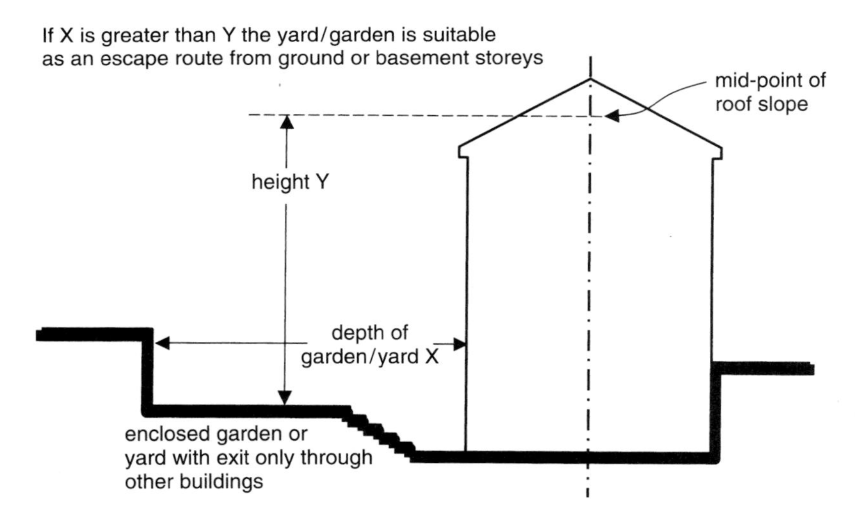

Fig. 7.12 Enclosed yard or garden suitable for escape purposes – dwellinghouses.

- be part of the same building from which escape is being made
- lead to a storey exit or external escape route
- be provided with 30 minutes fire resistance. This applies only to the part of the flat roof forming the escape route, its supporting structure and any opening within 3 m of the escape route.

Balconies and flat roofs provided for escape purposes should be guarded in accordance with the provisions of Approved Document K. (This relates to the provision of barriers at least 1100 mm high designed to prevent people falling from the escape route. The barriers should be capable of resisting at least the horizontal force given in BS 6399 *Loading for buildings,* Part 1: *1984 Code of practice for dead and imposed loads.*

7.12.5 Dwellinghouses with floors not more than 4.5 m above ground

For dwellings with floors not more than 4.5 m above ground, the means of escape measures outlined above need only to be augmented by the following provisions:

(1) All habitable rooms (except kitchens) in the upper storey(s) of a dwellinghouse served by only one stairway, should be provided with an external door or window which complies with Fig. 7.11. A single door or window can serve two rooms provided that they each have their own access to the stairs. A communicating door should be provided between the rooms so that it is possible to gain access to the door or window without entering the stair enclosure.
(2) All habitable rooms (except kitchens) in the ground storey should either:
  - open directly onto a hall which leads to an entrance or other suitable exit, or
  - be provided with a door or window which complies with Fig. 7.11.
(3) Where the dwelling contains a sleeping gallery (i.e. a floor which is used as a bedroom but which is open on at least one side to some other part of the dwelling) which is not more than 4.5 m above ground level:
  - The distance between the foot of the access stair to the gallery and the door to the room which contains the gallery should not be more than 3 m.
  - An alternative exit from the gallery complying with Fig. 7.11 should be provided, if the distance from the head of the access stair to any point in the gallery exceeds 7.5 m.
  - Unless they are enclosed with fire-resisting construction, any cooking facilities within a room containing a gallery should be remote from the stair to the gallery and positioned so that they do not prejudice the escape route from the gallery.

7.12.6 Houses with one floor more than 4.5 m above ground

Houses with one floor more than 4.5 m above ground are likely to consist of ground, first and second floors. Two alternative solutions are recommended in AD B1.

(1) Provide the first and second floors with a protected stairway which should either:
  - terminate directly in a final exit, or
  - give access to a minimum of two escape routes at ground level, separated from each other by self-closing fire doors and fire-resisting construction and each leading to a final exit (these alternatives are illustrated in Fig. 7.13) or
(2) Separate the top floor (and the main stairs leading to it) from the other floors by fire-resisting construction and provide it with an alternative means of escape leading to its own final exit. (A variation on this solution can be used where it is intended to convert the roof space of a two-storey dwelling, effectively making it three-storey. Further details of this are described below.)

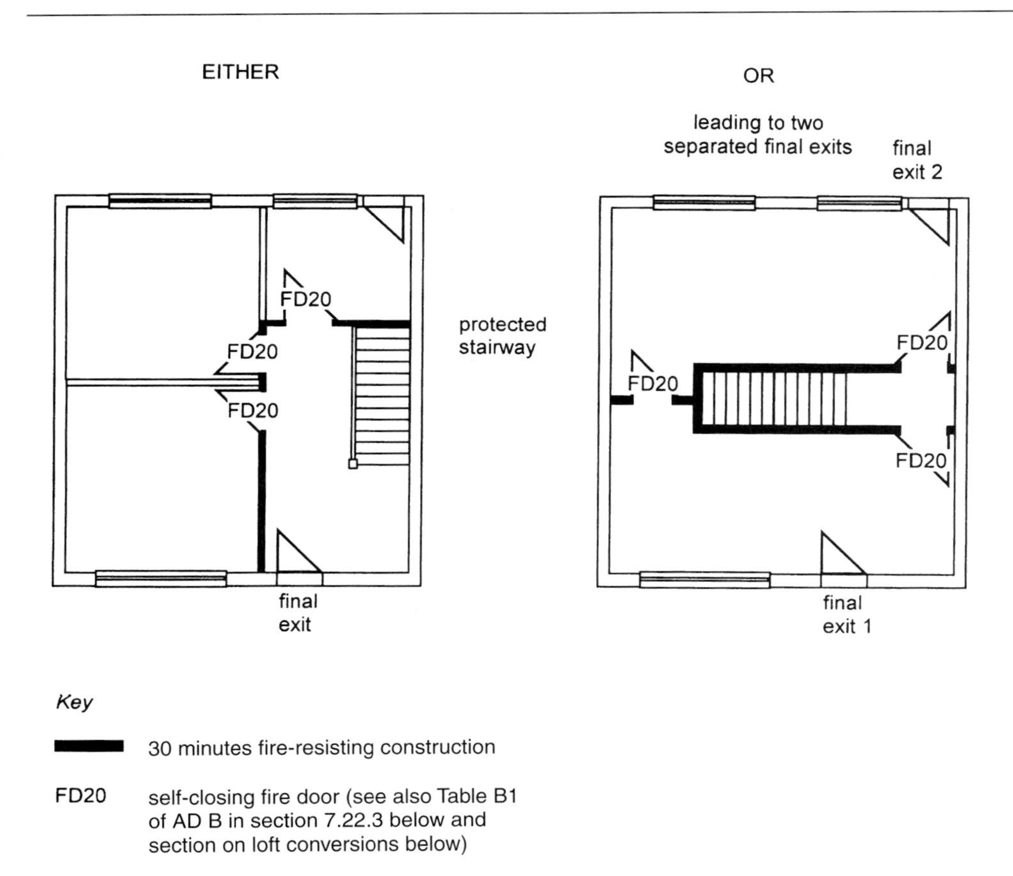

Fig. 7.13 Alternative final exit arrangements.

7.12.7 Houses with two or more floors more than 4.5 m above ground

Normally, this would include houses with three or more storeys above the ground floor. In most circumstances it will be necessary to provide an alternative means of escape for all floors which are 7.5 m or more above ground.

Where it is intended that the alternative escape route be accessed via the protected stairway (i.e. to an upper storey, see Fig. 7.14(a)) or via a landing within its enclosure (i.e. on the same storey, see Fig. 7.14(b)) it is possible a fire in the lower part of the dwellinghouse might make access to the alternative route impassable. Therefore, the protected stairway at or about 7.5 m above ground level should be separated from the lower storeys by fire-resisting construction.

7.12.8 Air circulation systems in houses with any floor more than 4.5 m above ground

Modern houses are often fitted with air circulation systems designed for heating, energy conservation or condensation control. With such systems, there is the possibility that fire or smoke may spread (possibly by forced convection, natural

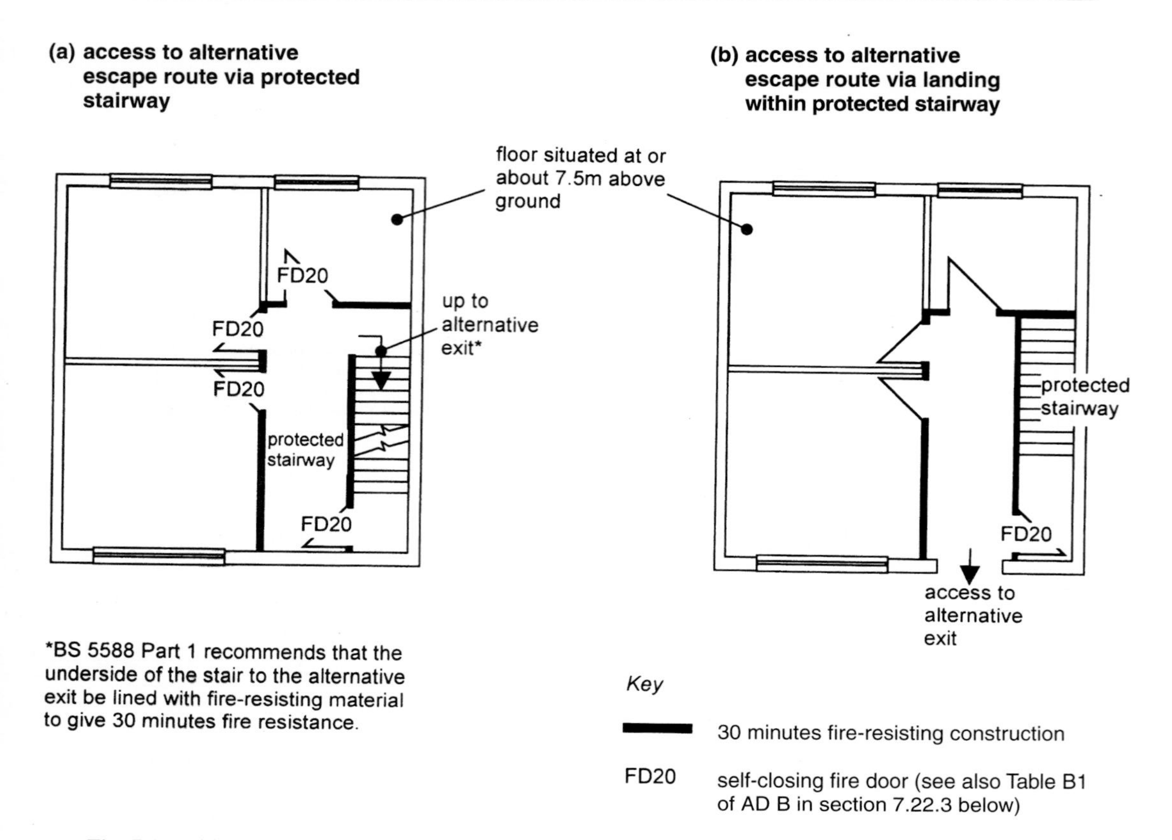

Fig. 7.14 Alternative exit arrangements in houses with more than one floor over 4.5 m above ground.

convection or fire-induced convection) from the room of origin of the fire to the protected stairway. Clearly, the risk to life is greater in dwellings where there are floors at high levels so AD B1 and BS 5588 Part 1 make the following recommendations for such buildings:

- The walls enclosing a protected stairway should not be fitted with transfer grilles.
- Where ductwork passes through the enclosure to a protected stairway, it should be fitted so that all joints between the ductwork and the enclosure are fire-stopped.
- Ductwork used to convey air into the protected stairway through its enclosure should be ducted back to the plant.
- Grilles or registers which supply or return air should be positioned not more than 450 mm above floor level.
- Any room thermostat for a ducted warm air heating system should be mounted in the living room at a height between 1370 mm and 1830 mm and its maximum setting should not be more than 27°C.
- Mechanical ventilation systems which recirculate air should comply with BS 5588 *Fire precautions in the design, construction and use of buildings,* Part 9: 1989 *Code of practice for ventilation and air conditioning ductwork.*

7.12.9 Passenger lifts in dwellinghouses

If a passenger lift is installed in a dwellinghouse which serves any floor situated more than 4.5 m above ground level, it should either be contained in a fire-resisting lift shaft or be located in the enclosure to a protected stairway.

7.12.10 Conversions to provide rooms in a roof space

The following provisions apply when it is proposed to convert the roof space of a two-storey house to provide living accommodation. The additional floor provided is most likely to be more than 4.5 m above ground and will therefore require additional protection to the means of escape or an alternative escape route. The measures described do not apply if:

- the new floor exceeds 50 m^2 in area; or
- it is proposed to provide more than two habitable rooms in the new storey.

In these cases, the loft conversion will have to meet the full provisions of the Building Regulations which apply to dwellings of three or more storeys.

The recommendations for means of escape are illustrated in Fig. 7.15(a) and 7.15(b), the general principles being:

- Every doorway within the existing stairway enclosure should be fitted with a door.
- Doors to habitable rooms in the existing stair enclosure should be self-closing to prevent the movement of smoke into the means of escape. For this purpose rising butt hinges are acceptable as self-closing devices in dwellinghouses.
- New doors to habitable rooms should be fire-resisting and self-closing, but existing doors will only need to be fitted with rising butt hinges.
- The stairway in the ground and first floors should be adequately protected with fire-resisting construction and should terminate as shown in Fig. 7.13 above.
- Glazing (whether new or existing) in the existing stair enclosure should be fire-resisting and retained by a suitable glazing system and beads compatible with the type of glass. This includes glazing in all doors (whether or not they need to be fire doors) except those to bathrooms or WCs. Fire-resisting glazing will usually consist of traditional annealed wired glass based on soda-lime-silica which is able to satisfy the integrity requirements of BS 476 Part 22. There are also some unwired glass products capable of satisfying the requirements for integrity. There is no limit on the area of such glass which can be incorporated in walls and doors above 100 mm from floor level.
- The new storey should be served by a stair which is located as shown in Fig. 7.15(a) and 7.15(b) and complies with Approved Document K. This could be an alternating tread stair.
- The new storey should be separated from the remainder of the dwelling by fire-resisting construction. Any openings should be protected by self-closing fire doors situated as shown in 7.15(a) and 7.15(b).

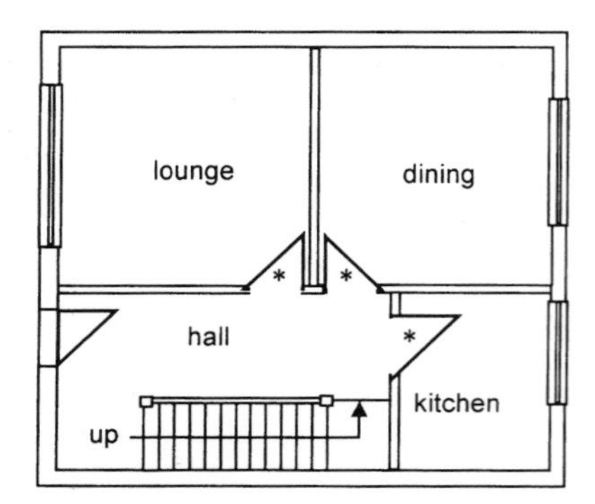

Existing ground floor common to all examples

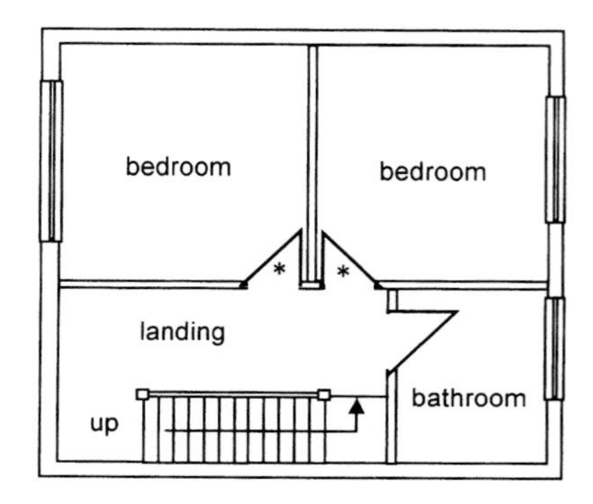

Proposed first floor - example A

Example A - new stairway rising in same staircase enclosure over existing stairway

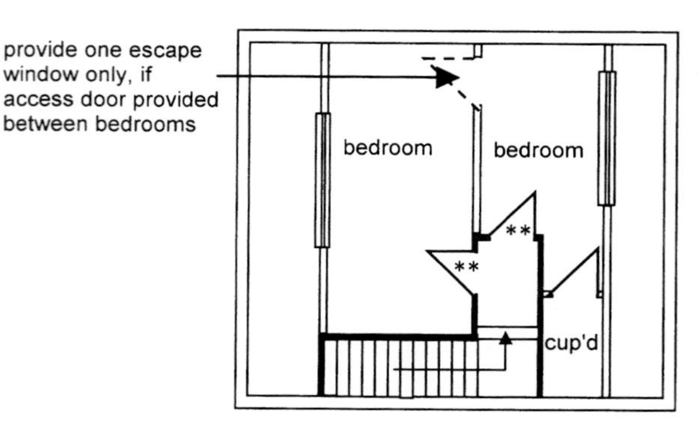

Proposed second floor - example A

Key

* self-closing door

** FD20 self-closing fire door (20 minutes integrity)

▬ 30 minutes fire-resisting construction

Fig. 7.15(a) Loft conversion to existing two-storey dwellinghouse.

- Windows and doors provided for emergency escape in the basement, ground and first floor of the existing dwelling provide a means of self-rescue. At higher levels, it is acceptable in loft conversions for escape to depend on a ladder being set up to a suitable escape window, or to a roof terrace, which is accessible by a door from the loft conversion. Any escape window should be large enough and suitably positioned to allow escape by means of a ladder to the ground. See Fig. 7.16.
- There should be suitable means of pedestrian access to the place at which the ladder would be set, to allow fire service personnel to carry a ladder from their vehicle (although it should not be assumed that only the fire service will effect a rescue). A ladder fixed to the slope of the roof is not recommended due to the danger inherent in using such ladders.
- Where the loft conversion consists of two rooms a single window can serve both rooms provided that they each have their own access to the stairs. A communicating door should be provided between the rooms so that it is possible to gain access to the window without entering the stair enclosure.

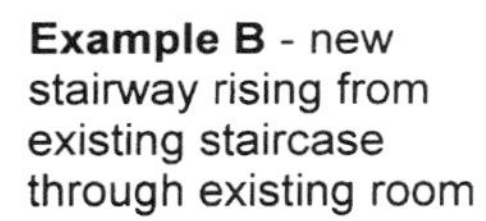

Example C - new
stairway separated
from existing staircase
at first floor level

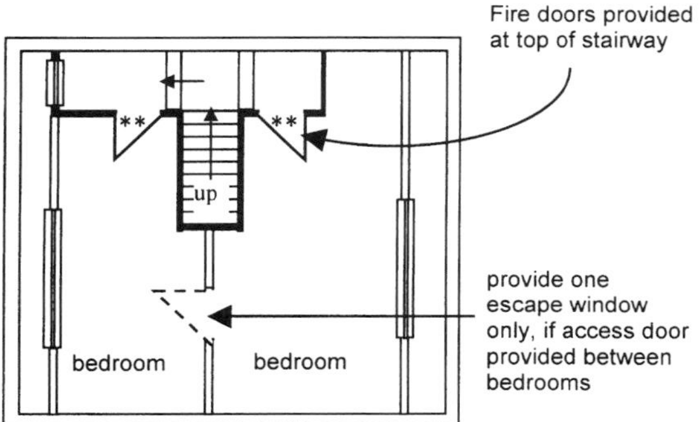

Proposed second floor - example B

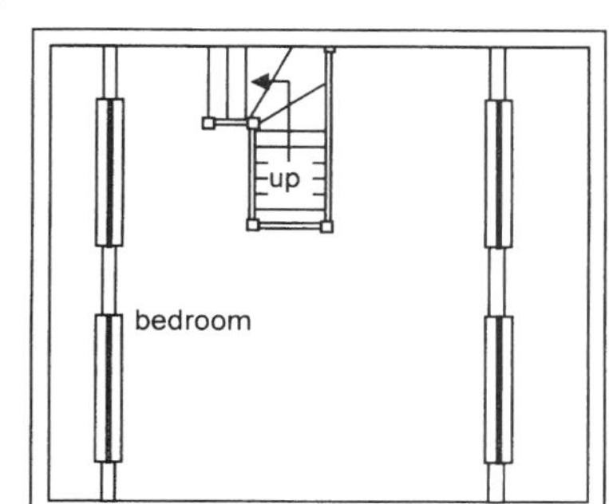

Proposed second floor - example C

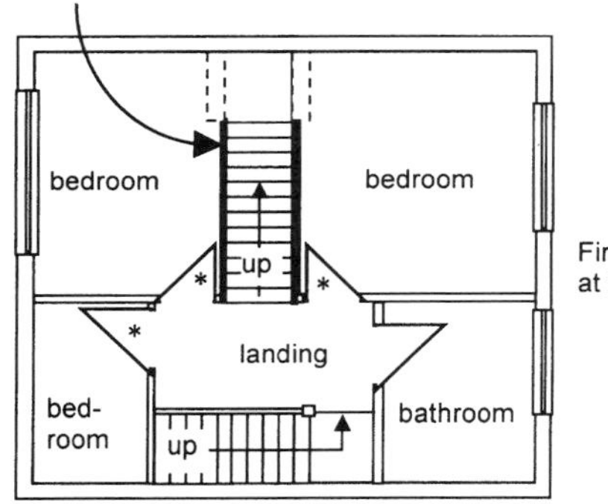

Proposed first floor - example B

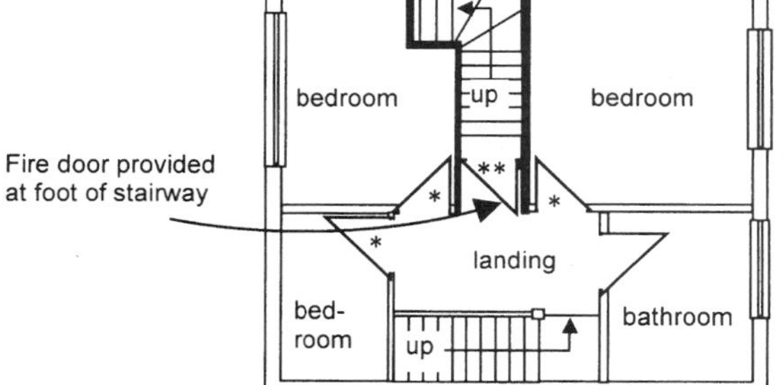

Proposed first floor - example C

Key

- * self-closing door
- ** FD20 self-closing fire door (20 minutes integrity)
- ▬ 30 minutes fire-resisting construction

Fig. 7.15(b) Loft conversion to existing two-storey dwellinghouse.

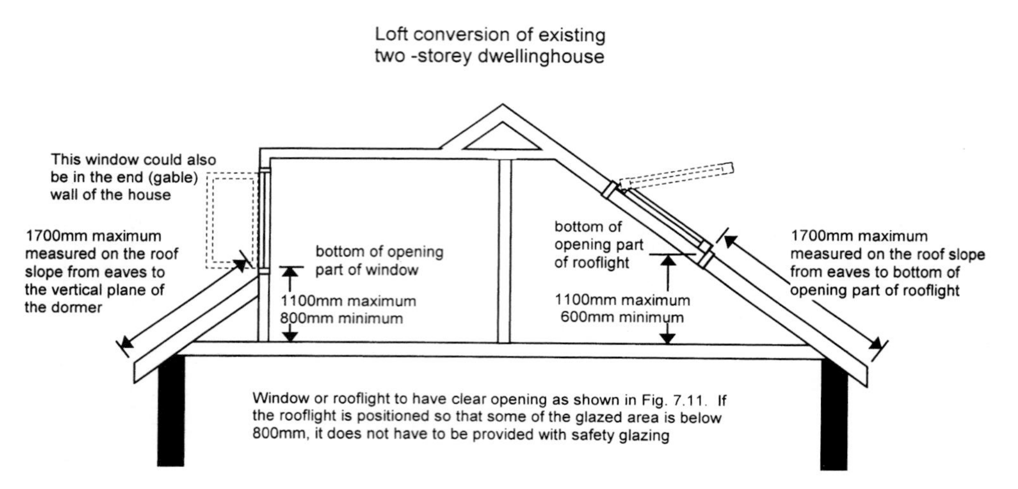

Fig. 7.16 Window positions for emergency escape dormers and rooflights in loft conversions.

- Escape across the roof of a ground storey extension is acceptable provided that the roof complies with the recommendations listed in section 7.12.4 above. Additionally, when it is proposed to erect an extension to a dwelling, the effect of the extension on the ability to escape from windows in other parts of the dwelling (especially from a loft conversion) should be carefully considered. This is particularly true when erecting a conservatory with a glazed roof.

The examples shown in Fig. 7.15(a) and 7.15(b) are typical illustrations of the principles outlined above for one particular form of dwellinghouse. They should not be taken to be definitive in any sense.

7.13 Flats and maisonettes

7.13.1 Introduction

Approved Document B1 deals with a limited range of common designs for flats and maisonettes. Where less common arrangements are desired (for example, where flats are entered above or below accommodation level, or flats contain sleeping galleries) the principles described in this chapter may still be applied or reference can be made to BS 5588: Part 1:1990, clauses 9 and 10.

The means of warning and escape recommendations listed above for dwellinghouses which consist of basement, ground and first floors only, apply equally to flats and maisonettes situated at these levels. At higher levels, escape through upper windows becomes more hazardous and more complex provisions are necessary, especially in maisonettes where internal stairs will need protection.

In addition to the general assumptions stated in section 7.8 above for all building types, the following assumptions are made when considering the means of warning and escape from flats and maisonettes.

- Fires generally originate in the dwelling.
- Rescue by ladders is not considered suitable.
- Fire spread beyond the dwelling of origin is unlikely due to the compartmentation recommendations in Approved Document B3, therefore simultaneous evacuation of the building should be unnecessary.
- Fires which occur in common areas are unlikely to spread beyond the immediate vicinity of the outbreak due to the materials and construction used there.

The guidance contained in this section is also applicable to flats and maisonettes when they are considered to be houses in multiple occupation. These are defined in section 7.12.1 above where references will be found to further guidance. Additionally, much of the following guidance will be applicable to flats used as sheltered housing although the nature of the occupancy may necessitate some additional fire protection measures. Guidance on these may be found in clause 17 of BS 5588: Part 1: 1990.

7.14 Means of escape in flats and maisonettes

There are two main components to the means of escape from flats and maisonettes:

- escape from within the dwelling itself; and
- escape from the dwelling to the final exit from the building (usually along a common escape route).

The following sections consider these two components.

7.14.1 Means of escape from within flats and maisonettes

The provisions relating to inner rooms, basements, balconies and flat roofs in dwellings (see sections 7.12.2 and 7.12.4 above) also apply to the inner parts of flats and maisonettes. However, where a balcony is provided as an alternative exit to a dwelling situated more than 4.5 m above ground it should be designed as a common balcony and should meet the conditions given in section 7.14.2 below for alternative exits.

Flats and maisonettes with floors not more than 4.5 m above ground

Generally, where the floor of a flat or maisonette is not more than 4.5 m above ground it should be planned so that any habitable room in a ground or upper storey is provided with a suitable escape window or door, as shown in Fig. 7.11 above.

A single door or window in an upper storey can serve two rooms provided that they each have their own access to the stair enclosure or entrance hall. A communicating door should be provided between the rooms so that it is possible to gain access to the door or window without entering the stair enclosure or entrance hall.

Alternatively, upper floors can be designed to follow the remainder of the guidance described below, in which case escape windows in upper storeys will be unnecessary.

Flats and maisonettes with floors more than 4.5 m above ground

Provided that the restrictions on inner rooms are observed, three possible solutions to the internal planning of flats are given in AD B1.

- All the habitable rooms in the flat are arranged to have direct access to a protected entrance hall. The maximum distance from the entrance door to the door of a habitable room should not exceed 9 m. (See Fig. 7.17(a)).
- In flats where a protected entrance hall is not provided, the 9 m distance referred to in (a) should be taken as the furthest distance from any point in the flat to the entrance door. Cooking facilities should be remote from the entrance door and positioned so that they do not prejudice the escape route from any point in the flat. (See Fig. 7.17(b)).
- Provide an alternative exit from the flat. In this case, the internal planning will be more flexible. An example of a typical flat plan where an alternative exit is provided, but not all the habitable rooms have direct access to the entrance hall, is shown in Fig. 7.17(c).

Where flats and maisonettes with floors more than 4.5m above ground are fitted with air circulation systems designed for heating, energy conservation or condensation control there is the possibility that fire or smoke may spread from the room of origin of the fire to the protected entrance hall or landing. Clearly, the risk to life is greater where there are floors at high levels so AD B1 and BS 5588 Part 1 make the following recommendations for such buildings:

- Any wall, door, floor or ceiling enclosing a protected entrance hall of a dwelling or protected stairway and landing of a maisonette should not be fitted with transfer grilles.
- Where ductwork passes through the enclosure to a protected entrance hall or protected stairway and landing, it should be fitted so that all joints between the ductwork and the enclosure are fire-stopped.
- Ductwork used to convey air into the protected entrance hall of a dwelling or protected stairway and landing of a maisonette through the enclosure of the protected hall or stairway, should be ducted back to the plant.
- Grilles or registers which supply or return air should be positioned not more than 450 mm above floor level.
- Any room thermostat for a ducted warm air heating system should be mounted in an area from which air is drawn directly to the heating unit at a height between 1370 mm and 1830 mm and its maximum setting should not be more than 27°C.
- Mechanical ventilation systems which recirculate air should comply with BS 5588 Part 9.

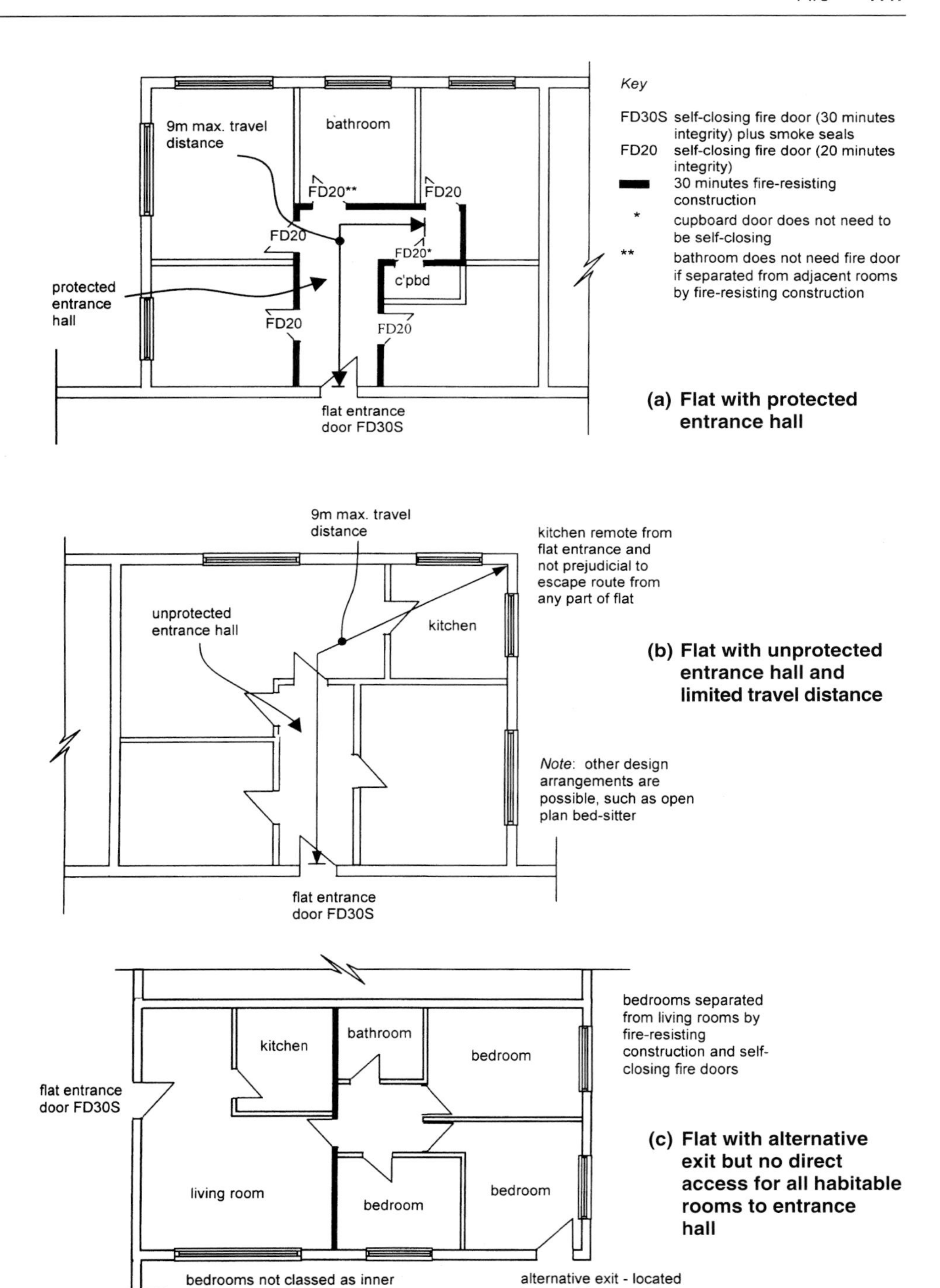

Fig. 7.17 Examples of alternative internal layouts to flats.

Maisonette with independent external entrance at ground level

A maisonette of this type is similar to a dwellinghouse and should have a means of escape which complies with the recommendations for dwellings described in section 7.12 above, depending on the height of the top storey above ground.

Maisonette with floor more than 4.5 m above ground and no external entrance at ground level

Two internal planning arrangements are described in AD B1 for maisonettes of this type. These are illustrated in Fig. 7.18(a) and 7.18(b), as follows:

- Provide each habitable room which is not on the entrance floor with an alternative exit, or
- Provide a protected entrance hall and/or landing entered directly from all the habitable rooms on that floor. Additionally, one alternative exit should be provided on each floor which is not the entrance floor.

7.14.2 Alternative exits from flats and maisonettes

That part of the means of escape from the entrance door of a flat or maisonette to a final exit from the building is often by way of a route which is common to all dwellings in the block. The provisions described below for means of escape in the common parts of flats and maisonettes are not applicable to such buildings where the top floor is not more than 4.5 m above ground level. However, they should be read in conjunction with the section below on 'Provision and protection of stairs' in section 7.15.8.

Reference has been made above, to the provision of alternative exits in certain planning arrangements for flats and maisonettes. Alternative exits will only be effective if they are remote from the main entrance door to the dwelling and lead to a common stair or final exit by means of:

- a door to an access corridor, access lobby or common balcony, or
- an internal private stairway leading to an access corridor, access lobby or common balcony on another level, or
- a door onto a common stair, or
- a door to an external stair, or
- a door to an escape route over a flat roof.

7.14.3 Means of escape in the common parts of flats and maisonettes

In general, flats and maisonettes should have access to an alternative means of escape. In this way, it will be possible to escape from a fire in a neighbouring flat by walking away from it. It is not always possible to provide alternative escape routes (which means providing two or more staircases) in all buildings containing flats and maisonettes, therefore single staircase buildings are permissible in certain circumstances. Typical examples of single and multi-stair buildings are described below.

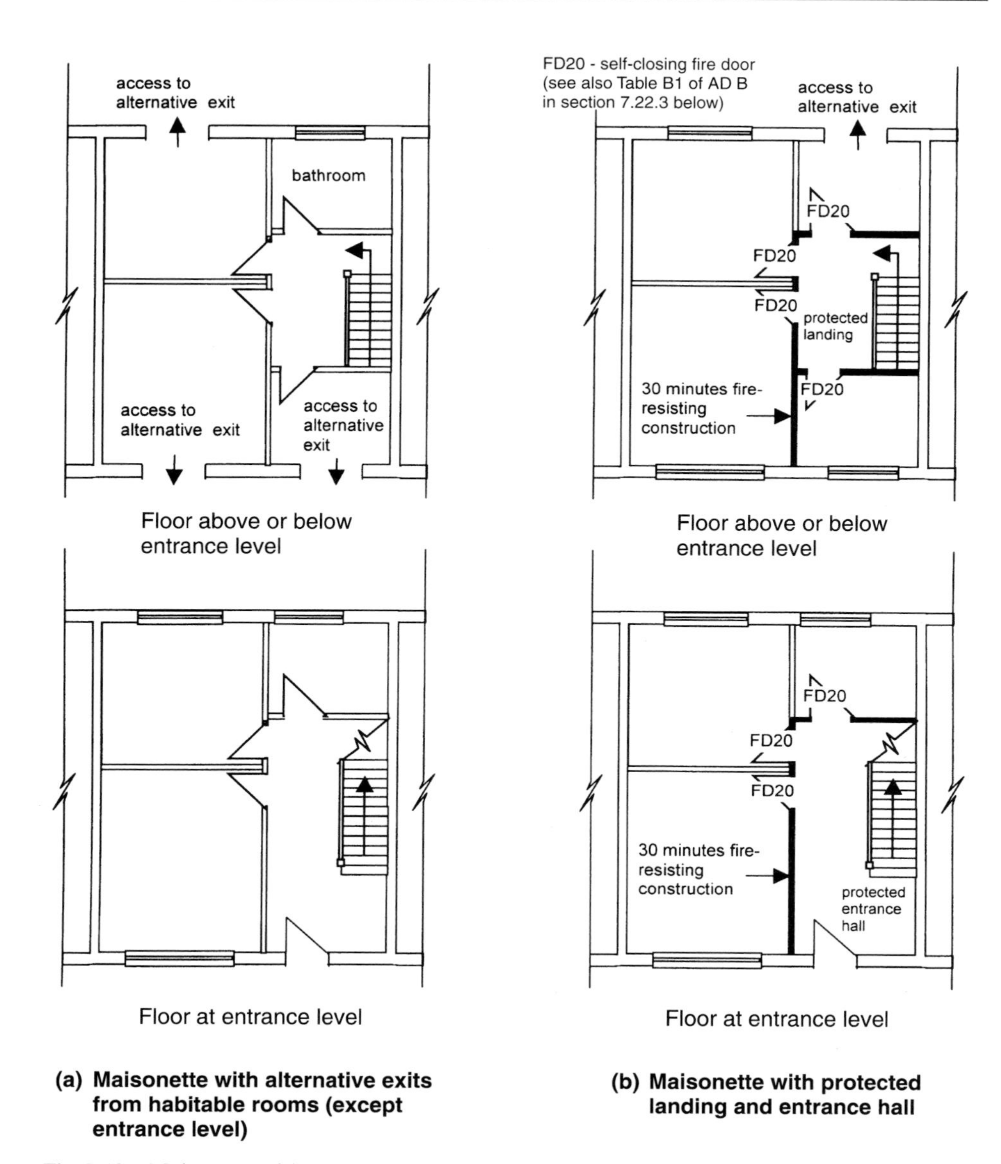

Fig. 7.18 Maisonette with no independent access at ground level and at least one storey more than 4.5 m above ground.

Flats and maisonettes with single common stairs

In larger buildings, it will be necessary to separate the entrance to each dwelling from the common stair by a protected lobby or common corridor. The maximum distance from any entrance door to the common stair or protected lobby should not exceed 7.5 m. (See Fig. 7.19(a) and 7.19(b).) These recommendations may be modified for smaller buildings in the following cases:

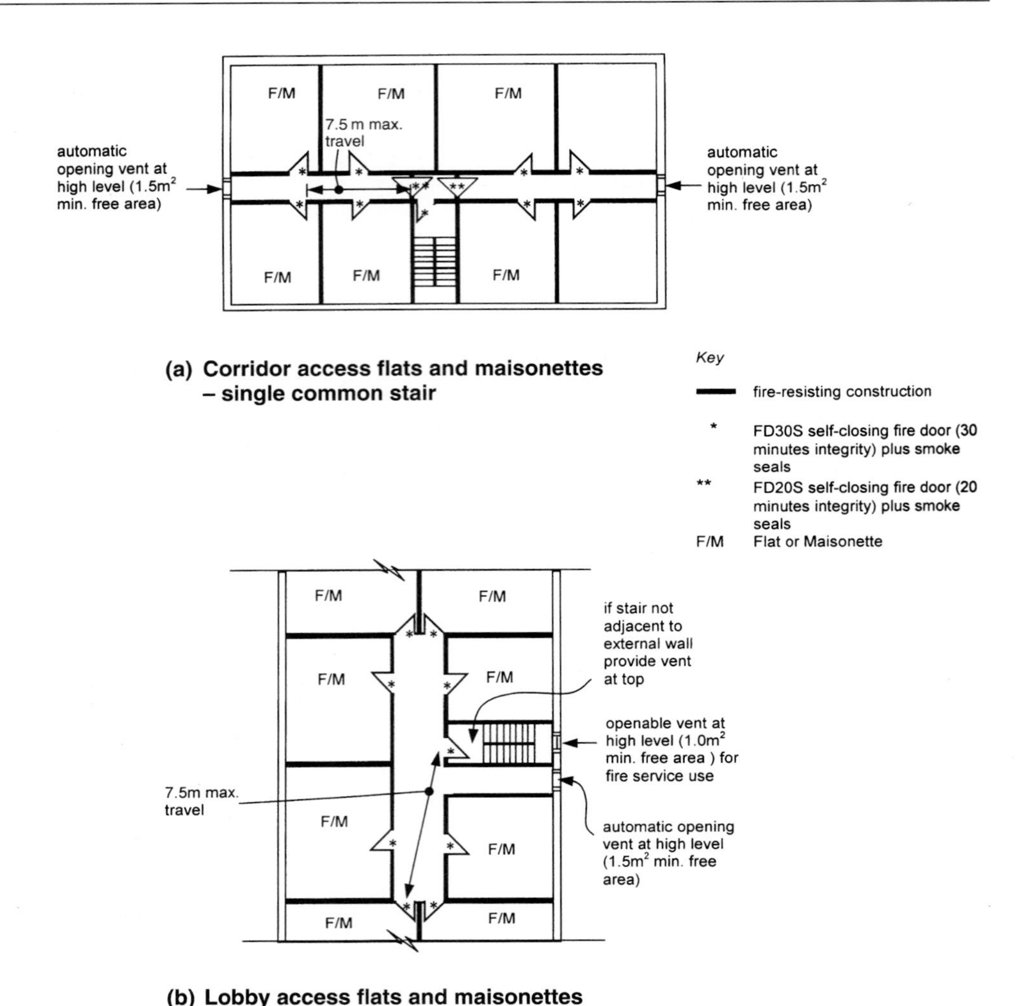

Fig. 7.19 Flats and maisonettes with single common stairs.

- The building consists of a ground storey and no more than three other storeys above the ground storey.
- The top floor does not exceed 11m above ground level.
- The stair does not connect to a covered car park unless it is open-sided.
- The stair does not also serve ancillary accommodation (see section 7.3 above), although this does not apply to ancillary accommodation:
  (a) in any storey which does not contain dwellings, and
  (b) which is separated from the stair by a ventilated protected lobby or ventilated protected corridor (i.e. provide permanent ventilation of not less than $0.4\,m^2$ or a mechanical smoke control system to prevent the ingress of smoke).

The modified recommendations are illustrated in Fig. 7.20(a). The maximum distance from the dwelling entrance door to the stair entrance should be reduced to 4.5 m. If the intervening lobby is provided with an automatic opening vent this distance may be increased to 7.5 m. Where the building contains only two flats per floor further simplifications as shown in Fig. 7.20(b) and (c) are possible.

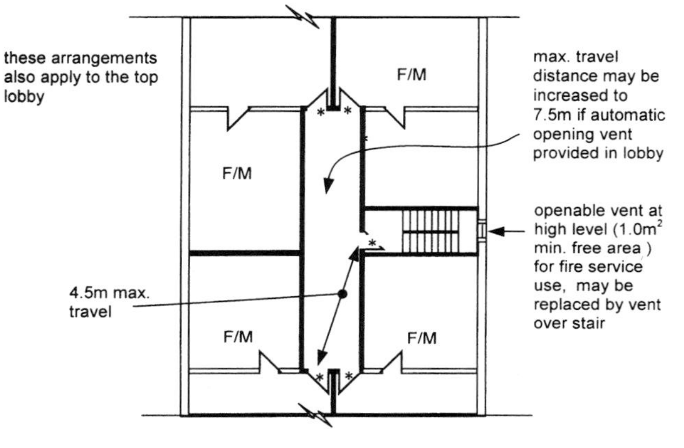

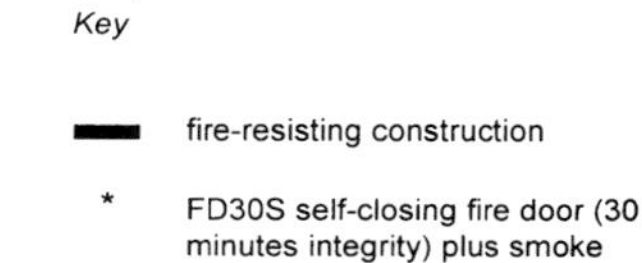

F/M Flat or Maisonette

(a) Common escape route for flats and maisonettes in small single stair building

Single stair access in small buildings shown in (a), (b) and (c) permitted if:

- maximum 4 storeys including ground storey.
- top floor not greater than 11m above ground level.
- stair does not connect to a covered car park unless it is open-sided.
- stair does not also serve certain types of ancillary accommodation (see section 7.3)

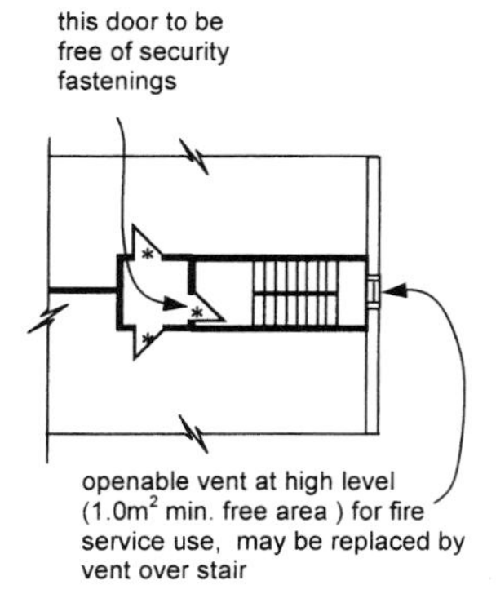

(b) Small single stair building – maximum two dwellings per floor

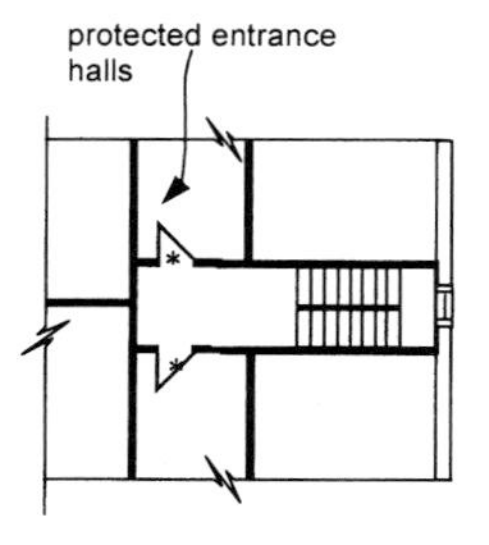

in (b) the lobby may be omitted if the dwellings have protected entrance halls

(c) Small single stair building – maximum two dwellings per floor with protected entrance halls

Fig. 7.20 Flats and maisonettes – small single stair buildings.

Flats and maisonettes with more than one common stair

Where escape is possible in two directions from the dwelling entrance door, the maximum escape distance to a storey exit may be increased to 30 m.

Furthermore, if all the dwellings on a storey have independent alternative means of escape, the maximum travel distance of 30 m does not apply. (There is still, however, the need to comply with the fire service access recommendations in Approved Document B5, where vehicle access for a pump appliance should be within 45 m of every dwelling entrance door.)

In buildings of this type it is possible to have a dead end situation where the stairs are not located at the extremities of each storey. This is permissible provided that the dead end portions of the corridor are fitted with automatic opening vents and the dwelling entrance doors are within 7.5 m of the common stair entrance.

Typical details of flats and maisonettes with more than one common stair are shown in Fig. 7.21.

Flats and maisonettes with balcony or deck approach

This is a fairly common arrangement whereby all dwellings are accessed by a continuous open balcony or deck on one or both sides of the block. AD B1 refers the reader to clause 13 of BS 5588 Part 1 on which the following notes are based.

The principal risk in such arrangements is smoke-logging of the balconies or decks.

This is less likely to occur when the balconies are relatively narrow, therefore the only considerations necessary are these.

- That vehicle access for a fire service pump appliance is within 45 m of every dwelling entrance door, and all parts of the building are within 60 m of a fire main.
- In the case of single stair buildings, that persons wishing to escape past the dwelling on fire can do so safely. This is usually achieved by ensuring that the external part of each dwelling facing the balcony is protected by 30 minutes fire-resisting construction for a distance of 1100 mm from the balcony floor level.

Smoke-logging is more likely to occur with the adoption of wider balconies or a deck approach. The provision of downstands from the soffit above a deck or balcony at right angles to the face of the building can reduce the possibility of smoke from any dwelling on fire spreading laterally along the deck. This would also reduce the chances of smoke logging on the decks above. Therefore, where the soffit above a deck or a balcony has a width of 2 m or more:

- it should be designed with downstands placed at 90° to the face of the building (on the line of separation between individual dwellings), and
- the down-stand should project 300 mm to 600 mm below any other beam or downstand parallel to the face of the building.

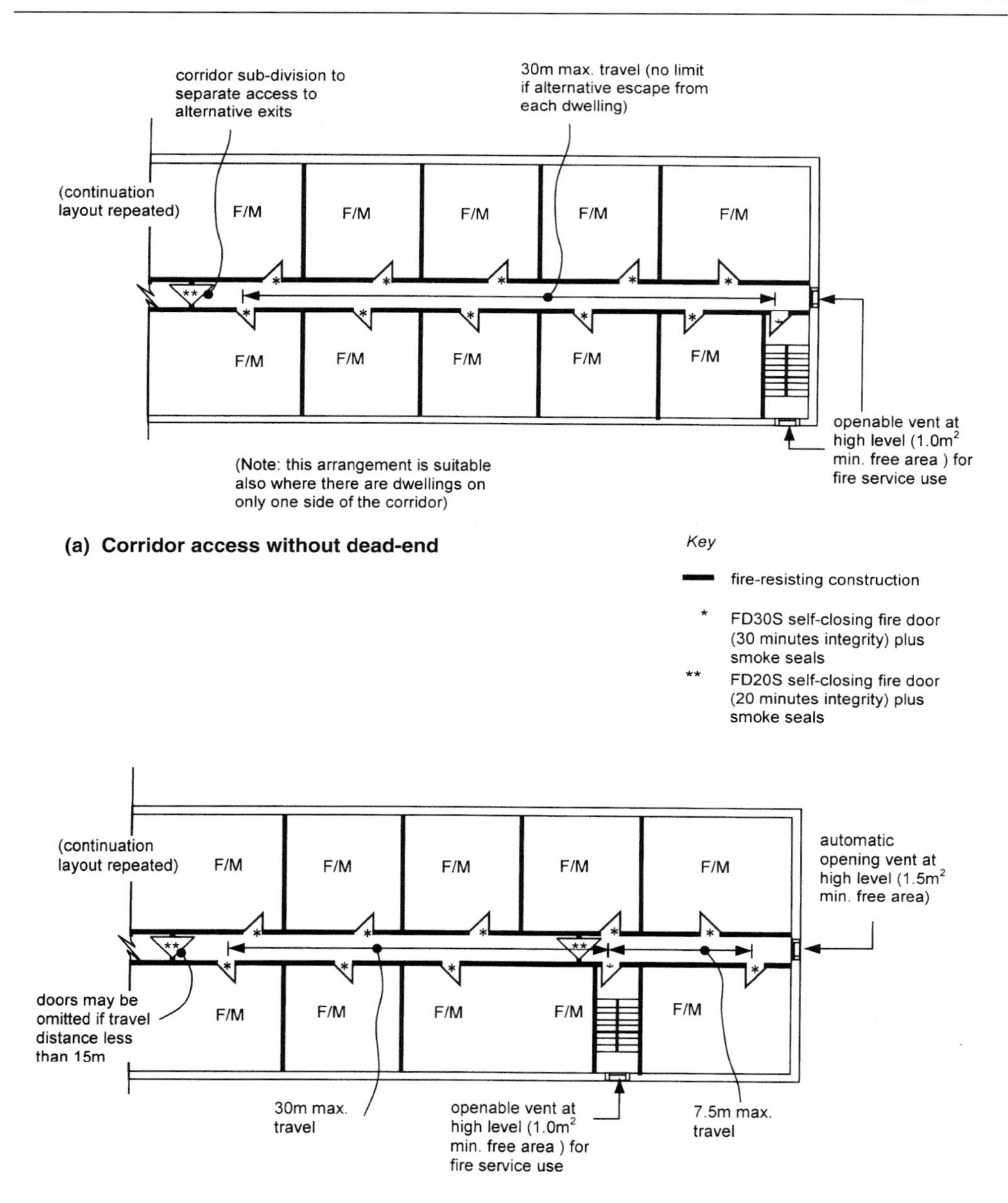

Fig. 7.21 Flats and maisonettes with more than one common stair.

There is a risk that occupants of dwellings with wider balconies or deck approach will use this opportunity of greater depth to erect 'external' stores or other fire risks. Therefore, no store or other fire risk should be erected externally on the balcony or deck.

Examples of common escape routes for dwellings with balcony or deck access are shown on Fig. 7.22 below.

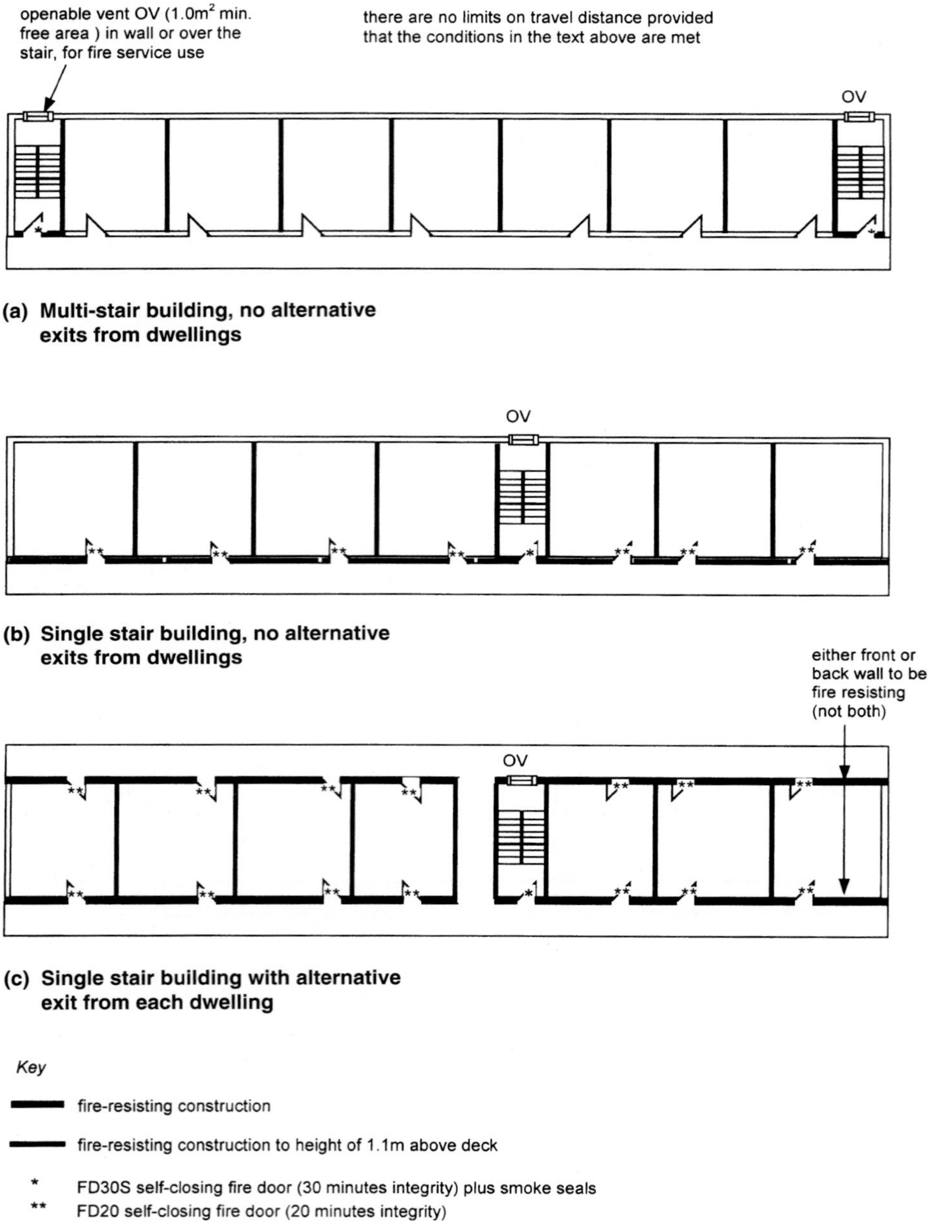

openable vent OV (1.0m^2 min. free area) in wall or over the stair, for fire service use

there are no limits on travel distance provided that the conditions in the text above are met

OV

(a) Multi-stair building, no alternative exits from dwellings

OV

(b) Single stair building, no alternative exits from dwellings

either front or back wall to be fire resisting (not both)

OV

(c) Single stair building with alternative exit from each dwelling

Key

fire-resisting construction

fire-resisting construction to height of 1.1m above deck

* FD30S self-closing fire door (30 minutes integrity) plus smoke seals

** FD20 self-closing fire door (20 minutes integrity)

Fig. 7.22 Flats and maisonettes with balcony/deck approaches.

Additional provisions for common escape routes

Common escape routes should be planned so that they are not put at risk by a fire in any of the dwellings or in any stores or ancillary accommodation. The following recommendations are designed to provide additional protection to these routes.

- It should not be necessary to pass through one stairway enclosure to reach another. Where this is unavoidable a protected lobby should be provided to the stairway. This lobby may be passed through in order to reach the other stair.
- Common corridors should be designed as protected corridors and should be constructed to be fire-resisting.
- The wall between each dwelling and the common corridor should be a compartment wall.
- A common corridor connecting two or more storey exits should be sub-divided with a self-closing fire door and/or fire-resistant screen, positioned so that smoke will not affect access to more than one storey exit.
- A dead end section of a common corridor should be separated in a similar manner from the rest of the corridor.
- Protected lobbies and corridors should not contain any stores, refuse chutes, refuse storage areas or other ancillary accommodation. See section 7.16.10 for for further details of the provision of refuse chutes and stores.

Ventilation of common escape routes

Although precautions can be taken to prevent the ingress of smoke onto common corridors and lobbies, it is almost inevitable that there will be some leakage since a flat entrance door must be opened in order that the occupants can escape. Provisions for ventilation of the common areas (which also provide some protection for common stairs) are therefore vital and may be summarised as follows.

- Subject to the variations shown in Fig. 7.19(a) and 7.19(b), common corridors or lobbies in larger, single-stair buildings should be provided with automatic opening ventilators, triggered by automatic smoke detection located in the space to be ventilated. These should be positioned as shown in the figure, should have a free area of at least 1.5 m^2 and be fitted with a manual override.
- Small single-stair buildings should conform to the guidance shown in Fig. 7.20(a), 7.20(b) and 7.20(c).
- Common corridors in multi-stair buildings should extend at both ends to the external face of the building where openable ventilators, or automatic opening ventilators, should be fitted for fire service use. They should have a free area of 1.0 m^2 at each end of the corridor (see Fig. 7.21).
- It is possible to protect escape stairways, corridors and lobbies by means of smoke control systems employing pressurisation. These systems should comply with BS 5588: *Fire precautions in the design, construction and use of buildings*, Part 4:1998 *Code of practice for smoke control using pressure differentials*. Where these are provided the cross corridor fire doors and the openable and automatic opening ventilators referred to above should be omitted.

Escape routes across flat roofs

Where more than one escape route exists from a storey or part of a building, one of those routes may be across a flat roof if the following conditions are observed.

- The flat roof should be part of the same building.
- The escape route over the flat roof should lead to a storey exit or external escape route.
- The roof and its structure forming the escape route should be fire-resisting.
- Any opening within 3 m of the route should be fire-resisting.
- The route should be adequately defined and guarded in accordance with Approved Document K. (This relates to the provision of barriers at least 1100 mm high designed to prevent people falling from the escape route. The barriers should be capable of resisting at least the horizontal force given in BS 6399: Part 1).

Provision of common stairs in flats and maisonettes

Stairs which are used for escape purposes should provide a reasonable degree of safety during evacuation of a building. Since they may also form a potential route for fire spread from floor to floor there are recommendations contained in Approved Document B3 which are designed to prevent this. Stairs may also be used for firefighting purposes. In this case reference should be made to the recommendations contained in Approved Document B5 (see section 7.30.2). The following recommendations are specifically for means of escape purposes.

- Each common stair should be situated in a fire-resisting enclosure with the appropriate level of fire resistance taken from Tables Al and A2 of Appendix A of Approved Document B.
- Each protected stair should discharge either:
  (a) direct to a final exit, or
  (b) by means of a protected exit passageway to a final exit.
- If two protected stairways or protected exit passageways leading to different final exits are adjacent, they should be separated by an imperforate enclosure.
- A protected stairway should not be used for anything else apart from a lift well or electricity meters.
- Openings in the external walls of protected stairways should be protected from fire in other parts of the building if they are situated where they might be at risk. (See section 7.14.3 'Additional provisions for common escape routes' above for details of protection measures.)
- A stair of acceptable width for everyday use will also be sufficient for escape purposes (BS 5588: Part 1 recommends a minimum width of 1 m); however if the stair is also a firefighting stair this should be increased to 1.1 m.
- Basement stairs will need to comply with special measures (see section 7.15.9 below).
- Gas service pipes and meters should only be installed in protected stairways if the installation complies with the requirements for installation and connection set out in the Pipelines Safety Regulations 1996, SI 1996/825 and the Gas Safety (Installation and Use) Regulations 1998 SI 1998/2451.
- A common stair which forms part of the only escape route from a flat or maisonette should not also serve any fire risk area such as a covered car park, boiler room, fuel storage space or other similar ancillary accommodation on the

same storey as that dwelling (but see the exceptions to this in section 7.14.3 'Flats and maisonettes with single common stairs' above).

- Where, in addition to the common stair, an alternative escape route is provided from a dwelling, it is permitted to serve ancillary accommodation from the common stair, provided that it is separated from that accommodation by a protected lobby or protected corridor.
- Where any stair serves an enclosed car park or place of special fire hazard (see section 7.2 above) it should be separated from that accommodation by a ventilated lobby or ventilated corridor (i.e. provide permanent ventilation of not less than $0.4\,m^2$ or a mechanical smoke control system to prevent the ingress of smoke).

7.14.4 External access and escape stairs

Where the building (or any part of it) is permitted to be served by a single access stair, that stair may be placed externally if it serves a floor which is not more than 6 m above ground level and it complies with the provisions listed at (2) to (6) below (see Fig. 7.23(a)).

Where there is more than one escape route available from a storey or part of a building some of the escape routes may be by way of an external escape stair if there is at least one internal escape stair serving every part of each storey (excluding plant areas) if the following provisions can be met.

(1) The stair should not serve any floors which are more than 6 m above the ground or a roof or podium. The roof or podium should itself be served by an independent protected stair.

(2) If it is more than 6 m in vertical extent, it is sufficiently protected from adverse weather. This does not necessarily mean that full enclosure will be necessary. The stair may be located so that protection may be obtained from the building itself. In deciding on the degree of protection it is necessary to consider the height of the stair, the familiarity of the occupants with the building, and the likelihood of the stair becoming impassable as a consequence of adverse weather conditions.

(3) Any part of the building (including windows and doors etc.) which is within 1.8 m of the escape route from the stair to a place of safety should be protected with fire-resisting construction. This does not apply if there is a choice of routes from the foot of the stair thereby enabling the people escaping to avoid the effects of fire in the adjoining building. Additionally, any part of an external wall which is within 1.8 m of an external escape route (other than a stair) should be of fire-resisting construction up to a height of 1.1 m from the paving level of the route.

(4) All the doors which lead onto the stair should be fire-resisting and self-closing. This does not apply to the only exit door to the landing at the head of a stair which leads downward.

(5) Any part of the external envelope of the building which is within 1.8 m of (and 9 m vertically below) the flights and landings of the stair, should be of fire-

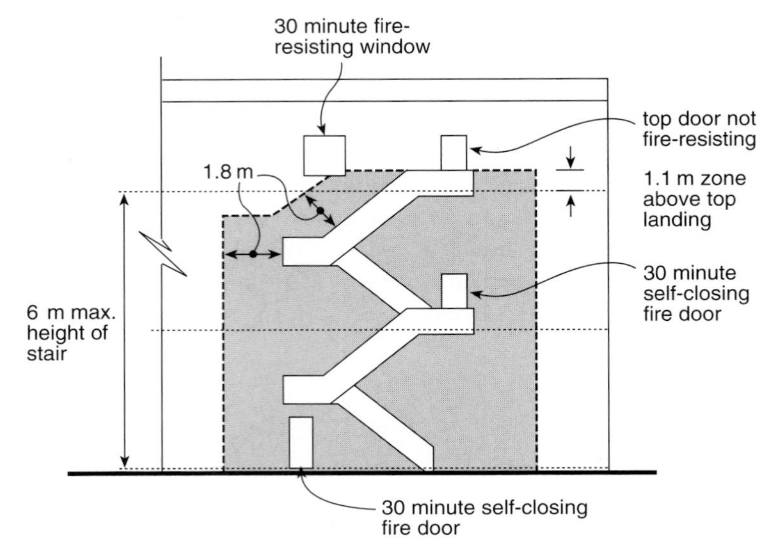

(a) Flats and maisonettes with external single access stair

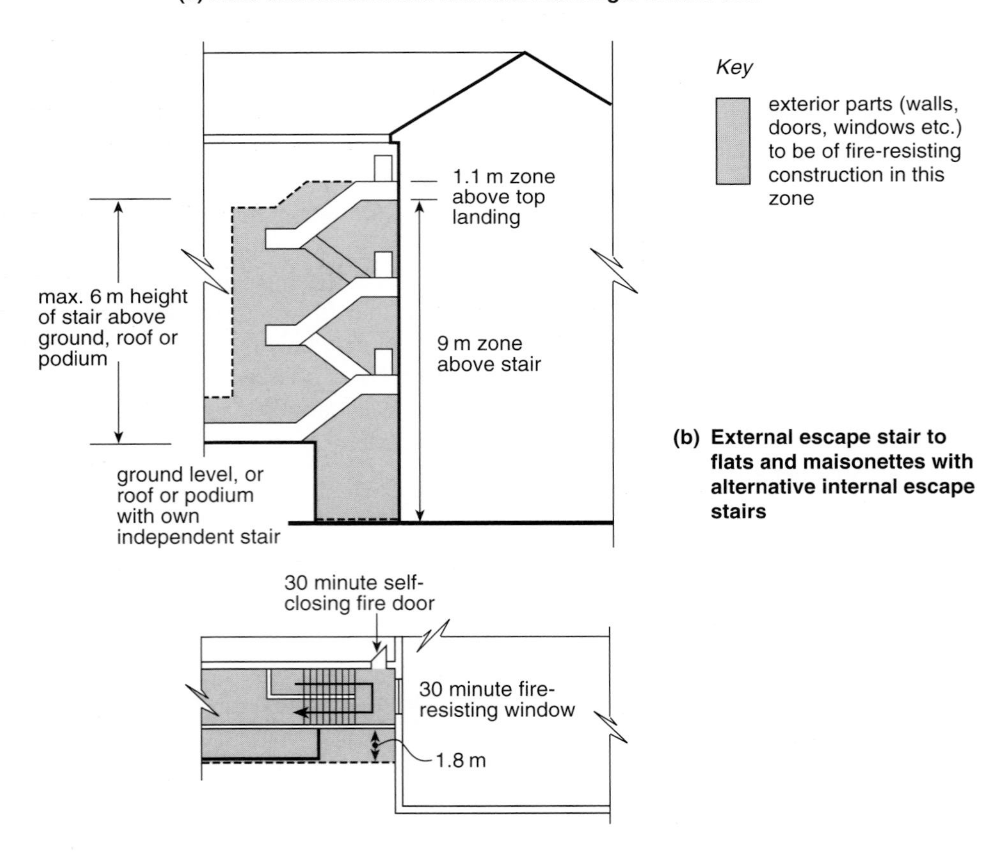

(b) External escape stair to flats and maisonettes with alternative internal escape stairs

Fig. 7.23 External escape stairs to flats and maisonettes.

resisting construction. This 1.8 m dimension may be reduced to 1.1 m above the top landing level provided that this is not the top of a stair up from basement level to ground.

(6) Any glazing which is contained within the fire-resisting areas mentioned above should also be fire-resisting in terms of maintaining its integrity in a fire, and be fixed shut. (For example, Georgian wired glass is adequate; it does not also have to meet the requirements for insulation.)

These provisions are illustrated in Fig. 7.23.

Stairs to dwellings in mixed use buildings

Many buildings consist of a mix of dwellings (i.e. flats and maisonettes) and other uses. Sometimes the dwellings are ancillary to the main use (such as a caretaker's flat in an office block), and sometimes they form a distinct separate use (as in the case of shops with flats over). Clearly, the degree of separation of the uses for means of escape purposes will depend on the height of the building and the extent to which the uses are interdependent.

Where a building has no more than three storeys above the ground storey, the stairs may serve both non-residential and dwelling uses, with the proviso that each occupancy is separated from the stairs by protected lobbies at all levels.

In larger buildings where there are more than three storeys above the ground storey, stairs may serve both the dwellings and the other occupancies if:

- the dwelling is ancillary to the main use of the building and is provided with an independent alternative escape route;
- the stair is separated from other occupancies in the building at lower storey levels by protected lobbies at those levels and has the same standard of fire resistance as that required by Approved Document B for the rest of the building (including any additional provisions if it is a firefighting stair);
- any automatic fire detection and alarm system fitted in the main part of the building is extended to the flat; and
- any security measures (usually installed for the benefit of the non-dwelling use) do not prevent escape at all material times.

Where fuels, such as petrol and liquid petroleum gas are stored, additional measures (such as an increase in the fire resistance period for the structure between the storage area and the dwelling) may be necessary.

7.15 Means of escape from buildings other than dwellinghouses, flats and maisonettes

7.15.1 Introduction

So far we have discussed the provisions contained in AD Bl for means of escape in buildings of Purpose Groups l(a), (b) and (c) i.e. dwellinghouses, flats and maisonettes. Although only a few of the more common arrangements for flats and maisonettes are covered in the Approved Document, the recommendations which are given are quite detailed and should form the basis for sound design guidance.

All other building types are grouped together in AD Bl and design recommendations are given for horizontal escape in Section 4, and vertical escape in Section 5. Of necessity, the guidance given is general in nature and is aimed at smaller, simpler types of buildings. For more complex or specialised buildings designers may well find that it is better to use other relevant design documents (such as the BS5588 series of codes) where more comprehensive guidance may be given.

7.15.2 Horizontal escape routes in buildings other than dwellings

Section 4 of AD B 1 deals with the provision of means of escape from any point in the floor of a building to the storey exit of that floor. It covers all buildings apart from dwellinghouses, flats and maisonettes. Whilst most of the guidance given in section 4 is related to general issues of design, the layouts of certain institutional buildings may warrant special provisions and some guidance on this is given in section 4.

The main decision that needs to be taken when designing the means of escape from a building is the number of escape routes and exits that are required. This will depend on:

- the maximum travel distance which is permitted to the nearest exit; and
- the number of occupants in the room, tier or storey under consideration.

Maximum travel distances and alternative escape routes in buildings other than dwellings

Ideally, there should be alternative escape routes provided from every part of the building. This is especially important in multi-storey buildings and in buildings where a mixture of Purpose Groups are present. In fact, if a mixed use building also contains Residential, or Assembly and Recreation purpose groups, these should be served by their own independent means of escape. (But see the exceptions to this for flats, described under 'Stairs to dwellings...' above.)

Where alternative escape routes are provided, escape will be possible in more than one direction. AD B1 places limits on the travel distance from any part of a room, tier or storey to a storey exit and these are shown in Table 3 from AD B1. The substance of Table 3 is summarised in Table 7.2 below. It should be read in conjunction with the following comments.

- The Table dimensions are actual travel distances and are measured along the shortest route taken by a person escaping in the event of a fire.
- Where there is fixed seating or there are other fixed obstructions, the travel distance is measured along the centre line of the seatways or gangways.
- Where the route of travel includes a stair it is measured along the pitch line on the centre line of travel.

- Where the layout of a room or storey is not known at the design stage, the direct distance measured in a straight line should be taken. Direct distances should be taken as two-thirds of the travel distance.

Once it has been established that *at least one exit* is within the distance limitations given in Table 7.2 the other exits may be further away than the distances given.

It will be observed from Table 7.2 that where escape is possible in one direction only the travel distances are much reduced. However, where a storey exit can be reached within these one-directional travel distances, it is not necessary to provide an alternative route except in the case of a room or storey that:

- has an occupant capacity exceeding 60 in the case of places of assembly or bars; or
- has an occupant capacity exceeding 30 if the building is in Purpose Group 2(a) Residential (institutional);
- is used for in-patient care in hospitals.

Similarly, it is often the case that there will not be alternative escape routes, especially at the beginning of an escape route. A room may have only one exit onto a corridor from where it may be possible to escape in two directions. This is permissible provided that:

- the overall distance from the furthest point in the room to the storey exit complies with the multi-directional travel distance from Table 7.2; and
- the single direction part of the route (in this case, in the room) complies with the 'one direction' travel distance specified in Table 7.2.

Although a choice of escape routes may be provided from a room or storey, it is possible that they may be so located, relative to one another, that a fire might disable them both. In order to consider them as true alternatives they should be positioned as shown in Fig. 7.24, i.e. the angle which is formed between the exits and any point in the space should be at least 45°. Where this angle cannot be achieved:

- the maximum travel distance for escape in one direction will apply; or
- the alternative escape routes should be separated from each other by fire-resisting construction.

Figure 7.24 illustrates these rules covering alternative escape routes.

Special rules apply where a dead-end situation exists in an open storey layout as shown in Fig. 7.25 as follows:

- XY should be within the one direction travel distance from Table 7.2
- whichever is the least distance of WXY and ZXY should be within the multi-direction travel distance from Table 7.2
- angle WXZ should be at least 45° plus 2.5° for each metre travelled in a single direction from Y to X.

Table 7.2 Travel distance limitations.

Purpose Group	Maximum Travel Distance (m) in:		Notes
	One direction	**Multi-direction**	
2(a) Institutional	9	18	In hospitals or other healthcare premises where the means of escape is being designed using the Department of Health's '*Firecode*' documents, the relevant travel distances recommended in those documents should be used
2(b) Other residential			
(i) in bedrooms	9	18	This is the maximum part of the travel distance within the bedroom but includes any associated dressing room, bathroom or sitting room etc. It is measured to the door onto the protected corridor serving the bedroom or suite
(ii) in bedroom corridors	9	35	This is the distance from the door onto the protected corridor serving the bedroom or suite to the storey exit
(iii) elsewhere	18	35	
3 Office	18	45	
4 Shop and commercial	18	34	For shopping malls see BS 5588: Part 10. This document applies more restrictive provisions to units with only one exit in covered shopping complexes. See also BRE Report (BR 368) *Design methodologies for smoke and heat exhaust ventilation* for guidance on associated smoke control measures
5 Assembly and recreation			
(i) buildings mainly for disabled people (not schools)	9	18	
(ii) schools	18	45	
(iii) areas with seating in rows	15	32	
(iv) elsewhere	18	45	

6 7	**Industrial and** **Storage & other non-residential**			
	(i) 'normal' fire risk	25	45	For 'normal' fire risk as defined in Home Office *Guide to fire precautions in existing places of work that require a fire certificate: Factories, Offices, Shops and Railway Premises*
	(ii) 'high' fire risk	12	25	For 'high' fire risk as defined in Home Office *Guide to fire precautions in existing places of work that require a fire certificate: Factories, Offices, Shops and Railway Premises*
2–7	**Places of special fire hazard**	9	18	This is the maximum part of the travel distance within the room or area. The travel distance outside such room or area should comply with the limits for the purpose group as shown above. Places of special fire hazard are: oil-filled transformer and switch gear rooms, boiler rooms, storage space for fuel or other highly flammable substances, and rooms housing fixed internal combustion engine. Plus, in schools: laboratories, technology rooms with open heat sources, kitchens and stores for PE mats or chemicals
2–7	**Plant room or rooftop plant:**			
	(i) distance within plant room	9	35	
	(ii) escape route not in open air	18	45	Overall travel distance
	(iii) escape route in open air	60	100	Overall travel distance

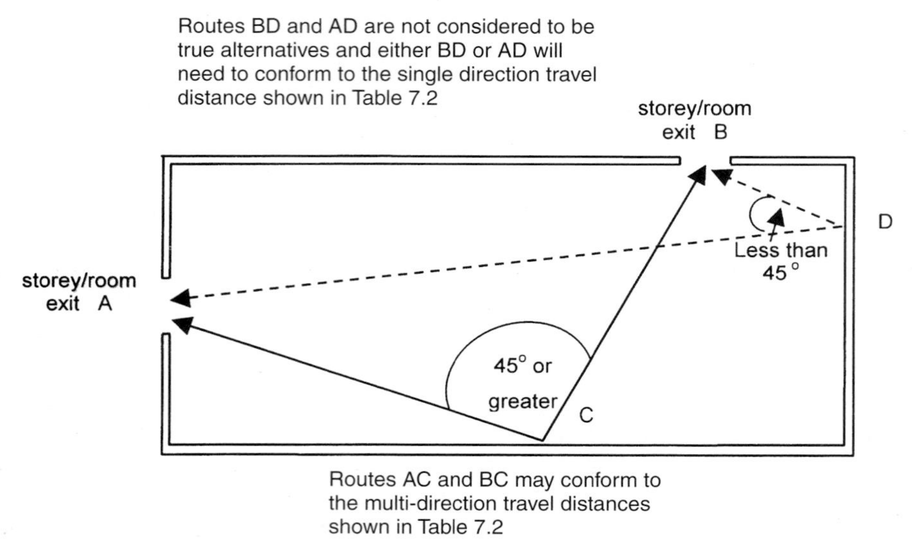

Fig. 7.24 Alternative escape routes.

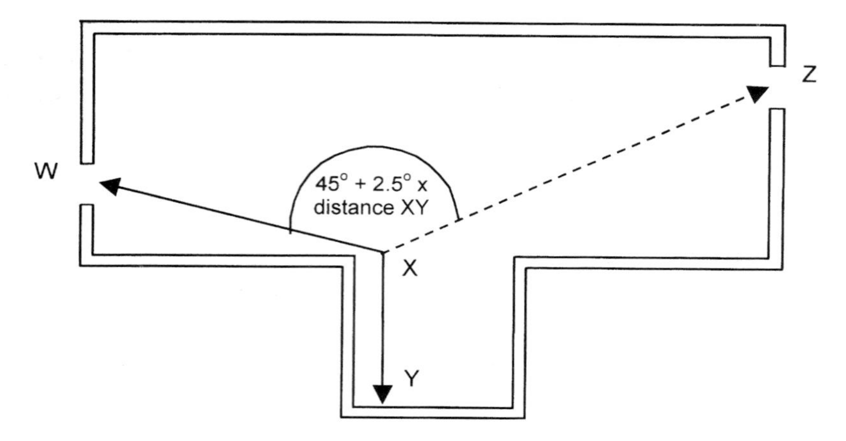

Fig. 7.25 Dead-end situation – single-storey layout.

Number and widths of exits related to the number of occupants in buildings other than dwellings

The number of occupants in a room, tier or storey influences the numbers and widths of the exits and escape routes that need to be provided from that area. Table 4 of AD B1 lists the minimum number(s) of escape routes or exits which should be provided relative to the maximum numbers of occupants as follows:

- up to 60 persons – 1 exit
- 61 to 600 persons – 2 exits
- over 600 persons – 3 exits

Realistically, the figures given above will only serve to define the absolute minimum number of exits for a means of escape. In practical terms the actual number of exits will be determined by travel distances and exit widths. The width of an escape route or exit may be determined by reference to Table 5 of AD B1. The information contained in this Table is restructured in Table 7.3 below to include relevant data from Approved Document M (Access and facilities for disabled people) since the minimum widths shown in Table 5 of AD B1 may not be adequate for disabled access.

Usually, the narrowest part of an escape route will be at the door openings which form the room or storey exits. These are measured as shown in Fig. 7.10 in section 7.9.3 above.

Where a storey has two or more exits it is assumed that one of them will be disabled by a fire. Therefore the remaining exits should have sufficient width to take the occupants safely and quickly. This means that the widest exit should be discounted and the remainder should be designed to take the occupants of the storey. Since stairs need to be as wide as the exit leading onto them, this recommendation for exit width may influence the width of the stairways. (Stairways may also need to be discounted and this is discussed in section 7.15.7 below.)

Except in doorways, all escape routes should have clear headroom of at least 2 m.

7.15.3 Horizontal escape routes – factors affecting internal planning

The efficacy of an escape route in a storey may be affected by a number of internal planning considerations, such as:

- the need for inner rooms
- the relationship between circulation routes and stairways
- the need for different occupancies in a building to use the same escape route
- the design and layout of means of escape corridors.

These are considered in more detail in the following paragraphs.

Provision of inner rooms

The rules governing the provision of inner rooms are more stringent than those for dwellings. Inner rooms are only acceptable under the following conditions.

- The occupant capacity of the inner room should not exceed 60 (or 30 for institutional buildings in Purpose Group 2(a)).
- The inner room should not be a bedroom.
- Only one access room should be passed through when escaping from the inner room.
- The maximum travel distance from the furthest point in the inner room to the exit from the access room should not exceed the appropriate limit given in Table 7.2 above.
- The access room should be in the control of the same occupier as the inner room.

Table 7.3 Widths of exits and escape routes.

Maximum number of persons	Min. width of exit*	Min. width of escape route*	Notes
Up to 50	750	750[(1)]	[(1)] Does not apply to: • schools where minimum width in corridors is 1050 (and 1600 in dead ends) • areas accessible to disabled people where minimum width in corridors is 1200 (or 1000 where lift access is not provided to the corridor or it is situated in an extension approached through an existing building) • gangways between fixed storage racking in Purpose Group 4 (Shop and Commercial) where minimum width may be 530 mm (but not in public areas) Widths of escape routes and exits less than 1050 should not be interpolated
51 to 110	850	850[(2)]	[(2)] Does not apply to: • schools where minimum width in corridors is 1050 (and 1600 in dead ends) • areas accessible to disabled people where minimum width in corridors is 1200 (or 1000 where lift access is not provided to the corridor or it is situated in an extension approached through an existing building) Widths of escape routes and exits less than 1050 should not be interpolated
111 to 220	1050	1050[(3)]	[(3)] Does not apply to: • schools where minimum width in corridor dead ends is 1600 mm • areas accessible to disabled people where minimum width in corridors is 1200
over 220	5 mm/person	5 mm/person	This method of calculation should not be used for any opening serving less than 220 persons [e.g. three exits each 850 mm wide will accommodate 3 × 110 = 330 people, not the 510 (i.e. 2550 ÷ 5) people that 3 × 850 = 2550 mm would accommodate]

* For method of measuring widths of escape routes and exits see section 7.9 above.

- The access room should not be a place of special fire hazard (e.g. a boiler room).

Where these conditions are met the inner room should be designed to conform to one of the following arrangements:

- the walls or partitions of the inner room should stop at least 500 mm from the ceiling, or
- a vision panel, which need not be more than 0.1 m^2 in area, should be situated in the walls or door of the inner room (this is to enable the occupiers to see if a fire has started in the access room), or
- a suitable automatic fire detection and alarm system should be fitted in the access room which will give warning of fire in that room to the occupiers of the inner room.

Horizontal escape routes and stairways

Care must be taken in the design of horizontal escape routes since they also form part of the normal circulation in a building and may jeopardise access to stairways unless the following points are considered.

- In any storey which has more than one escape stair, it should not be necessary to pass through one stairway to reach another. Where this is unavoidable a protected lobby should be provided to the stairway. This lobby may be passed through in order to reach the other stair.
- As part of the normal circulation in a building, it should not be necessary to pass through a stairway enclosure in order to reach another part of the building on the same level. (Such circulation patterns should be avoided since familiarity breeds contempt, and fire doors may become ineffective due to excessive use or misuse.) Such an arrangement is permissible if the doors to the protected stairway and any associated exit passageway are fitted with an automatic release mechanism (see below section 7.22.3).
- Where buildings are planned with more than one exit round a central core, these exits should be remote from each other and no two exits should be approached from the same lift hall, common lobby or undivided corridor, or linked together by any of these (see Fig. 7.26).

Use of common escape routes by different occupancies

It is common in mixed-use or multi-tenanted buildings for common escape routes to be used by all the occupants. There are restrictions on this and these have been referred to above in section 7.15.2. Where common escape routes are permitted the following rules should be observed.

- The means of escape from one occupancy should not pass through another.
- Common corridors or circulation spaces should either be fire-protected or fitted

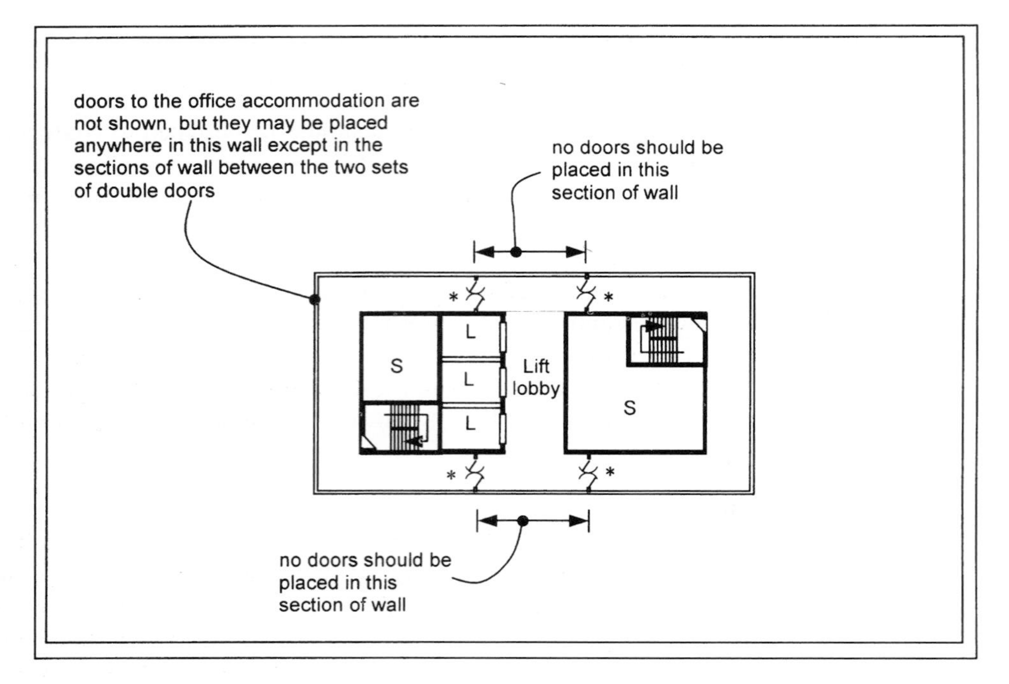

Key

* FD20 self-closing fire door (20 minutes integrity)

L Lift

S services, toilets etc.

Fig. 7.26 Corridor layout: building with central core.

with an automatic fire detection and alarm system which extends throughout the storey.

Storeys containing areas for consumption of food and/or drink by customers

In some buildings (such as department stores and shops) it may be desirable to provide an area for the consumption of food and/or drink by customers where this is ancillary to the main use of the building. Such an arrangement is permissible if the following conditions are met.

- At least two escape routes should be provided from each area (inner rooms which follow the guidance contained in section 7.15.3 'Provision of inner rooms' above are exempt from this); and
- Each escape route should lead directly to a storey exit without having to pass through a kitchen or similar area of high fire hazard.

Means of escape corridors – design factors

The following means of escape corridors should be fire-protected.

- Every dead-end corridor (although small recesses and extensions less than 2 m long and referred to in figures 10 and 11 of BS 5588: *Fire precautions in the design, construction and use of buildings:* Part 11: 1997 *Code of practice for shops, offices, industrial, storage and other similar buildings*, may be ignored).
- Every corridor serving bedrooms.
- Every corridor or circulation space common to two or more 'different occupancies' (i.e. where the premises are split into separate ownerships or tenancies of different organisations). In this case the need for fire protection may be omitted where an automatic fire detection and alarm system is installed throughout the storey. Even so, the means of escape from one occupancy should not pass through any other occupancy.

The way in which a storey layout is planned can have an effect on its means of escape characteristics. For example, whilst it is perfectly acceptable to have open plan floor areas, they offer no impediment to smoke spread but do have the advantage that occupants can become aware of a fire more quickly. On the other hand, the provision of a cellular layout, where the means of escape is enclosed by partitions, means that some defence is provided against smoke spread in the early stages of a fire even though the partitions may have no fire resistance rating.

To maintain the effectiveness of the partitions, they should be carried up to ceiling level (i.e. either the soffit of the structural floor above or to a suspended ceiling) and room openings should be fitted with doors (which do not need to be fire resisting).

Corridors which give access to alternative escape routes may become blocked by smoke before all the occupants of a building have escaped and may make both routes impassable. Additionally, the means of escape from any permitted dead-end corridors may be blocked. Therefore, corridors connecting two or more storey exits should be sub-divided by means of self-closing fire doors (and screens, if necessary) if they exceed 12 m in length. The doors (and screens) should be positioned so that:

- they are approximately mid-way between the two storey exits; and
- the route is protected from smoke, having regard to any adjacent fire risks and the layout of the corridor.

Unless the escape stairway and its associated corridors are protected by a pressurisation system complying with BS 5588: *Fire precautions in the design, construction and use of buildings:* Part 4: 1998 *Code of practice for smoke control using pressure differentials*, dead-end corridors exceeding 4.5 m in length giving access to a point from which alternative escape routes are available, should also be provided with fire doors so positioned that the dead-end is separated from any corridor which:

- provides two directions of escape; or
- continues past one storey exit to another.

These provisions for means of escape corridors are summarised in Fig. 7.27.

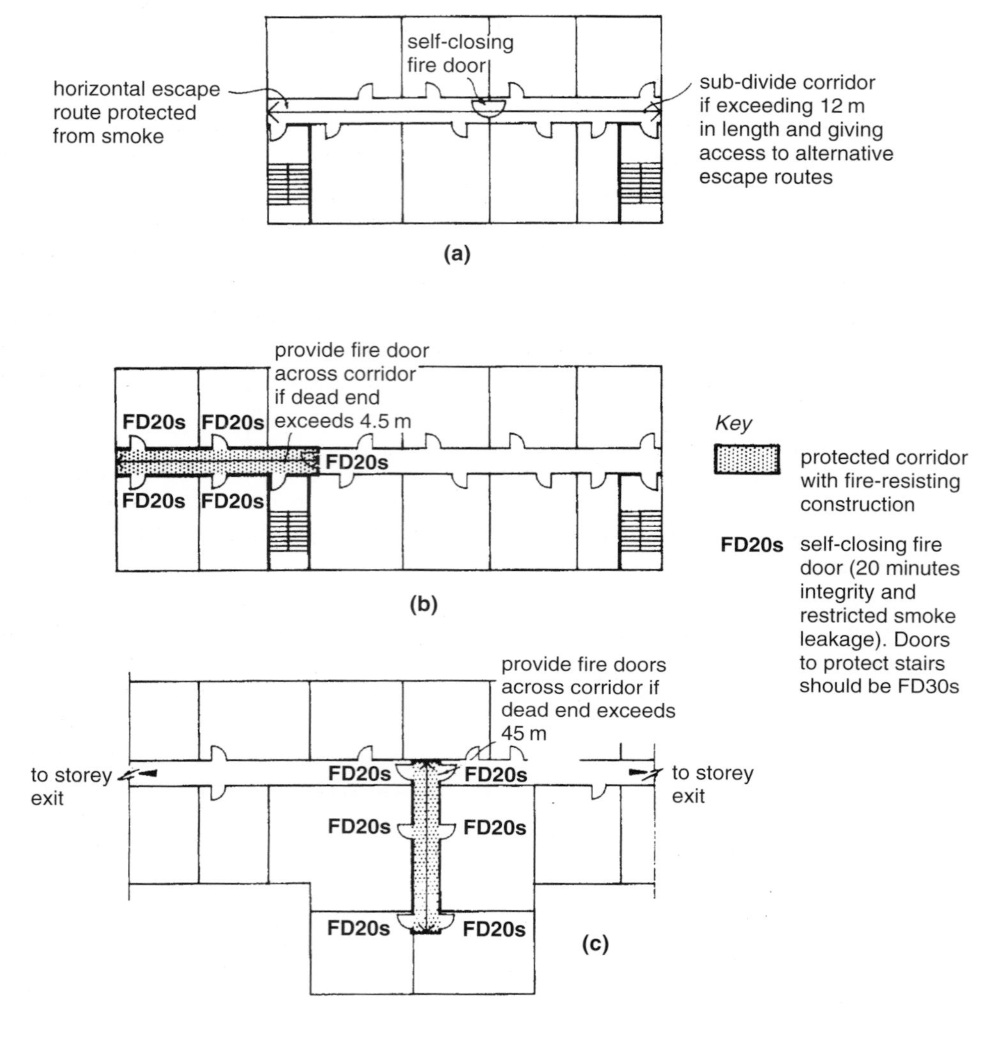

Fig. 7.27 Escape corridors – buildings other than dwellings. (a) Sub-division of corridors. (b) Dead end separation from continuation corridor. (c) Dead end separation from alternative escape routes.

7.15.4 Escape routes across flat roofs

The recommendations given above for escape over flat roofs from flats and maisonettes (section 7.14.3) also apply to all other building types except where the route serves an institutional building or part of a route used by members of the public.

7.15.5 Vertical escape routes in buildings other than dwellings

Section 5 of AD B1 deals with the provisions for vertical escape, by means of a sufficient number of adequately sized and protected escape stairs, for all buildings apart from dwellinghouses, flats and maisonettes.

The main decision to be taken when designing the vertical means of escape in a building is the number of stairways that need to be provided. It has already been shown above that alternative means of escape are required where horizontal constraints are imposed by travel distances, exit widths and numbers of occupants. Additionally, in buildings where there is a mix of different uses it may be the case that a fire in an unattended shop or office might have serious consequences for a residential or hotel use in the same building. Therefore, it is important to analyse the risks involved and to consider whether completely separate escape routes should be provided from each different use or whether other effective means of protecting common escape routes can be provided. (See above section 7.14 for examples of the use of common stairs in buildings which contain both dwellings and other uses).

Section 5 of AD B1 provides additional recommendations for assessing the number of stairways that are needed with regard to:

- the acceptability of single stairs for means of escape in a building; and
- the influence that adequate width of stairs may have on their provision whilst allowing for the fact that a stair may have to be discounted due to the effects of fire or smoke.

One further influence on the provision of stairs is the necessity to provide fire-fighting stairs in larger buildings. This is covered by AD B5 (see section 7.30) and may mean that extra stairways are required beyond those needed merely for means of escape purposes.

7.15.6 The provision of single stairs in buildings other than dwellings

Assuming that the building is not excluded from having a single escape route by virtue of the recommendations listed above, it may be served by a single escape stair in the following circumstances.

- Where it serves a basement which is allowed to have a single horizontal escape route (i.e. storey occupancy not exceeding 60, maximum travel distance within the limits for travel in one direction).
- Where it serves what are termed '*small premises*' and the recommendations of clause 10 of BS 5588 Part 11: 1997 *Code of practice for shops, offices, industrial, storage and other similar buildings* are followed (see section 7.15.8 below).
- Where it serves a building which has no floor more than 11m above ground and in which every floor is allowed to have a single horizontal escape route.

It should be noted that in schools where single stairs are provided the storeys above first floor level should only be occupied by adults. Additionally, where a two-storey school building (or part of a building) is provided with a single stair the following conditions apply.

- There should be no more than 120 pupils plus supervisors on the first floor.
- The first floor should not contain a place of special fire hazard.
- Classrooms and stores should not open onto the stairway.

7.15.7 Escape stair design – widths

Clearly, the width of escape stairs is related to the number of people that they can carry in an evacuation situation. AD B1 contains a number of provisions which enable the width of stairs to be calculated by reference to:

- the number of people who will use them,
- whether or not it will be necessary to discount any of the stairs, and
- their mode of use (i.e. simultaneous or phased evacuation).

Escape stairs should be at least as wide as any exits giving access to them and should not reduce in width as they approach the final exit. Additionally, if the exit route from a stair also picks up occupants of the ground and basement storeys, it may need to be increased in width accordingly. Although stairs need to be sufficiently wide for escape purposes, research has shown that people prefer to stay close to a handrail when making a long descent. Therefore the centre of a very wide stairway would be little used and might, in fact, be hazardous. For this reason, AD B1 puts a maximum limit of 1400 mm on the width of a stairway where its vertical extent exceeds 30 m unless it is centrally divided with a handrail. Where the design of the building calls for a stairway that is wider than 1400 mm, it should be at least 1800 mm wide and contain a central handrail. In this case, the stair width on either side of the central handrail will need to be considered separately when assessing stair capacity.

Minimum stair widths can, in the first instance, be assessed using Table 7.4 below. This is based on Table 6 from AD B1 and is suitable for most simple building designs where the maximum number of people served by the stair(s) does not exceed 220.

Where two or more stairways are provided, it is possible that one of the stairs may be inaccessible due to fire or smoke unless special precautions are taken. Therefore, it may necessary to discount each stair in turn in order to check that the remaining stairways are capable of coping with the demand. Discounting is unnecessary if:

- the escape stairs are approached through a protected lobby at each floor level (although a lobby is not needed for the top floor for the exception still to apply); or
- the stairs are protected by a pressurisation smoke control system designed in accordance with BS 5588: Part 4.

As Table 7.4 suggests, in multi-storey buildings where the number of occupants exceeds 220 it may be necessary to consider the mode of evacuation and use other methods to calculate stair widths.

Where it is assumed that all the occupants would be evacuated, this is termed 'simultaneous evacuation' and this should be the design approach for:

- all stairs which serve basements
- all stairs which serve buildings with open spatial planning (i.e. where the building is arranged internally so that two or more floors are contained within one undivided volume)
- all stairs which serve Assembly and Recreation buildings (PG 5) or Other Residential buildings (PG 2(b)).

Table 7.4 Minimum widths of escape stairs.

Description of stair	Numbers of people assessed as using stair in emergency[1]	Minimum width of stair (mm)
Escape stairs in any building (but see footnotes for exceptions)	Up to 50	800[2]
	51 to 150	1000
	151 to 220	1100
	Over 220	See note 3

Notes
(1) For methods of assessing occupancy see section 7.9.1 above.
(2) This minimum stair width does not apply:
 (a) in an Institutional buildings unless the stair will only be used by staff
 (b) in an assembly building unless the area served is less than 100 m^2 and/or is not for assembly purposes (e.g. office)
 (c) to any areas which are accessible to disabled people.
(3) See AD B1 Table 7 (and Formula 7.1) for simultaneous evacuation and AD B1 Table 8 (and Formula 7.2) for phased evacuation.

Using this approach the escape stairs should be wide enough to allow all the floors to be evacuated simultaneously. The calculations take into account the number of people temporarily housed in the stairways during evacuation.

A simple way of assessing the escape stair width is to use Table 7 from AD B1. This covers the capacity of stairs with widths from 1000 mm to 1800 mm for buildings up to ten storeys high (although the capacity of stairs serving more than ten storeys can be obtained from Table 7 by using linear extrapolation).

In fact, the capacities given in the table for stair widths of 1100 mm and greater are derived from the formula:

$$P = 200w + 50(w - 0.3)(n - 1) \quad \text{(Formula 7.1)}$$

where:
P = the number of people that can be served by the stair
w = the width of the stair in metres
n = the number of storeys in the building.

Formula 7.1 can be used for any size of building with no limit being placed on the occupant capacity or number of floors and it is probably advisable to use it for buildings which are larger than those covered by Table 7 from AD B1. It should be noted that separate calculations should be made for stairs serving basements and for those serving upper storeys.

The formula is particularly useful where the occupants of a building are not evenly distributed – either within a storey or between storeys. However, it cannot be used for stairs which are narrower than 1100 mm, so for stairs which are allowed to be 1000 mm wide, in buildings up to ten storeys high, the values have been extracted from Table 7 and are presented in Fig. 7.28.

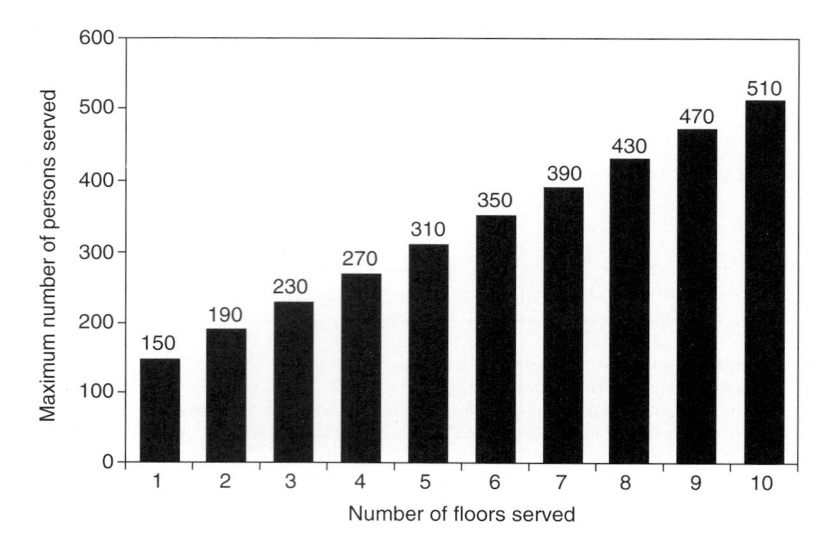

Fig. 7.28 Stair capacity for simultaneous evacuation – 1000 mm wide stairs.

In certain buildings it may be more advantageous to design stairs on the basis of 'phased evacuation'. Indeed, in high buildings it may be impractical or unnecessary to evacuate the building totally, especially if the recommendations regarding fire resistance, compartmentation and the installation of supporting facilities such as sprinklers and fire alarms are adhered to.

In phased evacuation, people with reduced mobility and those most immediately affected by the fire (i.e. those people on the floor of fire origin and the one above it) are evacuated first. After that, if the need arises, floors can be evacuated two at a time. Phased evacuation allows narrower stairs to be used and has the added advantage that it causes less disruption in large buildings than total evacuation.

Phased evacuation may be used for any buildings unless they are of the types listed above as needing simultaneous evacuation.

Where a building is designed for phased evacuation the following conditions should be met.

- The stairs should be approached through a protected lobby or protected corridor at each floor level (this does not apply to a top storey).
- The lifts should be approached through a protected lobby at each floor level.
- Each floor should be a compartment floor.
- If the building has a floor which is more than 30 m above ground, it should be protected throughout by an automatic sprinkler system which complies with the relevant requirements of BS 5306 *Fire extinguishing installations and equipment on premises,* Part 2:1990 *Specification for sprinkler systems* (i.e. the sections dealing with the relevant occupancy rating and the additional requirements for life safety). This provision does not apply to flats of PG 1(a) in a mixed use building.
- An appropriate fire warning system should be fitted which complies with

BS 5839 *Fire detection and alarm systems for buildings,* Part 1: 1988 *Code of practice for system design, installation and servicing*, to at least the L3 standard.

- An internal speech communication system (such as a telephone, intercom system or similar) should be provided so that conversation is possible between a fire warden at every floor level and a control point at the fire service access level.
- Where it is deemed appropriate to install a voice alarm, the recommendations regarding phased evacuation in BS 5839: Part 1 should be followed and the voice alarm system itself should conform to BS 5839: Part 8 1998 *Code of practice for the design, installation and servicing of voice alarm systems.*

When phased evacuation is used as the basis for design, the minimum stair width needed may be taken from Table 8 of AD B1 assuming phased evacuation of not more than two floors at a time. The data from Table 8, which has been reconfigured in Fig 7.29 below, is derived from the formula:

$w = [(P \times 10) - 100]$ mm..(Formula 7.2)

where:
w = the minimum width of stair (w must not be less than 1000 mm)
P = the number of people on the most heavily occupied storey.

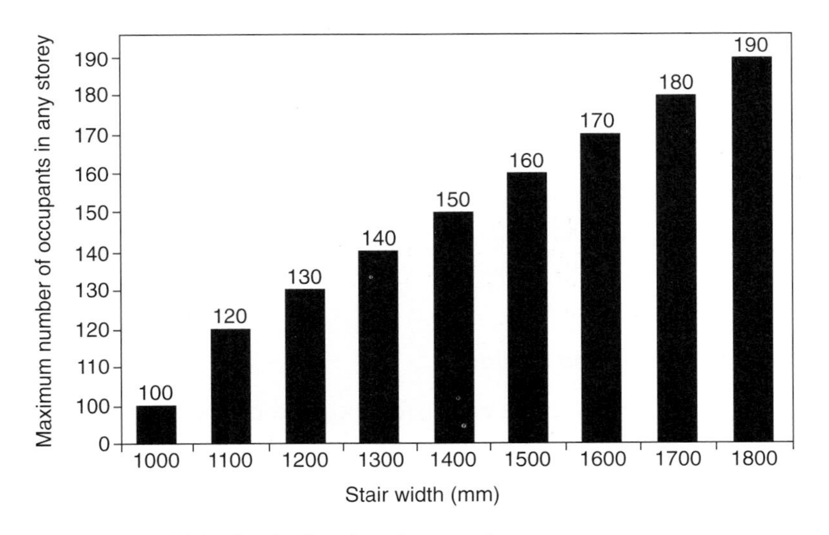

Fig. 7.29 Minimum width of stairs for phased evacuation.

7.15.8 Provision and protection of stairs

To be effective as an area of relative safety during a fire, escape stairs need to have an adequate standard of fire protection. This relates not only to the presence of fire-resisting enclosures but also to the provision of protected lobbies, corridors and final exits.

Fire-resisting enclosures

Each internal escape stair should be a protected stair situated in a fire-resisting enclosure. Additional measures may also be necessary for a stairway which is also a protected shaft (penetrating one or more compartment floors, see AD B section 9) or a firefighting shaft (see AD B Section 18). However, this does not preclude the provision of an accommodation stair (i.e. a stair which is provided for the convenience of occupants and is additional to those required for means of escape) if the design of the building so warrants it.

Exceptionally, an unprotected stair can form part of the internal escape route to a storey or final exit in low risk buildings if the number of people and the travel distance involved are very limited.

For example, BS 5588 Part 11 contains details in clause 10 of the use of an unprotected stair for means of escape in what are termed 'small premises'. Typically, the premises will be used as a small shop and will have two-storeys consisting of:

- ground and first floor; or
- basement and ground floor.

The following conditions will also need to be complied with.

- The maximum floor area in any storey must not exceed 90 m^2.
- The maximum direct travel distance from any point in the ground storey to the final exit must not exceed 18 m.
- The maximum direct travel distance from any point in the basement or first storey to the stair must not exceed 12 m.
- The stair must deliver into the ground storey not more than 3 m from the final exit.

It should be noted that restaurant or bar premises and those used for the sale, storage or use of highly flammable materials are barred from this arrangement.

Protected lobbies and corridors

Generally, protected lobbies or corridors should be provided at all levels including basements (but not at the top storey) where:

- the building has a single stair and there is more than one floor above or below the ground storey (except for small premises, see above)
- the building has a floor which is more than 18 m above ground
- the building is designed for phased evacuation
- the option has been taken to not discount one stairway when calculating stair widths (see section 7.15.7 above).

In these cases an alternative to a protected lobby or corridor is the use of a smoke control system designed in accordance with BS 5588: Part 4.

Protected lobbies are also needed where:

- the stairway is a firefighting stair (see AD B5)
- the stairway serves a place of special fire hazard (i.e. oil-filled transformer and switch gear rooms, boiler rooms, storage space for fuel or other highly flammable substances, rooms housing a fixed internal combustion engine, and in schools – laboratories, technology rooms with open heat sources, kitchens and stores for PE mats or chemicals). In this case, the lobby should be ventilated by permanent vents with an area of at least 0.4 m^2, or should be protected by a mechanical smoke control system.

Final exits from protected stairs

Ideally, every protected stairway should discharge directly to a final exit, i.e. it should be possible to leave the staircase enclosure and immediately reach a place of safety outside the building and away from the effects of fire.

Obviously, it is not always possible to achieve this ideal, especially where the building design calls for stairs to be remote from external walls. Therefore it is permissible for a protected stairway to discharge into a protected exit passageway, which in turn leads to a final exit from the building. Such a passageway can contain doors (e.g. to allow people on the ground floor to use the escape route) but they will need to be fire doors in order to maintain fire integrity and may need to be lobbied if the stairway needs to be served by lobbies. Therefore, if the exit route from a stair also picks up occupants of the ground and/or basement storeys, it may need to be increased in width accordingly. Thus, the width of the protected exit passageway will need to be designed in accordance with Table 7.3, for the estimated numbers of people that will use it in an emergency.

Sometimes the design of the building will call for two protected stairways or protected exit passageways to be adjacent to each other. Where this happens they should be separated by an imperforate enclosure.

Restrictions on the use of space in protected stairways

Since a protected stairway is considered to be a place of relative safety, it should be free of potential sources of fire. Therefore, the facilities that may be included in protected stairways are restricted to the following.

- Washrooms or sanitary accommodation provided that the accommodation is not used as a cloakroom. The only gas appliances that may be installed are water heaters or sanitary towel incinerators.
- A lift well, on condition that the stairway is not a firefighting stair.
- An enquiry office or reception desk area at ground or access level of not more than 10 m^2, provided that there is more than one stair serving the building.
- Fire-protected cupboards, provided that there is more than one stair serving the building.
- Gas service pipes and meters, but only if the gas installation is in accordance with

the requirements for installation and connection set out in the *Pipelines Safety Regulations 1996,* SI 1996/825 and the *Gas Safety (Installation and Use) Regulations 1998* SI 1998/2451.

Protection of external walls of protected stairways

If a protected stairway is situated on the external wall of a building it is not necessary for the external part of the enclosure to be fire-protected and in many cases it may be fully glazed. This is because fires are unlikely to start in protected stairways. Therefore these areas will not contribute to the radiant heat from a building fire which might put at risk another building.

In some building designs the stairway may be situated at an internal angle in the building façade (see Fig. 7.30) and may be jeopardised by smoke and flames coming from windows in the facing walls. This may also be the case if the stair projects from the face of the building (see Fig. 7.31) or is recessed into it. In these cases any windows or other unprotected areas in the face of the building and in the stairway should be separated by at least 1800 mm of fire-resisting construction. This provision also applies to flats and maisonettes.

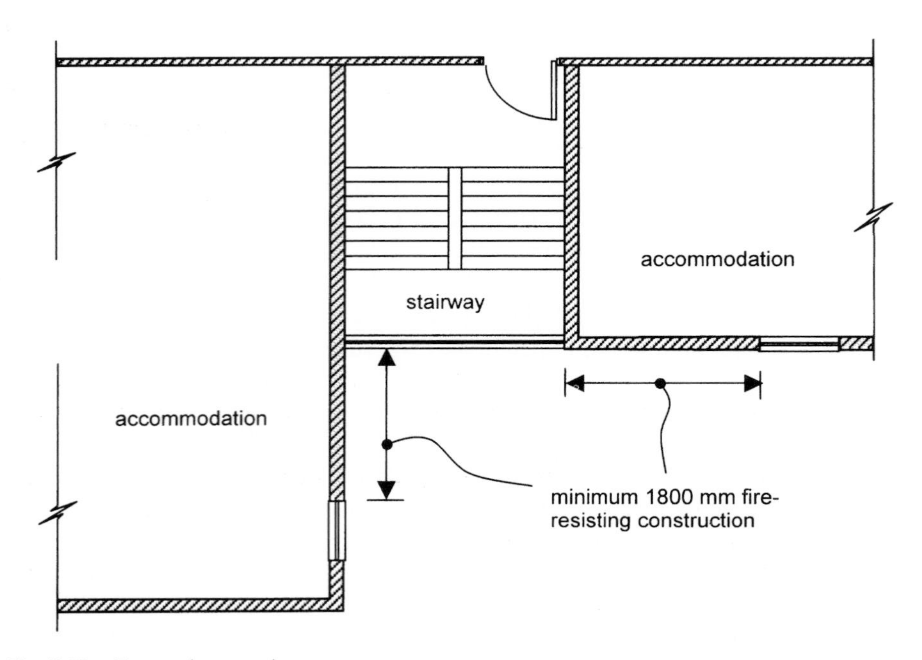

Fig. 7.30 External protection.

7.15.9 Basement stairs

Basement fires are particularly serious since combustion products tend to rise and find their way into stairways unless other smoke venting measures are taken. (See AD B5, section 7.30.13 below). Therefore, it is necessary to take additional pre-

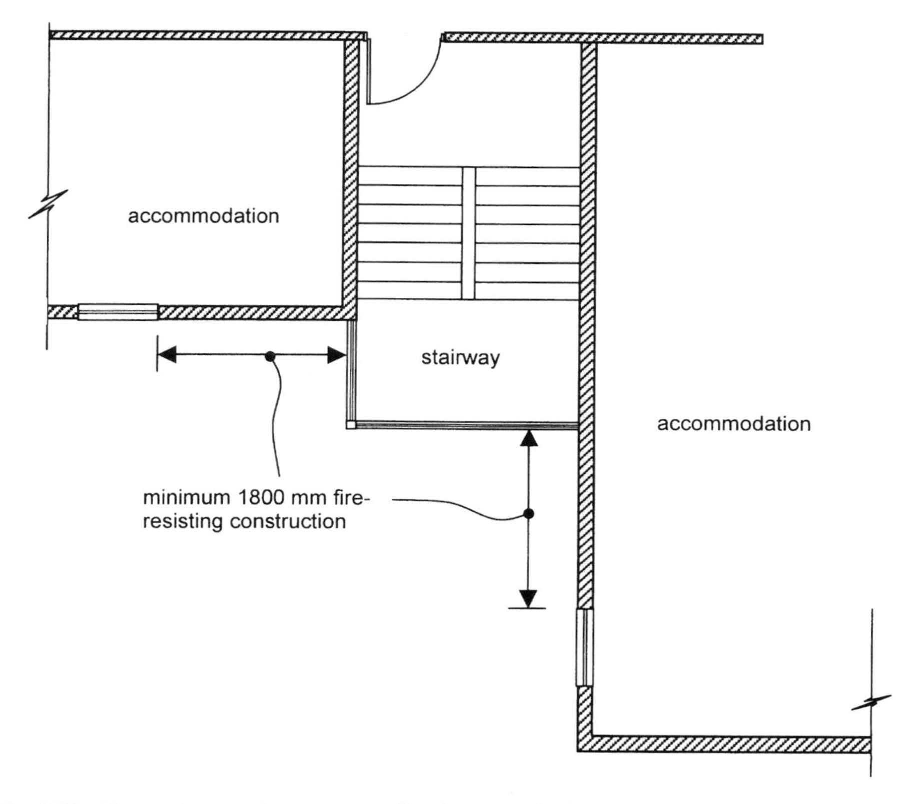

Fig. 7.31 External protection – protected stairway projecting beyond face of building.

cautions to prevent a basement fire endangering upper storeys in a building as follows.

- In most buildings with only one escape stair serving the upper storeys, this stair should not continue down to the basement, i.e. the basement should be served by a separate stair.
- In buildings containing more than one escape stair, at least one of the stairs should terminate at ground level and not continue down to the basement. The other stairs may terminate at basement level on condition that the basement accommodation is separated from the stair(s) by a protected lobby or corridor at basement level.

These provisions apply to all buildings, including flats and maisonettes.

7.15.10 External escape stairs

External escape stairs have long been used to provide additional vertical means of escape where parts of the building would otherwise be contained in long dead-ends or would be too distant from internal stairways. It is uncommon for such stairways to be fully protected from the elements therefore, they are not normally used for everyday access and egress around the building. Thus it can be argued that external escape stairs are not subject to the requirements of Part K of Schedule 1 to the Building Regulations 2000 because they do not form part of a building.

In buildings other than dwellings, an external escape stair may be used as an alternative means of escape provided that there is more than one escape route available from a storey, or part of a building, and the following conditions are met.

- There is at least one internal escape stair available from every part of each storey (plant areas excluded).
- It is not intended for use by members of the public if installed in assembly and recreation buildings (PG 5).
- It serves only office or residential staff accommodation if installed in an institutional building (PG 2(a)).
- If it is more than 6 m in vertical extent it is sufficiently protected from adverse weather. This does not necessarily mean that full enclosure will be necessary. The stair may be located so that protection may be obtained from the building itself. In deciding on the degree of protection it is necessary to consider the height of the stair, the familiarity of the occupants with the building, and the likelihood of the stair becoming impassable as a consequence of adverse weather conditions.
- Any part of the building (including windows and doors etc) which is within 1.8 m of the escape route from the stair to a place of safety should be protected with fire-resisting construction. This does not apply if there is a choice of routes from the foot of the stair thereby enabling the people escaping to avoid the effects of fire in the adjoining building. Additionally, any part of an external wall which is within 1.8 m of an external escape route (other than a stair) should be of fire-resisting construction up to a height of 1.1 m from the paving level of the route.
- All the doors which lead onto the stair should be fire-resisting and self-closing. This does not apply to the only exit door to the landing at the head of a stair which leads downward.
- Any part of the external envelope of the building which is within 1.8 m of (and 9 m vertically below) the flights and landings of the stair, should be of fire-resisting construction. This 1.8 m dimension may be reduced to 1.1 m above the top landing level provided that this is not the top of a stair up from basement level to ground.
- Any glazing which is contained within the fire-resisting areas mentioned above should also be fire-resisting in terms of maintaining its integrity in a fire, and be fixed shut. (For example, Georgian wired glass is adequate; it does not also have to meet the requirements for insulation).

7.16 General recommendations common to all buildings except dwellinghouses

Section 6 of AD B1 gives general guidance on a number of features of escape routes which apply to all buildings, except dwellinghouses, concerning:

- the standard of protection necessary for the elements enclosing the means of escape
- the provision of doors
- the construction of escape stairs
- the position and design of final exits
- lighting and signing
- mechanical services including lift installations
- protected circuits for the operation of equipment in the event of fire
- refuse chutes and storage
- the provision of fire safety signs.

These recommendations should be read in conjunction with the provisions described above for flats and maisonettes, and other buildings except dwellinghouses.

7.16.1 Protection of escape routes – standards and general constructional provisions

Those parts of a means of escape which are required by Part B1 to be fire-resisting should comply with the recommendations given in AD B3 (Internal fire spread – structure) or AD B5 (Access and facilities for the fire service) in addition to AD B1. In most cases 30 minutes fire protection is sufficient for the protection of a means of escape. The exceptions to this are when the element also performs a fire-separating function, or separates areas of different fire risk, such as:

- a compartment floor
- a compartment wall
- an external wall
- a protected shaft, or
- a firefighting shaft.

In these cases the element should achieve the standard of fire resistance given in Table A2 of Appendix A to Approved Document B. This may require considerably more than 30 minutes fire resistance (the fire resistance periods in Table A2 range from 30 minutes to 120 minutes).

The following general constructional provisions should also be met.

- Fully enclosed walk-in store rooms in shops should be separated from retail areas with 30 minute fire-resisting construction (see Table A1 of Appendix A of AD B) if they are sited so as to prejudice the means of escape. This does not apply if the store room is fitted with an automatic fire detection and alarm system or sprinklers.
- Glazed elements in fire-resisting enclosures and doors, which are only able to meet the requirements for integrity in the event of a fire, will be limited in area to the amounts shown in Table A4 of Appendix A of AD B (see section 7.22.3).
- There are no limitations on the use of glazed elements that can meet both the integrity and insulation performance recommendations of AD B1. However,

there may be some restrictions on the use of glass in firefighting stairs and lobbies in BS 5588: *Fire precautions in the design, construction and use of buildings* Part 5:1991 *Code of practice for firefighting stairways and lift* under the recommendations for robust construction. This is referred to in AD B5 and is mentioned below.

- Glazed elements may also need to comply with AD N (Chapter 18).
- All escape routes should have a minimum headroom of 2 m. The only projections allowed below this are for door frames.
- The floors of escape routes, including the surfaces of steps and ramps, should be chosen so that they are not unduly slippery when wet.
- Sloping floors or tiers should not have a pitch greater than 35° to the horizontal.
- Further guidance on the provision of ramps, stairs, aisles and gangways may be found in AD K (Chapter 15) and AD M (Chapter 17).

7.16.2 The provision of doors on escape routes

The time taken to pass through a closed door can be critical when escaping from a building in a fire situation. Doors on escape routes should be readily openable if undue delay in escaping from a building is to be avoided. They should also comply with the following general provisions.

- Doors on escape routes often need to be fire-resisting. This means that certain test criteria and performance standards as set out in Appendix B of Approved Document B (Table B1) will need to be met (see section 7.22.3 below).
- In general, escape doors should open in the direction of the means of escape where it is reasonably practicable to do so and should always do so if more than 60 people are likely to use the door in an emergency. However, for some industrial activities where there is a very high risk with potential for rapid fire growth it may be necessary for escape doors to open in the direction of escape for lower occupant numbers than 60. The exact figure will depend on the individual circumstances of the case and there is no specific guidance laid down in AD B.
- Ideally, doors on escape routes (whether or not they are fire doors) should not be fitted with fastenings unless these are simple to use and can be operated from the side of the door which is approached by people escaping. Any fastenings should be able to be operated without a key and without having to operate more than one mechanism, however this does not prevent doors being fitted with ironmongery which allows them to be locked when the rooms are empty. For example, this would permit a hotel bedroom to be fitted with a lock which could be operated from the outside with a key and from the inside by a knob or lever.
- Where security of final exit doors is important, as in Assembly and Recreation (PG 5) and Shop and Commercial (PG 4) buildings, panic bolts may be used. Additionally, it is accepted that in non-residential buildings it is appropriate for final exit doors to be locked when the building is empty. Clearly, a good deal of responsibility must be placed on management procedures for the safe use of these locks.

- Recommendations for self-closers and hold-open devices for fire doors are contained in Appendix B of AD B (see section 7.22.3 below).
- Doors on escape routes should swing through at least 90° to open, and should not reduce the effective width of any escape route across a landing. The swing should be clear of any changes in floor level, although a single step or threshold on the line of a door opening is permitted.
- Any door that opens towards a corridor or stairway should be recessed so that it does not encroach on or reduce the effective width of the corridor or stairway.
- Doors on escape routes which subdivide corridors, or are hung to swing in two directions, should contain vision panels. (See also Approved Document M and Chapter 17 of this book for vision panels in doors across accessible corridors).
- If revolving or automatic doors, or turnstiles, are placed across an escape route it is possible that they might obstruct the passage of people escaping. Therefore, they should not be placed across an escape route unless:
  - (a) in the case of automatic doors which are the correct width for the design of the route they:
    - (i) will fail safely to become outward opening from any position of opening, or
    - (ii) are provided with a monitored failsafe system for opening the doors in the event of mains power failure, or
    - (iii) fail safely in the open position in the event of mains power failure, or
  - (b) they have non-automatic swing doors of the required width adjacent to them which can provide an alternative exit.

7.16.3 The construction of escape stairs – conventional stairs

Escape stairs and their associated landings in certain high risk situations or buildings require the extra safeguard of being constructed in materials of limited combustibility. These are composite materials (such as plasterboard) which include combustible materials in their composition so that they cannot be classed as totally non-combustible. When exposed as linings to walls or ceilings they must achieve certain low flame spread ratings.

This recommendation applies in the following cases:

- where a building has only one stair serving it (this does not apply to two and three-storey flats and maisonettes)
- where a stair is located in a basement storey (except if it is a private stair in a maisonette)
- to any stair serving a storey in a building which is more than 18 m above ground or access level
- to any external stair (except where it connects the ground floor or paving level to a floor or flat roof which is not more than 6 m above ground)
- if the stair is a firefighting stair.

In all the above, except for the firefighting stair, it is permissible to add combustible materials to the upper surface of the stair.

Where possible, single steps should be avoided on escape routes unless prominently marked, since they can cause falls. It is permissible though, to have a single step on the line of a doorway.

7.16.4 The construction of escape stairs – special stairs and ladders

Although spiral and helical stairs, and fixed ladders are not as inherently safe as conventional stairs they may be used as part of a means of escape if the following restrictions are observed.

- Spiral and helical stairs should be designed in accordance with BS 5395 *Stairs, ladders and walkways,* Part 2:1984 *Code of practice for the design of helical and spiral stairs.* They are not suitable for use by pupils in schools, and if used by members of the public they should be type E (public) stair from the above standard.
- Fixed ladders are not suitable as a means of escape for members of the public. They should only be used to access areas which are not normally occupied, such as plant rooms, where it is not practical to provide a conventional stair. They should be constructed of non-combustible materials.

7.16.5 The position and design of final exits

Final exits should not be narrower than the escape routes they serve and should be positioned to facilitate evacuation of people out of and away from the building. This means that they should be:

- positioned so that rapid dispersal of people is facilitated to a street, passageway, walkway or open space clear of the effects of fire and smoke. The route from the building should be well defined and guarded if necessary;
- clearly apparent to users. This is very important where stairs continue up or down past the final exit level in a building; and
- sited so that they are clear of the effects of fire from risk areas in buildings such as basements (e.g. outlets for basement smoke vents), and openings to transformer chambers, refuse chambers, boiler rooms and other similar risk areas.

7.16.6 Lighting and signing

All escape routes should have adequate artificial lighting. In certain cases escape lighting which illuminates the route if the mains supply fails, should also be provided. These are listed in Table 7.5 below which is based on Table 9 to AD B1.

The lighting to escape stairs will also need to be on a separate circuit from that which supplies any other part of the escape route.

Standards for installation of escape lighting systems are given in BS 5266 *Emergency lighting,* Part 1:1988 *Code of practice for the emergency lighting of premises other than cinemas and certain other specified premises used for entertainment*, or CP 1007: 1955 *Maintained lighting for cinemas.*

Table 7.5 Provision of escape lighting.

Purpose Group	Description of building or part	Areas where escape lighting is required	Areas where escape lighting is *not* required
1(a) 2(a) 2(b)	Flat or maisonette Institutional Other Residential	All common escape routes (including external routes)	Common escape routes in two-storey flats. Dwellinghouses in PG 1(b) & 1(c)
3 4 6 7(a)	Office Shop and Commercial[1] Industrial Storage and other non-residential	(i) Underground or windowless accommodation (ii) Stairways in a central core or serving storey(s) over 18 m from ground level (iii) Internal corridors more than 30 m long (iv) Open-plan areas exceeding 60 m^2	
4 7(b)	Shop and Commercial[2] Car parks which admit the public	All escape routes (including external routes)	Escape routes in shops[3] of 3 or less storeys (with no sales floor exceeding 280 m^2)
5	Assembly and Recreation	All escape routes (including external routes), and accommodation	(i) Accommodation open on one side to view sport or entertainment during normal daylight hours (ii) Parts of school buildings with natural light and used only during normal school hours
All	All	(i) Windowless toilet accommodation with floor area not exceeding 8 m^2 (ii) All toilet accommodation with floor area exceeding 8 m^2 (iii) Electricity and generator rooms (iv) Switch room/battery room for emergency lighting system (v) Emergency control room	Dwellinghouses in PG 1(b) & 1(c)

Notes:
1. Those parts of the premises where the public are not admitted.
2. Those parts of the premises where the public are admitted.
3. Any 'shop' (see definition section 7.3 above) which is a restaurant or bar will require escape lighting as indicated in column 3.

Except in dwellinghouses, flats and maisonettes, emergency exit signs should be provided to every escape route. It is not necessary to sign exits which are in ordinary, daily use. The exit should be distinctively and conspicuously marked by a sign with letters of adequate size complying with the *Health and Safety (Safety signs and signals) Regulations 1996 (SI 1996/2341)*. In general, these regulations may be satisfied by signs containing symbols or pictograms which are in accordance with

BS 5499 *Fire safety signs, notices and graphic symbols,* Part 1: 1990 *Specification for fire safety signs.*

In some buildings other legislation may require additional signs.

7.16.7 Lift installations

Lifts are not normally used for means of escape since there is always the danger that they may become immobilised due to power failure and may trap the occupants. It is possible to provide lifts as part of a management plan for evacuating disabled people if the lift installation is appropriately sited and protected. It should also contain sufficient safety devices to ensure that it remains usable during a fire. Further details may be found in BS 5588: *Fire precautions in the design, construction and use of buildings* Part 8: 1988 *Code of practice for means of escape for disabled people.*

A further problem with lifts is that they connect floors and may act as a vertical conduit for smoke or flames thus prejudicing escape routes. This may be prevented if the following recommendations are observed.

- Lift wells should be enclosed throughout their height with fire-resisting construction if their siting would prejudice an escape route. Alternatively, they should be contained within the enclosure of a protected stairway.
- Any lift well which connects different compartments in a building should be constructed as a protected shaft.
- In buildings where escape is based on the principles of phased or progressive horizontal evacuation, if the lift well is not within the enclosure of a protected stairway, its entrance should be separated from the floor area on each storey by a protected lobby.
- Similarly, unless the lift is in a protected stairway enclosure, it should be approached through a protected lobby or corridor:
  - (a) if it is situated in a basement or enclosed car park, or
  - (b) where the lift delivers directly into corridors serving sleeping accommodation if any of the storeys also contain high fire risk areas such as kitchens, lounges or stores.
- A lift should not continue down to serve a basement if there is only one esca stairway in the building (since smoke from a basement fire might be able to prejudice the escape routes in the upper storeys) or if it is in an enclosure to a stairway which terminates at ground level.
- Lift machine rooms should be located over the lift shaft wherever possible. Where a lift is within the only protected stairway serving a building and the machine room cannot be located over the lift shaft, then it should be sited outside the protected stairway. This is to prevent smoke from a fire in the machine room from blocking the stair.
- Wall-climber and feature lifts often figure in large volume spaces such as open malls and atria. Such lifts do not have a conventional well and may place their occupants at risk if they pass through a smoke reservoir. Care will be needed in the design in order to maintain the integrity of the reservoir and protect the occupants of the lift.

7.16.8 Mechanical ventilation and air-conditioning services

Mechanical ventilation systems should be designed so that in a fire:

- air is drawn away from protected escape routes and exits, or
- the system (or the appropriate part of it) is closed down.

Systems which recirculate air should comply with BS 5588: *Fire precautions in the design, construction and use of buildings* Part 9:1989 *Code of practice for ventilation and air conditioning ductwork* for operation under fire conditions.

Recommendations for the use of mechanical ventilation in a place of assembly are given in BS 5588: *Fire precautions in the design, construction and use of buildings* Part 6:1991 *Code of practice for assembly buildings*.

Guidance on the design and installation of mechanical ventilation and air-conditioning plant is given in BS 5720: 1979 *Code of practice for mechanical ventilation and air conditioning in buildings*.

Where a pressure differential system is installed in a building (in order to keep smoke from entering the means of escape) it should be compatible with any ventilation or air-conditioning systems in the building, when operating under fire conditions.

7.16.9 Protected circuits for the operation of equipment in the event of fire

Protected power circuits are provided in situations where it is critical that the circuit should continue to function during a fire. For example, this will apply where the circuits provide power to:

- fire extinguishing systems
- smoke control systems
- sprinkler systems
- firefighting shaft systems (such as firefighting lifts)
- motorised fire shutters
- CCTV systems installed for monitoring means of escape
- data communications systems that link fire safety systems.

The cable used in a protected power circuit for operation of equipment in the event of fire should:

- meet the requirements for classification as CWZ in accordance with BS 6387: 1994 *Specification for performance requirements for cables required to maintain circuit integrity under fire conditions*;
- follow a route which passes through parts of the building in which there is negligible fire risk; and
- be separate from circuits which are provided for other purposes.

7.16.10 Refuse chutes and storage

Fires in refuse chute installations are extremely common and they are required to be built of non-combustible materials in Approved Document B3. So that escape routes are not jeopardised, refuse chutes and rooms for refuse storage should:

- be separated, by fire-resisting construction, from the rest of the building; and
- not be located in protected lobbies or stairways.

Rooms which store refuse or contain refuse chutes should:

- be approached directly from the open air; or
- be approached via a protected lobby provided with at least 0.2 m^2 of permanent ventilation.

Refuse storage chamber access points should be sited away from escape routes, final exits and windows to dwellings.

Refuse storage chambers, chutes and hoppers should be sited and constructed in accordance with BS 5906:1980 *Code of practice for storage and on-site treatment of solid waste from buildings*.

7.17 Alternative approach to the provision of means of escape in selected premises

Reference has been made throughout the text above to the use of design guides, other than AD B1, in the provision of means of escape. There are certain specialised types of premises where it is recommended that this other guidance be used in preference to the more general guidance in AD B1. Additionally, whilst AD M covers access and facilities for disabled people, there are no specific recommendations for means of escape for disabled people in AD B1. It may not be necessary to provide special structural measures to aid the escape of disabled people other than suitable management arrangements to cater for emergencies. Where it is felt that special provisions for means of escape are desirable, reference should be made to BS 5588: Part 8: 1988 *Code of practice for means of escape for disabled people*. Advice is given in the Code on refuges, evacuation lifts and the need for efficient management of escape.

7.17.1 Means of escape in health care premises and hospitals

Health care premises such as hospitals, nursing homes and homes for the elderly differ from other premises in that they contain people who are bed-ridden or who have very restricted mobility. In such buildings it is unrealistic to expect that the patients will be able to leave without assistance, or that total evacuation of the building is feasible.

Hence, the approach to the design of means of escape in these premises will demand a very different approach to that embodied in much of AD B1, and NHS Estates has prepared a set of guidance documents under the general title of *Firecode* for use in health care buildings. These documents are also applicable to non-National Health Service premises and are as follows.

- Means of escape in new hospitals – *Firecode. Health Technical Memorandum (HTM) 81, Fire precautions in new hospitals.* NHS Estates, HMSO, 1996.
- Work that affects the means of escape in existing hospitals – *Firecode. HTM 85 Fire precautions in existing hospitals.*
- Existing residential care premises – *Draft guide to fire precautions in existing residential care premises.* Home Office/Scottish Home and Health Department, 1983. This document is under review.

If an existing house of not more than two storeys is converted for use as an unsupervised Group Home for not more than seven mentally impaired or mentally ill people, it may be regarded as a dwellinghouse (PG 1(c)) if it has means of escape designed in accordance with *Firecode. HTM 88 Guide to fire precautions in NHS housing in the community for mentally handicapped (or mentally ill) people.* If the building is new, it might be better to regard it as being in PG 2(b) Residential (Other).

It should be noted that the *Firecode* documents contain managerial and other fire safety provisions which are outside the scope of the Building Regulations.

Progressive horizontal evacuation

Since total evacuation of health care premises is inappropriate in most cases, the *Firecode* documents contain guidance on progressive horizontal evacuation of premises. In-patients are evacuated, in the event of fire, to adjoining compartments or sub-divisions of compartments, the object being to provide a place of relative safety within a short distance. If necessary, further evacuation can be made from these safe places but under less pressure of time.

Section 4 of AD Bl gives a limited amount of guidance on progressive horizontal evacuation in some other residential buildings to which the *Firecode* documents do not apply, such as residential rest homes. When storeys are being planned for progressive horizontal evacuation, the following conditions should be considered when they are being divided into compartments.

- The compartment into which the evacuation is to take place should be large enough to take the occupants from the adjoining compartment and its own occupants. The design occupancy figures should be used to assess the total number of people involved.
- Each compartment should have an alternative escape route which is independent of the adjoining compartment. This may be through another compartment which also has an independent escape route.

7.17.2 Sheltered housing

Sheltered housing schemes which consist of specially adapted groups of houses, bungalows or two-storey flats with warden assistance and few communal facilities need not be treated differently from other one or two-storey houses or flats.

Sheltered accommodation in the institutional or other residential purpose groups would need to comply with the provisions for other buildings listed above. Additional guidance may be found in BS 5588: Part 1: 1990, clause 17.

7.17.3 Assembly buildings

A principal problem with assembly buildings is the difficulty in escaping from fixed seating. This occurs in theatres, concert halls, conference centres and at sports events. Specific guidance on this may be obtained from:

- sections 3 and 5 of BS 5588: Part 6: 1991 *Code of practice for places of assembly*, where guidance on the spacing of fixed seating and other aspects of means of escape may be found;
- *Guide to fire precautions in existing places of entertainment and like premises*. Home Office/Scottish Home and Health Department, HMSO, 1990; and
- *Guide to safety at sports grounds*, Department of National Heritage/Scottish Office. HMSO, 1997, for sports stadia, etc.

7.17.4 Schools and other educational buildings

As a result of the coming into force of the *Education (Schools and Further and Higher education)(Amendment) Regulations 2001* on 1 April 2001, schools in England are no longer exempt from the Building Regulations. Therefore, the fire safety objectives of the Department for Education and Skills may be met by following the guidance contained in AD B outlined in this chapter. It should be noted however that the Department for Education and Skills has issued guidance on constructional standards for schools (see *Guidance on the Constructional Standards for Schools*: DfES 142/2001-09-11), indicating where it is necessary to supplement the Approved Documents to take account of the specific requirements of schools. It is proposed that in due course the Approved Documents will be amended to take account of these. Further information on this can be obtained from the School Premises Team, Department for Education and Skills, Caxton House, Room 762, 6–12 Tothill Street, London SW1H 9NA, Tel 020 7273 6023, E-mail: premises.schools@dfes.gsi.gov.uk.

7.17.5 Shops and shopping complexes

British Standard BS 5588: Part 11: 1997 *Code of practice for shops, offices, industrial, storage and other similar buildings* may be used instead of AD B1.

Shopping complexes are not covered adequately by AD B1 and should be designed in accordance with section 4 of BS 5588: Part 10: 1991 *Code of practice for shopping complexes*. It should be noted that BS 5588: Part 10 applies more restrictive provisions to units with only one exit in covered shopping complexes than may be found in BS 5588: Part 11.

7.18 Internal fire spread (linings)

7.18.1 Introduction

Although the linings of walls and ceilings are unlikely to be the materials first ignited in a fire (this is more likely to occur in furniture and fittings), they can significantly affect the spread of fire and its rate of growth. This is especially true in circulation areas where rapid fire spread may prevent occupants from escaping. Part B2 of Schedule 1 to the 2000 Regulations seeks to control these surface linings.

7.18.2 Control of wall and ceiling linings

The spread of fire within a building may be inhibited by paying attention to the lining materials used on walls, ceilings, partitions and other internal structures. These linings must:

- offer adequate resistance to spread of flame over their surfaces; and
- if ignited, have a rate of heat release or fire growth which is reasonable in the circumstances.

It should be noted that floors and stairs are not covered by the requirements since they are not usually involved in a fire until it is well established. Consequently, they will not contribute to fire spread in the early stages of a fire.

7.18.3 Methods of test

In order to meet the requirements of AD B2 it is necessary for materials or products to meet certain levels of performance in appropriate tests.

Under National classifications the surface spread of flame characteristics of a material may be determined by testing it in accordance with the method specified in BS 476 *Fire tests on building materials and structures*: Part 7

- 1971 *Surface spread of flame tests for material; or*
- 1987 *Method for classification of the surface spread of flame of product; or*
- 1997 *Method of test to determine the classification of surface spread of flame of products.*

A strip of the material under test is placed with one end resting against a furnace and the rate at which flames spread along the material is measured.

Materials or products are thus placed in Classes 1, 2, 3 or 4, Class 1 representing a surface of very low flame spread. Class 4 (a surface of rapid flame spread) is not acceptable under the provisions of the approved document.

Under the European classifications, lining systems are classified in accordance with BS EN 1350: *Fire classification of construction products and building elements,* Part 1: 2002: *Classification using data from reaction to fire tests.* Materials or products are classified as A1, A2, B, C, D, E or F, with A1 being the highest. When a classification includes 's3, d2', it means that there is no limit set for smoke production and/or flaming droplets/particles. The relevant European test methods are specified as follows:

- BS EN ISO 1182:2002 *Reaction to fire tests for building products – Non combustibility test*;
- BS EN ISO 1716:2002 *Reaction to fire tests for building products – Determination of the gross calorific value*;
- BS EN 13823:2002 *Reaction to fire tests for building products – Building products excluding floorings exposed to the thermal attack by a single burning item*;
- BS EN ISO 11925 *Reaction to fire tests for building products,* Part: 2002 – *Ignitability when subjected to direct impingement of flame*; and
- BS EN 13238: 2001: *Reaction to fire tests for building products – conditioning procedures and general rules for selection of substrates.*

In the event of fire, some materials have a higher rate of heat release or ignite more easily than others. They are therefore more hazardous and this may mean a reduced time to flashover. In the National test, the way in which the rate of heat release may be assessed is contained in BS 476: Part 6: 1981 or 1989 *Method of test for fire propagation of products.*

The material or product is tested for a certain period of time in a furnace and is given two numerical indices related to its performance. The sub-index (i_1) is derived from the first three minutes of the test whilst the overall test performance is denoted by the index of performance (I).

7.18.4 Class 0 materials

In order to establish a high product performance classification for lining materials in high risk areas (such as circulation spaces), AD B2 recommends that materials in these areas should conform to either the Class 0 (National) standard or European Class B-s3, d2 or better. Class 0 is not a classification found in any British Standard test as such; however it is evident that it draws on BS 476 test results as the following definition shows.

The Class 0 standard will be achieved by any material or the surface of a composite product which:

(a) is composed of materials of limited combustibility (see below) throughout; or
(b) is a material of Class 1 which has an index of performance (I) of not more than 12 and a sub-index (i_1) of not more than 6.

7.18.5 Interpretation

The following terms are common to all parts of AD B but occur most frequently in AD B2, 3 and 4:

NON-COMBUSTIBLE MATERIAL – A material which has the highest level of reaction to fire performance when tested as follows:

- (National classes) to BS 476: Part 11; the material does not flame and there is no rise in temperature on either the centre (specimen) or furnace thermocouples; or
- (European classes) when classified as class A1 in accordance with BS EN13501 *Fire classification of construction products and building elements,* Part 1: 2002: *Classification using data from reaction to fire tests*:
  - (a) BS EN ISO 1182: 2002: *Reaction to fire tests for building products – Non-combustibility test*; and
  - (b) BS EN ISO 1716:2002 *Reaction to fire tests for building products – Determination of the gross calorific value.*

Table A6 from Appendix A of AD B (which is reproduced below) lists some examples of non-combustible materials and gives details of where they should be used.

MATERIALS OF LIMITED COMBUSTIBILITY – Materials in this group, whilst not regarded as non-combustible, would contribute little heat energy to a fire. Therefore, they can be used in situations where control of fire spread is essential, such as stairs to basements. They are defined in AD B, Table A7 (see below) when tested as follows:

- (National classes) to BS 476: Part 11, the material does not flame and there is no rise in temperature on either the centre (specimen) or furnace thermocouples; or
- (European classes) when classified as class A2-s3, d2 in accordance with BS EN13501 *Fire classification of construction products and building elements,* Part 1: 2002: *Classification using data from reaction to fire* test when tested to:
  - (a) BS EN ISO 1182: 2002: *Reaction to fire tests for building products – Non-combustibility test*, or
  - (b) BS EN ISO 1716: 2002 *Reaction to fire tests for building products – Determination of the gross calorific value*, and BS EN 13823: 2002, *Reaction to fire tests for building products – Building products excluding floorings exposed to the thermal attack by a single burning item.*

It should be noted that certain insulating materials in group (d) of the table are of a lower standard than the group (a), (b) or (c) materials and may only be used in the situations shown in items 9 and 10 of the table. It is of course permissible to use non-combustible materials whenever a recommendation for materials of limited combustibility is specified.

7.18.6 Materials for surface linings

As a guide to the materials which may be used for wall and ceiling linings, AD B lists in Table A8 (see below) the typical performance ratings for some generic materials and products. Test results for proprietary materials may be obtained from manufacturers and trade associations. However, small differences in detail (e.g. thickness, substrate, colour, form, fixings, adhesives, etc.) can significantly affect the rating. Therefore, the reference used to substantiate the spread of flame rating should be carefully checked to ensure that it is suitable, adequate and applicable to the construction to be used.

AD B, Appendix A

Table A6 Use and definitions of non-combustible materials

References in AD B guidance to situations where such materials should be used	Definitions of non-combustible materials	
	National class	**European class**
1. ladders referred to in the guidance to B1, paragraph 6.22 (see section 7.16.4) 2. refuse chutes meeting the provisions in the guidance to B3, paragraph 9.35c (see section 7.22.2) 3. suspended ceilings and their supports where there is provision in the guidance to B3, paragraph 10.13, for them to be constructed of non-combustible materials (see section 7.23.2) 4. pipes meeting the provisions in the guidance to B3, Table 15 (see section 7.24.1) 5. flue walls meeting the provisions in the guidance to B3, Diagram 39 (see section 7.24.3) 6. construction forming car parks referred to in the guidance to B3, paragraph 12.3 (see section 7.28.1)	a. Any material which when tested to BS 476: Part 11 does not flame nor cause any rise in temperature on either the centre (specimen) or furnace thermocouples b. Totally inorganic materials such as concrete, fired clay, ceramics, metals, plaster and masonry containing not more than 1% by weight or volume of organic material. (Use in buildings of combustible metals such as magnesium/aluminium alloys should be assessed in each individual case c. Concrete bricks or blocks meeting BS 6073: Part 1 d. Products classified as non-combustible under BS 476: Part 4	a. Any material classified as class A1 in accordance with BS EN 13501-1: 2002, Fire classification of construction products and building elements, Part 1 – Classification using data from reaction to fire tests b. Products made from one or more of the materials considered as Class A1 without the need for testing, as defined in Commission Decision 96/603/EC of 4th October 1996 establishing the list of products belonging to Class A1 'No contribution to fire' provided for in the Decision 94/611/EC implementing Article 20 of the Council Directive 89/106/EEC on construction products. None of the materials shall contain more than 1.0% by weight or volume (whichever is the lower) of homogeneously distributed organic material
		Note: The National classifications do not automatically equate with the equivalent classifications in the European column, therefore products cannot typically assume a European class unless they have been tested accordingly

7.18.7 Thermoplastic materials

A thermoplastic material is a synthetic polymer which has a softening point below 200°C when tested in accordance with BS 2782 Methods of testing plastics, Part 1 Thermal properties: Method 120A: 1990 Determination of the Vicat softening temperature of thermoplastics. If the thickness of the product to be tested is less than 2.5 mm then specimens for the test may be fabricated from the original polymer.

When used in isolation as a wall or ceiling lining, a thermoplastic material cannot be assumed to protect a substrate. The surface rating of both the substrate and the lining would need to meet the required classification. However, where the thermoplastic material is fully bonded to a non-thermoplastic substrate then only the surface rating of the composite would need to comply.

AD B, Appendix A

Table A7 Use and definitions of materials of limited combustibility.

References in AD.B guidance to situations where such materials should be used	Definitions of non-combustible materials	
	National class	**European class**
1. stairs where there is provision in the guidance to B1 for them to be constructed of materials of limited combustibility (see 6.19 and section 7.16.3) 2. materials above a suspended ceiling meeting the provision in the guidance to B3, paragraph 10.13 (see section 7.23.2) 3. reinforcement/support for fire-stopping referred to in the guidance to B3, see 11.13 (see section 7.24.4) 4. roof coverings meeting provisions: a. in the guidance to B3, paragraph 10.11 or (see section 7.23.2) b. in the guidance to B4, Table 17 or c. in the guidance to B4, Diagram 47 (see section 7.18.9) 5. roof deck meeting the provisions of the guidance to B3, Diagram 28a (see section 7.22.1) 6. class 0 materials meeting the provisions in Appendix A, paragraph 13(a) (see section 7.18.4) 7. ceiling tiles or panels of any fire protecting suspended ceiling (Type Z) in Table A3 (see section 7.19) 8. compartment walls and compartment floors in hospitals referred to in paragraph 9.32	a. Any non-combustible material listed in Table A6 b. Any material of density 300/kg/m^2 or more, which when tested to BS 476: Part 11, does not flame and the rise in temperature on the furnace thermocouple is not more than 20°C c. Any material with a non-combustible core at least 8 mm thick having combustible facings (on one or both sides) not more than 0.5 mm thick. (Where a flame spread rating is specified, these materials must also meet the appropriate test requirements)	a. Any material listed in Table A6 b. Any material/product classified as Class A2-s3, d2 or better in accordance with BS EN 13501-1: 2002, *Fire classification of construction products and building elements, Part 1 – Classification using data from reaction to fire tests*
9. insulation material in external wall construction referred to in paragraph 13.7 (see section 7.27) 10. Insulation above any fire-protecting suspended ceiling (Type Z) in Table A3 (see section 7.19)	Any of the materials (a), (b) or (c) above, or: d. Any material of density less than 300 kg/m^3, which when tested to BS 476: Part 11, does not flame for more than 10 seconds and the rise in temperature on the centre (specimen) thermocouple is not more than 35°C and on the furnace thermocouple is not more than 25°C	Any of the materials/products (a) or (b) above
		Notes: 1. The National classifications do not automatically equate with the equivalent classifications in the European column, therefore products cannot typically assume a European class unless they have been tested accordingly 2. When a classification includes 's3, d2', this means that there is no limit set for smoke production and/or flaming droplets/particles

AD B, Appendix A

Table A8 Typical performance ratings of some generic materials and products.

Rating	Material or product
Class 0 (National)	1. any non-combustible material or material of limited combustibility. (Composite products listed in Table A7 must meet test requirements given in Appendix A, paragraph 13(b)) 2. brickwork, blockwork, concrete and ceramic tiles 3. plasterboard (painted or not with a PVC facing not more than 0.5 mm thick) with or without an air gap or fibrous or cellular insulating material behind 4. woodwool cement slabs 5. mineral fibre tiles or sheets with cement or resin binding
Class 3 (National)	6. timber or plywood with a density more than 400 kg/m^3, painted or unpainted 7. wood particle board or hardboard, either untreated or painted 8. standard glass reinforced polyesters
Class A1 (European)	9. any material that achieves this class and is defined as 'classified without further test' in a published Commission Decision
Class A2-s3, d2 (European)	10. any material that achieves this class and is defined as 'classified without further test' in a published Commission Decision
Class B-s3, d2 (European)	11. any material that achieves this class and is defined as 'classified without further test' in a published Commission Decision
Class C-s3, d2 (European)	12. any material that achieves this class and is defined as 'classified without further test' in a published Commission Decision
Class D-s3, d2 (European)	13. any material that achieves this class and is defined as 'classified without further test' in a published Commission Decision

Notes (National):

1. Materials and products listed under Class 0 also meet Class 1.
2. Timber products listed under Class 3 can be brought up to Class 1 with appropriate proprietary treatments.
3. The following materials and products may achieve the ratings listed below. However, as the properties of different products with the same generic description vary, the ratings of these materials/products should be substantiated by test evidence.
   Class 0 – aluminium faced fibre insulating board, flame retardant decorative laminates on a calcium silicate board, thick polycarbonate sheet, phenolic sheet and UPVC;
   Class 1 – phenolic or melamine laminates on a calcium silicate substrate and flame retardant decorative laminates on a combustible substrate.

Notes (European):

For the purposes of the Building Regulations

1. Materials and products listed under Class A1 also meet Classes A2-s3, d2, B-s3, d2, C-s3, d2 and D-s3, d2.
2. Materials and products listed under Class A2-s3, d2 also meet Classes B-s3, d2, C-s3, d2 and D-s3, d2.
3. Materials and products listed under Class B-s3, d2 also meet Classes C-s3, d2 and D-s3, d2.
4. Materials and products listed under Class C-s3, d2 also meet Class D-s3, d2.
5. The performance of timber products listed under Class D-s3, d2 can be improved with appropriate proprietary treatments.
6. Materials covered by the CWFT process (classification without further testing) can be found by accessing the European Commission's website via the link on the ODPM's web site www.odpm.gov.uk/bregs/cpd/index.htm.
7. The National classifications do not automatically equate with the equivalent classifications in the European column, therefore products cannot typically assume a European class unless they have been tested accordingly.
8. When a classification includes 's3, d2', this means that there is no limit set for smoke production and/or flaming droplets/particles.

Some thermoplastic materials can be tested under BS 476: Parts 6 and 7, and can be used in accordance with their ratings as described above. Alternatively, they may be classified as TP(a) rigid, TP(a) flexible, or TP(b), as described below, but their use would be restricted to rooflights, lighting diffusers, suspended ceiling panels and external window glazing (except in circulation areas). These uses are described more fully below.

TP(a) rigid means:

- rigid solid PVC sheet
- solid (i.e. not double- or multi-skin) polycarbonate sheet at least 3 mm thick
- multi-skinned rigid sheet made from uPVC or polycarbonate with a BS 476: Part 7 rating of Class 1
- any other rigid thermoplastic product which, when tested in accordance with BS 2782: Part 5: 1970 (1974): Method 508A, performs so that the flame extinguishes before reaching the first mark, and the duration of the flame or afterglow after removal of the burner does not exceed five seconds.

TP(a) flexible means:

- flexible products not more than 1 mm thick complying with the Type C requirements of BS 5867 *Specification for fabrics for curtains and drapes,* Part 2: 1980 *Flammability requirements*, when tested to BS 5438: Test 2, 1989. In the BS 5438 test, the flame should be applied to the specimens for 5,15,20 and 30 seconds respectively, although it is not necessary to include the cleansing procedure described in the British Standard.

TP(b) means:

- rigid solid polycarbonate sheet products less than 3 mm thick, or multi-skin polycarbonate sheet products which do not qualify as TP(a) by test
- any other product, a specimen of which between 1.5 mm and 3 mm thick, when tested in accordance with BS 2782: Part 5: 1970 (1974): Method 508A, has a rate of burning not exceeding 50 mm/minute.

If it is not possible to cut or machine a 3 mm thick test specimen from the product, then it is permissible to mould one from the original material used for the product.

Currently, no new guidance is possible on the assessment or classification of thermoplastic materials under the European system since there is no generally accepted European test procedure and supporting comparative data.

7.18.8 Specific recommendations for wall and ceiling linings

Table 10 from Section 6 of AD B2 gives the recommended flame spread classifications for the surfaces of walls and ceilings in any room or circulation space, for all building types.

Different standards are set for 'small rooms', which are totally enclosed rooms with floor area of not more than 4 m^2 in residential buildings or 30 m^2 in non-residential buildings, and for other rooms and circulation spaces.

When considering the performance of wall linings, window glazing and ceilings which slope at more than 70° to the horizontal are treated as a wall. Certain vertical surfaces are excluded from this definition, such as:

- doors, door frames and glazing in doors
- window frames and other frames containing glazing
- narrow members, such as architraves, cover moulds, picture rails and skirtings
- fireplace surrounds, mantle shelves and fitted furniture.

Similarly, ceiling surfaces include glazing and walls which slope at 70° or less to the horizontal, but exclude:

- trap doors and frames
- window frames, rooflight frames and other frames in which glazing is fitted
- narrow members, such as architraves, cover moulds, exposed beams and picture rails.

Therefore, bearing in mind the above definitions and exclusions the linings of walls and ceilings should conform to the following classifications:

- in circulation spaces in buildings other than dwellings (but including the common areas of flats and maisonettes) – not lower than Class 0 (or European Class B-s3, d2);
- in circulation spaces within dwellings – not lower than Class 1 (or European Class C-s3, d2);
- in any other rooms in any building (other than in small rooms as defined above) – not lower than Class 1 (or European Class C-s3, d2);
- in small rooms (see definition above) – not lower than Class 3 (or European Class D-s3, d2); and
- in domestic garages of area not exceeding 40 m^2 – not lower than Class 3. Above this floor area – not lower than Class 1 (or European Class D-s3, d2).

(Note: the National classifications do not automatically equate with the equivalent European classifications; therefore products cannot typically assume a European class, unless they have been tested accordingly. When a classification includes 's3, d2', this means that there is no limit set for smoke production and/or flaming droplets/particles).

Approved Document B2 allows certain variations from the strict lining classifications shown above provided that no lining is lower than Class 3 (National Class) or D-s3, d2 (European Class), as follows.

- Wall linings in rooms may be lower than Class 1 if their area does not exceed half the floor area of the room subject to the following maxima:

(a) in residential buildings – 20 m^2
(b) in non-residential buildings – 60 m^2.

- Plastic rooflights, and lighting diffusers fitted in suspended ceilings, may be lower than Class 0 or Class 1 (but not lower than Class 3) if they comply with recommendations for thermoplastic rooflights or diffusers shown below. Rooflights of other materials should comply with the lining classifications shown above.
- External windows to rooms may be glazed with a TP(a) rigid thermoplastic product, but not in circulation areas. However, internal glazing should comply with the lining classifications shown above.
- Suspended or stretched-skin ceilings made from thermoplastic material with a TP(a) flexible classification are permitted provided they do not form part of a fire-resisting ceiling. Each panel should be supported on all its sides and should not exceed 5 m^2 in area.

7.18.9 Rooflights and lighting diffusers

Rooflights, and lighting diffusers which form an integral part of a ceiling (i.e. not attached to the soffit or suspended beneath the ceiling), may:

- comply with the classification shown above; or
- consist of thermoplastic materials with a TP(a) rigid or TP(b) classification or at least a Class 3 rating if used under the conditions described in Table 11 from AD B2, and limited in extent and layout as illustrated in Diagram 24 from AD B2. (It should be noted that no guidance is currently possible on the performance requirements in the European fire tests as there is no generally accepted test and classification procedure.) These conditions and limitations are combined in Fig. 7.32 below. Rooflights made from these materials must not be used in a protected stairway.

The space in the ceiling above a lighting diffuser should comply with the lining classifications shown above for flame spread, according to the type of space below the ceiling. Lighting diffusers should only be used in fire-resisting or fire-protecting ceilings if they have been satisfactorily tested as part of the ceiling system being used to provide the appropriate fire protection.

The external surfaces of rooflights may also need to follow the recommendations for roofs contained in AD B4. Tables 18 and 19 of AD B4 give details of the limitations on the use of plastic rooflights in roofs. The main recommendations of the tables are combined in Table 7.6 and are summarised below.

- TP(a) rigid thermoplastic material may be used for a rooflight over any space except a protected stairway. Unless located in any of the buildings listed in (a) and (b) immediately below, the rooflight should be at least 6 m from any point on a boundary,
- Rooflights serving:

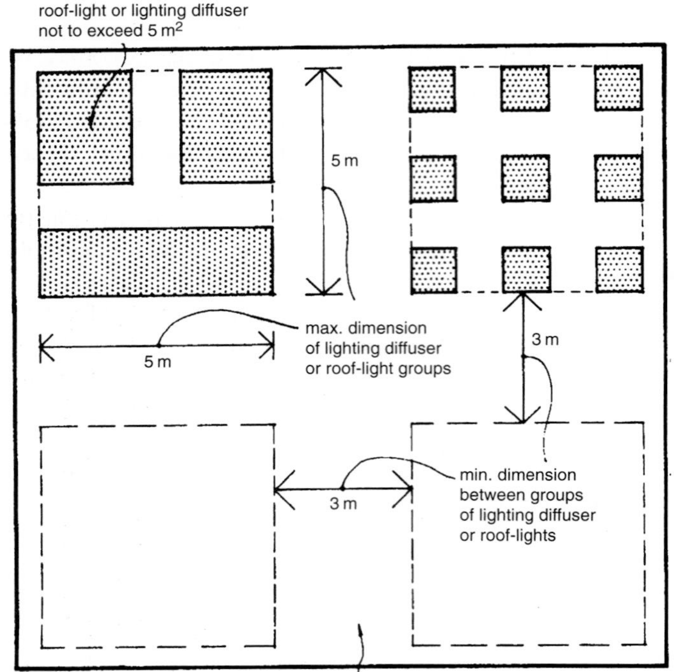

Notes:

1. For Class 3 or TP(b), total area of lighting diffusers or roof-lights should not exceed:
   - 50% of floor area if located in rooms; or
   - 15% of floor area if located in circulation spaces
2. Plastics roof-lights and lighting diffusers should not be used in protected stairways
3. There are no restrictions on the use of Class 3 materials in small rooms
4. There are no restrictions on the use of TP (a) materials except for note 2 above
5. Class 3 rooflights to rooms in industrial and other non-residential purpose groups may be spaced 1800mm apart provided the rooflights are evenly distributed and do not exceed 20% of the area of the room
6. The minimum 3m separation between each $5m^2$ must be maintained. Therefore, in some cases it may not be possible to use the maximum percentage quoted.

Upper and lower surface of suspended ceiling between roof-lights or diffusers to comply with AD B2, paragraph 7.1 regarding lining classification

Ceiling plan

Fig. 7.32 Limitations on use of Class 3 plastics rooflights, TP(b) rooflights and TP(b) lighting diffusers in suspended ceilings.

(a) balconies, verandas, carports, covered ways or loading bays with one longer side permanently open, or detached swimming pools; or

(b) garages, conservatories or outbuildings with a floor area not exceeding $40\,m^2$;

may consist of plastics materials with a lower surface of not less than Class 3 surface spread of flame or TP(b). The rooflight should be 6 m from any point on the boundary if the external surface is designated AD, BD, CA, CB, CC, CD or TP(b). (For details of designatory letters see section 7.27.9.) If the external surface is designated DA, DB, DC or DD the rooflight should be 20 m from any point on the boundary.

- The internal and external surface limits specified in (b) immediately above also apply to roof-lights in all other types of buildings. However, there are additional limitations on the area and spacing of the rooflights.

Table 7.6 Plastics rooflights: limitations on use and boundary distance.

Space which rooflight can serve	Min. classification on lower surface[1]	Minimum distance from any point on relevant boundary to rooflight with an external designation[2] of: AB BD CA CB CC CD	an external designation[2] of: DA DB DC DD	an external surface classification[1] of: TP(a)	an external surface classification[1] of: TP (b)
1. Any space except a protected stairway	TP(a) rigid	N/A	N/A	6 m	N/A
2. Balcony, veranda, carport, covered way or loading bay, which has at least one longer side wholly or permanently open	TP(b)	N/A	N/A	N/A	6 m
3. Detached swimming pool 4. Conservatory, garage or outbuilding, with a maximum floor area of 40 m^2	Class 3	6 m	20 m	N/A	N/A
5. Circulation space[3] (except a protected stairway)	TP(b)	N/A	N/A	N/A	6 m[4]
6. Room[3]	Class 3	6 m	20 m	N/A	N/A

Table References:

(1) See also the guidance to AD B2 (section 7.18.3).
(2) The designation of external roof surfaces is explained in Appendix A of AD B (see section 7.27.9).
(3) Single skin rooflight only, in the case of non-thermoplastic material.
(4) The rooflight should also meet the provision of Diagram 47 of AD B4 (see Fig. 7.32).

Notes

(a) N/A Not applicable.
(b) Polycarbonate and PVC rooflights which achieve a Class 1 rating by test, see paragraph 15.7 of AD B4 (section 7.27.9), may be regarded as having an AA rating.
(c) None of the above designations are suitable for protected stairways.
(d) Products may have upper and lower surfaces with different properties if they have double skins or are laminates of different materials. In which case the more onerous distance applies.
(e) Where the roof covering continues over a compartment wall, rooflights should be at least 1.5 m from the compartment wall (see section 7.22.1).

- Individual rooflights serving circulation spaces or rooms should not exceed 5 m^2 in area. They should be separated by roof covering materials of limited combustibility at least 3 m wide. If the rooflight is not thermoplastic it should consist of a single-skin material.

These recommendations for plastic rooflights and lighting diffusers are illustrated in Fig. 7.33.

D = 6 m if external surface of roof-light is AD, BD, CA, CB, CC, CD, TP(a) or TP(b)
D = 20 m if external surface of roof-light is DA, DB, DC, or DD

boundary

D

lower surface
of Class 3, TP(a) or TP(b)

(a) Balcony, verandah, carport,
detached swimming pool,
covered way or loading bay with one
longer side permantly open

OR: (b) Garage,
conservatory or
outbuilding not
exceeding 40 m^2
floor area

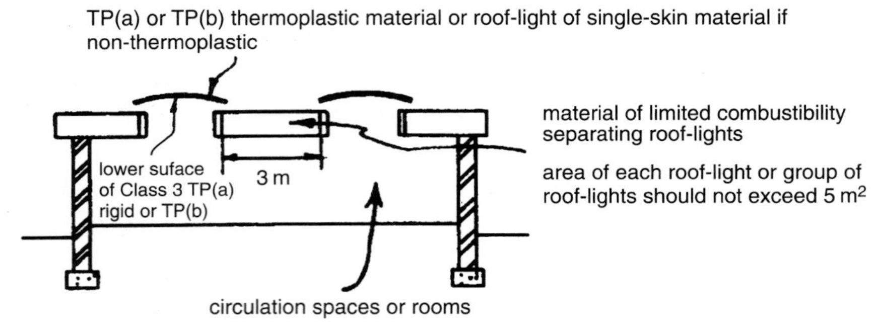

distance D as above except that for roof-lights of TP (a) material, D = 6 m

Note: see also Fig. 7.32 and Table 7.6 for more details of plastics roof-lights

Fig. 7.33 Plastics rooflights.

7.18.10 Flexible membranes and air supported structures

In recent years there has been a move towards the use of flexible membranes to provide the external envelope of a structure, often where this is to be of a temporary nature. The most famous example of this was the Millennium Dome which was supported by masts and cables. It is equally possible for the structure to be air supported. Since such materials are unlikely to satisfy the strict criteria laid down for the linings of walls and ceilings in AD B2, the Approved Document gives a number of alternative sources where guidance may be sought as follows.

- For any flexible membrane covering a structure (except an air supported structure) see Appendix A of BS 7157: 1989 *Method of test for ignitability of fabrics used in the construction of large tented structures.*
- Guidance on the use of PTFE-based materials for tension-membrane roof structures may be found in BRE Report 274: 1994 *Fire safety of PTFE-based materials used in buildings.*
- Air supported structures should follow the guidance given in BS 6661: 1986 *Guide for the design, construction and maintenance of single-skin air supported structures.*

7.19 Further requirements for ceilings

Fire-protecting suspended ceilings may be used to contribute towards the fire-resistance of a floor if they meet certain criteria. These are contained in Table A3 of Appendix A of AD B which is reproduced below. The highest grade of suspended ceiling is type Z. This (together with any insulating material) should be constructed of materials of limited combustibility (National) or Class A2-s3, d2 or better

AD B, Appendix A

Table A3 Limitations on fire-protecting suspended ceilings (see Table A1, Note 4).

Height of building or separated part (m)	Type of floor	Provision for fire resistance or floor (minutes)	Description of suspended ceiling
less than 18	not compartment	60 or less	Type W, X, Y or Z
	compartment	less than 60	
		60	Type X, Y or Z
18 or more	any	60 or less	Type Y or Z
no limit	any	more than 60	Type Z

Notes:

1. Ceiling type and description (the change from Types A–D to Types W–Z is to avoid confusion with Classes A–D (European)):
   W. Surface of ceiling exposed to the cavity should be Class 0 or Class 1 (National) or Class C-s3, d2 or better (European).
   X. Surface of ceiling exposed to the cavity should be Class 0 (National) or Class B-s3, d2 or better (European).
   Y. Surface of ceiling exposed to the cavity should be Class 0 (National) or Class B-s3, d2 or better (European). Ceiling should not contain easily openable access panels.
   Z. Ceiling should be a material of limited combustibility (National) or of Class A2-s3, d2 or better (European) and not contain easily openable access panels. Any insulation above the ceiling should be of a material of limited combustibility (National) or Class A2-s3, d2 or better (European).
2. Any access panels provided in fire protecting suspended ceilings of type Y or Z should be secured in position by releasing devices or screw fixings, and they should be shown to have been tested in the ceiling assembly in which they are incorporated.
3. European classifications.

The National classifications do not automatically equate with the equivalent European classifications, therefore products cannot typically assume a European class unless they have been tested accordingly.

When a classification includes 's3, d2', this means that there is no limit set for smoke production and/or flaming droplets/particles.

(European) and should not contain easily openable access panels. Type Z ceilings may be used anywhere under any conditions and should be used where the fire-resistance of the total floor/ceiling assembly exceeds 60 minutes. All the ceiling types need to comply with certain surface spread of flame requirements for the upper surface facing into the cavity, in addition to the Table 10 recommendations of AD B2 for the lower surface.

Similarly, fire-resisting ceilings may be used to reduce the need for cavity barriers in concealed void spaces in some floors and roofs. These are discussed in section 7.23.2 below, in the section on cavity barriers (see Fig. 7.41).

7.20 Internal fire spread (structure)

Paragraph B3 of Schedule 1 to the Building Regulations lists a number of factors which must be considered in order to reduce the effects of fire spread throughout the structure of a building as follows.

- The building must be so designed and constructed that its stability will be maintained for a reasonable period during a fire.
- Walls which are common to two or more buildings must be designed and constructed so that they resist the spread of fire between those buildings. Semi-detached and terraced houses are treated as separate buildings for the purposes of this requirement.
- The building must be subdivided by fire-resisting construction, depending on its size and intended use, where this is necessary to inhibit the spread of fire. (This requirement does not apply to material alterations to prisons provided under section 33 of the Prisons Act 1952.)
- Fire and smoke may spread unseen through concealed spaces in the structure and fabric of a building. The building must be designed and constructed so that this fire and smoke spread is inhibited.

7.21 Fire resistance and structural stability

If the structural elements of a building can be satisfactorily protected against the effects of fire for a reasonable period, it will be possible for the occupants to be evacuated safely and also the spread of fire throughout the building will be kept to a minimum. The risk to firefighters (who may have to search for or rescue people who are trapped) will be reduced and there will be less risk to people in the vicinity of the building from falling debris or as a result of an impact with an adjacent building from the collapsing structure.

One way to measure the standard of protection to be provided is by reference to the fire resistance of the elements under consideration. A number of factors which have a bearing on fire resistance are considered in Appendix A of AD B including the following.

- Fire severity – estimated from the purpose group (and therefore, the use) of the building. This assumes that the contents (which constitute the fire load) are the same for buildings of similar usage and that the contents of some building types will be more hazardous than others.
- Height of the top floor above ground – affects ease of escape, firefighting and the consequences of a large-scale collapse.
- Building occupancy – influences the speed of evacuation.
- The presence of basements – lack of venting may increase heat build-up and the duration of a fire, and hinder firefighting.
- The number of floors – escape from single-storey buildings is easier and a structural failure is unlikely to happen before evacuation has taken place.

It can be seen from the foregoing that an assessment of the standard of fire resistance in a building is a complicated matter. It is further complicated by the fact that there can be little control exercised over a building's future fire load unless a material change of use occurs. If a fire engineering approach is adopted for the assessment of fire severity based on fire load for a particular use, then future changes in use should also be borne in mind.

The method of assessment of fire resistance contained in AD B is based on the performance of an element of structure, door or other part of a building by reference to standard tests contained in:

- BS 476: Parts 20-24: 1987 (or to BS 476: Part 8: 1972, for items tested prior to 1 January 1988); or
- (European tests) Commission Decision 2000/367/EC of 3 May 2000 implementing Council Directive 89/106/EEC as regards the classification of the resistance to fire performance of construction products, construction works and parts thereof.

All products are classified in accordance with:

- BS EN 13501: *Fire classification of construction products and building elements,* Part 2: xxxx: *Classification using data from fire resistance tests (excluding products for use in ventilation systems)*
- BS EN 13501: *Fire classification of construction products and building elements,* Part 3: xxxx: *Classification using data from fire resistance tests on components of normal building service installations (other than smoke control systems)*
- BS EN 13501: *Fire classification of construction products and building elements,* Part 4: xxxx: *Classification using data from fire resistance tests on smoke control systems.*

(It should be noted that the designation of xxxx above is used for the year reference for standards that are not yet published. It is permissible to use the latest version of any standard provided that it continues to address the relevant requirements of the Regulations.)

The tests relate to the ability of the element:

- to resist a fire without collapse (loadbearing capacity), denoted R in the European classification of the resistance to fire performance;
- to resist fire penetration (integrity), denoted E in the European classification of the resistance to fire performance; and
- to resist excessive heat penetration so that fire is not spread by radiation or conduction (insulation), denoted I in the European classification of the resistance to fire performance.

Clearly the criteria of resistance to fire and heat penetration are applicable only to fire-separating elements, such as walls and floors. The criterion of resistance to collapse is applicable to all load-bearing elements, such as columns and beams, in addition to floors and load-bearing walls; however, it does not apply to external curtain walling or other claddings which transmit only self-weight and wind loads.

Table A1 to Appendix A of AD B (see below) shows the method of exposure required for the various elements of structure and other forms of construction, together with the BS 476 requirements which should be satisfied in terms of load-bearing capacity, integrity and insulation, and the minimum provisions when tested to the relevant European standard. For some items the table indicates the actual period of fire resistance recommended under each heading, but for others Table A2 of Appendix A gives the detailed recommendations in respect of fire resistance periods. The performance standards for doors are contained in Table B1 of AD B which is reproduced in the section on fire doors below.

In addition to the elements of structure defined in section 7.2 and illustrated in Fig. 7.3 above, there are requirements for some other elements of the building to be of fire-resisting construction. Included in this category are some doors, pipe casings and cavity barriers. These are considered later under the actual element references.

7.21.1 Minimum period of fire resistance

In order to establish the minimum period of fire resistance for the elements of structure of a building, it is necessary, first, to determine the building's use or Purpose Group. The fire resistance period will then depend on the height of the top storey of the building above ground or the depth of the lowest basement storey below ground.

It will be seen that the fire resistance recommendations for basements are generally more onerous than for ground floors in the same building. This reflects the greater difficulty experienced in dealing with a basement fire. However, it is sometimes the case that due to the slope of the ground, at least one side of a basement is accessible at ground level. This gives opportunities for smoke venting and firefighting and in these circumstances it may be reasonable to adopt the less onerous fire resistance provisions of the upper elements of the construction for the elements of structure in the basement.

The minimum periods of fire resistance recommended for the elements of structure in the basements, ground or upper storeys of a building are given in Table A2 from Appendix A of AD B which is reproduced below.

AD B, Appendix A

Table A1 Specific provisions of test for fire resistance of elements of structure etc.

Part of building	Minimum provisions when tested to the relevant part of BS 476(1) (minutes)			Minimum provisions when tested to the relevant European standard (minutes) (12)	Method of exposure
	Loadbearing capacity (2)	Integrity	Insulation		
1. **Structural** frame, beam or column	see Table A2	not applicable	not applicable	R see Table A2	exposed faces
2. **Loadbearing wall** (which is not also a wall described in any of the following items)	see Table A2	not applicable	not applicable	R see Table A2	each side separately
3. **Floors (3)**					
a. in upper storey of two-storey dwelling house (but not over garage or basement)	30	15	15	REI 30 (9)	from underside (4)
b. between a shop and flat above	60 or see Table A2 (whichever is greater)	60 or see Table A2 (whichever is greater)	60 or see Table A2 (whichever is greater)	REI 60 or see Table A2 (whichever is greater)	
c. any other floor, including compartment floors	see Table A2	see Table A2	see Table A2	REI see Table A2	
4. **Roofs**					
a. any part forming an escape route	30	30	30	REI 30	from underside (4)
b. any roof that performs the function of a floor	see Table A2	see Table A2	see Table A2	REI see Table A2	
5. **External walls**					
a. any part less than 1000 mm from any point on the relevant boundary	see Table A2	see Table A2	see Table A2	REI see Table A2	each side separately
b. any part 1000 mm or more from the relevant boundary (5)	see Table A2	see Table A2	15	REI see Table A2 (10)	from inside the building
c. any part adjacent to an external escape route (see Section 6, Diagram 22)	30	30	no provision (6)(7)	RE 30	from inside the building
6. **Compartment wall** Separating occupancies (see 9.20f)	60 or see Table A2 (whichever is less)	60 or see Table A2 (whichever is less)	60 or see Table A2 (whichever is less)	REI 60 or see Table A2 (whichever is less)	each side separately

(Contd)

Table A1 (*Contd*).

Part of building	Minimum provisions when tested to the relevant part of BS 476(1) (minutes)			Minimum provisions when tested to the relevant European standard (minutes) (12)	Method of exposure
	Loadbearing capacity (2)	Integrity	Insulation		
7. **Compartment walls** (other than in item 6)	see Table A2	see Table A2	see Table A2	REI see Table A2	each side separately
8. **Protected shafts**, excluding any firefighting shaft					
a. any glazing described in Section 9, Diagram 30	not applicable	30	no provision (7)	E 30	
b. any other part between the shaft and a protected lobby/corridor described in Diagram 30 above	30	30	30	REI 30	each side separately
c. any part not described in (a) or (b) above	see Table A2	see Table A2	see Table A2	REI see Table A2	
9. **Enclosure** (which does not form part of a compartment wall or a protected shaft) to a:					each side separately
a. protected stairway	30	30	30 (8)	REI 30 (8)	
b. lift shaft	30	30	30	REI 30	
10. **Firefighting shafts**					
a. construction separating firefighting shaft from rest of building;	120	120	120	REI 120	from side remote from shaft
	60	60	60	REI 60	from shaft side
b. construction separating firefighting stair, firefighting lift shaft and firefighting lobby	60	60	60	REI 60	each side separately
11. **Enclosure** (which is not a compartment wall or described in item 8) to a:					each side separately
a. protected lobby	30	30	30 (8)	REI 30 (8)	
b. protected corridor	30	30	30 (8)	REI 30 (8)	

(*Contd*)

Table A1 (*Contd*).

Part of building	Minimum provisions when tested to the relevant part of BS 476(1) (minutes)			Minimum provisions when tested to the relevant European standard (minutes) (12)	Method of exposure
	Loadbearing capacity (2)	Integrity	Insulation		
12. **Sub-division of a corridor**	30	30	30 (8)	REI 30 (8)	each side separately
13. **Wall separating** an attached or integral garage from a dwellinghouse	30	30	30 (8)	REI 30 (8)	from garage side
14. **Enclosure** in a flat or maisonette to a protected entrance hall, or to a protected landing	30	30	30 (8)	REI 30 (8)	each side separately
15. **Fire-resisting construction**					each side separately
a. in dwellings not described elsewhere	30	30	30 (8)	REI 30 (8)	
b. enclosing places of special fire hazard (see 9.12)	30	30	30	REI 30	
c. between store rooms and sales area in shops (see 6.54)	30	30	30	REI 30	
d. fire-resisting subdivision described in Section 10, Diagram 34(b)	30	30	30	REI 30	
16. **Cavity barrier**	not applicable	30	15	EI 30 (11)	each side separately
17. **Ceiling** described in Section 10, Diagram 33 or Diagram 35	not applicable	30	30	EI 30	from underside
18. **Duct** described in paragraph 10.14e	not applicable	30	no provision	E 30	from outside
19. **Casing** around a drainage system described in Section 11, Diagram 38	not applicable	30	no provision	E 30	from outside

(*Contd*)

Table A1 (*Contd*).

Part of building	Minimum provisions when tested to the relevant part of BS 476(1) (minutes)			Minimum provisions when tested to the relevant European standard (minutes) (12)	Method of exposure
	Loadbearing capacity (2)	Integrity	Insulation		
20. **Flue walls** described in Section 11, Diagram 39	not applicable	half the period specified in Table A2 for the compartment wall/floor	half the period specified in Table A2 for the compartment wall/floor	EI half the period specified in Table A2 for the compartment wall/floor	from outside
21. **Construction** described in Note (a) to paragraph 15.9	not applicable	30	30	EI 30	from underside
22. **Fire doors**		see Table B1		see Table B1	

Notes:

1. Part 21 for loadbearing elements, Part 22 for non-loadbearing elements, Part 23 for fire-protecting suspended ceilings, and Part 24 for ventilation ducts. BS 476: Part 8 results are acceptable for items tested or assessed before 1st January 1988.
2. Applies to loadbearing elements only (see B3.ii and Appendix E).
3. Guidance on increasing the fire resistance of existing timber floors is given in BRE Digest 208 increasing the fire resistance of existing timber floors (BRE 1988).
4. A suspended ceiling should only be relied on to contribute to the fire resistance of the floor if the ceiling meets the appropriate provisions given in Table A3.
5. The guidance in Section 14 allows such walls to contain areas which need not be fire-resisting (unprotected areas).
6. Unless needed as part of a wall in item 5a or 5b.
7. Except for any limitations on glazed elements given in Table A4.
8. See Table A4 for permitted extent of uninsulated glazed elements.
9. For the purposes of meeting the Building Regulations floors under item 3a will be deemed to have satisfied the provisions above, provided that they achieve loadbearing capacity of at least 30 minutes and integrity and insulation requirements of at least 15 minutes when tested in accordance with the relevant European test.
10. For the purposes of meeting the Building Regulations external walls under item 5b will be deemed to have satisfied the provisions above, provided that they achieve the loadbearing capacity and integrity requirements as defined in Table A2 and an insulation requirement of at least 15 minutes.
11. For the purposes of meeting the Building Regulations cavity barriers will be deemed to have satisfied the provisions above, provided that they achieve an integrity requirement of at least 30 minutes and an insulation requirement of at least 15 minutes.
12. The National classifications do not automatically equate with the equivalent classifications in the European column, therefore products cannot typically assume a European class unless they have been tested accordingly. 'R' is the European classification of the resistance to fire performance in respect of loadbearing capacity; 'E' is the European classification of the resistance to fire performance in respect of integrity; and 'I' is the European classification of the resistance to fire performance in respect of insulation.

The following points should also be taken into account when using Table A2.

- Any element of structure should have fire resistance at least equal to the fire resistance of any element which it carries, supports or to which it gives stability. This principle may be varied where:
  (a) the supporting structure is in the open air and would be unlikely to be affected by a fire in the building, or

AD B, Appendix A

Table A2 Minimum periods of fire resistance.

Purpose group of building	Minimum periods (minutes) for elements of structure in a:					
	Basement storey ($) including floor over		**Ground or upper storey**			
of a lowest basement	**Depth (m) of a lowest basement**		**Height (m) of top floor above ground, in a building or separated part of a building**			
	more than 10	**not more than 10**	**not more than 5**	**not more than 18**	**not more than 30**	**more than 30**
1. Residential (domestic):						
a. flats and maisonettes	90	60	30*	60**†	90**	120**
b. and c. dwellinghouses	not relevant	30*	30*	60@	not relevant	not relevant
2. Residential:						
a. institutional œ	90	60	30*	60	90	120#
b. other residential	90	60	30*	60	90	120#
3. Office:						
– not sprinklered	90	60	30*	60	90	not permitted
– sprinklered (2)	60	60	30*	30*	60	120#
4. Shop and commercial:						
– not sprinklered	90	60	60	60	90	not permitted
– sprinklered (2)	60	60	30*	60	60	120#
5. Assembly and recreation:						
– not sprinklered	90	60	60	60	90	not permitted
– sprinklered (2)	60	60	30*	60	60	120#
6. Industrial:						
– not sprinklered	120	90	60	90	120	not permitted
– sprinklered (2)	90	60	30*	60	90	120#
7. Storage and other non-residential:						
a. any building or part not described elsewhere:						
– not sprinklered	120	90	60	90	120	not permitted
– sprinklered (2)	90	60	30*	60	90	120#
b. car park for light vehicles:						
i. open sided car park (3)	not applicable	not applicable	15* + (4)	15* + (4)	15* + (4)	60
ii. any other car park	90	60	30*	60	90	120#

(Contd)

Table A2 (*Contd*).

Single storey buildings are subject to the periods under the heading 'not more than 5'. If they have basements, the basement storeys are subject to the period appropriate to their depth.

Modifications referred to in Table A2: (for application of the table notes in text above)

$ The floor over a basement (or if there is more than 1 basement, the floor over the topmost basement) should meet the provisions for the ground and upper storeys if that period is higher.

* Increased to a minimum of 60 minutes for compartment walls separating buildings.

** Reduced to 30 minutes for any floor within a maisonette, but not if the floor contributes to the support of the building.

œ Multi-storey hospitals designed in accordance with the NHS Firecode document should have a minimum 60 minutes standard.

Reduced to 90 minutes for elements not forming part of the structural frame.

+ Increased to 30 minutes for elements protecting the means of escape.

† Refer to p. 7.149 below regarding the acceptability of 30 minutes in flat conversions.

@ 30 minutes in the case of three storey dwellinghouses, increased to 60 minutes minimum for compartment walls separating buildings.

Notes:

1. Refer to Table A1 for the specific provisions of test.
2. 'Sprinklered' means that the building is fitted throughout with an automatic sprinkler system meeting the relevant recommendations of BS 5306 Fire extinguishing installations and equipment on premises. Part 2 Specification for sprinkler systems; ie the relevant occupancy rating together with the additional requirements for life safety.
3. The car park should comply with the relevant provisions in the guidance on requirement B3, Section 12.
4. For the purposes of meeting the Building Regulations, the following types of steel elements are deemed to have satisfied the minimum period of fire resistance of 15 minutes when tested to the European test method;
   (i) Beams supporting concrete floors, maximum Hp/A = 230m-1 operating under full design load.
   (ii) Free standing columns, maximum Hp/A = 180m-1 operating under full design load.
   (iii) Wind bracing and struts, maximum HP/A = 210m-1 operating under full design load.

Guidance is also available in BS 5950 Structural use of steelwork in building. Part 8 Code of practice for fire resistant design.

(b) where a roof top plant room needs a higher standard of fire resistance than the structure supporting it, or

(c) the supporting and supported structures are in different compartments (the separating element between the compartments would have to have the higher standard of fire resistance; see next item).

- If an element of structure forms part of more than one building or compartment, and is thus subject to two or more different fire resistances, it is the greater of these which applies.
- A structural frame, beam, column or load-bearing wall of a *single-storey building* (or which is part of the ground storey of a building that consists of a ground storey and one or more basement storeys) is generally not required to have fire resistance. (This reflects the view that, given satisfactory means of escape, and the restricted use of combustible materials as wall and ceiling linings, fire resistance in the elements of structure in the ground storey will contribute little to the safety of the occupants.) However, the above concession will only apply if the element of structure:

  (a) which is part of or supports an external wall, is sufficiently far from its relevant boundary to be regarded as a totally unprotected area;

  (b) is not part of and does not support a compartment wall or a wall which is common to two or more buildings;

   (c) is not a wall between a house and an attached or integral garage; or
   (d) does not support a gallery.
- Single-storey buildings should comply with the fire resistance periods under the heading 'not more than 5'. Where they have basements, these should, of course, comply with the recommendations for basement storeys depending on their depth below ground level.
- Further fire resistance provisions relating to the following elements may be found in the relevant sections below:
   (a) compartment walls, external walls and the wall between a dwelling-house and a domestic garage;
   (b) walls which enclose a firefighting shaft or protect a means of escape;
   (c) compartment floors.

7.21.2 Meeting the performance recommendations

Reference has been made above to the BS 476 tests, where the fire resistance period of a material, product, structure or system may be assessed. It should be realised that the aim of the standard fire tests is to measure or assess the response of the sample to one or more aspects of fire behaviour under standardised conditions. The tests cannot normally measure fire hazard and represent only one aspect of the total fire safety package.

In a real fire the conditions will not be standard and there is always the possibility of premature failure of a particular component due to faulty design or workmanship. Therefore, the periods stated in Table A2 should be used for guidance and not taken as 'cast in tablets of stone', there being little correlation between the period derived from the standard test and the performance in a real fire. They do however, enable different systems to be compared under similar circumstances and are useful in this sense.

Therefore, in order that a material, product or structure may be used in a building it should:

- be in accordance with a specification or design which has been shown by test to be capable of meeting a performance standard referred to in a relevant British or European Standard or European Technical Approval (British Standards may continue to be used for products or materials where European standards or approvals are not yet available). For this purpose, laboratories accredited by the United Kingdom Accreditation Service (UKAS) for conducting the relevant tests would be expected to have the necessary expertise; or
- be assessed from test evidence against appropriate standards, or by using relevant design guides, as meeting the relevant performance (suitably qualified fire safety engineers and laboratories accredited by UKAS, might be expected to have the necessary expertise to carry out the assessment. Additionally, any body notified to the UK Government by the Government of another member state of the European Union as capable of assessing materials and products against the relevant British Standards, may also be expected to have the necessary expertise); or

- comply with an appropriate specification given in relevant tables of notional performance included in AD B. (Over the years since the 1976 Regulations, there has been a tendency to reduce the practical guidance given on forms of construction which will satisfy the fire resistance requirements. The 1992 edition of AD B contains no such examples, the reader is merely referred to the publication listed in the next point); or
- for fire-resisting elements,
  (a) conform with an appropriate specification from Part II of the BRE Report *Guidelines for the construction of fire-resisting structural elements* (BR 128, BRE, 1988)
  (b) be designed in accordance with a relevant British Standard or Eurocode.

Where test evidence is used to substantiate a fire resistance rating of a construction, care should be taken to check that it demonstrates compliance which is adequate and applicable to the intended use. For example, small differences in detail (e.g. fixing method, joint details, dimensional variations etc.) may significantly affect the rating.

In order to provide some assistance to readers, Table 7.7 gives notional periods of fire resistance for some common floor and wall constructions. It is based on a selection of constructions from Table A3 of DOE's 1985 Approved Document B2/3/4, which in turn is based on the BRE Report. The BRE Report mentioned above also contains much information on fire protection to structural frameworks of beams and columns. Additionally, there are available various mineral based insulating boards, sprayed coatings and intumescent paint systems which are capable of providing differing degrees of fire protection depending on their thickness and method of fixing. Reference should be made to individual manufacturers or their trade associations for further details.

Information on tests on fire-resisting elements is also given in such publications as:

- Association for Specialist Fire Protection/Steel Construction Institute/Fire Test Study Group *Fire protection for structural steel in buildings*, second edition (revised, 1992 and available from the ASFP, Association House, 99 West Street, Farnham, Surrey GU9 7EN and the Steel Construction Institute, Silwood Park, Ascot, Berks SL5 7QN);
- PD 6520 *Guide to fire test methods for building materials and elements of construction* (available from the British Standards Institution);
- BS 6336 *Guide to development and presentation of fire tests and their use in hazard assessment*.

For the first two items in Table 7.7 it should be noted that the upper floor of a two-storey dwelling is regarded as a special case. Such a floor, when tested for fire resistance from the underside is required only to provide

- stability for 30 minutes
- integrity for 15 minutes
- insulation for 15 minutes.

Table 7.7 Notional periods of fire resistance of some common constructions.

These constructions are a selection from Table A3 of DOE's 1985 Approved Document B2/3/4.
A large number of constructions other than those shown are capable of providing the fire resistance looked for. For example, various mineral based insulating boards can be used. Because their performance varies and it dependent on their thickness, it is not possible to give specific thicknesses in this table. However, manufacturers will normally be able to say what thickness would be needed to achieve the particular performance.

Fire resistance	No.	Construction
Floors: timber joist		
Modified 30 minutes (stability 30 minutes) (integrity 15 minutes) (insulation 15 minutes)	**1**	any structurally suitable flooring: floor joists at least 37 mm wide ceiling: (a) 12.5 mm plasterboard[a] with joints taped and filled and backed by timber, or (b) 9.5 mm plasterboard[a] with 10 mm lightweight gypsum plaster finish
	2	at least 15 mm t&g boarding or sheets of plywood or wood chipboard, floor joists at least 37 mm wide ceiling: (a) 12.5 mm plasterboard[a] with joints taped and filled, or (b) 9.5 mm plasterboard[a] with at least 5 mm neat gypsum plaster finish
30 minutes	**3**	at least 15 mm t&g boarding or sheets of plywood or wood chipboard, floor joists at least 37 mm wide ceiling: 12.5 mm plasterboard[a] with at least 5 mm neat gypsum plaster finish
	4	at least 21 mm t&g boarding or sheets of plywood or wood chipboard, floor joists at least 37 mm wide ceiling: 12.5 mm plasterboard[a] with joints taped and filled
60 minutes	**5**	at least 15 mm t&g plywood or wood chipboard, floor joists at least 50 mm wide ceiling: not less than 30 mm plasterboard[a] with joints staggered and exposed joints taped and filled
Floors: concrete		
60 minutes	**6**	reinforced concrete floor not less than 95 mm thick, with not less than 20 mm cover on the lowest reinforcement

(*Contd*)

Table 7.7 (*Contd*).

Walls: internal		
30 minutes load-bearing	**7**	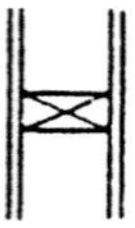 framing members at least 44 mm wide[b] and spaced at not more than 600 mm apart, with lining (both sides) of 12.5 mm plasterboard[a] with all joints taped and filled
	8	 100 mm reinforced concrete wall[c] with minimum cover to reinforcement of 25 mm
60 minutes load-bearing	**9**	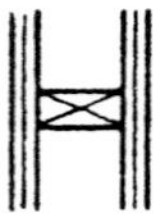 framing members at least 44 mm wide[b] and spaced at not more than 600 mm apart, with lining (both sides) at least 25 mm plasterboard[a] in 2 layers with joints staggered and exposed joints taped and filled
	10	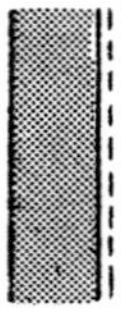 solid masonry wall (with or without plaster finish) at least 90 mm thick (75 mm if non-load-bearing) *Note:* for masonry cavity walls, the fire resistance may be taken as that for a single wall of the same construction, whichever leaf is exposed to fire
	11	 120 mm reinforced concrete wall[c] with at least 25 mm cover to the reinforcement
Walls: external		
Modified 30 minutes (stability 30 minutes) (integrity 30 minutes) (insulation 15 minutes) load-bearing wall 1 m or more from relevant boundary	**12**	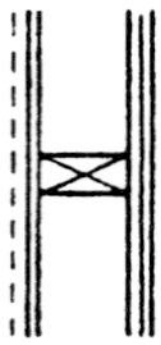 any external weathering system with at least 8 mm plywood sheathing, framing members at least 37 mm wide and spaced not more than 600 mm apart internal lining: 12.5 mm plasterboard[a] with at least 10 mm lightweight gypsum plaster finish
30 minutes load-bearing wall less than 1 m from the relevant boundary	**13**	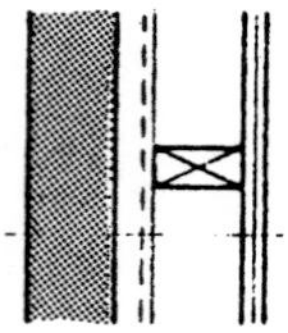 100 mm brickwork or blockwork external face (with, or without, a plywood backing); framing members at least 37 mm wide and spaced not more than 600 mm apart internal lining: 12.5 mm plasterboard[a] with at least 10 mm lightweight gypsum plaster finish

(*Contd*)

Table 7.7 (*Contd*).

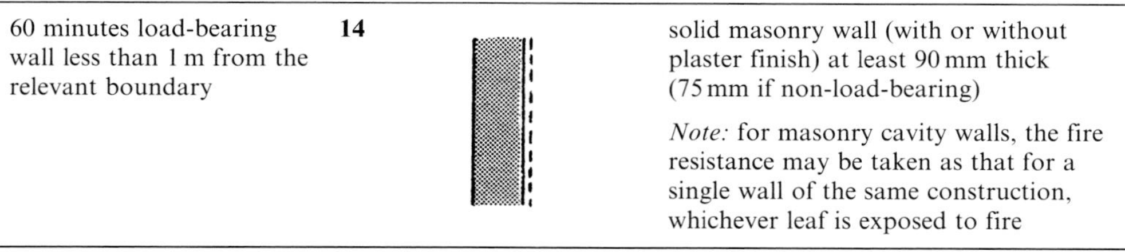

60 minutes load-bearing wall less than 1 m from the relevant boundary	**14**		solid masonry wall (with or without plaster finish) at least 90 mm thick (75 mm if non-load-bearing) *Note:* for masonry cavity walls, the fire resistance may be taken as that for a single wall of the same construction, whichever leaf is exposed to fire

Notes

[a] Whatever the lining material, it is important to use a method of fixing that the manufacturer says would be needed to achieve the particular performance. For example, if the lining is plasterboard, the fixings should be at 150 mm centres as follows (where two layers are being used each should be fixed separately):
9.5 mm thickness, use 30 mm galvanised nails
12.5 mm thickness, use 40 mm galvanised nails
19 mm–25 mm thickness, use 60 mm galvanised nails.

[b] Thinner framing members, such as 37 mm, may be suitable depending on the loading conditions.

[c] A thinner wall may be suitable depending on the density of the concrete and the amount of reinforcement. (See *Guidelines for the construction of fire-resisting structural elements* (BRE, 1988).)

This is termed 'modified 30 minutes' fire resistance in the BRE Report (see Table A1, Appendix A, item 3, 'Floors').

7.21.3 Compartmentation

In order to prevent the rapid spread of fire within buildings (which could trap the occupants) and to restrict the size of any fires which do occur, AD B3 contains provisions for subdividing a building into compartments separated from one another by fire-resisting walls and floors.

The extent of the subdivisions will depend on the same factors which were considered for fire resistance, namely:

- the severity of the fire
- the height of the top floor above ground
- the building occupancy
- the presence of basements
- the number of floors.

These items are considered more fully in section 7.21 above. Additionally, the provision of a sprinkler system can affect the growth rate of a fire and may suppress it altogether.

The subdivision is achieved by means of compartment walls and compartment floors and since these come under the definition of elements of structure they should be fire resisting (but not, of course, the lowest floor in the building).

Most multi-storey buildings should be compartmented because of the increased risk to life should a fire occur. However, in the case of a two storey building in Purpose Group 4 – Shop and Commercial, or Purpose Group 6 – Industrial, the ground storey may be treated as a single storey building for the purposes of fire compartmentation if the use of the upper storey is ancillary to the use of the ground storey. This concession is dependent on the following conditions:

- the upper storey floor area should not exceed the lesser of 20% of the ground storey area or 500 m^2;
- the upper storey should be compartmented from the ground storey; and
- there should be an independent means of escape from the upper storey which is separated from the ground storey escape routes.

In single storey buildings, where the risk to life is obviously less than in multi-storey buildings, compartmentation is recommended only for:

- single-storey hospitals with a floor area limit of 3000 m^2;
- schools, where the floor area limit is 800 m^2; and
- unsprinklered shops, limited to a maximum floor area of 2000 m^2.

Where atria (see section 7.7 above for definition of an atrium) are incorporated into the design of a building there are obvious implications for the integrity of any compartmentation in the building. Detailed advice on all issues relating to atria is given in BS 5588: Part 7: 1997 *Code of practice for the incorporation of atria in buildings*; however, the standard is only relevant where compartmentation is breached by the atria.

Generally, two main principles are adopted when deciding how to compartment a building:

(1) For non-residential buildings, the compartment sizes should be restricted to the maximum dimensions shown in Table 12 of AD B3, which is reproduced below, and the following points should be noted.
   - Limits are given by reference to maximum floor areas for differing top storey heights (for multi-storey buildings) for assembly and recreation, shop and commercial, and industrial buildings.
   - For storage and other non-residential buildings, the limits are according to maximum compartment volumes for differing top storey heights.
   - The installation of sprinklers allows the compartment size limits to be doubled.
   - If a building has a storey which is more than 30 m above ground or a basement more than 10 m below ground, all floors should be constructed as compartment floors, including any ground floor over a basement (except for small premises described in section 7.15.8 above).

(2) For all buildings, the following walls and floors which separate buildings into different ownerships, uses, or occupancies (or protect particular risk areas) should be constructed as compartment walls and floors:
   - Walls which are common to two or more buildings (including walls between semi-detached and terraced houses). These walls should run the full height of the building in a continuous vertical plane, the adjoining buildings being separated only by walls, not floors.
   - Walls and floors dividing parts of buildings used for different purposes. This does not apply where one use is ancillary to another (see section 7.3).

AD B3, Section 9

Table 12 Maximum dimensions of building or compartment (non-residential buildings).

Purpose Group of building or part	Height of floor of top storey above ground level (m)	Floor area of any one storey in the building or any one storey in a compartment (m^2)	
		in multi-storey buildings	in single storey buildings
Office	no limit	no limit	no limit
Assembly & Recreation Shop & Commercial: a. schools	no limit	800	800
b. shops – not sprinklered	no limit	2000	2000
shops – sprinklered (1)	no limit	4000	no limit
c. elsewhere – not sprinklered	no limit	2000	no limit
elsewhere – sprinklered (1)	no limit	4000	no limit
Industrial (2) not sprinklered	not more than 18 more than 18	7000 2000 (3)	no limit no limit
sprinklered (1)	not more than 18 more than 18	14000 4000 (3)	no limit no limit
	Height of floor of top storey above ground level (m)	**Maximum compartment volume (m^3)**	
		in multi-storey buildings	**in single storey buildings**
Storage (2) & other non-residential: a. car park for light vehicles	no limit	no limit	no limit
b. any other building or part: not sprinklered	not more than 18 more than 18	20000 4000 (3)	no limit no limit
sprinklered (1)	not more than 18 more than 18	40000 8000 (3)	no limit no limit

Notes:
1. 'Sprinklered' means that the building is fitted throughout with an automatic sprinkler system meeting the relevant recommendations of BS 5306: Part 2, ie the relevant occupancy rating together with the additional requirements for life safety.
2. There may be additional limitations on floor area and/or sprinkler provisions in certain industrial and storage uses under other legislation, for example in respect of storage of LPG and certain chemicals.
3. This reduced limit applies only to storeys that are more than 18 m above ground level. Below this height the higher limit applies.

- Walls dividing buildings into separated parts. These walls should also run the full height of the building in a continuous vertical plane, in a similar manner to common walls above (see section 7.10) and the two separated parts can have different standards of fire resistance.
- Walls and floors bounding a protected shaft. (Protected shafts are considered in more detail below.)
- Construction enclosing places of special fire hazard (see section 7.2).

Curiously, any walls or floors enclosing such places are not considered to be compartment walls and floors for the purposes of AD B3.

Finally, the following walls and floors in particular Purpose Groups should be constructed as compartment walls and floors:

- Any floor or wall separating an attached or integral garage from a dwellinghouse should be constructed as shown in Fig. 7.34.
- Any floor in an institutional or other residential building.
- Any floor in flats and maisonettes, except an internal floor in an individual dwelling.
- Any wall separating a flat or maisonette from any other part of the same building.
- Any wall enclosing a refuse storage chamber in flats and maisonettes.

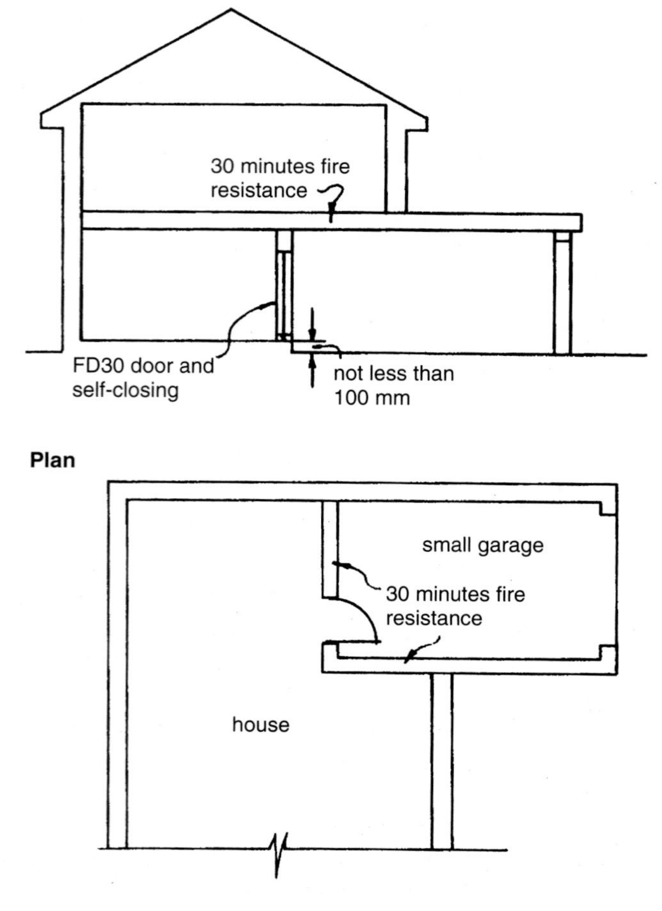

Fig. 7.34 Attached small garages.

- Any wall or floor in a shopping complex referred to in Section 5 of BS 5588: Part 10 as needing to be constructed to the compartmentation standard.
- Any wall or floor provided to divide a building into separate occupancies (i.e. parts of the building used by different organisations irrespective of whether or not they fall within the same Purpose Group) in Shop and Commercial, Industrial or Storage premises.
- Any wall between compartments in the upper storeys of health-care premises used for in-patients. The upper storeys of such buildings should be divided into at least two compartments (so that no compartments exceed 2000 m^2) so as to permit progressive horizontal evacuation of each compartment (see section 7.17.1 above).

7.22 Compartment walls and floors – construction details

Since the purpose of compartment walls and floors is to form a complete barrier to the passage of fire between the compartments which they separate, it follows that they should have the appropriate standards of fire resistance indicated in Tables A1 and A2 above.

Additionally, compartment walls and floors in hospitals designed on the basis of *Firecode* (see section 7.17.1 above) should be constructed of materials of limited combustibility if they have a fire resistance of 60 minutes or more. This does not apply if the building is fitted throughout with a suitable sprinkler system as detailed in *Firecode*.

Any points of weakness in compartment walls and floors should be adequately protected. These points of weakness occur:

- at junctions with other compartment walls, external walls and roofs
- where timber beams, joists, purlins and rafters are built into or pass through a compartment wall
- at openings for doors, pipes and ducts of various kinds, refuse chutes and protected shafts.

7.22.1 Junction details

Where a compartment wall and roof meet, the wall may be carried at least 375 mm above the roof covering surface, measured at right angles to the roof slope, or the wall may be taken up to meet the underside of the roof covering or deck, and the junction fire-stopped. Acceptable design solutions are illustrated in Fig. 7.35 for buildings in different purpose groups.

Generally, the covering in a 1.5 m wide zone on either side of the wall should be designated AA, AB or AC (see below for designations) and it should be carried on a substrate or deck consisting of a material of limited combustibility. The roof covering and deck could be of a composite structure such as profiled steel cladding. Where double skinned insulated roof sheeting is specified it should incorporate a band of material of limited combustibility at least 300 mm wide centred over the wall.

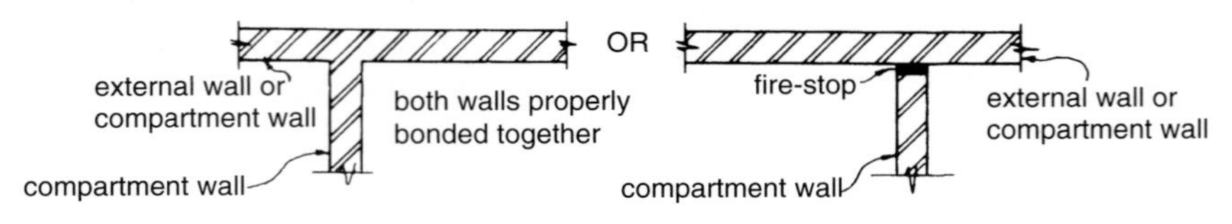

junction of external and compartment walls

EITHER

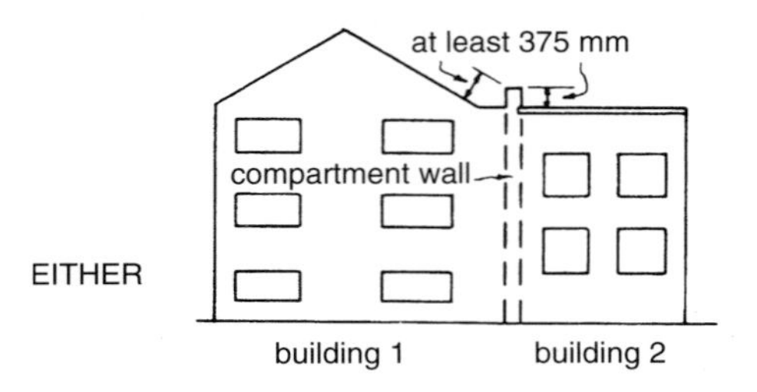

OR **(a) Building or compartment of any use or height**

- roof covering to be AA, AB or AC on a substrate or deck of material of limited combustibility
- roof covering and deck could be of a composite structure such as profiled steel cladding
- double skinned insulated roof sheeting should incorporate a band of material of limited combustibility at least 300 mm wide centred over the wall.

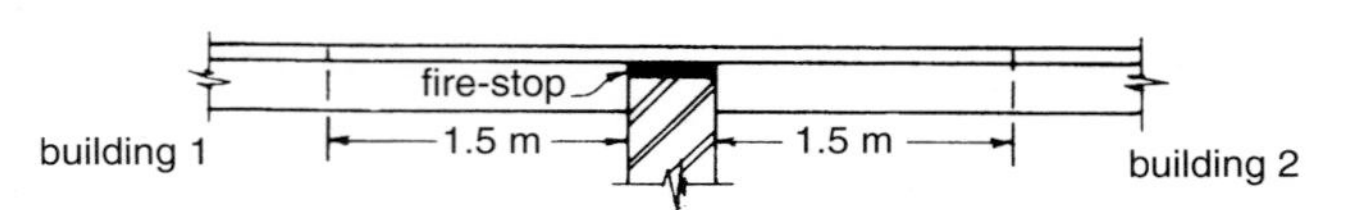

(b) Dwellinghouse, office, assembly and recreation, or residential (not institutional) nor more than 15 m high

roof covering: AA, AB or AC

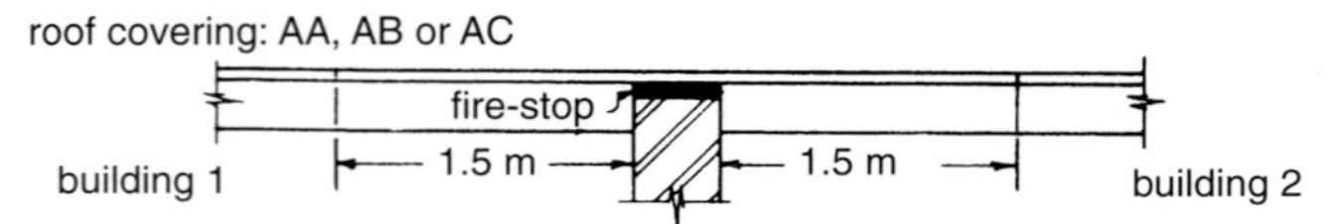

see below for materials which may be carried over or through compartment wall

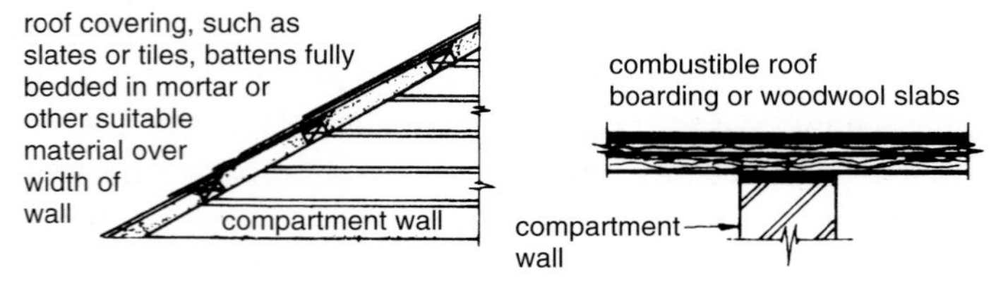

Combustible materials permitted over wall

Fig. 7.35 Junction details – compartment walls.

Any roof support members which pass through the wall may need to have fire protection on their undersides for at least 1.5 m on either side of the wall to delay distortion at the junction.

Some exceptions to the need for the substrate or deck to be in materials of limited combustibility are permitted in the case of certain roof constructions in buildings not more than 15 m high in dwellinghouses, offices, assembly and recreation buildings, and residential buildings (but not institutional). In the roof shown at (b) in Fig. 7.35, the following combustible materials may be carried over the top of the compartment wall:

- roof boarding serving as a base for roof covering; or
- woodwool slabs; or
- timber slating or tiling battens
- sarking felt.

The materials should be fully bedded in mortar or similar material.

Where a compartment wall or floor meets another compartment wall or an external wall, the junction should maintain the fire resistance of the compartmentation. This will normally mean that the various structures should be bonded together or the junction fire-stopped (see Fig. 7.35). The compartment wall should also be continued across any eaves cavity.

It is permissible for timber beams, joists, purlins and rafters to be built into or carried through a concrete or masonry compartment wall provided that the openings for them are kept as small as is practicable and are then fire-stopped.

If trussed rafters bridge a compartment wall they should be designed so that failure of any part of the truss caused by fire in one compartment does not lead to failure of any part of the truss in another compartment.

7.22.2 Openings in compartment walls and floors

Generally, the only openings permitted in compartment walls and floors are one or more of the following.

- An opening fitted with a door which has the appropriate fire resistance given in Table B1 of Appendix B to AD B and is fitted in accordance with Appendix B of AD B (see Fire doors below).
- An opening for a protected shaft (see Protected shafts in section 7.22.4).
- An opening for a refuse chute of non-combustible construction.
- An opening for a pipe, ventilation duct, chimney, appliance ventilation duct or duct encasing one or more flue pipes, provided it complies with the relevant parts of Section 11 of AD B (see Pipes, ventilation ducts and flues in section 7.24).
- Atria designed in accordance with BS 5588: Part 7.

In the case of compartment walls which are common to two or more buildings or which separate different occupancies in the same building, it would not be sensible or necessary to allow all of the above openings to exist. Therefore such walls should be imperforate except for:

- an opening for a door which is needed as a means of escape in case of fire. The door should have the same fire resistance as the wall and should be fixed in accordance with the provisions of Appendix B to AD B (see below); or
- an opening for a pipe complying with the provisions of Section 11 of AD B (see section 7.24).

7.22.3 Fire doors

All fire doors should be fitted with an automatic self-closing device unless they are to cupboards or service ducts which are normally kept locked shut. Rising butt hinges are not considered as automatic self-closing devices unless the door is:

- to or within a dwelling
- in a cavity barrier
- between a dwellinghouse and a garage.

As a general rule, no device should be provided to hold a door open. However, in some cases a self-closing device may be considered a hindrance to normal use. In such cases a fire-resisting door may be held open by:

- a fusible link device, provided that the door is not fitted in an opening used as a means of escape (this does not apply where two doors are provided in the opening as mentioned below);
- a door closure delay device; or
- an automatic release mechanism, actuated by an automatic fire detection and alarm system.

In this context an automatic release mechanism is one which automatically closes a door in the event of each or any one of:

- smoke detection by appropriate apparatus suitable in nature, quality and location
- manual operation by a suitably located switch
- failure of the electricity supply to the device, smoke apparatus, or switch
- operation of a fire alarm system, if fitted.

All fire-resisting doors should have the appropriate fire performance described in Table Bl of Appendix B (see below), either:

- in accordance with a rating given in terms of their performance under test to BS 476: Part 22. This rating relates to the ability of the door to maintain its integrity for a specified period in minutes, e.g. FD30. Doors should be tested from each side separately; however, since lift doors are only at risk from one side in the event of a fire, it is only necessary to test these from the landing side. The rating for some doors has the suffix S added where restricted smoke leakage is needed at ambient temperatures. The leakage rate should not exceed 3 m^3/m/hour (head and jambs only) when tested at 25 Pa to BS 476: Section 31.1, unless pressurisation techniques complying with BS 5588: Part 4 are used; or

- as determined with reference to Commission Decision 2000/367/EC of 3 May 2000 implementing Council Directive 89/106/EEC as regards the classification of the resistance to fire performance of construction products, construction works and parts thereof. All such fire doors should be classified in accordance with BS EN 13501 *Fire classification of construction products and building elements,* Part 2: xxxx: *Classification using data from fire resistance tests (excluding products for use in ventilation systems).* They are tested to the relevant European method from the following:
  - (a) BS EN 1634 *Fire resistance tests for door and shutter assemblies:* Part 1: 2000 *Fire doors and shutters;*
  - (b) BS EN 1634 *Fire resistance tests for door and shutter assemblies:* Part 2: xxxx *Fire door hardware;*
  - (c) BS EN 1634 *Fire resistance tests for door and shutter assemblies:* Part 3: xxxx *Smoke control doors.*

  The performance requirement is in terms of integrity (E) for a period of minutes. An additional classification of Sa is used for all doors where restricted smoke leakage at ambient temperatures is needed. The designation of xxxx is used for standards that are not yet published. The latest version of any standard may be used provided that it continues to address the relevant requirements of the Regulations. Until such time that the relevant harmonised product standards are published, for the purposes of meeting the Building Regulations, products tested in accordance with BS EN 1634: Part 1 (with or without pre-fire test mechanical conditioning) will be deemed to have satisfied the provisions provided that they achieve the minimum fire resistance in terms of integrity, as detailed in Table B1.

No fire door should be hung on hinges made of a material with a melting point less than 800°C, unless the hinges can be shown to be satisfactory when tested as part of a fire door assembly.

Although each fire door should have the appropriate period of fire resistance defined in Table Bl, it is permissible for two fire doors to be fitted in an opening if each door is capable of closing the opening and the required level of fire resistance can be achieved by the two doors together. However, if these two doors are fitted in an opening provided for a means of escape, both doors should be self-closing. One of them may be held open by a fusible link and be fitted with an automatic self-closing device if the other is easily openable by hand and has at least 30 minutes fire resistance.

Doors which are to be kept closed or locked when not in use, or which are held open by an automatic release mechanism, should be marked with the appropriate fire safety sign in accordance with BS 5499 *Fire safety signs, notices and graphic symbols,* Part 1: 1990 *Specification for fire safety signs.* The signs should be on both

AD B, Appendix B

Table B1 Provisions for fire doors.

Position of door	Minimum fire resistance of door in terms of integrity (minutes) when tested to BS 476 part 22(1)	Minimum fire resistance of door in terms of integrity (minutes) when tested to the relevant European standard
1. In a compartment wall separating buildings	As for the wall in which the door is fitted, but a minimum of 60	As for the wall in which the door is fitted, but a minimum of 60
2. In a compartment wall:		E30 S. (3)
a. if it separates a flat or maisonette from a space in common use;	FD 30S (2)	
b. enclosing a protected shaft forming a stairway situated wholly or partly above the adjoining ground in a building used for Flats, Other Residential, Assembly and Recreation, or Office purposes;	FD 30S (2)	E30 S. (3)
c. enclosing a protected shaft forming a stairway not described in (b) above;	Half the period of fire resistance of the wall in which it is fitted, but 30 minimum and with suffix S (2)	Half the period of fire resistance of the wall in which it is fitted, but 30 minimum and with suffix S. (3)
d. enclosing a protected shaft forming a lift or service shaft;	Half the period of fire resistance of the wall in which it is fitted, but 30 minimum	Half the period of fire resistance of the wall in which it is fitted, but 30 minimum
e. not described in (a), (b), (c) or (d) above	As for the wall it is fitted in, but add S (2) if the door is used for progressive horizontal evacuation under the guidance to B1	As for the wall it is fitted in, but add S. (3) if the door is used for progressive horizontal evacuation under the guidance to B1
3. In a compartment floor	As for the floor in which it is fitted	As for the floor in which it is fitted
4. Forming part of the enclosures of:		
a. a protected stairway (except where described in item 9); or	FD 30S (2)	E20 S. (3)
b. a lift shaft (see paragraph 6.42b); which does not form a protected shaft in 2(b), (c) or (d) above	FD 30	E30
5. Forming part of the enclosure of:		
a. a protected lobby approach (or protected corridor) to a stairway;	FD 30S (2)	E30 S. (3)
b. any other protected corridor; or	FD 20 (S)	E20 S. (3)
c. a protected lobby approach to a lift shaft (see paragraph 6.42)	FD 30S (2)	E30 S. (3)
6. Affording access to an external escape route	FD 30	E30

(*Contd*)

Table B1 (*Contd*).

Position of door	Minimum fire resistance of door in terms of integrity (minutes) when tested to BS 476 part 22(1)	Minimum fire resistance of door in terms of integrity (minutes) when tested to the relevant European standard
7. Sub-dividing:		
a. corridors connecting alternative exits;	FD 20S (2)	E20 S. (3)
b. dead-end portions of corridors from the remainder of the corridor	FD 20S (2)	E20 S. (3)
8. Any door:		
a. within a cavity barrier;	FD 30	E30
b. between a dwellinghouse and a garage	FD 30	E20
9. Any door:		
a. forming part of the enclosures to a protected stairway in a single family dwellinghouse;	FD 20	E20
b. forming part of the enclosure to a protected entrance hall or protected landing in a flat or maisonette;	FD 20	E20
c. within any other fire-resisting construction in a dwelling not described elsewhere in this table	FD 20	E20

Notes:
1. To BS 476: Part 22 (for BS 476: Part 8 subject to paragraph 5 in Appendix A).
2. Unless pressurization techniques complying with BS 5588: Part 4 *Fire precautions in the design, construction and use of buildings, Code of practice for smoke control using pressure differentials* are used, these doors should also either:
   (a) have a leakage rate not exceeding 3 m^3/m/hour (head and jambs only) when tested at 25 Pa under BS 476 *Fire tests on building materials and structures,* Section 31.1 *Methods for measuring smoke penetration through doorsets and shutter assemblies, Method of measurement under ambient temperature conditions*; or
   (b) meet the additional classification requirement of S. when tested to BS EN 1634-3:xxxx, *Fire resistance tests for door and shutter assemblies, Part 3 – Smoke control doors.*
3. The National classifications do not automatically equate with the equivalent classifications in the European column, therefore products cannot typically assume a European class unless they have been tested accordingly.

sides of the fire doors, except for cupboards and service ducts where it is only necessary to mark the doors on the outside. This recommendation does not apply to:

- fire doors within dwellinghouses
- fire doors to and within flats or maisonettes
- bedroom doors in other residential buildings (PG 2(b))
- lift entrance/landing doors.

Normal fire doors do not provide any significant amount of insulation. Therefore it is necessary to limit the proportion of doorway openings in compartment walls to 25% of the length of the wall, unless the doors provide both integrity and insulation (see Appendix A, Table A2).

Further sources of guidance

The following additional sources of guidance to fire doors and ironmongery are referred to in AD B and may be found useful.

- Recommendations for the specification, design, construction, installation and maintenance of fire doors constructed with non-metallic leaves may be found in BS 8214: *Code of practice for fire door assemblies with non-metallic leaves*, 1990.
- Guidance on timber fire-resisting doorsets, in relation to the new European test standard, may be found in *Timber Fire-Resisting Doorsets: maintaining performance under the new European test standard* published by TRADA.
- Guidance for metal doors is given in *Code of practice for fire-resisting metal doorsets* published by the DSMA (Door and Shutter Manufacturers' Association) in 1999.
- Ironmongery used on fire doors can significantly affect their performance in a fire. Further guidance is available in *Hardware for timber and escape doors* published by the Builders Hardware Industry Federation, 2000.

Rolling shutters

Rolling shutters are sometimes used to protect compartments and means of escape. They are usually held open by automatic release mechanisms (see above). Where rolling shutters are provided across a means of escape, they should only be released by a heat sensor (such as a fusible link or electric heat detector) situated in the immediate vicinity of the door. Closure initiated by smoke detectors or a fire alarm system should not be considered unless it is also the intention that the shutter will descend partially to form a boundary to a smoke reservoir.

All rolling shutters should be capable of manual operation for firefighting purposes.

Glazing to fire doors

It is often desirable to provide glazed vision panels in fire doors. Where the glazing can satisfy the relevant insulation criteria from Table Al of Appendix A, there are no limitations on its use in fire doors. Where this is not the case, the uninsulated glazing should comply with the recommendations of Table A4 of Appendix A, which is reproduced below. Table A4 also contains details of the use of uninsulated glazing in protected stairways, lobbies and corridors. This is described more fully under 'Protected shafts' below.

7.22.4 Protected shafts

Protected shafts are needed when it is necessary to pass persons, things or air between compartments. Therefore, they should only be used to accommodate:

AD B, Appendix A

Table A4 Limitations on the use of uninsulated glazed elements on escape routes. (These limitations do not apply to glazed elements which satisfy the relevant insulation criterion, see Table A1) (See BS 5588: Part 7 for glazing to atria; see BS 5588: Part 8 for glazing to refuges)

Position of glazed element	Maximum total glazed area in parts of a building with access to:			
	a single stairway		more than one stairway	
	walls	door leaf	walls	door leaf
Single family dwellinghouses 1. a. Within the enclosures of: i. a protected stairway, or within fire-resisting separation shown in Fig. 7.13; or	fixed fanlights only	unlimited	fixed fanlights only	unlimited
ii. an existing stair (see section 7.12.10)	unlimited	unlimited	unlimited	unlimited
b. Within fire-resisting separation: i. shown in Fig. 7.14; or ii. described in section 7.12.6	unlimited above 100 mm from floor	unlimited above 100 mm from floor	unlimited above 100 mm from floor	unlimited above 100 mm from floor
c. Existing window between an attached/integral garage and the house	unlimited	not applicable	unlimited	not applicable
Flats and maisonettes 2. Within the enclosures of a protected entrance hall or protected landing or within fire-resisting separation shown in Fig. 7.17	fixed fanlights only	unlimited above 1100 mm from floor	fixed fanlights only	unlimited above 1100 mm from floor
General (except dwellinghouses) 3. Between residential/sleeping accommodation and a common escape route (corridor, lobby or stair)	nil	nil	nil	nil
4. Between a protected stairway (1) and: a. the accommodation; or b. a corridor which is not a protected corridor. Other than in item 3 above	nil	25% of door area	unlimited above 1100 mm (2)	50% of door area
5. Between: a. a protected stairway (1) and a protected lobby or protected corridor; or b. accommodation and a protected lobby. Other than in item 3 above	unlimited above 1100 mm from floor	unlimited above 100 mm from floor	unlimited above 100 mm from floor	unlimited above 100 mm from floor

(Contd)

Table A4 (*Contd*).

Position of glazed element	Maximum total glazed area in parts of a building with access to:			
	a single stairway		more than one stairway	
	walls	door leaf	walls	door leaf
6. Between the accommodation and a protected corridor forming a dead end. Other than in item 3 above	unlimited above 1100 mm from floor	unlimited above 100 mm from floor	unlimited above 1100 mm from floor	unlimited above 100 mm from floor
7. Between accommodation and any other corridor; or subdividing corridors. Other than in item 3 above	not applicable	not applicable	unlimited above 100 mm from floor	unlimited above 100 mm from floor
8. Adjacent an external escape route described in section 7.15.10	unlimited above 1100 mm from paving	unlimited above 1100 mm from paving	unlimited above 1100 mm from paving	unlimited above 1100 mm from paving
9. Adjacent an external escape stair (see section 7.15.10 and Fig. 7.23) or roof escape (see section 7.15.4)	unlimited	unlimited	unlimited	unlimited

Notes:
1. If the protected stairway is also a protected shaft (see section 7.22.4) or a firefighting stair (see section 7.30.2) there may be further restrictions on the uses of glazed elements.
2. Measured vertically from the landing floor level or the stair pitch line.
3. The 100 mm limit is intended to reduce the risk of fire spread from a floor covering.
4. Items 1c, 3 and 6 apply also to single storey buildings.

- stairs, lifts and escalators
- pipes, ducts or chutes
- sanitary accommodation and/or washrooms.

They should form a complete barrier between the different compartments which they connect and, except for glazed screens which should meet the recommendations referred to below, should have fire resistance as specified in Table A1 of Appendix A (see above).

A protected shaft containing a stairway is often approached by way of a corridor or lobby. It is sometimes desirable to glaze the wall between the shaft and the corridor or lobby in order to allow light and visibility in both directions.

This glazing is permitted provided that it has at least 30 minutes fire resistance in terms of integrity and the following conditions are met:

- the stair enclosure is not required to have more than 60 minutes fire resistance
- the corridor or lobby has at least 30 minutes fire separation from the rest of the floor (including doors)
- the protected shaft is not a firefighting shaft.

These recommendations are illustrated in Fig. 7.36. Where these provisions cannot be met, the guidance shown in Table A4 of Appendix A relating to the limits on areas of uninsulated glazing will apply. There should be no oil pipe or ventilating

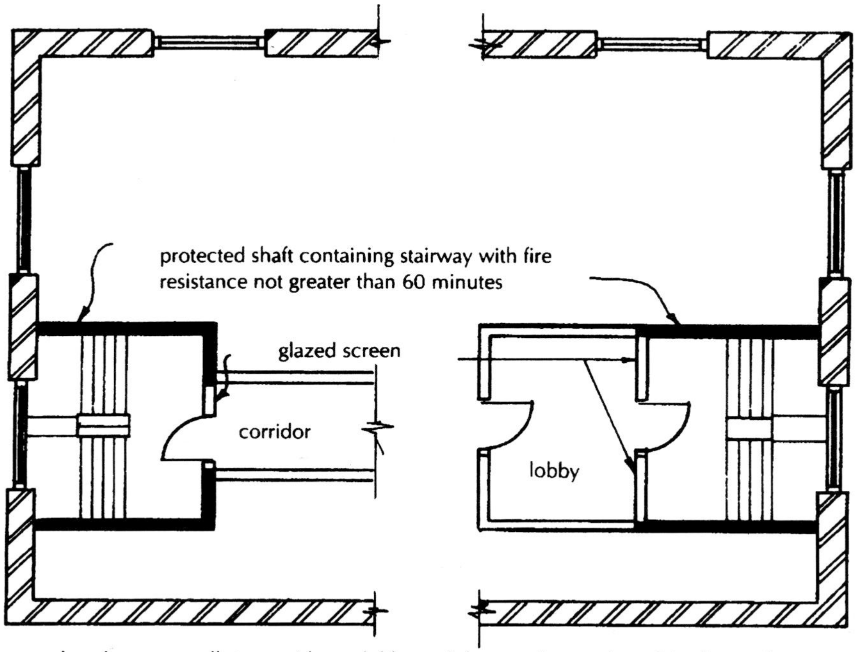

Fig. 7.36 Glazed screen separating protected shaft from lobby or corridor.

duct within any protected shaft which contains any stairway and/or lift (although pipes which convey oil for hydraulic lift mechanisms and ducts used in pressurisation systems aimed at keeping stairs smoke-free are permitted).

Where a protected shaft contains a pipe carrying natural gas or LPG, the pipe should be of screwed steel or of all welded steel construction in accordance with the Pipelines Safety Regulations 1996, S.I. 1996 No. 825, and Gas Safety (Installation & Use) Regulations 1998, SI 1998 No 2451. Where a pipe is completely separated from a protected shaft by fire-resisting construction it is not considered to be contained within a protected shaft. The shaft should be adequately ventilated direct to external air by ventilation openings at high and low level in the shaft and any extension of the floor of the storey into the shaft should not compromise the free movement of air over the entire length of the shaft. Guidance on shafts, such as those which convey piped flammable gas, may be found in BS 8313: 1989 *Code of practice for accommodation of building services in ducts.*

Ideally, protected shafts should be imperforate except for certain openings mentioned below. The number of openings permitted will depend, to a great extent, on the function of the wall surrounding the protected shaft.

Generally, external walls to protected shafts do not need to be fire-resisting and hence there are no restrictions on the number of openings in such walls. This would not be so if the external wall was part of a firefighting shaft; in this case, reference should be made to BS 5588: Part 5: 1991 *Code of practice for firefighting stairs and lifts*. See also *Protection of external walls of protected stairways* in section 7.15.8 above.

Where part of the enclosure to a protected shaft consists of a wall which is common to two or more buildings, the only openings permitted are those referred to in section 7.22.2 above.

Any other walls which make up the enclosure should have no openings other than those referred to below.

- A protected shaft containing one or more lifts may have openings to allow lift cables to pass to the lift motor room. If this is at the bottom of the shaft, the openings should be kept as small as possible.
- Where a protected shaft contains or is itself a ventilating duct, any inlets to, outlets from and openings for the duct should comply with the guidance in section 11 of AD B3 (see Pipes, ventilation ducts and flues in section 7.24).
- It is permissible to form an opening for pipes (other than those specifically forbidden above), provided that the pipe complies with section 11 of AD B3.
- Any opening, other than those detailed above, should be fitted with a fire door complying with Table B1 of Appendix B of AD B.

Figure 7.37 illustrates the principles of protected shafts.

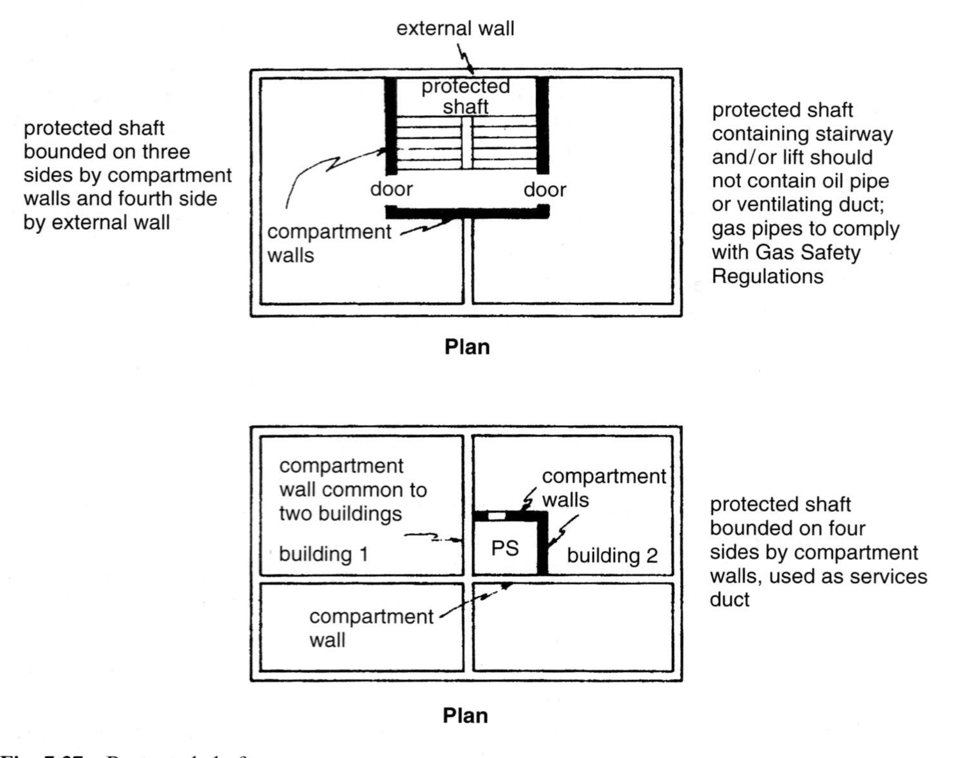

Fig. 7.37 Protected shafts.

7.23 Concealed spaces

Many buildings constructed today contain large hidden void spaces within floors, walls and roofs. This is particularly true of system-built housing, schools and other local authority buildings such as old people's homes.

These buildings may also contain combustible wall panels, frames and insulation, thereby increasing the risk of unseen smoke and flame spread through these concealed spaces.

Therefore, despite compartmentation and the use of fire-resistant construction, many buildings have been destroyed as a result of fire spreading through cavities formed by, or in, constructional elements, and by-passing compartment walls/ floors, etc.

Section 10 of AD B contains provisions designed to reduce the chance of hidden fire spread by making sure that the edges of openings are closed and that cavities are:

- interrupted if there is a chance that the cavity could form a route around a barrier to fire (such as a compartment wall or floor)
- subdivided if they are very large.

This interruption or sub-division of concealed spaces is achieved by using *cavity barriers*. These are defined in Appendix E of AD B as any form of construction (other than a smoke curtain) which is intended to close a cavity (concealed space) and prevent the penetration of smoke or flame, or is fitted inside a cavity in order to restrict the movement of smoke or flame within the cavity. Therefore, provided that it meets the requirements for cavity barriers, a form of construction designed for some other use (such as a compartment wall) may be acceptable as a cavity barrier.

7.23.1 Interruption of cavities

Where an element which is required to form a barrier to fire abuts another element containing a cavity, there is a risk that smoke and flame could by-pass the fire barrier via the cavity. Cavity barriers should, therefore, be provided to interrupt the cavity at the point of contact between the fire barrier and the element containing the cavity (see Fig. 7.38).

The degree to which cavity barriers should be provided will depend on the use (or Purpose Group) of the building concerned and is set out in Table 13 of AD B3. These provisions are summarised below.

Cavities in some elements are excluded from the above provisions regarding interruption, mainly on the grounds that they present little risk to unseen fire spread. These include any cavity:

- in an external wall of masonry construction which complies with the guidance shown in Fig. 7.39 below; and
- in a wall which requires fire resistance only because it is load-bearing (therefore it does not form a barrier to fire).

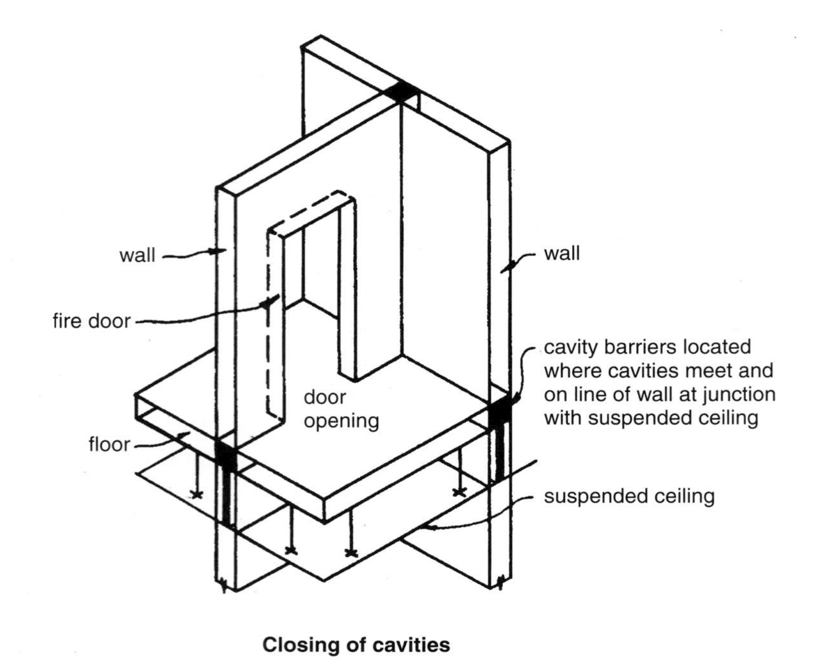

Closing of cavities

fire-resisting element

cavity barriers located in same plane as fire-resisting element (prevents fire by-passing element)

Includes: wall, floor, roof or other structure

element with cavity

Excludes: wall required to have fire resistance only because it is load-bearing

Interrupting cavities

Fig. 7.38 Cavity barriers.

7.23.2 Cavity subdivision

Any cavity, including a roof space, should generally be subdivided into separate sections by cavity barriers placed across the cavity at intervals not greater than the distances specified in Table 14 to section 10 of AD B3, which is reproduced below. These distances are measured along the members bounding the cavity and depend on the cavity location and the class of surface exposed within it (see Fig. 7.40).

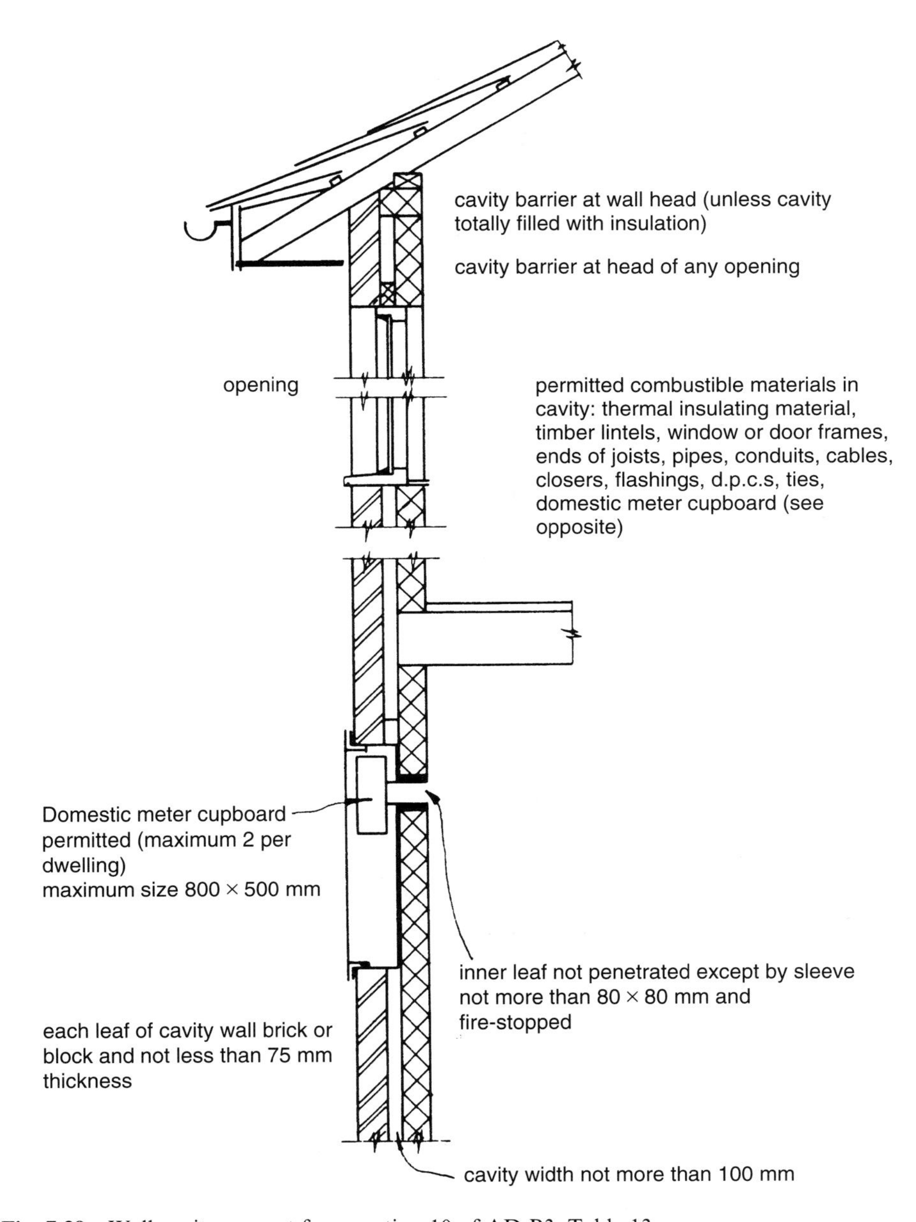

Fig. 7.39 Wall cavity exempt from section 10 of AD B3, Table 13.

Section 10 of AD B3 permits certain variations to the dimensions given in Table 14 as follows.

- If a room under a ceiling cavity exceeds the dimensions given in Table 14, then cavity barriers need only be placed on the line of the enclosing walls or partitions of that room, subject to a maximum cavity barrier spacing of 40 m, provided that

AD B3, Section 10

Table 14 Maximum dimensions of cavities in non-domestic buildings (Purpose Groups 2–7).

Locations of cavity	Class of surface/product exposed in cavity (excluding the surface of any pipe, cable or conduit, or any insulation to any pipe)		Maximum dimensions in any direction (m)
	National class	European class	
Between roof and a ceiling	Any	Any	20
Any other cavity	Class 0 or Class 1	Class A1 or Class A2-s3, d2 or Class B-s3, d2 or Class C-s3, d1	20
	Not Class 0 or Class 1	Not any of the above classes	10

Notes:
1. Exceptions to these provisions are given in paragraphs 10.11–10.13 of AD B3 and are summarised in section 7.23.2.
2. The National classifications do not automatically equate with the equivalent classifications in the European column, therefore products cannot typically assume a European class unless they have been tested accordingly.
3. When a classification includes 's3, d2', this means that there is no limit set for smoke production and/or flaming droplets/particles.

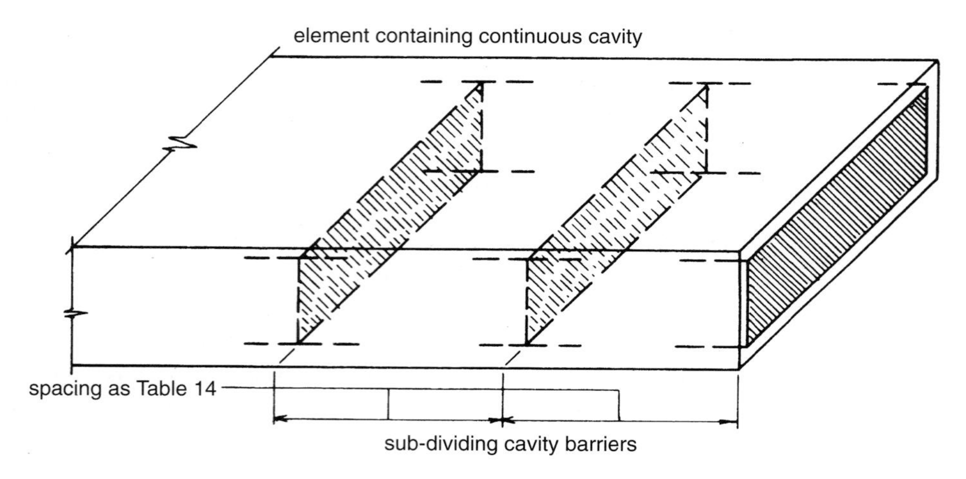

This requirement does not apply to:

- certain cavity walls in masonry construction (see Fig. 7.39)
- floors next to ground or oversite concrete if no access or not more than 1 m high space and no possibility of rubbish accumulating below floor
- cavity between non-combustible roof sheeting under certain conditions
- cavity in an underfloor service void

Fig. 7.40 Subdivision of cavities.

the surfaces exposed in the cavity are Class 0 or Class 1 (National class) or Class C-s3, d2 or better (European class).

- Cavities over undivided areas which exceed the 40 m limit mentioned above in both directions need not be divided with cavity barriers if the following conditions can be met:
  - (a) both room and cavity are compartmented from the rest of the building;
  - (b) an automatic fire detection and alarm system to BS 5839: Part 1: 1988 is fitted in the building (although detectors are not required in the cavity);
  - (c) if the cavity is used as a plenum for ventilating and air-conditioning ductwork the recommendations about recirculating air distribution systems in BS 5588: Part 9, should be followed;
  - (d) the ceiling surface exposed in the cavity is Class 0 (National class) or Class C-s3, d2 or better (European class) and any supports or fixings for the ceiling are non-combustible;
  - (e) any pipe insulation system should have a Class 1 flame spread rating;
  - (f) electrical wiring in the void should be laid in metal trays or metal conduit;
  - (g) any other materials in the cavity should be of limited combustibility.

Additionally, the following low-risk cavities are excluded from the provisions of Table 14 (as well as those specified above for cavity interruption):

- under a floor next to the ground or oversite concrete, provided that either:
  - (a) the height of the cavity is not more than 1 m, or
  - (b) there is no access to the cavity for persons;

This exclusion does not apply if it is possible for combustible material to accumulate in the cavity through openings in the floor (such as happened at the Bradford City Football Club fire). In this case the cavity should be accessible for cleaning and should contain cavity barriers as described in Table 14.

- between double-skinned corrugated or profiled insulated roof sheeting consisting of materials of limited combustibility, provided that the sheets are separated by insulating material having surfaces of Class 0 or Class 1(National class) or Class C-s3, d2 or better (European class), and that insulating material is in contact with both the inner and outer liner sheets. It should be noted that when a classification includes 's3, d2', this means that there is no limit set for smoke production and/or flaming droplets/particles;
- within an underfloor service void;
- within a floor or roof, or enclosed by a roof, provided the lower side of that cavity is enclosed by a ceiling which:
  - (a) extends throughout the whole building or compartment;
  - (b) is not designed to be demountable;
  - (c) has at least 30 minutes fire resistance;
  - (d) is imperforate except for any openings permitted in a cavity barrier (see below);

(e) has an upper surface facing the cavity of at least Class 1 surface spread of flame;
(f) has a lower surface of Class 0 (National class) or Class C-s3, d2 or better (European class).

Cavities above such fire-resisting ceilings are subject to an overall limit of 30 m in extent (see Fig. 7.41).

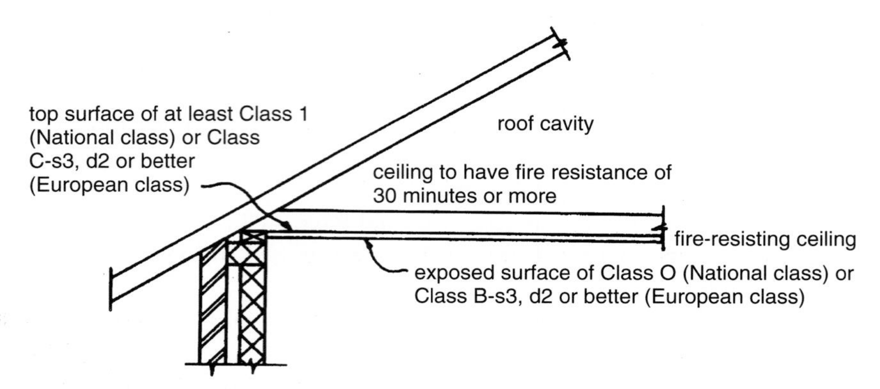

Fig. 7.41 Fire-resisting ceilings.

7.23.3 Provision of cavity barriers

Bearing in mind the exclusions already mentioned above, cavity barriers should be provided as indicated below:

In all buildings

(a) At the junction between a compartment wall which separates buildings and an external cavity wall. The top of the external wall in this case should also be closed with cavity barriers.
(b) At the edges of cavities (including around openings). Such cavity barriers may be formed by a window or door frame if this is appropriate.

In a dwellinghouse with a storey which is more than 4.5 m above ground level

- Above the enclosure to a protected stairway unless a fire-resisting ceiling is provided as shown in Figs. 7.41 and 7.42.

In flats and maisonettes

- At the junction between every compartment wall and compartment floor and an external cavity wall.

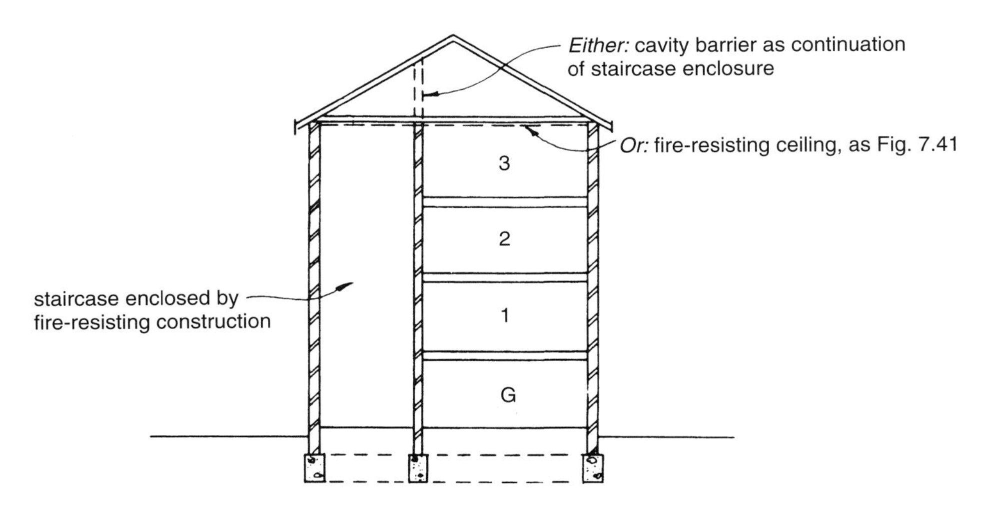

Fig. 7.42 Provision of cavity barriers in dwellinghouses of three or more storeys.

- At the junction between a cavity wall and every compartment wall, compartment floor or other wall or door assembly which forms a fire-resisting barrier.
- Above fire-resisting construction in a protected escape route if this is not carried up to the full storey height (or to the underside of the roof, if it is in the top storey).

In office, shop and commercial, assembly and recreation, industrial, storage and other non-residential buildings

- All as recommended for flats and maisonettes above.
- Above corridor enclosures where the corridor is subdivided to prevent smoke or fire affecting two escape routes simultaneously. The cavity barriers are not needed if the corridor enclosure is carried up to the full storey height (or to the underside of the roof, if it is in the top storey) or if the cavity is enclosed on the lower side by a fire-resisting ceiling complying with Fig. 7.41. (See also Fig.7.43.)
- To subdivide extensive cavities in floors, roofs, walls, etc. (but not in underfloor service voids) so that the distance between cavity barriers does not exceed the dimensions given in Table 14 (see above).

In other residential and institutional buildings:

- All as recommended for Office, etc., buildings above.
- Above any bedroom partitions if these are not carried up to the full storey height (or to the underside of the roof, if they are in the top storey).

Rain screen cladding

Rain screen cladding is a form of construction often used in refurbishment works to improve the weather resistance and thermal performance of the external walls of

To prevent both storey exits from becoming blocked by smoke:

EITHER

(a) sub-divide storey with fire-resisting construction

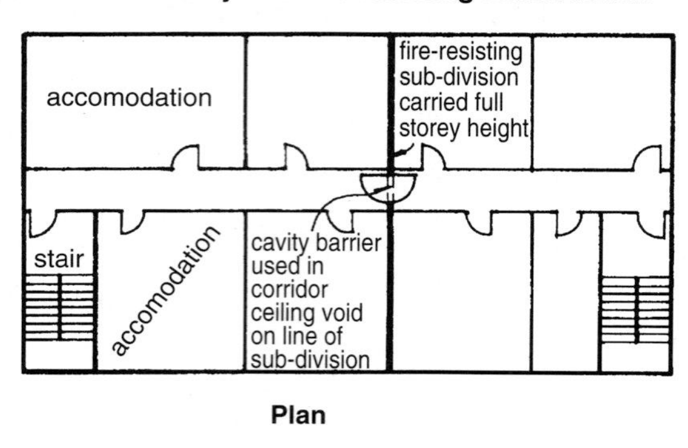

OR

(b) provide cavity barriers above corridor enclosure if enclosures are not carried up to full storey height

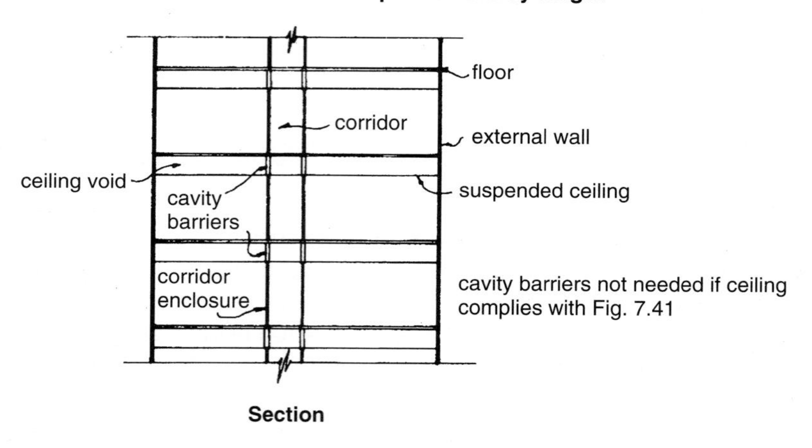

Fig. 7.43 Alternative forms of corridor enclosure.

buildings. It has been used regularly to upgrade the performance of blocks of high-rise flats (often built in the 1960s), where an external weatherproof cladding is lined internally with combustible insulation and this is fixed to the outer surface of the external walls. There have been occasions when fire has occurred within the void space containing the insulation and fire has spread upwards within the cladding, eventually affecting the interior of the building. Table 13 of AD B3 recommends that cavity barriers be provided within the void behind the external face of rain screen cladding at every floor level, and on the line of compartment walls abutting the external wall, where the building has any floor which is more than 18 m above ground level. This applies to Flats, Maisonettes, Other residential buildings and Institutional buildings. The spacing recommendations of Table 14 do not apply to such rain screen cladding, or to the over-cladding of an external masonry (or concrete) external wall, or to an over-clad concrete roof where the cavity does not contain combustible insulation and the cavity barriers are positioned as described above.

7.23.4 Construction of cavity barriers

A cavity barrier may be formed by construction provided for another purpose provided that it meets the recommendations for cavity barriers. However, where compartment walls are provided these should be taken up the full storey height to a compartment floor or roof and not completed by cavity barriers above them. This is because the fire resistance standards for compartment walls are higher than those for cavity barriers, and compartment walls should therefore be continued through the cavity to maintain the fire resistance standard.

Table A1 of Appendix A recommends that all cavity barriers should have a minimum standard of fire resistance of 30 minutes with regard to integrity and 15 minutes with regard to insulation. The only exception to this is in the case of a cavity barrier in a stud wall or partition which is permitted to be formed of:

- steel at least 0.5 mm thick
- timber at least 38 mm thick
- polythene sleeved mineral wool, or mineral wool slabs, in either case under compression when installed in the cavity
- calcium silicate, cement based or gypsum based boards at least 12 m thick.

Cavity barriers should be tightly fitted against rigid construction and mechanically fixed in position where possible. Where they abut against slates, tiles, corrugated sheeting and similar non-rigid construction, the junctions should be fire-stopped as described below.

Cavity barriers should also be fixed in such a way that their performance is unlikely to be affected by:

- building movements due to shrinkage, subsidence, thermal change, or the movement of the external envelope due to wind; or
- collapse in a fire of any services which penetrate them; or
- failure of their fixings, or any construction or material which they abut, due to fire. However, where cavity barriers are provided in roof spaces, the roof members to which they are fitted are not expected to have fire resistance.

7.23.5 Openings in cavity barriers

Cavity barriers should be imperforate except for one or more of the following openings:

- for a pipe which complies with Section 11 of AD B3
- for a cable, or for a conduit containing one or more cables

- if fitted with a suitably mounted fire damper
- for a duct (unless it is fire-resisting), fitted with an automatic fire damper where it passes through the barrier
- if fitted with a door which complies with Appendix B of AD B, and having at least 30 minutes fire resistance.

The above provisions do not apply to any cavity barrier provided above any bedroom partitions which are not carried up to the full storey height (or to the underside of the roof, if they are in the top storey).

7.24 Pipes, ventilation ducts and flues

It is impossible to construct a building without passing some pipes or ducts through the walls and floors, and such penetration of the fire separating elements of structure is a potential source of flame and smoke spread. Section 11 of AD B3 therefore attempts to control the specifications of such pipes and ducts and of their associated enclosing structures. The measures in section 11 are primarily designed to delay the passage of fire. They may also have the added benefit of retarding smoke spread, but the integrity test specified in Appendix A of AD B does not cover criteria for the passage of smoke as such.

7.24.1 Pipes

For the purposes of section 11, the term 'pipe' includes a ventilating pipe for an above-ground drainage system, but does not include any flue pipe or other form of ventilating pipe.

As is usual, the expression 'pipe' here may be read as 'pipeline', and should be taken to include all pipe fittings and accessories.

Requirements

Where a pipe as defined passes through an opening in:

- a compartment wall or compartment floor (unless the pipe is wholly enclosed within a protected shaft); or
- a cavity barrier;

then either a proprietary sealing system should be used which will maintain the fire resistance of the floor, wall or cavity barrier (and has been shown by test to do so) or, the nominal internal diameter of the pipe should not exceed the relevant dimension listed in Table 15 of section 11 of AD B3 (see below). The opening should be as small as practicable and fire-stopped around the pipe.

Where a pipe of specification (b) of Table 15 penetrates a structure, it is permissible to pass it through or connect it to a pipe or sleeve of specification (a) of Table 15 provided the pipe or sleeve of specification (a) extends on both sides of the

AD B3, Section 11

Table 15 Maximum nominal internal diameter of pipes passing through a compartment wall/floor (see para 11.5 et seq)

Situation	Pipe material and maximum nominal internal diameter (mm)		
	(a) Non-combustible material (1)	**(b) Lead, aluminium, aluminium alloy, uPVC (2), fibre cement**	**(c) Any other material**
1. Structure (but not a wall separating buildings) enclosing a protected shaft which is not a stairway or a lift shaft	160	110	40
2. Wall separating dwelling houses, or compartment wall or compartment floor between flats	160	160 (stack pipe) (3) 110 (branch pipe) (3)	40
3. Any other situation	160	40	40

Notes:
1. Any non-combustible material (such as cast iron, copper or steel) which if exposed to a temperature of 800°C, will not soften or fracture to the extent that flame or hot gas will pass through the wall of the pipe.
2. uPVC pipes complying with BS 4514 and uPVC pipes complying with BS 5255.
3. These diameters are only in relation to pipes forming part of an above-ground drainage system and enclosed as shown in Diagram 38. In other cases the maximum diameters against situation 3 apply.

structure for a minimum distance of 1m. The sleeve should be in contact with the pipe. (See Fig. 7.44.)

The following above-ground drainage system pipes complying with specification (b) of Table 15 may be passed through openings in a wall which separates houses, or through openings in a compartment wall or compartment floor in flats.

- A stack pipe of not more than 160 mm nominal internal diameter, provided it is contained within an enclosure in each storey; and
- a branch pipe of not more than 110 mm nominal internal diameter, provided it discharges into a stack pipe which is contained in an enclosure, the enclosure being partly formed by the wall penetrated by the branch pipe.

The enclosures referred to in both cases immediately above should comply with the following requirements.

- In any storey, the enclosure should extend from floor to ceiling or from floor to floor if the ceiling is suspended.
- Each side of the enclosure should be formed by a compartment wall or floor, external wall, intermediate floor or casing.
- The internal surface of the enclosure should meet the requirements of Class 0 (National class) or Class C-s3, d2 or better (European class), except for any supporting members.

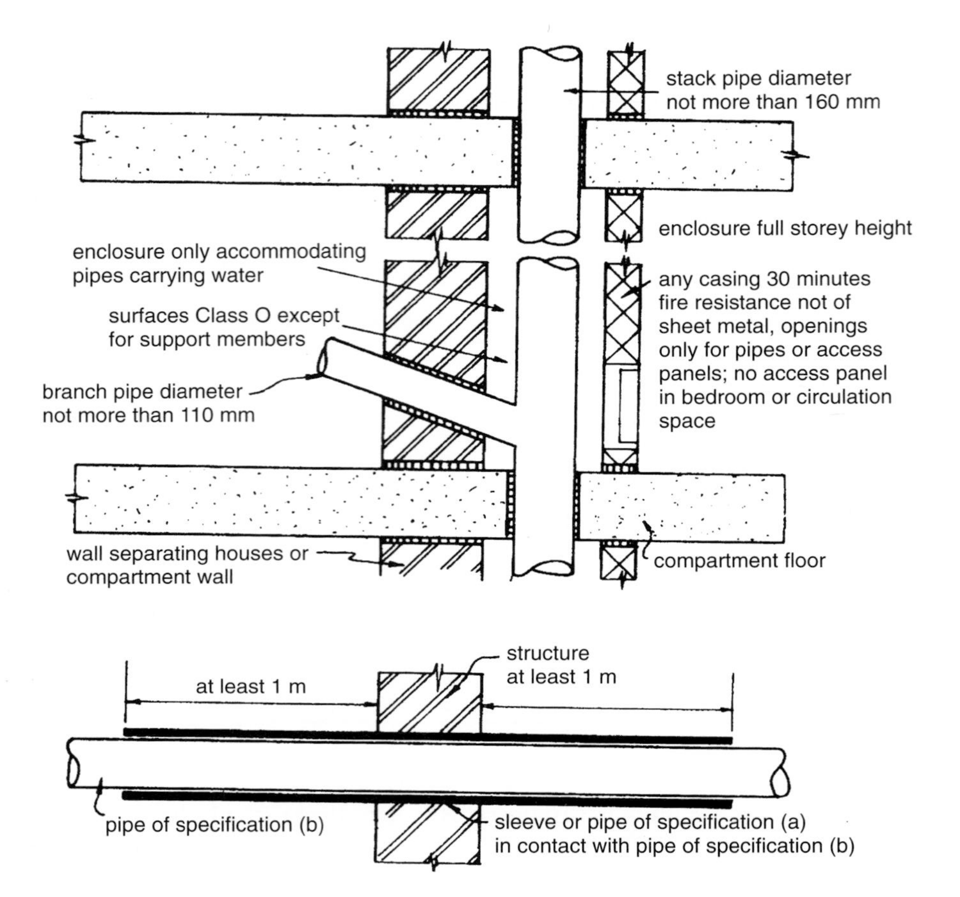

Fig. 7.44 Penetration of structure by pipes.

- No access panel to the enclosure should be fitted in any bedroom or circulation space.
- The enclosure should not be used for any purpose except to accommodate drainage or water supply pipes.

The 'casing' referred to in section 11, Diagram 34 and the second requirement immediately above, should provide at least 30 minutes fire resistance, including any access panel, and it should not be formed of sheet metal. The only openings permitted in a casing are openings for the passage of a pipe, or openings fitted with an access panel. The pipe opening, whether it be in the structure or the casing, should be as small as is practicable and fire-stopped around the pipe (see Fig. 7.44). A casing to the drainage or water supply pipes should always be provided if a wall which separates houses is penetrated by a branch pipe in the top storey.

7.24.2 Ventilating ducts

Ventilating ducts, normally forming part of an air-conditioning system, convey air to various parts of a building. It is, therefore, inevitable that they will need to pass through compartment walls and floors at some stage and it is important that the integrity of these fire separating elements is maintained. This can be achieved by following the guidance in Part 9 of BS 5588 where alternative ways of protecting compartmentation are described when air handling ducts pass from one compartment to another.

7.24.3 Flues

Where any flue, appliance ventilation duct or duct containing one or more flues:

- passes through a compartment wall or floor, or
- is built into a compartment wall,

the flue or duct walls should be separated from the compartment wall or floor by non-combustible construction of fire resistance equal to at least half that required for the compartment wall or floor (see Fig. 7.45).

For the purposes of the above, an appliance ventilation duct is a duct provided to convey combustion air to a gas appliance.

7.24.4 Fire stops

A fire stop is defined in Appendix E of AD B as a seal provided to close an imperfection of fit or design tolerance between elements or components, to restrict the passage of fire and smoke.

Therefore, fire stops should be provided:

- at junctions or joints between elements which are required to act as a barrier to fire; and
- where pipes, ducts, conduits or cables pass through openings in cavity barriers (see section 7.23.5 above), or elements which serve as a barrier to fire.

In the second case above, the openings should be kept as small and as few in number as possible, and the fire-stopping should not restrict the thermal movement of pipes or ducts.

Fire-stopping materials should be reinforced with or supported by materials of limited combustibility to prevent displacement if:

- the unsupported span exceeds 100 mm; and
- in any other case, non-rigid materials have been used (unless these have been shown by test to be satisfactory).

Suitable fire-stopping materials include:

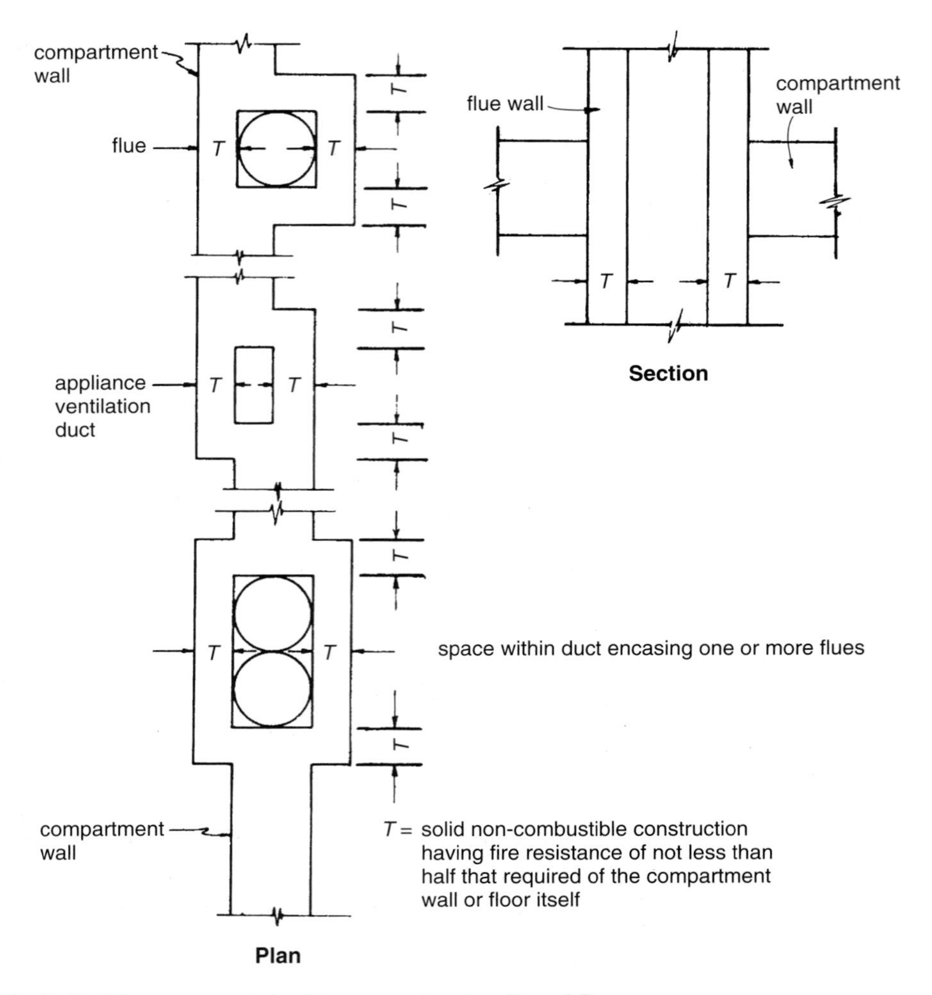

Fig. 7.45 Flues, etc., contained in compartment walls and floors.

- cement mortar
- gypsum based plaster
- cement or gypsum based vermiculite/perlite mixes
- glassfibre, crushed rock, blast furnace slag or ceramic based products (with or without resin binders)
- intumescent mastics
- any proprietary fire-stopping or sealing systems (including those designed for penetration by services) capable of maintaining the fire resistance of the element concerned (test results would be necessary to prove acceptability).

These materials should be used in appropriate situations, i.e. they may not all be suitable in every situation.

7.25 Variations to the provisions of Approved Document B3

As has been mentioned in the introduction to this chapter (see section 7.1 above), some difficulty may be encountered when trying to apply the provisions of AD B to existing buildings. Accordingly, AD B3 contains a number of specific recommendations related to raised storage areas, floors in domestic loft conversions and to the conversion of buildings to flats, where the 'normal' provisions are somewhat reduced.

7.25.1 Varying the provisions – raised storage areas

Sometimes, raised free standing floors (usually called mezzanine floors) are erected (often in single-storey industrial buildings) for the purposes of storage. They may be regarded merely as galleries or may be large enough to be considered as a floor forming an additional storey. In such cases, the normal recommendations regarding the fire resistance of elements of structure may prove to be unduly onerous if applied to raised storage areas.

It may be possible to reduce the level of fire resistance or even allow unprotected steelwork if the following precautions are taken.

(1) The structure should be used for storage purposes only and should contain only one tier.
(2) The number of people using the floor at any time should be limited. Members of the public should not be admitted.
(3) The floor is open both above and below to the space in which it is located.
(4) Means of escape is provided from the floor which meets the recommendations of Sections 4, 5 and 6 of AD B1.
(5) The floor should comply with the following parameters regarding its size:
- it should not be more than 10 m in length and/or width and should not be greater than half the floor area of the space in which it is located;
- the floor size may be increased to not more than 20 m in length and width where an automatic fire detection and alarm system is provided on the lower level, which complies with BS 5839: Part 1;
- there are no limits on the size of the floor if the building is fitted throughout with an automatic sprinkler system which meets the relevant recommendations of BS 5306: Part2. (This relates to the relevant occupancy rating together with the additional requirements for life safety).

7.25.2 Varying the provisions – dwellinghouses

Under the recommendations for means of escape in case of fire (AD Bl) certain special provisions apply where it is proposed to construct one or two rooms in the roof space of a two-storey dwelling thereby creating a three-storey dwelling (see section 7.12.10 above).

Floors in an existing two-storey dwelling may only be capable of achieving a modified 30 minutes fire resistance (see section 7.21.2 (Table 7.7) above). However,

AD B3 recommends that floors in a dwelling of three or more storeys should have a full 30 minutes fire resistance.

It is considered reasonable to relax the recommendation for the *existing* floor (thereby allowing the modified 30 minutes standard) provided the following provisions are complied with by way of compensation.

- Only one storey is being added with a floor area not exceeding 50 m^2.
- No more than two habitable rooms are provided in the new storey.
- The existing floor should only separate rooms (not circulation spaces).
- The means of escape provisions relevant to loft conversions in AD Bl (see section 7.12.10 above) should be complied with.

The relaxed recommendation will only apply to any floor which separates rooms (i.e. not circulation spaces). Therefore, the full 30 minute standard will need to be provided where the floor forms part of the enclosures to the circulation space between the loft conversion and the final exit.

It is sometimes the case that a floor is only capable of achieving a modified 30 minutes standard of fire resistance because it is constructed with plain edged boarding on the upper surface. The addition of a 3.2 mm thickness of standard hardboard nailed to the floor boards can usually upgrade the floor to the full 30 minutes standard (see Fig. 7.46).

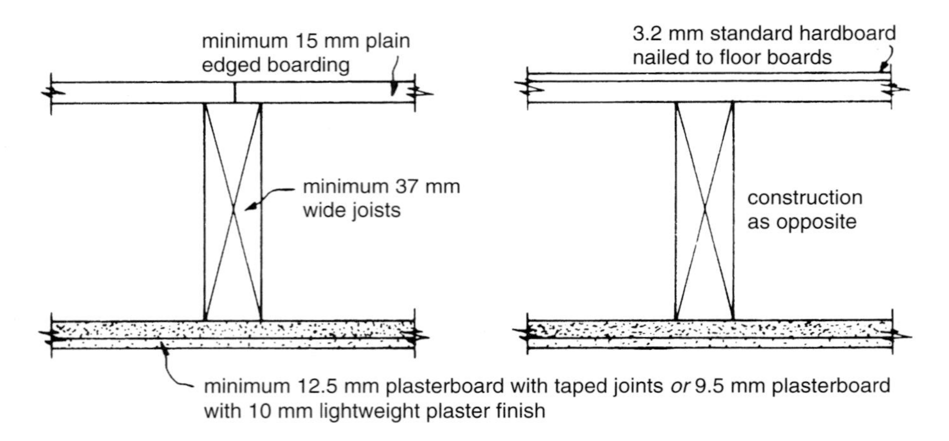

Fig. 7.46 Upgrading of existing floors in dwellings.

Other methods of upgrading existing timber floors can be found in *BRE Digest* 208.

7.25.3 Varying the provisions – conversion to flats

If it is proposed to convert a building into flats, Approved Document B of Schedule 1 of the 2000 Regulations will apply due to the material change of use.

It is often the case that the existing floors are of timber construction and have insufficient fire resistance for the proposed change of use.

The provision of an adequate, fully protected means of escape which complies with the recommendations of Section 3 of AD Bl will allow a 30 minute standard of fire resistance in the elements of structure in a building of not more than three storeys.

The full standard of fire resistance given in Table A2 of Appendix A would normally be required if the converted building contained four or more storeys.

7.26 External fire spread

The external walls of a building are required to adequately resist the spread of fire over their surfaces and from one building to another. In assessing the adequacy of resistance to fire spread, regard must be given to the height, use and position of the building.

The roof of a building must also offer adequate resistance to the spread of fire across its surface and from one building to another, having regard to the use and position of the building.

7.27 External walls

External walls serve to restrict the outward spread of fire to a building beyond the property boundary and also help resist fire from outside the building. This is achieved by ensuring that the walls have adequate fire resistance and external surfaces with restricted fire spread and low rates of heat release. Fire spread between buildings usually occurs by radiation through openings in the external walls (termed 'unprotected areas'). The risk of fire spread and its consequences are related to:

- the severity of the fire
- the fire resistance offered by the facing external walls including the number and disposition of the unprotected areas
- the distance between the buildings
- the risk presented to people in the opposite building.

In general, the severity of a fire will be related to the amount of combustible material contained in the building per unit of floor area (termed the 'fire load density'). Certain types of buildings, such as shops, industrial buildings and warehouses, may contain large quantities of combustible materials and are usually required to be sited further from their boundaries than other types of buildings.

7.27.1 External walls – general constructional recommendations

External walls are elements of structure and therefore they should have the relevant period of fire resistance specified in Appendix A of AD B. However, the provisions for space separation mentioned below allow increasingly large areas of the external walls of a building to be unprotected as the distance to the relevant boundary increases. A point will eventually be reached where the whole of a wall may be unprotected. In such a case only the load-bearing parts of the wall would need fire resistance. Similarly, where a wall is 1 m or more from the relevant boundary it only needs to resist fire from the inside and the insulation criteria of fire resistance are, in most cases, not applied.

The combustibility of the external envelope of a building is also controlled in certain circumstances. The limiting factors are the height of the building, its use and its distance from the relevant boundary. Table 7.8 sets out the recommendations for the external surfaces of walls and it is generally the case that buildings which are less than 1 m from the relevant boundary should have external surfaces of Class 0 (National class) or class B-s3, d2 or better (European class). For buildings which are 18 m or more in height, there are restrictions on the external surface materials irrespective of the distance to the boundary.

After the disastrous fire at the Summerland Leisure complex on the Isle of Man, special recommendations were introduced to prevent other assembly and recreation buildings from suffering a similar fate. The Summerland centre was constructed largely of plastics materials which extended to ground level. A fire was deliberately started adjacent to the building which, because of its rapid surface spread of flame characteristics, quickly became engulfed in flames. Therefore, any Assembly and Recreation Purpose Group building which has more than one storey (galleries counted, but not basements) should have only those external surfaces indicated in Fig. 7.47 below. This also applies in mixed use buildings which include Assembly and Recreation Purpose Group accommodation.

The provisions described above for the combustibility of external wall surfaces may, of course, be affected by the recommendations for space separation and the limits on unprotected areas contained in Section 14 of AD B and described below.

Mention has already been made of the risks involved with the use of combustible insulation in rain screen cladding of buildings (see section 7.23.3). In such a system the surface of the outer cladding which faces the cavity should comply with the provisions of Table 7.8 and Fig. 7.47. Furthermore, any insulation used in ventilated cavities in the external walls of a building over 18 m in height should be composed of materials of limited combustibility, although this restriction does not apply to insulation in masonry cavity walls which comply with Fig. 7.39 above. (Reference should also be made to the BRE Report *Fire performance of external thermal insulation for walls of multi-storey buildings*, BR 135, 1988.)

7.27.2 External walls and steel portal frames

Steel portal frames are commonly used in single-storey industrial and commercial buildings. Structurally, the portal frame acts as a single member. Therefore, where the column sections are built into the external walls, collapse of the roof sections may result in destruction of the walls.

If the building is so situated that the external walls cannot be totally unprotected,

Table 7.8 Limitations on external wall surfaces (all buildings).

Maximum height of building (m)	Distance of external surface from any point on the relevant boundary[a]		
	Less than 1 m	**1 m or more**	
Up to 18	Class 0[1]	No provision	
Over 18	Class 0[1]	Any surface less than 18 m above the ground	Timber at least 9 mm thick; or any material with an index of performance (*I*) not more than 20
		Any surface 18 m or more above the ground	Class 0

Notes:
1. Either class 0 (National class) or class B-s3, d2 or better (European class).
For meaning of class 0 and index of performance (*I*) see section 7.18.3 above.

Note: the provisions described in Table 7.8 and illustrated in Fig. 7.47 might also be met by following the guidance in BRE Fire Note 9 *Assessing the fire performance of external cladding systems: a test method* (BRE, 1999).

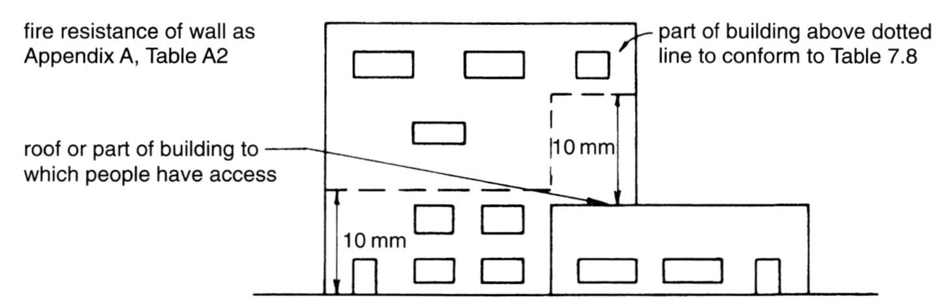

Fig. 7.47 External walls – special provisions for assembly and recreation buildings.

the provisions of AD B may recommend that both rafter and column sections be fire protected. This would result in an uneconomic building which would defeat the object of using a portal frame.

Investigations have been carried out into the behaviour of steel portal frames in fire. Provided that the connection between the frame and its foundation can be made sufficiently rigid to transfer the over-turning moment caused by collapse, in a fire, of the rafters, purlins and some of the roof cladding, it may be possible to remove the fire protection to the rafters and purlins while still allowing the external wall to perform its structural function.

Additional measures may be necessary in certain circumstances to ensure the stability of the external walls. Full details of the design method may be found in the publication *Fire and Steel Construction: The Behaviour of Steel Portal Frames in Boundary Conditions*, 1990 (2nd edition) which is available from the Steel Construction Institute, Silwood Park, Ascot, Berks, SL5 7QN. The publication also contains guidance on many aspects of portal frames including multi-storey types. The recommendations of this publication for designing the foundation to resist overturning need not be followed if the building is fitted with a sprinkler system that follows the relevant provisions in BS 5306: Part 2 (i.e. the relevant occupancy rating together with the additional requirements for life safety).

Normally, reinforced concrete portal frames can support external walls without specific measures at the base to resist overturning.

The following design guidance (which some existing buildings may already comply with) is also acceptable.

- To resist overturning, the column members should be rigidly fixed to a base of suitable size and depth.
- Brick, block or concrete protection should be provided up to a protected ring beam giving lateral support.
- Some form of roof venting should be provided to give early heat release (e.g. pvc roof lights covering at least 10% of the floor area evenly spaced out).

7.27.3 Space separation – permitted limits of unprotected areas

Unprotected areas in the external walls of a building are those areas which have less fire resistance than that recommended by Table A2 of Appendix A of AD B. Areas such as doors, windows, ventilators or combustible cladding are permitted in the external walls but their extent is limited depending on the use of the building and its distance from the relevant boundary. In order that a reasonable standard of space separation may be specified for buildings, the following basic assumptions are made in AD B, Section 14.

- A fire in a compartmented building will be restricted to that compartment and will not spread to adjoining compartments.
- The intensity of the fire is related to the purpose group of the building and it can be moderated by a sprinkler system.
- Residential, and Assembly and Recreation Purpose Groups represent a greater risk to life than other uses.
- Where buildings are on the same site, the spread of fire between them represents a low risk to life and can be discounted. This does not apply to buildings in the Residential, and Assembly and Recreation Purpose Groups.
- There is a building on the far side of the boundary situated an equal distance away with an identical elevation to the building in question.

- The amount of thermal radiation that passes through an external wall which has fire resistance may be discounted.
- A roof which is pitched at less than 70° to the horizontal does not need to comply with the recommendations in section 14 of AD B4. (See also definition of external wall in section 7.2 above).
- Vertical parts of a pitched roof (such as dormer windows) generally do not need to comply with the recommendations in section 14 of AD B4, unless they are part of a roof which is pitched at greater than 70° to the horizontal. However, a continuous run of dormer windows occupying most of a steeply pitched roof might need to be treated as a wall rather than a roof. This will be a matter of individual judgement.

It follows from the above that reduced separation distances (or increased amounts of unprotected areas) may be obtained by dividing a building into compartments.

7.27.4 Boundaries

It is clear from AD B4 that the separation distances referred to are those to the relevant boundaries of the site of the building in question. (Relevant boundary is defined in section 7.2 above and is illustrated in Fig. 7.6). Where the site boundary adjoins an area which is unlikely to be developed, such as a street, canal, railway or river, then the relevant boundary is usually taken as the centreline of that area.

Where buildings share the same site, the separation distance between them is usually discounted. However, if either or both of the buildings are in the Residential, or Assembly and Recreation Purpose Groups, then a notional boundary is assumed to exist between them such that they both comply with the space separation recommendations of AD B4. (Notional boundary is defined in section 7.2 and is illustrated in Fig. 7.5.)

7.27.5 Unprotected areas that can be discounted

Certain openings, etc. in walls have little effect on fire protection. Accordingly, AD B4 provides that four areas may be discounted when calculating the permitted limits of unprotected areas in the external walls of a building.

- Any unprotected area of not more than 0.1 m^2 which is at least 1.5 m away from any other unprotected area in the same side of the building or compartment, except an area of external wall forming part of a protected shaft.
- One or more unprotected areas, with a total area of not more than 1 m^2, which is at least 4 m away from any other unprotected area in the same side of the building or compartment, except a small area of not more than 0.1 m^2 as described above.
- Any unprotected area in an external wall of a stairway in a protected shaft. (But see Fig. 7.31 above for further provisions affecting stairways.)
- An unprotected area in the side of an uncompartmented building, if the area is at least 30 m above the ground adjoining the building.

Where part of an external wall is regarded as an unprotected area merely because of combustible cladding more than 1mm thick, the unprotected area presented by that cladding is to be calculated as only half the actual cladding area (see Fig. 7.48). Any cladding with a Class 0 (National class) or class B-s3, d2 or better (European class) surface spread of flame rating need not be counted as an unprotected area.

Therefore any wall which is situated within 1 m of the relevant boundary should contain only those unprotected areas listed above, and illustrated in Fig. 7.48. The

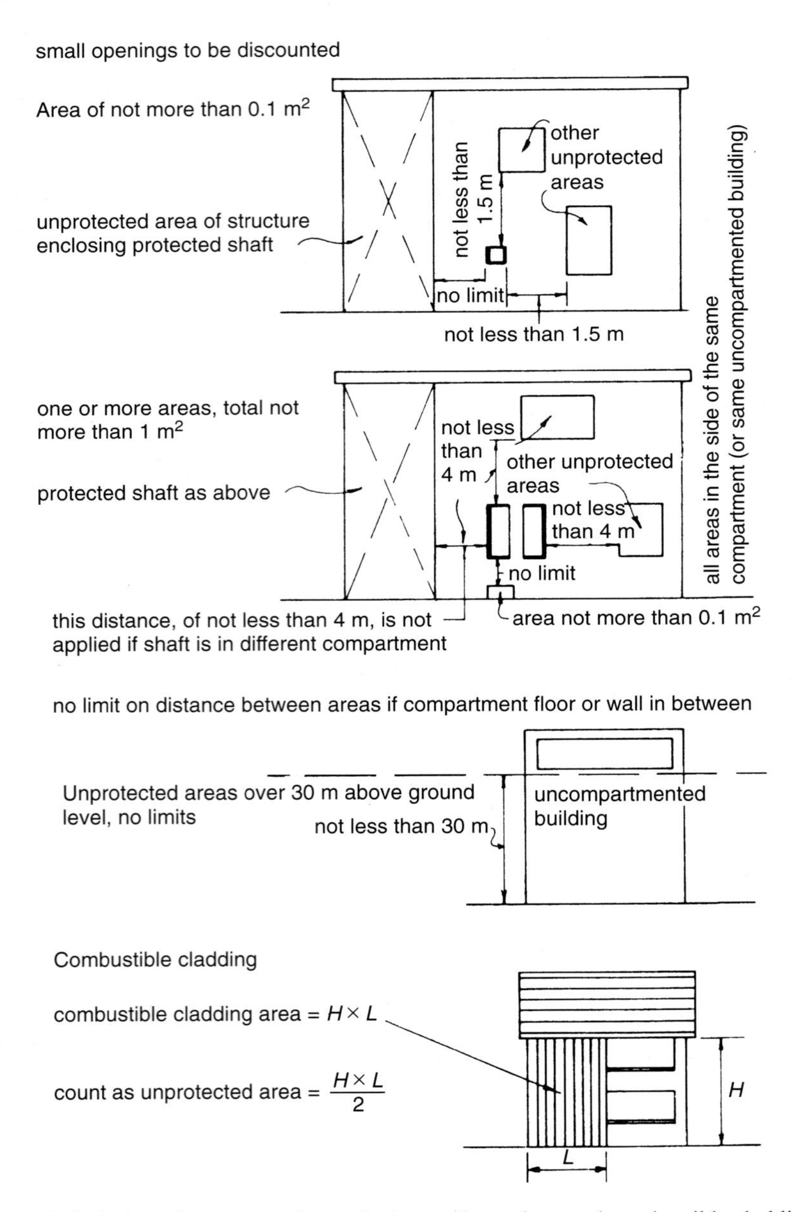

Fig. 7.48 Calculation of unprotected area limit, small openings and combustible cladding.

rest of the wall will need to meet the fire resistance requirements contained in Table A2 of Appendix A of AD B.

7.27.6 Unprotected areas – methods of calculation

Where a wall is situated 1m or more from the relevant boundary, the permitted limit of unprotected areas may be determined by either of two methods described in full in AD B4. The Approved Document also permits other methods, described in a BRE Report *External fire spread: Building separation and boundary distances*, BR 187, BRE, 1991. Part 1 of this report covers the 'Enclosing Rectangle' and 'Aggregate Notional Area' methods which were originally contained in the 1985 edition of AD B2/3/4 and which are also described below. An applicant may use whichever of these methods gives the most favourable result for his own building. Again, the rest of the wall should meet the fire resistance recommendations of Table 2 of Appendix A.

The basis of the two methods described in AD B4 is contained in the BRE Report mentioned above. The building should be separated from its boundary by at least half the distance at which the total thermal radiation intensity received from all unprotected areas in the wall would be 12.6 kW/m^2 in still air, assuming that the radiation intensity at each unprotected area is:

- 84kW/m^2 for buildings in the Residential, Office or Assembly and Recreation Purpose Groups, and
- 168 kW/m^2 for buildings in the Shop and Commercial, Industrial, Storage or Other non-residential Purpose Groups.

AD B, Section 14

Diagram 46 Permitted unprotected areas in small residential buildings.

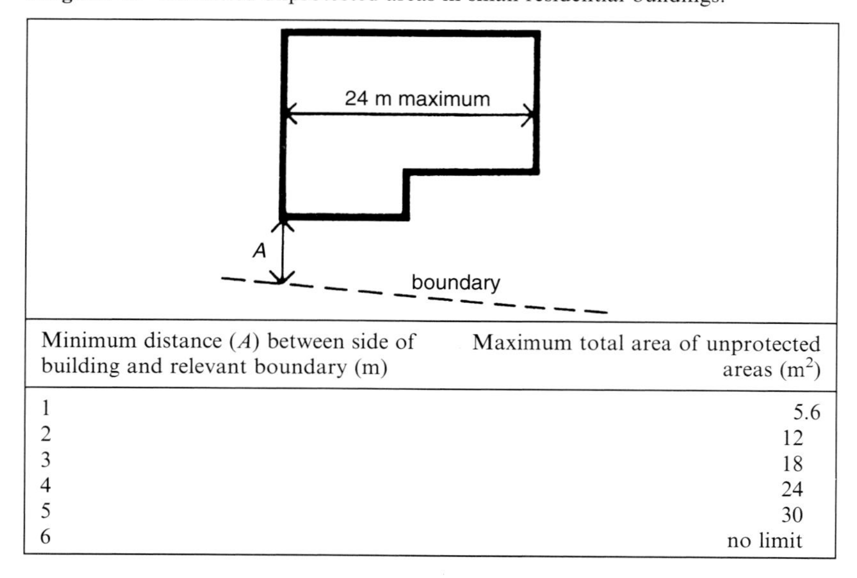

Minimum distance (A) between side of building and relevant boundary (m)	Maximum total area of unprotected areas (m^2)
1	5.6
2	12
3	18
4	24
5	30
6	no limit

This clearly illustrates the different fire load densities assumed for the two groups of buildings.

Where a sprinkler system complying with BS 5306: Part 2 is installed throughout a building, it is reasonable to assume that the extent and intensity of a fire will be reduced. In these circumstances the permitted boundary distances may be halved subject to a minimum distance of 1 m. Alternatively, if the boundary distance is kept the same, the amount of unprotected area can be doubled.

Method 1 – small residential buildings

Method 1 applies only to dwellinghouses, flats, maisonettes or other residential buildings (not institutional buildings) which:

- are not less than 1m from the relevant boundary
- are not more than three storeys high (basements not counted)
- have no side which exceeds 24 m in length.

The permitted limit of unprotected area in an external wall of any of these buildings is given in Diagram 46 of AD B4 which is reproduced above. It varies according to the size of the building and the distance of the side from the relevant boundary. Any parts of the side in excess of the maximum unprotected area should have the recommended fire resistance. The small areas referred to above may be discounted.

Method 2 – all buildings or compartments

This method applies to buildings and compartments in any purpose group which:

- are not less than 1m from the relevant boundary
- are not more than 10 m high (except open-sided car parks in Purpose Group 7(b)).

The permitted limits of unprotected areas given by this method are contained in Table 16 of AD B4 which is reproduced below. It should be noted that actual areas are not given. Column 3 of Table 16 expresses the permitted unprotected areas in percentage terms but fails to state what the percentage relates to. It is assumed that the percentages given refer to the total area of the side of the building in question. Thus, if a shop is at least 2 m from its relevant boundary then it is permitted to have 8% of its external wall area on that side as unprotected. Any other areas (except the small permitted areas) would need the requisite fire resistance.

For buildings or compartments that exceed 10m in height the methods set out in BRE Report, BR 187 can be applied. These are explained in the following paragraphs.

Additional methods – enclosing rectangles

This method of calculating the permitted limit of unprotected areas is based on the smallest rectangle of a height and width taken from Table 7.9 (which follows and is

AD B, Section 14

Table 16 Permitted unprotected areas in small buildings or compartments.

Minimum distance between side of building and relevant boundary (m)		Maximum total percentage of unprotected area %
Purpose groups		
Residential, Office, Assembly and Recreation (1)	Shop & Commercial, Industrial, Storage & other Non-residential (2)	(3)
n.a.	1	4
1	2	8
2.5	5	20
5	10	40
7.5	15	60
10	20	80
12.5	25	100

Notes:
n.a. = not applicable.
a. Intermediate values may be obtained by interpolation.
b. For buildings which are fitted throughout with an automatic sprinker system, see section 7.27.6.
c. In the case of open-sided car parks in Purpose Group 7(b), the distances set out in column (1) may be used instead of those in column (2).

based on Table J2 of the 1985 edition of AD B2/3/4), which would totally enclose all the relevant unprotected areas in the side of a building or compartment. This is referred to as the *enclosing rectangle* and is usually larger than the actual rectangle that would enclose these areas (see Fig. 7.49).

The unprotected areas are projected at right angles onto a *plane of reference* and Table 7.9 then gives the distance that the relevant boundary must be from the plane of reference according to the *unprotected percentage*, the height and width of the enclosing rectangle and the Purpose Group of the building. The figures in Table 7.9 relate to Shop and Commercial, Industrial and Storage and other non-residential buildings, whilst those in brackets relate to Residential, Office or Assembly and Recreation buildings.

The plane of reference is a vertical plane which touches some part of the outer surface of a building or compartment. It should not pass through any part of the building (except projections such as balconies or copings) and it should not cross the relevant boundary. It can be at any angle to the side of the building and in any position which is most favourable to the building designer, although it is usually best if roughly parallel to the relevant boundary. This method can be used to determine the maximum permitted unprotected areas for a given boundary position (Fig. 7.50) *or* how close to the boundary a particular design of building may be (Fig. 7.51).

It is permissible to calculate the enclosing rectangle separately for each compartment in a building. Therefore the provision of compartment walls and floors in a building can effectively reduce the enclosing rectangle thereby decreasing the

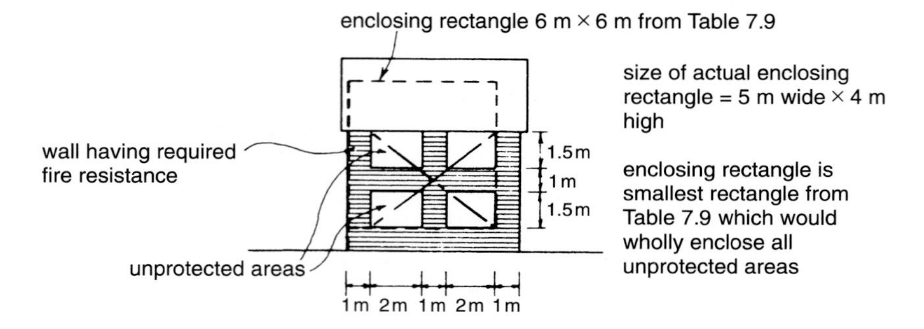

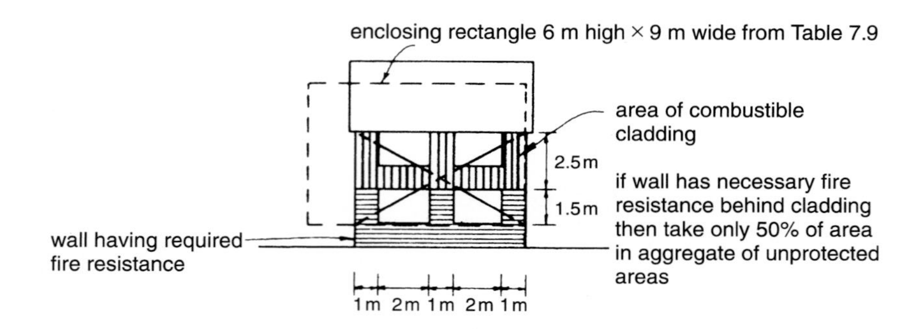

Note: plane of reference is taken to coincide with surface of wall

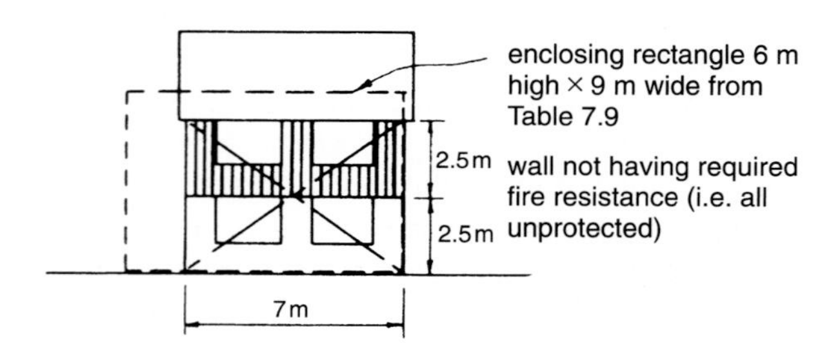

Fig. 7.49 Enclosing rectangles.

distance to the boundary without affecting the amount of unprotected areas provided. This technique is demonstrated in Fig. 7.52 which also shows how the enclosing rectangle method is applied in practice.

This method is quick and easy to use in practice but in certain circumstances it may give an uneconomical result with regard to the permitted distance from the boundary and it may unduly restrict the designer's freedom in choice of window areas, etc. It takes no account of the true distance from the boundary of unprotected areas in deeply indented buildings since all unprotected areas must be projected onto

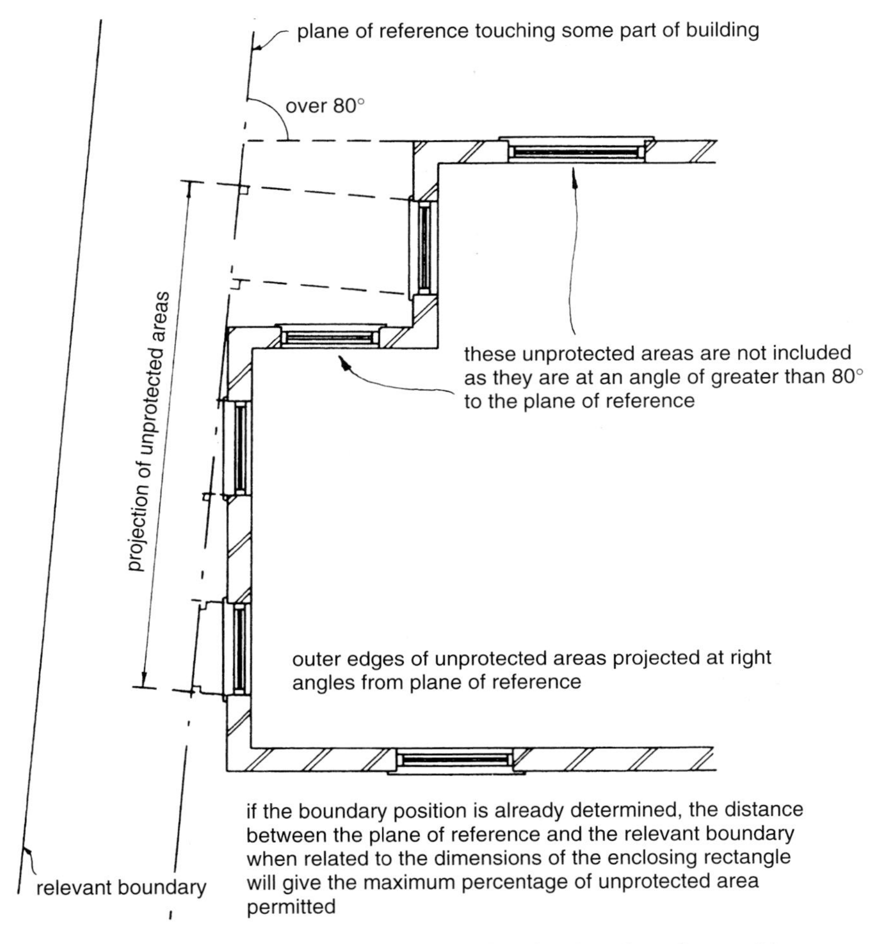

Fig. 7.50 Determination of maximum unprotected areas for given boundary position.

a single plane of reference. It also assumes that the effects of a fire will be equally felt at all points on the boundary from all unprotected areas despite the fact that some windows, for example, may be shielded from certain parts of the boundary by fire resistant walls. For these reasons it may be preferable to use the following method – the *aggregate notional areas* technique – as this method is usually more accurate in practice.

Additional methods – aggregate notional areas

The basis of the method of aggregate notional areas is to assess the effect of a building fire at a series of points 3 m apart on the relevant boundary.

A *vertical datum* of unlimited height is set at any position on the relevant boundary (see point P on Fig. 7.53). A *datum line* is drawn from this point to the nearest point on the building or compartment. A base line is then constructed at 90°

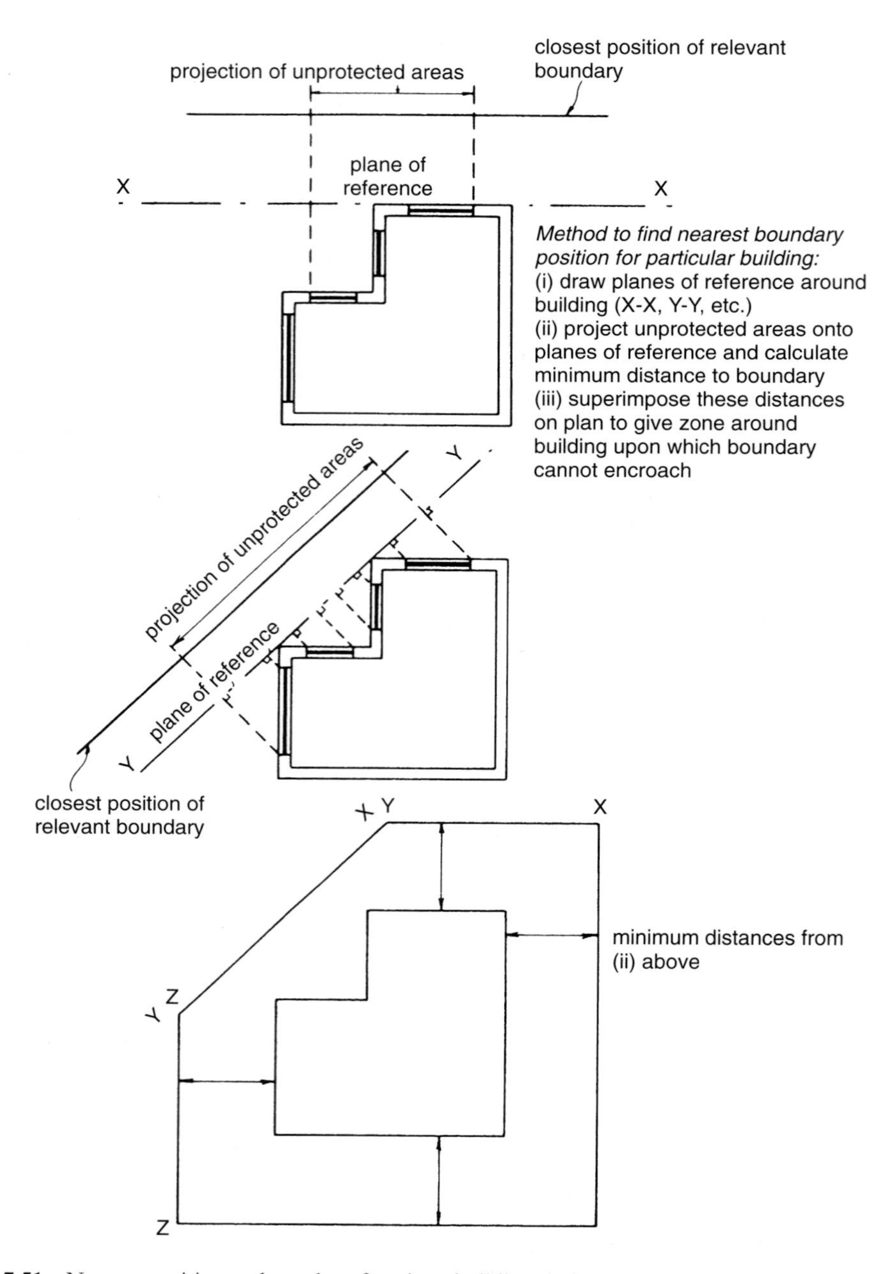

Fig. 7.51 Nearest position to boundary for given building design.

to the datum line and an arc of 50 m radius is drawn centred on the vertical datum to meet the base line.

Using this method it is possible to exclude certain unprotected areas that would have to be considered under the enclosing rectangles method (see Fig. 7.53).

For those unprotected areas which remain (that is, those that cannot be

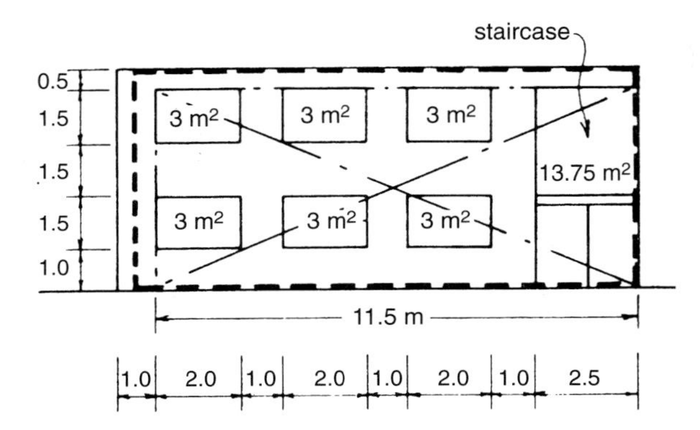

Uncompartmented building

(i) minimum size of rectangle enclosing unprotected areas = 11.5 m wide × 5.5 m high
(ii) from Table 7.9 enclosing rectangle = 12 m wide × 6 m high (take next highest values)
(iii) calculate aggregate of unprotected areas = 18 + 13.75 m^2 = 31.75 m^2
(iv) calculate unprotected percentage (aggregate of unprotected areas as percentage of enclosing rectangle) = $\frac{31.75}{12 \times 6} \times 100 = 44\% \therefore$ use 50% column in Table 7.9
(v) select distance from Table 7.9 (second part of table, fourth column, fourth row, figure in brackets) permitted distance to boundary = 3.5 m

Note:

(a) minimum size of rectangle indicated by diagonal lines
(b) enclosing rectangle indicated by dotted lines
(c) relevant boundary is parallel with wall face, plane of reference coincides with wall face (this need not be the case)

The situation can be improved if the staircase is enclosed in a protecting structure and a compartment floor is provided:

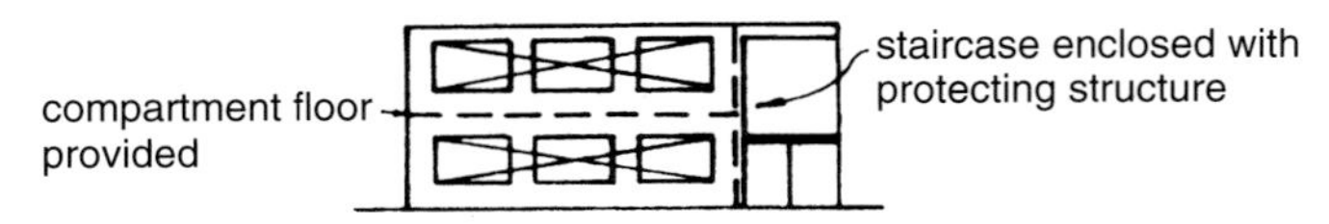

(i) each compartment now considered separately, the minimum rectangle being shown by diagonal lines above = 8 m × 1.5 m
(ii) enclosing rectangle from Table 7.9 = 9 m wide × 3 m high
(iii) aggregate of unprotected areas = 9 m^2
(iv) unprotected percentage = $\frac{9}{9 \times 3} \times 100 = 33\frac{1}{4}\%$ i.e. 40% in Table 7.9
(v) distance from Table 7.9 = 1.5 m

Compartment has therefore reduced the permitted distance to the boundary from 3.5 m to 1.5 m

Fig. 7.52 Enclosing rectangles – effects of compartmentation.

excluded), it is necessary to measure the distance of each from the vertical datum. Table J3 from the 1985 edition of AD B2/3/4 is reproduced below as Table 7.10. This contains a series of multiplication factors which are related to the distance from the vertical datum. The Table is based on the fact that the amount of heat caused by a fire issuing from an unprotected area will decrease in proportion to

Table 7.9 Permitted unprotected percentages in relation to enclosing rectangles.

Width of enclosing rectangle (m)	Distance from relevant boundary for unprotected percentage not exceeding								
	20%	30%	40%	50%	60%	70%	80%	90%	100%
	Minimum boundary distance (m); figures in brackets are for residential, office or assembly								
Enclosing rectangle 3 m high									
3	1.0 (1.0)	1.5 (1.0)	2.0 (1.0)	2.0 (1.5)	2.5 (1.5)	2.5 (1.5)	2.5 (2.0)	3.0 (2.0)	3.0 (2.0)
6	1.5 (1.0)	2.0 (1.0)	2.5 (1.5)	3.0 (2.0)	3.0 (2.0)	3.5 (2.0)	3.5 (2.5)	4.0 (2.5)	4.0 (3.0)
9	1.5 (1.0)	2.5 (1.0)	3.0 (1.5)	3.5 (2.0)	4.0 (2.5)	4.0 (2.5)	4.5 (3.0)	5.0 (3.0)	5.0 (3.5)
12	2.0 (1.0)	2.5 (1.5)	3.0 (2.0)	3.5 (2.0)	4.0 (2.5)	4.5 (3.0)	5.0 (3.0)	5.5 (3.5)	5.5 (3.5)
15	2.0 (1.0)	2.5 (1.5)	3.5 (2.0)	4.0 (2.5)	4.5 (2.5)	5.0 (3.0)	5.5 (3.5)	6.0 (3.5)	6.0 (4.0)
18	2.0 (1.0)	2.5 (1.5)	3.5 (2.0)	4.0 (2.5)	5.0 (2.5)	5.0 (3.0)	6.0 (3.5)	6.5 (4.0)	6.5 (4.0)
21	2.0 (1.0)	3.0 (1.5)	3.5 (2.0)	4.5 (2.5)	5.0 (3.0)	5.5 (3.0)	6.0 (3.5)	6.5 (4.0)	7.0 (4.5)
24	2.0 (1.0)	3.0 (1.5)	3.5 (2.0)	4.5 (2.5)	5.0 (3.0)	5.5 (3.5)	6.0 (3.5)	7.0 (4.0)	7.5 (4.5)
27	2.0 (1.0)	3.0 (1.5)	4.0 (2.0)	4.5 (2.5)	5.5 (3.0)	6.0 (3.5)	6.5 (4.0)	7.0 (4.0)	7.5 (4.5)
30	2.0 (1.0)	3.0 (1.5)	4.0 (2.0)	4.5 (2.5)	5.5 (3.0)	6.0 (3.5)	6.5 (4.0)	7.5 (4.0)	8.0 (4.5)
40	2.0 (1.0)	3.0 (1.5)	4.0 (2.0)	5.0 (2.5)	5.5 (3.0)	6.5 (3.5)	7.0 (4.0)	8.0 (4.0)	8.5 (5.0)
50	2.0 (1.0)	3.0 (1.5)	4.0 (2.0)	5.0 (2.5)	6.0 (3.0)	6.5 (3.5)	7.5 (4.0)	8.0 (4.0)	9.0 (5.0)
60	2.0 (1.0)	3.0 (1.5)	4.0 (2.0)	5.0 (2.5)	6.0 (3.0)	7.0 (3.5)	7.5 (4.0)	8.5 (4.0)	9.5 (5.0)
80	2.0 (1.0)	3.0 (1.5)	4.0 (2.0)	5.0 (2.5)	6.0 (3.0)	7.0 (3.5)	8.0 (4.0)	9.0 (4.0)	9.5 (5.0)
no limit	2.0 (1.0)	3.0 (1.5)	4.0 (2.0)	5.0 (2.5)	6.0 (3.0)	7.0 (3.5)	8.0 (4.0)	9.0 (4.0)	10.0 (5.0)
Enclosing rectangle 6 m high									
3	1.5 (1.0)	2.0 (1.0)	2.5 (1.5)	3.0 (2.0)	3.0 (2.0)	3.5 (2.0)	3.5 (2.5)	4.0 (2.5)	4.0 (3.0)
6	2.0 (1.0)	3.0 (1.5)	3.5 (2.0)	4.0 (2.5)	4.5 (3.0)	5.0 (3.0)	5.5 (3.5)	5.5 (4.0)	6.0 (4.0)
9	2.5 (1.0)	3.5 (2.0)	4.5 (2.5)	5.0 (3.0)	5.5 (3.5)	6.0 (4.0)	6.0 (4.5)	7.0 (4.5)	7.0 (5.0)
12	3.0 (1.5)	4.0 (2.5)	5.0 (3.0)	5.5 (3.5)	6.5 (4.0)	7.0 (4.5)	7.5 (5.0)	8.0 (5.0)	8.5 (5.5)
15	3.0 (1.5)	4.5 (2.5)	5.5 (3.0)	6.0 (4.0)	7.0 (4.5)	7.5 (5.0)	8.0 (5.5)	9.0 (5.5)	9.0 (6.0)
18	3.5 (1.5)	4.5 (2.5)	5.5 (3.5)	6.5 (4.0)	7.5 (4.5)	8.0 (5.0)	9.0 (5.5)	9.5 (6.0)	10.0 (6.5)
21	3.5 (1.5)	5.0 (2.5)	6.0 (3.5)	7.0 (4.0)	8.0 (5.0)	9.0 (5.5)	9.5 (6.0)	10.0 (6.5)	10.5 (7.0)
24	3.5 (1.5)	5.0 (2.5)	6.0 (3.5)	7.0 (4.5)	8.5 (5.0)	9.5 (5.5)	10.0 (6.0)	10.5 (7.0)	11.0 (7.0)
27	3.5 (1.5)	5.0 (2.5)	6.5 (3.5)	7.5 (4.5)	8.5 (5.0)	9.5 (6.0)	10.5 (7.6)	11.0 (7.0)	12.0 (7.5)

30	3.5 *(1.5)*	5.0 *(2.5)*	6.5 *(3.5)*	8.0 *(4.5)*	9.0 *(5.0)*	10.0 *(6.0)*	11.0 *(6.5)*	12.0 *(7.0)*	12.5 *(8.0)*
40	3.5 *(1.5)*	5.5 *(2.5)*	7.0 *(3.5)*	8.5 *(4.5)*	10.0 *(5.5)*	11.0 *(6.5)*	12.0 *(7.0)*	13.0 *(8.0)*	14.0 *(8.5)*
50	3.5 *(1.5)*	5.5 *(2.5)*	7.5 *(3.5)*	9.0 *(4.5)*	10.5 *(5.5)*	11.5 *(6.5)*	13.0 *(7.5)*	14.0 *(8.0)*	15.0 *(9.0)*
60	3.5 *(1.5)*	5.5 *(2.5)*	7.5 *(3.5)*	9.5 *(5.0)*	11.0 *(5.5)*	12.0 *(6.5)*	13.5 *(7.5)*	15.0 *(8.5)*	16.0 *(9.5)*
80	3.5 *(1.5)*	6.0 *(2.5)*	7.5 *(3.5)*	9.5 *(5.0)*	11.5 *(6.0)*	13.0 *(7.0)*	14.5 *(7.5)*	16.0 *(8.5)*	17.5 *(9.5)*
100	3.5 *(1.5)*	6.0 *(2.5)*	8.0 *(3.5)*	10.0 *(5.0)*	12.0 *(6.0)*	13.5 *(7.0)*	15.0 *(8.0)*	16.5 *(8.5)*	18.0 *(10.0)*
120	3.5 *(1.5)*	6.0 *(2.5)*	8.0 *(3.5)*	10.0 *(5.0)*	12.0 *(6.0)*	14.0 *(7.0)*	15.5 *(8.0)*	17.0 *(8.5)*	19.0 *(10.0)*
no limit	3.5 *(1.5)*	6.0 *(2.5)*	8.0 *(3.5)*	10.0 *(5.0)*	12.0 *(6.0)*	14.0 *(7.0)*	16.0 *(8.0)*	18.0 *(8.5)*	19.0 *(10.0)*
Enclosing rectangle 9 m high									
3	1.5 *(1.0)*	2.5 *(1.0)*	3.0 *(1.5)*	3.5 *(1.5)*	4.0 *(2.5)*	4.0 *(2.5)*	4.5 *(3.0)*	5.0 *(3.0)*	5.0 *(3.5)*
6	2.5 *(1.0)*	3.5 *(2.0)*	4.5 *(2.5)*	5.0 *(3.0)*	5.5 *(3.5)*	6.0 *(4.0)*	6.5 *(4.5)*	7.0 *(4.5)*	7.0 *(5.0)*
9	3.5 *(1.5)*	4.5 *(2.5)*	5.5 *(3.5)*	6.0 *(4.0)*	6.5 *(4.5)*	7.5 *(5.0)*	8.0 *(5.5)*	8.5 *(5.5)*	9.0 *(6.0)*
12	3.5 *(1.5)*	5.0 *(3.0)*	6.0 *(3.5)*	7.0 *(4.5)*	7.5 *(5.0)*	8.5 *(5.5)*	9.0 *(6.0)*	9.5 *(6.5)*	10.5 *(7.0)*
15	4.0 *(2.0)*	5.5 *(3.0)*	6.5 *(4.0)*	7.5 *(5.0)*	8.5 *(5.5)*	9.5 *(6.0)*	10.0 *(6.5)*	11.0 *(7.0)*	11.5 *(7.5)*
18	4.5 *(2.0)*	6.0 *(3.5)*	7.0 *(4.5)*	8.5 *(5.0)*	9.5 *(6.0)*	10.0 *(6.5)*	11.0 *(7.0)*	12.0 *(8.0)*	12.5 *(8.5)*
21	4.5 *(2.0)*	6.5 *(3.5)*	7.5 *(4.5)*	9.0 *(5.5)*	10.0 *(6.5)*	11.0 *(7.0)*	12.0 *(7.5)*	13.0 *(8.5)*	13.5 *(9.0)*
24	5.0 *(2.0)*	6.5 *(3.5)*	8.0 *(5.0)*	9.5 *(5.5)*	11.0 *(6.5)*	12.0 *(7.5)*	13.0 *(8.0)*	13.5 *(9.0)*	14.5 *(9.5)*
27	5.0 *(2.0)*	7.0 *(3.5)*	8.5 *(5.0)*	10.0 *(6.0)*	11.5 *(7.0)*	12.5 *(7.5)*	13.5 *(8.5)*	14.5 *(9.5)*	15.0 *(10.0)*
30	5.0 *(2.0)*	7.0 *(3.5)*	9.0 *(5.0)*	10.5 *(6.0)*	12.0 *(7.0)*	13.0 *(8.0)*	14.0 *(9.0)*	15.0 *(9.5)*	16.0 *(10.5)*
40	5.5 *(2.0)*	7.5 *(3.5)*	9.5 *(5.5)*	11.5 *(6.5)*	13.0 *(7.5)*	14.5 *(8.5)*	15.5 *(9.5)*	17.0 *(10.5)*	17.5 *(11.5)*
50	5.5 *(2.0)*	8.0 *(4.0)*	10.0 *(5.5)*	12.5 *(6.5)*	14.0 *(8.0)*	15.5 *(9.0)*	17.0 *(10.0)*	18.5 *(11.5)*	19.5 *(12.5)*
60	5.5 *(2.0)*	8.0 *(4.0)*	11.0 *(5.5)*	13.0 *(7.0)*	15.0 *(8.0)*	16.5 *(9.5)*	18.0 *(11.0)*	19.5 *(11.5)*	21.0 *(13.0)*
80	5.5 *(2.0)*	8.5 *(4.0)*	11.5 *(5.5)*	13.5 *(7.0)*	16.0 *(8.5)*	17.5 *(10.0)*	19.5 *(11.5)*	21.5 *(12.5)*	23.0 *(13.5)*
100	5.5 *(2.0)*	8.5 *(4.0)*	11.5 *(5.5)*	14.5 *(7.0)*	16.5 *(8.5)*	18.5 *(10.0)*	21.0 *(11.5)*	22.5 *(12.5)*	24.5 *(14.5)*
120	5.5 *(2.0)*	8.5 *(4.0)*	11.5 *(5.5)*	14.5 *(7.0)*	17.0 *(8.5)*	19.5 *(10.0)*	21.5 *(11.5)*	23.5 *(12.5)*	26.0 *(14.5)*
no limit	5.5 *(2.0)*	8.5 *(4.0)*	11.5 *(5.5)*	15.0 *(7.0)*	17.5 *(8.5)*	20.0 *(10.5)*	22.5 *(12.0)*	24.5 *(12.5)*	27.0 *(15.0)*

(Contd)

Table 7.9 (*Contd*).

Width of enclosing rectangle (m)	Distance from relevant boundary for unprotected percentage not exceeding								
	20%	30%	40%	50%	60%	70%	80%	90%	100%
	Minimum boundary distance (m); figures in brackets are for residential, office or assembly								
Enclosing rectangle 12 m high									
3	2.0 *(1.0)*	2.5 *(1.5)*	3.0 *(2.0)*	3.5 *(2.0)*	4.0 *(2.5)*	4.5 *(3.0)*	5.0 *(3.0)*	5.5 *(3.5)*	5.5 *(3.5)*
6	3.0 *(1.5)*	4.0 *(2.5)*	5.0 *(3.0)*	5.5 *(3.5)*	6.5 *(4.0)*	7.0 *(4.5)*	7.5 *(5.0)*	8.0 *(5.0)*	8.5 *(5.5)*
9	3.5 *(1.5)*	5.0 *(3.0)*	6.0 *(3.5)*	7.0 *(4.5)*	7.5 *(5.0)*	8.5 *(5.5)*	9.0 *(6.0)*	9.5 *(6.5)*	10.5 *(7.0)*
12	4.5 *(1.5)*	6.0 *(3.5)*	7.0 *(4.5)*	8.0 *(5.0)*	9.0 *(6.0)*	9.5 *(6.0)*	11.0 *(7.0)*	11.5 *(7.5)*	12.0 *(8.0)*
15	5.0 *(2.0)*	6.5 *(3.5)*	8.0 *(5.0)*	9.0 *(5.5)*	10.0 *(6.5)*	11.0 *(7.0)*	12.0 *(8.0)*	13.0 *(8.5)*	13.5 *(9.0)*
18	5.0 *(2.5)*	7.0 *(4.0)*	8.5 *(5.0)*	10.0 *(6.0)*	11.0 *(7.0)*	12.0 *(7.5)*	13.0 *(8.5)*	14.0 *(9.0)*	14.5 *(10.0)*
21	5.5 *(2.5)*	7.5 *(4.0)*	9.0 *(5.5)*	10.5 *(6.5)*	12.0 *(7.5)*	13.0 *(8.5)*	14.0 *(9.0)*	15.0 *(10.0)*	16.0 *(10.5)*
24	6.0 *(2.5)*	8.0 *(4.5)*	9.5 *(6.0)*	11.5 *(7.0)*	12.5 *(8.0)*	14.0 *(8.5)*	15.0 *(9.5)*	16.0 *(10.5)*	16.5 *(11.5)*
27	6.0 *(2.5)*	8.0 *(4.5)*	10.5 *(6.0)*	12.0 *(7.0)*	13.5 *(8.0)*	14.5 *(9.0)*	16.0 *(10.5)*	17.0 *(11.0)*	17.5 *(12.0)*
30	6.5 *(2.5)*	8.5 *(4.5)*	10.5 *(6.5)*	12.5 *(7.5)*	14.0 *(8.5)*	15.0 *(9.5)*	16.5 *(10.5)*	17.5 *(11.5)*	18.5 *(12.5)*
40	6.5 *(2.5)*	9.5 *(5.0)*	12.0 *(6.5)*	14.0 *(8.0)*	15.5 *(9.5)*	17.5 *(10.5)*	18.5 *(12.0)*	20.0 *(13.0)*	21.0 *(14.0)*
50	7.0 *(2.5)*	10.0 *(5.0)*	13.0 *(7.0)*	15.0 *(8.5)*	17.0 *(10.0)*	19.0 *(11.0)*	20.5 *(13.0)*	23.0 *(14.0)*	23.0 *(15.0)*
60	7.0 *(2.5)*	10.5 *(5.0)*	13.5 *(7.0)*	16.0 *(9.0)*	18.0 *(10.5)*	20.0 *(12.0)*	21.5 *(13.5)*	23.5 *(14.5)*	25.0 *(16.0)*
80	7.0 *(2.5)*	11.0 *(5.0)*	14.5 *(7.0)*	17.0 *(9.0)*	19.5 *(11.0)*	21.5 *(13.0)*	23.5 *(14.5)*	26.0 *(16.0)*	27.5 *(17.0)*
100	7.5 *(2.5)*	11.5 *(5.0)*	15.0 *(7.5)*	18.0 *(9.5)*	21.0 *(11.5)*	23.0 *(13.5)*	25.5 *(15.0)*	28.0 *(16.5)*	30.0 *(18.0)*
120	7.5 *(2.5)*	11.5 *(5.0)*	15.0 *(7.5)*	18.5 *(9.5)*	22.0 *(11.5)*	24.0 *(13.5)*	27.0 *(15.0)*	29.5 *(17.0)*	31.5 *(18.5)*
no limit	7.5 *(2.5)*	12.0 *(5.0)*	15.5 *(7.5)*	19.0 *(9.5)*	22.5 *(12.0)*	22.5 *(14.0)*	28.0 *(15.5)*	30.5 *(17.0)*	34.0 *(19.0)*
Enclosing rectangle 15 m high									
3	2.0 *(1.0)*	2.5 *(1.5)*	3.5 *(2.0)*	4.0 *(2.5)*	4.5 *(2.5)*	5.0 *(3.0)*	5.5 *(3.5)*	6.0 *(3.5)*	6.0 *(4.0)*
6	3.0 *(1.5)*	4.5 *(2.5)*	5.5 *(3.0)*	6.0 *(4.0)*	7.0 *(4.5)*	7.5 *(5.0)*	8.0 *(5.5)*	9.0 *(5.5)*	9.0 *(6.0)*
9	4.0 *(2.0)*	5.5 *(3.0)*	6.5 *(4.0)*	7.5 *(5.0)*	8.5 *(5.5)*	9.5 *(6.0)*	10.0 *(6.5)*	11.0 *(7.0)*	11.5 *(7.5)*
12	5.0 *(2.0)*	6.5 *(3.5)*	8.0 *(5.0)*	9.0 *(5.5)*	10.0 *(6.5)*	11.0 *(7.0)*	12.0 *(8.0)*	13.0 *(8.5)*	13.5 *(9.0)*
15	5.5 *(2.0)*	7.0 *(4.0)*	9.0 *(5.5)*	10.0 *(6.5)*	11.5 *(7.0)*	12.5 *(8.0)*	13.5 *(9.0)*	14.5 *(9.5)*	15.0 *(10.0)*
18	6.0 *(2.5)*	8.0 *(4.5)*	9.5 *(6.0)*	11.0 *(7.0)*	12.5 *(8.0)*	13.5 *(8.5)*	14.5 *(9.5)*	15.5 *(10.5)*	16.5 *(11.0)*

21	6.5 *(2.5)*	8.5 *(5.0)*	10.5 *(6.5)*	12.0 *(7.5)*	13.5 *(8.5)*	14.5 *(9.5)*	16.0 *(10.5)*	16.5 *(11.0)*	17.5 *(12.0)*
24	6.5 *(3.0)*	9.0 *(5.0)*	11.0 *(6.5)*	13.0 *(8.0)*	14.5 *(9.0)*	15.5 *(10.0)*	17.0 *(11.0)*	18.0 *(12.0)*	19.0 *(13.0)*
27	7.0 *(3.0)*	9.5 *(5.5)*	11.5 *(7.0)*	13.5 *(8.5)*	15.0 *(9.5)*	16.5 *(10.5)*	18.0 *(11.5)*	19.0 *(12.5)*	20.0 *(13.5)*
30	7.5 *(3.0)*	10.0 *(5.5)*	12.0 *(7.5)*	14.0 *(8.5)*	16.0 *(10.0)*	17.0 *(11.0)*	18.5 *(12.0)*	20.0 *(13.5)*	21.0 *(14.0)*
40	8.0 *(3.0)*	11.0 *(6.0)*	13.5 *(8.0)*	16.0 *(9.5)*	18.0 *(11.0)*	19.5 *(12.5)*	21.0 *(13.5)*	22.5 *(15.0)*	23.5 *(16.0)*
50	8.5 *(3.5)*	12.0 *(6.0)*	15.0 *(8.5)*	17.5 *(10.0)*	19.5 *(12.0)*	21.5 *(13.5)*	23.0 *(15.0)*	25.0 *(16.5)*	26.0 *(17.5)*
60	8.5 *(3.5)*	12.5 *(6.5)*	15.5 *(8.5)*	18.0 *(10.5)*	21.0 *(17.5)*	23.5 *(14.0)*	25.0 *(15.5)*	27.0 *(17.0)*	28.0 *(18.0)*
80	9.0 *(3.5)*	13.5 *(6.5)*	17.0 *(9.0)*	20.0 *(11.0)*	23.0 *(13.5)*	25.5 *(15.0)*	28.0 *(17.0)*	30.0 *(18.5)*	31.5 *(20.0)*
100	9.0 *(3.5)*	14.0 *(6.5)*	18.0 *(9.0)*	21.5 *(11.5)*	24.5 *(14.0)*	27.5 *(16.0)*	30.0 *(18.0)*	32.5 *(19.5)*	34.5 *(21.5)*
120	9.0 *(3.5)*	14.0 *(6.5)*	18.5 *(9.0)*	22.5 *(11.5)*	25.5 *(14.0)*	28.5 *(16.5)*	31.5 *(18.5)*	34.5 *(20.5)*	37.0 *(22.5)*
no limit	9.0 *(3.5)*	14.5 *(6.5)*	19.0 *(9.0)*	23.0 *(12.0)*	27.0 *(14.5)*	30.0 *(17.0)*	34.0 *(19.0)*	36.0 *(21.0)*	39.0 *(23.0)*
Enclosing rectangle 18 m high									
3	2.0 *(1.0)*	2.5 *(1.5)*	3.5 *(2.0)*	4.0 *(2.5)*	5.0 *(2.5)*	5.0 *(3.0)*	6.0 *(3.5)*	6.5 *(4.0)*	6.5 *(4.0)*
6	3.5 *(1.5)*	4.5 *(2.5)*	5.5 *(3.5)*	6.5 *(4.0)*	7.5 *(4.5)*	8.0 *(5.0)*	9.0 *(5.5)*	9.5 *(6.0)*	10.0 *(6.5)*
9	4.5 *(2.0)*	6.0 *(3.5)*	7.0 *(4.5)*	8.5 *(5.0)*	9.5 *(6.0)*	10.0 *(6.5)*	11.0 *(7.0)*	12.0 *(8.0)*	12.5 *(8.5)*
12	5.0 *(2.5)*	7.0 *(4.0)*	8.5 *(5.0)*	10.0 *(6.0)*	11.0 *(7.0)*	12.0 *(7.5)*	13.0 *(8.5)*	14.0 *(9.0)*	14.5 *(10.0)*
15	6.0 *(2.5)*	8.0 *(4.5)*	9.5 *(6.0)*	11.0 *(7.0)*	12.5 *(8.0)*	13.5 *(8.5)*	14.5 *(9.5)*	15.5 *(10.5)*	16.5 *(11.0)*
18	6.5 *(2.5)*	8.5 *(5.0)*	11.0 *(6.5)*	12.0 *(7.5)*	13.5 *(8.5)*	14.5 *(9.5)*	16.0 *(11.0)*	17.0 *(11.5)*	18.0 *(13.0)*
21	7.0 *(3.0)*	9.5 *(5.5)*	11.5 *(7.0)*	13.0 *(8.0)*	14.5 *(9.5)*	16.0 *(10.5)*	17.0 *(11.5)*	18.0 *(12.5)*	19.5 *(13.0)*
24	7.5 *(3.0)*	10.0 *(5.5)*	12.0 *(7.5)*	14.0 *(8.5)*	15.5 *(10.0)*	16.5 *(11.0)*	18.5 *(12.0)*	19.5 *(13.0)*	20.5 *(14.0)*
27	8.0 *(3.5)*	10.5 *(6.0)*	12.5 *(8.0)*	14.5 *(9.0)*	16.5 *(10.5)*	17.5 *(11.5)*	19.5 *(12.5)*	20.5 *(13.5)*	21.5 *(14.5)*
30	8.0 *(3.5)*	11.0 *(6.5)*	13.5 *(8.0)*	15.5 *(9.5)*	17.0 *(11.0)*	18.5 *(12.0)*	20.5 *(13.5)*	21.5 *(14.5)*	22.5 *(15.5)*
40	9.0 *(4.0)*	12.0 *(7.0)*	15.0 *(9.0)*	17.5 *(11.0)*	19.5 *(12.0)*	21.5 *(13.5)*	23.5 *(15.0)*	25.0 *(16.5)*	26.0 *(17.5)*
50	9.5 *(4.0)*	13.0 *(7.0)*	16.5 *(9.5)*	19.0 *(11.5)*	21.5 *(13.0)*	23.5 *(15.0)*	26.0 *(16.5)*	27.5 *(18.0)*	29.0 *(19.0)*
60	10.0 *(4.0)*	14.0 *(7.5)*	17.5 *(10.0)*	20.5 *(12.0)*	23.0 *(14.0)*	26.0 *(16.0)*	27.5 *(17.5)*	29.5 *(19.5)*	31.0 *(20.5)*
80	10.0 *(4.0)*	15.0 *(7.5)*	19.0 *(10.0)*	22.5 *(13.0)*	26.0 *(15.0)*	28.5 *(17.0)*	31.0 *(19.0)*	33.5 *(21.0)*	35.0 *(22.5)*
100	10.0 *(4.0)*	16.0 *(7.5)*	20.5 *(10.0)*	24.0 *(13.5)*	28.0 *(16.0)*	31.0 *(18.0)*	33.5 *(20.5)*	36.0 *(22.5)*	38.5 *(24.0)*
120	10.0 *(4.0)*	16.5 *(7.5)*	21.0 *(10.0)*	25.5 *(14.0)*	29.5 *(16.5)*	32.5 *(19.0)*	35.5 *(21.0)*	39.0 *(23.5)*	41.5 *(25.5)*
no limit	10.0 *(4.0)*	17.0 *(8.0)*	22.0 *(10.0)*	26.5 *(14.0)*	30.5 *(17.0)*	34.0 *(19.5)*	37.0 *(22.0)*	41.0 *(24.0)*	43.5 *(26.5)*

(Contd)

Table 7.9 (*Contd*).

Width of enclosing rectangle (m)	Distance from relevant boundary for unprotected percentage not exceeding								
	20%	30%	40%	50%	60%	70%	80%	90%	100%
	Minimum boundary distance (m); figures in brackets are for residential, office or assembly								
Enclosing rectangle 21 m high									
3	2.0 (*1.0*)	3.0 (*1.5*)	3.5 (*2.0*)	4.5 (*2.5*)	5.0 (*3.0*)	5.5 (*3.0*)	6.0 (*3.5*)	6.5 (*4.0*)	7.0 (*4.5*)
6	3.5 (*1.5*)	5.0 (*2.5*)	6.0 (*3.5*)	7.0 (*4.0*)	8.0 (*5.0*)	9.0 (*5.5*)	9.5 (*6.0*)	10.0 (*6.5*)	10.5 (*7.0*)
9	4.5 (*2.0*)	6.5 (*3.5*)	7.5 (*4.5*)	9.0 (*5.5*)	10.0 (*6.5*)	11.0 (*7.0*)	12.0 (*7.5*)	13.0 (*8.5*)	13.5 (*9.0*)
12	5.5 (*2.5*)	7.5 (*4.0*)	9.0 (*5.5*)	10.5 (*6.5*)	12.0 (*7.5*)	13.0 (*8.5*)	14.0 (*9.0*)	15.0 (*10.0*)	16.0 (*10.5*)
15	6.5 (*2.5*)	8.5 (*5.0*)	10.5 (*6.5*)	12.0 (*7.5*)	13.5 (*8.5*)	14.5 (*9.5*)	16.0 (*10.5*)	16.5 (*11.0*)	17.5 (*12.0*)
18	7.0 (*3.0*)	9.5 (*5.5*)	11.5 (*7.0*)	13.0 (*8.0*)	14.5 (*9.5*)	16.0 (*10.5*)	17.0 (*11.5*)	18.0 (*12.5*)	19.5 (*13.0*)
21	7.5 (*3.0*)	10.0 (*6.0*)	12.5 (*7.5*)	14.0 (*9.0*)	15.5 (*10.0*)	17.0 (*11.0*)	18.5 (*12.5*)	20.0 (*13.5*)	21.0 (*14.0*)
24	8.0 (*3.5*)	10.5 (*6.0*)	13.0 (*8.0*)	15.0 (*9.5*)	16.5 (*10.5*)	18.0 (*12.0*)	20.0 (*13.0*)	21.0 (*14.0*)	22.0 (*15.0*)
27	8.5 (*3.5*)	11.5 (*6.5*)	14.0 (*8.5*)	16.0 (*10.0*)	18.0 (*11.5*)	19.0 (*13.0*)	21.0 (*14.0*)	22.5 (*15.0*)	23.5 (*16.0*)
30	9.0 (*4.0*)	12.0 (*7.0*)	14.5 (*9.0*)	16.5 (*10.5*)	18.5 (*12.0*)	20.5 (*13.0*)	22.0 (*14.5*)	23.5 (*16.0*)	25.0 (*16.5*)
40	10.0 (*4.5*)	13.5 (*7.5*)	16.5 (*10.0*)	19.0 (*12.0*)	21.5 (*13.5*)	23.0 (*15.0*)	25.5 (*16.5*)	27.0 (*18.0*)	28.5 (*19.0*)
50	11.0 (*4.5*)	14.5 (*8.0*)	18.0 (*11.0*)	21.0 (*13.0*)	23.5 (*14.5*)	25.5 (*16.5*)	28.0 (*18.0*)	30.0 (*20.0*)	31.5 (*21.0*)
60	11.5 (*4.5*)	15.5 (*8.5*)	19.5 (*11.5*)	22.5 (*13.5*)	25.5 (*15.5*)	28.0 (*17.5*)	30.5 (*19.5*)	32.5 (*21.0*)	33.5 (*22.5*)
80	12.0 (*4.5*)	17.0 (*8.5*)	21.0 (*12.0*)	25.0 (*14.5*)	28.5 (*17.0*)	31.5 (*19.0*)	34.0 (*21.0*)	36.5 (*23.5*)	38.5 (*25.0*)
100	12.0 (*4.5*)	18.0 (*9.0*)	22.5 (*12.0*)	27.0 (*15.5*)	31.0 (*18.0*)	34.5 (*20.5*)	37.0 (*22.5*)	40.0 (*25.0*)	42.0 (*27.0*)
120	12.0 (*4.5*)	18.5 (*9.0*)	23.5 (*12.0*)	28.5 (*16.0*)	32.5 (*18.5*)	36.5 (*21.5*)	39.5 (*23.5*)	43.0 (*26.5*)	45.5 (*28.5*)
no limit	12.0 (*4.5*)	19.0 (*9.0*)	25.0 (*12.0*)	29.5 (*16.0*)	34.5 (*19.0*)	38.0 (*22.0*)	41.5 (*25.0*)	45.5 (*26.5*)	48.0 (*29.5*)
Enclosing rectangle 24 m high									
3	2.0 (*1.0*)	3.0 (*1.5*)	3.5 (*2.0*)	4.5 (*2.5*)	5.0 (*3.0*)	5.5 (*3.5*)	6.0 (*3.5*)	7.0 (*4.0*)	7.5 (*4.5*)
6	3.5 (*1.5*)	5.0 (*2.5*)	6.0 (*3.5*)	7.0 (*4.5*)	8.5 (*5.0*)	9.5 (*5.5*)	10.0 (*6.0*)	10.5 (*7.0*)	11.0 (*7.0*)
9	5.0 (*2.0*)	6.5 (*3.5*)	8.0 (*5.0*)	9.5 (*5.5*)	11.0 (*6.5*)	12.0 (*7.5*)	13.0 (*8.0*)	13.5 (*9.0*)	14.5 (*9.5*)
12	6.0 (*2.5*)	8.0 (*4.5*)	9.5 (*6.0*)	11.5 (*7.0*)	12.5 (*8.0*)	14.0 (*8.5*)	15.0 (*9.5*)	16.0 (*10.5*)	16.5 (*11.5*)
15	6.5 (*3.0*)	9.0 (*5.0*)	11.0 (*6.5*)	13.0 (*8.0*)	14.5 (*9.0*)	15.5 (*10.0*)	17.0 (*11.0*)	18.0 (*12.0*)	19.0 (*13.0*)
18	7.5 (*3.0*)	10.0 (*5.5*)	12.0 (*7.5*)	14.0 (*8.5*)	15.5 (*10.0*)	16.5 (*11.0*)	18.5 (*12.0*)	19.5 (*13.0*)	20.5 (*14.0*)

21	8.0 *(3.5)*	10.5 *(6.0)*	13.0 *(8.0)*	15.0 *(9.5)*	16.5 *(10.5)*	18.0 *(12.0)*	20.0 *(13.0)*	21.0 *(14.0)*	22.0 *(15.0)*
24	8.5 *(3.5)*	11.5 *(6.5)*	14.0 *(8.5)*	16.0 *(10.0)*	18.0 *(11.5)*	19.5 *(12.5)*	21.0 *(14.0)*	22.5 *(15.0)*	24.0 *(16.0)*
27	9.0 *(4.0)*	12.5 *(7.0)*	15.0 *(9.0)*	17.0 *(11.0)*	19.0 *(12.5)*	20.5 *(13.5)*	22.5 *(15.0)*	24.0 *(16.0)*	25.5 *(17.0)*
30	9.5 *(4.0)*	13.0 *(7.5)*	15.5 *(9.5)*	18.0 *(11.5)*	20.0 *(13.0)*	21.5 *(14.0)*	23.5 *(15.5)*	25.0 *(17.0)*	26.5 *(18.0)*
40	11.0 *(4.5)*	14.5 *(8.5)*	18.0 *(11.0)*	20.5 *(13.0)*	23.0 *(14.5)*	25.0 *(16.0)*	27.5 *(18.0)*	29.0 *(19.0)*	30.5 *(20.5)*
50	12.0 *(5.0)*	16.0 *(9.0)*	19.5 *(12.0)*	22.5 *(14.0)*	25.5 *(16.0)*	27.5 *(17.5)*	30.0 *(19.5)*	32.0 *(21.0)*	33.5 *(22.5)*
60	12.5 *(5.0)*	17.0 *(9.5)*	21.0 *(12.5)*	24.5 *(15.0)*	27.5 *(17.0)*	30.0 *(19.0)*	32.5 *(21.0)*	35.0 *(23.0)*	36.5 *(24.5)*
80	13.5 *(5.0)*	18.5 *(10.0)*	23.5 *(13.5)*	27.5 *(16.5)*	31.0 *(18.5)*	34.5 *(21.0)*	37.0 *(23.5)*	39.5 *(25.5)*	41.5 *(27.5)*
100	13.5 *(5.0)*	20.0 *(10.0)*	25.0 *(13.5)*	29.5 *(17.0)*	33.5 *(20.0)*	37.0 *(20.0)*	40.0 *(25.0)*	43.0 *(27.5)*	45.5 *(29.5)*
120	13.5 *(5.5)*	20.5 *(10.0)*	26.5 *(13.5)*	31.0 *(17.5)*	36.0 *(20.5)*	39.5 *(23.5)*	43.0 *(26.5)*	46.5 *(29.0)*	49.0 *(31.0)*
no limit	13.5 *(5.5)*	21.0 *(10.0)*	27.5 *(13.5)*	32.5 *(18.0)*	37.5 *(21.0)*	42.0 *(24.0)*	45.5 *(27.5)*	49.5 *(30.0)*	52.0 *(32.5)*
Enclosing rectangle 27 m high									
3	2.0 *(1.0)*	3.0 *(1.5)*	4.0 *(2.0)*	4.5 *(2.5)*	5.5 *(3.0)*	6.0 *(3.5)*	6.5 *(4.0)*	7.0 *(4.0)*	7.5 *(4.5)*
6	3.5 *(1.5)*	5.0 *(2.5)*	6.5 *(3.5)*	7.5 *(4.5)*	8.5 *(5.0)*	9.5 *(6.0)*	10.5 *(6.5)*	11.0 *(7.0)*	12.0 *(7.5)*
9	5.0 *(2.0)*	7.0 *(3.5)*	8.5 *(5.0)*	10.0 *(6.0)*	11.5 *(7.0)*	12.5 *(7.5)*	13.5 *(8.5)*	14.5 *(9.5)*	15.0 *(10.0)*
12	6.0 *(2.5)*	8.0 *(4.5)*	10.5 *(6.0)*	12.0 *(7.0)*	13.5 *(8.0)*	14.5 *(9.0)*	16.0 *(10.5)*	17.0 *(11.0)*	17.5 *(12.0)*
15	7.0 *(3.0)*	9.5 *(5.5)*	11.5 *(7.0)*	13.5 *(8.5)*	15.0 *(9.5)*	16.5 *(10.5)*	18.0 *(11.5)*	19.0 *(12.5)*	20.0 *(13.5)*
18	8.0 *(3.5)*	10.5 *(6.0)*	12.5 *(8.0)*	14.5 *(9.0)*	16.5 *(10.5)*	17.5 *(11.5)*	19.5 *(12.5)*	20.5 *(13.5)*	21.5 *(14.5)*
21	8.5 *(3.5)*	11.5 *(6.5)*	14.0 *(8.5)*	16.0 *(10.0)*	18.0 *(11.5)*	19.0 *(13.0)*	21.0 *(14.0)*	22.5 *(15.0)*	23.5 *(16.0)*
24	9.0 *(3.5)*	12.5 *(7.0)*	15.0 *(9.0)*	17.0 *(11.0)*	19.0 *(12.5)*	20.5 *(13.5)*	22.5 *(15.0)*	24.0 *(16.0)*	25.5 *(17.0)*
27	10.0 *(4.0)*	13.0 *(7.5)*	16.0 *(10.0)*	18.0 *(11.5)*	20.0 *(13.0)*	22.0 *(14.0)*	24.0 *(16.0)*	25.5 *(17.0)*	27.0 *(18.0)*
30	10.0 *(4.0)*	13.5 *(8.0)*	17.0 *(10.0)*	19.0 *(12.0)*	21.0 *(13.5)*	23.0 *(15.0)*	25.0 *(17.0)*	26.5 *(18.0)*	28.0 *(19.0)*
40	11.5 *(5.0)*	15.5 *(9.0)*	19.0 *(11.5)*	22.0 *(14.0)*	24.5 *(15.5)*	26.5 *(17.5)*	29.0 *(19.0)*	30.5 *(20.5)*	32.5 *(22.0)*
50	12.5 *(5.5)*	17.0 *(9.5)*	21.0 *(12.5)*	24.0 *(15.0)*	27.0 *(17.0)*	29.5 *(19.0)*	32.0 *(21.0)*	34.5 *(22.5)*	36.0 *(24.0)*
60	13.5 *(5.5)*	18.5 *(10.5)*	22.5 *(13.5)*	26.5 *(16.0)*	29.5 *(18.5)*	32.0 *(20.5)*	35.0 *(22.5)*	37.0 *(24.5)*	39.0 *(26.5)*
80	14.5 *(6.0)*	20.5 *(11.0)*	25.0 *(14.5)*	29.5 *(17.5)*	33.0 *(20.5)*	36.5 *(22.5)*	39.5 *(25.0)*	42.0 *(27.5)*	44.0 *(29.5)*
100	15.5 *(6.0)*	21.5 *(11.0)*	27.0 *(15.5)*	32.0 *(19.0)*	36.5 *(21.5)*	40.5 *(24.5)*	43.0 *(27.0)*	46.5 *(30.0)*	48.5 *(32.0)*
120	15.5 *(6.0)*	22.5 *(11.5)*	28.5 *(15.5)*	34.0 *(19.5)*	39.0 *(22.5)*	43.0 *(26.0)*	46.5 *(28.5)*	50.5 *(32.0)*	53.0 *(34.0)*
no limit	15.5 *(6.0)*	23.5 *(11.5)*	29.5 *(15.5)*	35.0 *(20.0)*	40.5 *(23.5)*	44.5 *(27.0)*	48.5 *(29.5)*	52.0 *(33.0)*	55.5 *(35.0)*

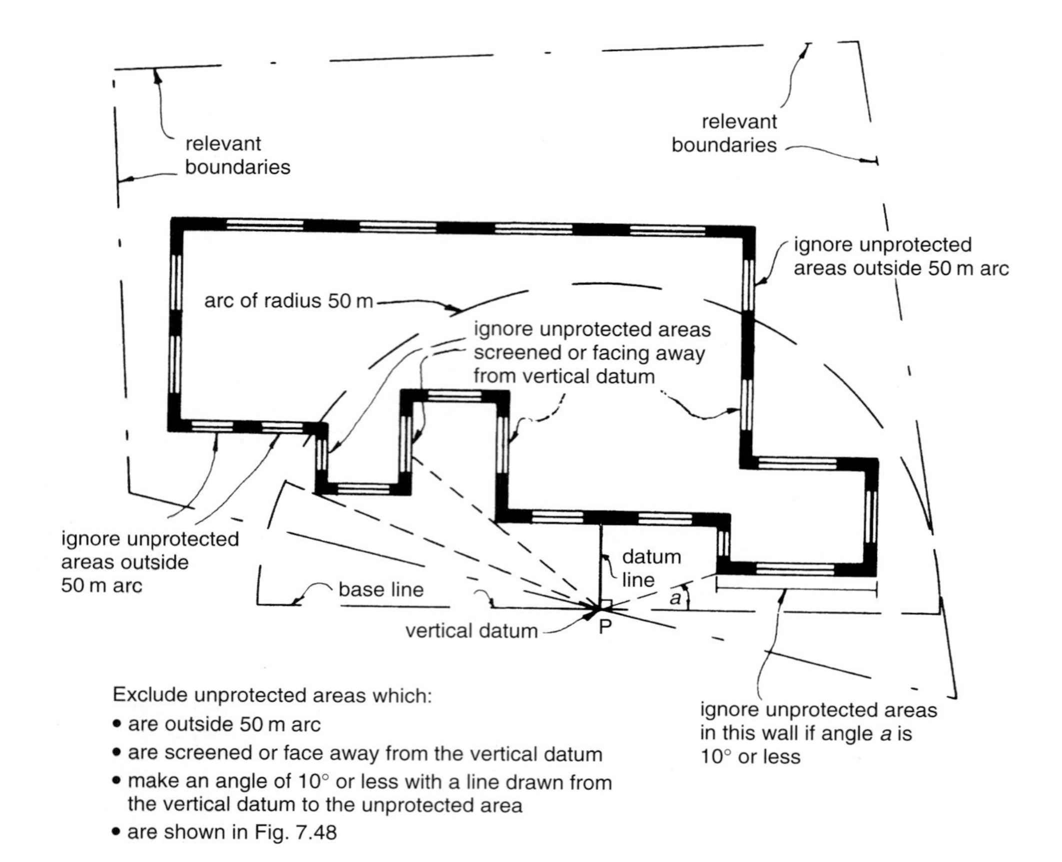

Fig. 7.53 Aggregate notional areas.

Table 7.10 Multiplication factors for aggregate notional area.

Distance of unprotected area from vertical datum (m)		
Not less than	**Less than**	**Multiplication factor**
1.0	1.2	80.0
1.2	1.8	40.0
1.8	2.7	20.0
2.7	4.3	10.0
4.3	6.0	4.0
6.0	8.5	2.0
8.5	12.0	1.0
12.0	18.5	0.5
18.5	27.5	0.25
27.5	50.0	0.1
50.0	no limit	0.0

its distance from the boundary (it does, in fact, correspond to an inverse square law of the type $y = 1/x^2$).

Therefore, each unprotected area is multiplied by its factor (which depends on its distance from the vertical datum) and these areas are then totalled to give the *aggregate notional area* for that particular vertical datum. The aggregate notional area thus achieved should not exceed:

- 210 m^2 for Residential, Assembly and Recreation or Office buildings; or
- 90 m^2 for Shop and Commercial, Industrial or Other non-residential buildings.

In order to confirm that the unprotected areas in the building comply at other points on the boundary, it is necessary to repeat the above calculations at a series of points 3 m apart starting from the original vertical datum. In practice it is usually possible, by observation, to place the first vertical datum at the worst position thereby obviating the need for further calculations.

The series of measurements and calculations mentioned above may be simplified if a number of protractors are made corresponding to different scales (i.e. 1:50, 1:100 and 1:200). A typical example is illustrated in Fig. 7.54.

7.27.7 Canopy structures and space separation

Since Building Regulations apply to the erection of a building, a canopy would need to comply with the provisions concerning space separation (referred to above) unless it falls into one of the exempt classes (see Class VI and Class VII in section 2.5 above). For free-standing canopies, this might prove unduly onerous (for example, canopies over petrol pumps), since the open sides would be regarded as unprotected areas. Paragraph 14.11 of AD B4 allows space separation recommendations to be disregarded for free-standing canopies which are more than 1m from the relevant boundary in view of the high degree of ventilation and heat dissipation achieved by the open sided construction. For canopies attached to the side of a building, provided that the edge of the canopy is at least 2 m from the boundary, the separation distance may be judged from the side of the building rather than the edge of the canopy. This exception would not apply if the canopy had side walls (such as in an enclosed loading bay).

7.27.8 Atrium buildings

An atrium is defined in section 7.7 as a space in a building (not necessarily vertically aligned) which passes through one or more structural floors. Clearly, the atrium effectively joins up all the relevant floors in the building to the extent that they can no longer be regarded as being compartmented from one another. The effect that this can have on space separation is explained above. In such buildings, AD B4 recommends the use of clause 28.2 of BS 5588: Part 7: 1997. This explains that, if the atrium building is fitted with a sprinkler system, the area affected in a fire will be sufficiently reduced so that the potential for fire spread to adjacent buildings will be comparable to that of an equivalent non-atrium building that is compartmented at each level and protected by a sprinkler system.

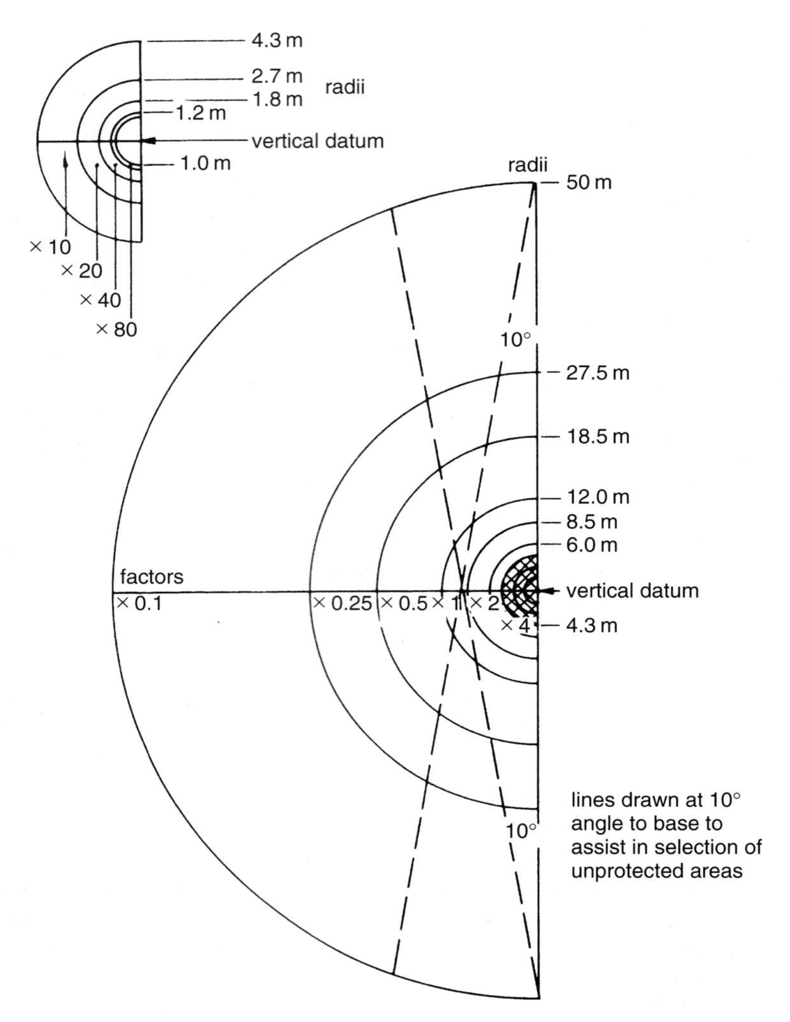

Note: the figure shown above is an enlargement to show radii and factors applicable near to the vertical datum (covering the portion shown hatched)

Fig. 7.54 Aggregate notional areas protractor.

Conversely, if the atrium building is unsprinklered, space separation needs to be calculated on the basis that all storeys not separated from the atrium by fire-resisting construction may be involved in the fire.

7.27.9 Roofs

Roofs are not required to provide fire resistance from the inside of a building but should resist fire penetration from outside and the spread of flame over their surfaces. The term roof covering means constructions which may contain one or more layers of material but it does not refer to the roof structure as a whole.

In addition to the recommendations for roof coverings contained in AD B4, reference should also be made to AD B 1 (Roofs as part of a means of escape), AD B2 (Internal surfaces of roof-lights) and AD B3 (Roofs used as part of a floor and roofs passing over the top of compartment walls).

The type of construction permitted for a roof depends on the purpose group and size of the building and its distance from the boundary.

Types of construction are specified by the two-letter designations from BS 476: Part 3: 1958 *External fire exposure roof test*. (It should be noted that this is not the most recent version of the standard but it is the one referred to in the latest edition of AD B. The current European standard for roofs is DD ENV 1187: 2002: *Test methods for external fire exposure to roofs*. Unfortunately, no guidance is possible on the performance in terms of the resistance of roofs to external fire exposure as determined by the methods specified in this document, since there is no accompanying classification procedure and no comparative supporting data.)

The first letter in the BS 476 designation method refers to flame penetration:

A – not penetrated within one hour
B – penetrated in not less than half an hour
C – penetrated in less than half an hour
D – penetrated in preliminary flame test.

The second letter in the BS 476 designation method refers to the surface spread of flame test:

A – no spread of flame
B – not more than 21 inches (533.4 mm) spread
C – more than 21 inches (533.4 mm) spread
D – those continuing to burn for five minutes after withdrawal of the test flame, or with a spread of more than 15 inches (381 mm) across the region of burning in the preliminary test.

Example

Roof surface classified AA. This means no fire penetration within one hour and no spread of flame.

Table A5 to AD B (reproduced below) gives a series of roof constructions together with their two-letter notional designations. In the example shown above, a roof constructed in accordance with Part 1 of Table A5 would satisfy the AA rating if it was of natural slates, fibre reinforced cement slates, clay tiles or concrete tiles and it was supported as shown in the table.

Table 17 of AD B4 is reproduced below and gives the notional two-letter designations for roofs in different buildings according to the distance of the roof from the relevant (or notional) boundary. Once the two-letter designation has been established a form of construction may be chosen from Table A5. Where it has been decided to use a different form of roof construction, the manufacturer's details should be consulted to confirm that the necessary designation will be achieved. It

AD B, Section 14

Table A5 Notional designations of roof coverings.

Part i: Pitched roofs covered with slates or tiles

Covering material	Supporting structure	Designation
1. Natural slates 2. Fibre reinforced cement slates 3. Clay tiles 4. Concrete tiles	timber rafters with or without underfelt, sarking, boarding, woodwool slabs, compressed straw slabs, plywood, wood chipboard, or fibre insulating board	AA

Note: Although the Table does not include guidance for roofs covered with bitumen felt, it should be noted that there is a wide range of materials on the market and information on specific products is readily available from manufacturers.

Part ii: Pitched roofs covered with self-supporting sheet

Roof covered material	Construction	Supporting structure	Designation
1. Profiled sheet of galvanised steel, aluminium, fibre reinforced cement, or pre-painted (coil coated) steel or aluminium with a pvc or pvf2 coating	single skin without underlay, or with underlay or plasterboard, fibre insulating board, or woodwool slab	structure of timber, steel or concrete	AA
2. Profiled sheet of galvanised steel, aluminium, fibre reinforced cement, or pre-painted (coil coated) steel or aluminium with a pvc or pvf2 coating	double skin without interlayer, or with interlayer of resin bonded glass fibre, mineral wool slab, polystyrene, or polyurethane	structure of timber, steel or concrete	AA

Part iii: Flat roofs covered with bitumen felt

A flat roof comprising of bitumen felt should (irrespective of the felt specification) be deemed to be of designation AA if the felt is laid on a deck constructed of 6 mm plywood, 12.5 mm wood chipboard, 16 mm (finished) plain edged timber boarding, compressed straw, slab, screeded wood wool slab, profiled fibre reinforced cement or steel deck (single or double skin) with or without fibre insulating board overlay, profiled aluminium deck (single or double skin) with or without fibre insulating board overlay, or concrete or clay pot slab (insitu or pre cast), and has a surface finish of:

a. bitumen-bedded stone chippings covering the whole surface to a depth of at least 12.5 mm;
b. bitumen-bedded tiles of a non-combustible material;
c. sand and cement screed; or
d. macadam.

Part iv. Pitched or flat roofs covered with fully supported material

Covering material	Supporting structure	Designation
1. Aluminium sheet 2. Copper sheet 3. Zinc sheet 4. Lead sheet 5. Mastic asphalt 6. Vitreous enamelled steel 7. Lead/tin alloy coated steel sheet 8. Zinc/aluminium alloy coated steel sheet 9. Pre-painted (coil coated) steel sheet including liquid-applied pvc coatings	timber joists and: tongued and grooved boarding, or plain edged boarding	AA*
	steel or timber joists with deck of: woodwool slabs, compressed straw slab, wood chipboard, fibre insulating board, or 9.5 mm plywood	AA
	concrete or clay pot slab (insitu or pre-cast) or non-combustible deck of steel, aluminium, or fibre cement (with or without insulation)	AA

Notes:
* Lead sheet supported by timber joists and plain edged boarding should be regarded as having a BA designation.

AD B4

Table 17 Limitations on roof coverings*

Designation† of covering of roof or part of roof	Minimum distance from any point on relevant boundary			
	Less than 6 m	At least 6 m	At least 12 m	At least 20 m
AA, AB or AC	●	●	●	●
BA, BB or BC	○	●	●	●
CA, CB or CC	○	● (1)(2)	● (1)	●
AD, BD or CD (1)	○	● (2)	●	●
DA, DB, DC or DD (1)	○	○	○	● (2)

Notes:

* See paragraph 15.8 for limitations on glass; paragraph 15.9 for limitations on thatch and wood shingles; and paragraphs 15.6 and 15.7 and Tables 18 and 19 for limitations on plastics rooflights.

† The designation of external roof surfaces is explained in Appendix A. (See Table A5, for notional designations of roof coverings.)

Separation distances do not apply to the boundary between roofs of a pair of semi-detached houses (see 15.5) and to enclosed/covered walkways. However, see Diagram 28 if the roof passes over the top of a compartment wall. Polycarbonate and PVC rooflights which achieve a Class 1 rating by test, see paragraph 15.7 may be regarded as having an AA designation.

● Acceptable.
○ Not acceptable.

1. Not acceptable on any of the following buildings:
   a. Houses in terraces of three or more houses;
   b. Industrial, Storage or Other non-residential purpose group buildings of any size;
   c. Any other buildings with a cubic capacity of more than 1500 m^3.
2. Acceptable on buildings not listed in Note 1, if part of the roof is no more than 3 m^2 in area and is at least 1500 mm from any similar part, with the roof between the parts covered with a material or limited combustibility.

should be noted that there are no restrictions on the use of roof coverings which are designated AA, AB or AC. Also, the boundary formed by the wall separating two semi-detached houses may be disregarded for the purposes of roof designations.

Where plastics rooflights form part of a roof structure they should comply with the provisions of AD B2, paragraph 7.13 and Tables 18 and 19 of AD B4 (see section 7.18.9 above) with regard to their separation, area and disposition. The following rooflight materials may be regarded as having an AA designation:

- rigid thermoplastic sheet made from polycarbonate or unplasticised PVC, which achieves a Class 1 surface spread of flame rating when tested to BS476: Part 7: 1971 or 1987 or 1997
- unwired glass at least 4 mm thick.

7.27.10 Thatch and wood shingles

Thatch and wood shingles that cannot achieve the performance specified in BS 476: Part 3: 1958 should be regarded as having a designation of AD/BD/CD in Table 17. However, it may be possible to locate thatch-roofed buildings closer to the boundary than the distances permitted by Table 17 if the following precautions (taken from *Thatched buildings. New properties and extensions* [the 'Dorset Model'], obtainable on www.dorset-technical-committee.org.uk) are incorporated into the design:

- the rafters are overdrawn with construction having at least 30 minutes fire resistance;
- the guidance in Approved Document J *Combustion appliances and fuel storage systems* is followed; and
- the smoke alarm system recommended in AD B1 is extended to the roof space.

7.28 Special provisions relating to shopping complexes and buildings used as car parks

Section 12 of AD B3 describes additional considerations which apply to the design and construction of buildings used as car parks, and to shopping complexes. Although these provisions are nominally placed in AD B3 (Internal fire spread – structure), most parts of AD B do have a bearing on these structures. Accordingly, the recommendations are dealt with in this separate section.

7.28.1 Car parks

A considerable amount of research has been carried out into the behaviour of fire in buildings used as parking for cars and light vans, with the following results.

- The fire load is not particularly high and is well defined.
- If the car park is well ventilated, there is a low risk of fire spread from one storey to another.

Because of the above, car parks are not normally sprinklered.

The best natural ventilation is achieved in open-sided car parks. Where this cannot be attained, heat and smoke will not be as readily dissipated and fewer concessions will apply.

Whatever standard of ventilation is achieved, certain provisions are common to all car parks as follows.

- The relevant provisions for means of escape in case of fire in AD Bl will apply.
- The recommendations of AD B5 regarding access and facilities for the fire service will apply.
- All materials used in the construction of the car park building should be non-combustible except for:
  - (a) any floor or roof surface finish,
  - (b) any fire door,
  - (c) any attendant's kiosk not greater than 15 m^2 in area,
  - (d) any shop mobility facility.
- Surface finishes in buildings, compartments or separated parts which are within the structure enclosing the car park do not need to be constructed from non-combustible materials but they should comply with the relevant provisions of AD B2 and AD B4.

7.28.2 Open-sided car parks

To be regarded as open-sided, the car park should comply with the following provisions.

- It should comply with the four recommendations immediately above.
- It should contain no basement storeys.
- Natural ventilation should be provided to each storey at each car parking level by permanent openings having an aggregate area of at least 5% of the floor area at that level, and at least half of the ventilation area (i.e. 2.5% of the floor area) should be equally provided in opposing walls.
- Where the building containing the car park is also used for other purposes, the part which contains the car park should be a separated part (for definition of 'separated part' see section 7.2 above).

If the above provisions can be met, the car park may be regarded as a small building or compartment for the purposes of space separation in Table 16 of AD B above (see section 7.27.6) and column (1) of that table may be used. Effectively, this halves the required distance to the relevant boundary (or doubles the permitted limit of unprotected areas). Additionally, the fire resistance recommendations of section 7b(i) of Table A2 of Appendix A (see section 7.21.1) will apply, reducing the fire resistance period to only 15 minutes in many cases.

7.28.3 Car parks not regarded as open-sided

If the ventilation recommendations mentioned above cannot be achieved, the car park cannot be regarded as open-sided and the fire resistance recommendations of section 7b(ii) of Table A2 of Appendix A (see section 7.21.1) will apply without any concessions. All car parks require some ventilation, therefore the provisions in the following section apply whatever standard of ventilation is provided.

7.28.4 Ventilation provisions

Ventilation may be provided by natural or mechanical means as follows.

- Natural ventilation to car parks that are not open-sided; provide either:
  - (a) permanent openings to each storey at each car parking level with an aggregate area of at least 2.5% of the floor area at that level. At least half of the ventilation should be equally provided between two opposing walls; or
  - (b) smoke vents at ceiling level with an aggregate area of permanent opening of at least 2.5% of the floor area, arranged to give a through draught.

(Reference should also be made to Approved Document F *Ventilation*, for additional guidance on normal ventilation of car parks.)

- Mechanical ventilation systems for basements and enclosed car parks should be:

(a) independent of any other ventilating system (other than any system providing normal ventilation to the car park);
(b) designed to operate at ten air changes per hour in a fire situation.
(c) designed to run in two parts, each capable of extracting half of the amount which would be extracted at the rates set out in (b) above and designed so that each part may operate singly or simultaneously;
(d) provided with an independent power supply for each part of the system, capable of operating in the event of failure of the main supply;
(e) provided with extract points arranged with half the points at high level and half the points at low level;
(f) provided with fans rated at 300°C for a minimum of 60 minutes;
(g) provided with ductwork and fixings constructed with materials having a melting point of at least 800°C.

(Further information on equipment for removing hot smoke may be obtained from BS 7346: Part 2 *Components for smoke and heat control systems, Specification for powered smoke and heat exhaust ventilators*. For an alternative method of providing smoke ventilation from enclosed car parks see BRE Report *Design methodologies for smoke and heat exhaust ventilation* (BR 368, 1999).)

7.29 Shopping complexes

Individual shops contained in single separate buildings should generally be capable of conforming to the recommendations of AD B. However, where a shop unit forms part of a covered shopping complex, certain difficulties may arise. Such complexes often include covered malls providing access to a number of shops and shared servicing areas. Clearly, it is not practical to compartment a shop from a mall serving it and provisions dealing with maximum compartment sizes may be difficult to meet. Certain other problems may arise concerning fire resistance, walls separating shop units, surfaces of walls and ceilings, and distances to boundaries.

In order to achieve a satisfactory standard of fire safety certain alternative arrangements and compensatory features to those set out in AD B may be appropriate. Reference should be made to sections 5 and 6 of BS 5588: Part 10: 1991 *Code of practice for enclosed shopping complexes* and the relevant recommendations of those sections should be followed.

7.30 Access and facilities for the fire service

7.30.1 Introduction

Part B5 of Schedule 1 to the 1991 Regulations introduced totally new requirements for dealing with access and facilities for the fire service. It reflected guidance produced by the Home Office Fire Department for the Fire Service which had been in use for many years, and replaced goodwill recommendations with statutory requirements. Many local authorities had, through the medium of local Acts of Parliament, applied means of access regulations in their own districts or boroughs for a considerable number of years. These regulations tended to vary from authority to authority so the new requirements brought consistency to this important area of control. In order to avoid duplication of control, most provisions in local Acts of Parliament contain a statutory bar which gives precedence to Building Regulations.

7.30.2 Interpretation

The following terms occur throughout AD B5:

FIRE SERVICE VEHICLE ACCESS LEVEL – The level at which the fire service gain access to a building. This may not always be at ground level, e.g. in a podium design the access may be above ground level.

FIREFIGHTING LIFT – A lift which is designed to have additional fire protection in which the controls may be overriden by the fire service so that they may control it directly for use in fighting fires (see fig. 7.55).

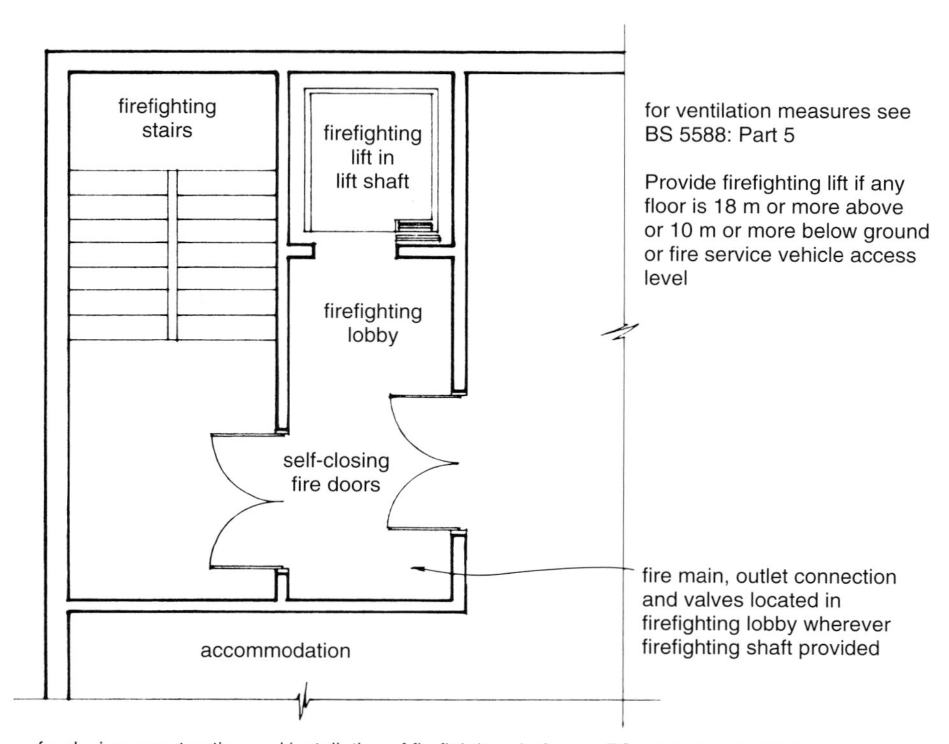

Fig. 7.55 Firefighting shafts and components.

FIREFIGHTING LOBBY – A protected lobby usually situated between a firefighting stair and the accommodation of a building which may also give access to any associated firefighting lift (see Fig. 7.55).

FIREFIGHTING SHAFT – A fire protected enclosure which contains a firefighting stair and firefighting lobbies. It may also contain a firefighting lift, if this is included in the building, together with its machine room. (See Fig. 7.55.)

FIREFIGHTING STAIR – A fire protected stair which is separated from the accommodation of the building by a firefighting lobby (see Fig. 7.55).

PERIMETER (of building) – The maximum aggregate plan perimeter. This is found by vertical projection of the building onto a horizontal plane and is illustrated in Fig. 7.56.

7.30.3 Access and facilities for the fire service – the statutory requirements

Buildings must be designed and constructed so as to provide reasonable facilities to assist firefighters in the protection of life. There must also be reasonable provision made within the site of the building to enable fire appliances to gain access to the building.

These requirements are interesting in that they apply provisions to the site of the building and not just to the building itself. This approach may be compared to that for access for disabled people in Chapter 17 of this book where site recommendations are also made. It should be noted that there is no definition of 'site' contained in the regulations or approved documents.

The main factor that determines the facilities which are needed to assist the fire service is the size of the building, since it is the philosophy of firefighting in the United Kingdom that this be carried out inside the building if it is to be effective. This philosophy ensures that the water used for firefighting actually reaches the fire and means that effective search and rescue can be carried out as close as possible to the source of the fire since it is at this point that trapped people will be in most peril.

Therefore, in order to meet these statutory requirements it is necessary to provide:

(1) in most buildings:
- sufficient means of vehicular access across the site of the building to enable fire appliances to be brought near to the building for effective use; and
- sufficient means of access for firefighting personnel into and within the building so that they may effect rescue and fight fire;

(2) in large buildings and/or buildings with basements:
- sufficient fire mains and other facilities, such as fire fighting shafts, to assist firefighters; and
- adequate means of venting heat and smoke from basement fires.

It should be noted that these arrangements for access and firefighting facilities are required in order to secure reasonable standards of health and safety for people

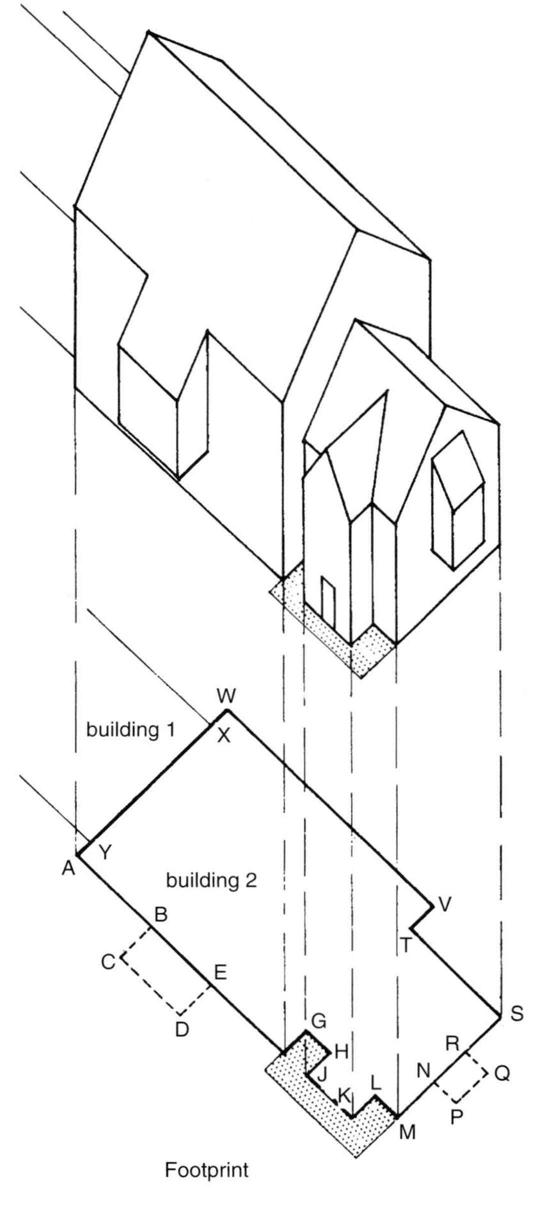

To calculate length of perimeter for purposes of Table 20 of AD B5, add together individual lengths (i.e. AB + BC + CD etc.). Do not include XY, as this is common to adjacent building and cannot be accessed anyway.

To find maximum aggregate plan perimeter (or footprint), project overhanging storeys onto ground floor plan, then footprint is outline denoted by letters opposite.

Example calculation

1. Total floor area of building = 600 m^2
2. Height of top storey above ground = 11 m
3. Perimeter length (less XY) = 60 m
4. From Table 20 provide access to 15% of perimeter i.e. 11 m shown by hatching on plan
5. Provide suitable door in this part of perimeter, min. 750 mm wide

Fig. 7.56 Calculation of perimeter.

(including firefighters) in or about buildings and for the purposes of protecting life by assisting the fire service.

7.30.4 Access facilities for fire appliances

Vehicle access to the exterior of a building is needed:

- to enable high-reach appliances (i.e. turntable ladders and hydraulic platforms) to be used; and
- to enable pumping appliances to supply water and equipment for rescue activities and firefighting.

Clearly, the requirements for access to buildings increase with building size and height. In large buildings it may be necessary to provide firefighting shafts and fire mains. Fire mains are provided in buildings to enable firefighters to connect their hoses to a convenient water supply at the floor level of the fire. Therefore, where these are fitted, it will be necessary for pumping appliances to gain access to the perimeter of the building at points near to the mains. This is especially so in the case of dry mains since these will need to be connected by hose to the pumping appliance. The provision of fire mains and firefighting shafts is described more fully below.

7.30.5 Buildings not fitted with fire mains

Fire mains need only be provided where there is a necessity to provide a firefighting shaft (see section 7.30.9 below). Therefore, in buildings not fitted with fire mains, access for fire service vehicles should be provided in accordance with Table 20 of AD B5 which is reproduced below. It should be noted that Table 20 does not apply to buildings with fire mains, or to blocks of flats and maisonettes, because every dwelling entrance door in such buildings should be within 45 m of a pump appliance.

Buildings with a total aggregate floor area of up to 2000 m^2 and a top storey less

AD B5

Table 20 Fire service vehicle access to buildings (excluding blocks of flats) not fitted with fire mains.

Total floor area (1) of building m^2	Height of floor of top storey above ground (2)	Provide vehicle access (3)(4) to:	Type of appliance
up to 2000	up to 11	see paragraph 17.2	pump
	over 11	15% of perimeter (5)	high reach
2000–8000	up to 11	15% of perimeter (5)	pump
	over 11	50% of perimeter (5)	high reach
8000–16,000	up to 11	50% of perimeter (5)	pump
	over 11	50% of perimeter (5)	high reach
16,000–24,000	up to 11	75% of perimeter (5)	pump
	over 11	75% of perimeter (5)	high reach
over 24,000	up to 11	100% of perimeter (5)	pump
	over 11	100% of perimeter (5)	high reach

Notes:

1. The total floor area is the aggregate of all floors in the building (excluding basements).
2. In the case of Purpose Group 7(a) (storage) buildings, height should be measured to mean roof level, see Methods of Measurement in Appendix C of AD B
3. An access door is required to each such elevation (see paragraph 17.5).
4. See paragraph 17.9 for meaning of access.
5. Perimeter is described in Fig. 7.56.

than 11 m above ground level are referred to as 'small buildings' in AD B5 (see the first row of Table 20). There should be vehicle access to 15% of, or to within 45 m of every point on, the maximum aggregate plan perimeter (or footprint, see Fig. 7.56) of such buildings, whichever figure is the less onerous. For single family dwelling houses, the 45 m may be measured to a door to the dwelling.

In Table 20, the key figure to remember is 11 m above ground level for the top storey of the building. Buildings above this height not fitted with fire mains will need access for high-reach appliances as well as pumping appliances. There should be a suitable door giving access to the interior of the building, at least 750 mm wide, situated in any elevations which are required to be accessed by virtue of Table 20. A typical example of this is shown in Fig. 7.56.

7.30.6 Buildings fitted with fire mains

As mentioned above, buildings provided with firefighting shafts should also have fire mains.

Where dry fire mains are fitted:

- access should be provided for a pumping appliance to within 18 m of each fire main inlet connection point
- the inlet should be visible from the appliance.

Where wet fire mains are fitted:

- access should be provided for a pumping appliance to within 18 m of a suitable entrance giving access to the main
- the entrance should be visible from the appliance
- the inlet for the emergency replenishment of the suction tank for the main should be visible from the appliance.

Sometimes, fire mains are fitted in buildings even though there is no recommendation in AD B5 that they should have firefighting shafts. In such cases the provisions for access listed above may be used instead of Table 20.

7.30.7 Access routes and hardstandings

In order to provide access for fire service vehicles across the site of a building, it is necessary to design a road or other route which is wide enough and has sufficient load-carrying capacity (including manhole covers, etc.) to take the necessary vehicles. Unfortunately, fire appliances are not standardised and so it is a wise precaution to check with the local fire service in a particular area in order to establish their weight and size requirements. Some design guidance is given in Table 21 of AD B5 (reproduced below), where typical vehicle access route specifications are shown. It should be noted that the typical minimum carrying capacity for a high-reach appliance is 17 tonnes. A roadbase designed to take 12.5 tonnes should be satisfactory for high-reach vehicles since the use would be infrequent and the weight of the vehicle is distributed over a number of axles. However, structures such as bridges should still be designed to take the full 17 tonnes.

In general, where access is provided to an elevation in accordance with Table 20, the access routes and hardstandings should be designed in accordance with the following provisions:

- for buildings up to 11 m high, access should be provided for a pump appliance adjacent to the building, for the percentage of the total perimeter specified in Table 20 (this does not apply where a pump appliance can get to within 45 m of every point on the perimeter of a building or to within 45 m of every dwelling entrance door in blocks of flats or maisonettes)
- for buildings over 11 m high, a zone should be established in accordance with Fig. 7.57, which should be kept clear of overhead obstructions, since it is possible that overhead obstructions such as cables and branches might interfere with the setting of ladders or the swing of high-reach appliances.

Any dead-end access route which is more than 20 m long should be provided with a turning point, such as a hammerhead or turning circle, designed in accordance with dimensions given in Table 21 of AD B5 (see below). This is to ensure that fire service vehicles do not have to reverse more than 20 m from the end of an access road.

In all the above considerations for site access it should be borne in mind that requirements cannot be made under the Building Regulations for work to be done outside the site of the works shown on the submitted plans, building notice or initial

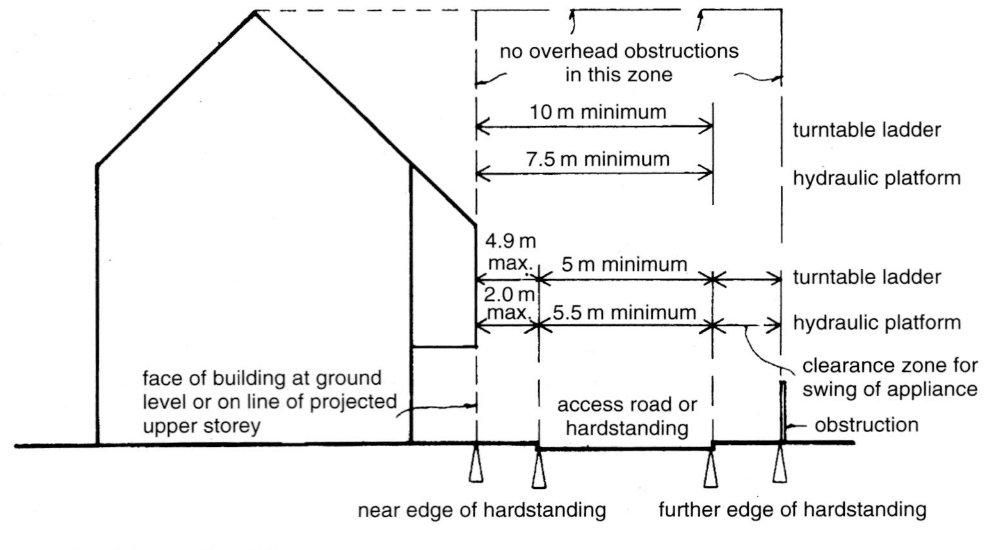

Fig. 7.57 Overhead obstructions – access dimensions for high-reach appliances.

AD B5

Table 21 Typical vehicle access route specification.

Appliance type	Minimum width of road between kerbs (m)	Minimum width of gateways (m)	Minimum turning circle between kerbs (m)	Minimum turning circle between walls (m)	Minimum clearance height (m)	Minimum carrying capacity (tonnes)
Pump	3.7	3.1	16.8	19.2	3.7	12.5
High-reach	3.7	3.1	26.0	29.0	4.0	17.0

notice. Therefore, it may not always be reasonable to upgrade an existing route across a site to a small building, such as a single dwelling house. The reasonableness of any proposals to carry out upgrading of certain features (such as the removal of a sharp bend) should be considered by the local authority or approved inspector in consultation with the fire service.

7.30.8 Access for firefighting personnel to buildings

As has been mentioned above, it is important that fire service personnel are able to gain access to buildings in order to reach the seat of the fire and to carry out effective search and rescue.

This is especially true in tall buildings and in those with deep basements. In such buildings firefighters will need additional facilities contained within a protected firefighting shaft, such as firefighting lifts, firefighting stairs and firefighting lobbies, equipped with fire mains. Additionally, fire appliances will need access to entry points near the fire mains. These facilities are necessary in order to avoid delay in tackling the fire and to provide a safe working base from which effective action may be taken.

In other buildings, the normal means of escape will offer personnel access and this, when combined with the ability to work from ladders and appliances on the perimeter, will generally mean that special internal arrangements are unnecessary. Vehicle access will usually be needed to some or all of the perimeter, but the extent of this will depend on the size of the building.

Dwellings and other small buildings should be sufficiently close to as point accessible to fire brigade vehicles – no other provisions are necessary.

In taller blocks of flats, the high degree of compartmentation means that the access facilities may be simpler than in other types of tall buildings, however, fire brigade personnel access facilities will still be needed within the building.

In basement fires, products of combustion tend to escape via stairways. This can make it difficult for fire service personnel to gain access to the seat of the fire. The problem can be reduced by providing smoke vents, which improve visibility, reduce temperatures and make search, rescue and firefighting less difficult.

7.30.9 Firefighting shafts

Firefighting shafts should be provided to serve the storeys indicated in the following list.

(1) Buildings which contain two or more basement storeys each exceeding 900 m^2 in area.
(2) Buildings in Purpose Groups 4, 6 and 7(a) (Shop and Commercial, Industrial, and Storage and other non-residential) with any storey of 900 m^2 or more in area situated more than 7.5 m above ground or fire service access level.
(3) Buildings with any floor more than 18 m above ground or fire service vehicle access level.
(4) Buildings with any floor more than 10 m below ground or fire service vehicle access level.
(5) Shopping complexes, in accordance with the recommendations of Section 3 of BS 5588: Part 10: 1991 *Code of practice for enclosed shopping complexes.*

Firefighting shafts should conform to the following recommendations.

- Those provided in (3) and (4) above should also contain firefighting lifts.
- Where provided to serve a basement under (1) or (4) above, there is no need to serve the upper floors also unless they qualify in their own right because of the size or height of the building.
- Similarly, where a shaft is provided to serve upper floors in (2) or (3) above, it need not also serve a basement which is not large or deep enough to qualify on its own.
- Where they are provided, firefighting shafts and lifts should serve all intermediate floors between the lowest and the highest in the building.

7.30.10 Standard of provision of firefighting shafts

The number of firefighting shafts which needs to be provided in a building may be obtained from Table 7.11 below. It can be seen that if the building is fitted with a sprinkler system meeting the relevant recommendations of BS 5306: Part 2: 1990 *Specification for sprinkler systems*, then it is possible to reduce the number of firefighting shafts that are provided.

Firefighting shafts should be located so that every part of each storey is within 60 m of the entrance to a firefighting lobby measured along a route which is suitable for laying fire hoses. This figure is reduced to 40 m where the internal layout of the building is not known at the design stage, the 40 m being measured in a straight line from every point in the storey to the entrance of the firefighting lobby. These distance recommendations do not apply to accommodation situated at fire service access level.

7.30.11 Layout and construction of firefighting shafts

Firefighting shafts should be designed and constructed to encompass the following recommendations.

(1) Except in blocks of flats and maisonettes, access to the accommodation in a building from a firefighting lift or stair should be through a firefighting lobby.

Table 7.11 Provision of firefighting shafts in buildings.

Area of largest qualifying floor (m^2)[a]	Number of firefighting shafts to be provided
A. With sprinklers fitted in the building (except basements)	
Under 900	1
900 to 2000	2
Over 2000	2 (plus 1 shaft for every extra 1500 m^2 or part thereof)
B. Without sprinklers and in any qualifying basement	
Up to 900	1
Over 900	2 (plus 1 shaft for every extra 900 m^2 or part thereof)

Notes:
[a] This is the largest floor area which is situated:
(a) Over 18 m above fire service vehicle access level, or
(b) Over 7.5 m or more above fire service vehicle access levels and 900 m^2 in area, or
(c) In any qualifying basement (see section 7.30.9).

(2) Every firefighting shaft should be equipped with a fire main which should have outlet connections and valves situated in fire-fighting lobbies.
(3) Firefighting shafts should comply with the following parts of BS 5588: Part 5: 1991 *Code of practice for fire fighting stairs and lifts*:
- section 2: *Planning and construction*
- section 3: *Firefighting lift installation*
- section 4: *Electrical services.*

The various components of a firefighting shaft are illustrated in Fig. 7.55 above.

7.30.12 Provision of fire mains

Fire mains are provided to enable the fire service to fight fires inside the building. They are equipped with valves which permit direct connection of fire hoses. This assists firefighters by making it unnecessary to take hoses up stairways from the pumping appliance at ground or access level, thus saving time and avoiding blockage of the escape route.

Fire mains which serve floors above ground or access level are commonly known as rising mains and those which serve floors below ground or access level (such as basements) are usually referred to as falling mains.

Where it is necessary to provide firefighting shafts in a building (see section 7.30.10 above), each shaft should contain a fire main.

There are two types of fire main.

- Wet mains (often called 'wet risers') are usually kept full of water by header tanks and pumps in the building. Since there is the danger that the water supply may run out in a serious fire, there should be a facility to replenish the wet main from the pumping appliance in an emergency. Wet risers should be provided in any building which has a floor situated more than 60 m above ground or access level. They may, of course, serve lower floors if so desired.

- Dry mains ('dry risers') are normally kept empty and are charged with water from a fire service pumping appliance in the event of a fire. Where provided, they may serve any floor which is less than 60 m above ground or access level.

The outlets from fire mains at each floor level in the building should be situated in fire fighting lobbies giving access to the accommodation from a firefighting shaft.

Further guidance on the design and construction of fire mains may be obtained from sections 2 and 3 of BS 5306 *Fire extinguishing installations and equipment on premises,* Part 1: 1976 (1988) *Hydrant systems, hose reels and foam inlets.*

7.30.13 Smoke and heat venting of basements

A basement fire differs from a fire in another part of a building in that it is difficult for heat and smoke to be adequately vented. Products of combustion from basement fires tend to escape via stairways making it difficult for fire service personnel to gain access to the fire. Consequently, visibility will be reduced and temperatures will tend to be higher in a basement fire making search, rescue and firefighting more difficult. If smoke outlets (or smoke vents) are installed they can provide a route for heat and smoke, enabling it to escape direct to outside air from the basement and permitting the ingress of cooler air. Two typical designs for smoke outlet shafts are shown in Fig. 7.58.

7.30.14 Standard of provision for smoke outlets

Basement storeys which exceed 200 m^2 in area or are more than 3 m below ground level should be provided with smoke outlets connected directly to outside air. Some basements are excepted from this rule as follows:

- any basement in a single family dwellinghouse (PG l(b) or l(c))
- any strong room.

If possible each basement room or space should have one or more smoke outlets. In some basements the plan may be too deep, or there may be insufficient external wall areas to permit this. An acceptable solution might be to vent perimeter spaces directly and to allow internal spaces to be vented indirectly by the fire service by means of connecting doors. This solution is not acceptable if the basement is compartmented since each compartment should have direct access to venting without the use of intervening doors between compartments.

7.30.15 Means of venting

Smoke venting may be by natural or mechanical means.

Natural smoke venting may be achieved by providing smoke outlets which conform to the following recommendations.

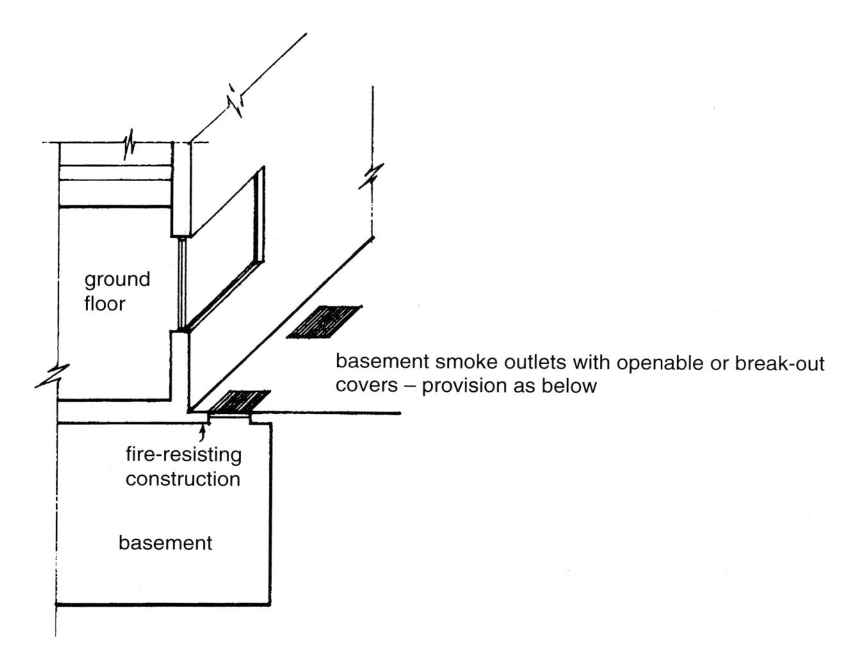

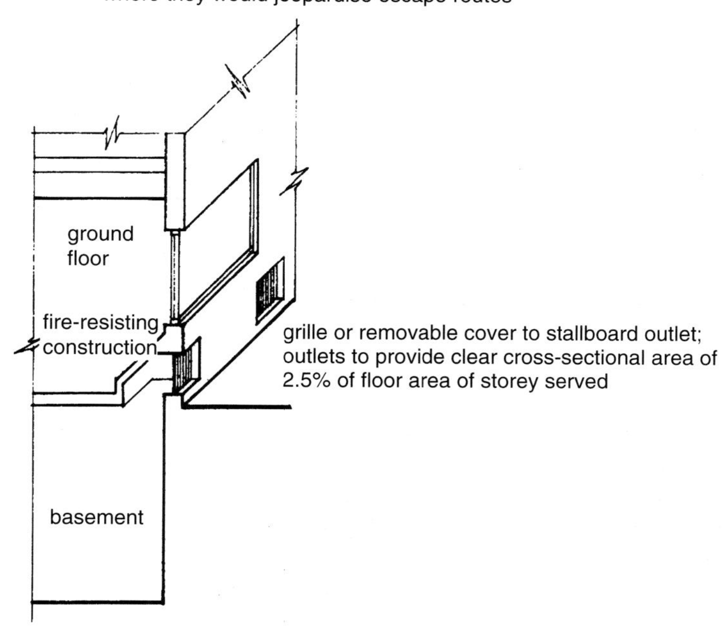

Fig. 7.58 Construction of smoke outlets.

- The total clear cross-sectional area of all the smoke outlets should be at least 2.5% of the floor area of the storey they serve.
- Places of special fire risk should be provided with separate outlets.
- Outlets should be positioned so that they do not compromise escape routes from the building.
- Outlets which terminate in readily accessible positions may be covered by pavement lights, stallboards or panels which can be broken out or opened. They should be suitably marked to indicate their position.
- Outlets which terminate in less accessible positions should be kept unobstructed and should only be covered by a louvre or grille which is non-combustible.
- Smoke outlets should be sited at high level (i.e. in the ceiling or wall of the space they serve) and should be distributed evenly around the perimeter of the building so as to discharge into open air outside the building.

Mechanical smoke extraction may be used as an alternative to natural venting if a sprinkler system conforming to BS 5306: Part 2 is installed in the basement. Unless needed for other reasons, it is not necessary to install sprinklers on the other storeys in the building merely because they are provided to allow mechanical smoke extraction in the basement to be used.

Any mechanical smoke extraction system should:

- achieve at least ten air changes per hour
- be capable of handling gas temperatures of 400°C for at least one hour
- come into operation automatically on activation of the sprinkler system, or be activated by an automatic fire detection system conforming to BS 5839: Part 1: 1988 (at least L3 standard).

7.30.16 Construction of smoke vent outlet ducts and shafts

Outlet ducts and shafts for smoke and heat venting should be enclosed in non-combustible fire-resisting construction. This applies equally to any bulkheads over the ducts or shafts as indicated in Fig. 7.58.

Natural smoke outlets from different compartments in the same basement storey or from different basement storeys should also be separated from each other by non-combustible fire-resisting construction.

7.30.17 Ventilation of basement car parks

The provisions contained in section 12 of AD B (see section 7.28.3 above) regarding the ventilation of basement car parks, may be regarded as satisfying the requirements contained in AD B5 for the smoke venting of any basement which is used as a car park.

7.31 Firefighting and the use of insulating core panels for internal structures

7.31.1 Introduction

Appendix F of Approved Document B provides a limited amount of guidance on the fire behaviour of insulating core panels when used for internal structures, where particular fire spread problems have been known to occur.

Internal insulating core panels are commonly used to provide chilled or subzero environment enclosures for the production, preservation, storage and distribution of perishable foodstuffs. They are also used where it is necessary to provide a hygienic environment.

A typical panel might consist of an inner insulating core sandwiched between, and bonded to, facing sheets of galvanised steel (sometimes bonded with a PVC facing where a hygienic finish is needed). The panels can be formed into a structure which can be free standing or can be attached to the building by lightweight fixings and hangers.

Common materials for the insulating core include:

- expanded or extruded polystyrene
- polyurethane
- mineral fibre
- polyisocyanurate
- modified phenolic.

7.31.2 The behaviour in fire of the core materials and fixing systems

When exposed to radiated or conducted heat from a fire, polymeric materials can be expected to degrade (and will produce large quantities of smoke). Fires involving mineral fibre cores produce less potential problems than those with polymeric cores.

It is also known that panels will tend to delaminate (the core separating from the facings because of expansion of the membrane and softening of the bond line) when exposed to the high temperatures of a developed fire, irrespective of the type of core material used.

Once it is involved in a fire, the panel will lose most of its structural integrity and the stability of the system will then depend on the residual structural strength of the non-exposed facing, the joint between panels and the fixing system. Because most panel jointing or fixing systems have an extremely limited structural integrity performance in fire conditions there is a real chance of total and unexpected collapse of the panel system (together with any associated equipment) if the fire starts to heat up the support fixings or structure to which they are attached. Fire can also spread behind the panels, where it may be hidden from the occupants of the building and this can prove to be a particular problem to fire fighters as, due to the insulating properties of the cores, it may not be possible to track the spread of fire, even using infra red detection equipment.

7.31.3 Problems for firefighters

When encountering a fire involving insulating core panel systems, firefighters may be confronted by the following problems:

- fire spread hidden within the panels;
- large quantities of black toxic smoke being produced; and
- rapid fire spread leading to flashover.

These characteristics are common to both polyurethane and polystyrene cored panels (although the rate of fire spread in polyurethane cores is significantly less than that of polystyrene cores) especially when any external heat source is removed.

Additionally, all systems are susceptible to the following problems, irrespective of the type of panel core used:

- delamination of the steel facing;
- collapse of the system; and
- hidden fire spread behind the system.

7.31.4 Design solutions

Risk assessment techniques should be used to identify the appropriate solution. The following design strategies can be adopted once the potential fire risks within the panel system enclosures have been identified:

- remove the risk
- provide an appropriate separation distance between the risk and the panels
- provide a fire suppression system for the risk
- provide a fire suppression system for the enclosure
- provide panels with suitable fire resistance
- specify appropriate materials, fixing and jointing systems.

Overall, the fire performance of the panel system should not be seen in isolation but should be considered in relation to the fire performance of the building as a whole, including the insulating envelope, the superstructure, the substructure etc.

7.31.5 Specification of panel core materials

Panels should be specified with core materials, which are appropriate to the application under consideration. This will help to ensure an acceptable level of performance for panel systems, under fire conditions. Table 7.12 below gives examples of core materials and appropriate applications.

7.31.6 Specification of materials/fixing and jointing systems

The aim of the specification should be to improve the stability of the panel system in the event of fire. Similarly, the aim of construction detailing should be to prevent the

Table 7.12 Insulating core panels – specification of core materials by application.

Core material	Applications[1]
Mineral fibre cores	Cooking areas, bakeries and other hot areas
	Fire breaks in combustible panels, fire stop panels and general fire protection
All core materials	Chill stores, cold stores and blast freezers
	Food factories
	Clean rooms

Notes: (1) Core materials may be used in other circumstances where a risk assessment has been made and other appropriate fire precautions have been put in place.

core materials from becoming exposed to the fire and contributing to the fire load. Therefore, it might be appropriate to consider the following.

- Design insulating envelopes, support systems, and supporting structure to maintain structural stability following failure of the bond line between insulant core and facing materials, by using alternative methods, such as catenary action. In a typical case this could require positive attachment of the lower faces of the insulant panels to supports.
- Where a supplementary support method is provided to support the panels, this should remain stable under fire conditions for an appropriate time. Therefore, consider fire protecting the building superstructure, together with any elements providing support to the insulating envelope, to prevent early collapse of the structure or the envelope.
- Light gauge steel members such as purlins and sheeting rails, which provide stability to building superstructures, may be compromised at an early stage of a fire. Although it is not practical to fire protect such members it may be possible to provide supplementary fire protected heavier gauge steelwork members at wider intervals than purlins to provide restraint in the event of a fire.
- In designated high risk areas, fire propagation through the insulant can be prevented by incorporating barriers consisting of, for example, non-combustible insulant cored panels into wall and ceiling construction at intervals, or strips of non-combustible material into specified wall and ceiling panels.
- The combustible insulant should be fully encapsulated by non-combustible facing materials that remain in place during a fire.
- Service penetration should be catered for by incorporating pre-finished and sealed areas in the panels.
- Do not allow panels or panel systems to support machinery or other permanent loads.
- Any cavity created by the arrangement of panels, their supporting structure or other building elements should be provided with suitable cavity barriers.

7.31.7 Further guidance on insulating core panels

Reference should be made to *Design, construction, specification and fire management of insulated envelopes for temperature controlled environments* published by the International Association of Cold Storage Contractors (European Division). This document contains examples of possible solutions and general guidance on insulating core panels construction. Chapter 8 of the document is of particular relevance since it gives guidance on the design, construction and management of insulated structures and is considered to be appropriate for most insulating core panel applications.

8 Materials, workmanship, site preparation and moisture exclusion

8.1 Materials and workmanship

8.1.1 Introduction

Regulation 7 of the 2000 Regulations is concerned with the fitness and use of the materials necessary for carrying out building work. It is supported by its own Approved Document entitled, rather aptly, *Approved Document to support Regulation 7* (AD Regulation 7). Apart from dealing generally with the standards of materials and workmanship needed for building work, AD Regulation 7 is also concerned with:

- the use of materials which are susceptible to changes in their properties;
- resistance to moisture and deleterious substances in the subsoil; and
- short-lived materials.

The use of materials which are unsuitable for permanent buildings is covered by section 19 of the Building Act 1984. Local authorities are enabled to reject plans for the construction of buildings of short-lived or otherwise unsuitable materials, or to impose a limit on their period of use. The Secretary of State may, by Building Regulations, prescribe materials which are considered unfit for particular purposes. Tables 1 and 2 of the 1976 Regulations listed materials which were considered unfit for the weather-resisting part of any external wall or roof. Neither the 2000 regulations nor AD Regulation 7 prescribe any materials as unfit for particular purposes as yet; however the AD does lay down some general criteria against which materials may be judged (see sections 8.4.5 and 8.4.6). Bearing this in mind, it is unlikely that section 19 can be used by local authorities to proscribe certain materials. Since it is now possible to use materials and methods of workmanship which comply with European Standards or Technical Approvals, all references to Agrément Certificates in the 1992 edition of AD Regulation 7 were replaced in the 1999 edition by references to national or European certificates issued by a European Technical Approvals issuing body.

8.1.2 Interpretation

A large number of terms and abbreviations appear in AD Regulation 7 as follows:

BS – British Standard, issued by the British Standards Institution (BSI). To achieve British Standard status a draft document is prepared by relevant experts and is submitted for public consultation. Comments received are considered and consensus is reached before the proposed document is issued. More information on British Standards may be obtained from the BSI at 389 Chiswick High Road, London W4 4AL (Internet: www.bsi.org.uk).

BUILDING CONTROL BODY – Includes both local authorities and approved inspectors.

CE MARKING – Materials bearing the CE mark (see Fig. 8.1) are presumed to comply with the minimum legal requirements as set out in the Construction Products Regulations 1991 (see Chapter 5). This is described more fully in section 8.4.2.

Fig. 8.1 The CE mark.

CEN – Comité Européen de Normalisation. This is the body recognised by the European Commission (see below) to prepare harmonised standards to support the Construction Products Directive (CPD) (see section 8.3). The committee comprises representatives of the standards bodies of participating members of the EU and EFTA (European Free Trade Association).

CPD – Construction Products Directive (see section 8.3).

EEA (EUROPEAN ECONOMIC AREA) – Those states which signed the Agreement at Oporto on 2 May 1992 plus the Protocol adjusting that Agreement signed in Brussels on 17 March 1993. The states are Austria, Belgium, Denmark, Finland, France, Germany, Greece, Iceland, Ireland, Italy, Luxembourg, Liechtenstein, Netherlands, Norway, Portugal, Spain, Sweden, and United Kingdom.

EOTA – European Organisation for Technical Approvals. This is the umbrella organisation for bodies which issue European Technical Approvals for individual products. Whilst EOTA operates over the same area as CEN, it complements their work by producing guidelines for innovative and other products for which standards do not exist.

EN – These letters indicate that a European standard has been implemented in a particular Member State. Thus, in the United Kingdom the designation will be BS EN, followed by the relevant standard number. A British Standard shown in this way will be identical to the standards of other Member States, but will also include additional guidance regarding its use and its relationship with other standards in the same group. An EN does not have a separate existence as a formally published document.

EUROPEAN COMMISSION – Based in Brussels, this is the executive organisation of the EU. The Commission ensures that Community rules are implemented and observed and it alone has power to propose legislation based on the Treaties. It also executes decisions taken by the Council of Ministers.

EUROPEAN TECHNICAL APPROVAL – A technical assessment which specifies that a construction product is fit for its intended use. It is issued for the purposes of the Construction Products Directive (CPD) and the issuing body must be authorised by a Member State and be notified to the European Commission under section 10 of the CPD. Details of the approval issuing bodies are published in the 'C' series of the Official Journal of the European Communities. For the United Kingdom, in addition to the British Board of Agrément the current listing also includes BRE Certification Ltd, Garston, Watford, Herts WD25 9XX. A current list can be found on the Building Regulations pages of the ODPM website at:
http://www.odpm.gov.uk

EU – The 15 countries of the European Union, *viz.* Austria, Belgium, Denmark, Finland, France, Germany, Greece, Ireland, Italy, Luxembourg, Netherlands, Portugal, Spain, Sweden, and United Kingdom.

ISO – International Organisation for Standardisation. This is the worldwide standards organisation and it is likely that some ISO standards may be adapted for use with the CPD. Such standards are identified by 'ISO' and a number and in the UK they may appear as BS ISO or, if they are adopted as European standards, as BS EN ISO. Unlike ENs, ISOs are published separately.

TECHNICAL SPECIFICATION – A standard or European Technical Approval Guide. A document against which compliance can be shown (e.g. for standards) and against which an assessment is made in order to deliver the European Technical Approval.

UKAS – United Kingdom Accreditation Service. An accreditation body for quality assurance and management schemes. Further details may be obtained from UKAS, 21–47 High Street, Feltham, Middlesex, TW3 4UN.

8.2 The influence of European standards

In order to understand the changes which are coming about with the advent of the Single European Market, the following brief summary sets out the context of these changes and their influence on the building control system in England and Wales.

A main goal of the European Community is to allow free movement among the Member States of goods, services, people and capital. Free movement of goods may be hampered by physical, technical or fiscal barriers.

Significant technical barriers arise from the use of different technical requirements or regulations in Member States. This results in the necessity to produce slightly different versions of the same product to satisfy different markets and different methods of test for suitability must be used, resulting in undue expense and waste of resources by manufacturers and suppliers.

To overcome these difficulties, early directives were issued for which it was necessary to resolve technical issues with unanimous agreement by all Member States. This turned out to be a slow and cumbersome process.

8.2.1 The New Approach

The Single European Act in 1986 declared an agreement to establish the Single European Market by the end of 1992. Progress to the single market has been aided by a radical change in the approach to the writing of directives and European standards – the so-called New Approach. This recognises that EU legislation should only apply to areas already subject to existing national laws or regulations and also allows qualified majority voting into the decision-making process.

The New Approach Directives express requirements in broad terms, called the Essential Requirements. Member States presume conformity with these requirements where a product satisfies a harmonised European technical specification or, as an interim measure, a national standard accepted by the Commission. The advent of European standards will prevent Member States from using their own standards to protect their own markets.

A number of new approach product directives have been adopted which are relevant to the construction industry, the most significant being the Construction Products Directive 89/106 EEC.

8.3 The Construction Products Directive (CPD)

The full title of this directive reflects its objectives:

> 'Council Directive of 21 December 1988 on the approximation of laws, regulations and the administrative provisions of the Member States relating to construction products.'

The CPD was implemented in the UK on 27 December 1991 by the Construction Products Regulations 1991 (see also DOE circular 13/91).

Construction products are defined as:

Those produced for incorporation in a permanent manner in construction works, in so far as the Essential Requirements (ERs) relate to them.

There are six Essential Requirements as follows:

(1) Mechanical resistance and stability.
(2) Safety in case of fire.
(3) Hygiene, health and the environment.
(4) Safety in use.
(5) Protection against noise.
(6) Energy economy and heat retention.

The link between the Essential Requirements and the product on the market is made in the Interpretative Documents (IDs).

8.4 Interpretative Documents

There is one ID for each Essential Requirement. These interpret the ERs more fully and indicate:

- appropriate product characteristics
- appropriate topics for harmonised technical specifications
- the need for different levels or classes of performance to allow for different regulation requirements in different Member States.

The UK is represented by experts from the DETR and BRE on the technical committees and drafting panels for the IDs. The IDs are intended mainly for standards' writers and enforcement authorities and will be of less use to manufacturers and suppliers.

8.4.1 Technical specifications

Sometimes referred to as harmonised technical specifications, these are deemed to comply with the Essential Requirements and so products meeting the specifications will immediately demonstrate their fitness to be placed on the market.

Three types exist:

(1) *Harmonised Standards* – ideally the best route for demonstrating compliance for a product. These standards are developed mainly by CEN on the basis of standardisation requests (called mandates) from the Commission. Only those parts of standards which relate to the ERs are mandated and these are the parts which support fixing of the CE mark to products.
(2) *European Technical Approvals* – these have replaced the Agrément certificate

form of product assessment and are carried out by bodies designated by individual Member States which are then notified by that Member State to the European Commission. All such bodies are members of the European Organisation for Technical Approvals (EOTA), and they operate under a common set of rules.

(3) *National Specifications* – although a possibility, the recognition of national specifications at community level is likely to be rare. The procedures have to be initiated by the Commission and their use may be expected only in situations where a barrier to trade has been demonstrated and the production of a Technical Specification is some way off.

8.4.2 The CE mark and attestation of conformity

The purpose of attestation of conformity with technical specifications is to assure purchasers and regulators that products placed on the market comply with the ERs. Such products may carry the CE mark.

Each Member State is required to maintain a register of designated or notified bodies which identifies:

- Test laboratories – for testing samples;
- Inspection bodies – for inspecting factories and processes;
- Certification authorities – to interpret results;

so as to allow the manufacturer to affix the CE mark to his product. The marking may be placed on the product itself, a label, the packaging or on the accompanying commercial information. It will be accompanied by a reference to the technical specification to which it conforms and, where appropriate, by an indication of its characteristics. In the UK the register of designated bodies is maintained by the ODPM.

It is important to appreciate that the CE mark is not a quality mark. It signifies only that the product satisfies the requirements. The CE mark is therefore primarily intended for enforcement officers and only states that the product may legally be placed on the market.

8.4.3 Interim procedures

It will be apparent that in order to harmonise the standards of all the countries in the European Union it is necessary for there to be a transitional period during which existing standards will have to co-exist with new standards.

With specific reference to British Standards, nearly all those which are related to construction products will be revised to become the British 'transposition' of the new European Standards (ENs) which are currently being drafted. In the past it has been the practice to adopt the old British Standard number when transposing an EN which is based on substantially the same material. Current BSI numbering policy is to adopt the CEN numbering, prefaced with BS.

Although British Standards are normally withdrawn when their equivalent

European Standards are published the following circumstances may require a deferred withdrawal of the British Standard:

- when it is necessary for the BS to remain available for work which has already commenced.
- when a BS is called up in an Approved Document which has yet to be revised.

In practice it may be necessary for some BSs to remain valid for a number of years, fully maintained alongside the new transposed standards of European origin. Therefore, it will be necessary for controllers, designers, etc., to check the applicability of the standard in each context. Where an old standard is retained it is reasonable to presume that it will satisfy the requirements of Regulation 7, and where a new standard is introduced it must be checked for applicability during the transitional period and if found suitable, compliance may reasonably also be presumed.

Standards of European origin have clauses specifically identified which relate to the 'harmonised' requirements relevant to the Building Regulations (i.e. those dealing with health and safety matters) and 'non-harmonised' requirements which contain additional material relating to trading requirements of the construction industry. The non-harmonised requirements are not the concern of Regulation 7 and are not covered by the AD Regulation 7 recommendations.

Interestingly, it is possible for a product to be tested and certified as complying with a British Standard by an approved body in another Member State of the European Community under the procedure covered by article 16 of the CPD. In this case it should normally be accepted by the building control body as complying with that standard. If there are reasons to doubt its acceptance then the burden of proof is on the controlling body which is obliged to notify the Trading Standards Officer so that the UK Government can notify the Commission.

With regard to CE marking the UK Construction Product Regulations state only that products must be 'fit for intended use', without reference to how this is demonstrated. The Regulations also apply only to products supplied after 27 December 1991. Therefore, any product which was legally usable before that date can continue to be sold.

Where a manufacturer identifies an existing barrier to trade with another Member State, a procedure exists under article 16 of the CPD to overcome this. This involves the testing of products against the requirements of the importing country by a notified body in the exporting country. This procedure will not lead to a CE mark.

Not all materials will necessarily be CE marked under the CPD and, from a practical viewpoint, it will not be possible for all products to be CE marked until all the relevant technical specifications are available. However, it should be noted that for some products CE marking is compulsory under other Directives (such as gas boilers).

8.4.4 Materials and workmanship generally

Building work must be carried out:

- with proper and adequate materials which are:
  (1) appropriate for the circumstances in which they are used,
  (2) adequately mixed and prepared, and
  (3) applied, used or fixed so as adequately to perform the functions for which they are designed; and
- in a workmanlike manner.

Guidance on the choice and use of materials, and on ways of establishing the adequacy of workmanship, is given in AD Regulation 7. It should be noted however, that materials and workmanship are controlled only to the extent of:

- securing reasonable standards of health and safety for persons in or about buildings for Parts A to K and N of Schedule 1 (except for paragraphs H2 and J6);
- conserving fuel and power in Part L; and
- providing access and facilities for disabled people in Part M.

Therefore, although it may be desirable for reasons of consumer satisfaction or protection to require higher standards, this cannot be required under building regulations.

In order to achieve a satisfactory standard of performance **materials** should be:

- suitable in nature and quality in relation to the purposes for which, and the conditions in which, they are used.

Additionally, **workmanship** should be such that, where relevant, materials are:

- adequately mixed and prepared (for example, in concrete mixes, the correct proportions must be used, there should be an appropriate water/cement ratio, mixing should be thorough, etc.); and
- applied, used or fixed so as to adequately perform their intended functions (for example, for reinforced concrete this would give control over the actual placing of the concrete, positioning of reinforcement, curing, etc.).

The definition of materials is quite broad and covers products, components, fittings, naturally occurring materials (such as timber, stone and thatch), items of equipment, and materials used in the backfilling of excavations in connection with building work. It should be noted that Building Regulations do not seek to control the use of materials after completion of the building work.

In order to reduce the environmental impact of building work careful thought should be given to the choice of materials, and where appropriate, recycled or recyclable materials should be considered. Obviously, the use of such materials must not have an adverse effect on the health and safety standards of the building work.

8.4.5 Fitness of materials

A number of ways of establishing the fitness of materials are dealt with in AD Regulation 7 and whilst this is mostly by reference to British Standards or certificates issued by European Technical Approvals issuing bodies other materials or products may be suitable in the particular circumstances. The following aids to establishing the fitness of materials are given in the approved document.

- A material may conform to the relevant provisions of an appropriate British Standard (but see the notes under interim procedures in section 8.4.3).
- A material may conform to the national technical specifications of other Member States which are contracting parties to the European Economic Area. It should be noted that where a person intends to use a product which complies with a national technical specification of another Member State, the onus is on that person to show that the product is equivalent to the relevant British Standard (and it would be necessary to provide a translation).
- A material may be covered by a national or European certificate issued by a European Technical Approvals issuing body. The conditions of use must be in accordance with the terms of the certificate and again, it will be up to the person intending to use the product to demonstrate equivalence and provide a translation.
- A material may bear a CE marking confirming that it conforms with a harmonised European standard or European Technical Approval together with the appropriate attestation procedure. Materials bearing the CE marking must be accepted if they are appropriate for the circumstances in which they are used. AD Regulation 7 qualifies this by saying that a CE marked product can only be rejected by a building control body on the basis that:
  - its performance is not in accordance with its technical specification, *or*
  - where a particular declared value or class of performance is stated for a product, the resultant value does not meet Building Regulation requirements.

  The burden of proof is on the controlling body which is obliged to notify the Trading Standards Officer so that the UK Government can notify the Commission.
- Independent certification schemes, (e.g. the kitemark scheme operated by the British Standards Institution), may also serve to show that a material is suitable for its purpose. However, some materials which are not so certified may still conform to a relevant standard. In the UK, many certification bodies which approve such schemes are accredited by UKAS.
- A material may be shown to be capable of performing its function by the use of tests, calculations or other means. It is important to ensure that tests are carried out in accordance with recognised criteria. UKAS run an accreditation scheme for testing laboratories, and together with similar schemes run by equivalent certification bodies (including accreditation schemes operated by other Member States of the EU), this ensures that standards of testing are maintained.
- In some cases past experience of a material in use in a building may be relied upon to ensure that it is capable of adequately performing its function.

- Local authorities are entitled under regulation 19 to take and test samples of materials in order to confirm compliance. Approved inspectors have similar powers under Regulation 11 of the Building (Approved Inspectors etc.) Regulations 2000 (as amended).

8.4.6 Short-lived materials

Only general guidance is given on the use of short-lived materials. These are materials which may be considered unsuitable due to their rapid deterioration when compared to the life of the building.

The main criteria to be considered are:

- accessibility for inspection, maintenance and replacement
- the effects of failure on public health and safety.

Clearly, if a material or component is inaccessible and its failure would create a serious health risk, it is unlikely that the material or component would be suitable. (See also the reference to section 19 of the Building Act 1984 in section 8.1.1.)

8.4.7 Materials subject to changes in their properties

Under certain environmental conditions, some materials may undergo a change in their properties which may affect their performance over time. A notable example of this occurred during the 1970s to structures constructed using high alumina cement. The subsequent deterioration of the concrete led to the collapse of a number of long span roof structures and the use of HAC was banned for all work except when the material was used as a heat-resisting material. It is known that a number of other materials, (such as certain stainless steels, structural silicone sealants and intumescent paints) may also be susceptible to changes in their properties under certain environmental conditions. In order to use such materials it will be necessary to estimate their final residual properties (including their structural properties) at the time the materials are incorporated into the work. It should then be shown that these residual properties will be adequate for the building to perform its intended function for its expected life.

8.4.8 Resistance to moisture and soil contaminants

Materials which are likely to suffer from the adverse effects of condensation, ground water or rain and snow may be satisfactory if:

- the construction of the building is such as to prevent moisture from reaching the materials; or
- the materials are suitably treated or otherwise protected from moisture.

Similarly, materials which are in contact with the ground will only be satisfactory if they can adequately resist the effects of deleterious substances, such as sulphates (see Site preparation and moisture exclusion in section 8.5).

8.4.9 Special treatment against house longhorn beetle infestation

The previous section in the 1992 edition of AD Regulation 7 on the house longhorn beetle was omitted from the 1999 edition. It is intended to incorporate the information into a revised Approved Document A which is currently under development and is expected to be published in 2004. In the interim, the advice given in the 1992 edition is still relevant and is described below.

In specified areas in the south of England all softwood roof timbers, including ceiling joists, should be treated with a suitable preservative against the house longhorn beetle.

The specified areas are as follows:

- The Borough of Bracknell Forest
- The Borough of Elmbridge
- The Borough of Guildford, other than the area of the former borough of Guildford
- The District of Hart other than the area of the former Urban District of Fleet
- The District of Runnymede
- The Borough of Spelthorne
- The Borough of Surrey Heath
- The Borough of Woking
- In the Borough of Rushmoor, the area of the former district of Farnborough
- The Borough of Waverley, other than the parishes of Godalming and Haslemere
- In the Royal Borough of Windsor and Maidenhead, the Parishes of Old Windsor, Sunningdale and Sunninghill.

No specific forms of treatment are recommended; however most reputable timber treatment companies (and members of such organisations as the British Wood Preserving and Dampproofing Association) have been providing treatment for house longhorn beetle for many years and will be able to recommend suitable treatments.

8.4.10 Adequacy of workmanship

It should be remembered that Building Regulations set different standards of workmanship to those imposed by, for example, a building specification. Building Regulations are not concerned with quality or value for money; they are concerned with public health and safety, the conservation of fuel and power, and access and facilities for disabled people.

Adequacy of workmanship, like that of materials, may be established in a number of ways.

- A British Standard Code of Practice or other equivalent technical specification (e.g. of Member States which are contracting parties to the European Economic Area) may be used. In this context, BS 8000: *Workmanship on Building Sites* may be useful since it gathers together guidance from a number of other BSI Codes and Standards.

- Technical approvals, such as national or European certificates issued by European Technical Approvals issuing bodies, often contain workmanship recommendations. Additionally, it may be possible to use an equivalent technical approval (such as those of a member of EOTA) if this provides an equivalent level of protection and performance. The onus of proof of acceptability rests with the user in this case.
- Workmanship which is covered by a scheme complying with BS EN ISO 9000: *Quality management and quality assurance standards,* will demonstrate an acceptable standard since these schemes relate to processes and products for which there may also be a suitable British or other technical standard. A number of such schemes have been accredited by UKAS. There are also a number of independent schemes for accreditation and registration of installers of materials, products and services and these ensure that work has been carried out to appropriate standards by knowledgeable contractors.
- In some cases past experience of a method of workmanship such as a building in use, may be relied upon to ensure that the method is capable of producing the intended standard of performance.
- Local authorities are empowered under Regulation 18 to make such tests of any building work to enable it to establish if the work complies with Regulation 7 or any other applicable requirements of Schedule 1. Approved inspectors have similar powers under Regulation 11 of the Building (Approved Inspectors etc.) Regulations 2000 (as amended).

8.5 Site preparation and moisture exclusion

8.5.1 Introduction

Part C of Schedule 1 to the 2000 Regulations is concerned with site preparation and resistance to moisture. In addition to moisture exclusion, paragraph C2 contains provisions controlling sites containing dangerous or offensive substances. This replaces section 29 of the Building Act 1984.

The supporting Approved Document C now contains recommendations relating to the control of radon gas in certain areas of England and Wales, and landfill gases on certain sites near waste disposal tips, etc. The guidance on the control of dampness in buildings in AD C does not extend to that concerning damage from condensation. For this, reference should be made to AD F, Ventilation (see Chapter 11), AD L, Conservation of fuel and power (see Chapter 16), and the BRE publication entitled *Thermal insulation: avoiding risks.*

Certain provisions regarding the damp-proofing and weather resistance of floors and walls do not apply to buildings used solely for storage of plant or machinery in which the only persons habitually employed are storemen, etc. Other similar types of buildings where the air is so moisture-laden that any increase would not adversely affect the health of the occupants are also excluded. These buildings are referred to as 'excepted buildings' throughout this chapter.

8.5.2 Preparation of site

The ground to be covered by the building is required to be reasonably free from vegetable matter.

Decaying vegetable matter could be a danger to health and it could also cause a building to become unstable if it occurred under foundations. AD C1/3, therefore, recommends that the site should be cleared of all turf and vegetable matter at least to a depth to prevent future growth. This does not apply to excepted buildings.

Below-ground services (such as foul or surface water drainage) should be designed to resist the effects of tree roots. This can be achieved by making services sufficiently robust or flexible and with joints that cannot be penetrated by roots.

8.5.3 Dangerous and offensive substances

Precautions must be taken to prevent any substances found on or in the ground from causing a danger to health and safety. This is, of course, the ground covered by the building and includes the area covered by the foundations.

There is a special definition of *contaminant* for the purposes of AD C – any material (including faecal or animal matter) and any substance which is or could become toxic, corrosive, explosive, flammable or radioactive and therefore likely to be a danger to health or safety. This material must be in or on the ground to be covered by the building.

Contaminants can be liquids or solids arising out of a previous use of land, or they may be gases. In recent years problems have arisen from the emission of landfill gas from waste disposal sites. Additionally, in certain parts of the country, contamination by the naturally-occurring radioactive gas radon and its products of decay has lead to concern over the long-term health of the occupants of affected buildings.

Where a site is being redeveloped, knowledge of its previous use, from planning or other local records, may indicate a possible source of contamination. Table 1 to AD

AD C2

Table 1 Sites likely to contain contaminants.

Asbestos works
Chemical works
Gas works, coal carbonisation plants and ancillary byproduct works
Industries making or using wood preservatives
Landfill and other waste disposal sites
Metal mines, smelters, foundries, steel workers and metal finishing works
Munitions production and testing sites
Nuclear installations
Oil storage and distribution sites
Paper and printing works
Railway land, especially the larger sidings and depots
Scrap yards
Sewage works, sewage farms and sludge disposal sites
Tanneries

C2 (reproduced below) lists a number of site uses that are likely to contain contaminants.

Where the presence of contaminants has not been identified at an early stage, a later site survey may indicate possible contamination. Table 2 to AD C2 (reproduced below) shows the signs to be looked for, indicates which materials may be responsible and suggests relevant actions that may be taken. The Environmental

AD C2

Table 2 Possible contaminants and actions.

Signs of possible contaminants	Possible contaminant	Relevant action
Vegetation (absence, poor or unnatural growth)	Metals Metal compounds*	None
	Organic compounds Gases	Removal[1]
Surface materials (unusual colours and contours may indicate wastes and residues)	Metals Metal compounds*	None
	Oily and tarry wastes	Removal, filling or sealing
	Asbestos (loose)	Filling[2] or sealing[3]
	Other mineral fibres	None
	Organic compounds including phenols	Removal or filling
	Combustible material including coal and coke dust	Removal or filling
	Refuse and waste	Total removal or see guidance
Fumes and odours (may indicate organic chemicals at very low concentrations)	Flammable explosive and asphyxiating gases including methane and carbon dioxide	Removal
	Corrosive liquids	Removal, filling or sealing
	Faecal animal and vegetable matter (biologically active)	Removal or filling
Drums and containers (whether full or empty)	Various	Removal with all contaminated ground

Notes:
Liquid and gaseous contaminants are mobile and the ground covered by the building can be affected by such contaminants from elsewhere. Some guidance on landfill gas and radon is given in AD C; other liquids and gases should be referred to a specialist.
*Special cement may be needed with sulphates.
Actions assume that ground will be covered with at least 100 mm in-situ concrete.
[1] Removal – the contaminant and any contaminated ground removed to a depth of 1 m below lowest floor (or less if local authority agrees) to place named by local authority.
[2] Filling – Area of building covered to a depth of 1 m (or less if local authority agrees) with suitable material. Filling material and ground floor design considered together. Combustible materials adequately compacted to avoid combustion.
[3] Sealing – Imperforate barrier between contaminant and building sealed at joints, edges and service entries. Polythene may not always be suitable if contaminants is tarry waste or organic solvent.

Health Officer should always be notified if contamination is suspected. He may then agree on the remedial measures necessary to make the site safe.

Some contaminants present such hazardous conditions that only complete removal of the offending substances can provide a complete remedy, while in other cases the risks may be reduced to acceptable levels by remedial measures. This may necessitate the removal of large quantities of material and such works or remedial actions should only be undertaken with the benefit of expert advice.

For guidance on site investigation, reference should be made to BS Draft for Development DD 175: 1988 *Code of practice for the identification of potentially contaminated land and its investigation*. Further information on sites in general may be obtained from BS 5930: 1981 *Code of practice for site investigations*.

8.5.4 Radon gas contamination

Radon is a naturally-occurring, colourless and odourless gas which is radioactive. It is formed in small quantities by the radioactive decay of uranium and radium, and thus travels through cracks and fissures in the subsoil until it reaches the atmosphere or enters spaces under or in buildings.

It is recognised that radon gas occurs in all buildings; however the concentration may vary from below 20 Bq/m^3 (the national average for houses in the UK) to more than 100 times this value. The National Radiological Protection Board (NRPB) has recommended an action level of 200 Bq/m^3 for houses. The lifetime risk of contracting lung or other related cancers at the action level is about 3%. Geographical distribution of houses at or above the action level is very uneven with about two-thirds of the total being in Devon and Cornwall.

The ODPM is reviewing the areas where preventative measures should be taken as information becomes available from the NRPB. This information has been placed in the BRE guidance document *Radon: guidance on protective measures for new dwellings* (BRE Report BR 211, obtainable from Building Research Establishment, Garston, Watford, WD2 7JR) which will be updated as necessary. The report identifies those areas where either basic or full radon protection is needed by reference to a series of maps derived from:

- statistical analysis of radon measurements of existing houses carried out by the NRPB (Annex A of BR 211), and
- an assessment of geological radon potential prepared by the British Geological Survey (Annex B of BR 211).

Use of the maps in accordance with directions given in BR 211 will determine whether basic, full or no protection is needed.

Areas most at risk include parts of:

- Devon
- Cornwall
- Somerset
- Oxfordshire

- Northamptonshire
- Leicestershire and Rutland
- Lincolnshire
- Staffordshire
- Derbyshire
- West Yorkshire
- Northumberland and parts of southern Cumbria
- South west, north east and mid Wales

As more information becomes available from NRPB it is likely that further areas will be covered by the need for radon precautions. The local building control authority should be contacted for confirmation.

8.5.5 Basic protection against radon

Basic protection may be provided by an airtight, and therefore radon-proof, barrier across the whole of the building including the floor and walls. This could consist of:

- polyethylene (polythene) sheet membrane of at least 300 micrometre (1200 gauge) thickness
- flexible sheet roofing materials
- prefabricated welded barriers
- liquid coatings
- self-adhesive bituminous-coated sheet products
- asphalt tanking.

It is important to have adequately sealed joints and the membrane must not be damaged during construction. Where possible, penetration of the membrane by service entries should be avoided. With careful design it may be possible for the barrier to serve the dual purpose of damp-proofing and radon protection. Some typical details are shown in Fig. 8.2.

8.5.6 Full protection against radon

In practical terms, a totally radon-proof barrier may be difficult to achieve. Therefore, in high-risk areas it is necessary to provide additional secondary protection. This might consist of:

- natural ventilation of an underfloor space by airbricks or ventilators on at least two sides
- the addition of an electrically operated fan in place of one of the airbricks to provide enhanced subfloor ventilation
- a subfloor depressurisation system comprising a sump located beneath the floor slab, joined by pipework to a fan. It may only be necessary to provide the sump and underfloor pipework during construction thus giving the owner the option of connecting a fan at a later stage if necessary.

Examples of these methods are shown in Figs. 8.2 and 8.3.

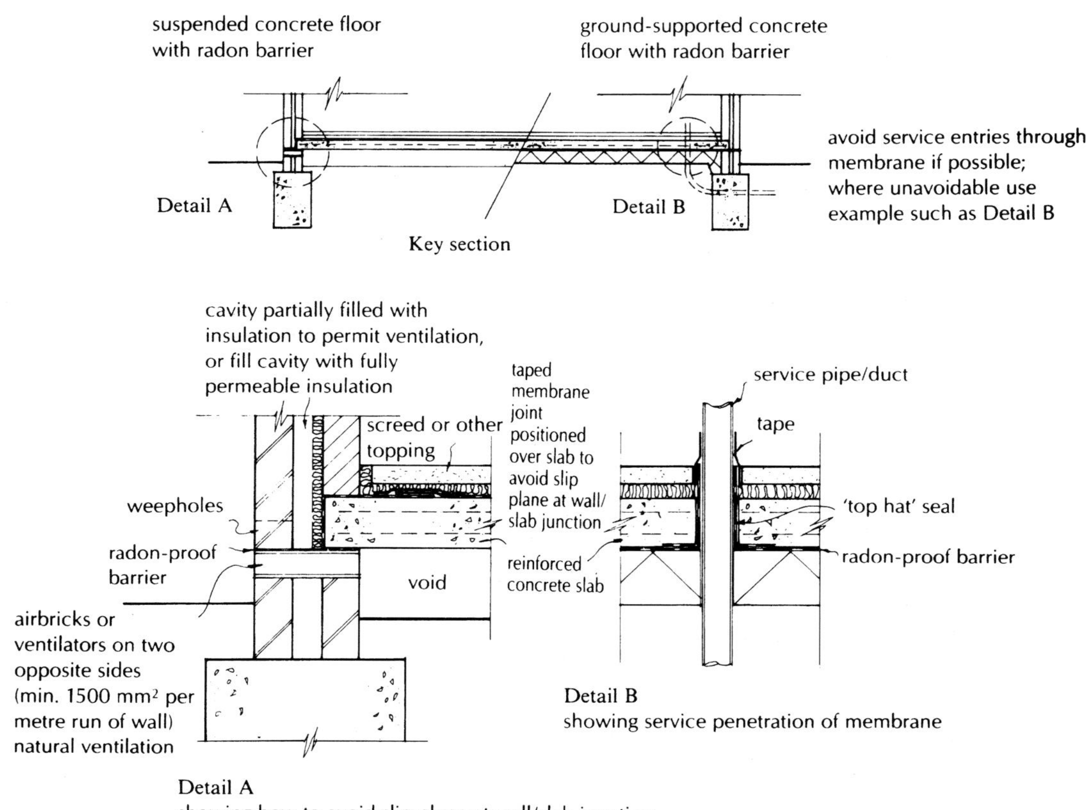

Fig. 8.2 Basic protection against radon.

It should be noted that the above brief notes on BR 211 are intended to give an idea of the content of that document. Designers of buildings in the delimited areas should consult the full report and any other relevant references, such as those listed in section 8.5.10 below.

8.5.7 Contamination of landfill gas

Landfill gas is typically made up of 60% methane and 40% carbon dioxide, although small quantities of other gases such as hydrogen, hydrogen sulphide and a wide range of trace organic vapours, may also be present. The gas is produced by the breakdown of organic material by micro-organisms under oxygen-free (anaerobic) conditions. Gases similar to landfill gas can also arise naturally from coal strata, river silt, sewage and peat.

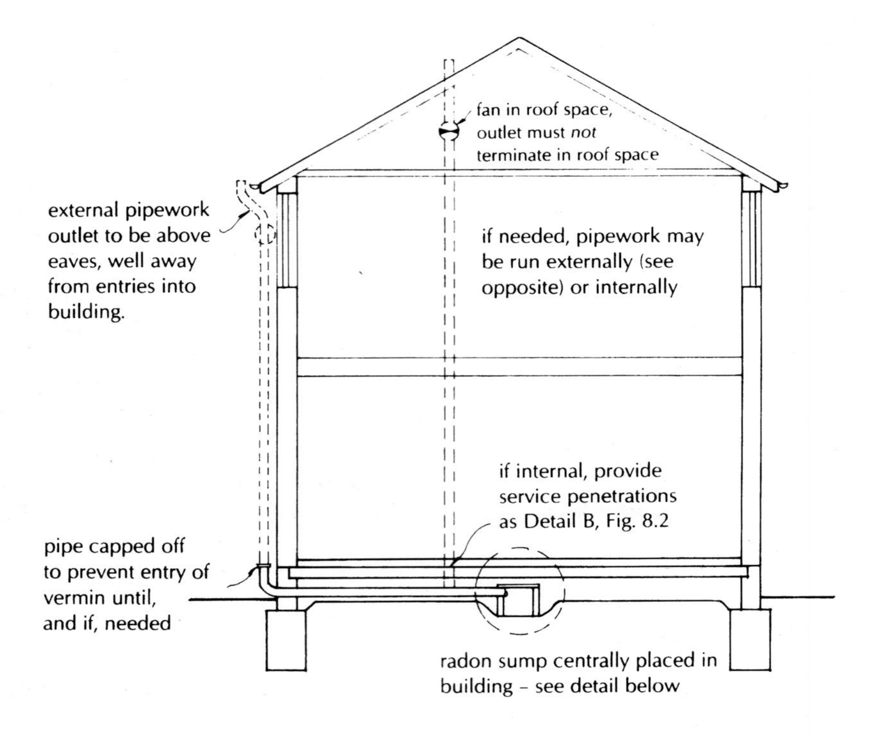

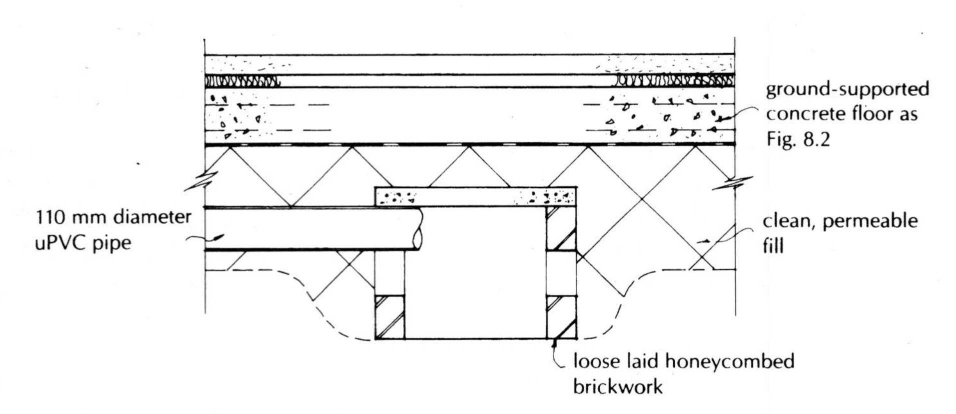

Fig. 8.3 Additional protection against radon.

8.5.8 Properties of landfill gases

The largest component of landfill gas, methane, is a flammable, asphyxiating gas with a flammable range between 5% and 15% by volume in air. If such a concentration occurs within a building and the gas is ignited it will explode. Methane is lighter than air.

The other major component, carbon dioxide, is a non-flammable, toxic gas which has a long-term exposure limit of 0.5% by volume and a short-term exposure limit of 1.5%. It is heavier than air.

8.5.9 Movement of landfill gases

The proportions of these two main gases and the amount of air mixed with them will largely determine the properties of the landfill gas since they remain mixed and do not separate, although the mixture can remain separate from surrounding air. These landfill gases will migrate from a landfill site as a result of diffusion through the ground and this migration may be increased by rainfall or freezing temperatures as these conditions tend to seal the ground surface. The gases will also follow cracks, cavities, pipelines and tunnels, etc., as these form ideal pathways. Landfill gas emissions can be increased by rapid falls in atmospheric pressure and by a rising water table. Thus, landfill gas may enter buildings and may collect in underfloor voids, drains and soakaways.

8.5.10 Building near landfill sites

AD C2 suggests that further investigations should be made to determine whether protective measures are necessary if:

- the ground to be covered by a building is on, or within 250 m of, a landfill; or
- there is reason to suspect that there may be gaseous contamination of the ground; or
- the building will be within the likely sphere of influence of a landfill.

Practical guidance on construction methods to prevent the ingress of landfill gas in buildings is not given in ADC2; instead, reference is made to the BRE Report *Construction of new buildings on gas-contaminated land* (BRE Report BR 212, obtainable from Building Research Establishment, Garston, Watford, WD2 7JR).

The report gives examples of ground floor construction and venting details for houses and other similar small buildings. These are suitable where the concentration of methane in the ground is unlikely to exceed 1% by volume.

With regard to carbon dioxide, a concentration greater than 1.5% by volume in the ground indicates that there is a need to consider using the construction details described in the report to prevent the ingress of gas. Where this is as high as 5% then the floor constructions are required.

Figure 8.4 gives typical examples of some of the constructional details contained in the report. It should be noted that the use of continuous mechanical ventilation for the removal of landfill gases in dwellings is not recommended since maintenance of the system cannot be guaranteed and a failure might result in a sharp increase in indoor methane concentration with the possibility of an explosion occurring.

In buildings other than dwellings, expert advice should be sought. This might include:

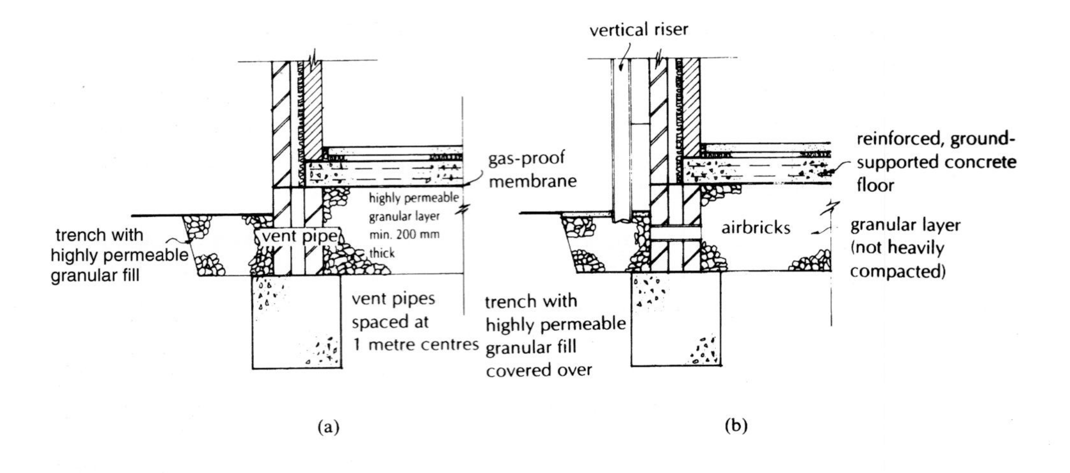

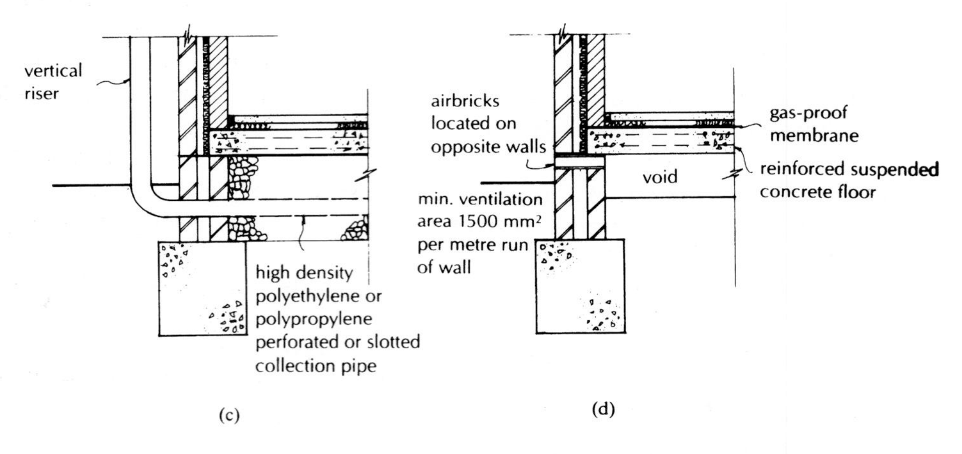

Fig. 8.4 Landfill gas protection details. (a) External trench ventilation. (b) Sealed trench with vertical riser. (c) Perforated pipe ventilation. (d) Suspended floor – airbrick ventilation.

- a complete investigation into the source and nature of any hazardous gases present;
- the potential for future gas generation from the landfill site;
- an assessment of the present and future risk posed by the gas, including the need for extended monitoring;
- the design of protective measures incorporated into the overall building design, including arrangements for maintenance and monitoring.

The BRE Report mentioned above also includes practical details for buildings other than dwellings.

It should be noted that the above brief notes on BR 212 are intended to give an idea of the content of that document. Designers of buildings which are likely to be affected by landfill gas should consult the full report and any of the references in the following list, which is reproduced from AD C2, as this gives further information regarding documents concerned with site investigations and the development of contaminated land.

Reference list to sources on site investigations

(a) Department of Environment (Her Majesty's Inspectorate of Pollution). *The control of landfill gas*, waste management paper no. 27 (1989), HMSO.
(b) British Standards Institution Draft for Development: *Code of Practice for the Identification of Potentially Contaminated Land and its Investigation* DD 175: 1988.
(c) Crowhurst, D. *Measurement of gas emissions from contaminated land.* BRE Report 1987. HMSO, ISBN 0 85125 246 X.
(d) ICRCL 17/78 *Notes on the development and after-use of landfill sites*, 8th edition, December 1990. Interdepartmental Committee on the Redevelopment of Contaminated Land.
(e) ICRCL 59/83 *Guidance on the assessment and redevelopment of contaminated land*, 2nd edition, July 1987. Interdepartmental Committee on the Redevelopment of Contaminated Land.
(f) Institute of Wastes Management. *Monitoring of landfill gas.* September 1989.
(g) BS 5930: 1981 *Code of practice for site investigations.*
(h) BRE. *Radon: guidance on protective measures for new dwellings.* BRE Report. Garston, BRE, 1991. ISBN 0 85125 511 6.
(i) BRE. *Construction of new buildings on gas contaminated land.* BRE Report. Garston, BRE, 1991. ISBN 0 85125 513 2.

(d) and (e) are obtainable from: Department of Environment Publication Sales Unit, Building 1, Victoria Road, South Ruislip, Middlesex HA4 0NZ.

8.5.11 Subsoil drainage

Subsoil drainage must be provided *if* it is necessary to avoid:

- the passage of moisture from the ground to the inside of the building; or
- damage to the fabric of the building.

There are no provisions in AD C1/3 concerning the flooding of sites. If the site of a building is subject to flooding it merely assumes that suitable steps are being taken.

Subsoil water may cause problems in the following cases.

- Where there is a high water table (i.e. within 0.25 m of the lowest floor in the building).
- Where surface water may enter or damage the building fabric.
- Where an active subsoil drain is severed during excavations.

Where problems are anticipated it will usually be necessary either to drain the site of the building or to design and construct it to resist moisture penetration.

Severed subsoil drains should be intercepted and continued in such a way that moisture is not directed into the building. Figure 8.5 illustrates a number of possible solutions.

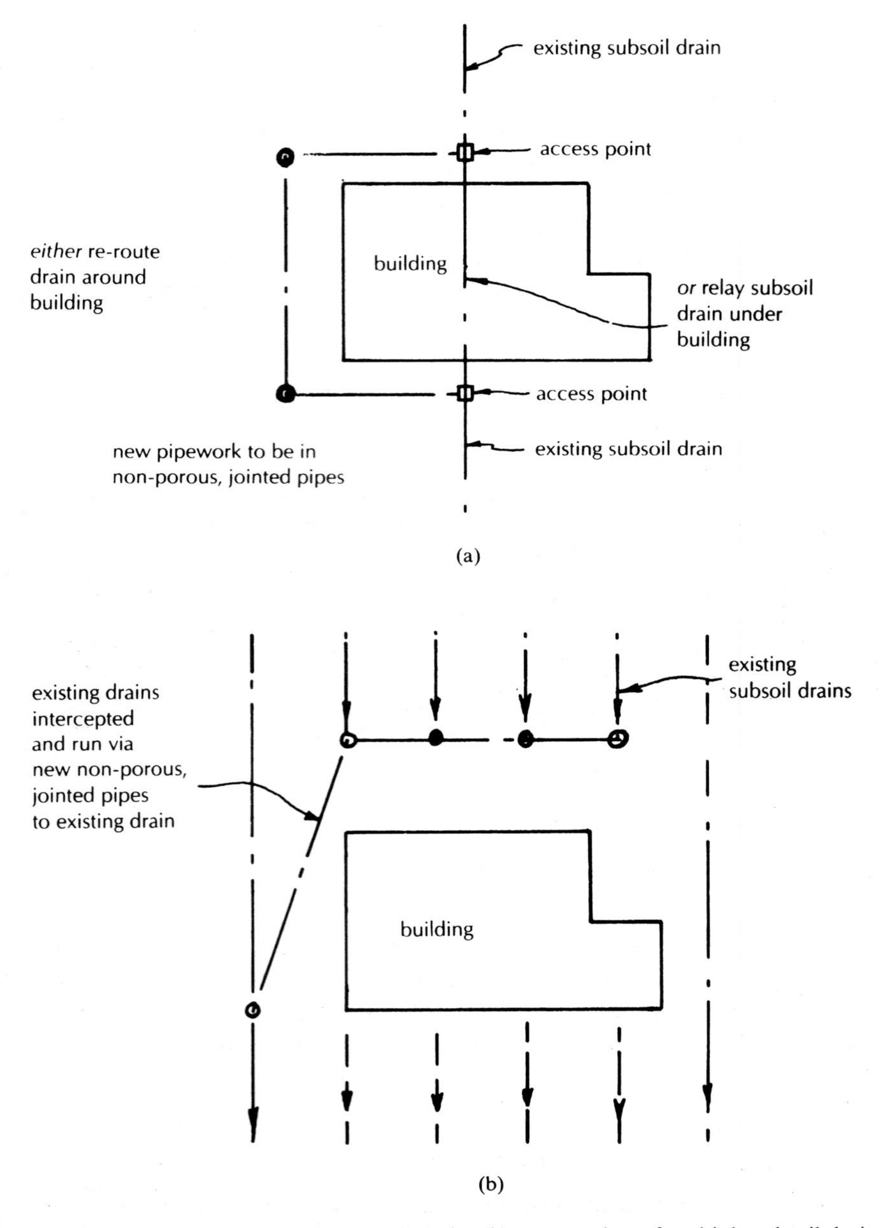

Fig. 8.5 Subsoil drainage. (a) Single subsoil drain. (b) Interception of multiple subsoil drains.

8.5.12 Resistance to weather and ground moisture

The floors, walls and roof of a building are required to resist the passage of moisture to the inside of the building.

8.5.13 Protection of floors next to the ground

The term *floor* means the lower surface of any space in a building and includes any surface finish which is laid as part of the permanent construction. This would, presumably, exclude carpets, lino, tiles, etc., but would include screeds and granolithic finishes.

A ground floor should be constructed so that:

- the passage of moisture to the upper surface of the floor is resisted (this does not apply to excepted buildings);
- it will not be adversely affected by moisture from the ground;
- it will not transmit moisture to another part of the building that might suffer damage (see Fig. 8.6).

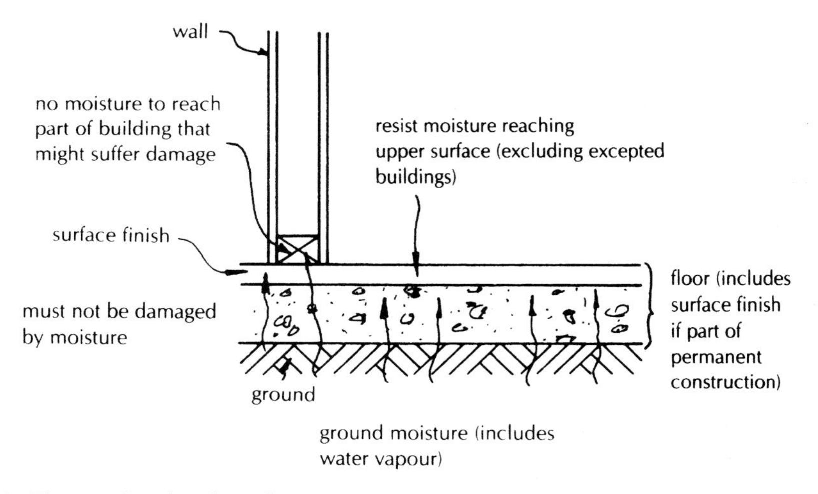

Fig. 8.6 Floors – functional requirements.

The term *moisture* is taken to include water vapour as well as liquid water. Damage caused by moisture is only significant if it would cause a material or structure to deteriorate to such a point that it would present an imminent danger to public health and safety or it would permanently reduce the performance of an insulating material.

8.5.14 Floors supported directly by the ground

The requirements mentioned above can be met, for ground supported floors, by covering the ground with dense concrete incorporating a damp-proof membrane, laid on a hardcore bed. If required, insulation may also be incorporated in the floor construction.

This form of construction is illustrated in Fig. 8.7 below. However, the following points should also be considered.

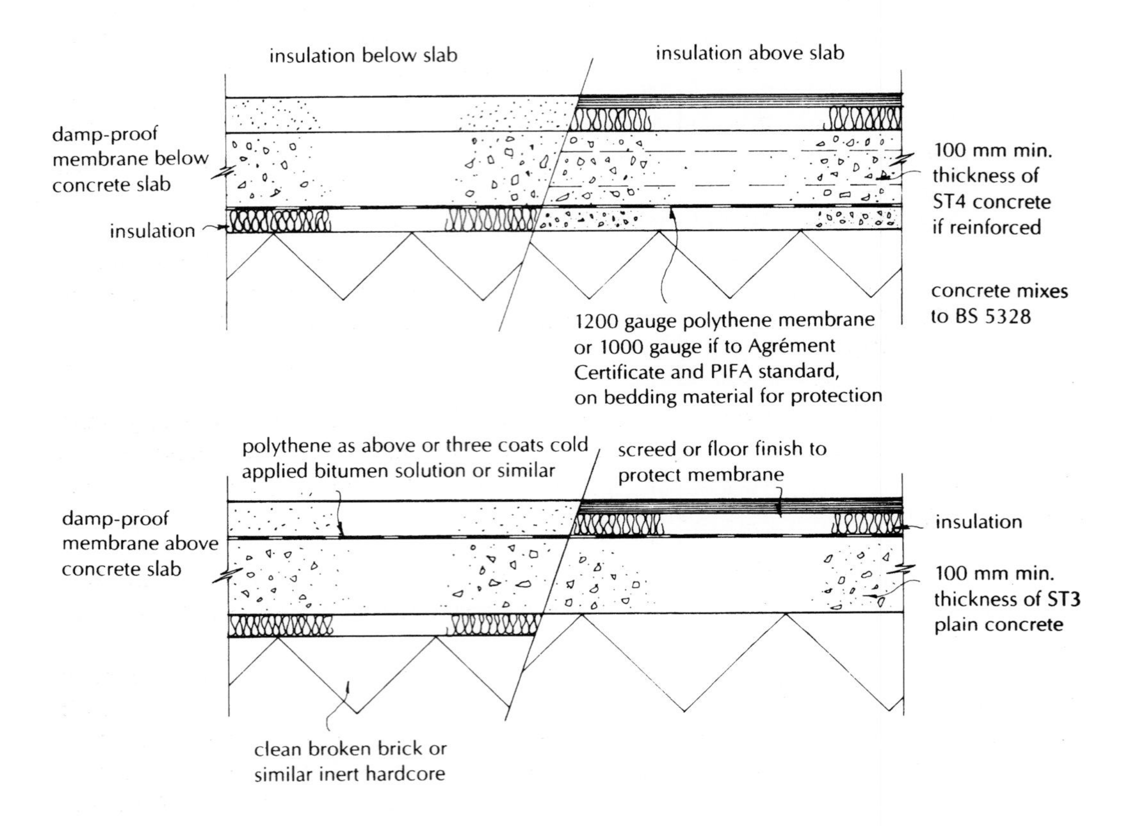

Fig. 8.7 Ground supported floor.

- Hardcore laid under the floor next to the ground should not contain water-soluble sulphates or deleterious matter in such quantities as might cause damage to the floor.

  Broken brick or stone are the best hardcore materials. Clinker is dangerous unless it can be shown that the actual material proposed is free from sulphates, etc., and colliery shales should likewise be avoided. In any event, the builder might well be liable for breach of an implied common law warranty of fitness of materials (see *Hancock* v. *B. W. Brazier (Anerley) Ltd* [1966] 2 All ER 901), where builders were held liable for subsequent damage caused by the use of hardcore containing sulphates.
- A damp-proof membrane (DPM) may be provided above or below the concrete floor slab and should be laid continuous with the damp-proof courses in walls, piers, etc.

If laid below the concrete, the DPM should be at least equivalent to 300 μm (1200 gauge) polyethylene (e.g. polythene). It should have sealed joints and should be supported by a layer of material that will not cause damage to the polythene. Polythene of 250 μm (1000 gauge) is also satisfactory if it is to the PIFA standard or is used in accordance with an appropriate BBA certificate.

If a polyethylene sheet membrane is laid above the concrete, there is no need to provide the bedding material. It is also possible to use a three-coat layer of cold applied bitumen emulsion in this position. These materials should be protected by a suitable floor finish or screed. Surface protection does not need to be provided where the membrane consists of pitchmastic or similar material which also serves as a floor finish.

- The minimum thickness for the concrete slab is 100 mm, although the structural design for the slab may require it to be thicker. Unreinforced concrete should be composed of 50 kg cement to maximum 0.11 m^3 fine aggregate to maximum 0.16 m^3 coarse aggregate or BS 5328 mix ST3. Reinforced concrete should be composed of 50 kg cement to maximum 0.08 m^3 fine aggregate to maximum 0.13 m^3 coarse aggregate or BS 5328 mix ST4.
- AD C4 gives no guidance on the position of the floor relative to outside ground level. Since this type of floor is unsuitable if subjected to water pressure, it is reasonable to assume that the top surface of the slab should not be below outside ground level unless special precautions are taken.

If it is proposed to lay a timber floor finish directly on the concrete slab, it is permissible to bed the timber in a material that would also serve as a damp-proof membrane.

No guidance is given regarding suitable DPM materials. However 12.5 mm of asphalt or pitchmastic will usually be satisfactory for most timber finishes and it may be possible to lay wood blocks in a suitable adhesive DPM. If a timber floor finish is fixed to wooden fillets embedded in the concrete, the fillets should be treated with a suitable preservative unless they are above the DPM (see BS 1282: 1975 *Guide to the choice, use and application of wood preservatives* for suitable preservative treatments).

Clause 11 of CP 102: 1973 *Code of practice for protection of buildings against water from the ground* may be used as an alternative to the above. Where ground water pressure is evident, recommendations may be found in BS 8102: 1990 *Code of practice for protection of structures against water from the ground.*

8.5.15 Suspended timber floors

The performance requirements mentioned above may be met for suspended timber ground floors by:

- covering the ground with suitable material to resist moisture and deter plant growth;
- providing a ventilated space between the top surface of the ground covering and the timber;

- isolating timber from moisture-carrying materials by means of damp-proof courses.

A suitable form of construction is shown in Fig. 8.8 and is summarised below.

- The ground surface should be covered with at least 100 mm of concrete composed of 50 kg cement to maximum 0.13 m^3 fine aggregate to maximum 0.18 m^3 coarse aggregate or BS 5328 mix ST1, if unreinforced. It should be laid on clean broken brick or similar inert hardcore not containing harmful quantities of water-soluble sulphates or other materials which might damage the concrete. (The Building Research Establishment suggests that over 0.5% of water-soluble sulphates would be a harmful quantity.)

  Alternatively, the ground surface may be covered with at least 50 mm of concrete, as described above, or inert fine aggregate, laid on a polythene DPM as described for ground supported floors above. The joints should be sealed and the membrane should be laid on a protective bed.

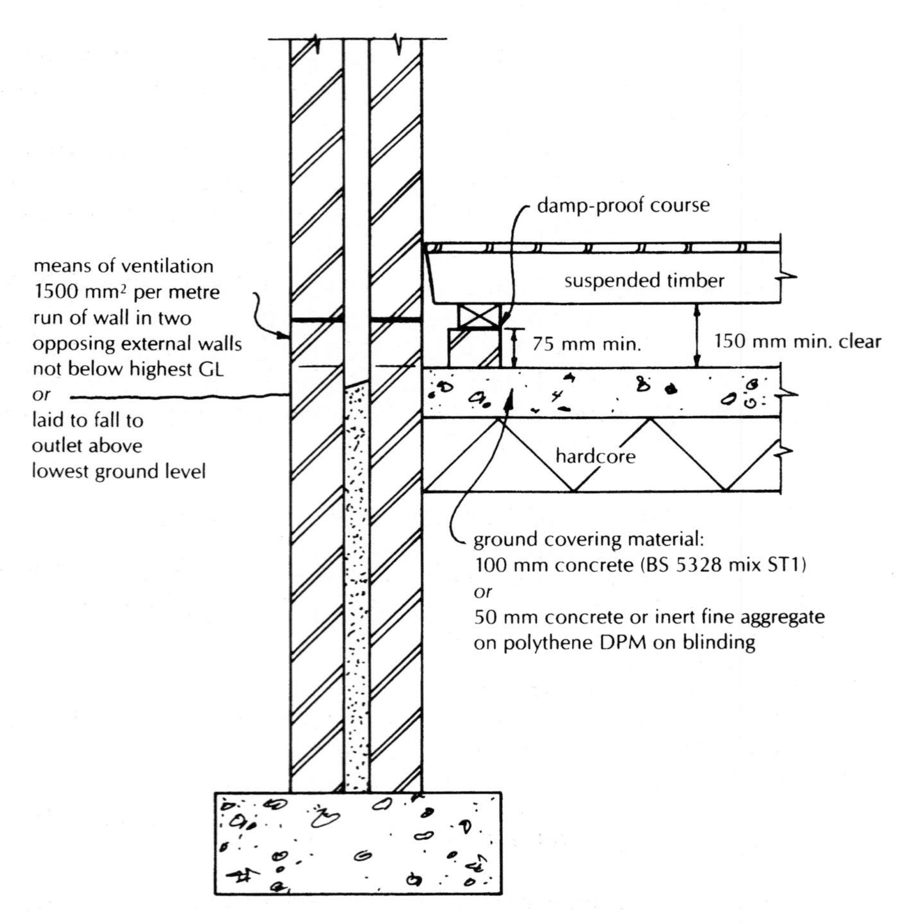

Fig. 8.8 Suspended timber floor.

- The ground covering material should be laid so that *either* its top surface is not below the highest level of the ground adjoining the building *or* it falls to an outlet above the lowest level of the adjoining ground.
- There should be a space above the top of the concrete of at least 75 mm to any wall-plate and 150 mm to any suspended timber. There should be ventilation openings in two opposing external walls allowing free ventilation to all parts of the floor. An actual ventilation area equivalent to 1500 mm^2 per metre run of wall should be provided and any ducts needed to convey ventilating air should be at least 100 mm in diameter.
- Damp-proof courses of sheet materials, slates or engineering bricks bedded in cement mortar should be provided between timber members and supporting structures to prevent transmission of moisture from the ground.

Again, the recommendations of Clause 11 of CP 102: 1973 may be used instead of the above, especially if the floor has a highly vapour-resistant finish.

8.5.16 Suspended concrete ground floors

Moisture should be prevented from reaching the upper surface of the floor and the reinforcement should be protected against moisture if the construction is to be considered satisfactory.

Suspended concrete ground floors may be of pre-cast construction with or without infilling slabs or they may be cast in-situ. A damp-proof membrane should be provided if the ground below the floor has been excavated so that it is lower than outside ground level and it is not effectively drained.

The space between the underside of the floor and the ground should be ventilated where there is a risk that an accumulation of gas could cause an explosion. The space should be at least 150 mm in depth (measured from the ground surface to the underside of the floor or insulation, if provided) and the ventilation recommendations should be as for suspended timber floors (see bottom of previous page). These recommendations are summarised in Fig. 8.9.

8.5.17 Protection of walls against moisture from the ground

The term *wall* means vertical construction, which includes piers, columns and parapets and may include chimneys if they are attached to the building. Windows, doors and other openings are not included.

Walls should be constructed so that:

- the passage of moisture from the ground to the inside of the building is resisted (this does not apply to excepted buildings);
- they will not be adversely affected by moisture from the ground;
- they will not transmit moisture from the ground to another part of the building that might be damaged.

The requirements mentioned above can be met for internal and external walls by providing a damp-proof course of suitable materials in the required position.

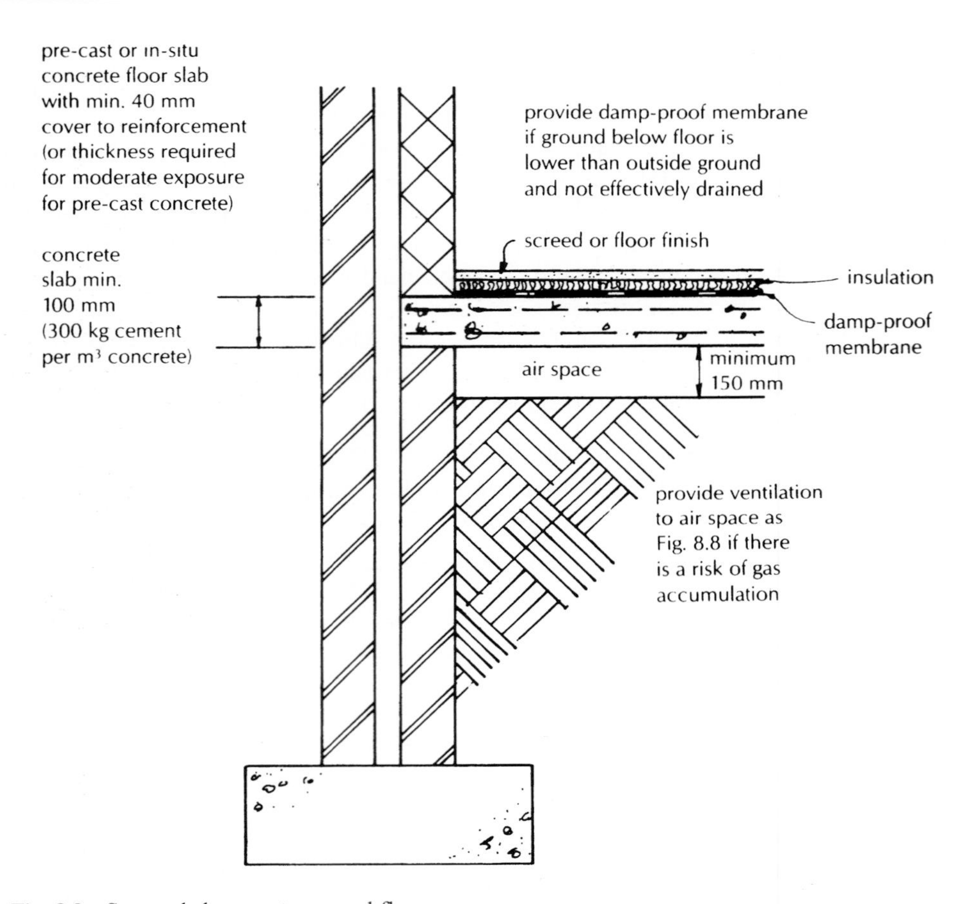

Fig. 8.9 Suspended concrete ground floors.

Figure 8.10 illustrates the main provisions, which are summarised below.

- The damp-proof course may be of any material that will prevent moisture movement. This would include bituminous sheet materials, engineering bricks or slates laid in cement mortar, polythene or pitch polymer materials.
- The damp-proof course and any damp-proof membrane in the floor should be continuous.
- Unless an external wall is suitably protected by another part of the building, the damp-proof course should be at least 150 mm above outside ground level.
- Where a damp-proof course is inserted in an external cavity wall, the cavity should extend at least 150 mm below the lowest level of the damp-proof course. However, where a cavity wall is built directly off a raft foundation, ground beam or similar supporting structure, it is impractical to continue the cavity down 150 mm. The supporting structure should therefore be regarded as bridging the cavity, and protection be provided by a flashing or damp-proof course as required (see Fig. 8.10).

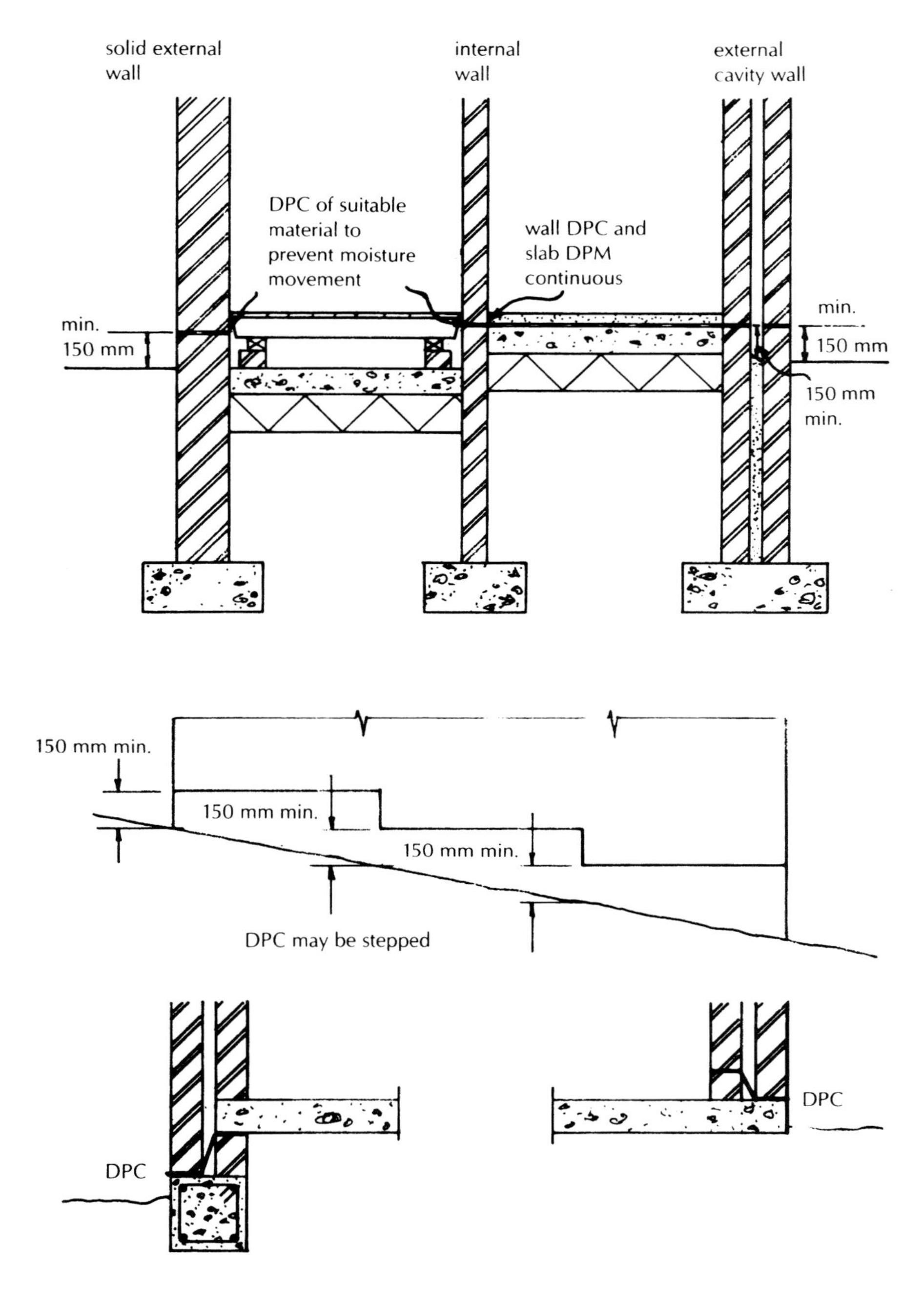

Cavity wall on ground beam or raft

Fig. 8.10 Protection of walls against moisture from the ground.

Alternatively, the provisions for protection of walls against moisture from the ground may be met by following the relevant recommendations of Clauses 4 and 5 of BS 8215: 1991 *Code of practice for design and installation of damp-proof courses in masonry construction*. BS 8102: 1990 *Code of practice for protection of structures against moisture from the ground* may be followed especially in the case of walls (including basement walls) subjected to ground water pressure.

8.5.18 Weather resistance of external walls

In addition to resisting ground moisture, external walls should:

- resist the passage of rain or snow to the inside of the building (this does not apply to excepted buildings)
- not be damaged by rain or snow
- not transmit moisture due to rain or snow to another part of the building that might be damaged.

There are a number of forms of wall construction which will satisfy the above requirements:

- a solid wall of sufficient thickness holds moisture during bad weather until it can be released in the next dry spell
- an impervious cladding prevents moisture from penetrating the outside face of the wall
- the outside leaf of a cavity wall holds moisture in a similar manner to a solid wall, the cavity preventing any penetration to the inside leaf.

These principles are illustrated in Fig. 8.11.

8.5.19 Solid external walls

The construction of a solid external wall will depend on the severity of exposure to wind-driven rain. This may be assessed for a building in a given area by using BSI Draft for Development DD93: 1984: *Method for assessing exposure to wind-driven rain*. Reference may also be made to BS 5628 *Code of practice for use of masonry*, Part 3: 1985 *Materials and components, design and workmanship* and the publication by the BRE entitled *Thermal insulation – avoiding the risks*. (See also Chapter 16, section 16.1.2.)

In conditions of *very* severe exposure it may be necessary to use an external cladding. However, in conditions of severe exposure a solid wall may be constructed as shown in Fig. 8.12. The following points should also be considered.

- The brickwork or blockwork should be rendered or given an equivalent form of protection.
- Rendering should have a textured finish and be at least 20 mm thick in two coats. This permits easier evaporation of moisture from the wall.
- The bricks or blocks and mortar should be matched for strength to prevent

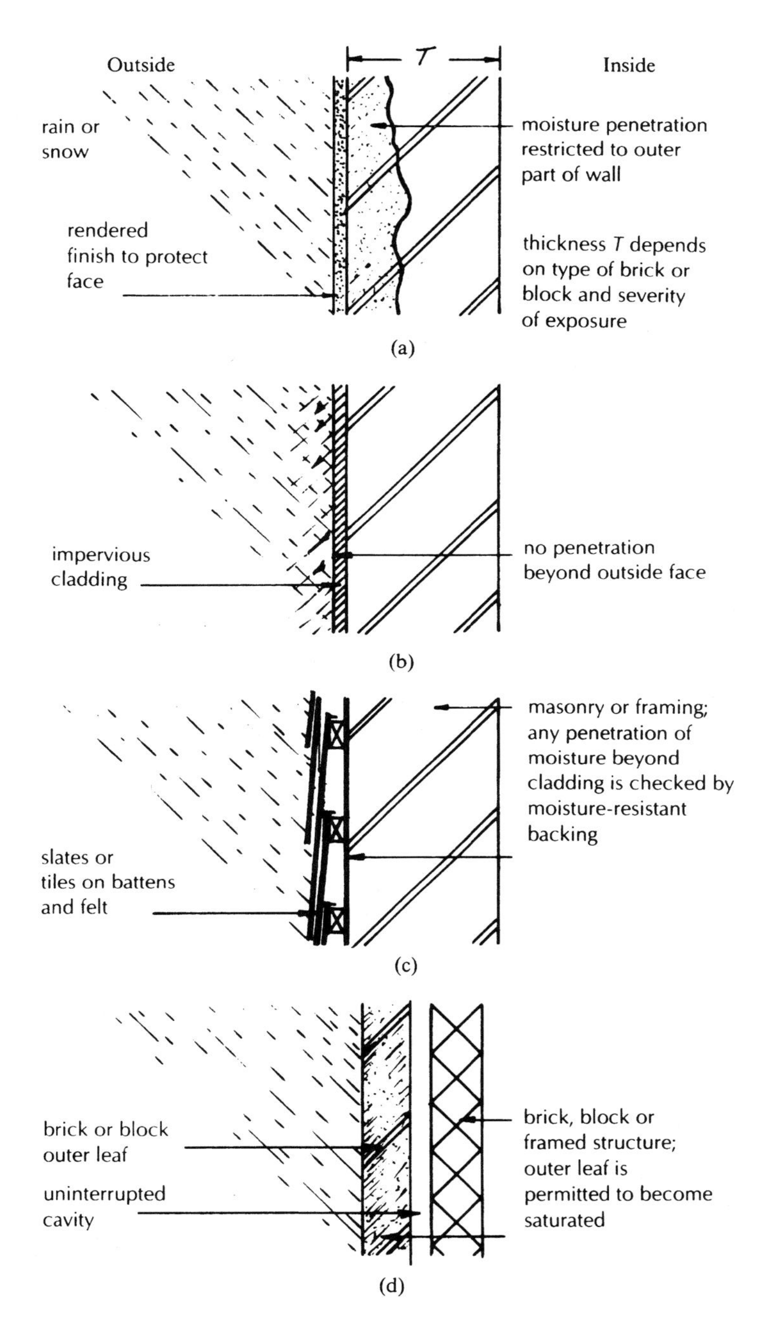

Fig. 8.11 Weather resistance of external walls – principles. (a) Solid external wall. (b) Impervious cladding. (c) Weather-resistant cladding. (d) Cavity wall.

cracking of joints or bricks and joints should be raked out to a depth of at least 10 mm in order to provide a key for the render.

- The render mix should not be too strong or cracking may occur. A mix of 1:1:6 cement:lime:sand is recommended for all walls except those constructed of dense concrete blocks where 1:½:4 should prove satisfactory.

  Further details of a wide range of render mixes may be obtained from BS 5262: 1976 *Code of practice. External rendered finishes.*
- Where the top of a wall is unprotected by the building structure it should be protected to resist moisture from rain or snow. Unless the protection and joints form a complete barrier to moisture, a damp-proof course should also be provided.
- Damp-proof courses should be provided to direct moisture towards the outside face of the wall in the positions shown in Fig. 8.12.
- Insulation to solid external walls may be provided on the inside or outside of the wall. Externally placed insulation should be protected unless it is able to offer resistance to moisture ingress so that the wall may remain reasonably dry (and the insulation value may not be reduced). Internal insulation should be separated from the wall construction by a cavity to give a break in the path for moisture. (Some examples of external wall insulation are given in Fig. 8.14.)

The performance requirements for solid and cavity external walls can also be met by complying with BS 5628 *Code of practice for use of masonry*, Part 3: 1985 *Materials and components, design and workmanship* or BS 5390: 1976 *Code of practice for stone masonry*.

8.5.20 External cavity walls

In order to meet the performance requirements, an external cavity wall should consist of an internal leaf which is separated from the external leaf by:

- a drained air space; or
- some other method of preventing moisture from rain or snow reaching the inner leaf.

An external cavity wall may consist of the following.

- An outside leaf of masonry (brick, block, natural or reconstructed stone).
- Minimum 50 mm uninterrupted cavity. Where a cavity is bridged (by a lintel, etc.) a damp-proof course or tray should be inserted in the wall so that the passage of moisture from the outer to the inner leaf is prevented. This is not necessary where the cavity is bridged by a wall tie, or where the bridging occurs, presumably, at the top of a wall and is then protected by the roof. Where an opening is formed in a cavity wall, the jambs should have a suitable vertical damp-proof course or the cavity should be closed so as to prevent the passage of moisture.
- An inside leaf of masonry or framing with suitable lining.

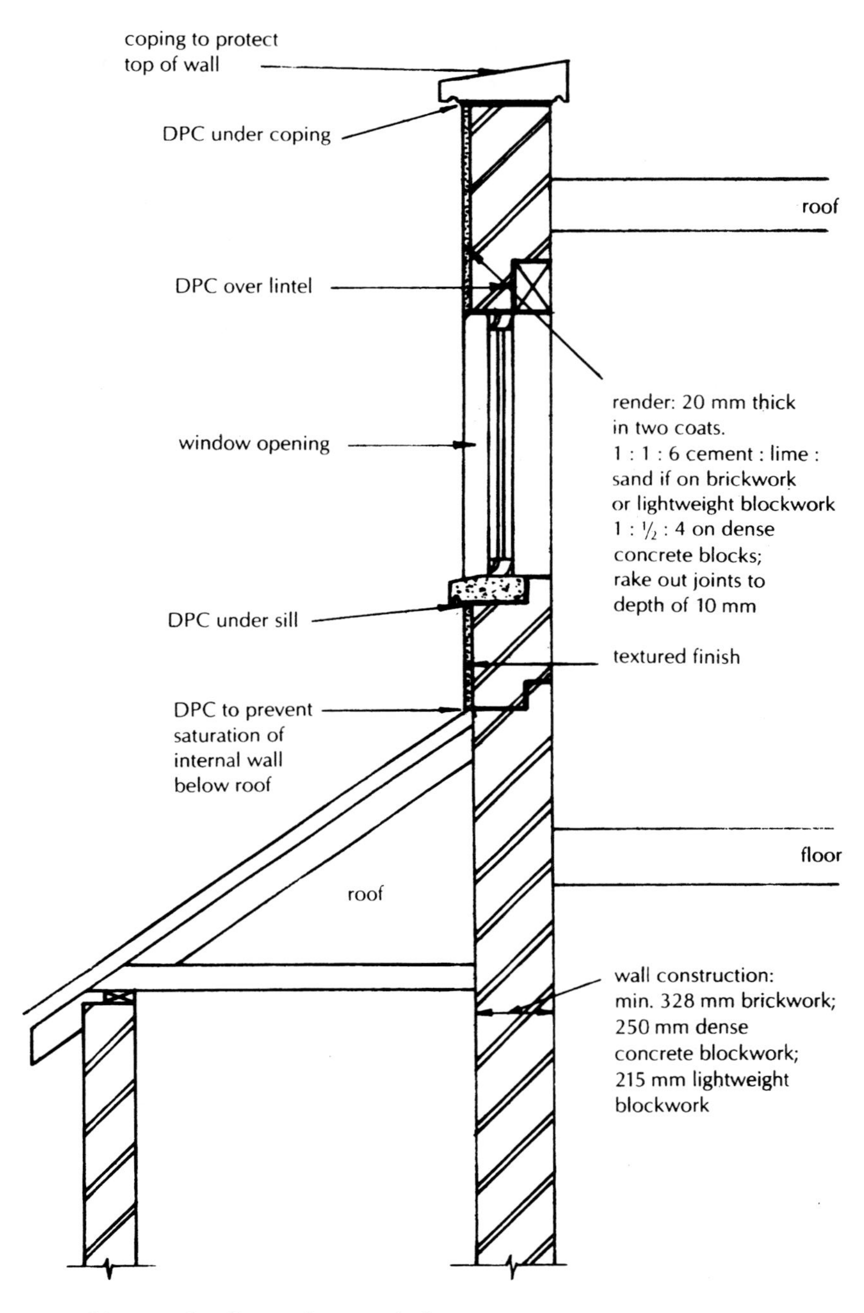

Fig. 8.12 Solid external walls – moisture exclusion.

These features are illustrated in Fig. 8.13.

- Where a cavity is only partially filled with insulation, the remaining cavity should be at least 50mm wide (see Fig. 8.14).

Alternatively, the relevant recommendations of BS 5628 *Code of practice for use of masonry*, Part 3: 1985 *Materials and components, design and workmanship* may be followed. Factors affecting rain penetration of cavity walls are indicated in the Code.

8.5.21 Weather resistance and cavity insulation

Since the installation of cavity insulation effectively bridges the cavity of a cavity wall and could give rise to moisture penetration to the inner leaf, it is most important that it be carried out correctly and efficiently. AD C4 lists a number of British Standards, Codes of Practice and other documents that cover the various materials that may be incorporated into a cavity wall.

- Rigid materials which are built in as the wall is constructed should be the subject of a current British Board of Agrément Certificate or European Technical Approval and the work should be carried out to meet the requirements of that document.
- Urea-formaldehyde foam inserted after the wall has been constructed should comply with BS 5617: 1985 *Specification for urea-formaldehyde (UF) foam systems, etc.* and should be installed in accordance with BS 5618: 1985 *Code of practice for thermal insulation of cavity walls, etc.*
- Other insulating materials inserted after the wall has been constructed should comply with BS 6232 *Thermal insulation of cavity walls by filling with blown man-made mineral fibre*, Parts 1 and 2: 1982, or be the subject of a current British Board of Agrément Certificate, or European Technical Approval. The work should be carried out to meet the requirements of the relevant document by operatives directly employed by the document holder. Alternatively, they may be employed by an installer approved to operate under the document.
- Where materials are inserted into a cavity after the wall has been constructed, the suitability of the wall for filling should be assessed, before installation, in accordance with BS 8208 *Guide to assessment of suitability of external cavity walls for filling with thermal insulants*, Part 1: 1995 *Existing traditional cavity construction*, and the person carrying out the work should operate under a current BSI Certificate of Registration of Assessed Capability or a similar document issued by an equivalent body.

(See Fig. 8.14 for further information regarding the insulation of external walls.)

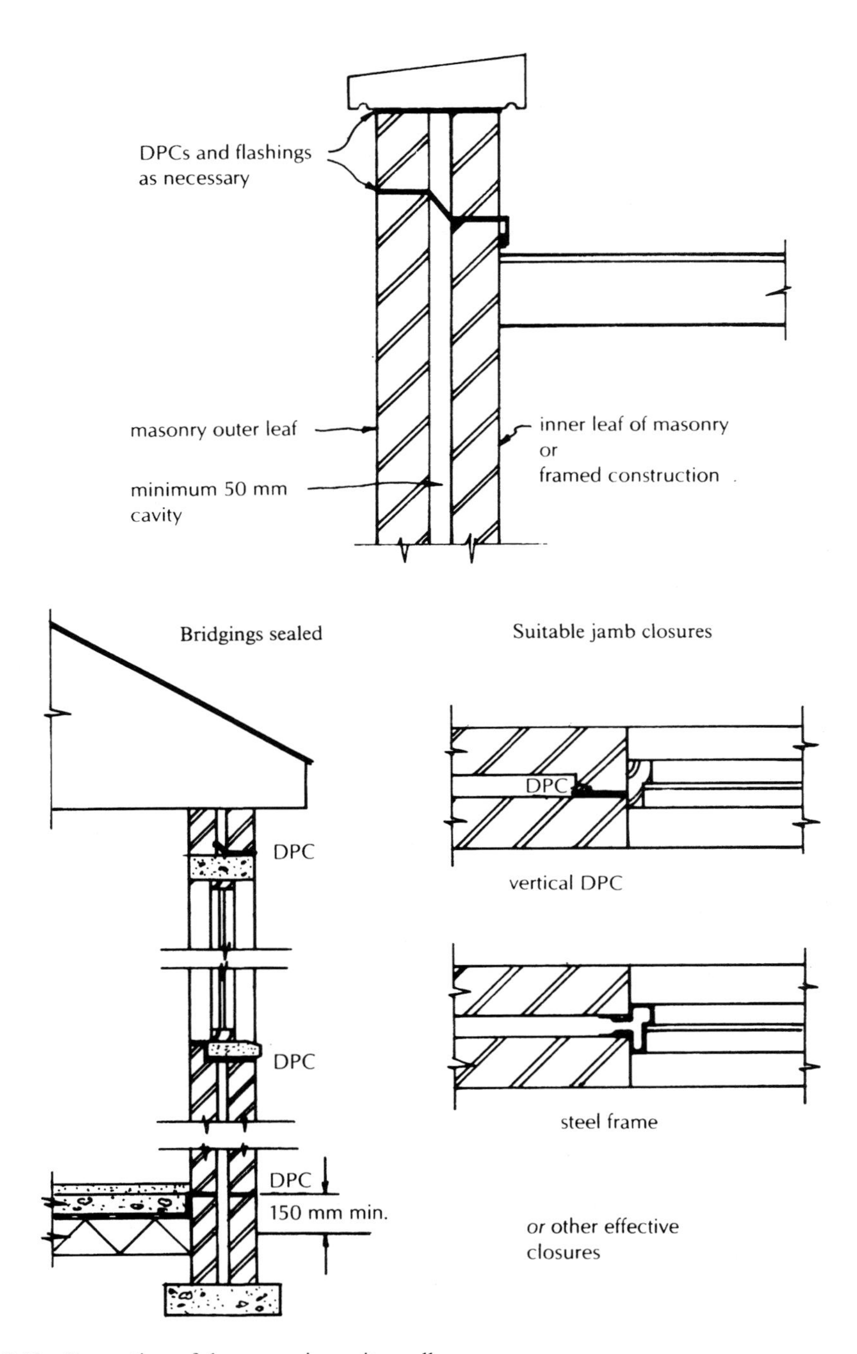

Fig. 8.13 Prevention of dampness in cavity walls.

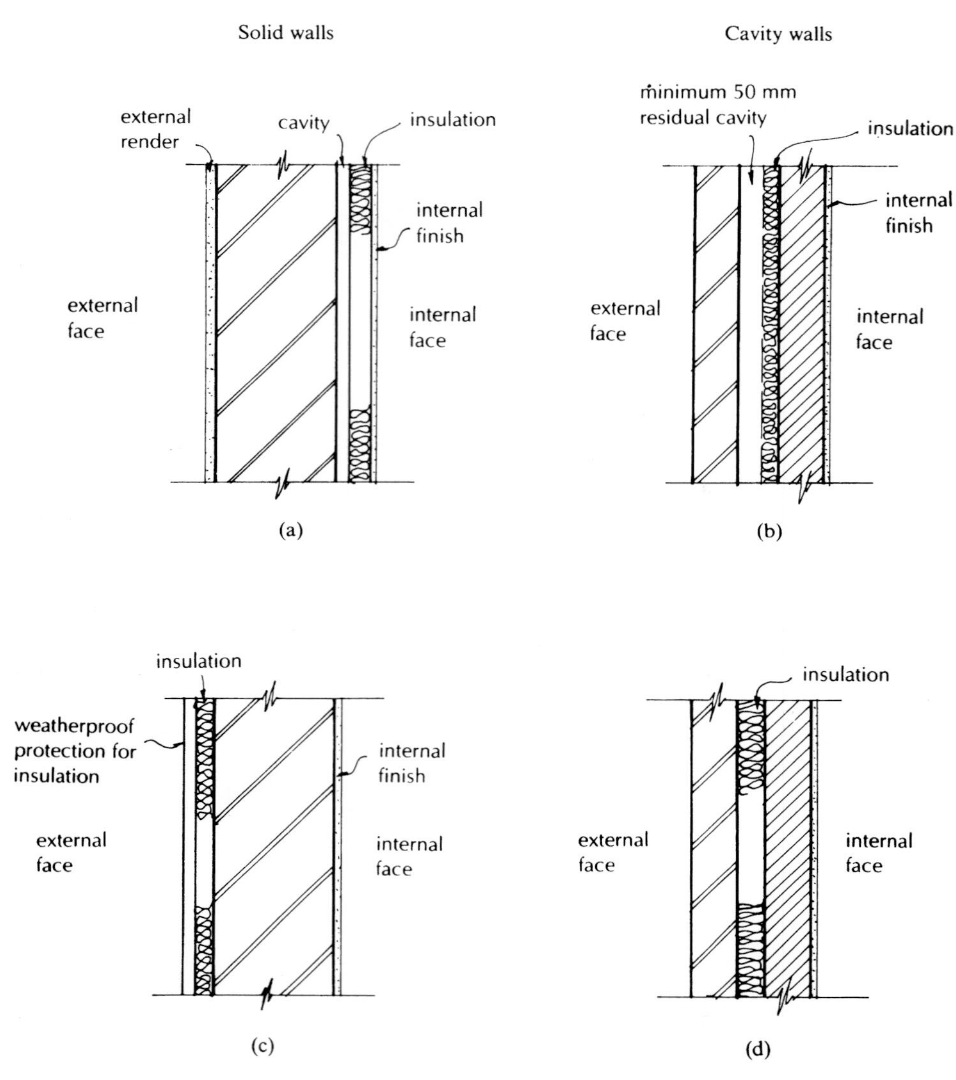

Fig. 8.14 Weather resistance and insulation of external walls. (a) Internal insulation. (b) Partially filled cavity. (c) External insulation. (d) Fully-filled cavity.

8.5.22 Claddings for external walls

The principles of external claddings are illustrated in Fig. 8.11(b) and (c) above. Therefore the cladding should be either:

(a) jointless (or have sealed joints) and be impervious to moisture (such as sheets of metal, glass, plastic or bituminous materials); or
(b) have overlapping dry joints and consist of impervious or weather-resisting materials (such as natural stone or slate, cement based products, fired clay or wood).

Dry jointed claddings should be backed by a material (such as sarking felt) which will direct any penetrating moisture to the outside surface of the structure.

Moisture-resisting materials consisting of bituminous or plastic products lapped at the joints are permitted but they should be permeable to water vapour unless there is a ventilated space behind the cladding.

Materials that are jointless or have sealed joints should be designed to accommodate structural and thermal movement.

Dry joints between cladding units should be designed either to resist moisture penetration or to direct any moisture entering them to the outside face of the structure. The suitability of dry joints will depend on the design of the joint and cladding and the severity of exposure of the building.

All external claddings should be securely fixed.

Some materials, such as timber claddings, are subject to rapid deterioration unless properly treated. These materials should only be used as the weather-resisting part of a roof or wall if they can meet the conditions specified in AD Regulation 7 described earlier in this chapter. It should be noted that the weather-resisting part of a roof or wall does not include paint or any surface rendering or coating which does not of itself provide all the weather resistance.

Insulation may be incorporated into the roof or wall cladding provided that it is protected from moisture (or is unaffected by it). Possible problems may arise due to interstitial condensation and cold bridges in the construction. Further guidance on this may be found in Ventilation (Chapter 11) and in the BRE publication *Thermal insulation – avoiding risks*.

8.5.23 Weather resistance of roofs

The roof of a building should:

- resist the passage of rain or snow to the inside of the building
- not be damaged by rain or snow
- not transmit moisture due to rain or snow to another part of the building that might be damaged.

The requirements for external wall claddings mentioned above apply equally to roof covering materials.

The performance requirements for external wall and roof claddings can also be met if they comply with:

- British Standard Code of Practice 143 *Code of practice sheet roof and wall coverings* (this includes recommendations for aluminium, zinc, galvanised corrugated steel, copper and semi-rigid asbestos bitumen sheet).
- BS 6915: 1988 *Specification for design and construction of fully supported lead sheet roof and wall coverings*.
- BS 5247 *Code of practice for sheet roof and wall coverings*, Part 14: 1975 *Corrugated asbestos-cement*.
- BS 8200: 1985 *Code of practice for design of non-loadbearing external vertical enclosures of buildings*.

The following codes refer to walls only.

- British Standard Code of Practice 297: 1972 *Precast concrete cladding (non-loadbearing).*
- BS 8298: 1989 *Code of practice for design and installation of natural stone cladding and lining.*

The above documents describe the materials to be used and contain design guidance including fixing recommendations.

9 Toxic substances

9.1 Introduction

In recent years there has been evidence to suggest that fumes from urea-formaldehyde foam, when used as a cavity wall filling, can have an adverse effect on the health of people occupying the building. This is still a contentious issue, but has nevertheless become a subject for building control.

9.2 Cavity insulation

Where insulating material is inserted into a cavity in a cavity wall reasonable precautions must be taken to prevent toxic fumes from penetrating occupied parts of the building.

It should be noted that Approved Document D1 does not require total exclusion of formaldehyde fumes from buildings but merely that these should not increase to an irritant concentration.

The inner leaf of the cavity wall should provide a continuous barrier to the passage of fumes and for this purpose it should be of brick or block construction.

Before work is commenced the wall should be assessed for suitability in accordance with BS 8208 *Guide to assessment of suitability of external walls for filling with thermal insulants* Part 1: 1985 *Existing traditional cavity construction.*

The work should be carried out by a person holding a current BSI Certificate of Registration of Assessed Capability for this particular type of work.

The urea-formaldehyde foam should comply with the requirements of BS 5617: 1985 *Specification for urea-formaldehyde (UF) foam systems etc.* and the installation with BS 5618: 1985 *Code of practice for thermal insulation of cavity walls etc.*

10 Sound insulation

M.W. Simons

10.1 Introduction

Over recent years, living standards have improved, life styles have altered and as a consequence, sources of the noise which people commonly encounter have changed. Householders are now subjected to higher levels of noise from traffic and neighbourhood sources and this, coupled with higher expectations of indoor environmental quality and reduced tolerance to unwanted noise, has led to increasing dissatisfaction with the standard of sound insulation provided in dwellings. Part E of Schedule 1 to the Building Regulations 2000 (as amended) is aimed at targeting widespread dissatisfaction of occupants living in properties constructed under the preceding legislation that derived from studies undertaken prior to the 1965 Regulations. In addition to annoyance, it should be recognised that domestic noise is also a potential risk to health and whilst difficult to quantify, it is known to be a source of loss of sleep, stress and in extreme cases contributes to assaults on neighbours.

Initial proposals for amending Requirement E included the insulation of the envelope of the building against noise from external sources, but this has not been included in Part E of the 2000 Regulations. Requirements E1 and E2 relate to resistance to the passage of sound between dwellings constructed as part of the same building and within dwellings, and E3 is aimed at limiting reverberation of noise in the common parts of multi-occupancy buildings. The only requirement which does not relate to buildings providing living accommodation is E4, the subject of which is acoustic conditions in schools. This requirement is not addressed in Approved Document E and designers are referred to the Stationery Office publication Building Bulletin 93, *The acoustic design of schools*.

In Approved Document E 1992 it was stated that one way of satisfying the requirements of Part E of Schedule 1 to the 1991 Building Regulations was to build components for which sound insulation was required in accordance with specific 'deemed to satisfy' constructions referred to in the document. Whilst elements constructed to a standard equal to the prescribed constructions may well have provided adequate sound insulation, a fundamental shortcoming of such a procedure is that it makes no allowance for the consequences of bad workmanship other than inspection by building control authorities. Unfortunately, sound insulation is seriously affected by localised construction imperfections and details which on completion may be hidden from view may render a construction unsatisfactory. For this reason, specific performance standards are stipulated in Approved Document E 2003 and there is a requirement for pre-completion testing, by the controlling authority, of a sample of completed units to ensure.compliance with the standards. This is a very important aspect of the control procedure, and it places a requirement on the contractor to maintain quality control.

The Building Regulations 2000 (as amended) extend the requirement to provide sound insulation in dwellings beyond the circumstances stipulated in the 1991 Regulations. Prior to the current legislation, the requirement to provide sound insulation was restricted to walls separating houses together with walls and floors separating flats. This has now been extended to include similar levels of insulation between 'rooms for residential purposes' which include hotel rooms, rooms in residential homes and student halls of residence accommodation. In addition, there is now a requirement to restrict the passage of sound between certain walls and between floors within dwelling houses and other types of accommodation. An additional source of annoyance in multi-occupancy buildings containing dwelling units is noise generated in the common parts of such buildings which tend to be constructed using hard durable noise-reflecting surfaces. As a consequence, the regulations now include a requirement to restrict the level of reverberant sound in such areas.

For convenience, dwelling houses, flats and rooms for residential purposes will, for the remainder of this chapter be referred to collectively as 'dwelling places'. Also, following each section heading within this chapter is a list of the Approved Document (2003) E sections to which it relates. However, since material relating to a specific subject may occur in a number of places within the AD, these reference lists may not be totally complete or comprehensive. Section 10.14 contains important information relating to possible future developments.

10.2 Performance requirements (Approved Document E Section 0: Performance)

10.2.1 Sound insulation between dwelling places AD E: 0.1–0.8

Requirement E1 of Schedule 1 to the Building Regulations 2000 (as amended) states that the design and construction of houses, flats and rooms for residential purposes should be such that there is reasonable resistance to sound transmission between them and other parts of the same building and other buildings with which they are in direct physical contact. It should be noted that this requirement applies not only to purpose-built units but also to those resulting from the material change of use from some other type of building. Rooms for residential purposes include (see definitions in Chapter 2, section 2.4) rooms in residential homes, student halls of residence and hotels in which people may be expected to sleep.

In circumstances where Requirement E1 applies, Regulation 20A of the Building Regulations 2000 (as amended) and Regulation 12A of the Building (Approved Inspectors etc.) Regulations 2000 (as amended) apply and these regulations impose a requirement to undertake an approved programme of sound insulation testing and provide the appropriate local authority with a copy of the test results. In the case of both regulations there are specific time limits within which the results must be presented to the building control body. It is stated in Approved Document E that the normal way of satisfying regulation 20A or 12A will be to undertake a programme of testing in accordance with section 1 of the approved document. The nature and extent of this pre-completion testing is considered in sections 10.3 and 10.11 of this chapter.

Requirement E1 will be satisfied for separating walls and floors together with stairs which separate dwelling units if they are constructed such that they satisfy the sound insulation values given in Tables 1a and 1b in the Approved Document and also in Table 10.1 below. Separating floors are defined as floors which separate flats or rooms for residential purposes and separating walls are those which separate adjoining dwelling houses, flats or rooms for residential purposes. It is important to note that sound will not only be transmitted directly through the element in question but also through adjacent, or flanking, elements and that when assessing compliance with Table 10.1, flanking as well as direct transmission must be taken into account. See section 10.13 for explanations of these terms.

Table 10.1 Performance standards for separating walls, separating floors and stairs that have a separating function.

	Minimum Value of Airborne Sound Insulation $D_{nT,w} + C_{tr}$ dB	**Maximum Value of Impact Sound Insulation $L'_{nT,w}$ dB**
Purpose built dwelling houses and flats		
Walls (inc. walls that separate houses and flats from rooms for residential purposes)	45	–
Floors and stairs	45	62
Dwelling houses, flats and rooms for residential purposes formed by material change of use		
Walls	43	–
Floors and stairs	43	64
Purpose built rooms for residential purposes		
Walls	43	–
Floors and stairs	45	62

Requirement E1 does not relate only to walls and floors which separate dwelling units. If a wall or floor separates a dwelling unit from a space used for communal or non-domestic purposes, e.g. a refuse chute, the level of sound insulation required may be greater than that given in Table 10.1 for separating walls and floors. In such circumstances, the required value of sound insulation will depend on the level of noise generated in the non-domestic space and expert advice should be sought to assess the anticipated level and, if necessary, recommend higher levels of sound insulation. Figure 10.1 summarises the obligations of E1 in terms of elements requiring insulation and the nature of insulation required.

It will be seen from Table 10.1 that the criterion applied to assess airborne sound insulation is a single figure index, $D_{nT,w} + C_{tr}$, which is obtained from a frequency

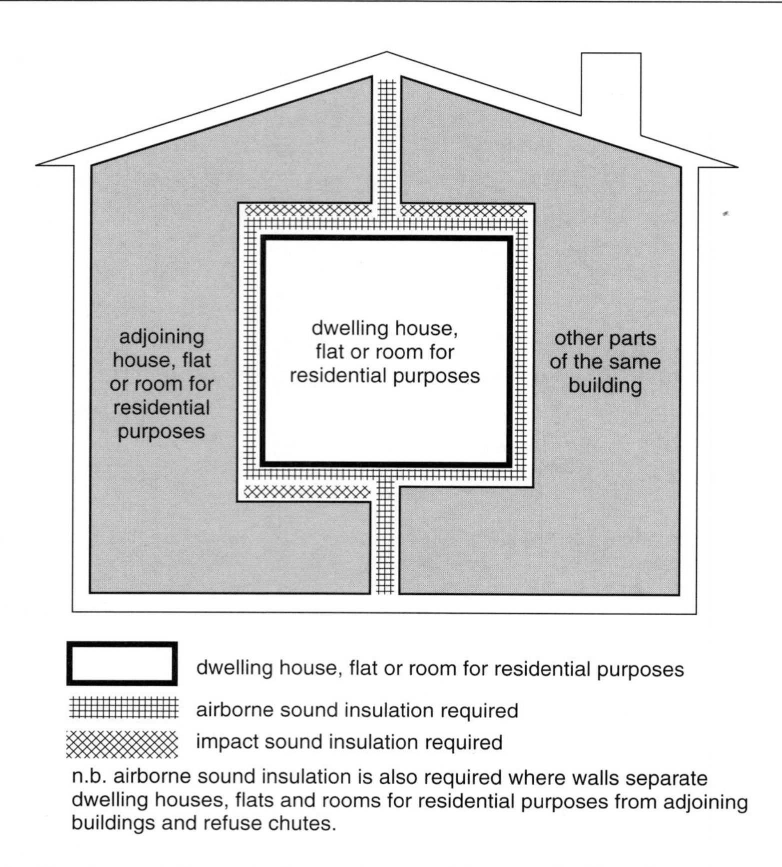

Fig. 10.1 Requirements for protection against sound between dwelling places.

band analysis of the noise. The term 'airborne sound insulation' is explained in section 10.13 and the procedure required for measurement of $D_{nT,w} + C_{tr}$ is described in section 10.11. This index replaces $D_{nT,w}$ which applied in the 1992 approved document by the addition of the C_{tr} term. C_{tr} modifies the index by increasing the significance of low frequency noise. The addition of C_{tr} is considered important since the powerful low frequency output of modern musical entertainment systems is such that sound transmission at low frequencies is now a frequent source of considerable annoyance.

The value of 45dB for airborne sound insulation between purpose-built dwellings, which appears in Table 10.1, appears at first sight to represent a reduction in standard set when compared with the equivalent value of 49dB which appears in the preceding code of practice. However, the effect of C_{tr} will be to reduce the measured value by approximately 5dB and, in addition, a reduction of 2dB is made to take into account the accuracy of the field measurement procedure. Consequently, the new value may be considered as an increase in the standard of 3dB, i.e:

$$49\text{dB} - 5\text{dB} - 2\text{dB} + 3\text{dB} = 45\text{dB}$$

In the case of a construction with a particularly poor low frequency performance the overall improvement in performance would need to be higher than 3dB in order to compensate for the lowering effect of C_{tr}.

Required values of the index used to assess the standard of impact sound insulation, $L'_{nT,w}$, which remains unchanged from the 1992 approved document, are shown in Table 10.1. See section 10.13 for an explanation of the term 'impact sound insulation', and section 10.11 for an explanation of the index $L'_{nT,w}$ and a description of the associated measurement procedure.

There are two sets of circumstances when it may not be necessary to fully comply with the requirement E1. Firstly, buildings constructed from prefabricated assemblies which are newly made or delivered from stock are subject to the full requirements of Schedule 1 of the building regulations but in some cases, e.g. temporary dwelling places, the requirement to provide reasonable sound insulation may vary. Examples referred to in the approved document are buildings:

- formed by dismantling, transporting and reassembling sub-assemblies on the same premises
- constructed from sub-assemblies from other premises or stock manufactured before 1 July 2003.

These buildings would normally be acceptable if they satisfied the requirements of Approved Document E 1992 or, in the case of school buildings, the relevant provisions of the 1997 edition of Building Bulletin 87.

Secondly, in the case of the material change of use of some historic buildings it may not be viable to fully comply with E1 whilst still conserving their special historical features. In such cases the aim should be to improve sound insulation as much as is practically possible without adversely effecting the character of the building or its long term rate of deterioration. When this approach is adopted, and E1 is not fully complied with, a prominent notice should be displayed in the building indicating the level of insulation obtained by tests carried out using the appropriate procedure. For the purposes of Approved Document E 'historic buildings' include:

- listed buildings
- buildings in conservation areas
- buildings of architectural or historic interest and referred to as a material consideration in local authority development plans or in national parks, areas of outstanding national beauty or world heritage sites.

It is suggested in the AD that when assessing the relative merits of reduced sound transmission and conservation, advice should be obtained from the local planning authority's conservation officer.

10.2.2 Sound insulation within dwelling places AD E: 0.9–0.10

Requirement E2 of Schedule 1 to the Building Regulations 2000 (as amended) states that the design and construction of houses, flats and rooms for residential purposes should be such that:

- internal walls between bedrooms and any other rooms
- internal walls between rooms containing water closets and any other rooms
- all internal floors

provide reasonable resistance to the transmission of sound. These requirements are shown diagrammatically in Fig. 10.2.

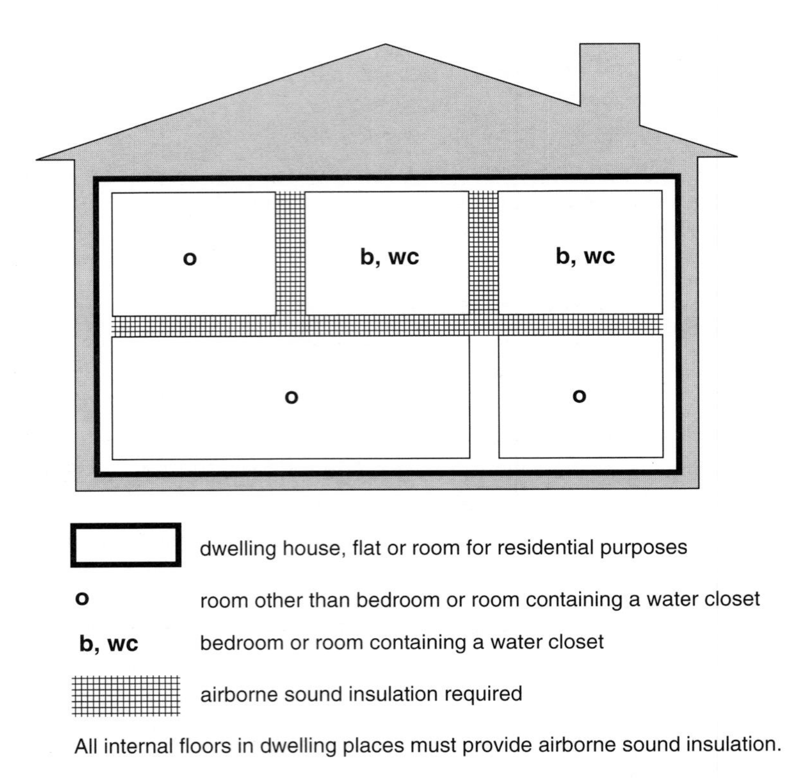

Fig. 10.2 Requirements for protection against sound within dwelling places.

Requirement E2 will be satisfied if the above elements meet the insulation values quoted in Table 2 of the AD. This simply states that new internal walls and floors in dwelling houses, flats and rooms for residential purposes, whether purpose-built or formed by material change of use must have a value of airborne sound insulation, R_w = 40dB or above. R_w, the weighted sound reduction index, is a single figure index which results from a frequency band analysis carried out in a laboratory of a sample of the element of construction. It follows therefore that pre-completion testing of internal elements is not necessary but that builders will need to show that elements used satisfy the above criterion. For laboratory results to be acceptable it will be necessary to demonstrate that the organisation undertaking the work has appropriate accreditation, preferably by the United Kingdom Accreditation Service (UKAS), or a European equivalent, for laboratory measurements. The laboratory measurement of sound insulation is discussed later in section 10.11. There is no requirement to provide impact sound insulation in order to satisfy E2.

As with E1, Requirement E2 extends to dwelling places resulting from the material change of use of a building but unlike E1 the same value applies to both categories of accommodation. Figure 10.2 summarises the obligations of E2 in terms of elements requiring insulation and the nature of insulation required.

10.2.3 Reverberation in common spaces and acoustic conditions in schools AD E: 0.11-0.12

Requirement E3 of the Building Regulations 2000 (as amended) states that the construction of the common internal spaces in buildings which contain flats or rooms for residential purposes shall be such as to prevent more reverberation than is reasonable. This requirement will be satisfied if sound absorptive measures described in section 7 of the approved document are applied. The provision of absorption in common spaces is considered in section 10.9 of this chapter.

Requirement E4 states that the design and construction of all rooms and other places in school buildings shall be such that their acoustic conditions and insulation against disturbance by noise is appropriate to their intended use. It is stated in AD E that the normal way of satisfying these conditions will be to meet the requirements given in section 1 of Building Bulletin 93, *The acoustic design of schools*. See section 10.10 of this chapter.

10.3 Pre-completion testing
(Approved Document E Section 1: Pre-completion testing)

10.3.1 Requirements AD E: 1.1–1.10

AD E provides guidance on the nature and extent of the pre-completion testing that should be carried out in order to comply with Regulation 20A, or 12A which states that where Requirement E1 applies, sound insulation testing must be carried out in accordance with an approved procedure.

It is written in the AD that sound insulation testing should now be considered as part of the construction process, the work being carried out at the cost of the builder. Regulations 12A and 20A state that it is the responsibility of the builder to ensure that:

- the work is carried out in an approved manner
- the results are recorded in an approved manner
- the results are given to the approved authority or inspector within five days of the completion of the work.

However, it is expected that the building control body will actually select which properties to test and the nature of the tests in each case. Whilst Regulations 12A and 20A state that the testing is the responsibility of the person carrying out the building work it is recognised that the testing will be carried out by an appropriate testing body. Indeed, for field measurements of sound insulation to be acceptable it will be necessary to demonstrate that the organisation undertaking the work has appropriate accreditation for field measurements, preferably by the United Kingdom Accreditation Service (UKAS), or a European equivalent.

Testing must be carried out on:

- purpose-built dwelling houses, flats and rooms for residential purposes
- dwelling houses, flats and rooms for residential purposes formed by material change of use

with the standard to be achieved being that given in Table 10.1 . Guidance on appropriate construction details is given in Approved Document E and developers are recommended in the AD that this, or other suitable guidance, is closely followed in order to ensure that the standards set in Table 10.1 are achieved. This is very important since there is no tolerance for measurement uncertainty, which has been built into the values given in the table; and if a test does not reach the prescribed value, by even a very small margin, it will be deemed to be a failure. Where there is doubt as to the required form of construction, specialist advice should be obtained at design stage. The exception to the standards defined in Table 10.1 is when historic buildings are undergoing a material change of use and it is not practicable to achieve the prescribed standard. In this case, appropriate testing is still required with, as mentioned earlier, a notice of the standard achieved posted in a prominent position in the building.

The tests are designed to be carried out on the common wall which separates specific types of rooms and to be undertaken as soon as the dwelling place is, with the exception of decoration, complete. In the case of impact tests, with certain exceptions which are referred to in section 10.11.3, it is important that these are carried out before soft coverings are installed.

10.3.2 Grouping for tests AD E: 1.11-1.18

Not all constructions have to be tested, but those that are must allow inferences of the performance of all constructions in the particular development to be drawn. It is therefore necessary to classify the units into a number of notional groups which are representative of the whole. Houses, flats and rooms for residential purposes must form three different groups.

If there are any significant differences in the method of construction used within each of these groups, then each must be classified as a sub-group. The AD provides guidance, as follows, as to how sub-groups should be determined.

- Dwelling houses (including bungalows) by type of separating wall, and flats and rooms for residential purposes by type of separating wall and separating floor
- Where there are significant differences in flanking elements, e.g. walls, floors and cavities
- Units with specific features which are acoustically unfavourable.

It should be noted that sub-grouping is not necessary if houses, flats and rooms for residential purposes are built with the same separating and flanking constructions and have similar room layouts and dimensions. However, whilst the same principles apply in the case of material change of use as for new buildings, in this case it is probable that significant differences between separating walls, separating floors and flanking constructions will occur more frequently. It follows that more sub groups will be necessary than in the case of new build work.

10.3.3 Sets of tests AD E: 1.19–1.28

The nature of the tests required between dwelling places depends on the type of accommodation they offer, i.e. there are different requirements for houses and flats. The tests required on a specific type of dwelling place are referred to collectively as a 'set of tests' and the specific requirements for sets of tests between different types of dwelling place are shown in Table 10.2. Note that:

- it is not appropriate to undertake impact tests between houses and bungalows;
- tests are only required in the circumstances identified in the table; and
- the tests referred to in Table 10.2 should each be carried out between a pair of adjacent rooms on opposite sides of the separating wall or floor between them.

Each set of tests should preferably contain individual tests between bedrooms and between living rooms. However, although it is recognised that this may not always be possible when, for example, room types are not mirrored on each side of a separating wall or floor, at least one of the rooms in one pair should be a bedroom and at least one in the other pair should be a living room. Tests should not be carried out between living spaces and corridors, stairwells or halls even though the walls between them may need to be constructed to separating wall standard. When the layout of a property is such that there is only one pair of rooms opposite to each other over the complete area of a separating wall or floor, the number of sound insulation tests may be reduced accordingly.

In the case of rooms for residential purposes, whether in new buildings or constructed by material change of use, the sound insulation of their walls and floors should be tested by application of the measurement principles applicable to dwelling houses and flats but adapted to suit the individual circumstances. It is sometimes the case that dwelling places are sold before fitting out with internal walls and internal fitments, the example of loft apartments being quoted in the approved document. In these situations the dimensions and locations of the principal room are not available but tests should still be undertaken, again applying the principles outlined for dwelling houses and flats. It is stated in the approved document that it is necessary in such cases to make sure that the subsequent work will not be detrimental to sound insulation by application of guidance provided on relevant construction details, see following sections.

Table 10.2 Requirements for a 'set of tests' between different types of dwelling places.

	Dwelling houses (including bungalows)	**Flats with separating floors but without separating walls**	**Flats with separating floors and separating walls**
Airborne test of walls Where possible between rooms suitable for use as living rooms	✓		✓
Airborne test of walls Where possible between rooms suitable for use as bedrooms	✓		✓
Airborne test of floors Where possible between rooms suitable for use as living rooms		✓	✓
Airborne test of floors Where possible between rooms suitable for use as bedrooms		✓	✓
Impact test of floors Where possible between rooms suitable for use as living rooms		✓	✓
Impact test of floors Where possible between rooms suitable for use as bedrooms		✓	✓
Total number of tests in a set	2	4	6

10.3.4 Testing programmes AD E: 1.29–1.31

It is expected that the building control authority will determine the nature and extent of the testing programme which is required. However, the approved document lays down the overall strategy that should be adopted.

Tests should be carried out between the first pair of properties scheduled for completion/sale in each group or sub-group irrespective of the size of the development, thus providing all parties with an early indication as to the likelihood of any problems. To this end, the frequency of testing should be higher in the early stages of completion than towards its end although in the case of large developments testing should continue for a substantial part of the time that construction is taking place. Notwithstanding the fact that the rate of testing may change during the construction period, the approved document states that, assuming that there are no failures, the building control body should require at least one set of tests for every ten dwelling units completed in each group or sub-group. It should be recognised that each set of tests relates to the sound insulation between at least two properties, three in the case of flats with separating walls and floors.

10.3.5 Failure of tests AD E: 1.32–1.39

A set of tests is deemed to have failed if one or more of its individual tests do not satisfy the requirements of Table 10.1. In the event of a failure, remedial action must be taken to rectify the failure between the actual rooms which have failed and also, since the separating element between those rooms may be defective, the developer must demonstrate that other rooms which share the element in question satisfy the performance criteria by:

- further testing between the other rooms; and/or
- application of remedial measures to the other rooms; and/or
- showing that the failure is limited to the rooms initially tested.

If a test between two rooms fails, careful consideration must be given to other properties which incorporate similar elements and forms of flanking construction, and developers will be expected to take the action necessary to satisfy building control authorities that the required standards are being achieved in those properties. Clearly, this will present problems if, by then, some of the properties are occupied. In any event, it is stated in the approved document that following the failure of a set of tests, the rate of testing should be increased from what had previously been agreed until the building control body is satisfied that the problems have been overcome.

The AD recommends reference to BRE Information Paper IP 14/02 for guidance relating to the failure of sound insulation tests. It should be recognised that failure may be due not to the construction of the separating wall but rather to excessive flanking transmission, the two causes requiring totally different types of remedial action. The reason for the failure will also indicate which of the other rooms that have not been tested are likely to fail to meet the standard. Such rooms will require remedial treatment. Whenever remedial treatment is applied, the building control body will normally call for further sound insulation testing in order that they can be satisfied with its effectiveness.

10.3.6 Recording of pre-completion test results AD E: 1.41

Regulations 20A and 12A state that the results of sound insulation testing shall be recorded in a manner approved of by the Secretary of State. In order to comply with this requirement, the report on a set of tests must contain at least the following information in the order in which it is listed. The following list is quoted directly from the AD:

(1) Address of Building.
(2) Type(s) of property. Use the definitions in Regulation 2: dwelling house, flat, room for residential purposes. State if the building is a historic building (see definition in the section on Requirements of this Approved Document).

(3) Date(s) of testing.
(4) Organisation carrying out testing, including:
   (a) name and address;
   (b) third party accreditation number (e.g. UKAS or European equivalent);
   (c) name(s) of person(s) in charge of test;
   (d) name(s) of client(s).
(5) A statement (preferably in a table) giving the following information:
   (a) the rooms used for each test within the set of tests;
   (b) the measured single-number quantity ($D_{nT,w} + C_{tr}$ for airborne sound insulation and $L'_{nT,w}$ for impact sound insulation) for each test within the set of tests;
   (c) the sound insulation values that should be achieved according to the values set out in Section 0: Performance – Table 1a or 1b, and
   (d) an entry stating 'Pass' or 'Fail' for each test within the set of tests according to the sound insulation values set out in Section 0 Performance – Table 1a or 1b.
(6) Brief details of the test, including:
   (a) equipment;
   (b) a statement that the test procedures in Annex B have been followed. If the procedure could not be followed exactly then the exceptions should be described and reasons given;
   (c) source and receiver room volumes (including a statement on which rooms were used as source rooms);
   (d) results of tests shown in tabular and graphical form for third octave bands according to the relevant part of the BS EN ISO 140 series and BS EN ISO 717 series, including:
      (i) single-number quantities and the spectrum adaptation terms;
      (ii) D_{nT} and L'_{nT} data from which the single-number quantities are calculated.

10.4 Separating walls and their flanking constructions – new buildings

(Approved Document E, Section 2: Separating walls and associated flanking constructions for new buildings)

10.4.1 Construction forms AD E: 2.1–2.17

Section 2 of the Approved Document E gives examples of walls which should achieve the performance standards, for houses and flats, tabulated in section 0: Performance – Table 1a of the approved document and in section 10.2 of this chapter. However, as is stated in the document, the actual performance of a wall will depend on the quality of its construction. The information in this section of the AD is provided only for guidance; it is not intended to be exhaustive and alternative constructions may achieve the required standard. Designers are also advised to seek advice from other sources, such as manufacturers, regarding alternative designs, materials or products. The details in no way override the requirement to undergo pre-completion testing. For the provision of walls in 'rooms for residential purposes', see section 10.8 of this chapter.

The separating wall constructions described in the approved document are divided into four types.

(1) **Type 1 Solid masonry**
Walls consisting of one leaf of brick or concrete block, plastered on both sides.
(2) **Type 2 Cavity masonry**
Walls consisting of two leaves of brick or concrete block separated by a cavity and usually plastered on both sides.
(3) **Type 3 Masonry between independent panels**
Walls consisting of a layer of brick or block with independent panels on each side.
(4) **Type 4 Framed walls with absorbent material**
Walls consisting of independent frames of, for example, timber with panels, typically of plasterboard, fixed to the outside of each.

Within each wall type, alternatives are presented in a ranking order such that the one which should offer the best sound insulation is described first.

It is very important that the junctions between a separating wall and the elements to which it is attached are carefully designed because this will have a significant influence on its overall acoustic effectiveness. Since the level of insulation achieved will be strongly dependent on the extent to which flanking transmission exists, considerable advice is given in the AD on this issue. Guidance is provided on the junctions between separating walls and the following other elements:

- external cavity walls with masonry inner leaf
- external cavity wall with timber framed inner leaf
- solid masonry external walls
- internal walls – masonry
- internal walls – timber
- internal floors – timber
- internal floors – concrete
- ground floors – timber
- ground floors – concrete
- ceiling and roof spaces
- separating floors Types 1, 2 and 3 (these are considered with separating floors).

However, not all wall/junction combinations are considered. It is stated in the AD that if any element has a separating function, e.g. a ground floor which is also a separating floor for a basement flat, then the separating element requirements take priority.

The following notes should be borne in mind when considering the alternatives which are presented.

(1) The mass per unit area of walls is quoted in kilograms per square metre (kg/m^2) and procedures for calculating it, as described in Annex A of the AD, are outlined in section 10.12. Densities are quoted in kilograms per cubic metre and when used for calculating the mass per unit area of brick and blockwork they should take into account moisture content by using data available in CIBSE Guide A (1999). Alternatively, manufacturers data relating to mass per unit area may be used where this is available.
(2) The guidance assumes solid blocks which do not have voids within them. The presence of voids may affect sound transmission and relevant information should be sought from manufacturers.
(3) Drylining laminates of plasterboard with mineral wool may be used whenever plasterboard is recommended. Manufacturer's advice should be obtained for all other drylining laminates. Plasterboard lining should always be fixed in accordance with manufacturer's instructions.
(4) The cavity widths recommended in the guidance are minimum values.
(5) The requirements of Building Regulations Part C (Site preparation and resistance to moisture) and Part L (Conservation of fuel and power) must be considered when applying Part E to junctions between separating walls and ground floors.
(6) The walls described in Tables 10.3, 10.5, 10.6 and 10.7 are only example constructions.

Wall ties in separating and external masonry walls AD E: 2.18–2.24

The sound insulating properties of cavity walls are effected by the coupling effect of wall ties and so it is important that they are used correctly. In AD E, wall ties are defined, in accordance with BS1243:1978, *Metal ties for cavity wall construction*, as Type A (butterfly ties) and Type B (double triangle ties) and their layout should be as described in BS 5628-3:2001, *Code of practice for use of masonry*. However, BS 5628-3 limits the above tie types and spacing to prescribed combinations of cavity widths and minimum masonry leaf thicknesses. It is stated in the approved document that tie Type B should be used only in external cavity walls where tie Type A does not satisfy the requirements of Building Regulation Part A – Structure, and that tie Type B may decrease airborne sound insulation, due to flanking transmission via the external leaf, compared with tie Type A.

An option, which is of value in situations outside the above limitations, is to select wall ties with an appropriate dynamic stiffness for the cavity width as explained in AD E. The procedure for measuring the dynamic stiffness of wall ties is explained in BRE IP 3/01, *Dynamic stiffness of wall ties used in masonry cavity walls*. It is essential to ensure that, if wall ties are selected on the basis of satisfying dynamic stiffness criteria, they also comply with the requirements of Building Regulation A – Structure.

10.4.2 Design of wall Type 1 – solid masonry AD E: 2.29–2.35

The insulation offered against airborne sound transmission by solid walls is primarily determined by their mass per unit area. Three forms of construction based on solid walls, i.e. wall Type 1, are described in the AD, and also in Table 10.3, and the document states that these should, if built correctly, comply with Requirement E1, see Table 10.1. Examples of each of the three types of wall are provided in the approved document and these are also described in Table 10.3 and shown in Fig. 10.3.

Table 10.3 Types and examples of solid masonry separating wall constructions.

	Wall Type		
	Type 1.1	**Type 1.2**	**Type 1.3**
Form			
Description	Dense aggregate concrete block, plastered on both faces	Dense aggregate cast in situ concrete, plastered on both faces	Brickwork, plastered on both faces
Minimum Mass per unit area	415 kg/m^2 including plaster	415 kg/m^2 including plaster	375 kg/m^2 including plaster
Surface finish	13 mm plaster on both room faces	Plaster on both room faces	13 mm plaster on both room faces
Other factors	Blocks laid flat to full thickness of the wall		Bricks laid 'frog up', coursed with headers
Example construction			
Thickness	215 mm block	190 mm of concrete	215 mm brick
Detail of structure	Dense aggregate concrete block, density 1840 kg/m^3	Density of concrete 2200 kg/m^3	Density of brick 1610 kg/m^3
Coursing	110 mm	n.a.	75 mm
Surface Finish	13 mm lightweight plaster with a minimum mass per unit area of 10kg/m^2, on both faces	13 mm lightweight plaster with a minimum mass per unit area of 10kg/m^2, on both faces	13 mm lightweight plaster with a minimum mass per unit area of 10kg/m^2, on both faces
Further information	215 mm blocks laid flat		

In order for the performance of solid separating walls to be satisfactory, the junctions between them and their surrounding construction elements must be designed correctly. The requirements detailed in the AD are explained below. See also section 10.4.7.

Junctions with external cavity walls which have masonry inner leaves AD E: 2.36–2.39

- There are no restrictions on the outer leaf of the wall.
- Unless the cavity is fully filled with mineral wool, expanded polystyrene beads or some other suitable insulating material (the AD states that manufacturer's advice should be sought regarding alternative suitable materials), the cavity should be stopped with a flexible closer as shown in Fig. 10.4.

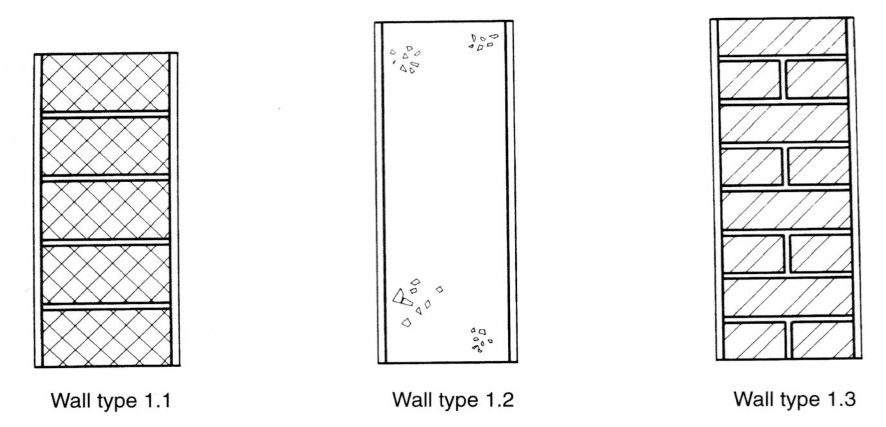

Fig. 10.3 Wall Types 1.1, 1.2 and 1.3.

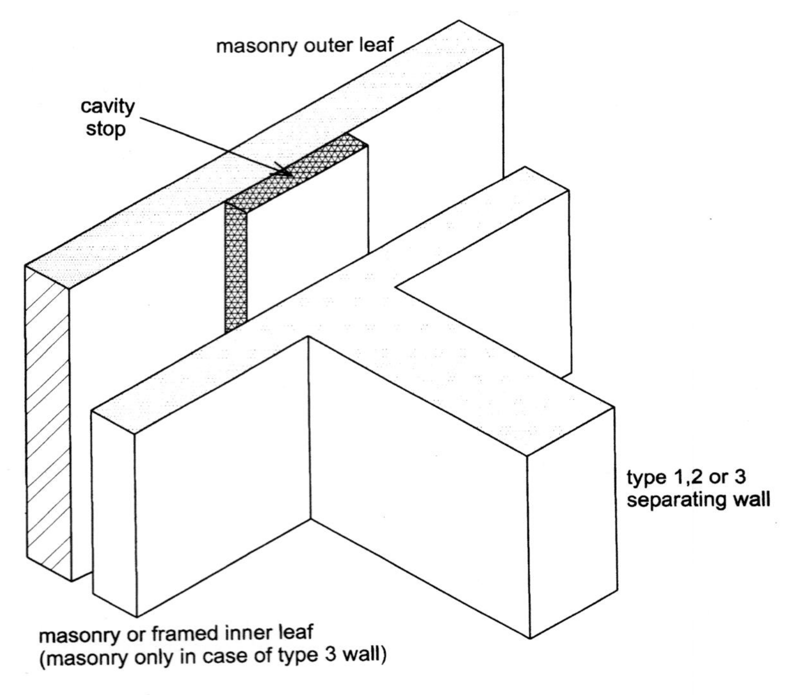

Fig. 10.4 Provision of a flexible cavity stop at junction of a Type 1, 2 or 3 separating wall and an external cavity wall.

- The separating wall should be joined to the inner leaf of the external wall either by bonding with at least half of the bond provided by the separating wall, or by using tied construction with the inner leaf abutting the separating wall see Fig. 10.5. See, also, Building Regulation Part A – Structure.
- The mass per unit area of the inner leaf of the external wall should be at least $120 kg/m^2$ excluding its finish in order to reduce the effects of flanking trans-

mission. This minimum mass requirement is waived if there is no separating floor and there are openings in the external wall which are at least 1 m high, within 0.7 m of the separating wall and on both sides of the separating wall at every storey.

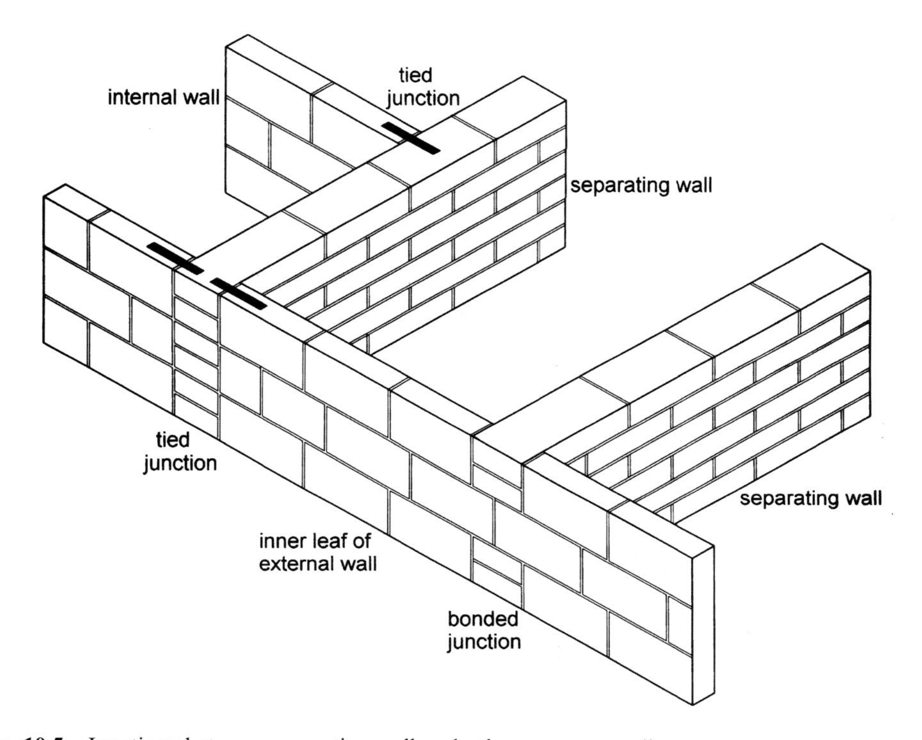

Fig. 10.5 Junctions between separating wall and other masonry walls.

Junctions with external cavity walls which have timber framed inner leaves AD E: 2.40–2.41

- There are no restrictions on the outer leaf of the wall.
- The cavity should be stopped with a flexible closer as shown in Fig. 10.4.
- The inner leaf frame of the external wall should abut the separating wall and be tied to it vertically with ties at not greater than 300 mm centres.
- The finish of the inner leaf of the external wall should be one layer of plasterboard with a mass per unit area of at least 10 kg/m^2 and having all joints sealed with tape or caulked with appropriate sealant. Two layers of plasterboard, each with a mass per unit area of at least 10 kg/m^2, are required if there is a separating floor.

Junctions with solid external masonry walls AD E: 2.42

No guidance is provided in the AD. Designers are advised to seek specialist advice.

Junctions with internal framed walls AD E: 2.43

No restrictions are imposed.

Junctions with internal masonry walls AD E: 2.44

Internal masonry walls should abut the separating wall and have a mass per unit area of at least 120 kg/m^2 excluding finish, see Fig. 10.5.

Junctions with internal timber floors AD E: 2.45

Floor joists should not be built into this type of wall. Instead, they should be supported on joist hangers.

Junctions with internal concrete floors AD E: 2.46–2.48

- The requirements for solid concrete slabs are shown in Fig. 10.6.
- Hollow core concrete plank floors and concrete beam with infill block floors should not be continuous through Type 1 separating walls.
- If concrete beam with infill block floors are built into separating walls, then the blocks in the floor must fill the space between the beams where they penetrate the wall.

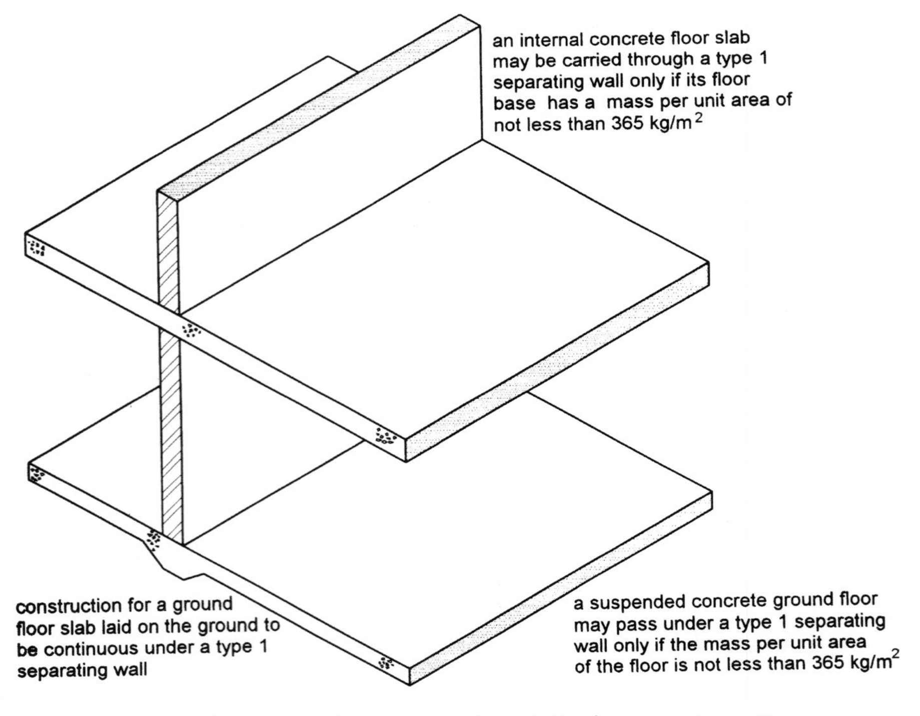

Fig. 10.6 Conditions for concrete floors to pass through Type 1 separating walls.

Junctions with timber ground floors AD E: 2.49

Ground floor joists should not be built into this type of wall. Instead, they should be supported on joist hangers.

Junctions with concrete ground floors AD E: 2.51–2.53

- For suspended concrete ground floors and solid slabs laid directly onto the ground, see Fig. 10.6.
- Hollow core concrete plank floors and concrete beam with infill block floors should not be continuous under Type 1 separating walls.

Junctions with ceilings and roofs AD E: 2.55–2.59

- The separating wall should continue to the underside of the roof with a flexible closer, which is suitable for use as a fire stop, at its junction with the roof, see Fig. 10.7.
- If the roof or loft space is not a habitable room, then the mass per unit area of the separating wall may be reduced to 150 kg/m^2 above the ceiling which separates the roof or loft space from the uppermost habitable rooms. For this reduction to apply, the mass per unit area of the ceiling must be at least 10kg/m^2 with sealed joints around the perimeter of the ceiling, see Fig. 10.7.
- If lightweight aggregate blocks with a density of less than 1200kg/m^3 are used above ceiling level, one side of the wall should be sealed, see Fig. 10.7.
- Cavities in external walls should be closed at their eaves since this will have the

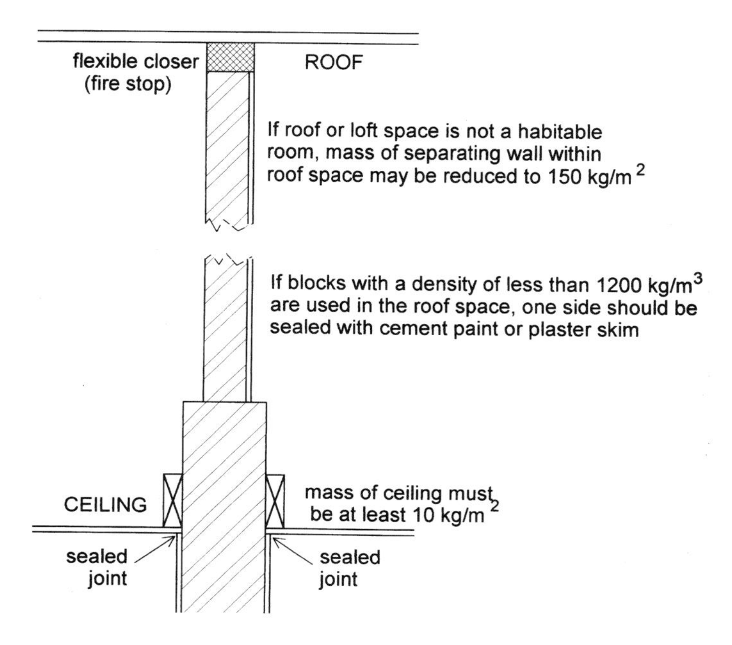

Fig. 10.7 Roof junction using wall Type 1, assuming roof space is not a habitable room.

effect of reducing any sound transmission through the cavity from the roof space. A flexible material should be used for the closure with no rigid joint between the two leaves. It is important to avoid a rigid connection between the two leaves and if a rigid material is used it should only be rigidly attached to one leaf, see BRE BR 262, *Thermal insulation: avoiding risks*, section 2.3.

Junctions with separating floors AD E: 2.60

See section 3 of the AD (section 10.5 of this chapter) for details of junctions between separating floors and Type 1 separating walls.

10.4.3 Design of wall Type 2 – cavity masonry AD E: 2.61–2.72

The insulation offered against airborne sound transmission by cavity masonry walls depends on the mass per unit area of the masonry, the cavity width and also the nature of the connection there is between the two leaves by, for example, wall ties and foundations.

Four types of cavity masonry separating wall are described in Approved Document E, which according to the document should, if built correctly, comply with Requirement E1, see Table 10.4. However, only two of these are appropriate for general application, two being restricted to situations where there is a step or stagger in the separating wall. Examples of each of the four types of wall are provided in the AD and these are described in Table 10.5. For each of these constructions it is important to note that:

- cavity widths are minimum values
- Type A wall ties should be used to connect the leaves of the walls
- blocks without voids are used (manufacturers advice should be sought regarding the use of blocks with voids).

In order for the performance of cavity masonry separating walls to be satisfactory, the junctions between them and their surrounding construction elements must be designed correctly. The requirements detailed in the approved document are explained below. See also section 10.4.7.

Junctions with external cavity walls which have masonry inner leaves AD E: 2.73–2.76

- There are no restrictions on the outer leaf of the wall.
- Unless the cavity is fully filled with mineral wool, expanded polystyrene beads or some other suitable insulating material (the approved document states that manufacturer's advice should be sought regarding alternative suitable materials), the cavity should be stopped with a flexible closer as shown in Figs. 10.4 and 10.8. Note the nature of the cavity closer for staggered wall construction in Fig. 10.8.
- The separating wall should be joined to the inner leaf of the external wall either

Table 10.4 Types of cavity masonry separating wall constructions.

	Wall Type			
	Type 2.1	**Type 2.2**	**Type 2.3**	**Type 2.4**
Description of wall	Two leaves of dense aggregate concrete block with 50 mm cavity, plaster on both room faces	Two leaves of lightweight aggregate block with 75 mm cavity, plaster on both room faces	Two leaves of lightweight aggregate block with 75 mm cavity and step/stagger. Plasterboard on both room faces	Two leaves of aircrete block with 75 mm cavity and step/stagger. Plasterboard or plaster on both room faces
Limitations			**Should only be used where there is a step and/or stagger of at least 300 mm**	**Should only be used in constructions without separating floors and where there is a step and/or stagger of at least 300 mm**
Minimum mass per unit area	415 kg/m^2 including plaster	300 kg/m^2 including plaster	290 kg/m^2 including plasterboard	150 kg/m^2 including finish
Minimum cavity width	50 mm	75 mm	75 mm	75 mm
Surface finish	13 mm plaster on both room faces	13 mm plaster on both room faces	Plasterboard on both room faces, minimum mass per unit area of each sheet 10 kg/m^2	Plasterboard on both room faces, minimum mass per unit area of each sheet 10 kg/m^2 or, 13 mm plaster on both room faces.
Other factors			The lightweight blocks should have a density in the range 1350 to 1600 kg/m^3 The composition of the lightweight aggregate blocks contributes to performance of this construction with a plasterboard finish. Denser blocks may not give equivalent performance Increasing size of step or stagger in the separating wall tends to increase airborne sound insulation	Increasing size of step or stagger in the separating wall tends to increase airborne insulation

Table 10.5 Examples of cavity masonry separating wall constructions.

	Wall Type			
	Type 2.1	**Type 2.2**	**Type 2.3**	**Type 2.4**
Description of wall	Two leaves of dense aggregate concrete block with 50 mm cavity, plaster on both room faces	Two leaves of lightweight aggregate block with 75 mm cavity, plaster on both room faces	Two leaves of lightweight aggregate block with 75 mm cavity and step/stagger. Plasterboard on both room faces	Two leaves of aircrete block with 75 mm cavity and step/stagger. Plasterboard or plaster on both room faces
Limitations			**Should only be used where there is a step and/or stagger of at least 300 mm**	**Should only be used in constructions without separating floors and where there is a step and/or stagger of at least 300 mm**
Thickness of block leaves	100 mm	100 mm	100 mm	100 mm
Description of blocks	Dense aggregate concrete block, density 1990 kg/m^3	Lightweight aggregate block, density1375 kg/m^3	Lightweight aggregate block, density 1375 kg/m^3	Aircrete block, density 650 kg/m^3
Coursing	225 mm	225 mm	225 mm	225 mm
Surface finish	13 mm lightweight plaster with a minimum mass per unit area of 10kg/m^2, on both room faces	13 mm lightweight plaster with a minimum mass per unit area of 10kg/m^2, on both room faces	Plasterboard on both room faces. Minimum mass per unit area of each sheet 10 kg/m^2	Plasterboard on both room faces. Minimum mass per unit area of each sheet 10 kg/m^2

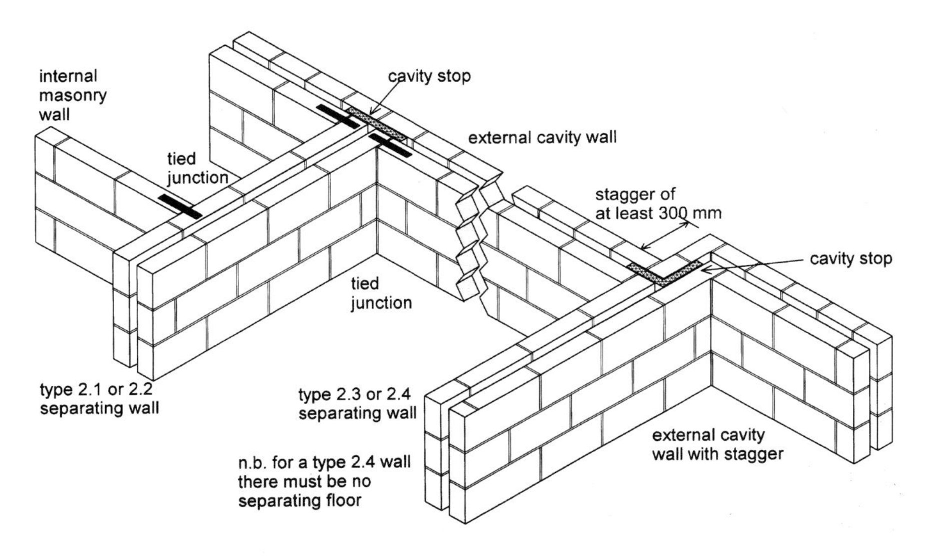

Fig. 10.8 Junctions of cavity separating walls.

by bonding with at least half of the bond provided by the separating wall, or by using tied construction with the inner leaf abutting the separating wall. See Building Regulation Part A – Structure.

- The mass per unit area of the inner leaf of the external wall should be at least 120kg/m^2 excluding its finish for Type 2.2 separating walls in order to reduce the effects of flanking transmission. This requirement does not apply to Type 2.1, 2.3 and 2.4 walls unless there is also a separating floor in which case it applies to all Type 2 separating walls.

Junctions with external cavity walls which have timber framed inner leaves AD E: 2.77–2.78

- There are no restrictions on the outer leaf of the wall.
- The cavity should be stopped with a flexible closer as shown in Fig. 10.4.
- The inner leaf frame of the external wall should abut the separating wall and be tied to it vertically with ties at not greater than 300 mm centres.
- The finish of the inner leaf of the external wall should be one layer of plasterboard with a mass per unit area of at least 10 kg/m^2 with all joints sealed with tape or caulked with appropriate sealant. Two layers of plasterboard, each with a mass per unit area of at least 10 kg/m^2, are required if there is a separating floor.

Junctions with solid external masonry walls AD E: 2.79

No guidance is provided in the AD. Designers are advised to seek specialist advice.

Junctions with internal framed walls AD E: 2.80

No restrictions are imposed.

Junctions with internal masonry walls AD E: 2.81–2.83

Internal masonry walls which abut a Type 2 separating wall, see Fig. 10.8, should have a mass per unit area of at least 120 kg/m^2 excluding finish. Also, where there is a separating floor, internal masonry walls should have a mass of at least 120 kg/m^2 excluding finish. In the case of Type 2.3 and 2.4 separating walls (which are only applicable if there is a step or stagger), and when there is no separating floor, there is no minimum mass requirement for internal masonry walls.

Junctions with internal timber floors AD E: 2.84

Floor joists should not be built into this type of separating wall. They should be supported on joist hangers.

Junctions with internal concrete floors AD E: 2.85

Internal concrete floors should generally be built into Type 2 separating walls and continue to the face of the cavity, see Fig. 10.9. Such floors should not bridge the cavity.

Junctions with timber ground floors AD E: 2.86

Ground floor joists should not be built into this type of separating wall. They should be supported on joist hangers.

Junctions with concrete ground floors AD E: 2.88–2.89

- Concrete floor slabs laid directly onto the ground should not be continuous beneath Type 2 separating walls but should abut them as shown in Fig. 10.9.
- Suspended concrete ground floors should not be continuous beneath Type 2 separating walls but should be built in and continue to the face of the cavity, see Fig. 10.9. Floors should not bridge the cavity.

Junctions with ceilings and roofs AD E: 2.91–2.95

- The separating wall should continue to the underside of the roof with a flexible closer, which is suitable for use as a fire stop, at its junction with the roof, see Fig. 10.10.
- If the roof or loft space is not a habitable room, then the mass per unit area of the separating wall may be reduced to 150 kg/m^2 above the ceiling which separates the roof or loft space from the uppermost habitable rooms. For this reduction to apply, the mass per unit area of the ceiling must be at least 10kg/m^2 with sealed

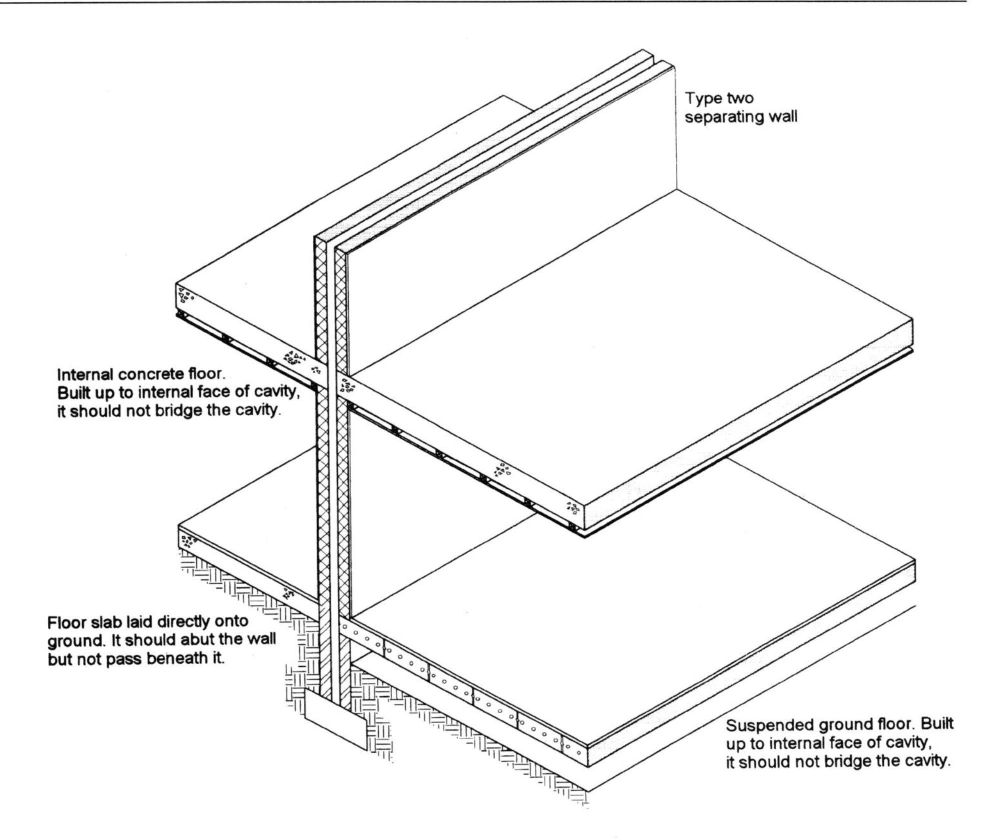

Fig. 10.9 Conditions for concrete floors and Type 2 separating walls.

joints around the perimeter of the ceiling and the wall must remain a cavity wall, see Fig. 10.10.

- If lightweight aggregate blocks with a density of less than 1200 kg/m^3 are used above ceiling level, one side of the wall should be sealed, see Fig. 10.10.
- Cavities in external walls should be closed at their eaves since this will have the effect of reducing any sound transmission through the cavity from the roof space. A flexible material should be used for the closure with no rigid joint between the two leaves. It is important to avoid a rigid connection between the two leaves and if a rigid material is used it should only be rigidly attached to one leaf, see BRE BR 262, *Thermal insulation: avoiding risks*, section 2.3.

Junctions with separating floors AD E: 2.96

See section 3 of the AD (section 10.5 of this chapter) for details of junctions between separating floors and Type 2 separating walls.

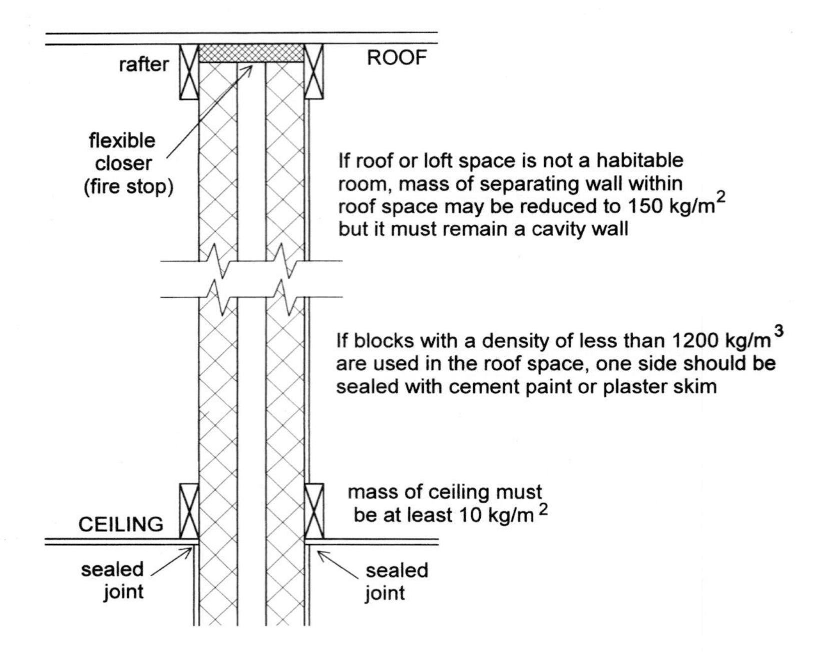

Fig. 10.10 Roof junction using wall Type 2, assuming roof space is not a habitable room.

10.4.4 Design of wall Type 3 – masonry walls between independent panels AD E: 2.97–2.107

The independent panels used in this type of construction are typically of plasterboard fixed to timber studs and the insulation such walls offer to airborne sound transmission depends on the mass per unit area of the masonry and the independent panels. It is also influenced by the degree of isolation between the masonry core and the panels. Separating walls of this type are capable of providing very good airborne and impact sound insulation.

Three types of masonry/independent panel walls are described in Approved Document E, which according to the document should, if built correctly, comply with Requirement E1, see Table 10.6. Examples of each of the three types of wall are provided in the approved document and these are also described in Table 10.6 . The essential components of this form of construction are a masonry core of either solid or cavity construction with independent panels on each side of it. The panels, and any framing which is associated with them, must not be in contact with the masonry core.

For each of these constructions:

- cavity widths are minimum values;
- Type A wall ties should be used to connect the leaves of cavity masonry cores; and

Table 10.6 Types and examples of masonry with independent panels separating wall constructions.

	Wall Type		
	Type 3.1	**Type 3.2**	**Type 3.3**
Form			
Description	Solid masonry core of dense aggregate concrete block, independent panels on both room faces	Solid masonry core of lightweight concrete block, independent panels on both room faces	Cavity masonry core of brickwork or blockwork, 50 mm cavity, independent panels on both room faces
Minimum mass per unit area of core	300 kg/m^2	150 kg/m^2	No minimum specified
Cavity width	n.a.	n.a.	50 mm minimum
Panels	Independent panels on both room faces	Independent panels on both room faces	Independent panels on both room faces
Other factors	Structural requirements determine minimum core width. Ref: Part A Structure	Structural requirements determine minimum core width. Ref: Part A Structure	Structural requirements determine minimum core width. Ref: Part A Structure
Example construction			
Thickness of core	140 mm	140 mm	Two leaves, each one at least 100 mm thick
Cavity width	n.a.	n.a.	50 mm minimum
Detail of core	Dense aggregate concrete block, density 2200 kg/m^3	Lightweight concrete block, density 1400 kg/m^3	Concrete block, density not specified
Coursing	110 mm	225 mm	Not specified
Form of panels	Independent panels, each consisting of two sheets of plasterboard with staggered joints Mass per unit area of each panel 20 kg/m^2	Independent panels, each consisting of two sheets of plasterboard joined by a cellular core Mass per unit area of each panel 20 kg/m^2	Independent panels, each consisting of two sheets of plasterboard joined by a cellular core Mass per unit area of each panel 20 kg/m^2

- a stringent specification for the construction of the independent panels should be adhered to as follows:
  (a) the independent panels must each have a mass per unit area of at least 20 kg/m^2 excluding the mass of any supporting framework;
  (b) panels to consist of two or more layers of plasterboard with staggered joints, or composite panels formed by two sheets of plasterboard separated by a cellular core; and,
  (c) a gap of at least 35 mm between panels and masonry core if the panels are not supported on a frame, or a gap of at least 10 mm between the frame and the masonry core if the panels are supported on a frame.

In order for the performance of masonry with independent panel separating walls to be satisfactory, the junctions between them and their surrounding construction elements must be designed correctly. The requirements detailed in the AD are explained below. See also section 10.4.7.

Junctions with external cavity walls which have masonry inner leaves AD E: 2.108–2.112

- There are no restrictions on the outer leaf of the wall.
- Unless the cavity is fully filled with mineral wool, expanded polystyrene beads or some other suitable insulating material (the approved document states that manufacturers' advice should be sought regarding alternative suitable materials), the cavity should be stopped with a flexible closer as shown in Figs. 10.4 and 10.11.
- There should be a bonded or tied connection between the core of the separating wall and the inner leaf of the external wall. See Fig. 10.11 for detail of tied connection.
- The inner leaf of the external wall should be lined in the same way as the separating wall, see Fig. 10.11.
- The mass per unit area of the inner leaf of the external wall should be at least $120 kg/m^2$ excluding its finish, but if there is no separating floor and the inner leaf of the external wall is lined with independent panels in the same way as the separating wall there is no minimum mass requirement for the inner leaf.
- Where the mass per unit area of the inner leaf of the external wall is at least $120 kg/m^2$ excluding its finish, there is no separating floor and wall Type 3.1 or 3.3 is used, the inner leaf of the external wall may be finished with plaster or plasterboard with a mass per unit area of not less than $10 kg/m^2$.

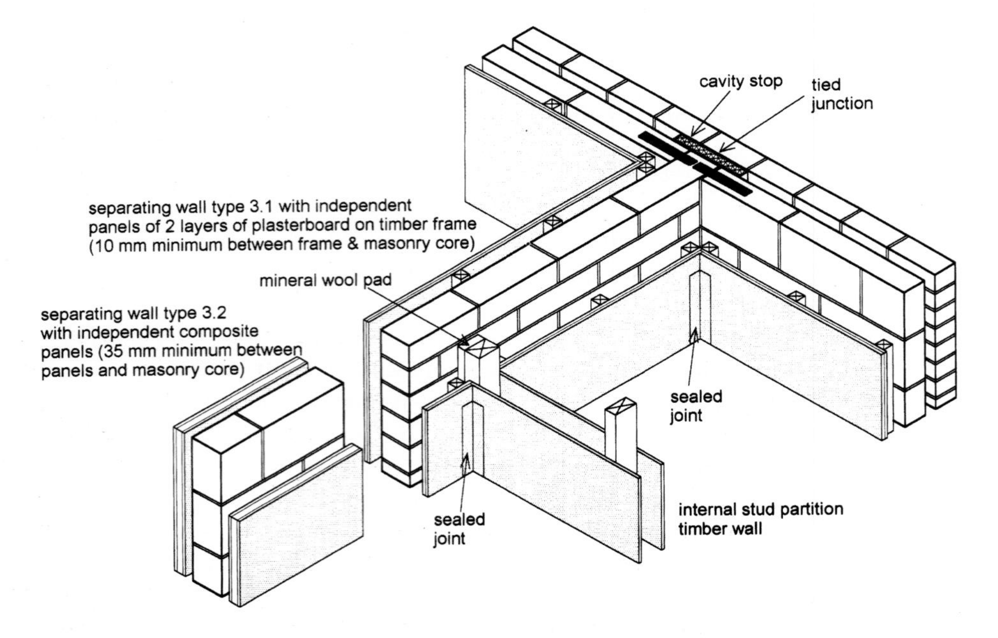

Fig. 10.11 Junctions using Type 3 separating walls.

Junctions with external cavity walls which have timber framed inner leaves AD E: 2.113

No guidance is provided in the AD. Designers are advised to seek specialist advice.

Junctions with solid external masonry walls AD E: 2.114

No guidance is provided in the AD. Designers are advised to seek specialist advice.

Junctions with internal framed walls AD E: 2.115–2.117

Load bearing internal framed walls should be fixed to the masonry core using a continuous pad of mineral wool as shown in Fig. 10.11, whereas non-load bearing internal walls should abut the independent panels. The joints between all internal wall panels should be either sealed with tape or caulked with appropriate sealant.

Junctions with internal masonry walls AD E: 2.118

Internal walls of masonry construction should not abut Type 3 separating walls.

Junctions with internal timber floors AD E: 2.119–2.120

Floor joists should not be built into this type of wall. They should be supported on joist hangers and the spaces at the wall surface between the joists should be sealed to the full depth of the joists with timber blocking.

Junctions with internal concrete floors AD E: 2.121–2.122

For separating wall Types 3.1 and 3.2 (solid), internal concrete floor slabs may only be carried through the solid masonry core if the floor base has a mass per unit area of 365 kg/m^2 or more. For separating wall Type 3.3 (cavity), internal concrete floors should be built into the wall and continue to the face of the cavity. Such floors should not bridge the cavity.

Junctions with timber ground floors AD E: 2.123–2.124

Floor joists should not be built into this type of wall. They should be supported on joist hangers and the spaces at the wall surface between the joists should be sealed to the full depth of the joists with timber blocking.

Junctions with concrete ground floors AD E: 2.127–2.131

- A concrete floor slab laid on the ground may be continuous under the solid core of a Type 3.1 or 3.2 (solid core) separating wall.
- A suspended concrete floor may pass under the masonry core of a Type 3.1 or 3.2 (solid core) separating wall only if the mass per unit area of the floor is 365 kg/m^2 or more.

- Hollow core concrete plank floors and concrete beam with infill block floors should not be continuous under the masonry core of a Type 3.1 or 3.2 (solid core) separating wall.
- Concrete floor slabs laid directly onto the ground should not be continuous beneath Type 3.3 (cavity core) separating walls but should abut them in a manner similar to that shown in Fig. 10.9 for cavity walls.
- Suspended concrete ground floors should not be continuous beneath Type 3.3 (cavity core) separating walls but should be built in and continue to the face of the cavity in a manner similar to that shown in Fig. 10.9 for cavity walls. The floor should not bridge the cavity.

Junctions with ceilings and roofs AD E: 2.133–2.139

- The masonry core should continue to the underside of the roof with a flexible closer, which is also suitable for use as a fire stop, at its junction with the roof. Also, the independent panels should be either sealed with tape or caulked with appropriate sealant at their junction with the ceiling, see Fig. 10.12.

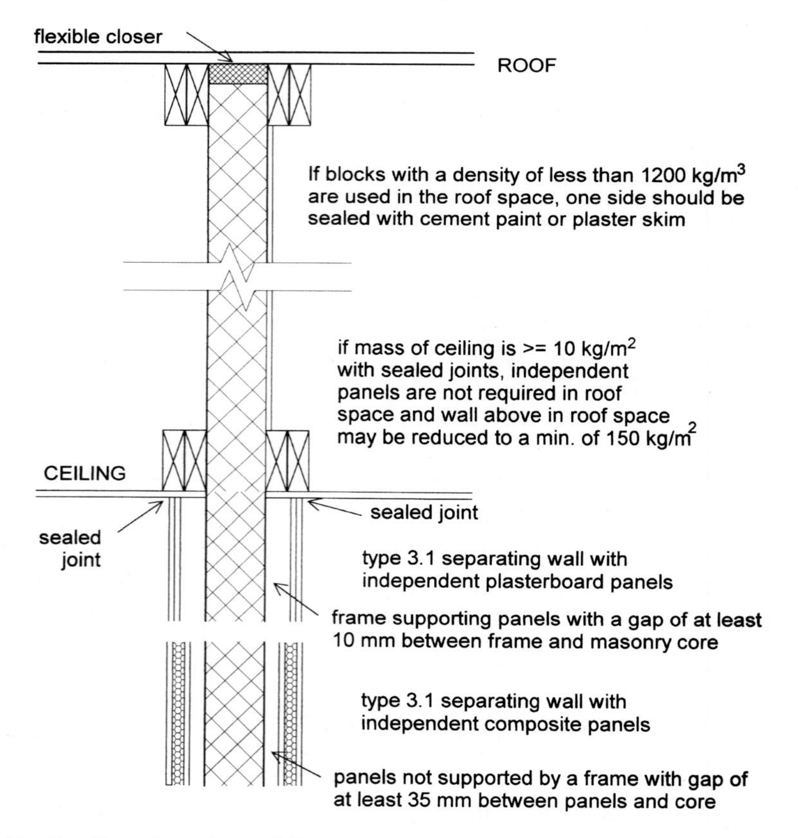

Fig. 10.12 Roof junction using wall Type 3.1 or 3.2, assuming roof space is not a habitable room.

- Cavities in external walls should be closed at their eaves since this will have the effect of reducing any sound transmission through the cavity from the roof space. A flexible material should be used for the closure with no rigid joint between the two leaves. It is important to avoid a rigid connection between the two leaves and if a rigid material is used it should only be rigidly attached to one leaf, see BRE BR 262, *Thermal insulation: avoiding risks*, section 2.3.
- If the roof or loft space is not a habitable room and the mass per unit area of the ceiling is at least 10kg/m^2 and it has sealed joints, then the independent panels may be omitted in the roof space. Also, in the case of wall Types 3.1 and 3.2 the mass per unit area of the separating wall may be reduced to 150 kg/m^2 above the ceiling, see Fig. 10.12. However, in the case of wall Type 3.3, the cavity masonry core must continue to the underside of the roof.
- If lightweight aggregate blocks with a density of less than 1200 kg/m^3 are used above ceiling level, one side of the wall should be sealed with cement paint or plaster skim.

Junctions with separating floors AD E: 2.140

See section 3 of the AD (section 10.5 of this chapter) for details of junctions between separating floors and Type 3 separating walls.

10.4.5 Design of wall Type 4 – framed walls with absorbent material AD E: 2.141–2.147

Walls of this type consist of two independent frames of timber or steel with panels fixed to the outside of each and sound absorbent material such as mineral wool placed in the void between the two panels. The insulation they offer to airborne sound transmission depends on the mass per unit area of the panels, the extent to which the frames are isolated from each other and on the absorption properties of the void between them.

One type of framed wall consisting of timber frames and plasterboard lining is described in Approved Document E, see Table 10.7, and its construction form is shown in Fig. 10.13. According to the document this wall should, if built correctly, comply with Requirement E1. Designers are advised in the AD to seek advice from manufacturers for steel framed alternatives.

In order for the performance of frame and absorbent infill separating walls to be satisfactory, the junctions between them and their surrounding construction elements must be designed correctly. The requirements detailed in the AD are explained below. See also section 10.4.7.

Junctions with external cavity walls which have masonry inner leaves AD E: 2.148

No guidance is provided in the approved document. Designers are advised to seek specialist advice.

Table 10.7 One type of framed wall with absorbent material.

Form	Wall Type 4.1
Structure	A two leaf frame
Lining	Each lining to consist of two or more sheets of plasterboard with staggered joints. Minimum mass per unit area of each sheet: 10kg/m^2
Width	Distance between inside lining faces: 200 mm minimum
Form of absorbent material	Unfaced mineral wool batts or quilt with a minimum density of 10kg/m^3. These may be wire reinforced
Thickness of absorbent material	25 mm min. if suspended in cavity between frames 50 mm min. if fixed to one frame 25 mm min. per batt or quilt if one is fixed to each frame
Other information	Plywood sheathing may be used in the cavity as necessary for structural reasons

Notes: 1 See Fig. 10.13 for details associated with this type of wall.
2 A masonry core may be incorporated, if this is necessary for structural reasons, but it may only be connected to one frame.

Junctions with external cavity walls which have timber framed inner leaves AD E: 2.149–2.150

- No restrictions are imposed on the form of the outer leaf.
- The cavities in the external wall and the separating wall should be stopped with flexible closers as shown in Fig. 10.13.
- The finish of the inner leaf of the external wall, assuming there is no separating floor, should consist of one layer of plasterboard with a mass per unit area of at least 10 kg/m^2, all joints being sealed with tape or caulked with appropriate sealant. If there is a separating floor, two layers of plasterboard, each with a mass per unit area of at least 10 kg/m^2, should be provided.

Junctions with solid external masonry walls AD E: 2.151

No guidance is provided in the approved document. Designers are advised to seek specialist advice.

Junctions with internal framed walls AD E: 2.152

No restrictions are imposed.

Junctions with internal masonry walls AD E: 2.153

No restrictions are imposed.

Junctions with internal timber floors AD E: 2.154

In order to prevent flanking transmission, air paths into the cavity from floor voids must be blocked by solid timber blockings, continuous ring beams or joists.

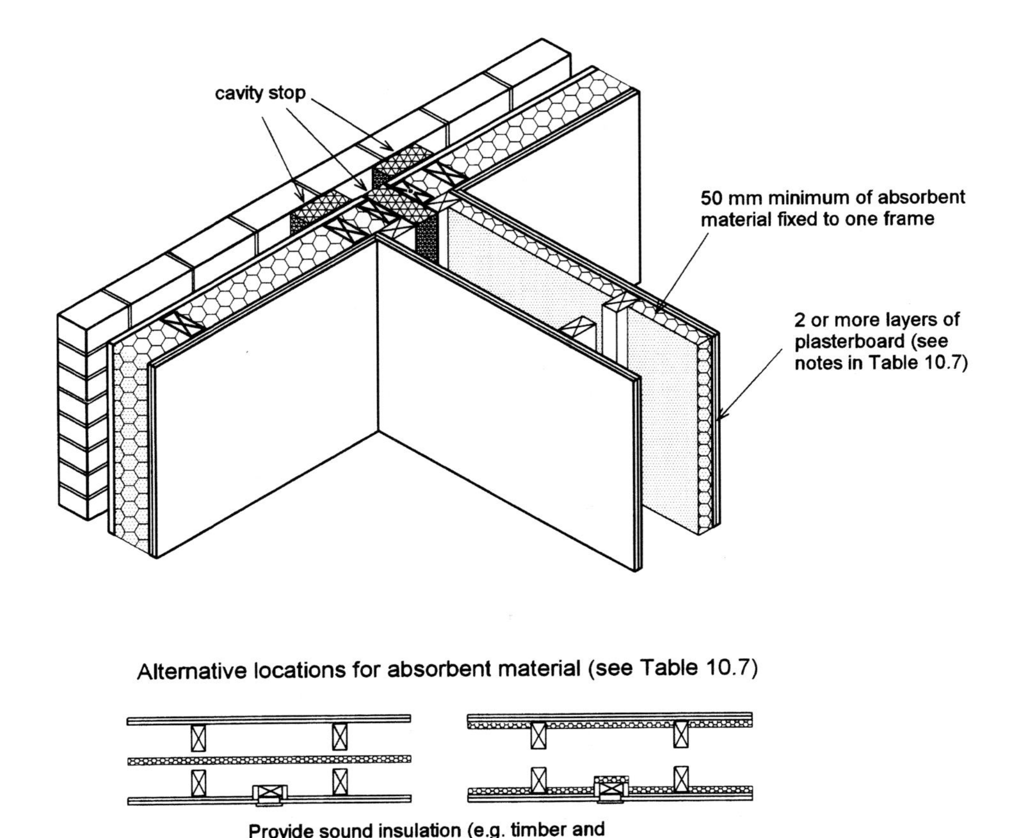

Fig. 10.13 Details of timber framed separating wall with absorbent infill.

Junctions with internal concrete floors AD E: 2.155

No guidance is provided in the approved document. Designers are advised to seek specialist advice.

Junctions with timber ground floors AD E: 2.156

In order to prevent flanking transmission, air paths into the cavity from floor voids must be blocked by solid timber blockings, continuous ring beams or joists.

Junctions with concrete ground floors AD E: 2.158

A concrete ground floor slab laid directly on the ground may be of continuous construction under a Type 4 separating wall whereas a suspended concrete ground floor may only pass beneath a Type 4 wall if the floor has a mass per unit area of not less than 365 kg/m^2.

Junctions with ceilings and roofs AD E: 2.160–2.163

- The separating wall should preferably continue to the underside of the roof, a flexible closer should be provided at its junction with the roof and the separating wall linings should be sealed with tape or caulked with appropriate sealant at their junction with the ceiling.
- If the roof or loft space is not a habitable room and the ceiling has sealed joints and a mass per unit area of at least 10kg/m^2, then within the roof space, either:
  (a) the lining on each of the two frames may be reduced to two layers of plasterboard, each sheet with a mass per unit area of at least 10 kg/m^2; or
  (b) the wall may be reduced to one frame with two layers of plasterboard, each sheet with a mass per unit area of at least 10 kg/m^2, on each side of the frame. In this case, the cavity must be closed at ceiling level in such a way that the two frames are not rigidly connected together.
- Cavities in external walls should be closed at their eaves. A flexible material should be used for the closure with no rigid joint between the two leaves. It is important to avoid a rigid connection between the two leaves and if a rigid material is used it should only be rigidly attached to one leaf, see BRE BR 262, *Thermal insulation: avoiding risks*, section 2.3.

Junctions with separating floors AD E: 2.164

See section 3 of the AD (section 10.5 of this chapter) for details of junctions between separating floors and Type 4 separating walls.

10.4.6 Walls separating habitable rooms from other parts of a building AD E: 2.25–2.28

Buildings containing flats and rooms for residential purposes often contain non-habitable spaces juxtaposed to those which are occupied. The approved document stipulates the sound insulation provided by the walls of refuse chutes in terms of mass, including finishes, per unit area. If the refuse chute is separated from a habitable room or kitchen its mass per unit area should be 1320kg/m^2 or more, whereas if it is a non-habitable room the mass per unit area may be reduced to 220kg/m^2.

Another common source of noise in multi-occupancy buildings is that transmitted from corridors. The AD states that in order to control noise from this source, walls between corridors and flats should be constructed to the standard used for separating walls, see sections 10.4.2 to 10.4.5. A weak point in the sound insulation provided from corridor noise is that transmitted through doors. For this reason, the document states that doors to corridors should have good sealing around their perimeters (including, if practical, their thresholds), and a mass per unit area of at least 25kg/m^2, or a weighted sound reduction index of at least 29 dB. The term sound reduction index is described in section 10.13 and the relevant measurement standards are listed in section 10.11.4 (AD E B3.9). In 'noisy' parts of a building, noise should be contained by a lobby, two doors in series or a high performance

Table 10.8 Construction procedures relating to the sound insulation of separating walls.

Separating Wall Type	Correct Procedure	Procedures which must be avoided
1, 2, 3, 4	Control flanking transmission between separating walls and connected walls and floors as described in the text Ensure external cavity walls are stopped with flexible closers at junctions with separating wall. See text	
1, 2, 3	Fill and seal all masonry joints with mortar Ensure flue blocks will not adversely affect sound insulation. Ensure that a suitable finish is used over flue blocks (see BS 1289-1:1986 and obtain manufacturer's advice)	
1, 2, 4	Stagger the positions of sockets on opposite sides of separating walls. In the case of Type 4 walls, use a similar thickness of cladding material behind socket boxes	In the case of Type 1 and 2 walls, do not use deep sockets and chases in the wall and do not locate them back to back In the case of Type 4 walls do not locate sockets back to back or chase plasterboard and a minimum of 150 mm edge to edge stagger is recommended between sockets
1, 2		Do not create a junction between Type 1 and 2 walls in which the cavity is bridged by the solid wall Do not attempt to convert a Type 2 separating wall into a Type 1 wall by inserting mortar or concrete into the cavity
1	Lay bricks 'frog up' Use bricks and blocks which extend to the full thickness of the wall	
2	Keep cavity leaves separate below ground floor level	Do not change to a solid wall in the roof space. A rigid connection between the leaves will adversely affect performance Do not build a cavity wall off a continuous concrete floor slab
3	Fix panels or supporting frames to ceiling and floor only Tape and seal all joints	Do not fix, tie or otherwise connect the free standing panels to the masonry core wall
4	Ensure that fire stops in the cavity between frames are either flexible or fixed to one frame only Ensure that each layer of plasterboard is independently fixed to the frame	If the two leaves have to be connected for structural purposes, do not use ties of greater cross section than 40 mm by 3 mm. Fix them to the studwork at or just below ceiling level. Do not set the ties at closer than 1.2 m centres

door set and if this is impossible, flats in the vicinity should be provided with this type of protection at their entrances. It is very important when considering the sound insulation of doors to bear in mind also the requirements of Building Regulations Part B – Fire safety and Part M – Access and facilities for disabled people.

10.4.7 Construction procedures

The AD provides information regarding correct and incorrect construction procedures which will influence the sound insulation offered by separating walls. This is summarised in Table 10.8.

10.5 Separating floors and their flanking constructions – new buildings (Approved Document E, Section 3: Separating floors and associated flanking constructions for new buildings)

10.5.1 Construction forms AD E: 3.1–3.7 & 3.9

Section 3 of the Approved Document E gives examples of floors which should achieve the performance standards for houses and flats tabulated in section 0: Performance – Table 1a of the AD and in section 10.2 of this chapter. However, as is stated in the document, the actual performance of a floor will depend on the quality of its construction. The information in this section of the AD is provided only for guidance; it is not intended to be exhaustive and alternative constructions may achieve the required standard. Designers are also advised to seek advice from other sources such as manufacturers regarding alternative designs, materials or products. The details in no way override the requirement to undergo pre-completion testing. For the provision of floors in 'rooms for residential purposes', see section 10.8 of this chapter.

The separating floor constructions described in the AD are divided into three types:

- **Type 1 Concrete base with ceiling and soft floor covering**
- **Type 2 Concrete base with ceiling and floating floor**
  Floors of this type require one of three types of floating floor. These are defined by the terms (a), (b) and (c).
- **Type 3 Timber frame base with ceiling and platform floor**

The three floor types are shown in Fig. 10.14 and within each floor type, alternatives are presented in a ranking order such that the one which should offer the best sound insulation is described first.

The following notes should be borne in mind when considering the alternatives which are presented.

- The mass per unit area of floors is quoted in kilograms per square metre (kg/m^2) and this may be obtained from manufacturers' data. Alternatively it may be obtained using a procedure described in Annex A of the AD, and outlined in

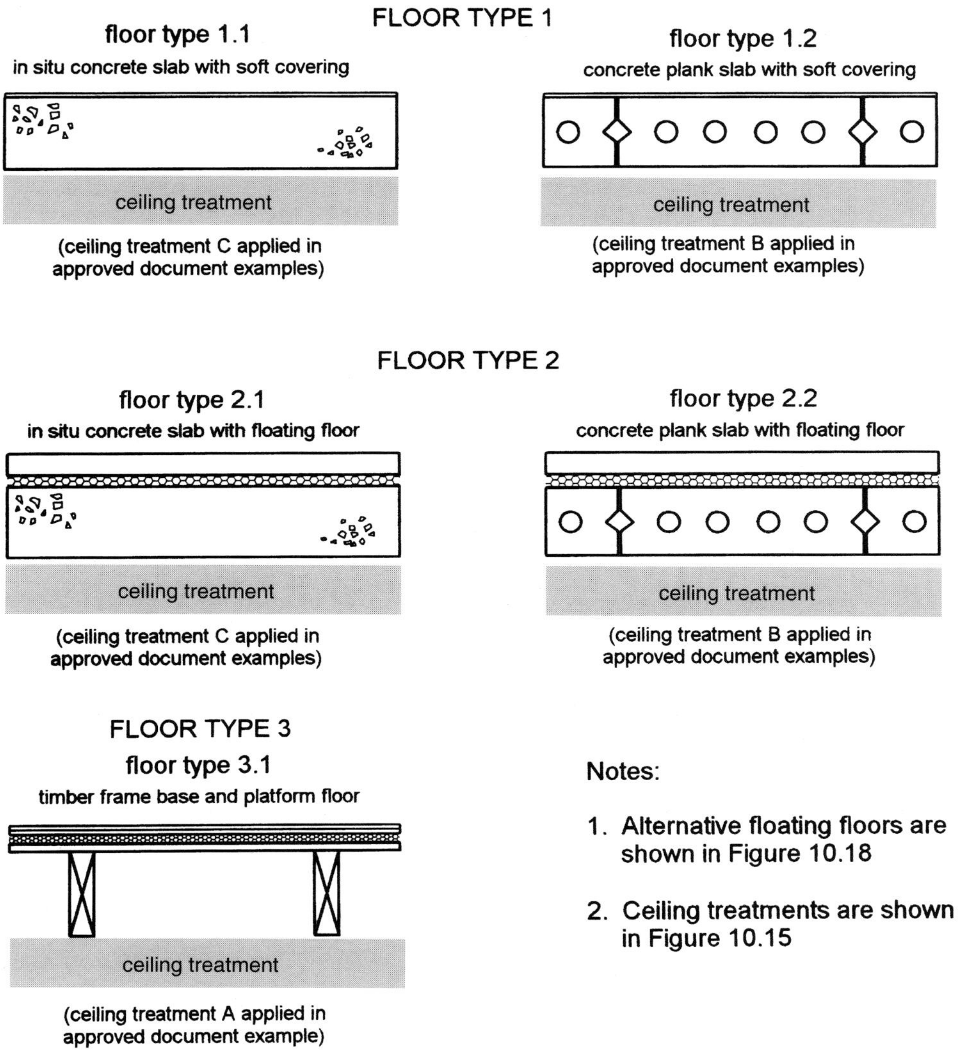

FLOOR TYPE 1

floor type 1.1
in situ concrete slab with soft covering
ceiling treatment
(ceiling treatment C applied in approved document examples)

floor type 1.2
concrete plank slab with soft covering
ceiling treatment
(ceiling treatment B applied in approved document examples)

FLOOR TYPE 2

floor type 2.1
in situ concrete slab with floating floor
ceiling treatment
(ceiling treatment C applied in approved document examples)

floor type 2.2
concrete plank slab with floating floor
ceiling treatment
(ceiling treatment B applied in approved document examples)

FLOOR TYPE 3

floor type 3.1
timber frame base and platform floor
ceiling treatment
(ceiling treatment A applied in approved document example)

Notes:

1. Alternative floating floors are shown in Figure 10.18
2. Ceiling treatments are shown in Figure 10.15

Fig. 10.14 Alternative types of floor construction.

section 10.12 of this chapter. Densities of materials are quoted in kilograms per cubic metre (kg/m^3).

- Acoustically, the mass of a bonded screed acts with the slab on to which it is laid and therefore, where appropriate, may be taken into account when calculating the mass per square metre of a floor, but the mass of a floating screed does not, and therefore must not be taken into account.
- Solid in situ concrete and hollow plank floors are considered in the guidance notes but beam and block floors are not. It is suggested that designers take advice from manufacturers regarding the sound insulation properties of this latter type of floor.

Floor penetrations AD E: 3.41–3.43 & 3.79–3.82 & 3.117–3.120

Particular attention must be given to pipes or ducts which pass through separating floors. The points in the AD are summarised as follows.

- Pipes and ducts penetrating a floor separating habitable rooms in different dwelling places should be enclosed for their full height, i.e. for the full height of the room if they pass from floor to ceiling, with material which has a mass per unit area of at least 15 kg/m^2 and either: the pipe or duct should be wrapped or the enclosure lined with 25 mm of unfaced mineral fibre.
- If the floor incorporates a floating layer, a gap of approximately 5 mm should be left between this and the enclosure with the gap filled with sealant or neoprene. The enclosure may extend down to the floor base but if so it must be isolated from the floating layer.
- There are fire safety implications where separating floors are penetrated and fire protection to satisfy the requirements of Building Regulation Part B must be provided. Flexible fire stopping should be used and, in order to prevent structure-borne transmission, it is important that there is no rigid contact between pipes and floor.
- Ducts containing gas pipes must be ventilated at each floor. Gas pipes may be housed in separate ventilated ducts or, alternatively, gas pipes need not be enclosed. It is essential that relevant codes and standards are complied with to ensure gas safety. Reference: The Gas Safety (Installation and Use) Regulations 1998, SI 1998 No. 2451.

Ceiling construction AD E: 3.17–3.22

The type of ceiling provided has a significant bearing on the acoustic performance of a floor and so, since alternative ceiling treatments may be combined with each type of floor, the three ceiling treatments described in the AD are defined by the letter A, B or C and each floor type is qualified by the letter A, B or C depending on its ceiling treatment. The AD states that the ceiling treatments, as described in Table 10.9 and shown in Fig. 10.15, are ranked such that A should provide the highest level of sound insulation and, further, that if a ceiling treatment of a higher ranking than one applied in a guidance example is used, then providing that there is no significant flanking transmission, the result should be improved sound insulation from the complete floor.

Junctions and flanking transmission AD E: 3.10

It is very important that the junctions between a separating floor and the elements to which it is attached are carefully designed because this will have a significant influence on its overall effectiveness. Since the level of sound insulation achieved will be strongly dependent on the extent to which flanking transmission exists, considerable advice is given in the AD on this issue. Guidance is provided on the junctions between separating floors and the following other elements:

Table 10.9 Ceiling treatments for separating wall constructions.

Ceiling treatment	Form	Specification
A	Independent ceiling with absorbent material	At least two layers of plasterboard with staggered joints and total mass per unit area of at least 20 kg/m^2 Mineral wool laid in cavity above ceiling to provide sound absorption. Minimum thickness 100 mm, minimum density 10 kg/m^3 See note below regarding fixings
B	Plasterboard on proprietary resilient bars with absorbent material to fill ceiling void	Single layer of plasterboard with mass per unit area of at least 10 kg/m^2 Plasterboard fixed using proprietary resilient bar. For concrete floor, fix bar to timber battens. Bar should be fixed in accordance with manufacturers instructions Fill ceiling void with mineral wool with density of at least 10 kg/m^3
C	Plasterboard on timber battens, or plasterboard on resilient channels with absorbent material to fill ceiling void	Single layer of plasterboard with mass per unit area of at least 10 kg/m^2 Plasterboard should be fixed using proprietary resilient channels or timber battens Fill ceiling void with mineral wool with density of at least 10 kg/m^3 if resilient channels are used

Notes:
(a) If using ceiling treatment A with floor types 1, 2 the ceiling should be attached to independent joists supported by surrounding walls with a clearance of a minimum of 100 mm between top of ceiling plasterboard and underside of base floor. If using floor type 3, the ceiling should be attached to independent joists supported by surrounding walls with extra support from resilient hangers attached directly to the floor. In this case there should be a clearance of a minimum of 100 mm between the top of the ceiling joists and underside of base floor.
(b) Light fittings recessed into ceilings may reduce sound insulation.
(c) See BRE BR 262, *Thermal Insulation: Avoiding Risks*, section 2.4, regarding heat emission from electrical cables which may be covered by absorbent material.

- external cavity walls with masonry inner leaf
- external cavity wall with timber framed inner leaf
- solid masonry external walls
- internal walls – masonry
- internal walls – timber
- floor penetrations (see above)
- separating walls Type 1, 2, 3 and 4 (these being relevant to the design of flats).

However, not all floor/junction combinations are considered. It is stated in the AD that if any element has a separating function, e.g. a ground floor which is also a separating floor for a basement flat, then the separating element requirements take precedence.

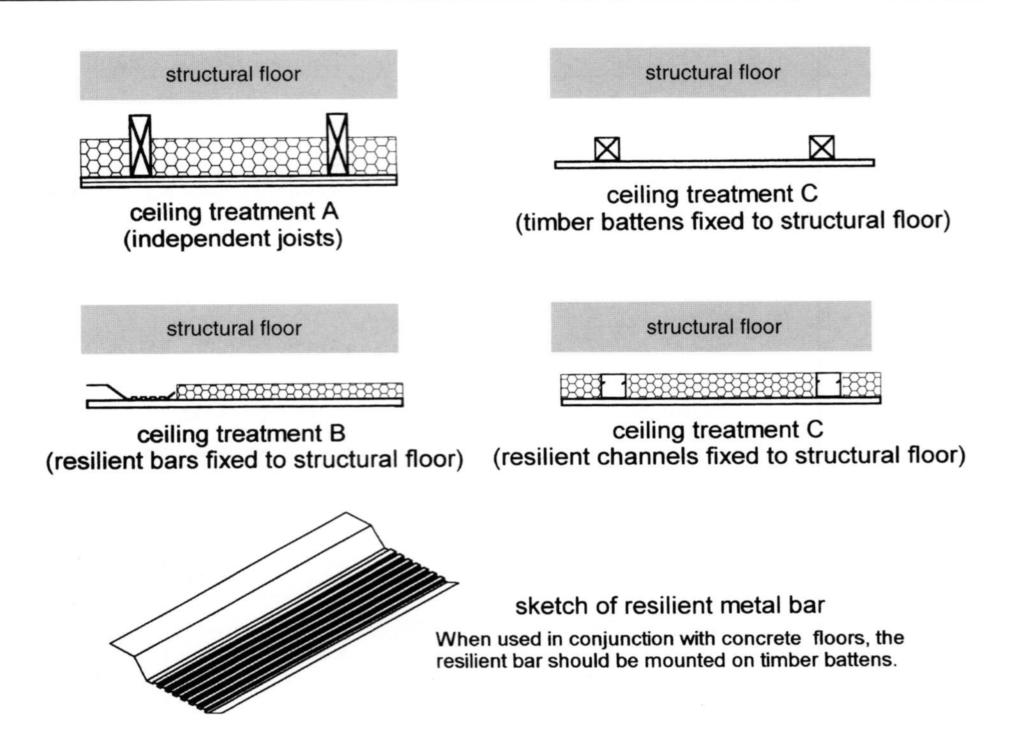

Fig. 10.15 Alternative ceiling treatments.

10.5.2 Design of floor Type 1 – concrete base with ceiling and soft floor covering AD E: 3.23–3.30

It is the large mass of the concrete slab together with that of the ceiling which provides airborne sound insulation from this type of floor. For impact sound insulation it relies on the soft covering preventing the contact of hard surfaces.

To be acceptable, a soft covering should be a resilient material, or have a resilient base, with a thickness, when uncompressed, of 4.5 mm or more. An alternative approach is to measure the 'weighted reduction of impact sound pressure level', ΔL_w, of the floor covering in accordance with the procedure described in Appendix B3 of the approved document, see section 10.11.4 of this chapter (Floor coverings and floating floors). To be acceptable the measured value of ΔL_w should be 17dB or more.

Two variations of floor Type 1 are described in Approved Document E, which according to the document should, if built correctly, comply with Requirement E1, see Table 10.10. The first floor has ceiling treatment C and hence is referenced 1.1C, whereas the second has ceiling treatment B and hence is referenced 1.2B.

In order for the performance of Type 1 floors to be satisfactory, the junctions between them and their surrounding elements of constructions must be designed correctly. The requirements detailed in the approved document are explained below. See also section 10.5.5.

Table 10.10 Examples of concrete base with ceiling and soft covering floor constructions.

	Floor Type	
	1.1C	**1.2B**
Description	Solid concrete slab, cast in situ with or without permanent shuttering. Soft floor covering and ceiling treatment C	Hollow or solid concrete planks. Soft floor covering and ceiling treatment B
Minimum mass per unit area of concrete base	365 kg/m^2 including any bonded screed. Permanent shuttering of solid concrete or metal may be also be included	365 kg/m^2 including any bonded screed
Floor covering	Soft floor covering. This is essential	Soft floor covering. This is essential
Ceiling treatment	C or better is essential	B or better is essential
Other requirements		Use regulating floor screed All joints between and around planks to be fully grouted to ensure complete air tightness

Junctions with external cavity walls which have masonry inner leaves AD E: 3.31–3.35

- There are no restrictions on the outer leaf of the wall.
- Unless the cavity is fully filled with mineral wool, expanded polystyrene beads or some other suitable insulating material (the AD states that manufacturers' advice should be sought regarding alternative suitable materials), the cavity should be stopped with a flexible closer as shown in Fig. 10.16. The flexible closer must be protected from the effects of moisture and it is essential that adequate drainage is provided (note the provision of a flexible cavity tray, or an equivalent, as shown in Fig. 10.16).
- The mass per unit area of the inner leaf of the external wall should be at least 120 kg/m^2 excluding its finish in order to reduce the effects of flanking transmission.
- The floor base, but not its screed, should continue to the face of the cavity without bridging the cavity and if floor type 1.2B is used, with the planks laid parallel to the wall, the first joint between planks should be 300 mm or more from the cavity face, as shown in Fig. 10.16.
- The use of wall ties in external masonry walls is considered in Section 2 of the approved document. See section 10.4.1 (wall ties) of this chapter.

Junctions with external cavity walls which have timber framed inner leaves AD E: 3.36

- There are no restrictions on the outer leaf of the wall.
- It is assumed that the cavity will not be filled and so the cavity should be stopped with a flexible closer.

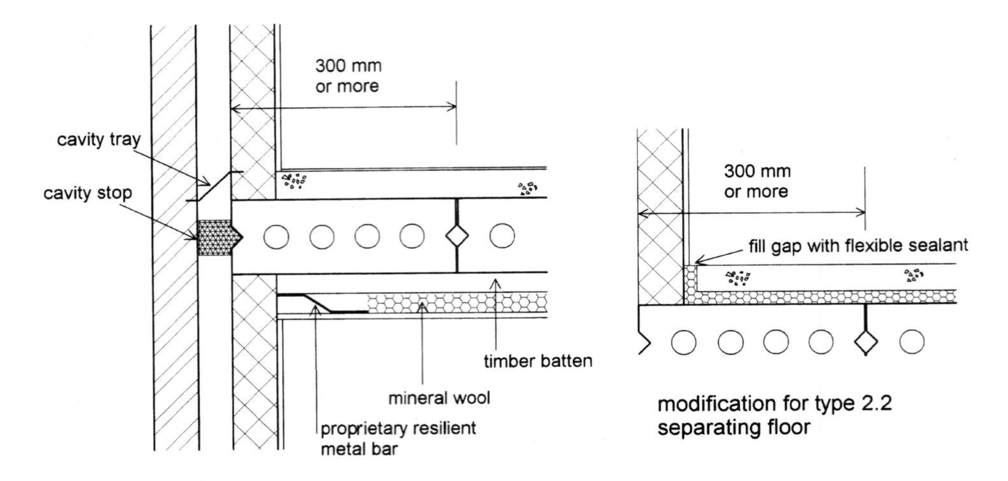

Fig. 10.16 Details of junctions between Type 1 and 2 separating floors and external cavity walls.

- The finish of the inner leaf should consist of two layers of plasterboard each with a mass per unit area of at least 10 kg/m^2 with all joints sealed with tape or caulked with appropriate sealant.

Junctions with solid external masonry walls AD E: 3.37

No guidance is provided in the AD. Designers are advised to seek specialist advice.

Junctions with internal framed walls AD E: 3.38

No restrictions are imposed.

Junctions with internal masonry walls AD E: 3.39–3.40

The floor base should be continuous through or above such walls and any internal load-bearing wall or any other wall rigidly connected to a separating floor should have a mass per unit area of at least 120 kg/m^2 excluding finish.

Junctions with solid masonry (Type 1) separating walls AD E: 3.44–3.45

Floors bases of Type 1.1C should pass through Type 1 separating walls. Floor bases of Type 1.2B should not be continuous through Type 1 separating walls. In neither case should any screed penetrate the separating wall. See Fig. 10.17 for examples and requirements of both types of construction.

Junctions with cavity masonry (Type 2) separating walls AD E: 3.46–3.48

- The mass per unit area of any leaf, excluding finish, that supports or adjoins the floor should be 120 kg/m^2 or more.
- The floor base, but not its screed, should continue to the face of the cavity without bridging the cavity and if floor Type 1.2B is used, with the planks laid parallel to the wall, the first joint between planks should be 300 mm or more from the cavity face as shown in Fig. 10.17.

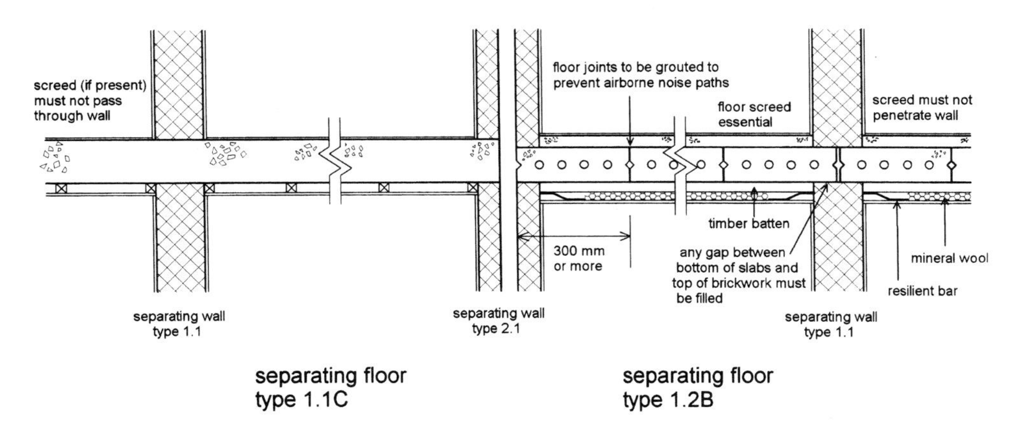

Fig. 10.17 Details of junctions between Type 1 separating floors and Type 1 and 2 separating walls.

Junctions with masonry between independent panels (Type 3) separating walls AD E: 3.49–3.54

Assuming the separating wall has a solid core, i.e. wall Types 3.1 and 3.2:

- Floor bases of Type 1.1C should pass through the wall. Floor bases of Type 1.2B should not be continuous through this type of separating wall. In neither case should any screed penetrate the separating wall. Construction should be similar to that shown in Fig. 10.17 for solid masonry separating walls.
- If floor Type 1.2B is used in conjunction with wall Type 3.2, with the planks laid parallel to the wall, the first joint between planks should be 300 mm or more from the centreline of the masonry core.

Assuming the separating wall has a cavity core, i.e. wall Type 3.3:

- The mass per unit area of any leaf that is supporting or adjoining the floor, excluding finish, should be 120 kg/m^2 or more.
- The floor base, but not its screed should continue to the face of the cavity without bridging the cavity and if floor Type 1.2B is used, with the planks laid parallel to the wall, the first joint between planks should be 300 mm or more from the cavity face of the adjacent leaf of the masonry core. Construction

should be similar to that shown in Fig. 10.17 for cavity masonry separating walls.

Junctions with framed with absorbent material (Type 4) separating walls AD E: 3.55

No guidance is provided in the AD. Designers are advised to seek specialist advice.

10.5.3 Design of floor Type 2 – concrete base with ceiling and floating floor AD E: 3.56–3.60

The complete floor construction consists of a concrete floor base on top of which is a resilient layer. On top of the resilient layer there is a solid 'floating' layer and below the concrete floor base is a ceiling which, as has previously been defined, may be classified as Type A, B or C.

The sound insulation offered by floors of this type depends on the mass of the concrete base, the floating layer and the ceiling. It is also improved by the isolation of the floating layer and the ceiling. Impact sound is reduced at source by the floating layer. However, even if resistance to airborne sound only is required, the full construction should still be used.

Floating floors AD E: 3.62–3.66

Examples of three types of floating floors are presented in the AD. These are described in Table 10.11 and shown in Fig. 10.18.

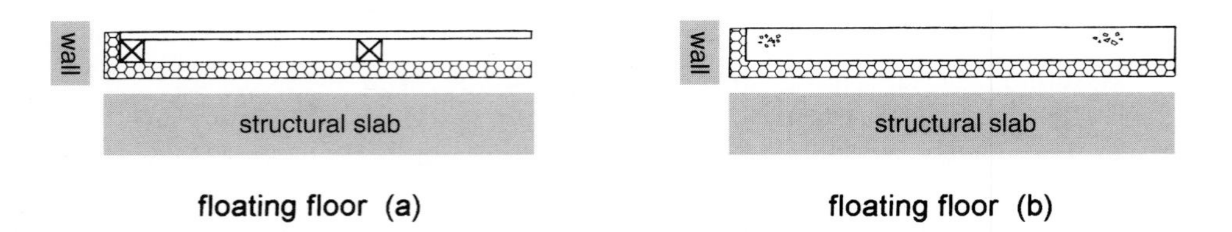

Fig. 10.18 Alternative floating floor constructions.

Floor, floating floor, ceiling combinations AD E: 3.67–3.68

Two variations of floor Type 2 are described in Approved Document E, which according to the document should, if built correctly, comply with Requirement E1, see Table 10.12 .The first floor has ceiling treatment C and hence is referenced 2.1C, whereas the second has ceiling treatment B and hence is referenced 2.2B.

When considering the alternatives in Table 10.12, it should be borne in mind that in the approved document, alternatives are ranked such that the one which should give the best performance appears first. Also, the ceiling treatment specified is that which is required to meet the requirements of the AD. Since, as already stated, the ceilings are ranked in descending order of performance from A to C, substitution of

Table 10.11 Alternative types of floating floors as described in the approved document.

	Floating Floor Type		
	(a)	(b)	(c)
Description	Timber raft floating layer supported on a resilient layer	Sand and cement screed floating layer supported on a resilient layer	A floating floor designed to satisfy a performance criterion
Floating layer	Raft of board material with bonded edges, such as t. & g. timber with mass per unit area of not less than 12 kg/m^2, fixed to 45 mm × 45 mm battens Raft laid on resilient layer without any fixings. Do not lay battens along joints in resilient layer	65 mm of sand and cement or suitable proprietary screed product Screed to have mass per unit area of 80 kg/m^2 or more Resilient layer must be protected while screed is laid, e.g. by using a 20–50 mm wire mesh	The floating floor should consist of a rigid board and under layer which provides a weighted reduction in impact sound pressure level, ΔL_w, of 29 dB or more ΔL_w is the improvement in impact sound insulation obtained in a laboratory by installing a floating floor over a test floor
Resilient layer	Mineral wool with density of 36 kg/m^3 and thickness of at least 25 mm. The layer of mineral wool may have a paper faced under side	Mineral wool with density of 36 kg/m^3 and thickness of at least 25 mm. The layer of mineral wool to be paper faced on upper side to prevent wet screed entering the resilient layer, or, a layer meeting the specific dynamic stiffness criterion of 15MN/m^3 and min thickness of 5mm under specified load. See BS 29052-1:1992	Laboratory measurement should be in accordance with the procedure described in section 10.11.4

Designers are advised to take advice from manufacturers on proprietary screed products and the performance and installation of proprietary floating floors.

Type A or B ceilings in Floor 2.1 or a Type A ceiling in Floor 2.2 should give improved sound insulation.

In order for the performance of Type 2 floors to be satisfactory, the junctions between them and their surrounding elements of constructions must be designed correctly. The requirements detailed in the approved document are explained below. See also section 10.5.5.

Junctions with external cavity walls which have masonry inner leaves AD E: 3.69–3.73

- There are no restrictions on the outer leaf of the wall.
- Unless the cavity is fully filled with mineral wool, expanded polystyrene beads or some other suitable insulating material (the AD states that manufacturers' advice should be sought regarding alternative suitable materials), the cavity

Table 10.12 Examples of concrete base with ceiling and floating floor constructions.

	Floor Type	
	2.1C	**2.2B**
Description	Floating floor Solid concrete slab, cast in situ with or without permanent shuttering Ceiling treatment C	Floating floor Hollow or solid concrete planks Ceiling treatment B
Minimum mass per unit area of concrete base	300 kg/m² including any bonded screed. Permanent shuttering of solid concrete or metal may be also be included	300 kg/m² including any bonded screed
Floor covering	Floating floor (a), (b) or (c) is essential	Floating floor (a), (b) or (c) is essential
Ceiling treatment	C or better is essential	B or better is essential
Other requirements	Regulating floor screed is optional	Use regulating floor screed All joints between and around planks to be fully grouted to ensure complete air tightness

should be stopped with a flexible closer as shown in Fig. 10.16. The flexible closer must be protected from the effects of moisture and it is essential that adequate drainage is provided (note the provision of a flexible cavity tray, or an equivalent, as shown in Fig. 10.16).

- The mass per unit area of the inner leaf of the external wall should be at least 120 kg/m² excluding its finish in order to reduce the effects of flanking transmission.
- The floor base, but not its screed, should continue to the face of the cavity without bridging the cavity and if floor Type 2.2B is used, with the planks laid parallel to the wall, the first joint between planks should be 300 mm or more from the cavity face, as shown in Fig. 10.16.
- Treatment of the intersection of the floating layer and the wall should be as shown in Figs. 10.16 and 10.19.
- The use of wall ties in external masonry walls is considered in section 2 of the approved document. See section 10.4.1 of this chapter.

Junctions with external cavity walls which have timber framed inner leaves AD E: 3.74

- There are no restrictions on the outer leaf of the wall.
- It is assumed that the cavity will not be filled and so the cavity should be stopped with a flexible closer.
- The finish of the inner leaf should consist of two layers of plasterboard each with a mass per unit area of at least 10 kg/m² with all joints sealed with tape or caulked with appropriate sealant.

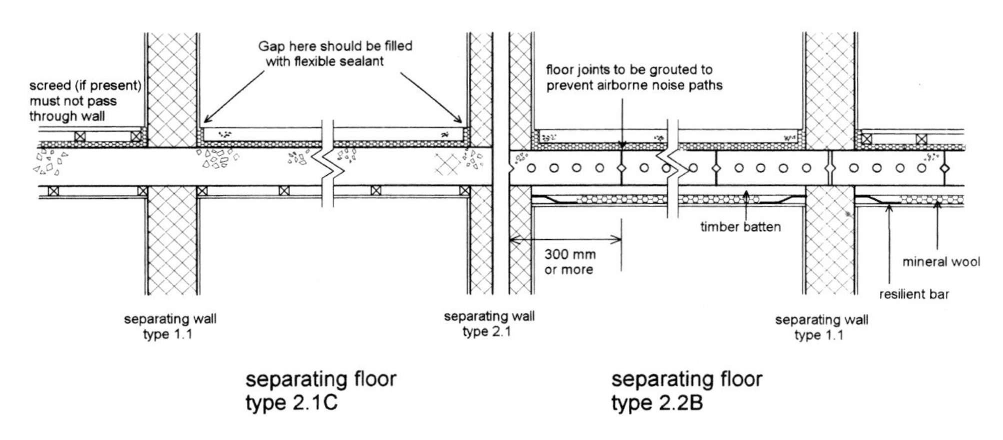

Fig. 10.19 Details of functions between Type 2 separating floors and Type 1 and 2 separating walls.

Junctions with solid external masonry walls AD E: 3.75

No guidance is provided in the AD. Designers are advised to seek specialist advice.

Junctions with internal framed walls AD E: 3.76

No restrictions are imposed.

Junctions with internal masonry walls AD E: 3.77–3.78

The floor base should be continuous through or above such walls and any internal load bearing wall or any other wall rigidly connected to a separating floor should have a mass per unit area of at least 120 kg/m^2 excluding finish.

Junctions with solid masonry (Type 1) separating walls AD E: 3.83–3.84

Floors bases of Type 2.1C should pass through Type 1 separating walls. Floor bases of Type 2.2B should not be continuous through Type 1 separating walls. In neither case should any screed penetrate the separating wall. See Fig. 10.19 for examples and requirements of both types of construction.

Junctions with cavity masonry (Type 2) separating walls AD E: 3.85–3.86

The floor base, but not its screed, should continue to the face of the cavity without bridging the cavity and if floor Type 2.2B is used, with the planks laid parallel to the wall, the first joint between planks should be 300 mm or more from the cavity face as shown in Fig. 10.19.

Junctions with masonry between independent panels (Type 3) separating walls AD E: 3.87–3.92

Assuming the separating wall has a solid core, i.e. wall Types 3.1 and 3.2:

- Floor bases of Type 2.1C should pass through the wall. Floor bases of Type 2.2B should not be continuous through this type of separating wall. In neither case should any screed penetrate the separating wall. Construction should be as shown in Fig. 10.20.
- If floor Type 2.2B is used in conjunction with wall Type 3.2, with the planks laid parallel to the wall, the first joint between planks should be 300 mm or more from the centreline of the masonry core.

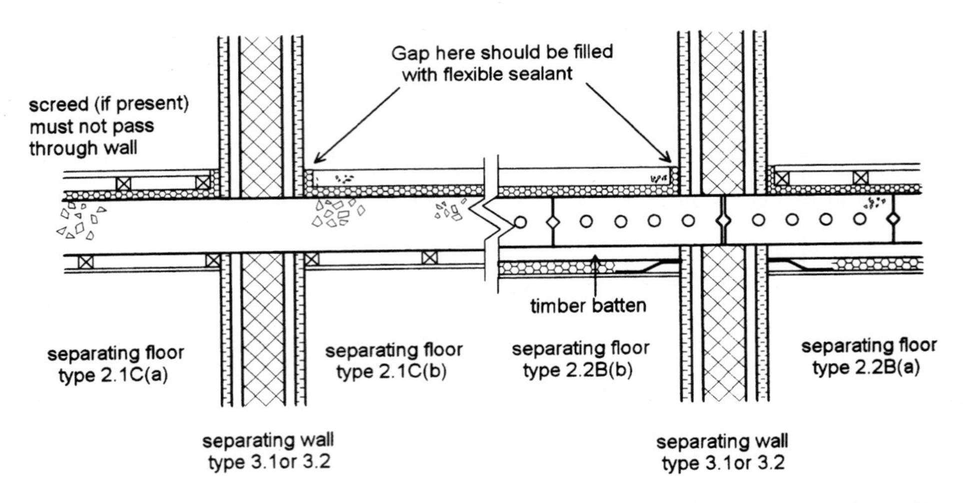

Fig. 10.20 Details of junctions between Type 2 separating floors and Type 3 separating walls.

Assuming the separating wall has a cavity core, i.e. wall Type 3.3:

- The mass per unit area of any leaf that is supporting or adjoining the floor, excluding finish, should be 120 kg/m^2 or more.
- The floor base, but not its screed should continue to the face of the cavity without bridging the cavity and if floor type 2.2B is used, with the planks laid parallel to the wall, the first joint between planks should be 300 mm or more from the cavity face of the adjacent leaf of the masonry core.

Junctions framed with absorbent material (Type 4) separating walls AD E: 3.93

No guidance is provided in the AD. Designers are advised to seek specialist advice.

10.5.4 Design of floor Type 3 – timber frame base and platform floor with ceiling treatment AD E: 3.94–3.102

Floors of this type consist of a structural timber base comprising of boarding supported on timber joists upon which is a floating layer resting on a resilient layer. The platform floor consists of the floating layer and the resilient layer. Beneath the structural floor is a Type A ceiling treatment. The construction form is shown in Fig. 10.21. In order to obtain good insulation against both airborne and impact sound transmission it is essential that there is good acoustic isolation between the platform floor and the structural base, and between the structural base and the ceiling. The platform floor is important because it reduces the effects of impact noise at source. However, even if resistance to airborne sound only is required, the full construction should still be used.

There are fewer variations of this floor than there are of floor Types 1 and 2 and only one construction form is described in the AD. According to the document this should, if built correctly, comply with Requirement E1. Since this example, which is described in Table 10.13 and shown in Fig. 10.21, utilises ceiling treatment A it is referred to as separating floor Type 3.1A.

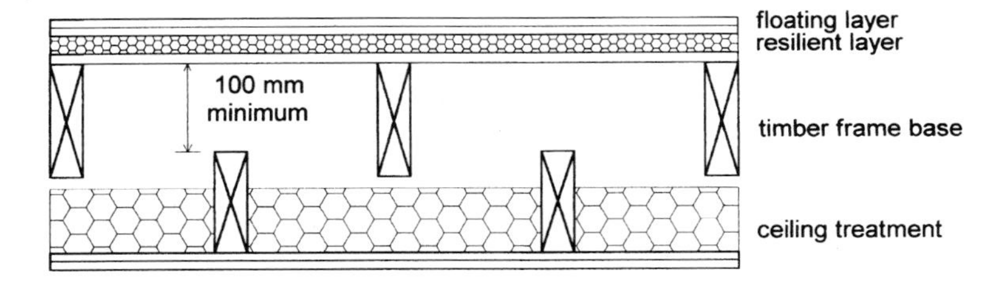

Fig. 10.21 Construction form of floor Type 3.1A.

In order for the performance of Type 3 floors to be satisfactory, the junctions between them and their surrounding elements of constructions must be designed correctly. The requirements detailed in the approved document are explained below. See also section 10.5.5.

Junctions with external cavity walls which have masonry inner leaves AD E: 3.103–3.108

- There are no restrictions on the outer leaf of the wall.
- Unless the cavity is fully filled with mineral wool, expanded polystyrene beads or some other suitable insulating material (the approved document states that manufacturers' advice should be sought regarding alternative suitable materials), the cavity should be stopped with a flexible closer.
- The AD states that it is necessary to line the internal face of the masonry wall with independent panels as is described for Type 3 separating walls However, if the mass per unit area of the inner leaf is greater than 375 kg/m^2, the independent panels need not be provided.

Table 10.13 Example of timber frame base with ceiling and platform floor.

	Floor Type 3.1A
Description	Timber joists supporting a deck and platform floor (consisting of resilient layer and floating layer) together with ceiling treatment A beneath the structural floor
Structural floor	Timber joists selected to satisfy structural requirements together with a deck which has a minimum mass per unit area of 20 kg/m^2
Floating layer	A minimum of two layers of board each with a minimum thickness of 8 mm to provide a total mass per unit area of at least 25 kg/m^2. The layers should be fixed together ensuring staggered joints and laid loose on the resilient layer
	Two example platform floor constructions are provided in Approved Document E. These are as follows:
	1. 18 mm of timber or wood based board with tongued and grooved edges and glued joints. These should be spot bonded to a substrate of 19 mm thick plasterboard with staggered joints to give a total mass per unit area of at least 25 kg/m^2
	2. Two layers of cement bonded particle board glued and screwed together with staggered joints. The resulting platform must have a total thickness of 24 mm and a mass per unit area of at least 25 kg/m^2
Resilient layer	Mineral wool, which may be paper faced on its underside, with a thickness of at least 25 mm and density in the range 60 to 100 kg/m^3
	If the material chosen is towards the lower end of the above range, impact sound insulation is improved but it may result in what is described in the document as a 'soft' floor, i.e. one which may be considered to respond excessively to fluctuating loads. It is suggested in the document that this may be overcome by providing additional support via a timber batten which is fixed to the walls and has a foam strip along its top
Ceiling treatment	Ceiling treatment A must be used in conjunction with this floor

- The ceiling, which consists of independent joists supporting plasterboard, must continue to the masonry of the inner leaf. However, where the ceiling passes above the independent panels it should be sealed with tape or caulked with appropriate sealant.
- The method of connecting the floor base, i.e. the floor joists, to the external wall is not prescribed but it is emphasised that there must be no air paths connecting the floor to the wall cavity. This would pose a source of flanking transmission which it would be difficult to remedy at a later date.
- The use of wall ties in external masonry walls is considered in section 2 of the AD. See section 10.4.1 of this chapter.

Junctions with external cavity walls which have timber framed inner leaves AD E: 3.109–3.112

- There are no restrictions on the outer leaf of the wall.
- It is assumed that the cavity will not be filled and so the cavity should be stopped with a flexible closer.

- The finish of the inner leaf should consist of two layers of plasterboard each with a mass per unit area of at least 10 kg/m^2 with all joints sealed with tape or caulked with appropriate sealant.
- The method of connecting the floor base, i.e. the floor joists, to the external wall is not prescribed but where floor joists are perpendicular to the wall, the spaces between the floor joists should be sealed to the full depth of the floor with timber blocking.
- The junction between the ceiling and wall lining should be sealed with tape or caulked with appropriate sealant.

Junctions with solid external masonry walls AD E: 3.113

No guidance is provided in the AD. Designers are advised to seek specialist advice.

Junctions with internal framed walls AD E: 3.114–3.115

- Where floor joists are perpendicular to the internal framed wall, the spaces between the floor joists should be sealed to the full depth of the floor with timber blocking.
- The junction between the ceiling and the wall should be sealed with tape or caulked with appropriate sealant.

Junctions with internal masonry walls AD E: 3.116

No guidance is provided in the AD. Designers are advised to seek specialist advice.

Junctions with solid masonry (Type 1) separating walls AD E: 3.121–3.122

- When floor joists are to be supported on this type of wall, they should not be built into the wall but must be supported on joist hangers.
- The junction between the ceiling and the wall should be sealed with tape or caulked with appropriate sealant.

Junctions with cavity masonry (Type 2) separating walls AD E: 3.123–3.126

- When floor joists are to be supported on this type of wall, they should not be built into the wall but must be supported on joist hangers.
- The AD states that it is necessary to line the leaf of the cavity wall which is nearest to the room in question with independent panels as is described for Type 3 separating walls. However, if the mass per unit area of the nearest leaf to the room in question is greater than 375 kg/m^2, the independent panels need not be provided.
- The ceiling, which consists of independent joists supporting plasterboard, should continue to the masonry of the wall. However, where the ceiling passes above the

Table 10.14 Construction procedures relating to the sound insulation of separating floors.

Element	Correct procedure	Procedures which must be avoided
Floor Types 1, 2 and 3	Seal the perimeter of independent ceilings with tape or sealant Give extra attention to workmanship and detailing at perimeter and where there are floor penetrations, to reduce flanking transmission and prevent air paths Ensure notes on junction construction are complied with to reduce flanking effects	Do not create a rigid or direct connection between an independent ceiling and its floor base
Floor Types 1 and 2	Fill all joints between floor components to avoid air paths between floors Build concrete separating floors (or floor bases) into all of their masonry perimeter walls Ensure that all gaps between heads of masonry walls and undersides of concrete floors are filled with mortar	Do not allow a floor base to bridge across a cavity in a masonry cavity wall
Floor Type 1	Fix or glue the soft covering to the floor	A soft floor covering must be used. Do not use ceramic floor tiles, wood blocks or any other non-resilient floor finish rigidly connected to the floor base
Floor Type 2	Leave small gap of size recommended by manufacturers between floating layer and wall at room edge and fill with flexible sealant Leave gap of about 5 mm between skirting and floating layer and fill with flexible sealant Lay rolls or sheets of resilient materials with lapped joints or tightly butted and taped joints Prevent screed entering resilient layers by laying fibrous materials with paper facing uppermost	Do not bridge between the floating layer and the base or surrounding walls, e.g. with services or fixings which penetrate the resilient layer between floating layer and floor base or walls Do not allow the floating layer to bridge to the floor base or walls, e.g. through a void in the resilient layer
Floor Type 3	With respect to the platform floor: (a) Ensure that the density of the resilient layer is correct and adequate to carry the applied load (b) Use a resilient material, e.g. expanded or extruded polystyrene strip around the perimeter to ensure that a gap is maintained between wall and floating layer during construction. The strip should be about 4 mm higher than top surface of the floating layer. The gap may be filled with a flexible sealant (c) Lay sheets of resilient materials with tightly butted and taped joints	Do not bridge between the floating layer and the timber frame base or surrounding walls, e.g. with services or fixings which penetrate the resilient layer between platform floor and timber base or walls

independent panels it should be sealed with tape or caulked with appropriate sealant.

Junctions with masonry between independent panels (Type 3) separating walls AD E: 3.127–3.128

- When floor joists are to be supported on this type of wall, they should not be built into the wall but must be supported on joist hangers.
- The ceiling, which is supported on independent joists, should pass through the independent panels to the masonry core. The junction where the ceiling passes over the independent panels should be sealed with tape or caulked with appropriate sealant.

Junctions framed with absorbent material (Type 4) separating walls AD E: 3.129–3.130

- Where the floor joists are perpendicular to the separating wall, the spaces between the floor joists should be sealed to the full depth of the floor with timber blocking.
- The junction between the ceiling and the wall lining should be sealed with tape or caulked with appropriate sealant.

10.5.5 Construction procedures

The AD provides information regarding correct and incorrect construction procedures which will influence the sound insulation offered by separating floors. This is summarised in Table 10.14.

10.6 Dwelling houses and flats formed by material change of use (Approved Document E, Section 4: Dwelling-houses and flats formed by material change of use)

10.6.1 General requirements AD E: 4.1–4.21

Section 4 of Approved Document E describes forms of treatment to walls, floors, stairs and their associated junctions which it may be appropriate to apply when converting existing buildings by material change of use to provide dwelling houses and flats. Rooms for residential purposes are considered in section 6 of the AD, see section 10.8 of this chapter. The point is made in the AD that existing components may already satisfy the requirements of Table 1a of the document, which are outlined in section 10.2, and it is suggested that this would be so if the construction of a component, its junction detailing and flanking construction was sufficiently similar to that of one of the example walls or floors presented for new buildings. Alternatively, the examples provided for the construction of new buildings may be used to provide guidance as to what work may be necessary to bring existing components into line with the required standard. The importance of the control of flanking transmission, see section 10.6.6, in instances of change of use is also emphasised in the AD.

The construction of an existing component may be such that a designer has little idea as to whether it will comply with the requirements of Table 1a of the AD. To help in this case, treatments to existing components are provided as follows:

- walls, one form of treatment
- floors, two forms of treatments
- stairs, one form of treatment.

The nature of each of these is explained below. It is important to note that the AD states that the example treatments 'can be used to increase sound insulation', but not that if constructed correctly, treated components will achieve the required standard. This is because the level of insulation reached will depend heavily on the existing construction. It is also stated in the AD that the information provided is only guidance and that other designs, materials or products may be used to achieve the required performance standard. Designers are recommended to seek advice from manufacturers and/or other expert sources.

It is suggested in the AD that, with respect to requirements for mass, for example, an existing component may be sufficiently similar to comply with the required performance requirements if its mass was within 15% of that of an equivalent component recommended for the construction of new buildings. This may be the case, but due to uncertainty about the workmanship and density of material used throughout the component, additional treatment may still be advisable.

In view of the complex nature of the construction forms resulting from conversion work, it may be necessary to seek specialist acoustic advice. The AD cites, as an example of a construction which will require special attention, the consequences of constructing a wall or floor across an existing continuous floor or wall such that the original floor or wall becomes a flanking element. The nature of the building may make the simple isolation of such flanking components technically difficult and in such circumstances and when significant additional loads are imposed on the building as a result of floor and wall treatment, structural advice should be sought.

The AD recommends that the following work should be undertaken to existing floors before the recommended treatments to improve sound insulation are applied:

- Since gaps in existing floorboards are a potential source of airborne sound transmission they should be covered with hardboard or filled with sealant.
- Replacement of floorboards. If required, these should be at least 12 mm thick and mineral wool with a density of not less than 10 kg/m^2 should be laid between joists to a depth of not less than 100 mm.
- The mass per unit area of concrete floors should be increased to at least 300 kg/m^2, air gaps within them should be sealed and a regulating screed provided if required.
- Existing lath and plaster ceilings should be retained as long as they comply with Building Regulation B – Fire Safety, whereas other ceilings should be modified

such that they consist of at least two layers of plasterboard with staggered joints and a total mass per unit area of 20 kg/m^2.

Sound transmission from corridors and, particularly, through doors in corridors is often a serious source of annoyance in dwellings formed by material change of use. It is suggested in the AD that the separating walls described in this section, i.e. section 4 of the AD, should be used to control sound insulation and flanking transmission between houses and flats formed by material change of use and corridors. Corridor doors should be constructed as corridor doors for new buildings, as described in section 10.4.6 of this chapter.

10.6.2 Wall Treatment 1: independent panel(s) with absorbent material AD E: 4.22–4.24

This treatment consists of adding a panel, or panels, to an existing wall which has poor sound insulating properties and the result may be similar to the 'masonry wall with independent panels' as described for new constructions but with insulation in the void between the two. The existing wall will offer some sound insulation and this will be enhanced by the provision of an independent panel, the isolation of that panel and the provision of absorbent material between the two.

The guidance suggests that if the original wall is of masonry construction, is 100 mm or more thick and is plastered on both sides, one independent panel will suffice, but in the case of different types of existing wall, a panel should be built on each side of the wall.

The specification of Wall Treatment 1 is as described in Table 10.15 and shown in Fig. 10.22.

Table 10.15 Construction form of Wall Treatment 1.

	Description
Construction form	Plasterboard supported on a timber frame or freestanding panels consisting of plasterboard sandwiching a cellular core
Panels	The mass per unit area of each panel should be at least 20 kg/m^2 excluding the mass of any framework
	If supported on a frame, panels should consist of at least two layers of plasterboard with staggered joints. Alternatively, freestanding panels, e.g. two sheets of plasterboard separated by a cellular core, may be used
Spacing	If panels are supported on a frame there should be a gap of at least 10 mm between the frame and the existing wall. If freestanding panels are used, there should be a gap of at least 35 mm between them and the existing wall
Mineral wool	Mineral wool with a minimum density of 10 kg/m^2 and minimum thickness of 35 mm to be located in the cavity formed between the panel and the existing wall

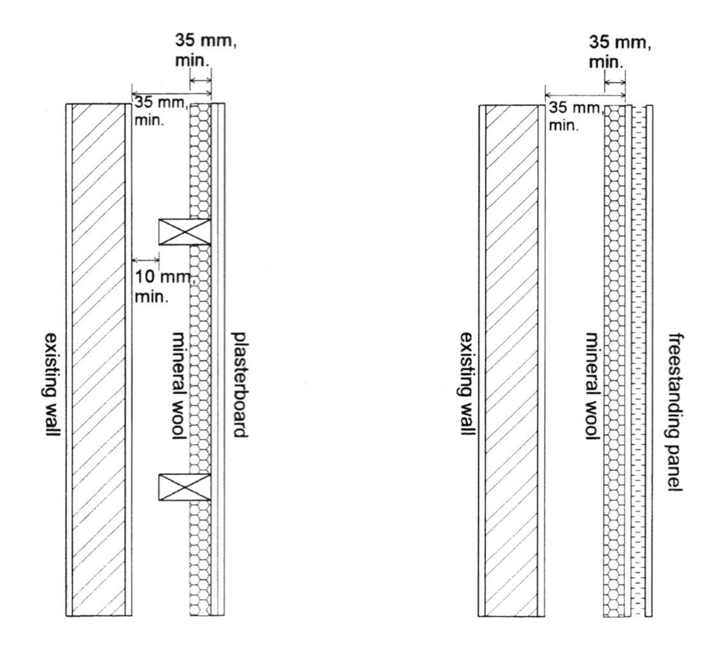

Fig. 10.22 Construction form of Wall Treatment 1.

10.6.3 Floor Treatment 1: independent ceiling with absorbent material AD E: 4.26–4.29

This treatment consists of adding an additional independent ceiling comprising plasterboard, joists and absorbent infill to an existing conventional floor comprising joists, boarding and plasterboard. The mass of the new ceiling, its absorbent infill together with its acoustic isolation will add significantly to the airborne and impact sound insulation offered by the original ceiling. Further, for good insulation, the construction should be made as airtight as possible.

The specification of Floor Treatment 1 is as described in Table 10.16.

The point is made in the AD that adoption of this procedure significantly reduces the floor to ceiling height in the converted building and that this should be borne in mind at design stage, i.e. the reduction will be a minimum of 125 mm plus the thickness of the plasterboard and even more if the joists need to be deeper than 100 mm. This can present a problem if window heads are close to the height of the original ceiling and a method of solution is proposed. This, together with the wall junction treatment that should be used when this procedure is employed are shown in Fig. 10.23.

10.6.4 Floor Treatment 2: platform floor with absorbent material AD E: 4.31–4.33

This treatment consists of adding a platform floor, i.e. a floating layer supported on a resilient layer, to an existing conventional floor comprising timber joists, boarding

Table 10.16 Construction form of Floor Treatment 1.

	Description
Construction form	Independent joist and plasterboard ceiling with absorbent mineral wool infill constructed below a conventional timber joist floor
Independent ceiling	Two or more layers of plasterboard having staggered joints and a mass per unit area of at least 20 kg/m^2
Mineral wool	Mineral wool with a minimum density of 10 kg/m^3 and minimum thickness of 100 mm to be located between the joists in the void between the new and old ceilings
Ceiling support	Independent joists fixed to surrounding walls There should be either: (a) no further support with a clearance of 25 mm or more between top of the joists and underside of the existing floor; or (b) additional support by resilient hangers fixed directly to the underside of the existing floor base
Existing ceiling	Upgrade as necessary to provide two or more layers of plasterboard having staggered joints and a mass per unit area of at least 20 kg/m^2

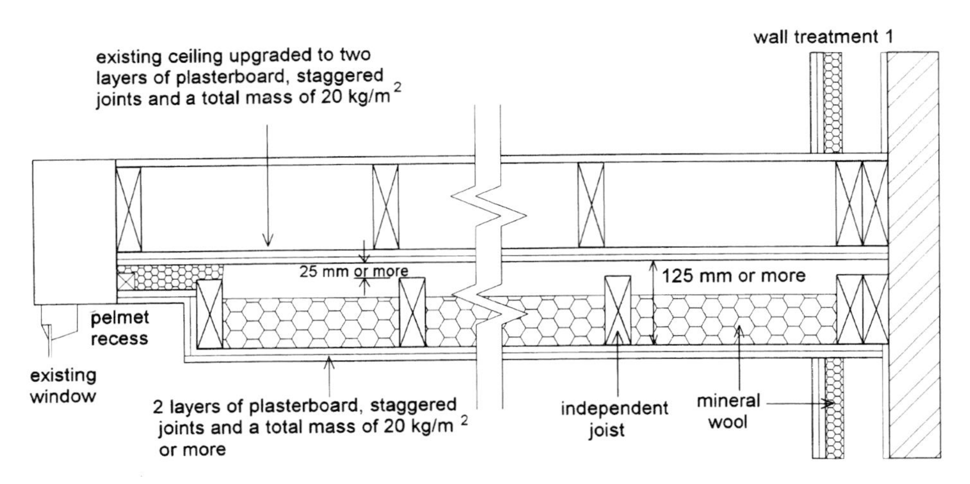

Fig. 10.23 Construction details of Floor Treatment 1 showing junction with Wall Treatment 1.

and plasterboard ceiling. The increase in the total mass of the floor, the action of the resilient layer and the absorbent material should all serve to increase the airborne and impact sound resistance of the floor.

The specification of Floor Treatment 2 should be as described in Table 10.17 and the construction form together with a junction treatment recommended in the AD are shown in Fig. 10.24.

Table 10.17 Construction form of Floor Treatment 2.

	Description
Construction form	A floating layer laid on a resilient layer which has first been laid onto the boarding of a conventional timber joist floor
Floating layer	A minimum of two layers of board each with a minimum thickness of 8 mm to provide a total mass per unit area of at least 25 kg/m^2. The layers should be fixed together, for example by spot bonding or glueing and screwing, ensuring staggered joints and then laid loose on the resilient layer
	Two examples of how a platform floor/floating layer of this type could be constructed are provided in the approved document, see new floor Type 3.1A
Resilient layer	Mineral wool, which may be paper faced on its underside, with a thickness of at least 25 mm and density in the range 60 to 100 kg/m^3
	If the material chosen is towards the lower end of the above range, impact sound insulation is improved but it may result in what is described in the document as a 'soft' floor, i.e. one which may be considered to respond excessively to fluctuating loads. It is suggested in the document that this may be overcome by providing additional support via a timber batten which is fixed to the walls and has a foam strip along its top
Mineral wool	Mineral wool with a minimum density of 10 kg/m^3 and minimum thickness of 100 mm to be located between the joists in the floor cavity
Existing ceiling	Upgrade as necessary to provide two or more layers of plasterboard having staggered joints and a mass per unit area of at least 20 kg/m^2

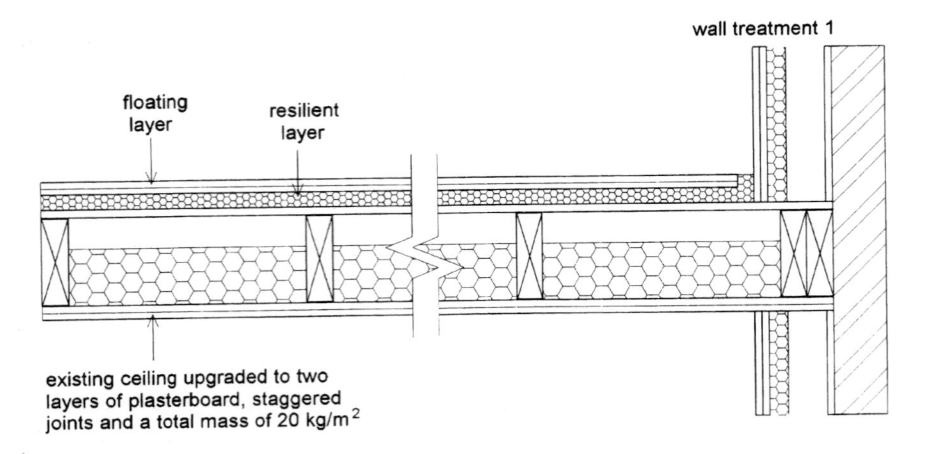

Fig. 10.24 Construction details of Floor Treatment 2 showing junction with Wall Treatment 1.

10.6.5 Stair treatment: stair covering and independent ceiling with absorbent material AD E: 4.35–4.38

This treatment consists of adding a soft layer over the treads, which reduces impact noise at source, and constructing an independent ceiling lined with absorbent material beneath the stairs. It is the mass of the stair and independent ceiling, the isolation of the independent ceiling and the presence of the absorbent material which provides sound insulation. This is improved by constructing a cupboard enclosure beneath the stairs. It may well be necessary to consider the influence of stairs when constructing dwellings by material change of use since when stairs provide a separating function they must make the same contribution to sound insulation as separating floors.

The specification of the stair treatment is as described in Table 10.18 and its construction form, together with associated treatment of the space beneath the stair as recommended in the AD, are shown in Fig. 10.25. If a staircase performs a separating function, reference should be made to Building Regulation Part B – Fire Safety.

Table 10.18 Construction form of stair treatment.

	Description
Construction form	The addition of a soft layer over the existing treads and an independent ceiling lined with absorbent material beneath the stairs A cupboard enclosure may be provided beneath the stairs
Soft covering	At least 6 mm thick, the soft covering must be securely fixed in order that it does not become a safety hazard
If there is an under stairs cupboard	The stair within the cupboard should be lined with plasterboard which has a mass per unit area of at least 10 kg/m^2. This should be mounted on battens, the space between the battens being filled with mineral wool with a density of at least 10 kg/m^3. See Fig. 10.25 The cupboard should be built with a small heavy well fitted door and walls consisting of two layers of plasterboard, or its equivalent, each layer of which has a mass per unit area of 10 kg/m^2
If there is NOT an under stairs cupboard	In this case an independent ceiling as recommended in Floor Treatment 1 should be constructed beneath the stair

10.6.6 Specific junction requirements in the event of change of use AD E: 4.39–4.50

- Advice is provided in the document for abutments of floor treatments:
  (a) In the case of floating floors, e.g. Floor Treatment 2, the resilient layer should be turned vertically upwards at room edges to ensure that there is no contact between the floating layer and the wall and a gap of about 5 mm, filled with flexible sealant, should be left between the floating layer and the skirting.
  (b) The junctions between new ceilings and walls should be sealed with tape or caulked with appropriate sealant, and detailed as indicated in Figs. 10.23 and 10.24.

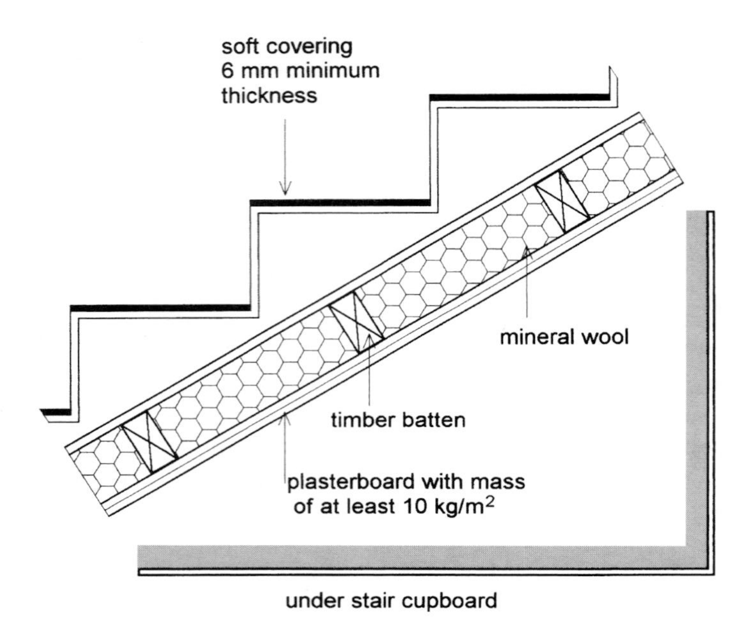

Fig. 10.25 Stair treatment assuming an under stair cupboard.

- Junctions of separating walls and floors with other elements, e.g. external or other load bearing walls are a potential source of significant flanking transmission and it is stated in the approved document that it may be necessary to seek expert advice. As also stated in the approved document, if there is significant flanking transmission via adjoining walls, this can be reduced by lining all adjoining masonry walls with:
  (a) an independent layer of plasterboard; or
  (b) a laminate of plasterboard and mineral wool.
  The approved document suggests seeking manufacturers' advice for other dry-lining laminates and does not provide guidance as to the thickness of plasterboard which is appropriate since this will depend on the structure of the adjoining wall. Indeed, the document states that if the adjoining masonry wall has a mass per unit area of more than 375 kg/m^2, the lining may not significantly improve the insulation and so its use may not be appropriate.
- The requirements in the AD for penetrating services, i.e. piped services and ducts, passing through floors that separate habitable rooms in different flats in conversion buildings are essentially the same as for penetrating services in new buildings which are explained in section 10.5.1.

10.6.7 Construction procedures

The AD provides information regarding correct and incorrect construction procedures which will influence the sound insulation provided when undertaking material change of use. This is summarised in Table 10.19.

Table 10.19 Construction procedures relating to material change of use.

Element	Correct procedure	Procedures which must be avoided
Wall Treatment 1	Ensure independent panel (and frame if present) have no contact with existing wall Seal perimeter of independent panel with tape or sealant	Do not tightly compress absorbent material as this may bridge the cavity
Floor Treatments 1 and 2	Apply appropriate remedial work to the existing construction Seal perimeter of any new or independent ceiling with tape or sealant	
Floor Treatment 1		Do not create a rigid or direct connection between an independent ceiling and its floor base Do not tightly compress the absorbent material as doing so may bridge the cavity
Floor Treatment 2	Use correct density of resilient layer and ensure it can carry anticipated load Allow for movement of materials, e.g. expansion of chipboard after laying, to maintain isolation Carry resilient layer up to room edges to isolate floating layer from wall surfaces Leave a gap of approximately 5 mm between skirting and floating layer and fill with flexible sealant Lay resilient layers in sheets with joints tightly butted and taped	Do not bridge between the floating layer and the timber base or surrounding walls, e.g. with services or fixings which penetrate the resilient layer between the floating layer and timber base or walls
Junctions with floor penetrations	Seal joints between casings and ceilings with tape or sealant Leave a gap of approximately 5 mm between casing and floating layer and fill with flexible sealant.	

10.7 Internal walls and floors for new buildings (Approved Document E, Section 5: Internal walls and floors for new buildings)

10.7.1 General requirements AD E: 5.1, 5.2 & 5.9–5.12

Requirement E2 of Schedule 1 to the Building Regulations 2000 (as amended) relates to protection against noise *within* dwelling places. Unlike Requirement E1 which relates to protection against noise *between* dwelling places, the performance of components within dwelling places is not assessed by pre-completion testing but simply by demonstrating that they meet the laboratory sound insulation values which are tabulated in Table 2 of section 0: Performance of the AD, see section 10.2.2 of this chapter. This requirement relates only to airborne sound insulation and advice to occupiers wishing to improve impact sound insulation should be to provide carpets or other soft coverings to floor surfaces. The examples in the AD are for guidance, they are not exhaustive and other designs, materials or products may be used to provide the desired performance.

The following points should be considered when designing internal walls and floors.

- If a door assembly in an internal wall offers a lower level of insulation than the wall in which it is located this will reduce the overall insulation offered. The avoidance of air paths by good perimeter sealing is an essential way of reducing sound transmission, and in addition the AD recommends the use of door sets.
- Stairs provide an important route of sound transmission between floors within dwellings, and if not enclosed the overall airborne sound insulation will be such that the potential of the associated floor will not be realised. However, as stated in the AD, the floor must still be constructed to comply with Requirement E2.
- The requirements for internal noise control should be borne in mind by designers who ought to ensure that noise sensitive rooms such as bedrooms are not constructed immediately adjacent to noise source rooms. The AD refers designers to BS 8233:1999, *Sound Insulation and Noise Reduction for Buildings – Code of Practice*.
- The sealing of walls, floors and gaps around doors has the potential to reduce air supply and in this context the requirements of Building Regulation Part F – Ventilation and Building Regulation Part J – Combustion Appliances and fuel storage systems must be taken into account.

10.7.2 Internal walls AD E: 5.4, 5.5 & 5.8

Four examples of different types of construction which should meet the requirements for internal walls are described in the AD. These are presented in the document, as follows, as wall Types A to D in a ranking order such that, as far as possible, the ones which give the best sound insulation appear first.

- Type A Timber or metal frames with plasterboard linings on each side.
- Type B Timber or metal frames with plasterboard linings on each side and absorbent infill material.
- Type C Concrete block wall with plaster or plasterboard finish on both sides.
- Type D Aircrete block wall with plaster or plasterboard finish on both sides.

Type A and B internal walls AD E: 5.17–5.18

These types of walls are similar, not only in construction form but in the ways in which they insulate against airborne sound transmission. Increasing the mass of the panels and the provision of absorbent infill between them increases their resistance to airborne sound transmission which is also influenced by cavity width and material with which the frame is constructed. The construction of the examples described in the AD is as outlined in Table 10.20.

Table 10.20 Internal walls Type A and B.

Element	Wall Type A	Wall Type B
Frame	Timber or metal	Timber or metal
Linings	Linings to be provided on each side of the frame Each lining to consist of two or more layers of plasterboard, each sheet of which has a mass per unit area of 10 kg/m^2 or more	Linings to be provided on each side of the frame Each lining to consist of a single layer of plasterboard which has a mass per unit area of 10 kg/m^2 or more
Distance between linings	A minimum of 75 mm if fixed to a timber frame A minimum of 45 mm if fixed to a metal frame	A minimum of 75 mm if fixed to a timber frame A minimum of 45 mm if fixed to a metal frame
Absorbent layer	Not required	Unfaced wool batts or quilt with: • thickness of 25 mm or more; and • density of 10 kg/m^3 or more suspended in the cavity The absorbent layer may be wire reinforced
Other requirements	All joints to be well sealed	All joints to be well sealed

Type C and D internal walls AD E: 5.19–5.20

These types are similar, not only in construction form but in the ways in which they insulate against airborne sound transmission. Increasing the mass of the panels influences their resistance to airborne sound transmission. The construction of the examples described in the approved document is as outlined in Table 10.21.

10.7.3 Internal floors AD E: 5.6, 5.7 & 5.8

Three examples of different types of construction which should meet the requirements for internal floors are described in the AD. These are presented in the document, as follows, as floor Types A to C in a ranking order such that, as far as possible, the ones which give the best sound insulation appear first.

- Type A Concrete planks.
- Type B Concrete beams with infilling blocks.
- Type C Timber or metal joists with board and plasterboard surfaces.

Table 10.21 Internal walls Type C and D.

Element	Wall Type C	Wall Type D
Core structure	Concrete block wall	Aircrete block wall
Required mass per unit area	A mass per unit area, excluding finish, of 120 kg/m^2 or more	A mass per unit area, including finish, of • 90 kg/m^2 or more for plaster finish • 75 kg/m^2 or more for plasterboard finish
Surface finish	Plaster or plasterboard on both sides	Plaster or plasterboard on both sides
Other requirements	All joints to be well sealed	All joints to be well sealed
Restrictions	No specific restrictions	This type of wall should: • not be used as a load-bearing wall • not be rigidly connected to the separating floors described in the AD (see guidance relating to separating floors) • only be used with the separating walls described in the document where there is no minimum mass requirement on internal masonry walls, e.g. walls Type 2.3 and 2.4 when there is no separating floor (see guidance relating to separating walls)

Type A and B internal floors AD E: 5.21–5.22

These types of floors are similar, not only in construction form but in the ways in which they insulate against airborne sound transmission. Increasing the mass of the planks (Type A) or beams and infilling blocks (Type B) and screed influences their resistance to airborne sound transmission. The provision of a soft covering, such as carpet, will improve impact sound insulation by reducing impact noise at its source. The construction of the examples described in the document is as outlined in Table 10.22 and shown in Fig. 10.26.

Type C internal floor AD E: 5.23

Resistance to airborne sound transmission is determined by the joist and board construction, the ceiling and the absorbent material used. The provision of a soft covering, such as carpet, will improve impact sound insulation by reducing impact noise at its source. The construction form of the example described in the document is as outlined in Table 10.23 and shown in Fig. 10.26.

10.7.4 Internal wall and floor junctions AD E: 5.13–16

Guidance on the form of junctions between separating walls and internal floors, and separating floors and internal walls is provided in sections 2 and 3 respec-

Table 10.22 Internal floors Type A and B.

Element	Floor Type A	Floor Type B
Structure	Concrete planks	Concrete beams with infilling blocks, bonded screed and ceiling
Required mass per unit area	Mass per unit area of 180 kg/m^2 or more	Mass per unit area of concrete beams and blocks to be 220 kg/m^2 or more
Screed	The provision of a regulating screed is optional	A bonded screed is required. If a sand and cement screed is used, it should have minimum thickness of 40 mm. If a proprietary product is used, manufacturers advice should be sought regarding its thickness
Ceiling finish	The provision of a ceiling finish is optional	Ceiling finish C or better, as defined in section 3, of the approved document, is required. See section 10.5.1 of this chapter for details
Other requirements	Although not stated in the approved document, floor joints should be fully grouted to ensure air tightness	

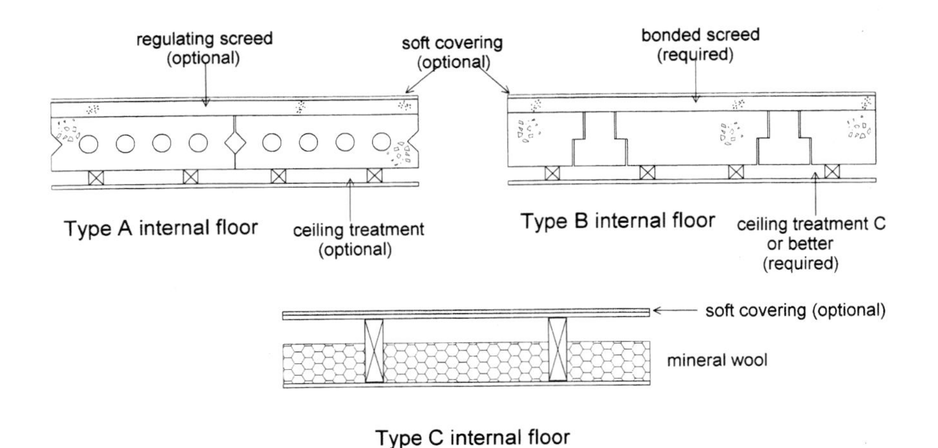

Fig. 10.26 Internal floors Type A, B and C.

tively of the approved document and this is considered in sections 10.4 and 10.5 of this chapter. When separating elements are present, the junctions between them and internal walls and floors should always be constructed in accordance with this guidance.

Whenever internal walls or floors are constructed, care should be taken to ensure that there are no air paths, i.e. direct or indirect routes of air passage, between rooms by filling all gaps around the wall or floor.

Table 10.23 Internal floor Type C.

Element	Floor Type C
Structure	Timber or metal joists supporting timber or wood based boarding, a plasterboard ceiling and absorbent material between the joists
Required mass per unit area	The timber or wood based boarding should have a mass per unit area of not less than 15 kg/m^2
Ceiling finish	A ceiling consisting of a single layer of plasterboard. Mass per unit area of not less than 10 kg/m^2. Normal method of fixing
Absorbent layer	Mineral wool laid in cavity between joists. Thickness not less than 100 mm and density not less than 10 kg/m^3
Other requirements	See BRE BR 262 Thermal Insulation: Avoiding risks, section 2.4 regarding heat emission from electrical cables which may be covered by absorbent material

10.8 Rooms for residential purposes (Approved Document E, Section 6: Rooms for residential purposes)

10.8.1 General requirements AD E: 6.1–6.3 & 6.16–6.18

The expression 'rooms for residential purposes', which is defined in Regulation 2, is used to refer to rooms such as those in hotels, hostels, boarding houses, halls of residence and residential homes but not those in, for example, hospitals.

Section 6 of the AD gives examples of walls and floors which it states should achieve the performance standards laid out in section 0: Performance – Table 1b of the AD and section 10.2 of this chapter although, as is stated in the document, the actual performance of a wall or floor will depend on the quality of its construction. The information in this section of the AD is provided only for guidance. It is not exhaustive and the details in no way override the requirement to undergo pre-completion testing. Designers are recommended also to seek advice from other sources such as manufacturers regarding alternative designs, materials or products.

It is not only sound insulation but room layout and building services which determine internal noise levels and these factors may be particularly significant in the confined spaces of 'rooms for residential purposes'. The requirements for internal noise control should be borne in mind by designers who ought to endeavour to ensure that noise sensitive rooms such as bedrooms are not constructed immediately adjacent to noise source rooms. The AD refers designers to BS 8233:1999 *Sound insulation and noise reduction for buildings – code of practice* and the BRE/CIRIA Report: *Sound control for homes*.

10.8.2 New buildings containing rooms for residential purposes

Separating walls AD E: 6.4

Examples of alternative types of separating walls suitable for new buildings are presented in section 2 of the AD and described in section 10.4 of this chapter. It is stated in the AD that where buildings contain rooms for residential purposes, of the examples given, the most appropriate choices are:

- wall Type 1 (solid masonry), where the in situ plastered finishes associated with each of the variants 1.1, 1.2 and 1.3 may be substituted for by a single sheet of plasterboard on each face provided that the sheets each have a mass per unit area of 10 kg/m^2 or more
- wall Type 3 (masonry between independent panels) but only Types 3.1 and 3.2 which have solid masonry cores.

The AD states that wall Type 2 (cavity masonry) and wall Type 4 (framed wall with absorbent material) may be used but particular care must be taken to maintain isolation between the leaves and specialist advice may be required.

Separating floors AD E: 6.8

Examples of alternative types of separating floors suitable for new buildings are presented in section 3 of the approved document and described in section 10.5 of this chapter. It is stated in the AD that where the buildings contain rooms for residential purposes, of the examples given, the most appropriate choice is one of the following subgroups of floor Type 1 (concrete base with soft covering):

- floor Type 1.1C Solid concrete slab, cast in situ with or without permanent shuttering. Soft floor covering and ceiling treatment C
- floor Type 1.2B Hollow or solid concrete planks. Soft floor covering and ceiling treatment B.

The AD states that floor Type 2 (concrete base with ceiling and floating floor) and floor Type 3 (timber frame base with ceiling and platform floor) may be used but their floating floors and ceilings must not be continuous across the walls which separate rooms for residential purposes. Designers are advised that this type of construction may require specialist advice.

10.8.3 Corridor walls and doors AD E: 6.5–6.7

The walls between rooms for residential purposes and corridors should be built to the specification, as described in section 6 of the AD, of the separating walls and their junction details. However, a weak point in the sound insulation provided from corridor noise is that transmitted through doors. For this reason, the document states that doors to corridors should have good sealing around their perimeters (including, if practical, their thresholds), and a mass per unit area of at least 25kg/m^2, or alternatively, a doorset with a weighted sound reduction index of at least 29 dB. The term sound reduction index is described in section 10.13 and the relevant measurement standards are listed in section 10.11.4.

Particular attention should be paid to potentially noisy places such as bars, which ideally should be separated from the rest of the building by a lobby, two doors in series or a high performance doorset. If this provision can not be made to the noisy room, it should be applied to rooms for residential purposes in its vicinity.

The requirements of Building Regulation Part B – Fire safety and Building regulation Part M – Access and facilities for disabled people must be taken into account when selecting doors for buildings of this type.

10.8.4 Material change of use – rooms for residential purposes AD E: 6.9–6.10

Section 6 of the AD considers forms of treatment to walls, floors, stairs and their associated junctions which it may be appropriate to apply when converting existing buildings by material change of use to provide rooms for residential purposes. The point is made in the AD that existing components may already satisfy the requirements of Table 1b of section 0: Performance (see section 10.2 of this chapter) and it is suggested that this would be so if the construction of a component, its junction detailing and flanking construction was sufficiently similar to that of one of the example walls or floors described above for 'rooms for residential purposes' created in new buildings. It is suggested in the AD that, with respect to requirements for mass, for example, an existing component may be sufficiently similar to comply with the required performance requirements if its mass was within 15% of that of an equivalent component recommended for the construction of new buildings. This may be the case, but due to uncertainty about the workmanship and density of material used throughout the component, additional treatment may still be advisable.

The construction of an existing component may be such that it is not possible to demonstrate whether or not it will comply with the requirements of Table 1b of section 0: Performance. In this case, use may be made of the floor, wall and stair treatments recommended in section 4 of the AD for houses and flats resulting from change of use, see section 10.6 of this chapter. The design of separating components and their associated junctions to provide adequate sound insulation in such circumstances may be of a complex nature and developers are advised in the approved document that specialist advice may be required when undertaking this type of work.

10.8.5 Junctions AD E: 6.11–6.15

As with other forms of dwelling place, in order for separating elements to fulfil their full potential it is essential that flanking transmission is restricted by careful junction design and selection of flanking elements. The AD states that in the case of new buildings, the guidance relating to junction design and flanking transmission provided in sections 2 and 3 of the document for separating walls and floors respectively of houses and flats should be adhered to when building 'rooms for residential purposes'. This guidance is described in sections 10.4 and 10.5 of this chapter.

Similarly, the guidance relating to junction and flanking details for houses and flats formed by material change of use in section 4 of the AD should be applied when creating 'rooms for residential purposes' by material change of use. This guidance is explained in section 10.6 of this chapter.

A relaxation to the guidance for houses and flats is that where a solid masonry Type 1 separating wall is used, the wall need not be continuous through a ceiling void or roof space to the underside of a structural floor or roof, subject to the following conditions:

- there is a ceiling consisting of two or more layers of plasterboard with a total mass per unit area of not less than 20 kg/m^2;
- there is a layer of mineral wool in the roof void which is not less than 200 mm thick and has a density of not les than 10 kg/m^3; and
- the ceiling is not perforated.

Also, the structure of the building should be such that neither the ceiling joists or plasterboard sheets are continuous across the wall separating 'rooms for residential purposes' with sealed joints as shown in Fig. 10.27. As stated in the approved document, this construction detail may only be used if Building Regulation Part B – Fire Safety and Building Regulation Part L – Conservation of Fuel and Power are satisfied.

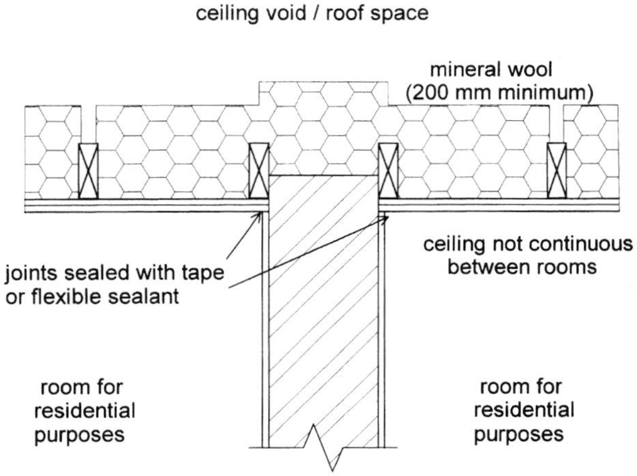

Fig. 10.27 Ceiling void /roof space detail – rooms for residential purposes only.

10.9 Reverberation in common parts of buildings
(Approved Document E, Section 7: Reverberation in the common internal parts of buildings containing flats or rooms for residential purposes)

10.9.1 General requirements AD E: 7.1–7.9

The common parts of buildings containing flats and/or 'rooms for residential purposes' tend to be constructed with hard durable surface finishes, which are easily maintained. Unfortunately, such surfaces lack the soft open texture which efficiently absorbs sound and so the level of reflected, or reverberated, sound tends to be high in such places. Requirement E3 of Part E of the Building Regulations 2000 (as amended) states that the design and construction of the common parts of such buildings shall be such as to prevent more reverberation around them than is reasonable.

Fortunately, it is relatively easy to increase sound absorption and hence reduce reverberant noise levels by surface treatment. Procedures for determining the amount of additional absorption which is required in corridors, hallways, stairwells and entrance halls providing access to flats and 'rooms for residential purposes' places are described in section 7 of Approved Document E. The document differentiates between different types of common space as shown in Table 10.24.

Table 10.24 Limiting dimensions of common areas.

Type of space	Ratio of longest to shortest floor dimensions
corridor, hallway	greater than three
entrance hall	three or less
stairwell	not defined in terms of ratio of dimensions

The types of space referred to in Table 10.24 are frequently interconnected and in such circumstance the guidance for each space should be followed individually.

The AD describes two methods, A and B, for calculating how much sound absorption is required and the procedure for the use of each method is explained in sections 10.9.3 and 10.9.4 with an example of their application in section 10.9.5. However, whichever procedure is adopted, it is essential that any material chosen to line the internal surfaces complies with the requirements of Building Regulation Part B – Fire Safety and it is recommended in the AD that if designers require guidance regarding the provision of additional sound absorbent material, this should be sought at an early stage of the design process.

10.9.2 Sound absorption properties of materials

When sound energy strikes a surface, some passes through it, some is absorbed by it and some is reflected. The absorbent coefficient, α, of a material is defined as:

$$\alpha = \frac{\text{sound energy not reflected}}{\text{sound energy incident}}$$

It follows that if no sound is reflected $\alpha = 1$ and if none is absorbed or transmitted $\alpha = 0$. In theory, all building materials have absorption coefficients between zero and one, values close to zero representing reflective surfaces with low absorption power and vice versa. Sound absorption is highly dependent on frequency. Some of the types of material used to line rooms have good absorption power in the mid to high frequencies of the audible spectrum whereas for others it is highest at the lower end of the spectrum. For this reason it is usual to quote one third octave or one octave band values of absorption coefficients measured across the audible spectrum. Octave-band values of absorption coefficient for a range of materials are shown in Table 10.27 below.

10.9.3 Calculation of additional absorption – Method A AD E: 7.10–7.12

This method makes use of a procedure for classifying sound absorbers which is described in EN ISO 11654:1997. To apply this procedure, absorption coefficients are measured at one-third-octave bands and the arithmetic mean of the three one-third-octave bands within each octave band is taken to the obtain the 'practical sound absorption coefficient', α_{pi}, a one-octave band average value where i is the ith octave band. The one-octave band values are then compared with a reference curve which is 'shifted' in increments of 0.05dB until the sum of the adverse deviations does not exceed 0.01dB. The value of the reference curve at 500 Hz, α_w, is then taken as a single figure weighted average absorption coefficient of the material in question. The material is then given a classification letter according to its value of α_w as shown in Table 10.25 which is an extract from EN ISO 11654:1997.

Table 10.25 Sound absorption classes defined in EN ISO 11654:1997.

Sound absorption class	α_w
A	0.90; 0.95; 1.00
B	0.80; 0.85
C	0.60; 0.65; 0.70; 0.75
D	0.30; 0.35; 0.40; 0.45; 0.50; 0.55
E	0.25; 0.20; 0.15
Not classified	0.10; 0.05; 0.00

Method A involves covering a prescribed area with an absorbent material of a defined classification as follows.

Entrance halls, corridors, hallways

Cover an area, at least as great as the floor area with an absorber which is Class C or better. The document states that it will normally be expedient to cover the ceiling with the additional absorptive material.

Stairwells and stair enclosures

Calculate the combined area of:

- stair treads
- upper surface of intermediate landings
- upper surface of landings (excluding ground floor) and
- ceiling area on the top floor

and cover an area which is equal to or larger than the calculated area with a Class D absorber, or cover an area which is at least half of the calculated area with an absorber of Class C or better. In either case the sound absorbing material should be distributed equally between floor levels. It is stated in the approved document that it will normally be appropriate to cover the:

- underside of intermediate landings
- underside of the other landings and
- ceiling area on the top floor.

The materials used to increase sound absorption tends to be light and porous, lacking the robustness which is required to withstand day-to-day wear and tear and vandalism in common areas where they may come into contact with human beings. This is why it is recommended in the AD that they are used to cover ceilings and the undersides of landings despite the fact that they would be equally effective on other surfaces. The AD states that the use of proprietary acoustic ceilings is appropriate. These have the benefits of being available with a variety of surface finishes and are lightweight and very efficient absorbers of sound.

10.9.4 Calculation of additional absorption – Method B AD E: 7.13–7.21

This method takes into account the actual absorption power of the surfaces of the enclosure prior to the provision of additional absorbent material and allows the amount of additional material which is required to be calculated. It is stated in the approved document that it is only intended for corridors, hallways and entrance halls (not stairwells) but that it offers a more flexible approach which may require less additional absorption than Method A. It is also a more efficient approach in that the additional absorption can be directed at the frequencies at which it is most needed.

For a particular surface of area S m^2 and absorption coefficient α, its absorption area A is defined as the product of S and α and it has units of m^2. This may be considered as the hypothetical area of a material which is a perfect absorber, i.e. one with $\alpha = 1$, which would provide the same sound absorption as the actual material with area S.

If an enclosure has n internal surfaces, its total absorption area, A_T is defined as the sum of the absorption areas of its individual components, such that:

$$A_T = \alpha_1 S_1 + \alpha_2 S_2 + \alpha_3 S_3 + \ldots\ldots + \alpha_n S_n$$

It follows that A_T may be considered as the hypothetical area of a material which is a perfect absorber which would provide the same total sound absorption as the sum of the actual materials which are present.

Requirement E3 will be satisfied when the total absorption area stipulated in Table 10.26 for different types of common area has been provided at each octave band from 250 Hz to 4000 Hz.

Absorption coefficient data which should be accurate to two decimal places may be obtained from a variety of sources. For generic materials, data is provided in Table 7.1 of the AD which is reproduced as Table 10.27. It is stated in the AD that this may be supplemented by other published data. Whilst published data will be necessary and appropriate for generic surface materials not provided specifically to

Table 10.26 Total absorption areas for common areas using Method B.

Type of common area	Minimum total absorption area required	Location of additional absorptive material
Entrance halls	0.20 m^2 per cubic metre of volume	To be distributed over the available surfaces
Corridors, hallways	0.25 m^2 per cubic metre of volume	To be distributed over one or more of the surfaces

Table 10.27 Copy of Table 7.1 from Approved Document E providing absorption coefficients for commonly used materials.

	Sound absorption coefficient, α in octave frequency bands (Hz)				
Material	**250**	**500**	**1000**	**2000**	**4000**
Fair-faced concrete or plastered masonry	0.01	0.01	0.02	0.02	0.03
Fair-faced brick	0.02	0.03	0.04	0.05	0.07
Painted concrete block	0.05	0.06	0.07	0.09	0.08
Windows, glass facade	0.08	0.05	0.04	0.03	0.02
Doors (timber)	0.10	0.08	0.08	0.08	0.08
Glazed tile/marble	0.01	0.01	0.01	0.02	0.02
Hard floor coverings (e.g. lino, parquet) on concrete floor	0.03	0.04	0.05	0.05	0.06
Soft floor coverings (e.g. carpet) on concrete floor	0.03	0.06	0.15	0.30	0.40
Suspended plaster or plasterboard ceiling (with large airspace behind)	0.15	0.10	0.05	0.05	0.05

increase the absorption area of the enclosure, materials which are selected for their absorption properties will usually be trade products and their absorption coefficients should be available from their manufacturers or suppliers. These should be established by measuring one-third-octave band values of absorption coefficients in accordance with BS EN 20354:1993 and then converting them to one-octave band values of 'practical sound absorption coefficient', α_{pi}, using the procedure described EN ISO 11654:1997, see section 10.9.3.

When calculating the area and absorption coefficient of the material necessary to satisfy the requirements of Table 10.26, calculation steps should be rounded to two decimal places.

10.9.5 Example of absorption calculation

There follows an example of the application of Methods A and B to calculate the extra absorption required for an entrance hall, see Fig. 10.28.

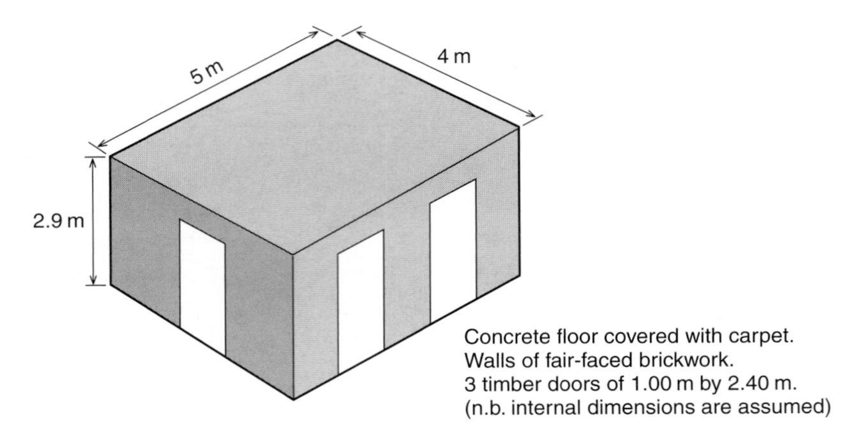

Fig. 10.28 Entrance hall relating to example question.

Application of Method A

This requires covering an area equal or greater than the floor area with a Class C, or better, absorber.

An appropriate solution would therefore be to cover $4.0 \times 5.0 = 20\ m^2$, e.g. the ceiling with an appropriate material, probably acoustic ceiling tiles, of Class C or better.

Application of Method B

Assuming that the designer has decided to cover the entire ceiling with absorptive material the necessary calculations are as described below.

Stage 1

Calculate area of each internal surface and establish the absorption coefficient from Table 10.27 for each surface which is not being selected for its sound absorbing properties.

Surface	area (m^2)	Absorption coefficient for each octave band				
		250 (Hz)	500 (Hz)	1000 (Hz)	2000 (Hz)	4000 (Hz)
floor (concrete, carpet covered)	20	0.03	0.06	0.15	0.30	0.40
doors (timber)	7.2	0.10	0.08	0.08	0.08	0.08
walls (exc. doors) (fair-faced brick)	45	0.02	0.03	0.04	0.05	0.07
ceiling (yet to be determined)	20					

Note that the carpeted floor provides good absorption at the higher frequency octave bands. This reduces the absorption required at these frequencies from the ceiling covering.

Stage 2
Calculate the absorption area of floor, doors and walls (the product of area and absorption coefficient) and then establish the overall absorption area at each octave band from these surfaces by summing.

	Absorption area (m^2)				
Surface	**250 (Hz)**	**500 (Hz)**	**1000 (Hz)**	**2000 (Hz)**	**4000 (Hz)**
floor	0.60	1.20	3.00	6.00	8.00
doors	0.72	0.58	0.58	0.58	0.58
walls (exc. doors)	0.90	1.35	1.80	2.25	3.15
sum of absorption areas	2.22	3.13	5.38	8.83	11.73

Stage 3
Calculate the total absorption area, A_T, required at each octave band, i.e. $0.2 \times \text{volume} = 0.2 \times 5 \times 4 \times 2.9 = 11.6\ m^2$

Stage 4
Calculate the additional absorption area required from ceiling by subtracting the values found at Stage 2 from that found at Stage 3:

	Absorption area (m^2)				
	250 (Hz)	**500 (Hz)**	**1000 (Hz)**	**2000 (Hz)**	**4000 (Hz)**
Additional area required from ceiling	9.38	8.47	6.22	2.77	−0.13

The negative value at 4000 Hz indicates that sufficient absorption is provided at this octave band by the floor, doors and walls.

Stage 5
Calculate the absorption coefficient required for the ceiling by dividing the required absorption area by the area of the ceiling.

	Absorption area (m^2)				
	250 (Hz)	**500 (Hz)**	**1000 (Hz)**	**2000 (Hz)**	**4000 (Hz)**
Required absorption coefficient of ceiling material	0.47	0.42	0.31	0.14	any value

Select a patent ceiling product which has absorption coefficients greater than those in the above table.

10.9.6 Report on compliance AD E: 7.22

Demonstration of compliance with Requirement E3 may take the form of an annotated drawing or a report. The information which should be included, quoted directly from the approved document, is as follows.

(1) A description of the enclosed space (entrance hall, corridor, stairwell etc.).
(2) The approach used to satisfy Requirement E3, Method A or B.
- With Method A, state the absorber class and the area to be covered.
- With Method B, state the total absorption area of additional absorptive material used to satisfy the Requirement.

(3) Plans indicating the assignment of the absorptive material in the enclosed space.

10.10 School acoustics
(Approved Document E, Section 8: Acoustic conditions in schools)

Requirement E4 states that the design and construction of all rooms and other places in school buildings shall be such that their acoustic conditions and insulation against disturbance by noise is appropriate to their intended use. It is stated in Approved Document E that the normal way of satisfying these conditions will be to meet the requirements given in section 1 of Building Bulletin 93, *The acoustic design of schools* with respect to sound insulation, reverberation time and internal ambient noise levels. This document is produced by DfES and published by The Stationery Office, ISBN No. 0 11 271105 7.

10.11 Calculation of sound transmission indices
(Approved Document E, Annex E: Procedures for sound insulation testing)

10.11.1 Overview of procedures for establishing sound insulation values from field measurements

Field measurement of airborne sound insulation for compliance with Requirement E1 requires calculation of the 'weighted standardised level difference' $D_{nT,w}$ and the 'spectrum adaptation term' C_{tr}. Initially, a noise source is located in one of two rooms between which there is a separating element and microphones are located in this and the room on the other side of the separating element as shown in Fig. 10.29. On activation of the noise source, noise passes from the source room to the receiving room by the direct and indirect routes indicated on the figure and the level difference between them, D, is measured in accordance with BS EN ISO 140-4:1998 and as described in section 10.11.3. The level measured at the microphone in the receiving room will be determined not only by the energy passing directly to the microphone but also by noise energy reflected within the room which in turn will influence the calculated level difference D.

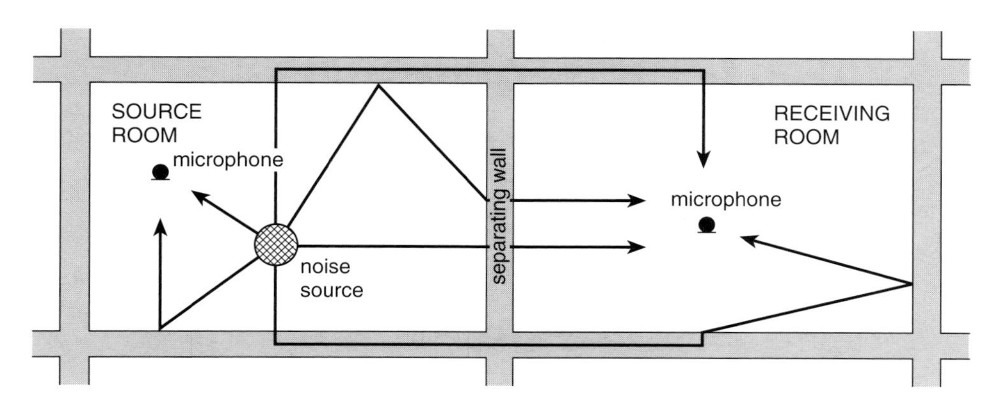

Fig. 10.29 Measurement of airborne sound insulation.

To take account of the reflected energy in the receiving room, a noise source is activated in the receiving room and then switched off. The sound pressure then falls, and the time taken for the sound to decay by 60 dB is established. This is known as the reverberation time and the standards relevant to reverberation time measurement are referred to in section 10.11.3. The level difference is then normalised by calculating the value D_{nT}:

$$D_{nT} = D + 10\ \mathrm{Log}\left(\frac{T}{T_0}\right)\mathrm{dB} \quad \text{with } T_0 = 0.5\ \mathrm{s}.$$

where T is the reverberation time of the receiving room , and T_0 is a reference value. D_{nT} is known as the 'standardised level difference' and is the level difference which would have been measured if the receiving room had a reverberation time of 0.5 s. Hence all level differences are normalised to a receiving room reverberation time of 0.5 s. D_{nT} is measured using one-third-octave bands and these values are combined to give a single value, the weighted standardised level difference $D_{nT,w}$, using a procedure described in BS EN ISO 717-1:1997.

The spectrum adaptation term, C_{tr}, is also a single number value measured in decibels and calculated from the spectrum D_{nT} using BS EN ISO 717-1:1997. If a separating element provides poor insulation at low frequencies, C_{tr} will be large and negative and so when added to $D_{nT,w}$ for the purposes of requirement E1, its effect will be to reduce the overall magnitude of the insulation value.

Field measurement of impact sound insulation for compliance with Requirement E1 requires calculation of the 'weighted standardised impact sound pressure level', $L'_{nT,w}$. Initially a tapping machine is located on top of the separating floor and a microphone(s) is located in the room beneath it, see Fig. 10.30. A tapping machine is a device with five small hammers which consecutively strike the floor producing impacts of predetermined momentum. On activation of the tapping machine, the impact sound pressure level, L_i is measured in the room beneath the floor (the receiving room) in accordance with BS EN ISO 140-7:1998. As with airborne sound transmission, the level measured at the microphone in the receiving room will be

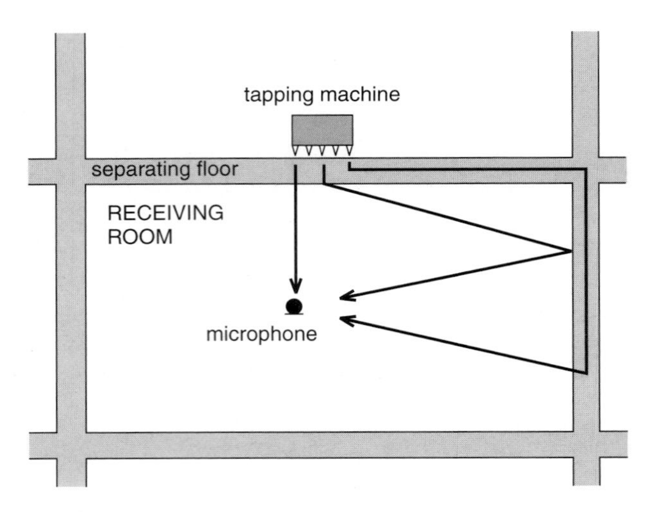

Fig. 10.30 Measurement of impact sound transmission.

influenced by noise energy reflected within the room, and so reverberation time is also measured when assessing impact sound insulation. In this case, the standardisation is also to the equivalent of a room with a reverberation time of 0.5 s. to give the 'standardised impact sound pressure level' L'_{nT}:

$$L'_{nT} = L_i + 10 \text{ Log}\left(\frac{T}{T_0}\right) \text{dB} \quad \text{with } T_0 = 0.5 \text{ s.}$$

L'_{nT} is measured using one-third-octave bands and these values are combined to give a single value, the weighted standardised impact sound pressure level, $L'_{nT,w}$ using a procedure described in BS EN ISO 717-2:1997.

10.11.2 General requirements AD E: B1.4, B2.1 & B3.2

It is the responsibility of the person undertaking the building work to arrange for the necessary sound insulation testing to be carried out by a testing organisation with appropriate third party accreditation. The AD states that the organisation undertaking the testing should preferably have United Kingdom Accreditation Service (UKAS) accreditation, or a European equivalent. It is also a requirement that a valid traceable calibration certificate exists for the instrumentation used for the measurement and that it has been tested within the preceding two years.

Measurements are based on the BS EN ISO 140 series of standards and rating is based on the BS EN ISO 717 series of standards and as stated in the AD it is important when calculating test results that rounding does not occur until required by the standard in question.

10.11.3 Field measurement of the sound insulation of separating walls and floors AD E: 0.1

Requirement E1 is satisfied if separating elements comply with Tables 1a and 1b of the AD following pre-completion testing, see section 10.2 of this chapter.

Regulations 20A and 12A apply to building work to which E1 applies and these regulations state that pre-completion testing must be carried out in accordance with an approved procedure. The approved procedure is described in Annex B2 of Approved Document E and its content is outlined below.

Airborne insulation of separating walls or floors AD E: B2.2–B2.8

Procedure:
Measure in accordance with BS EN ISO140-4:1998
(all measurements and calculations carried out in one-third-octave bands).
Rate performance in accordance with BS EN ISO 717-1:1997
(in terms of the weighted standardised level difference $D_{nT,w}$ and spectrum adaptation term C_{tr}).

(1) Measurements using a single sound source
The average sound pressure level (SPL) is measured in one-third-octave-bands in source and receiving rooms for each source position using either fixed microphone positions, and averaging values on an energy basis, or a moving microphone.

The differences between average source room SPL measurements in adjacent one-third-octave bands should not exceed 6dB. If this criterion is not satisfied, the instrumentation should be adjusted and measurement repeated until it is. When it is satisfied, the average SPL in the receiving room and hence the level difference should be established.

The sound source must not be moved or its output level adjusted until measurements in the source and receiving rooms are complete.

The procedure should be repeated with the sound source moved to another position in the source room. At least two source positions should be used and the level differences obtained at each position should be averaged to obtain the level difference D defined in BS EN ISO 140-4:1998.

(2) Measurements using simultaneously operating multiple sound sources
The average SPL is measured in one-third-octave-bands in source and receiving rooms with the multiple sources operating simultaneously. Measurement is undertaken using either fixed microphone positions, and averaging values on an energy basis, or a moving microphone.

The differences between average source room SPL measurements in adjacent one-third-octave bands should not exceed 6dB. If this criterion is not satisfied, the instrumentation should be adjusted and measurement repeated until it is. When it is satisfied, the average SPL in the receiving room and hence the level difference, D defined in BS EN ISO 140-4:1998 may be established.

Impact sound transmission through separating floors AD E: B2.9

Procedure:
Measure in accordance with BS EN ISO140-7:1998
(all measurements and calculations carried out in one-third-octave bands).
Rate performance in accordance with BS EN ISO 717-2:1997
(in terms of the weighted standardised impact sound pressure level, $L'_{nT,w}$).

Measurement of reverberation time AD E: B2.10

In order to calculate the weighted standardised level difference $D_{nT,w}$, and the weighted standardised impact sound pressure level, $L'_{nT,w}$ measurements have to be corrected to take into account the degree of reverberation in the receiving room. This requires measurement of its reverberation time which is defined as the time taken for the SPL to decrease by 60dB after a sound source is switched off. The measurement standards referenced above refer to ISO 354 for the method of measuring reverberation time. Guidance in BS EN ISO 140-7:1998 with respect to positioning of equipment and number of measurements should be adhered to.

Room types and sizes AD E: B2.11

The types of rooms to be used for testing are considered in section 10.3. They should have volumes of $25m^3$ or more and if not, the volumes of the rooms used should be reported.

Room conditions for tests AD E: B2.12–B2.17

The following conditions should apply.

(1) Rooms (or available spaces in the case of properties sold before fitting out) used for testing should be completed but unfurnished.
(2) Floors without soft coverings should be used for impact tests. Exceptions are separating floors Type 1 and structural concrete floor bases which have integral soft coverings. If soft coverings have been laid on other types of floor, these should be removed and if that is not possible at least half of the floor should be exposed, with the tapping machine (this is required for impact transmission measurement) used on the exposed part of the floor.
(3) When a pair of rooms used for airborne sound insulation measurement are of different volumes, the sound source should be activated in the larger room.
(4) All doors and windows should be closed and the doors of fitments such as kitchen units and cupboards on all walls should be empty and have open doors.

Accuracy of measurement AD E: B2.18–B2.19

SPL and reverberation time measurements should be to an accuracy of 0.1 dB and 0.01 s respectively.

Measurement using moving microphones AD E: B2.20–B2.21

If a moving microphone is employed, measurements should be taken with it centred at two or more positions. Due to the transient nature of the reverberation time measurement process, fixed microphone positions as opposed to moving microphones should be used when measuring reverberation times.

10.11.4 Laboratory measurement procedures AD E: B1.2 & B3.1–B3.2

Laboratory testing may be required to establish the performance of building elements in order to comply with Requirement E2, components such as wall ties and floating floors to which Requirement E1 applies as well as to assess the performance of proposed novel constructions. Appropriate laboratory testing procedures are described in Annex B3 of the AD and are outlined below.

No rounding in calculations associated with sound insulation test results should take place until required by the relevant BS EN ISO 140 and BS EN ISO 717 series standards.

Floor coverings and floating floors: AD E: B3.3–B3.6

Procedure:
Test in accordance with BS EN ISO 140-8:1998 (using a test floor with a thickness of 140 mm).
Rate performance in accordance with BS EN ISO 717-2:1997.

The AD states that text has been omitted from BS EN ISO 140-8:1998 and that for the purposes of the document, section 6.2.1 of BS EN ISO 140-8:1998 should be disregarded and section 5.3.3 of BS EN ISO 140-7:1998 referred to instead.

BS EN ISO-8:1998 refers to ISO 354 for the method of measuring reverberation time. It is stated in the AD that guidance in BS EN ISO 140-8:1998 with respect to positioning of equipment and number of required decay measurements should be adhered to.

When assessing Category II specimens, defined in BS EN ISO 140-8:1998 as large specimens including rigid homogeneous surface materials (e.g. floating floors), the value ΔL_w should be measured with and without the floor carrying a defined load. ΔL_w is a measure of the improvement in impact sound insulation provided by installing a floor covering or floating floor over a laboratory test floor.

Dynamic stiffness AD E: B3.7–B3.8

Procedure for resilient layer:
Measure in accordance with BS EN 29052-1:1992 (use sinusoidal signals method with no pre-compression of specimens).
Procedure for wall ties:
Measure in accordance with BRE Information Paper IP 3/01.

Airborne sound insulation of wall and floor elements AD E: B3.9

Procedure:
Measure in accordance with BS EN ISO 140-3:1995.
Rate performance in accordance with BS EN ISO 717-1:1997
(determine weighted sound reduction index R_w).
Sound reduction index and weighted sound reduction index are described in section 10.13.

Flanking laboratory measurement AD E: B3.10–B3.14

Although tests in flanking laboratories are very useful because they include the effects of transmission through direct and flanking routes, they may not be used to indicate compliance with Requirement E1 since this relates to field performance.

Flanking laboratory measurements are a useful way of assessing the probable performance in practice of novel or alternative constructions and it is stated in the AD that if a test construction provides airborne insulation, $D_{nT,w} + C_{tr}$, value of 49 dB or more when measured in a flanking laboratory, this may be taken as an indication that a construction which is identical in all significant details may achieve a $D_{nT,w} + C_{tr}$ value of 45 dB or more if built in the field. Note first paragraph of this section.

It is also stated in the AD that if a test construction provides impact sound transmission , $L'_{nT,w}$, of 58 dB or less measured in a flanking laboratory, this may be taken as an indication that a construction which is identical in all significant details may achieve an $L'_{nT,w}$ value of 62 dB or less if built in the field. Note first paragraph of this section.

A standard for laboratory measurement of flanking transmission is being developed and construction details of a flanking laboratory are available from The Acoustics Centre, BRE, Garston, Watford WD25 9XX.

10.11.5 Test report information

Field test report information AD E: B4.1

The information to be provided in a test report relating to tests done in compliance with Regulations 20A or 12A is itemised in section 1 of Approved Document E, see section 10.3.6 of this chapter. It is suggested in the AD that, although not mandatory, it may be useful to also provide the following information which is quoted directly from the document:

(1) sketches showing the layout and dimensions of rooms tested;
(2) description of separating walls, external walls, separating floors, and internal walls and floors including details of materials used for their construction and finishes;
(3) mass per unit area in kg/m^2 of separating walls, external walls, separating floors, and internal walls and floors;

(4) dimensions of any step and/or stagger between rooms tested;
(5) dimensions and position of any windows or doors in external walls.

Laboratory test report information AD E: B4.2

It is stated that the following information, which is quoted directly from the AD, should be provided in test reports:

(1) Organisation conducting test, including:
  (a) name and address;
  (b) third party accreditation number (e.g. UKAS or European equivalent);
  (c) Name(s) of person(s) in charge of test.
(2) Name(s) of client(s).
(3) Date of test.
(4) Brief details of test, including:
  (a) equipment;
  (b) test procedures.
(5) Full details of the construction under test and the mounting conditions.
(6) Results of test shown in tabular and graphical form for third octave bands according to the relevant part of the BS EN ISO 140 series and BS EN ISO 717 series, including:
  (a) single-number quantity and the spectrum adaptation terms;
  (b) data from which the single-number quantity is calculated.

10.12 The calculation of mass
(Approved Document E, Annex A: Method for calculating mass per unit area)

The mass per unit area, expressed in kilograms per square metre may be obtained from manufacturers' data. Alternatively, it can be calculated for walls and floors by the following methods as described in Annex A of Approved Document E.

10.12.1 Walls AD E: A2–A4

In the case of a brick and block wall, mortar beds and perpends may constitute a considerable proportion of the total volume and, since the density of mortar may be significantly different to that of the blocks, account must be taken of this when calculating the mass per unit area. The procedure used is to define the 'co-ordinating area' as shown in Fig. 10.31, and to calculate the mass per unit area, M_A from:

$$M_A = \frac{\text{mass of co-ordinating area}}{\text{co-ordinating area}} = \frac{M_B + \rho_m(T\,d\,(L + H - d) + V)}{LH}\,\text{kg/m}^2$$

where, in the above equation:

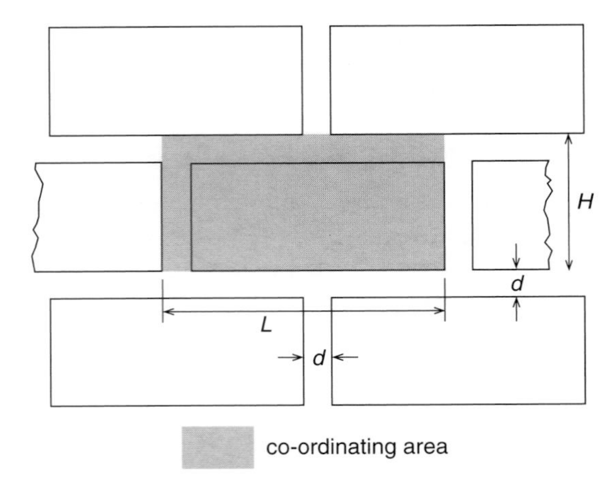

Fig. 10.31 Co-ordinating area used for calculation of mass per unit area.

M_B = brick or block mass (kg)
ρ_m = density of mortar (kg/m^3)
T = brick or block thickness (m) (excluding any finish)
d = mortar joint thickness (m)
L = co-ordinating length (m)
H = co-ordinating height (m)
V = volume of any mortar filled void, e.g. a frog

In using the above equation the following points should be noted.

- Since density is a function of moisture content, the mass of brick/block and density of mortar should be taken at the appropriate moisture content. The AD makes reference to Table 3.2 of CIBSE Guide A (1999) for this information.
- The above formula calculates the mass per unit area of a single leaf without surface finish. If a finish, e.g. plaster, is to be applied and/or if it is a cavity wall with another leaf, the mass per unit area of each component must be added together to obtain the total mass per unit area of the wall.
- Manufacturers' data should be used to obtain the mass per unit area of surface finishes.

Example calculation

Assume that a concrete block has the following dimensions:

thickness = 0.120 m
length = 0.440 m
height = 0.290 m
mass of block = 14 kg
and it is to be used on edge with mortar which has a density, of 1800 kg/m^3 and a thickness, of 0.010 m (10mm).

In this case:

$$\begin{aligned} M_B &= 14 \text{ kg} \\ \rho_m &= 1800 \text{ kg/m}^3 \\ T &= 0.120 \text{ m} \\ d &= 0.010 \text{ m} \\ L &= 0.450 \text{ m} \\ H &= 0.300 \text{ m} \\ V &= 0 \text{ (it is assumed that there is no frog)} \end{aligned}$$

and substituting into the above equation gives a mass of 115.5 kg/m^2.

The AD makes reference to 'simplified equations'. These, however, are simply representations of the equation shown above with substitution for all variables for the particular case with the exception of the mass of the block. In this format the above equation would be written as:

$$\text{Mass per unit area} = 7.4\ M_B + 11.8 \text{ kg/m}^2$$

Various values of M_B may now be substituted into the equation to establish the mass per unit area.

10.12.2 Floors AD E: A5

The mass per unit area of a flat solid concrete slab floor, M_F, may be obtained by multiplying its density, ρ_c, by its thickness, T:

$$M_F = \rho_c\ T$$

In the case of a conventional beam and block floor, as shown in Fig. 10.32, the mass per unit area, M_F, can be obtained from:

$$M_F = \frac{(M_{beam,1m} + M_{block,1m})}{L_B}$$

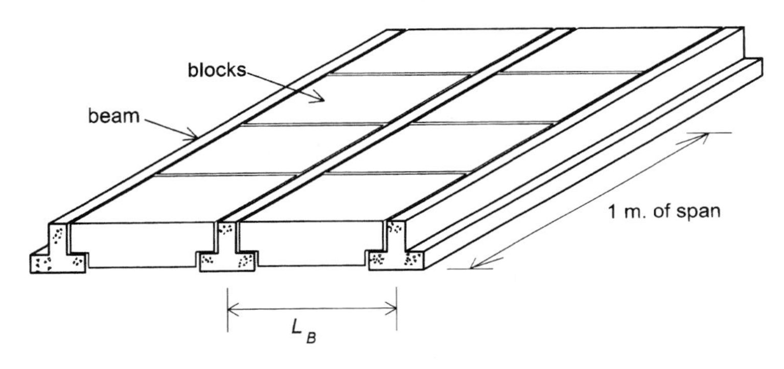

Fig. 10.32 Beam and block floor.

where, in the above equation:

$M_{beam,1m}$	=	the mass of 1 m of the length of the beam
$M_{block,1m}$	=	the mass of the blocks spanning between two consecutive beams for 1 m of the length of the beam
L_B	=	the dimension defined in Fig. 10.32.

No other examples are given in the AD. Designers are advised in the AD to seek advice from manufacturers on the mass per unit area of other floor types.

10.13 Explanation of important terms

Absorption of sound

When sound strikes a surface, a fraction is reflected back from the surface and the remainder is absorbed at the surface. Sound absorption, which may be thought of as the conversion of sound energy to heat, may be caused by a number of mechanisms. The absorption coefficient is the fraction of the incident energy which is not reflected from the surface and has a value between zero and one, the extreme values representing perfectly reflecting and perfectly absorbing surfaces respectively.

Airborne sound insulation

Airborne sound is sound, from a source such a speech or loudspeakers, which travels through air. Elements reducing its transmission are therefore providing airborne sound insulation. Airborne sound may pass from one room to another by:

- direct transmission: sound which passes directly through the element separating two rooms
- flanking transmission: sound which passes indirectly between rooms, e.g. via elements which abut the separating element.

Dwelling places

Within this chapter, dwelling houses, flats and rooms for residential purposes are, for convenience, referred to collectively as ‘dwelling places’.

Frequency

Sound results from cyclic variations in air pressure, the frequency of which is measured in the units of hertz and given the symbol Hz. The frequency in hertz refers to the number of pressure variation cycles per second. The human ear can detect sound within a range of frequencies from about 20 Hz to 20,000 Hz. An octave band contains all frequencies from a lower limiting frequency to an upper limiting frequency of twice the lower limit. One-third-octave bands result from dividing an octave band into three contiguous bands in which the upper limiting frequency in each sub band is its lower limiting frequency multiplied by $2^{1/3}$.

Impact sound insulation

Impact sound is sound created by the impact of objects, e.g. footsteps, directly with part of the building structure. If a component is provided which reduces impact sound transmission it is providing impact sound insulation. When sound is transmitted via the structural components of a building it is known as structure borne sound.

Impact sound pressure level

The standardised impact sound pressure level, L'_{nT}, is obtained from frequency band measurements of the impact sound generated by a standard test. The weighted standardised impact sound pressure level, $L'_{nT,w}$, is a single number expression of impact sound level measured in decibels and obtained from L'_{nT}. These indices are described in section 10.11.

Level difference

The standardised level difference, D_{nT}, is obtained from frequency band measurements of the sound pressure level difference between two rooms, and the weighted standardised level difference, $D_{nT,w}$, is a single number expression of level difference obtained from D_{nT}. The spectrum adaptation term, C_{tr}, is a single number modification to $D_{nT,w}$. These indices are described in section 10.11.

Separating floors and walls

These are the elements between dwelling places. Separating floors separate flats and 'rooms for residential purposes'. Separating walls separate dwelling houses, flats and 'rooms for residential purposes'.

Sound reduction index

The sound reduction index, R, of a building element is a laboratory measured value relating to its airborne sound insulating properties. Sound reduction index is measured in accordance with BS EN ISO 140-3:1995 in one-third-octave bands across the audible frequency range and is expressed in decibels. The weighted sound reduction index, R_w, is a single number quantity which expresses the airborne sound insulation of a building element. It is derived from the sound reduction index in accordance with BS EN ISO 717-1:1997.

Units of measurement

The decibel, given the symbol dB, is a convenient unit for representing magnitude of sound. Sound pressure level, which is used to measure the magnitude of sound in building acoustics, is measured in decibels.

10.14 Robust standard details and application

Requirement E1 of Schedule 1 to the Building Regulations 2000 stipulates that a predefined level of acoustic performance must be provided between adjoining dwelling places and, via Regulations 12A and 20A, that this should be demonstrated by way of a programme of pre-completion testing, as described in section 10.3. This procedure ensures that buildings are constructed not only to a high notional standard of sound insulation but also that compliance with the standard will be verified by routine testing of a sample of completed buildings. Whilst attractive to purchasers such a scheme is less appealing to the house building industry since there is no guarantee that dwellings will not require retrospective acoustic treatment following adverse pre-completion test measurements.

The principal reason for introducing pre-completion testing was that for various reasons, such as design detailing and construction site practices, components have not in the past been achieving their full performance potential. It follows that if there could be a sufficiently high level of confidence that the performance levels specified in the Building Regulations were being achieved in practice, such that the pass rate in pre-completion testing approached 100%, the need for the pre-completion test programme would be reduced.

The House Builders Federation have proposed the introduction of a series of robust standard details (RSDs) for use as separating walls and separating floors which would not require routine pre-completion testing. The justification which is proposed for not applying the pre-completion tests is that the RSDs would be designed such that the level of insulation provided would be consistently greater than the standard required to satisfy requirement E1.

To guarantee that an installed separating wall or floor will provide a performance, in terms of sound insulation, which is greater than the prescribed values of $D_{nT,w} + C_{tr}$ in Approved Document E is demanding since there are inherent variations in the composition of the materials forming the components as well as quality of workmanship. Due allowance has had to be made for variables such as these in selecting target values for the sound insulation offered by an RSD.

Shortly before the publication of Approved Document E 2003, a project was initiated by the House Builders Federation targeted at providing a series of RSDs for new houses and apartments as an alternative to pre-completion testing. The impetus for this work was a ministerial statement to the effect that the Building Regulations Advisory Committee will give consideration as to whether or not the RSDs produced by the House Builders Federation achieve their objective. The results of this project were presented to the Building Regulations Advisory Committee during the summer of 2003 and a consultation document:

> 'Amendment of the Building Regulations to allow Robust Standard Details to be used as an alternative to pre-completion testing'

was produced by the Office of the Deputy Prime Minister in August 2003.

It is not known whether or not the concept of RSDs as an alternative to pre-completion testing will be accepted by government. If RSDs are accepted it is likely that Regulations 20A and 12A will require amendment. House builders will then have a choice between two approaches to acoustic design.

The subject of robust standard details is not discussed in this chapter. The aim of the chapter is to provide a guide to Approved Document E 2003.

11 Ventilation

11.1 Introduction

The need to provide adequate ventilation to buildings has long been recognised in building control legislation. Formerly, however, it was restricted to dwellings, and to bathrooms and rooms containing sanitary conveniences in buildings other than dwellings. The building regulations now extend the requirement for adequate means of ventilation to all building types.

It should be noted that if the provisions of Approved Document F are followed, then it would prevent the service of an improvement notice under Section 23(3) of the Health and Safety at Work etc Act 1974. This relates to the requirements for ventilation contained in regulation 6 (1) of the Workplace (Health, Safety and Welfare) Regulations 1992 and is the first time that such a connection between the Building Regulations and the Workplace Regulations has been made.

The 1995 edition of Approved Document F introduced a number of significant changes to the way in which ventilation is treated in domestic buildings including:

- The use of passive stack ventilation or open-flued heating appliances as an alternative to mechanical extract ventilation.
- The need to ventilate utility rooms.
- Removal of the restrictions on enclosed courtyards.
- Removal of the need to ventilate common spaces in flats.

11.2 Means of ventilation

In general, there must be adequate means of ventilation provided for people in buildings. The following are exempted from this rule because providing ventilation in them would not serve to protect the health of the users:

- Buildings or spaces within buildings where people do not normally go.
- Buildings or spaces within buildings used solely for storage.
- Garages used solely in connection with a single dwelling.

The provisions of Approved Document F1 are designed to ensure that suitable air quality is maintained in buildings.

Without adequate ventilation, moisture (leading to mould growth) and pollutants (originating inside a building) may accumulate to such levels that they become a hazard to the health of users of the building. For these reasons AD F1 recommends the following methods of ventilation which may be adopted for use in buildings:

- Extract ventilation (either natural or mechanical)

This is used to remove water vapour or pollutants from areas where they are produced in significant quantities and before they become widespread. Clearly, this would apply to kitchens, utility rooms and bathrooms in the case of water vapour. Interestingly, with regard to the extraction of pollutants, AD F1 includes not only rooms containing processes which produce harmful contaminants but also rest rooms where smoking is permitted. This would appear to be yet another victory for the anti-smoking lobby!

- Rapid dilution

Normally, this would be achieved by providing a door or window which could be thrown open as required. In sanitary accommodation which is not within a bathroom a similar level of rapid dilution may be obtained by mechanical extraction.

- Background ventilation

The guiding principle here is that a minimum supply of fresh air should be available over a long period of time to disperse residual water vapour as necessary. It is important that the means of ventilation should not compromise security or comfort and should resist rain penetration.

In non-domestic buildings, it is often the case that ventilation is provided by mechanical means or by or air-conditioning systems. These are permissible provided that they achieve the performance listed above, and they are:

(a) designed, installed and commisioned so that their performance will not put at risk the health of people in the building, and
(b) designed to permit necessary maintenance so that all the objectives outlined above may continue to be achieved.

11.2.1 Interpretation

Special definitions apply to AD F1.

- VENTILATION OPENING – Includes any permanent or closeable means of ventilation which opens directly to external air as follows:
  - (a) opening lights in windows
  - (b) louvres
  - (c) airbricks
  - (d) progressively openable ventilators, window trickle ventilators
  - (e) doors.

Undoubtedly, the most common way of providing background ventilation is via a trickle ventilator located in or above a window frame and it is also possible to obtain glazing systems containing this facility.

Airbricks, ducted through a wall and finished internally with a 'hit and miss' ventilator are also permissible provided that the main air passages are large enough to minimise resistance to airflow. Therefore, slots should have a minimum dimension of 5 mm and any square or circular holes should be at least 8 mm across (excluding any insect screens or baffles, etc.).

The two methods mentioned above rely on providing means of ventilation which are additional to the windows or external doors of a room. In fact, the windows themselves can be used for background ventilation provided that they are of a suitable type and can be secured in the open position to provide the amount of ventilation recommended in Table 11.1. Vertical sliding sash windows are ideal for this purpose, the background ventilation being provided by opening the top sash and locking it in the required position. Top hung opening casement windows may also be suitable provided that the light can be locked in at least two opening positions. Since these windows are more easily forced they should be restricted to use above ground floor level.

However background ventilation is provided, it should always be located so that it does not cause discomfort from cold draughts. Additionally, it is unlikely to be effective if it is less than 1.75 m above floor level.

Where windows with adjustable locking positions are used to supply background ventilation there is a danger that they would be unusable as a means of escape in case of fire. Approved Document B1 contains details of windows which would need to be used for escape purposes and it is unlikely that these could be used for background ventilation as described above. (See Chapter 7, section 7.12.3 for details of means of escape windows.)

- HABITABLE ROOM – A room used for dwelling purposes which is not solely a kitchen.
- BATHROOM – A room containing a bath or shower with or without sanitary accommodation.
- SANITARY ACCOMMODATION – A room which contains one or more closets or urinals. If sanitary accommodation contains one or more cubicles it is not necessary to provide separate ventilation to each if air is free to circulate throughout the space.
- UTILITY ROOM – A room in which water vapour is likely to be produced in significant quantities because it is designed or intended to be used to contain clothes washing or similar equipment such as a sink, washing machine or tumble drier, etc. It should be noted that ventilation does not need to be provided under Building Regulations if the utility room can be entered solely from outside the building.
- OCCUPIABLE ROOM – Includes rooms occupied by people in non-domestic buildings such as offices, workrooms, classrooms, hotel bedrooms etc.

  Excluded from this definition are:

Table 11.1 Ventilation recommendations.

Ventilation recommendations for rooms capable of containing openable windows				Ventilation recommendations for rooms not containing openable windows	
1 Room or space	**2 Rapid ventilation**	**3 Background ventilation**	**4 Extract ventilation (fan rates)**	**5 Mechanical extract ventilation (fan rates)**	**6 Notes**
1 Domestic buildings					
(a) Habitable rooms	Ventilation opening equal to at least $\frac{1}{20}$th room floor area	8000 mm^2	* See note column 6	For mechanical ventilation see BS 5720: 1979 and BRE Digest 398	* No recommendation given in AD F1
(b) Kitchens	Opening window (any size)	4000 mm^2	30 litres/sec in or adjacent to hob. 60 litres/sec elsewhere	Mechanical extract as column 4, with 15 minute overrun on fan connected to light switch for rooms without natural light	See also text page 11.7 and note at foot of column 4 below for alternatives to mechanical extract
(c) Utility room	Opening window (any size)	4000 mm^2	30 litres/sec	Mechanical extract as column 4, with 15 minute overrun on fan connected to light switch for rooms without natural light	No ventilation provisions necessary if room entered only from outside
(d) Bathroom	Opening window (any size)	4000 mm^2	15 litres/sec	Mechanical extract as column 4, with 15 minute overrun on fan connected to light switch for rooms without natural light	Bathroom may or may not contain WC
(e) Sanitary accommodation if separate from bathroom	$\frac{1}{20}$th room floor area as habitable room in (a) above	4000 mm^2	* See note column 6		* See also BS 5720: 1979 and BRE Digest 398
2 Non-domestic buildings					
(f) Occupiable room	$\frac{1}{20}$th room floor area as habitable room in (a) above	4000 mm^2 for room floor areas up to 10 m^2 400 mm^2/m^2 for room floor areas over 10 m^2	* See note column 6	For mechanical ventilation allow 8 litres/sec per occupant of fresh air for rooms where no smoking is permitted and 16 litres/sec per occupant for rooms where light smoking is permitted. For rooms designed for heavy smoking, such as rest rooms where smoking is allowed see Table 11.2	* No table recommendation given in AD F1

(g) Kitchen (For type see note)	Opening window (any size)	4000 mm^2	30 litres/sec in or adjacent to hob. 60 litres/sec elsewhere	Mechanical extract as column 4, with 15 minute overrun on fan connected to light switch or occupant detecting sensor for rooms without natural light	Recommendations are for kitchens similar to domestic i.e. *not* commercial kitchens. See also text page 11.11 for alternatives to mechanical extract
(h) Bathroom	Opening window (any size)	4000 mm^2 per bath or shower	15 litres/sec per bath or shower	Mechanical extract as column 4, with 15 minute overrun on fan connected to light switch or occupant detecting sensor for rooms without natural light	(Includes shower rooms)
(j) Sanitary accommodation (with or without washing facilities)	$\frac{1}{20}$th room floor area as habitable room in (a) above	4000 mm^2 per WC	* See note column 6	6 litres/sec per WC mechanical extract or 3 air changes/hour, with 15 minute overrun on fan. Fan may be connected to light switch or occupant detector for rooms without natural light	* See also BS 5720: 1979 and CIBSE Guides A and B
(k) Common spaces	No recommendations for rapid, background or extract ventilation in AD F1. Instead provide either natural ventilation equal to $\frac{1}{50}$th of floor area of common space *or* mechanical ventilation at rate of 1 litre/sec per m^2 floor area				Applies to spaces where large numbers of people gather, e.g. shopping malls and foyers *not* spaces used principally for circulation
Notes		In domestic buildings the areas given in (a) to (e) above may be varied to give an average of 6000 mm^2 for the dwelling with at least 4000 mm^2 per room	Extract ventilation may also be by passive stack ventilation or appropriate open-flued heating appliance – see text page 11.7	In domestic buildings in (b) to (e) and non-domestic buildings in (g) to (j) above, an air inlet should be provided to each room such as a 10 mm gap under the door	

(a) Bathrooms, sanitary accommodation and utility rooms, and
(b) Rooms or spaces used solely or mainly for circulation, building services plant and storage.

- DOMESTIC BUILDINGS – Buildings used for dwelling purposes such as dwelling houses, flats, residential accommodation and student hostels.
- NON-DOMESTIC – All buildings not contained in the definition of domestic buildings above. To avoid confusion, buildings where people reside only temporarily such as hotels, are regarded as non-domestic buildings.
- PASSIVE STACK VENTILATION (PSV) – A system of ventilation which relies on the natural stack effect (in which warm air rises due to the difference in temperature between the inside and outside of a building) and the effect of wind passing over the roof. The system consists of a series of ceiling outlets connected by ducts to terminals on the roof of a building (see Fig. 11.1).

  No positive guidance on PSV is given in AD F1 apart from referring the reader to BRE Information Paper 13/94 (*Passive stack ventilation systems: design and installation*). Additionally, any system with appropriate third party certification (e.g. a British Board of Agrément certificate) would also be acceptable.

It should be noted that where rooms have a double function the individual provisions for rapid, background and extract ventilation shown in Table 11.1 need not be duplicated. Instead, the room will need to be provided with the greater provision for each of the individual functions listed in the Table. Therefore, if a room is a kitchen-diner then the extract ventilation recommendations for the kitchen will need

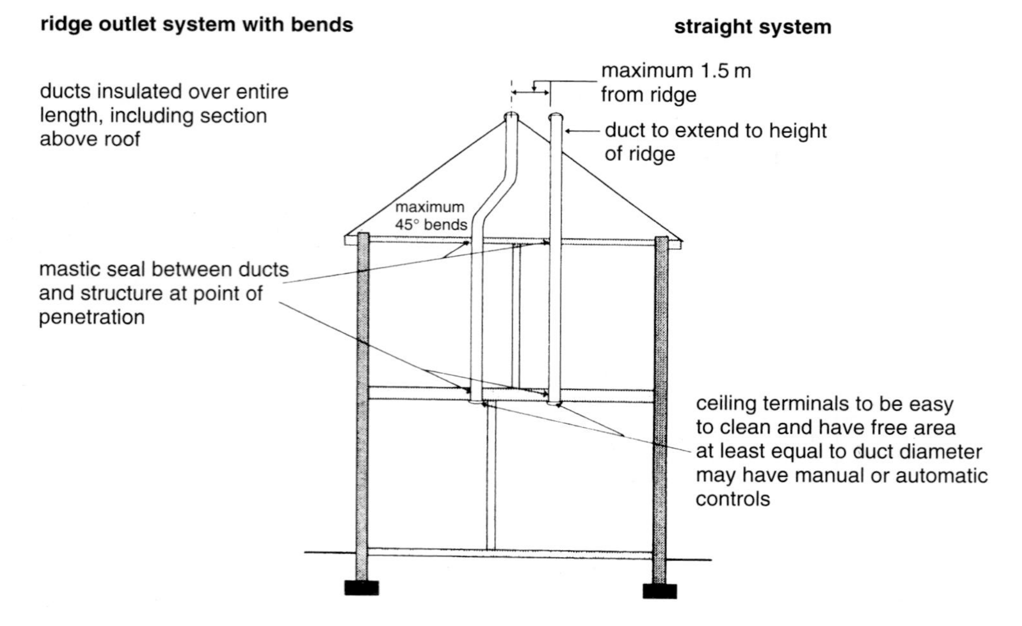

Fig. 11.1 Passive stack ventilation – principles.

to be included in addition to those for rapid and background ventilation for the dining area.

11.3 Ventilation of domestic buildings

11.3.1 Background and rapid ventilation

Kitchens, habitable rooms, utility rooms, bathrooms and sanitary accommodation in domestic buildings should be provided with *background ventilation* and *rapid ventilation*.

Both are achieved by providing a ventilation opening (or openings), some part of which should be at high level (typically 1.75 m above floor level) and with a total area as shown in Table 11.1.

Additionally, background ventilation should be controllable, secure and located to avoid draughts and rain ingress. It might consist of an airbrick, trickle ventilator or suitable opening window as described on page 11.3 above. It should be noted that it is permissible to vary the recommendations of Table 11.1 provided that an average of 6000 mm^2 per room for background ventilation can be achieved with an absolute minimum of 4000 mm^2 in each room.

11.3.2 Extract ventilation

In addition to the recommendations for background and rapid ventilation mentioned above, kitchens, utility rooms and bathrooms in domestic buildings should also be provided with *extract ventilation*. Extract ventilation may be achieved using any of the following:

- mechanical extract ventilation; or
- passive stack ventilation; or
- a suitable open-flued heating appliance.

Mechanical extract ventilation can be operated manually and/or automatically by a controller or sensor. The recommendations are shown in Table 11.1 and fans should be rated at not less than:

- 15 litres/second for a bathroom,
- 30 litres/second for a utility room,
- 30 litres/second for a kitchen where the fan is located within a cooker hood or is not less than 300 mm from the centreline of the hob space, is under humidistat control and is located near the ceiling, or
- 60 litres/second if the fan is located elsewhere in the kitchen.

Passive stack ventilation may be provided by a manual and/or automatic system which uses controllers or sensors to close the system when moisture has been removed (see Fig. 11.1 above).

Open-flued heating appliances can be used as a means of extract ventilation when they are in operation because they take air which they need for combustion from the room or space in which they are installed. When not in operation the appliance should still be capable of providing adequate extract ventilation. Most solid fuel open-flued appliances are acceptable provided that they are used as a primary source of heating, cooking or hot water production. Appliances which burn other fuels may have control dampers which block the air flow when they are not in use or they may have a flue diameter which is insufficient to allow a free flow of air. In these circumstances it is necessary to check that:

- the appliance has a flue with a free area which is at least equivalent to a 125 mm diameter duct, and
- the appliance has combustion and dilution air inlets which are permanently open when it is not in use so that the ventilation path is unrestricted. These recommendations for rapid, background and extract ventilation are illustrated in Fig. 11.2.

11.3.3 Ventilation of rooms not containing opening windows

In order to make best use of available space in a dwelling, *non-habitable* rooms, such as kitchens, utility rooms, bathrooms and sanitary accommodation are often positioned away from the external walls. This is true especially in flats and since it is not possible to have opening windows in these situations AD F1 permits the use of mechanical and other forms of ventilation as follows:

- mechanical extract ventilation rated as in Table 11.1 operated by a fan, (controlled either automatically using an occupant detector or manually by connection to the light switch) which should continue to run for 15 minutes after the room has been left or the light switched off; or
- passive stack ventilation as in section 11.2.1 and Fig. 11.1; or
- a suitable open-flued heating appliance.

An air inlet should always be provided to internal rooms ventilated as above and this could be, for example, a 10 mm gap under the door to the room.

An internal *habitable* room may be ventilated through an adjoining room if there is a permanent opening between them with an area equal to $\frac{1}{20}$th of their combined floor areas. Additionally, the ventilation recommendations in Table 11.1 should be based on the combined floor areas of the two rooms.

A habitable room opening onto a conservatory or similar space may be treated as one with the conservatory for the purposes of ventilation. The opening between the room and conservatory (which may contain a door or window, for example) should have an area equal to $\frac{1}{20}$th of the combined floor area of the two rooms (some part of which should be at least 1.75 m above floor level), for rapid ventilation. Background ventilation of 8000 mm^2 should also be provided between the two rooms. Additionally, these provisions for both rapid and background ventilation should be made from the conservatory to outside air (see Fig. 11.2).

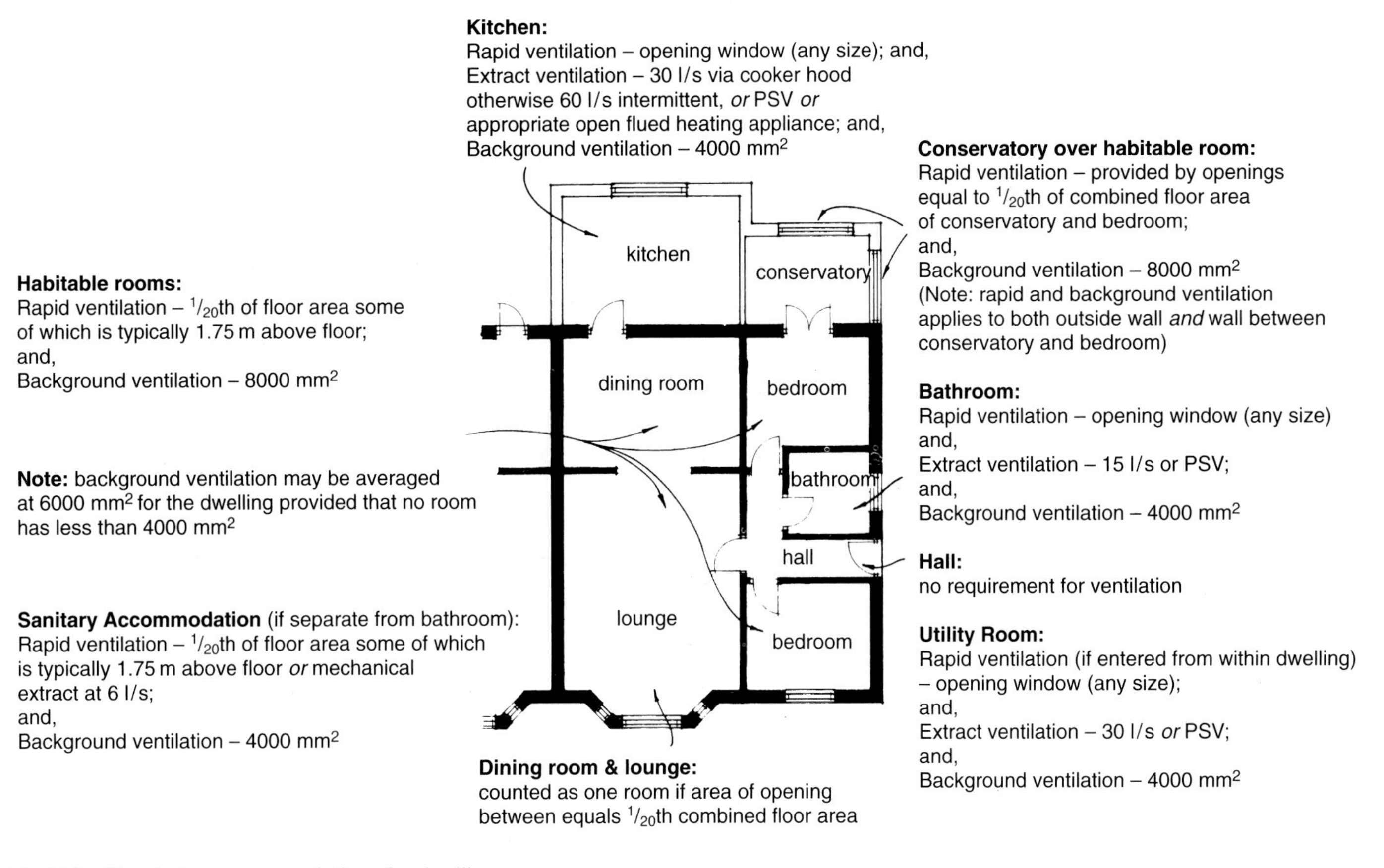

Fig 11.2 Ventilation recommendations for dwellings.

11.3.4 Mechanical ventilation

As an alternative to the foregoing methods of ventilation, the requirements of regulation F1 may also be satisfied by:

- following the recommendations of BRE Digest 398 *Continuous mechanical ventilation in dwellings; design installation and operation.* The Digest describes two approaches to the provision of continuous mechanical ventilation which may be applied either to the entire dwelling using a balanced (supply and extract) system or may only apply to the kitchen, utility room, bathroom and sanitary accommodation; or,
- following clauses 2.3.2.1, 2.3.3.1, 2.5.2.9, 3.1.1.1, 3.1.1.3 and 3.2.6 of BS 5720: 1979 *Code of practice for mechanical ventilation and air-conditioning in buildings.*

11.3.5 Mechanical extract ventilation and open-flued heating appliances

Mechanical extract ventilation can create dangerous conditions where open-flued appliances are also present due to the spillage of flue gases, whether or not the fans and appliances are in the same room. Where this form of ventilation is provided merely to comply with the recommendations for extract ventilation shown in column 4 of Table 11.1, it is perfectly feasible to use the flue of the combustion appliance for extract ventilation provided it complies with the recommendations shown in section 11.3.2 above (see *Open-flued heating appliances*).

However, there may still be occasions when it is thought desirable to install open-flued appliances in conjunction with mechanical extract ventilation. In these circumstances it is essential that the appliance is able to operate safely whether or not the fan is running. The risk of danger from the spillage of flue gases will vary according to the type of fuel being burnt. Therefore, whereas mechanical extract ventilation should never be provided in the same room as an open-flued appliance burning *solid* fuel (but for further advice contact the Heating Equipment Testing and Approval Scheme, PO Box 37, Bishop's Cleeve, Gloucestershire, GL52 4TB), it may be possible to use *gas* or *oil* burning appliances in conjunction with mechanical extract ventilation as follows:

- with gas appliances that are located in a kitchen which is mechanically ventilated, it has been found that an extract rate of not more than 20 litres/sec will be unlikely to cause spillage of flue gases, although it will be necessary to carry out a spillage test in accordance with BS 5440: Part 1, Clause 4.3.2.3. This should be done even though the appliance may be located in a different room to the fan. If this causes spillage then it may be necessary to reduce the extract rate still further until the problem is cured.
- advice on the installation of *oil-fired* appliances is contained in Technical Information Note T1/112 which may be obtained from the Oil Firing Technical Association for the Petroleum Industry (OFTEC), Century House, 100 High Street, Banstead Surrey, SM7 2NN.

General information and advice on the subject of the interaction between mechanical extract ventilation and open-flued appliances, including details of the spillage test, may be found in BRE Information Paper 21/92, *Spillage of flue gases from open-flued combustion appliances*.

See also Approved Document J (section 14.5.1 below) for details of the provision of combustion air to fuel burning appliances.

11.3.6 Ventilation of domestic buildings – alternatives to the AD F1 recommendations

Since AD F1 is not a mandatory document, it is possible to use other advice when providing ventilation in domestic buildings. It is not possible to summarise these other sources of information in this book but the reader may find the following to be of interest:

- BS 5925:1991 *Code of practice for ventilation principles and designing for natural ventilation*, especially clauses 4.4, 4.5, 4.6.1, 4.6.2, 5.1, 6.1, 6.2, 7.2, 7.3, 12 and 13, or,
- BS 5250: 1989 *Code of practice for the control of condensation in buildings*, especially clauses 6, 7, 8, 9.1, 9.8, 9.9.1, 9.9.2, 9.9.3 and Appendix C.

11.4 Ventilation of non-domestic buildings – general activities

The ventilation recommendations for general activities in non-domestic buildings in AD F1 follow a similar pattern to those already described for domestic buildings. Provision should be made for rapid, background and extract ventilation and the guidance summarised in Table 11.1 should be read with the following comments:

- An occupiable room in which heavy smoking is to take place (such as a rest room designed for this purpose) should comply with the recommendations shown in Table 11.2.
- The kitchens referred to in the Table are of the domestic type and are not to be construed as commercial kitchens. For further guidance on these see Table 11.2.
- Bathrooms include shower-rooms.
- Sanitary accommodation includes rooms which also contain washing facilities or rooms containing solely washing facilities.
- Extract ventilation can be provided by mechanical means operated manually and/or automatically by a controller or sensor, or by passive stack ventilation for domestic type facilities. The use of open-flued combustion appliances to provide extract ventilation is not mentioned in this part of AD F1 although it is permitted in domestic buildings. Even so, the approved document still recommends that caution be exercised with regard to the use of mechanical extract in a building containing open-flued appliances.
- Background ventilation may be provided by the same means as is described for domestic buildings in section 11.2.1 above.

Table 11.2 Ventilation recommendations for specialist activities – non-domestic buildings.

1 **Use of building or room**	2 **Approved Document F1-specific recommendations**	3 **Alternative further guidance documents**	4 **Notes**
(a) School/ educational establishment	General areas as Table 11.1. Sanitary accommodation at rate of 6 air changes/hour	See Education (School Premises) Regulations	Fume cupboards complying with Dept of Education Design Note 29 may be needed for areas where noxious fumes generated
(b) Workplaces	—	See Health and Safety Executive Guidance Note EH 22 *Ventilation of the workplace*	
(c) Hospitals	—	See DHSS *Activity Data Base*. For general guidance and standard of provision see individual Dept of Health Building Notes for specific departmental areas	Ventilation needs of different types of accommodation vary with use and may vary throughout year
(d) Building services plant rooms	—	See BS 4434: 1989 *Specification for safety aspects in the design, construction and installation of refrigeration appliances and systems*	Provision may be necessary for emergency ventilation to control dispersal of contaminating gas releases, such as refrigerant leaks. See HSE Guidance Note EH 22 *Ventilation of the workplace*, paragraphs 25 to 27
(e) Rest rooms where smoking allowed	If natural ventilation possible provide both: • air supply to Table 11.1 for occupiable room, and • local extraction to remove tobacco smoke. If mechanical ventilation provided allow extract rate of 16 litres/sec per person	—	Workplace (Health and Safety) Regulations 1992 require rest rooms and rest areas to have suitable arrangements to protect non-smokers from discomfort caused by tobacco smoke

(f) Car parks	If *naturally* ventilated provide well distributed permanent ventilation at each level equivalent to $\frac{1}{20}$th floor area at that level with at least 50% in opposing walls. If *mechanically* ventilated provide *either*: • *both* natural permanent vents not less than $\frac{1}{40}$th of floor area *and* mechanical ventilation of min. 3 air changes/hour, *or* • 6 air changes per hour for basement car parks and local ventilation at rate of 10 air changes per hour on ramps and exits wherer cars queue inside building with engines running.	See Association for Petroleum and Explosives Administration publication entitled *Code of practice for ground floor, multi-storey and underground car parks*, or CIBSE Guide B, Section B2.6 and Table B2.7.	Recommendations apply to car parks which are: • below ground level, or • enclosed, or • multi-storey. Instead of provisions in columns 2 and 3 it is also possible to calculate mean predicted pollutant levels and design ventilation system to limit carbon monoxide concentration to: • not exceeding 50 parts per million average over 8 hour period, and • not exceeding 100 parts per million for periods not exceeding 15 minutes of peak concentration on ramps and exits.
(g) Commercial Kitchens	—	See Chartered Institution of Building Services Engineers Guide B, Tables B2.3 and B2.11	

11.4.1 Ventilation of communal areas in non-domestic buildings

Many non-domestic buildings have areas where large numbers of people gather, such as foyers in cinemas and theatres, or enclosed shopping malls. Clearly, such spaces need to be ventilated or the air in them will become stale and unhealthy conditions might arise. AD F1 recommends that common spaces should be ventilated either:

- Naturally by means of suitably positioned ventilation opening(s) sufficient to give an opening area equivalent to $\frac{1}{50}$th of the floor area of the common space, or
- Mechanically so that fresh air may be provided at a rate of 1 litre/sec per m^2 of floor area.

It should be noted that the above recommendations do not apply to common spaces used solely or principally for circulation, although AD B1 (means of escape) should also be consulted since it contains certain recommendations regarding ventilation of such spaces (see Chapter 7 section 7.14.3 above).

11.4.2 Ventilation of non-domestic buildings – alternatives to the AD F1 recommendations

As was the case with domestic buildings, it is possible to use certain alternative recommendations to those contained in AD F1 when providing ventilation in non-domestic buildings. The recommendations for rapid, background and extract ventilation shown in Table 11.1, and the guidance given on ventilation to common spaces in non-domestic buildings may also be satisfied by following the advice given in:

- BS 5925: 1991 *Code of practice for ventilation principles and designing for natural ventilation*, clauses 5.1, 5.2, 6.1, 6.2, 7.3, 12 and 13; or
- Chartered Institution of Building Services Engineers (CIBSE) Guide A: *Design data*, section A4 *Air infiltration and natural ventilation*, and CIBSE Guide B: *Installation and equipment data*, section B2 *Ventilation and air-conditioning (requirements)*.

11.4.3 Mechanical ventilation

Many non-domestic buildings contain non-habitable rooms such as, kitchens, bathrooms and sanitary accommodation which are situated away from external walls and are unable to be provided with windows or other ventilation openings. These rooms can be fitted with mechanical extract ventilation at the rates shown in Table 11.1 operated by connection to the light switch or an occupant detector. The fan should have a 15 minute overrun facility and some form of air inlet should be provided to the room, such as a 10 mm gap under the door.

Mechanical ventilation to occupiable rooms should be provided at a rate of:

- 8 litres/sec per occupant for rooms where smoking is not permitted, or
- 16 litres/sec per occupant for rooms designed for light smoking.
- For rooms which are specifically designed for heavy smoking, such as rest rooms where smoking is allowed see Table 11.2.

11.4.4 Mechanical ventilation and air-conditioning plant – design, maintenance and commissioning

Since the air in a building cannot be continuously recycled, at some point in the design of a mechanical ventilation or air-conditioning system it is necessary to introduce fresh air to replace stale air which is being exhausted. Unless care is taken in the siting of the inlets and outlets to the system it is possible that contaminants which are injurious to health may be introduced into the system. Therefore, fresh air inlets should be situated away from areas such as:

- flues
- exhaust outlets from ventilation systems
- evaporative cooling towers
- areas where vehicles manoeuvre.

General guidance on how to deal with recirculated air in mechanical ventilation and air-conditioning systems may be found in paragraph 32 of the *Approved Code of Practice and Guidance* L24, issued under the Workplace (Health, Safety and Welfare) Regulations 1992, by the Health and Safety Executive.

Further guidance on the design of mechanical ventilation and air-conditioning systems may be found in BS 5720: 1979 by following clauses 2.3.2, 2.3.3, 2.4.2, 2.4.3, 2.5, 3.2.6, 3.2.8, and 5.5.6 *or* CIBSE Guide B sections B2 and B3.

There is one particular disease which is associated with air-conditioning systems known as legionnaires' disease. This is a form of lung infection caused by the bacteria legionella pneumophilia and is named after an epidemic which affected 182 people attending an American Legion Convention in 1976. In the original outbreak the germ was found to have been transmitted through the cooling and evaporating elements of a large, central air-conditioning system. Since a large number of subsequent outbreaks of the disease have been traced to similar sources it is essential that cooling and heating systems are cleaned regularly and filters are changed often. Further information may be obtained from the guide issued by the Health and Safety Executive, *The control of legionellosis including legionnaires' disease*, paragraphs 71 to 89.

In order to be able to carry out the regular maintenence of cleaning the system and replacing filters it is essential that all parts are available for access and sufficient space is provided, especially in central plant rooms. Normally, special provision for access will be made in the design of the system. Where this is not the case, AD F1 recommends the following minimum dimensions for access passageways and cleaning points in central plant rooms:

- 600 mm wide by 2000 mm high general access passageways for walking between plant, and

- 1100 mm wide by 1400 mm high kneeling spaces for routine cleaning and maintenance of equipment. Additionally, a 690 mm high space should be available for access to low level equipment.

Since these figures are the minimum recommended they do not necessarily include for access doors which may need additional space.

This approved document guidance is very limited and it may prove more useful to consult Building Services Research and Information Association Technical Note TN 10/92: *Space allowances for building services distribution systems – detailed design stage*, Sections A5 and D2. (BSRIA 1992, ISBN 0 86226 350 7).

Mechanical ventilation and air-conditioning systems are complex and need to be commissioned and tested to ensure that they are performing in accordance with their design specifications. Therefore, the local building control authority will need to be satisfied that the installed systems have been commissioned and tested so that they are performing their ventilation functions effectively. This recommendation applies only to a system:

- which is installed in a building to serve a floor area greater than 200 m^2, and
- in which the other provisions for mechanical ventilation and air-conditioning mentioned above have been followed.

Compliance may be demonstrated to the local authority by presenting them with test reports and commissioning certificates showing that commissioning and testing has been carried out in accordance with the CIBSE commisioning codes.

11.4.5 Ventilation of non-domestic buildings – specialist activities

The recommendations listed in Table 11.1 refer to the ventilation of rooms which are used for activities of a general nature where the production of water vapour or small amounts of tobacco smoke are the main problems, and therefore the activities are not dissimilar to those encountered in domestic premises.

Many non-domestic buildings have rooms or spaces where large quantities of water vapour are produced or where noxious fumes may be generated and the provisions of Table 11.1 may be inadequate in these circumstances. Comprehensive recommendations for the ventilation of these specialist activities is beyond the scope of AD F1 so the reader is referred to a number of additional guidance documents where more detailed guidance may be sought. These are summarised in Table 11.2 and it should be noted that the recommendations for ventilation of car parks shown in the table relate to the provision of air to ensure normal healthy conditions are maintained. Reference should be made also to Approved Document B (see section 7.28.1 above) for guidance on the design of mechanical ventilation and air-conditioning systems for the purposes of fire safety.

11.5 Condensation in roofs

In buildings, adequate provision must be made to prevent excessive condensation in roofs and roof voids over insulated ceilings.

When condensation occurs in roof spaces it can have two main effects:

- the thermal performance of the insulant materials may be reduced by the presence of the water; and
- the structural performance of the roof may be affected due to increased risk of fungal attack.

Approved Document F2 recommends that, under normal conditions, condensation in roofs and in spaces above insulated ceilings should be limited such that the thermal and structural performance of the roof will not be substantially and permanently reduced.

AD F2 applies only to roofs where the insulation is placed at ceiling level (cold roofs) irrespective of whether the ceiling is flat or pitched. Warm roofs where the insulation is placed above the structural system and roof void do not present the same risks and, therefore, are not covered.

It should be noted that the provisions of AD F2 apply to roofs of any pitch even though a roof which exceeds 70° in pitch is required to be insulated as if it were a wall.

Small roofs over porches or bay windows, etc., may sometimes be excluded from the requirements of regulation F2 if there is no risk to health or safety.

11.5.1 Roofs with a pitch of 15° or more

Pitched roofs should be cross-ventilated by permanent vents at eaves level on the two opposite sides of the roof, the vent areas being equivalent in area to a continuous gap along each side of 10 mm width.

Mono-pitch or lean-to roofs should have ventilation at eaves level as above and also at high level either at the point of junction or through the roof covering at the highest practicable point. The high level ventilation should be equivalent in area to a continuous gap 5 mm wide (see Fig. 11.3).

11.5.2 Roofs with a pitch of less than 15°

In low-pitched roofs the volume of air contained in the void is less and therefore the risk of saturation is greater.

This also applies to roofs with pitch greater than 15° where the ceiling follows the pitch of the roof. High level ventilation should be provided as in 1.4 above.

Cross-ventilation should again be provided at eaves level but the ventilation gap should be increased to 25 mm width.

Where the roof span exceeds 10 m or the roof plan is other than a simple rectangle, more ventilation, totalling 0.6% of the roof area, may be needed.

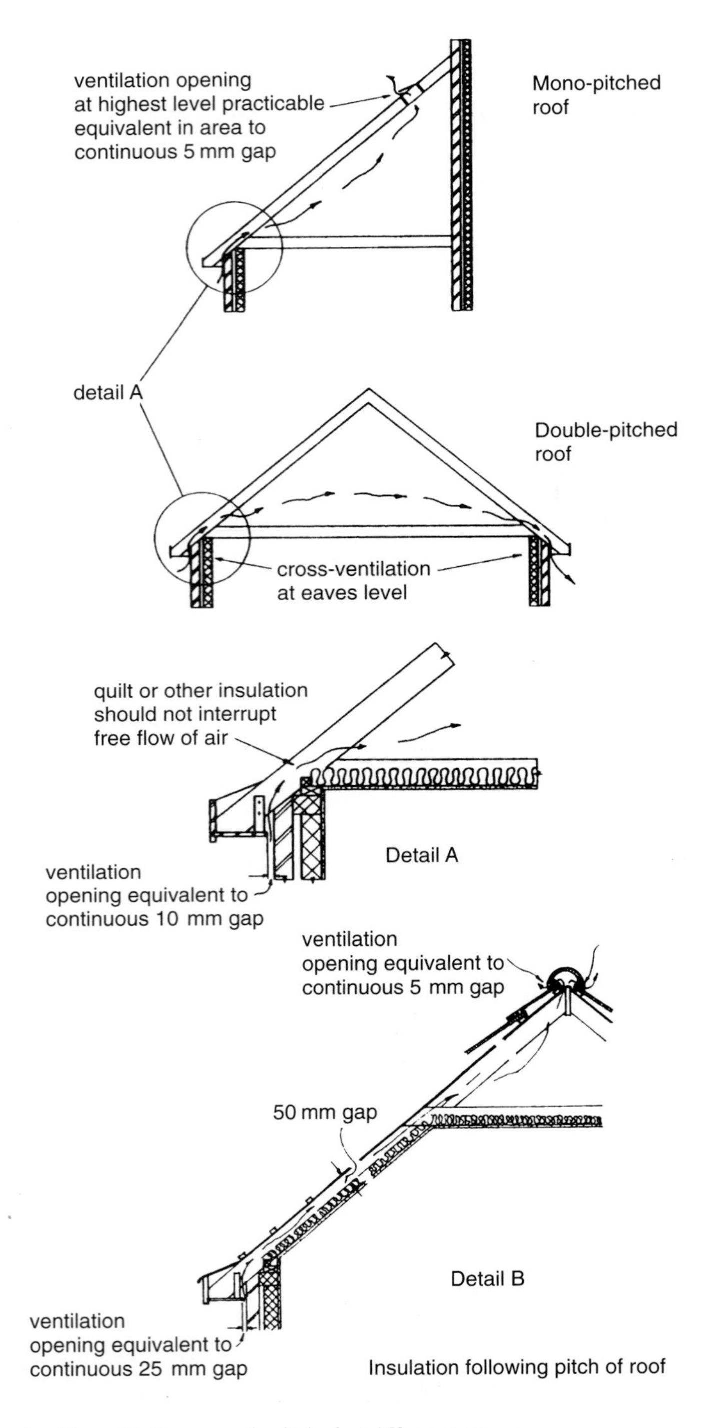

Fig. 11.3 Roof void ventilation – roofs pitched at 15° or more.

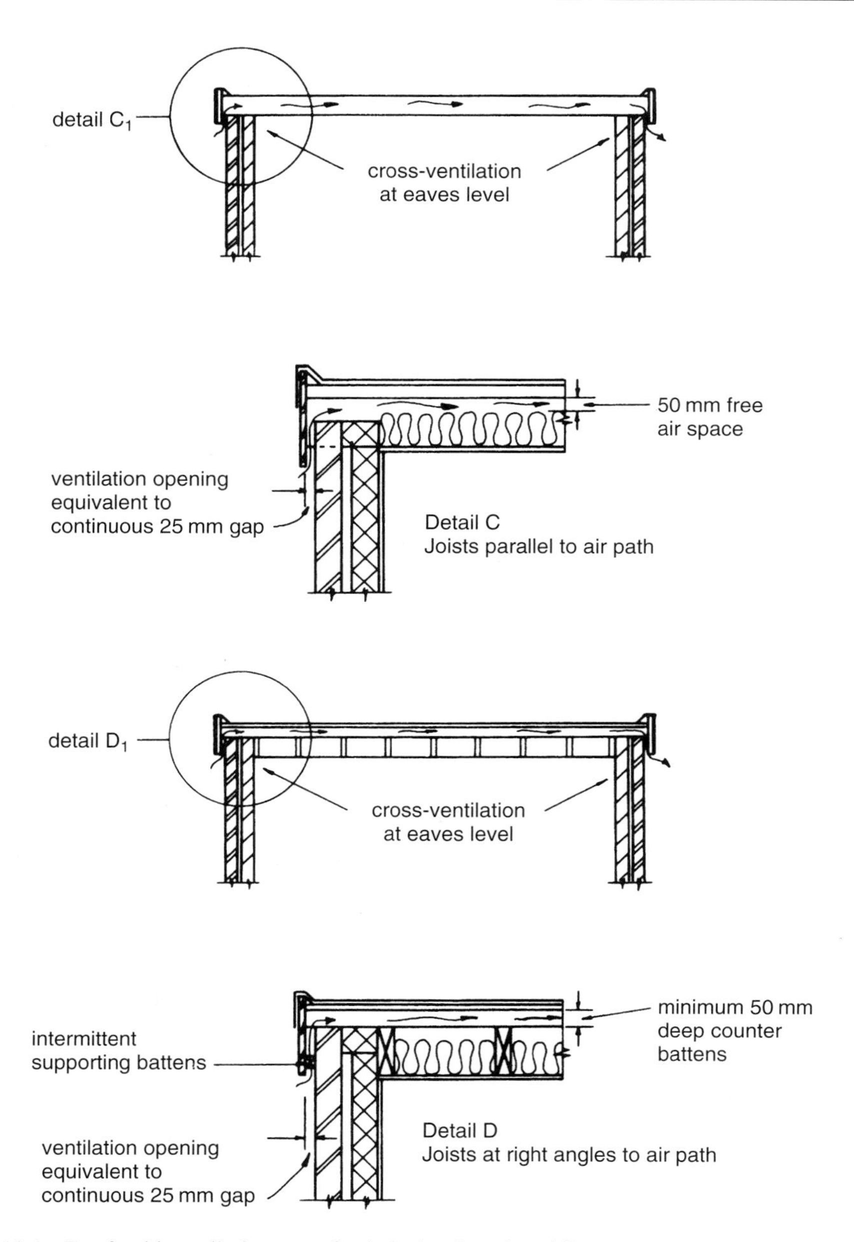

Fig. 11.4 Roof void ventilation – roofs pitched at less than 15°.

A free airspace of at least 50 mm should be provided between the roof deck and the insulation. This may need to be formed using counter-battens if the joists run at right angles to the flow of air (see Fig. 11.4).

Where it is not possible to provide proper cross-ventilation an alternative form of roof construction should be considered.

It is possible to install vapour checks (called vapour control layers in BS 5250) at ceiling level using polythene or foil-backed plasterboard, etc., to reduce the amount of moisture reaching the roof void. This is not acceptable as an alternative to ventilation unless a complete vapour barrier is installed.

The requirements can also be met for both flat and pitched roofs by following the relevant recommendations of BS 5250: 1989 *Code of practice: the control of condensation in buildings*, Clauses 9.1, 9.2 and 9.4. Further guidance may also be found in the 1994 edition of BRE Report BR 262 *Thermal insulation: avoiding the risks*.

12 Hygiene

12.1 Introduction

When first introduced, Part G of Schedule 1 to the Building Regulations 1985 consisted of four requirements grouped under the title 'Hygiene'.

The first of these requirements (G1 – Food storage) required that dwellings be provided with adequate food storage accommodation. Since most people have refrigerators or deep freezers today this regulation has become outdated and food storage is no longer controlled by the building regulations.

Consequently, the former regulation G4 (Sanitary conveniences and washing facilities) was renumbered G1 in the 1991 Regulations to fill the void left by the now defunct food storage requirements. This convention has been continued in the 2000 Regulations

The remaining regulations in Part G relate to Bathrooms (G2) and Hot water storage (G3).

12.2 Sanitary conveniences and washing facilities

Adequate sanitary conveniences (i.e. closets and urinals) situated in purpose-built accommodation or bathrooms, must be provided in buildings. This requirement replaces section 26 of the Building Act 1984.

Additionally, adequate washbasins with suitable hot and cold water supplies must be provided in rooms containing water-closets or in adjacent rooms or spaces.

These sanitary conveniences and washbasins must be separated from places where food is prepared and must be designed and installed so that they can be cleaned effectively.

It may be noted that section 66 of the 1984 Act enables the local authority to serve a notice on an occupier requiring him to replace any closet provided for his building which is not a water-closet. The notice can only be served where the building has a sufficient water supply and a sewer available. Where a notice requiring closet conversion is served, the local authority must bear half the cost of carrying out the work.

A satisfactory level of performance will be achieved if:

- sufficient numbers of the appropriate type of sanitary convenience are provided depending on the sex and age of the users of the building
- washbasins with hot and cold water supply are provided either in or adjacent to rooms containing water-closets.

Both sanitary conveniences and washbasins should be sited, designed and installed so as not to be a health risk.

12.2.1 Provision of sanitary conveniences and washbasins

The following definitions apply in AD G1.

SANITARY CONVENIENCE – Closets and urinals.

SANITARY ACCOMMODATION – A room containing closets or urinals. Other sanitary fittings may also be present. Sanitary accommodation containing more than one cubicle may be treated as a single room provided there is free air circulation throughout the room.

WATER-CLOSET – Defined by section 126 of the Building Act 1984 as a closet which has a separate fixed receptacle connected to a drainage system and separate provision for flushing from a supply of clean water, either by the operation of mechanism or by automatic action.

AD G1 also permits the use of a chemical or other means of treatment where drains and water supply are not available. It is not clear whether earth-closets would be permitted, but on normal principles of interpretation it is unlikely that they would be.

Houses, flats and maisonettes should have at least one closet and one washbasin. This also applies to houses in multiple occupation (houses where the occupants are not part of a single household), if the facilities are available for the use of all the occupants.

In other types of buildings the scale of provision and the siting of appliances may be the subject of other legislation as follows:

- Workplace (Health, Safety & Welfare) Regulations 1992. Approved Code of Practice & Guidance.

(This document is not referred to in the current edition of ADGI because it was published after that document. However, it repeals those provisions of the Offices, Shops and Railway Premises Act 1963 and the Factories Act 1961 which are referred to in ADGI.)

- The Food Hygiene (General) Regulations 1970
- Part M of Schedule 1 to the 2000 Regulations (Access and facilities for disabled people).

The requirement to provide satisfactory sanitary conveniences can also be met, subject to other legislation, by referring to the relevant clauses of BS 6465 *Sanitary installations*, Part 1: 1984 which contains details of the scale of provision, selection and installation of sanitary appliances.

A room or space containing closets or urinals should be separated by a door from any area in which food is prepared or washing up done. AD G1 makes it clear, therefore, that a separate lobby is not required.

Additionally, washbasins should be placed:

- in the room containing the closet; or
- in the room or space immediately leading to the room containing the closet provided it is not used for food preparation; or
- in the case of dwellings, in the room or space adjacent to the room containing the closet.

In this last case it is unclear whether or not the space may be used for the preparation of food, but in all probability it is not.

Closets, urinals and washbasins should have smooth, readily-cleaned, non-absorbent surfaces.

Any flushing apparatus should be capable of cleansing the receptacle effectively. The receptacle should only be connected to a flush pipe or branch discharge pipe.

Any washbasins required by the provisions of regulation G4 should have a supply of hot water from a central source or unit water heater and a piped cold water supply.

12.2.2 Discharge from sanitary conveniences and washbasins

Water-closets should discharge via a trap and branch pipe to a soil stack pipe or foul drain.

In recent years a system of waste disposal has been developed in which the discharge from a waste appliance is fed into a macerator. The liquified contents are then pumped via a small bore pipe to the normal foul drainage system. A closet is permitted to be connected to such a system provided:

- a closet discharging directly to a gravity system is also available; and
- the macerator system is the subject of a current European Technical Approval issued by a member body of the European Organisation for Technical Approvals e.g. the British Board of Agrément. The conditions of use must be in accordance with the terms of the ETA.

Urinals which are fitted with flushing apparatus should have an outlet fitted with an effective grating and trap and should discharge via a branch pipe to a soil stack pipe or foul drain (see Approved Document H1 and Chapter 13 for details of drainage).

Washbasins should discharge via a trap and branch discharge pipe to a soil stack. If on the ground floor, it is permissible to discharge the basin to a gulley or direct to a drain.

12.3 Bathrooms

Dwellings are required to be provided with a bathroom containing a fixed bath or shower. Hot and cold water must also be supplied to the bath or shower. This requirement replaces section 27 of the Building Act 1984.

The foregoing requirements apply to dwellings (i.e. houses, flats and maisonettes) and houses in multiple occupation (houses where the occupants are not part of a single household). In the latter case the facility should be available to all the occupants.

The hot and cold water supplies should be piped to the bath or shower and hot water may come from a central source such as a hot water cylinder or from a unit water heater.

The discharge from the bath or shower should be via a trap and waste pipe to a gulley, soil stack pipe or foul drain direct (see Approved Document H1 and Chapter 13 for details of drainage).

A bath or shower may be connected via a macerator system provided it complies with a current European Technical Approval.

12.4 Hot water storage

A hot water storage system incorporating a hot water storage vessel which is not vented to the atmosphere must be installed by a competent person and adequate precautions must be taken to:

- prevent the water temperature exceeding 100°C; and
- ensure that any hot water discharged from safety devices is conveyed safely to a disposal point where it is visible but will not be a danger to users of the building.

The above requirements do not apply to space heating systems, systems which heat or store water for industrial processes and systems which store 15 litres or less of water.

Approved Document G3 describes the provisions for an unvented hot water storage system. In such a system, the stored hot water is heated in a closed vessel. Without adequate safety devices an uncontrolled heat input would cause the water temperature to rise above the boiling point of water at atmospheric pressure (100°C). At the same time the pressure would increase until the vessel burst. This would result in an almost instantaneous conversion of water to steam with the large increase in volume producing a steam explosion.

Water for domestic use is required at temperatures below 100°C, therefore, an explosion cannot occur if the water is released at these temperatures, however great the pressure, hence the precautions required by regulation G3 to prevent the water temperature exceeding 100°C.

The term 'domestic hot water' is defined in AD G3 as water which has been heated for washing, cooking and cleaning purposes. The term is used irrespective of the type of building in which an unvented hot water storage system is installed.

Figure 12.1 illustrates the three independent levels of protection which should be provided for each source of energy supply to the stored water. These are:

- Thermostatic control (see Part L, Chapter 16).
- Non self-resetting thermal cut-outs to BS 3955: 1986 (electrical controls) or BS 4201: 1979 (for gas burning appliances).
- One or more temperature operated relief valves to BS 6283 *Safety devices for use in hot water systems* Part 2: 1991 or Part 3: 1991.

These safety devices are required for both directly and indirectly heated unvented hot water storage systems. The safety devices are designed to work in sequence as the temperature rises. All three means of protection would have to fail for the water temperature to exceed 100°C.

AD G3 provides separate recommendations for smaller (usually domestic) systems (not exceeding 500 litres capacity with a heat input below 45 kW) in section 3. Systems which exceed 500 litres capacity or have a heat input in excess of 45 kW are dealt with in section 4.

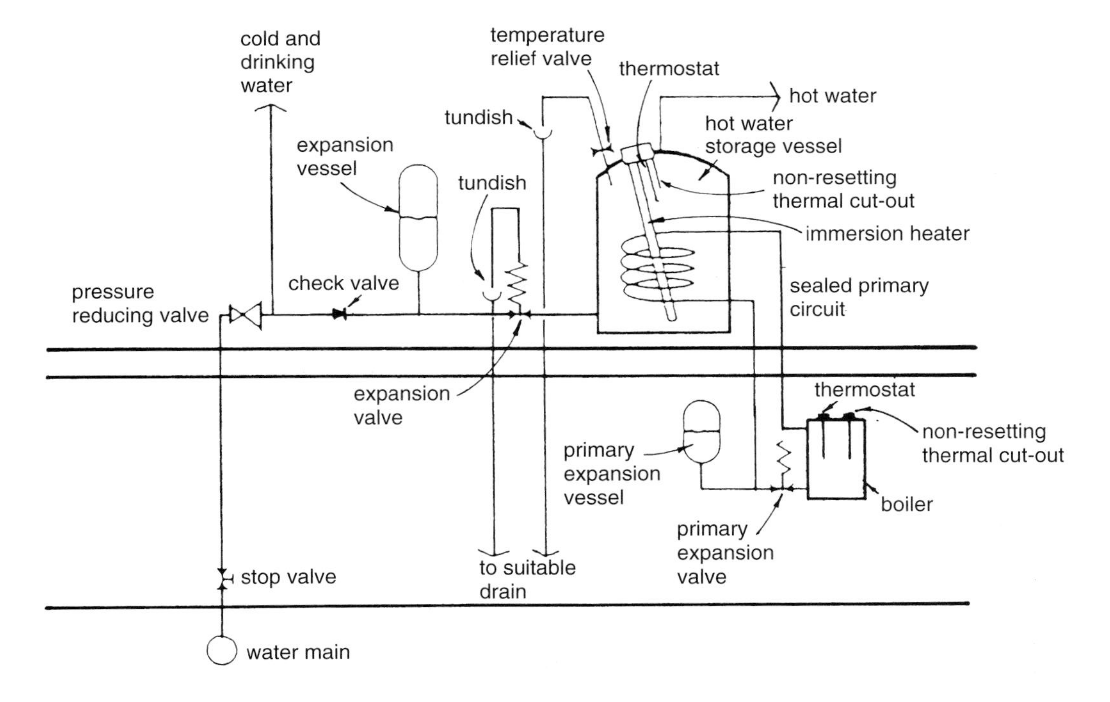

Fig. 12.1 Directly and indirectly heated unvented hot water storage system.

12.4.1 Section 3 hot water storage systems

Generally, a system covered by section 3 of AD G3 should be in the form of a unit or package which is:

- approved by a member body of the European Organisation for Technical Approvals (EOTA) operating a technical approvals scheme which ensures that the relevant requirements of regulation G3 will be met (e.g. the British Board of Agrément); or
- approved by a certification body having accreditation from the National Accreditation Council for Certification Bodies (NACCB). This would include testing to the requirements of an appropriate standard to ensure compliance with regulation G3 (e.g. BS 7206: 1990 *Specification for unvented hot water storage units and packages*); or
- independently assessed to clearly demonstrate an equivalent level of verification and performance to those above.

This means that the system should be factory made and supplied either as a *unit* (fitted with all the safety protection devices mentioned above and incorporating any other operating devices to stop primary flow, prevent backflow, control working pressure, relieve excess pressure and accommodate expansion fitted to the unit by the manufacturer) or as a *package* in which the safety devices are fitted by the manufacturer but the operating devices are supplied in kit form to be fitted by the installer.

This approach ensures that the design and installation of the safety and operating devices are carried out by the manufacturer who is conversant with his own equipment and can control the training and supervision of his staff.

It should be noted that where a system is subject to the above approvals it is unlikely to need site inspection by the building control authorities.

This may not be the case in other situations.

The recommendations for approval mentioned above ensure that the system is fit for its purpose and that the information regarding installation, maintenance and use of the system is made available to all concerned.

12.4.2 Provision of non self-resetting thermal cut-outs

Storage systems may be heated directly or indirectly. In an unvented, indirectly heated system (see Fig. 12.1) the non self-resetting thermal cut-out should be wired up to a motorised valve or other approved device or should shut off the flow to the primary heater. These devices should be subject to the same approvals as the units or packages.

Sometimes a unit system may incorporate a boiler. In this case the thermal cut-out may be located on the boiler.

In many cases an indirect system will also contain an alternative direct method of water heating (such as an immersion heater). This alternative heating source will also need to be fitted with a non self-resetting thermal cut-out. The non self-resetting thermal cut-out should be connected to the direct heat source or the indirect primary flow control device in accordance with BS 7671 Requirements for Electrical Installations. IEE Wiring Regulations.

12.4.3 Provision of temperature relief valves

Whether the unit or package is directly or indirectly heated the temperature relief valve should be situated directly on the storage vessel in order to prevent the stored water exceeding 100°C.

BS 6283 requires that each valve be marked with a discharge rating (in kW). This rating should never be less than the maximum power input to the vessel which the valve protects. More than one valve may be needed.

Valves should also comply with the following.

- They should not be disconnected except for replacement.
- They should not be relocated in any other position.
- The valve connecting boss should not be used to connect any other devices or fittings.
- They should discharge through a short length of metal pipe (D1) which is of at least the same bore as the valve's nominal outlet size.

  The discharge should either be direct or by way of a manifold which is large enough to take the total discharge of all the pipes connected to it. It should then continue via an air break to a tundish which is located vertically as near as possible to the valve.

It may be possible to provide an equivalent degree of safety using other safety devices but these would need to be assessed in a similar manner to units or packages (see above).

12.4.4 Installation

The installation of the system should be carried out by a competent person.

This means a person who holds a current Registered Operative Identity Card for the installation of unvented domestic hot water storage systems issued by:

- the Construction Industry Training Board (CITB); or
- the Institute of Plumbing; or
- the Association of Installers of Unvented Hot Water Systems (Scotland and Northern Ireland); or
- designated Registered Operatives who are employed by companies included on the list of Approved Installers published by the BBA up to 31 December 1991; or
- an equivalent body.

12.4.5 Discharge pipes

Discharge pipe D1 (see above and Fig. 12.2) is usually supplied by the storage system manufacturer. (If not it should be fitted by the installer of the system.)

In either case the tundish should be:

- vertical;
- located in the same space as the unvented hot water storage system; and
- fitted within 500 mm of the safety device.

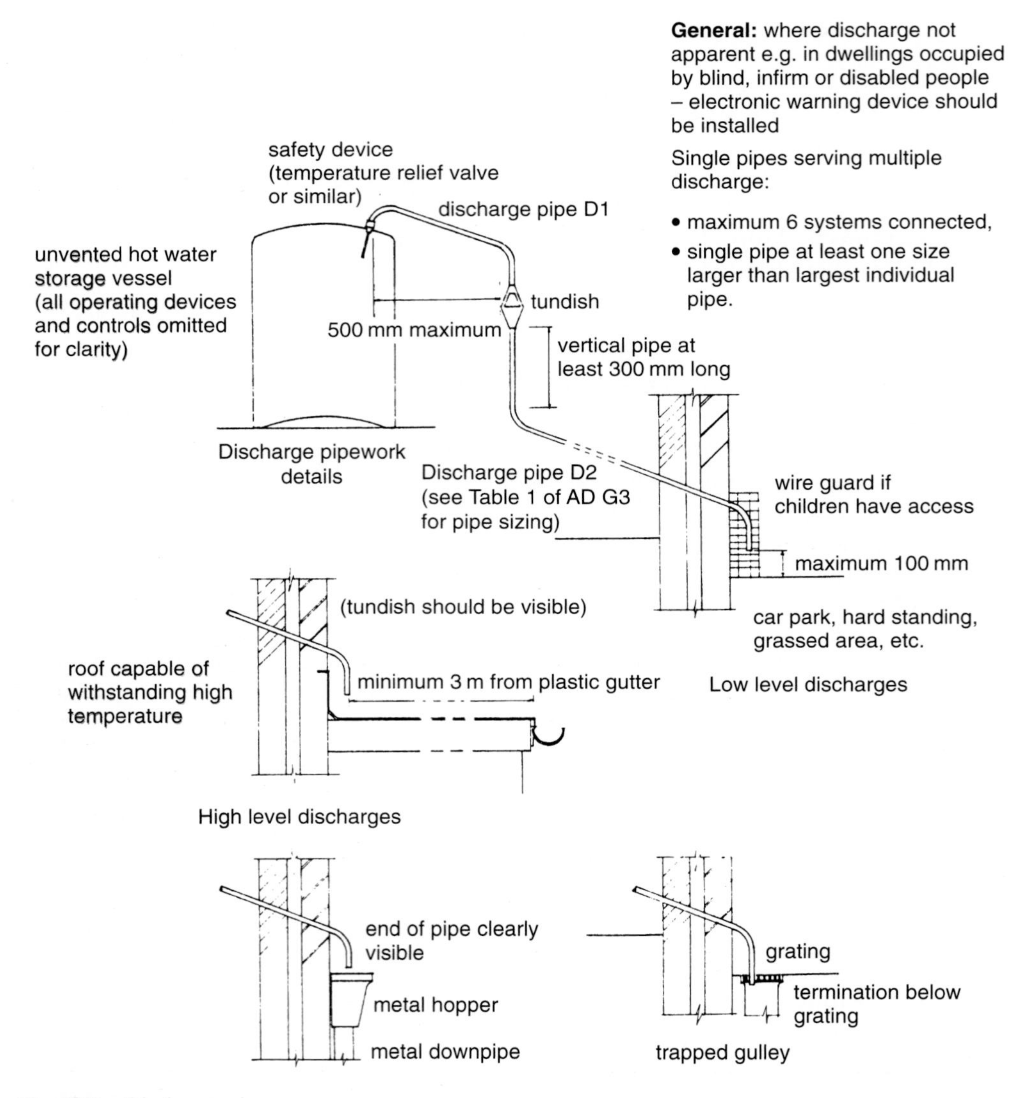

Fig. 12.2 Discharge pipes.

Discharge pipe D2 (see Fig. 12.2) from the tundish should comply with the following:

- it should terminate in a safe place where it cannot present a risk of contact to users of the building;
- it should be laid to a continuous fall;
- the discharge should be visible at either tundish or final outlet, but preferably at both of these locations (Fig. 12.2 shows possible discharge arrangements);

- it should be at least one pipe size larger than the nominal outlet size of the safety device unless its total equivalent hydraulic resistance exceeds that of a straight pipe 9 m long (bends will increase flow resistance, therefore Table 1 from AD G3 is reproduced below and shows how to calculate the minimum size of discharge pipe D2); and
- it should have a vertical section of pipework at least 300 mm long below the tundish before any changes in direction.

12.4.6 Alternative approach

Discharge pipes may also be sized in accordance with BS 6700: 1987 *Specification for design, installation, testing and maintenance of services supplying water for domestic use within buildings and their curtilages*, Appendix E, section E2 and Table 21.

AD G3, Section 3

Table 1 Sizing of copper discharge pipe 'D2' for common temperature relief valve outlet sizes

Valve outlet size	Minimum size of discharge pipe D1*	Minimum size of discharge pipe D2* from tundish	Maximum resistance allowed, expressed as a length of straight pipe (i.e. no elbows or bends)	Resistance created by each elbow or bend
G½	15 mm	22 mm	up to 9 m	0.8 m
		28 mm	up to 18 m	1.0 m
		35 mm	up to 27 m	1.4 m
G¾	22 mm	28 mm	up to 9 m	1.0 m
		35 mm	up to 18 m	1.4 m
		42 mm	up to 27 m	1.7 m
G1	28 mm	35 mm	up to 9 m	1.4 m
		42 mm	up to 18 m	1.7 m
		54 mm	up to 27 m	2.3 m

* see 3.5, 3.9, 3.9(a) and Fig. 12.2

Worked example:
The example below is for a G½ temperature relief valve with a discharge pipe (D2) having 4 No. elbows and length of 7 m from the tundish to the point of discharge.

From Table 1:
Maximum resistance allowed for a straight length of 22 mm copper discharge pipe (D2) from a G½ temperature relief valve is: 9.0 m
Subtract the resistance for 4 No. 22 mm elbows at 0.8 m each = 3.2 m

Therefore the maximum permitted length equates to: 5.8 m

5.8 m is less than the actual length of 7 m therefore calculate the next largest size.

Maximum resistance allowed for a straight length of 28 mm pipe (D2) from a G½ temperature relief valve equates to: 18 m

Subtract the resistance for 4 No. 28 mm elbows at 1.0 m each = 4 m

Therefore the maximum permitted length equates to: 14 m

As the actual length is 7 m, a 28 mm (D2) copper pipe will be satisfactory.

12.4.7 Section 4 hot water storage systems

Systems within the scope of section 4 exceed 500 litres in capacity or have a power input of more than 45 kW. Generally they will be individual designs for specific projects and therefore, not systems appropriate for EOTA or NACCB certification. Nevertheless, these systems should still conform to the same general safety recommendations as in section 3 including design by an appropriately qualified engineer and installation by a competent person.

Systems with a storage vessel of more than 500 litres capacity but with a power input of not more than 45kW should have safety devices conforming to BS 6700: 1987 (section two, clause 7) or other equivalent practice specifications which recommend a similar operating sequence for the safety devices to prevent the stored water temperature exceeding 100°C.

An unvented hot water storage vessel with a power input which exceeds 45 kW should also have an appropriate number of temperature relief valves which:

- either comply with BS 6283: Parts 2 or 3 or equivalent giving a combined discharge rating at least equivalent to the power input; or
- are equally suitable and marked with the set temperature in °C and a discharge rating marked in kW, measured in accordance with Appendix F of BS 6283: Part 2: 1991 or Appendix G of BS 6283: Part 3: 1991 and certified by a member of EOTA such as BBA or another recognised testing body (e.g. the Associated Offices Technical Committee, AOTC).

The temperature relief valves should be factory fitted to the storage vessel and the sensing element located as described in paragraph 3.5 of AD G3.

The non self-resetting thermal cut-outs should be installed in the system as described in Section 3 and the discharge pipes should also comply with that section.

13 Drainage and waste disposal

13.1 Introduction

This chapter describes Part H of Schedule 1 to the Building Regulations 2000 (as amended) and the associated Approved Document H. Together, these documents cover:

- Foul water drainage (H1);
- Wastewater treatment systems and cesspools (H2);
- Rainwater drainage (H3);
- Building over existing sewers (H4);
- Separate systems of drainage (H5); and
- Solid waste storage (H6).

Part H was substantially revised in 2002 and the current edition of Approved Document H draws together not only guidance on the drainage items listed above but also a certain amount of information on legislation related to drainage and waste disposal under the following headings:

- Repairs, alterations and discontinued use of drains and sewers (Appendix H1-B)
- Adoption of sewers and connection to public sewers (Appendix H1-C)
- Maintenance of wastewater treatment systems and cesspools (Appendix H2-A)
- Relevant waste collection legislation (Appendix H6-A).

13.2 Repairs, alterations and discontinued use of drains and sewers

Reconstruction and alteration to existing drains and sewers is deemed to constitute a material alteration of a controlled service under the Building Regulations and should be carried out to the same standards as new drains and sewers. Therefore, where new drainage is connected to existing pipework, the following points should be considered.

- Existing pipework should not be damaged, (e.g. use proper cutting equipment when breaking into existing drain runs).
- The resulting joint should be watertight, (e.g. by making use of purpose made repair couplings).

- Care should be taken to avoid differential settlement between the existing and new pipework, (e.g. by providing proper bedding of the pipework).

Even though the Building Regulations do not cover requirements for ongoing maintenance or repair of drains or sewers, sewerage undertakers and local authorities have a variety of powers under other legislation to make sure that drains, sewers, cesspools, septic tanks and settlement tanks do not deteriorate to the extent that they become a risk to public health and safety. This includes powers to ensure that:

- adequate maintenance is carried out
- repairs and alterations are properly carried out
- disused drains and sewers are sealed.

Requirements for inspection, maintenance, repairs and alterations

Section 48 of the Public Health Act 1936 (*power of relevant authority to examine and test drains etc. believed to be defective*) enables a local authority to examine and test any sanitary convenience, drain, private sewer or cesspool where it feels that it has reasonable grounds for believing that the drain is in such a condition:

- as to be prejudicial to health or a nuisance; or
- is so defective as to admit subsoil water (where the drain or private sewer connects indirectly with a public sewer).

Similar powers exist to enable sewerage undertakers to examine and test drains and private sewers under section 114 of the Water Industry Act 1991 (*power to investigate defective drain or sewer*).

Section 59 of the Building Act 1984 (*drainage of building*) allows a local authority to require a building owner to carry out remedial works on soil pipes, drains, cesspools or private sewers where these are deemed to be:

- insufficient for adequately draining the building
- prejudicial to health or a nuisance
- so defective as to admit subsoil water.

Section 59 also applies to disused cesspools, septic tanks or settlement tanks where these are considered to be prejudicial to health or a nuisance. The local authority can require the owner or occupier to fill or remove the tank or otherwise render it innocuous.

Under section 60 of the of the Building Act 1984 a pipe for conveying rainwater from a roof may not be used for conveying soil or drainage from a sanitary convenience, or as a ventilating shaft to a foul drain. The practical effect of this provision is that all rainwater pipes must be trapped before entering a foul drain.

Section 61 of the Building Act 1984 (*Repair etc. of drain*), requires any person intending to repair, reconstruct or alter a drain to give 24 hours notice to the local authority of their intention to carry out the works. This does not apply in an emergency; however such work must not be subsequently covered over without giving 24 hours notice. Free access must also be given to the local authority to inspect the works.

Section 17 of the Public Health Act 1961 (*power to repair drains etc. and to remedy stopped-up drains*) provides a swift procedure whereby local authorities may repair or clear blockages on drains or private sewers which have not been properly maintained. The repairs etc. must not cost more than £250 and can only be carried out after a notice has been served on the owner from whom costs can be recovered.

Section 50 of the Public Health Act 1936 (*overflowing and leaking cesspools*) allows the local authority to take action against any person who has caused by their action, default or sufferance, a septic tank, settlement tank or cesspool to leak or overflow. The person can be required to carry out repairs or to periodically empty the tank. This does not apply to the overflow of treated effluent or flow from a septic tank into a drainage field, provided the overflow is not prejudicial to health or a nuisance. It should be noted that under this section action can be taken against a builder who had caused the problem, as well as against the owner.

Sealing and/or removal of disused drains and sewers

Disused drains and sewers can be prejudicial to health in that they harbour rats, allow them to move between sewers and the surface, and may collapse causing possible subsidence. Therefore local authorities have a number of powers to control the sealing and removal of such drains and sewers as follows.

- Where a person carries out work which results in any part of a drain becoming permanently disused, under section 62 of the Building Act 1984 (*disconnection of drain*) a local authority may require the drain to be sealed at such points as it directs.
- Section 82 of the Building Act 1984 (*notices about demolition*), allows the local authority to require any person demolishing a building to remove or seal any sewer or drain to which the building was connected (see Chapter 1, section 1.6).
- A local authority can also use its powers under section 59 of the Building Act 1984 (see above) to require an owner of a building to remove or otherwise render innocuous any disused drain or sewer which is a health risk.

Disused drains or sewers should be disconnected from the sewer system as near as possible to the point of connection. Care should be taken not damage any pipe which is still in use and to ensure that the sewer system remains watertight. Disconnection is usually carried out by removing the pipe from a junction and placing a stopper in the branch of the junction fitting. If the connection is to a public sewer the sewerage undertaker should be consulted.

Shallow drains or sewers (i.e. less than 1.5 m deep) in open ground should, where possible, be removed. To ensure that rats cannot gain access, other pipes should be grout filled and sealed at both ends and at any point of connection. Larger pipes

(225 mm diameter or greater) should be grout filled to prevent subsidence or damage to buildings or services in the event of collapse.

Pollution of watercourses and ground water

Under Section 85 (*offences of polluting controlled waters*) of the Water Resources Act 1991 the Environment Agency have powers to prosecute anyone causing or knowingly permitting pollution of any stream, river, lake etc. or any groundwater. They also have powers under section 161A (*notices requiring persons to carry out anti-pollution works and operations*) of the Water Resources Act 1991 (as amended by the Environment Act 1995) to take action against any person causing or knowingly permitting a situation in which pollution of a stream, river, lake etc. or groundwater, is likely. Such a person can be required to carry out works to prevent the pollution.

Control over solid waste storage

With regard to solid waste storage, all dwellings are now required to have satisfactory means of storing solid waste and the provision of sections 23(1) and (2) of the Building Act 1984 which required satisfactory means of access for removal of refuse have been replaced by paragraph H4 of Schedule 1 to the Building Regulations 2000 (as amended). This paragraph of the regulations must be read in light of other legislative provisions in respect of refuse disposal. In particular, sections 45 to 47 of the Environmental Protection Act 1990 should be referred to (see Chapter 5) since those sections deal with the removal of refuse and allied matters. Thus, under section 45 of the 1990 Act a duty is placed on the local authority to collect all household waste in their area, while sections 46 and 47 make provision for the removal of trade and other refuse. Section 23(3) of the Building Act 1984 requires the local authority's consent to close or obstruct the means of access by which refuse is removed from a house.

13.3 Sanitary pipework and drainage

Paragraph H1 of Schedule 1 to the Building Regulations 2000 (as amended) requires that an adequate system of drainage must be provided to carry foul water from appliances in a building to one of the following, listed in order of priority:

- a public sewer; or
- a private sewer communicating with a public sewer; or
- a septic tank which has an appropriate form of secondary treatment or another wastewater treatment system; or
- a cesspool.

Movement to a lower level in the order of priority may only be on the grounds of reasonable practicability. For example, if no public or private sewer was available within a reasonable distance then a septic tank might be a suitable alternative.

FOUL WATER is defined as waste water which comprises or includes:

- waste from a sanitary convenience, bidet or appliance used for washing receptacles for foul waste, or
- water which has been used for food preparation, cooking or washing.

Where it is proposed to divert water that has been used for personal washing or for the washing of clothes, linen or other articles to a collection system for reuse, then the provisions of requirement H1 will not apply.

Further guidance on the meaning of SANITARY CONVENIENCE is given in the guidance to Approved Document G4 where it is defined as a closet or urinal.

FOUL WATER OUTFALL may be a foul or combined sewer, cesspool, septic tank or holding tank. This term is not specifically defined in AD H1; however the term is inferred from the description of Performance on page 6.

The requirements of Paragraph H1 may be met by any foul water drainage system which:

- conveys the flow of foul water to a suitable foul water outfall;
- reduces to a minimum the risk of leakage or blockage;
- prevents the entry of foul air from the drainage system to the building, under working conditions;
- is ventilated;
- is accessible for clearing blockages, and
- does not increase the vulnerability of the building to flooding.

AD H1 sets out detailed provisions in two sections. Section 1 deals with sanitary pipework (i.e. above ground foul drainage) and is applicable to domestic buildings and small non-domestic buildings. Section 2 deals with foul drainage (i.e. below ground foul drainage). There is also an appendix (H1-A) which contains additional guidance for large buildings. Complex systems in larger buildings should follow the guidance in BS EN 12056 *Gravity drainage systems inside buildings*.

13.3.1 Above-ground foul drainage

A number of terms are used throughout AD H1. These are defined below and illustrated in Fig. 13.1. It should be noted that these definitions do not appear in the AD.

DISCHARGE STACK – A ventilated vertical pipe which carries soil and waste water directly to a drain.

VENTILATING STACK – A ventilated vertical pipe which ventilates a drainage system either by connection to a drain or to a discharge stack or branch ventilating pipe.

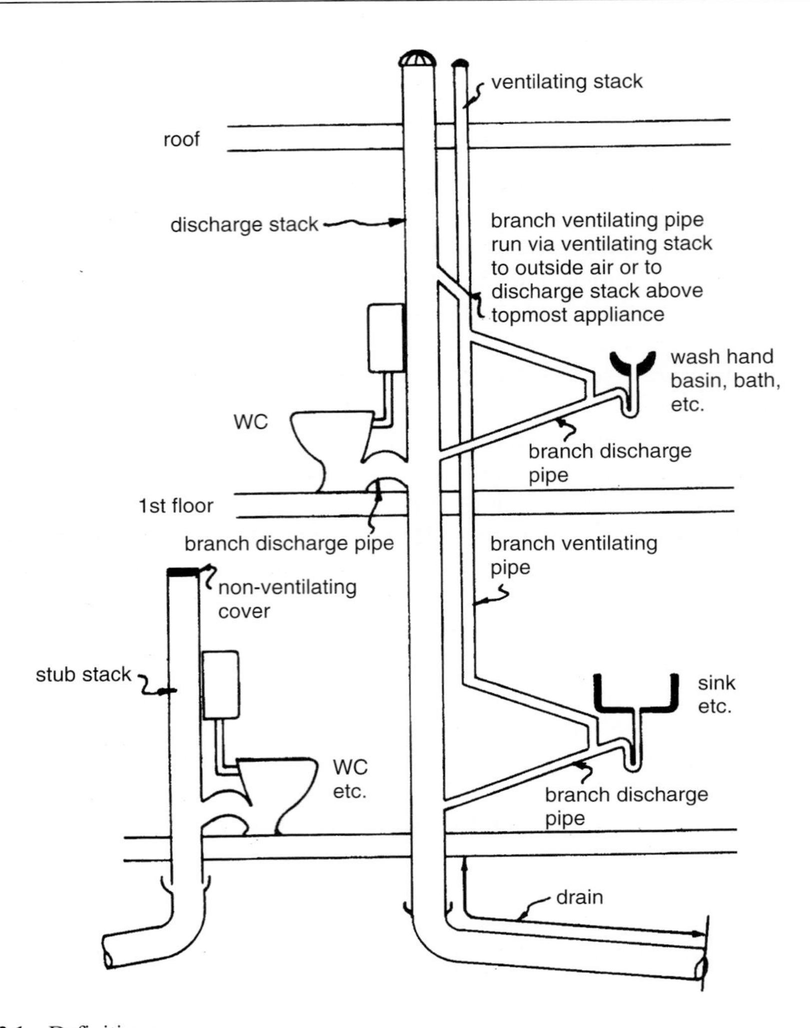

Fig. 13.1 Definitions.

BRANCH DISCHARGE PIPE (sometimes referred to as a BRANCH PIPE) – The section of pipework which connects an appliance to another branch pipe or a discharge stack if above the ground floor, or to a gully, drain or discharge stack if on the ground floor.

BRANCH VENTILATING PIPE – The section of pipework which allows a branch discharge pipe to be separately ventilated.

STUB STACK – An unventilated discharge stack.

A drainage system, whether above or below ground, should have sufficient capacity to carry the anticipated flow at any point. The capacity of the system, therefore, will depend on the size and gradient of the pipes whereas the flow will depend on the type, number and grouping of appliances. Table 13.1 below is based on information from BS EN 12056 and Table A2 of AD HI, and gives the expected flow rates for a range of appliances.

Since sanitary appliances are seldom used simultaneously, the normal size of discharge stack or drain will be able to take the flow from quite a large number of appliances. Table A1 of AD H1 is reproduced below and is derived from BS EN 12056. It shows the approximate flow rates from dwellings and is based on an appliance grouping per household of 1 WC, 1 bath, 1 or 2 washbasins and 1 sink.

The guidance given in section 1 of AD H1 is applicable for WCs with major flush volumes of 5 litres or more. WCs with flush volumes of less than 5 litres may give rise to an increased risk of blockages, however BS EN 12056 contains guidance on the design of sanitary pipework suitable for WCs with flush volumes as low as 4 litres.

13.3.2 Pipe sizes

Since individual manufacturer's pipe sizes will vary, the sizes quoted in AD H1 are nominal and give a numerical designation in convenient round numbers. Similarly, equivalent pipe sizes for individual pipe standards are given in the standards listed in AD H Tables 4, 7 and 14 reproduced below.

Table 13.1 Appliance flow rates.

Appliance	Flow rate (litres/sec)
WC (9 litre washdown)	2.3
Washbasin	0.6
Sink	0.9
Bath	1.1
Shower	0.1
Washing machine	0.7
Urinal (per person unit)	0.15
Spray tap basin	0.06
Dishwashing machine	0.25

AD H1, section 1

Table 1 Flow rates from dwellings

Number of dwellings	Flow rate (litres/sec)
1	2.5
5	3.5
10	4.1
15	4.6
20	5.1
25	5.4
30	5.8

13.3.3 Trap water seals

Trap water seals are provided in drainage systems to prevent foul air from the system entering the building. All discharge points into the system should be fitted with traps and these should retain a minimum seal of 25 mm or equivalent under test and working conditions.

Traditionally the 'one pipe' and 'two pipe' systems of plumbing have required the provision of branch ventilating pipes and ventilating stacks unless special forms of trap are used. The 'single-stack' system of plumbing obviates the need for these ventilating pipes and is illustrated in Fig. 13.2. Table 13.2 below, which is based on Table 1 and Table A3 of AD H1, gives minimum dimensions of pipes and traps where it is proposed to use appliances other than those shown in Fig. 13.2.

It is permissible to reduce the depth of trap seal to 38mm where washing machines, dishwashers, baths or showers discharge directly to a gully. Additionally, traps used on appliances with flat bottom (trailing waste) discharge which discharge to a gully with a grating may also have a water seal of not less than 38 mm.

It should be stressed that the minimum pipe sizes given above relate to branch pipes serving a single appliance. Where a number of appliances are served by a single branch pipe which is unventilated, the diameter of the pipe should be at least the size given in Table 2 to section 1 of AD HI, which is reproduced below.

If it is not possible to comply with the figures given in Table 13.1, Fig. 13.2 or Table 2, then the branch discharge pipe should be ventilated in order to prevent loss of trap seals. This is facilitated by means of a *branch ventilating pipe* which is connected to the discharge pipe within 750 mm of the appliance trap. The branch ventilating pipe may be run direct to outside air, where it should finish at least 900 mm above any opening into the building which is nearer than 3m, or, it may be connected to the ventilating stack or stack vent above the 'spillover' level of the highest appliance served. In this case it should have a continuous incline from the branch discharge pipe to the point of connection with the stack (see Fig. 13.3).

Where a branch ventilating pipe serves only one appliance it should have a minimum diameter of 25 mm. This should be increased to 32 mm diameter if the branch ventilating pipe is longer than 15 m or contains more than five bends.

Table 13.2 Minimum dimensions of branch pipes and traps.

Appliance	Minimum diameter of pipe and trap (mm)	Depth of trap seal (mm)
Bidet	32	75
Shower Food waste disposal unit Urinal bowl Sanitary towel macerator Washing machine Dishwashing machine	40	75
Industrial food waste disposal unit	50	75
Urinal stall (1 to 6 person position)	65	50

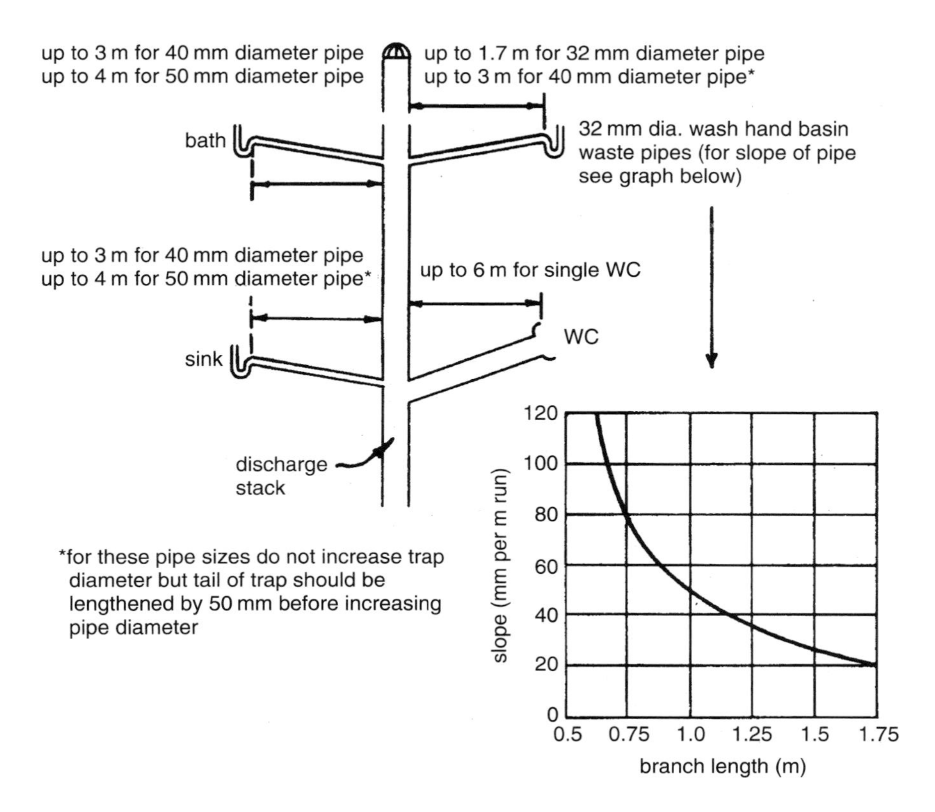

Appliance	*Minimum diameter of pipe and trap (mm)*	*Depth of trap seal*	*Slope (mm/m)*
Sink	40	75	18–90
Bath	40	50	18–90
WC – outlet < 80 mm	75	50	18
WC – outlet > 80 mm	100	50	18
washbasin	32	75*	See graph above

*Depth of seal may be reduced to 50 mm only with flush grated wastes without plugs on spray tap basins

Fig. 13.2 Single stack system – design limits.

As appliance traps present an obstacle to the normal flow in a pipe they may be subject to periodic blockages. It is important, therefore, that they be fitted immediately after an appliance and either be removable or be fitted with a cleaning eye. Where a trap forms an integral part of an appliance (such as in a WC pan), the appliance should be removable.

AD H1 Section 1

Table 2 Common Branch discharge pipes (unventilated).

Appliance	Max no. to be connected	Max length of branch pipe (m)	Min size of pipe (mm)	Gradient limits (mm fall per metre)
WC outlet > 80 mm	8	15	100	18[2] to 90
WC outlet < 80 mm	1	15	75[3]	18 to 90
Urinal – bowl		3[1]	50	
Urinal – trough		3[1]	65	18 to 90
Urinal – slab[4]		3[1]		
Washbasin or bidet	3	1.7	30	18 to 22

Notes:
[1] Should be as short as possible to prevent deposition
[2] May be reduced to 9 mm on long drain runs where space is restricted, but only if more than one WC is connected
[3] Not recommended where disposal of sanitary towels may take place via the WC, as there is an increased risk of blockages
[4] Slab urinals longer than seven, persons should have more than one outlet.

13.3.4 Branch discharge pipes – design recommendations

In addition to size and gradient there are other design recommendations for branch discharge pipes that should be adhered to for efficient operation, and in order to prevent loss of trap seals.

Branch pipes should only discharge into another branch pipe, a discharge stack or a gully. Gullies are usually at ground floor level but may be situated in a basement and are only permitted to take wastewater. It is not permissible to discharge a branch pipe into an open hopper. Branch pipes to ground floor appliances may also discharge into a stub stack or directly to a drain.

In high buildings especially, back-pressure may build up at the foot of a discharge stack and may cause loss of trap seal in ground floor appliances. Therefore, the following recommendations should be followed.

- For multi-storey buildings up to five storeys high there should be a minimum distance of 750 mm between the point of junction of the lowest branch discharge pipe connection and the invert of the tail of the bend at the foot of the discharge stack. This is reduced to 450 mm for discharge stacks in single dwellings up to three storeys high (see Fig. 13.4).
- For appliances above ground floor level the branch pipe should only be run to a discharge stack, to another branch pipe or to a stub stack (but see also section 13.3.8 below for more information on stub stacks).
- Ground floor appliances may be run to a separate drain, gully or stub stack. (A gully connection should be restricted to pipes carrying waste water only.) They may also be run to a discharge stack in the following circumstances:
  (a) in buildings up to five storeys high – without restriction;
  (b) in buildings with six to twenty storeys – to their own separate discharge stack;

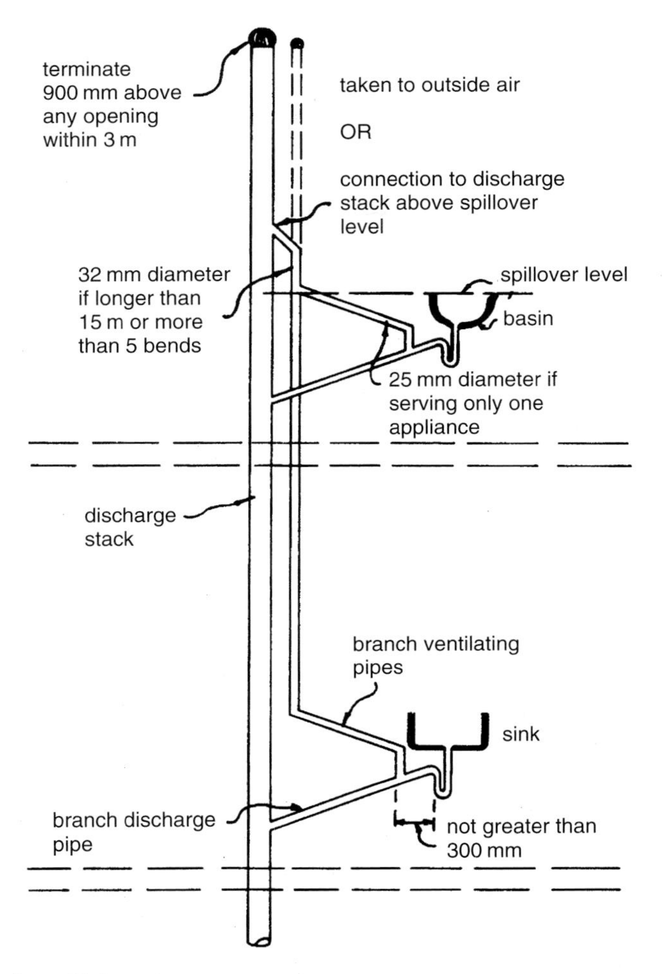

Fig. 13.3 Branch ventilating pipes.

(c) in buildings over 20 storeys – ground and first floor appliances to their own separate discharge stack; (see Fig. 13.5).

Back-pressure and blockages may occur where branches are connected so as to be almost opposite one another. This is most likely to occur where bath and WC branch connections are at or about the same level. Figure 13.6 illustrates ways in which possible cross flows may be avoided.

Additionally, a long vertical drop from a ground floor water closet to a drain may cause self-syphonage of the WC trap. To prevent this the drop should not exceed 1.3 m from floor level to invert of drain (see Fig. 13.7).

Similarly, there is a chance of syphonage where a branch discharge pipe connects with a gully. This can be avoided by terminating the branch pipe above the water level but below the gully grating or sealing plate (see Fig. 13.7).

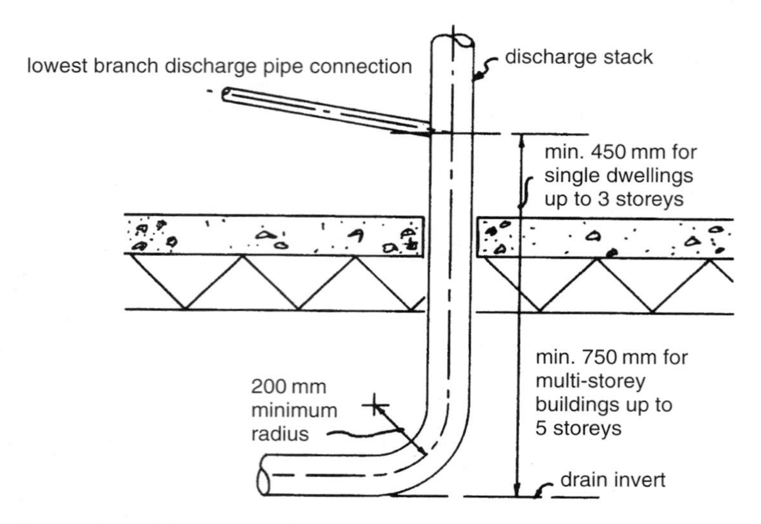

Fig. 13.4 Connection of lowest branch to discharge stack.

Self-syphonage can also be prevented by ensuring that bends in branch discharge pipes are kept to a minimum. Where bends are unavoidable they should be made with as large a radius as possible. Junctions on branches should be swept in the direction of flow with a minimum radius of 25 mm or should make an angle of 45° with the discharge stack. Where a branch diameter is 75 mm or more the sweep radius should be increased to 50 mm (see Fig. 13.6). Branch pipes up to 40 mm diameter joining other branch pipes which are 100 mm diameter or greater should, where possible, connect to the upper part of the pipe wall of the larger branch.

Branch discharge pipes should be fully accessible for clearing blockages. Additionally rodding points should be provided so that access may be gained to any part of a branch discharge pipe which cannot be reached by removing a trap or an appliance with an integral trap.

13.3.5 Drainage of condensate from boilers

It is permissible to connect condensate drainage from boilers to sanitary pipework. The connecting pipework should have a minimum diameter of 22 mm and should pass through a 75 mm condensate trap. This can be by means of an additional trap provided externally to the boiler to achieve the 75 mm seal. If this is the case, an air gap should be provided between the boiler and the trap. The following recommendations should also be observed.

- For preference, the connection should be made to an internal stack with a 75 mm condensate trap.
- Any connection made to a branch discharge pipe should be downstream of any sink waste connection.
- All sanitary pipework receiving condensate should be made of materials which can resist a pH value of 6.5 and lower.

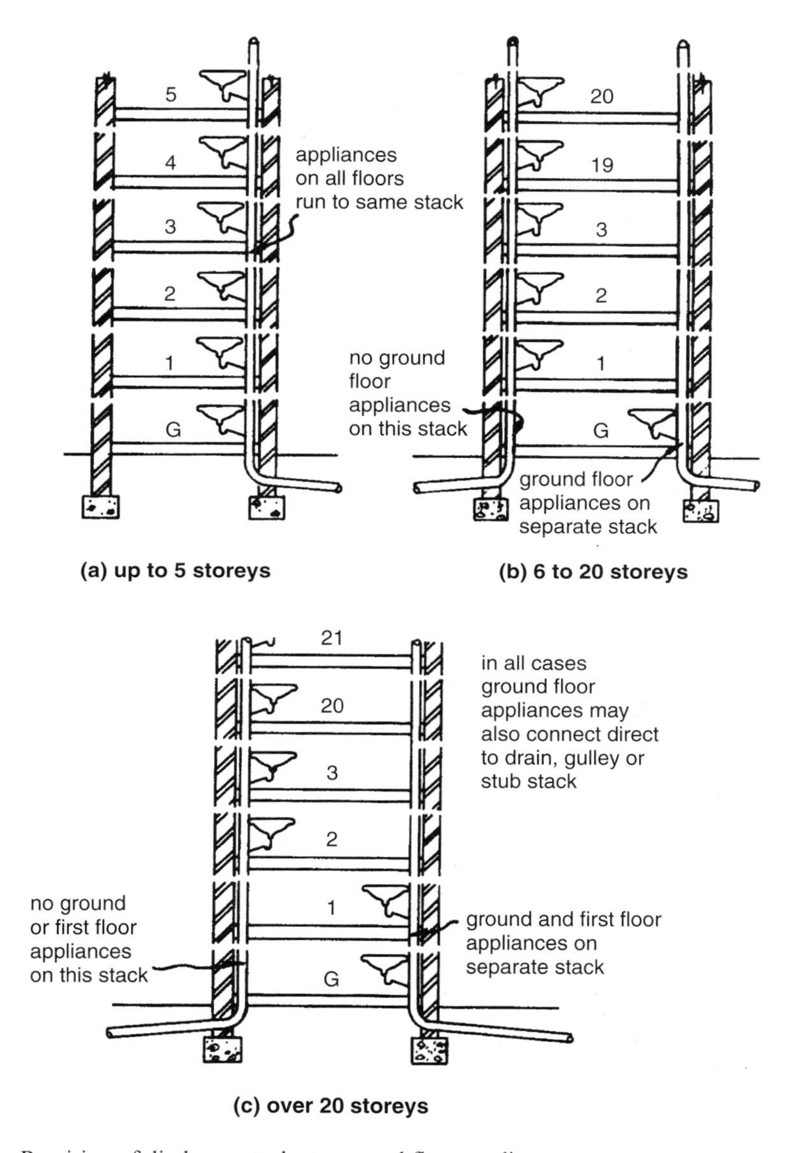

Fig. 13.5 Provision of discharge stacks to ground floor appliances.

- The installation should follow the guidance in BS 6798 *Specification for installation of gas-fired hot water boilers of rated input not exceeding 60 kW.*

13.3.6 Discharge stacks – design recommendations

The satisfactory performance of a discharge stack will be ensured if it complies with the following rules.

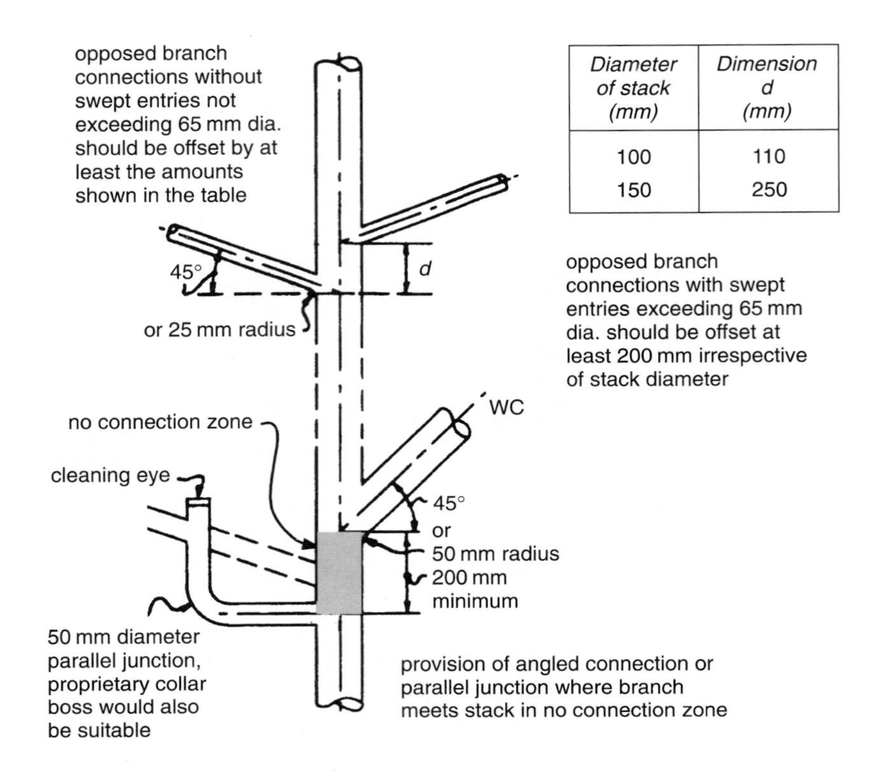

Diameter of stack (mm)	*Dimension d (mm)*
100	110
150	250

Fig. 13.6 Avoidance of cross flows in discharge stacks.

- The foot of the stack should only connect with a drain and should have as large a radius as possible (at least 200 mm at the centreline).
- Ideally, there should be no offsets in the wet part of a stack (i.e. below the highest branch connection).
- If offsets are unavoidable then:
  (a) buildings over three storeys should have a separate ventilation stack connected above and below the offset; and
  (b) buildings up to three storeys should have no branch connection within 750 mm of the offset.
- The stack should be placed inside a building, unless the building has not more than three storeys. This rule is intended to prevent frost damage to discharge stacks and branch pipes.
- The stack should comply with the minimum diameters given in Table 3 to section 1 of AD Hl (see below). Additionally, the following minimum internal diameters for discharge stacks also apply:
  (a) serving urinals – 50 mm,
  (b) serving closets with outlets less than 80 mm – 75 mm, and
  (c) serving closets with outlets greater than 80 mm – 100 mm.

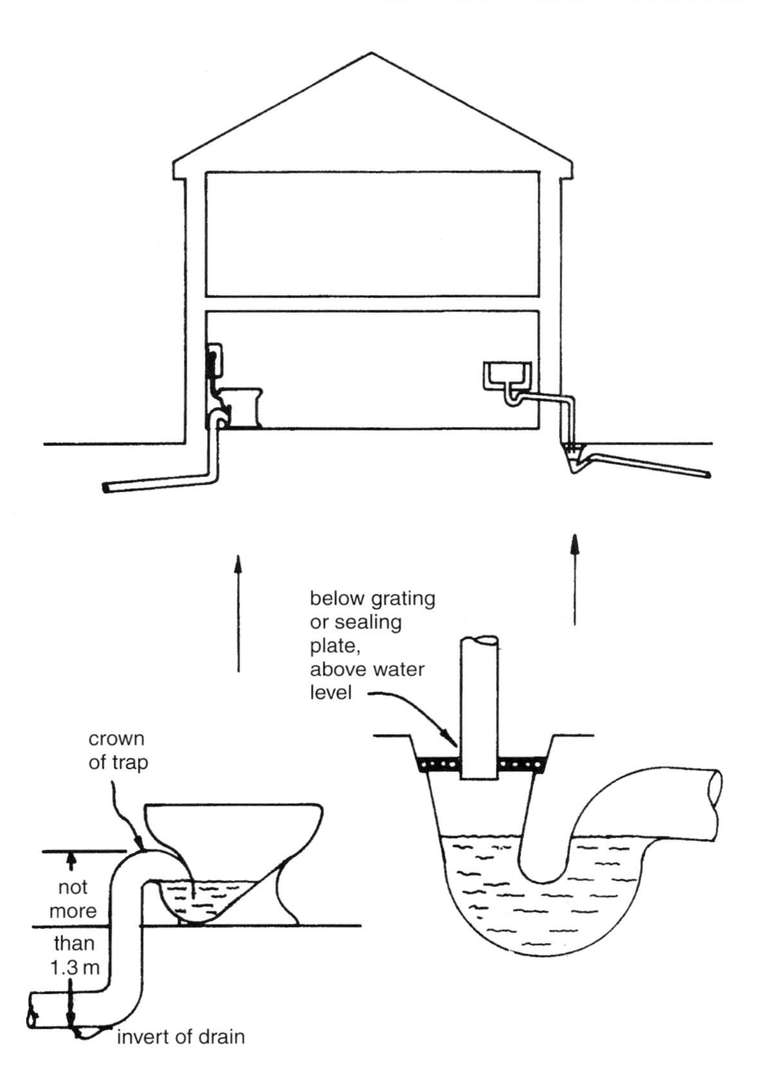

Fig. 13.7 Ground floor connections for water closets and gullies.

- The diameter of a discharge stack should not reduce in the direction of flow and the internal diameter of the stack should not be less than that of the largest trap or branch discharge pipe.
- Adequate access points for clearing blockages should be provided and all pipes should be reasonably accessible for repairs. Rodding points in stacks should be above the spillover level of appliances.

13.3.7 Discharge stacks – ventilation recommendations

In order to prevent the loss of trap seals it is essential that the air pressure in a discharge stack remains reasonably constant. Therefore, the stack should be

AD H1, section 1

Table 3 Minimum diameters for discharge stacks.

Stack size (mm)	Max capacity (litres/sec)
50*	1.2
65*	2.1
75†	3.4
90	5.3
100	7.2

Note
* No wcs.
† Not more than 1 syphonic wc with 75 mm outlet.

ventilated to outside air. For this purpose it should be carried up to such a height that its open end will not cause danger to health or a nuisance. AD HI recommends that the pipe should finish at least 900 mm above the top of any opening into the building within 3 m. The open end should be fitted with a durable ventilating cover (see Fig. 13.8). In areas where rodent control is a problem the cover should be metallic.

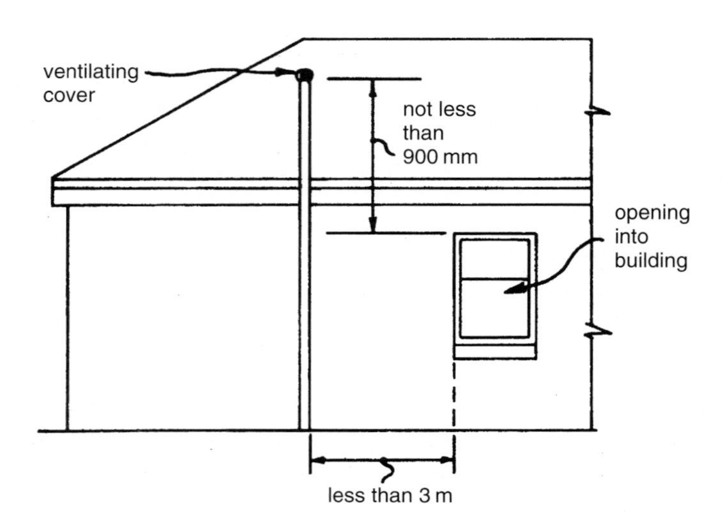

Fig. 13.8 Termination of discharge stacks.

The dry part of a discharge stack above the topmost branch, which serves only for ventilation, may be reduced in size in one and two storey houses to 75 mm diameter.

It is permissible to terminate a discharge stack inside a building if it is fitted with an air admittance valve. This valve allows air to enter the pipe but does not allow foul air to escape. It should comply with prEN 12380 *Ventilating pipework, air admittance valves* and should not adversely affect the operation of the underground drainage system which normally relies on ventilation from the open stacks to the sanitary pipework.

Air admittance valves should also be:

- located in areas which have adequate ventilation
- accessible for maintenance
- removable to give access for clearing blockages.

Air admittance valves should not be used:

- in dust laden atmospheres
- outside buildings
- where there is no open ventilation on a drainage system or through connected drains – other means to relieve positive pressures should be considered.

Some underground drains are subject to surcharging. Where this is the case the discharge stack should be ventilated by a pipe of not less than 50 mm diameter connected at the base of the stack above the expected flood level. This would also apply where a discharge pipe is connected to a drain near an intercepting trap.

13.3.8 Stub stacks

There is one exception to the general rule that discharge stacks should be ventilated. This involves the use of an unvented stack (or *stub stack*). A stub stack should connect to a ventilated discharge stack or a ventilated drain which is not subject to surcharging and should comply with the dimensions given in Fig. 13.9. It is permissible for more than one ground floor appliance to connect to a stub stack.

13.3.9 Dry ventilating stacks

Where an installation requires a large number of branch ventilating pipes and the distance to a discharge stack is also large it may be necessary to use a dry ventilating stack.

It is normal to connect the lower end of a ventilating stack to a ventilated discharge stack below the lowest branch discharge pipe and above the bend at the foot of the stack or to the crown of the lowest branch discharge pipe connection provided that it is at least 75 mm diameter.

Ventilating stacks should be at least 32 mm in diameter if serving a building containing dwellings not more than ten storeys high. For all other buildings reference should be made to BS EN 12056 *Gravity drainage systems inside buildings*.

13.3.10 Greywater recovery systems

Greywater is defined in the Water Regulations Advisory Scheme leaflet No. 09-02-04 *Reclaimed water systems. Information about installing, modifying or maintaining*

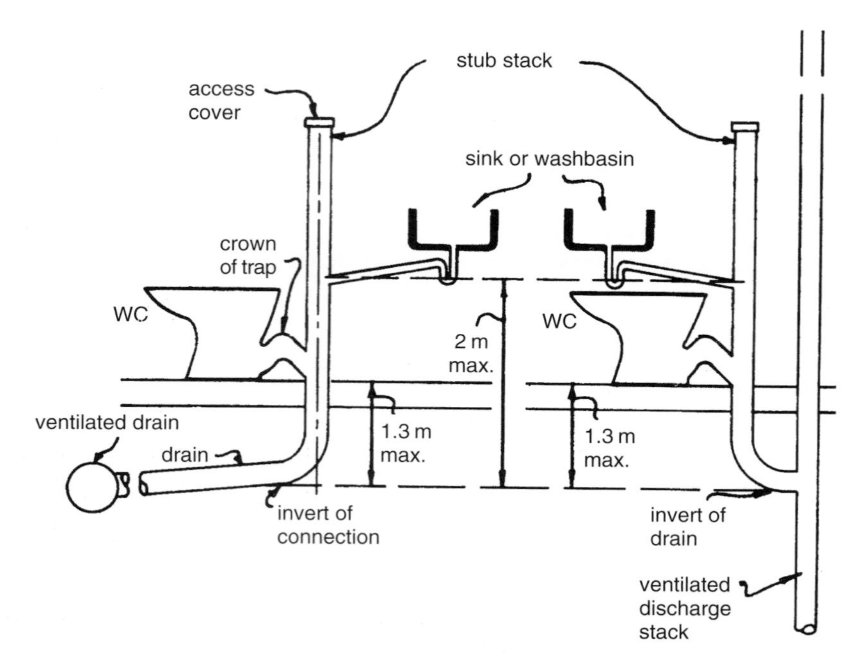

Fig. 13.9 Stub stacks.

reclaimed water systems as – 'water originating from the mains potable water supply that has been used for bathing or washing, washing dishes or laundering clothes'.

Such water can be used for irrigation and for other purposes such as toilet flushing or car washing; however care must be taken to prevent contamination of potable water supplies or accidental misuse due to the greywater being mistaken for potable water.

Approved Document H1 gives very little guidance on the use of greywater other than a passing reference to the leaflet mentioned above. It is more concerned with the identification of the pipes conveying the greywater and the suitability of storage systems. Accordingly, it recommends that all sanitary pipework carrying greywater for reuse should be clearly marked with the word 'GREYWATER' in accordance with the Water Regulations Advisory Scheme leaflet No. 09-02-05 *Marking and identification of pipework for reclaimed greywater systems.*

Guidance on the provision of external tanks for the storage of greywater is given in section 13.5 below.

13.3.11 Materials for above-ground drainage systems

Table 4 to section 1 of AD H1, which is reproduced below, gives details of the materials that may be used for pipes, fittings and joints in above ground drainage systems. The following matters should also be addressed when considering which materials to use in a system of sanitary pipework:

AD H1, section 1

Table 4 Materials for sanitary pipework.

Material	British Standard
Pipes	
Cast Iron	BS 416, BS EN 877
Copper	BC EN 1254, BS EN 1057
Galvanised steel	BS 3868
PVC-U	BS EN 1329
Polypropylene (PP)	BS EN 1451
ABS	BS EN 1455
Polyethylene (PE)	BS EN 1519
Styrene Copolymer blends (PVC + SAN)	BS EN 1565
PVC-C	BS EN 1566
Traps	BS EN 274, BS 3943
Note: Some of these materials may not be suitable for carrying trade effluent or condensate from boilers	

- pipes of different metals should be separated where necessary by non-metallic material to prevent electrolytic corrosion;
- pipes should be adequately supported without restricting thermal movement;
- care should be taken to ensure continuity of any electrical earth bonding;
- care should be taken where pipes pass through fire separating elements (see Part B of Schedule 1 to the Building Regulations 2000 and Approved Document B);
- light should not be visible through the pipe wall when sanitary pipework is connected to WCs as this is believed to encourage damage by rodents.

13.3.12 Workmanship

Workmanship should be in accordance with BS 8000 *Workmanship on Building Sites* Part 13: *Code of practice for above ground drainage*.

13.3.13 Test for airtightness

In order to ensure that a completed installation is airtight it should be subjected to a pressure test of air or smoke of at least 38 mm water gauge for a maximum of three minutes. A satisfactory installation will maintain a 25 mm water seal in every trap. PVC-U pipes should not be smoke tested.

13.3.14 Alternative method of design

The requirements of the 2000 Regulations for above-ground drainage can also be met by following the relevant recommendations of BS EN 12056 *Gravity drainage systems inside buildings*. These are:

- in Part 1 *General and performance requirements* – clauses 3 to 6;
- in Part 2 *Sanitary pipework, layout and calculation*, clauses 3 to 6 and national annexes NA to NG (System III is traditionally in use in the UK);
- in Part 5 *Installation and testing, instructions for operation, maintenance and use*, clauses 4 to 6, 8, 9 & 11.

For vacuum drainage systems, designers should follow the guidance in BS EN 12109 *Vacuum drainage systems inside buildings*.

13.3.15 Below-ground foul drainage

Section 2 of AD H1 gives guidance on the construction of underground drains and sewers from buildings to the point of connection to a suitable outfall. This may be an existing sewer, a wastewater treatment system or a cesspool and includes any drains or sewers outside the curtilage of the building.

Section 2 also gives guidance in Appendix H1-B on the repair, alteration and discontinued use of drains and sewers and in Appendix H1-C on the adoption of sewers and connection to public sewers.

In most modern systems of underground drainage foul water and rainwater are carried separately. However, some public sewers are on the combined system taking foul and rainwater in the same pipe. The provisions of AD HI will apply equally to combined systems although pipe gradients and sizes may have to be adjusted to take the increased flows. In some circumstances separate drainage should still be provided on a development even though the outfall of the drainage system is to a combined sewer (see Requirement H5 in section 13.9.2 below). Combined systems should never discharge to a cesspool or septic tank.

13.3.16 Foul water outfalls and connections with sewers

Ideally, foul drainage from a development should connect to a public foul or combined sewer. Section 106 of the Water Industry Act 1991 gives the owner or occupier of a building the right to connect to a public sewer subject to the following conditions:

- where separate foul and surface water sewers are provided the connections must match this appropriately and proof of connectivity will be needed by the Building Control Body;
- the manner of the connection must not prejudice the public sewer system; and
- 21 days notice of intention to connect must be given to the sewerage undertaker.

Section 107 of the Water Industry Act 1991 allows the sewerage undertaker to make the connection and recover reasonable costs from the developer. Alternatively, the sewerage undertaker may allow the developer to carry out the work under its supervision.

Drain connections (drains to drains, drains to public or private sewers, and private sewers to public sewers) should be made obliquely, or in the direction of flow. Connections should be made using prefabricated components and where holes are cut to make the connection, these should be drilled to avoid damaging the pipe. Sometimes, in making a connection, it is preferable to remove a section of pipe and insert a junction. Repair couplings should be used for this to ensure a watertight joint. The coupling should be carefully packed to avoid differential settlement with adjacent pipes.

Where a sewer serves more than one property it should be kept as far away as is practicable from the position where a future extension might be built.

The degree to which it is possible to connect to a public sewer may, to a certain extent, depend on the size of the development. For example, for a small development, it may be reasonable to connect to a public sewer up to 30 m from the development provided that the developer has the right to construct drainage over any intervening private land. This might necessitate the provision of a pumping installation where the levels do not permit drainage by gravity (see below section 13.3.28). The economies of larger developments may make it feasible to connect to a public sewer which is some distance away.

It is also possible, for developments which comprise more than one curtilage, for the developer to requisition a sewer from the sewerage undertaker. This may be done under section 98 of the Water Industry act 1991. In constructing the sewer, the sewerage undertaker may use its rights of access to private land, however the person requisitioning the sewer may be required to contribute towards its cost over a 12 year period.

It may be possible to connect to an existing private sewer that connects with a public sewer where it is not reasonably practicable to connect directly to a public sewer. In such a case permission will need to be granted by the owner(s) of the private sewer and it should be in a satisfactory condition and have sufficient capacity to take the increased flows.

A wastewater treatment system or cesspool should only be provided where it is not reasonably practicable to connect to a sewer as described above.

13.3.17 Design and performance factors

The performance of a below ground foul drainage system depends on the drainage layout, provision for ventilation, the pipe cover and bedding, the pipe sizes and gradients, the materials used and the provisions for clearing blockages.

DRAINAGE LAYOUT – The drainage layout should be kept as simple as possible with pipes laid in straight lines and to even gradients. The number of access points provided should be limited to those essential for clearing blockages. If possible, changes of gradient and direction should be combined with access points, inspection chambers or manholes.

A slight curve in a length of otherwise straight pipework is permissible provided the line can still be adequately rodded. Bends should only be used in or close to inspection chambers and manholes, or at the foot of discharge or ventilating stacks. The radius of any bend should be as large as practicable.

In commercial hot food premises, drains serving kitchens should be fitted with a grease separator in compliance with prEN 1825: *Installations for separation of grease*, Part 1: *Principles of design, performance and testing, marking and quality control* and Part 2: *Selection of nominal size, installation and maintenance*, unless other effective means of grease removal are provided.

VENTILATION – It is important to ventilate an underground foul drainage system with a flow of air. Ventilated discharge pipes may be used for this purpose and should be positioned at or near the head of each main run. An open ventilating pipe (without an air admittance valve) should be fitted on any drain run fitted with an intercepting trap (especially on sealed systems) and on drains subject to surcharge. Ventilating pipes should not finish near openings in buildings (see section 13.3.3 above).

PIPE COVER AND BEDDING – The degree of pipe cover to be provided will usually depend on:

- the invert level of the connections to the drainage system
- the slope and level of the ground
- the necessary pipe gradients
- the necessity for protection to pipes.

In order to protect pipes from damage it is essential that they are bedded and backfilled correctly. The choice of materials for this purpose will depend mainly on the depth, size and strength of the pipes used. If the limits of cover cannot be attained it may be possible to choose another pipe strength and bedding class (see also BS EN 1295-1:1998 *Structural design of buried pipelines under various conditions of loading*) or provide special protection (see section 13.3.20 below).

Pipes used for underground drainage may be classed as rigid or flexible. Flexible pipes will be subject to deformation under load and will therefore need more support than rigid pipes so that the deformation may be limited.

13.3.18 Rigid pipes

Tables 8 and 9 of AD HI are set out below and contain details of the limits of cover that need to be provided for rigid clay and concrete pipes in any width of trench. For details of the bedding classes referred to in the Tables, see Figs. 13.10 and 13.11.

The backfilling materials should comply with the following.

(1) Granular material for rigid pipes should conform to BS EN 1610 Annex B Table B.15. The granular material should be single sized or graded from 5 mm up to:
   - 10 mm for 100 mm pipes
   - 14 mm for 150 mm pipes
   - 20 mm for pipes from 150 mm to 600 mm diameter
   - 40 mm for pipes more than 600 mm diameter.

The compaction fraction maximum should be 0.3 for class N or B and 0.15 for class F.

(2) Selected fill should be free from stones larger than 40 mm, lumps of clay over 100 mm, timber, frozen material or vegetable matter.

(3) It is possible that groundwater may flow in trenches with granular bedding. Provisions may be required to prevent this.

(4) Socketed pipes used with class D bedding should have holes formed in the trench bottom under the sockets to give a clearance of at least 50 mm. The holes should be as short as possible.

(5) Sockets for pipes used with class F or N bedding should be at least 50 mm above the floor of the trench.

AD H1, section 2

Table 8 Limits of cover for class 120 Clayware pipes in any width of trench.

Nominal size	Laid in fields	Laid in light roads	Laid in main roads
100 mm	0.6 m–8 + m	1.2 m–8 + m	1.2 m–8 m
225 mm	0.6 m–5 m	1.2 m–5 m	1.2 m–4.5 m
400 mm	0.6 m–4.5 m	1.2 m–4.5 m	1.2 m–4 m
600 mm	0.6 m–4.5 m	1.2 m–4.5 m	1.2 m–4 m

Notes:

1. All pipes assumed to be Class 120 to BS EN 295, other strengths and sizes of pipe are available, consult manufacturers;
2. Bedding assumed to be Class B with bedding factor of 1.9, guidance is available on use of higher bedding factors with clayware pipes
3. Alternative designs using different pipe strengths and/or bedding types may offer more appropriate or economic options using the procedures set out in BS EN 1295;
4. Minimum depth in roads set to 1.2 m irrespective of pipe strength.

AD H1, section 2

Table 9 Limits of cover for class M Concrete pipes in any width of trench.

Nominal size	Laid in fields	Laid in light roads	Laid in main roads
300 mm	0.6 mm–3 m	1.2 m–3 m	1.2 m–2.5 m
450 mm	0.6 m–3.5 m	1.2 m–3.5 m	1.2 m–2.5 m
600 mm	0.6 m–3.5 m	1.2 m–3.5 m	1.2 m–3 m

Notes:

1. All pipes assumed to be Class M to BS 5911, other strengths and sizes of pipe are available, consult manufacturers;
2. Bedding assumed to be Class B with bedding factor of 1.9;
3. Alternative designs using different pipe strengths and/or bedding types may offer more appropriate or economic options using the procedures set out in BS EN 1295;
4. Minimum depth in roads set to 1.2 m irrespective of pipe strength.

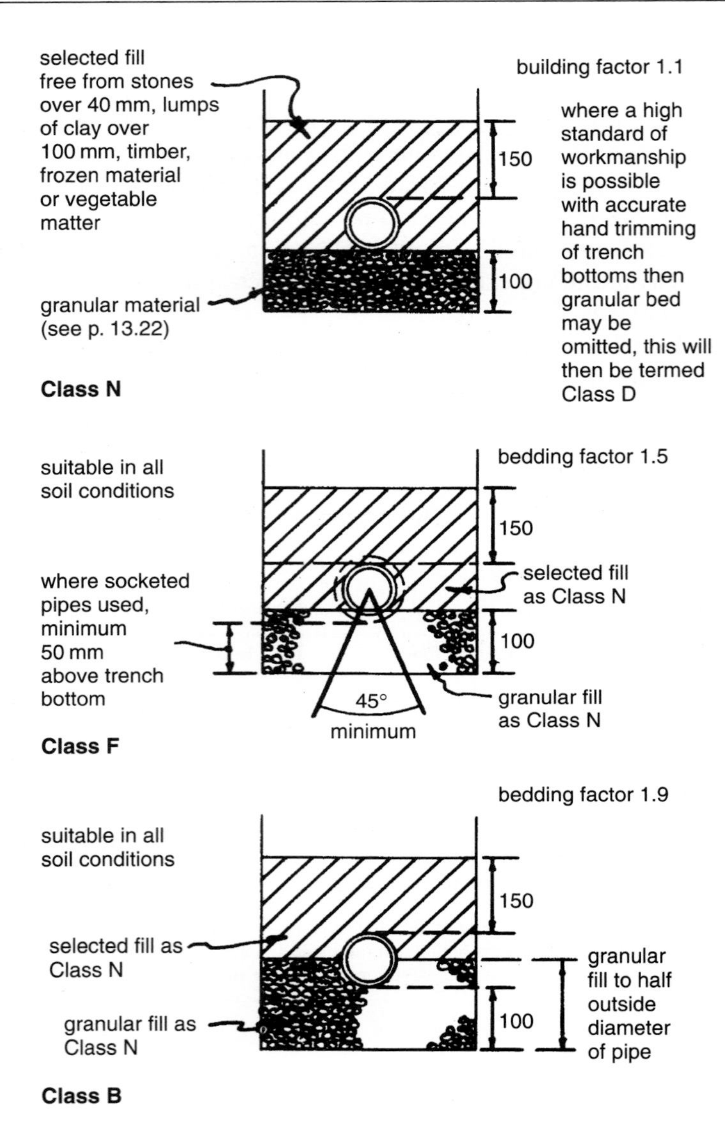

Fig. 13.10 Bedding classes for rigid pipes.

13.3.19 Flexible pipes

Flexible pipes should be provided with a minimum depth of cover of 900 mm under any road. This may be reduced to 600 mm in fields and gardens. The maximum permissible depth of cover is 7 m. Figure 13.11 shows typical bedding and backfilling details for flexible pipes. Table 10 of AD H1 gives limits of cover for thermoplastics pipes in any width of trench. Where flexible pipes have less than the minimum cover depths given in the tables they should be protected where necessary, as shown in Fig. 13.12.

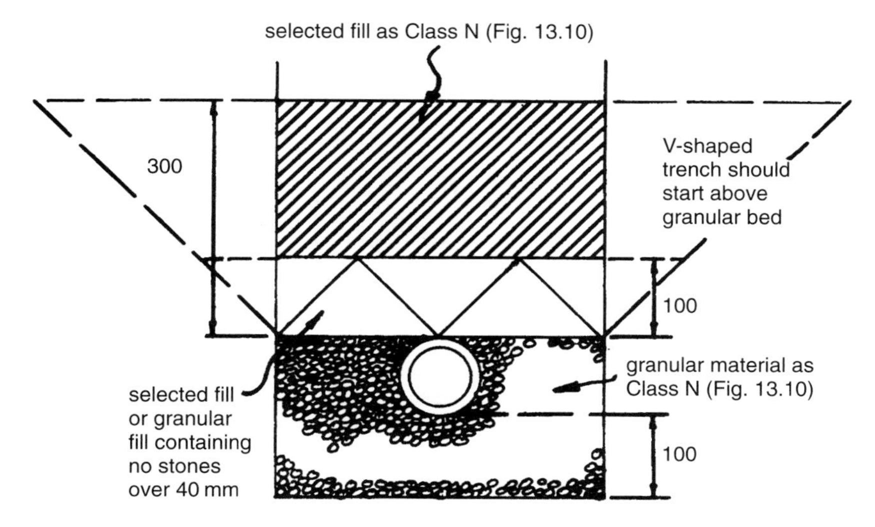

Fig. 13.11 Bedding for flexible pipes.

13.3.20 Special protection to pipes

Where pipes have less than the minimum cover recommended in Tables 8, 9 or 10 they should be bridged by reinforced concrete cover slabs resting on a flexible filler with at least 75 mm of granular fill between the top of the pipe and the underside of the flexible filler (see Fig. 13.12). Where it is necessary to backfill a trench with concrete to protect adjacent foundations (see section 13.3.25 below), the pipes should be surrounded in concrete to a thickness of at least 100 mm. Expansion joints should also be provided at each socket or sleeve joint face (see Fig. 13.12).

PIPE SIZES AND GRADIENTS – Drains should be laid to falls and should be large enough to carry the expected flow. The rate of flow will depend on the type,

AD H1, section 2

Table 10 Limits of cover for thermoplastics (nominal ring stiffness SN4) pipes in any width of trench

Nominal size	Laid in fields	Laid in light roads	Laid in main roads
100 mm–300 mm	0.6 m–7 m	0.9 m–7 m	0.9 m–7 m

Notes:

1. For drains and sewers less than 1.5 m deep where there is a risk of excavation adjacent to the drain, a special calculation is necessary, see BS EN 1295, paragraph NA. 6.2.3
2. All pipes assumed to be in accordance with the relevant standard listed in Table 7 of AD H1 section 2 with nominal ring stiffness SN4, other strengths and sizes of pipe are available, consult manufacturers;
3. Bedding assumed to be Class S2 with 80% compaction and average soil conditions;
4. Alternative designs using different pipe strengths and/or bedding types may offer more appropriate or economic options using the procedures set out in BS EN 1295;
5. Minimum depth in roads is set to 1.5 m irrespective of pipe strength, to cover loss of side support from parallel excavations.

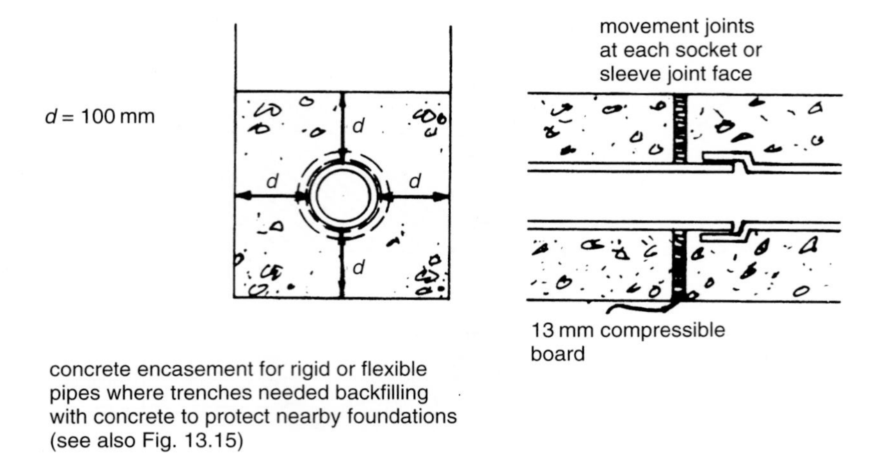

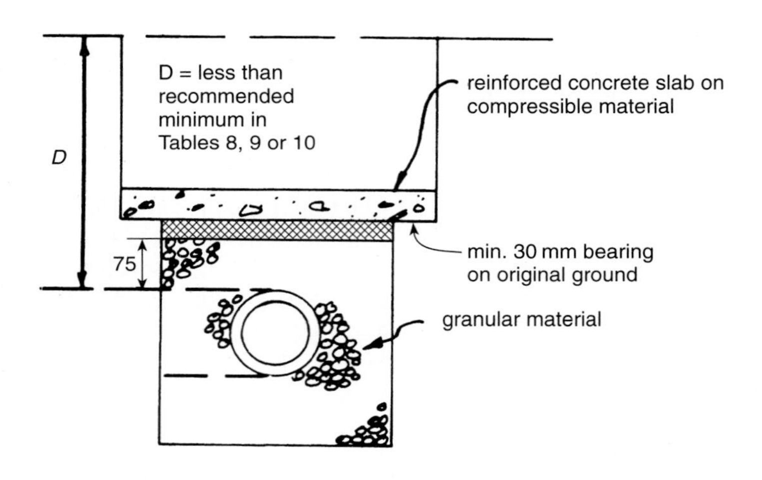

Fig. 13.12 Special protection to pipes.

number and grouping of appliances that are connected to the drain (see Table A1, and Table 13.1, section 13.3.1). The capacity will depend on the diameter and gradient of the pipes.

Table 6 to section 2 of AD HI gives recommended minimum gradients for different sized foul drains and shows the maximum capacities they are capable of carrying. The Table is set out below.

As a further design guide Diagram 9 from AD HI is reproduced below. This gives discharge capacities for foul drains running at 0.75 proportional depth.

A drain serving more than one property (i.e. a sewer) should normally have a minimum diameter of 100 mm if serving no more than ten dwellings. Sewers serving more than ten dwellings should normally have a diameter of at least 150 mm.

AD H1, section 2

Table 6 Recommended minimum gradients for foul drains.

Peak flow (litres/sec)	Pipe size (mm)	Minimum gradient (l:...)	Maximum capacity (litres/sec)
< 1	75	1:40	4.1
	100	1:40	9.2
> 1	75	1:80	2.8
	100	1:80*	6.3
	150	1:150†	15.0

Notes:
* Minimum of 1 wc.
† Minimum of 5 wcs.

If a drain is carrying only foul water it may have a minimum diameter of 75 mm. This is increased to a minimum of 100 mm if the drain is carrying effluent from a WC or trade effluent. Where foul and rainwater drainage systems are combined, the capacity of the system should be large enough to take the combined peak flow (see Rainwater drainage, section 13.6 below).

MATERIALS – Table 7 to section 2 of AD HI, which is reproduced below, gives details of the materials that may be used for pipes, fittings and joints in below-ground foul drainage systems. Joints should remain watertight under working and test conditions and nothing in the joints, pipes or fittings should form an obstruction inside the pipeline. To avoid damage by differential settlement pipes should have flexible joints appropriate to the material of the pipes.

To prevent electrolytic corrosion, pipes of different metals should be separated where necessary by non-metallic material.

PROVISIONS FOR CLEARING BLOCKAGES – Every part of a drainage system should be accessible for clearing blockages. The type of access point chosen and its siting and spacing will depend on the layout of the drainage system and the depth and size of the drain runs. A drainage system designed in accordance with the provisions of AD H1 should be capable of being rodded by normal means (i.e. not by mechanical methods).

13.3.21 Access points

Four types of access points are described in AD H1.

- Rodding eyes (or points). These are extensions of the drainage system to ground level where the open end of the pipe is capped with a sealing plate.
- Access fittings. Small chambers situated at the invert level of a pipe and without any real area of open channel.
- Inspection chambers. Chambers having working space at ground level.
- Manholes. Chambers large enough to admit persons to work at drain level.

AD HI, section 2

Diagram 9 Discharge capacities of foul drains running 0.75 proportional depth.

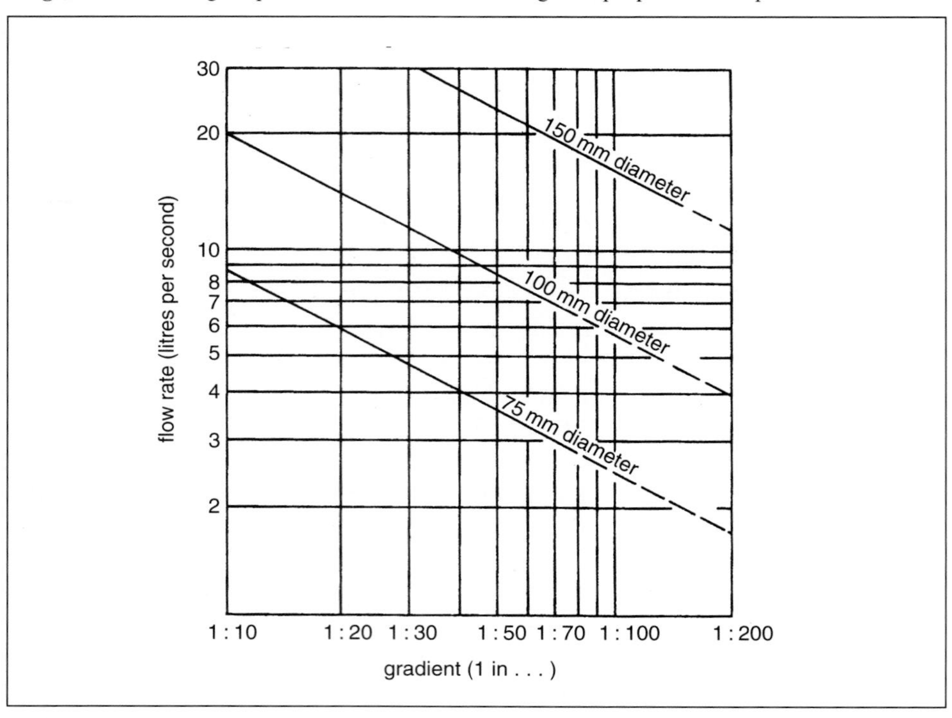

AD H1, section 2

Table 7 Materials for below ground gravity drainage.

Material	**British Standard**
Rigid pipes	
Vitrified clay	BS 65, BS EN 295
concrete	BS 5911
grey iron	BS 437
ductile iron	BS EN 598
Flexible pipes	
UPVC	BS EN 1401 +
PP	BS EN 1852 +
Structure Walled Plastic pipes	BS EN 13476

+ Application area code UD should normally be specified
Note: Some of these materials may not be suitable for conveying trade effluent

Some typical access point details are illustrated in Fig. 13.13.

Whatever form of access point is used it should be of sufficient size to enable the drain run to be adequately rodded. Tables 11 and 12 to section 2 of AD HI set out the maximum depths and minimum internal dimensions for

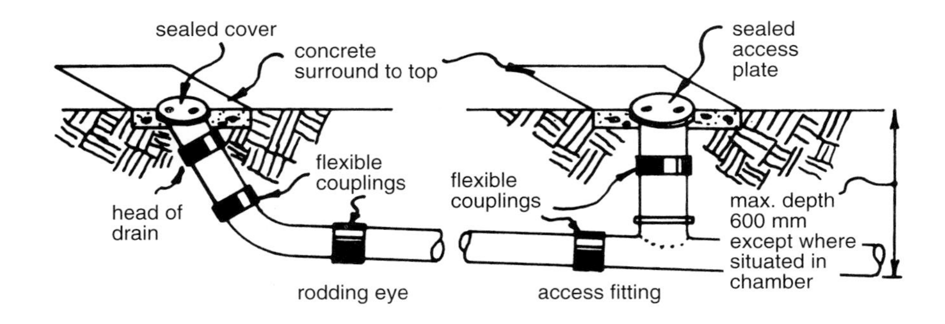

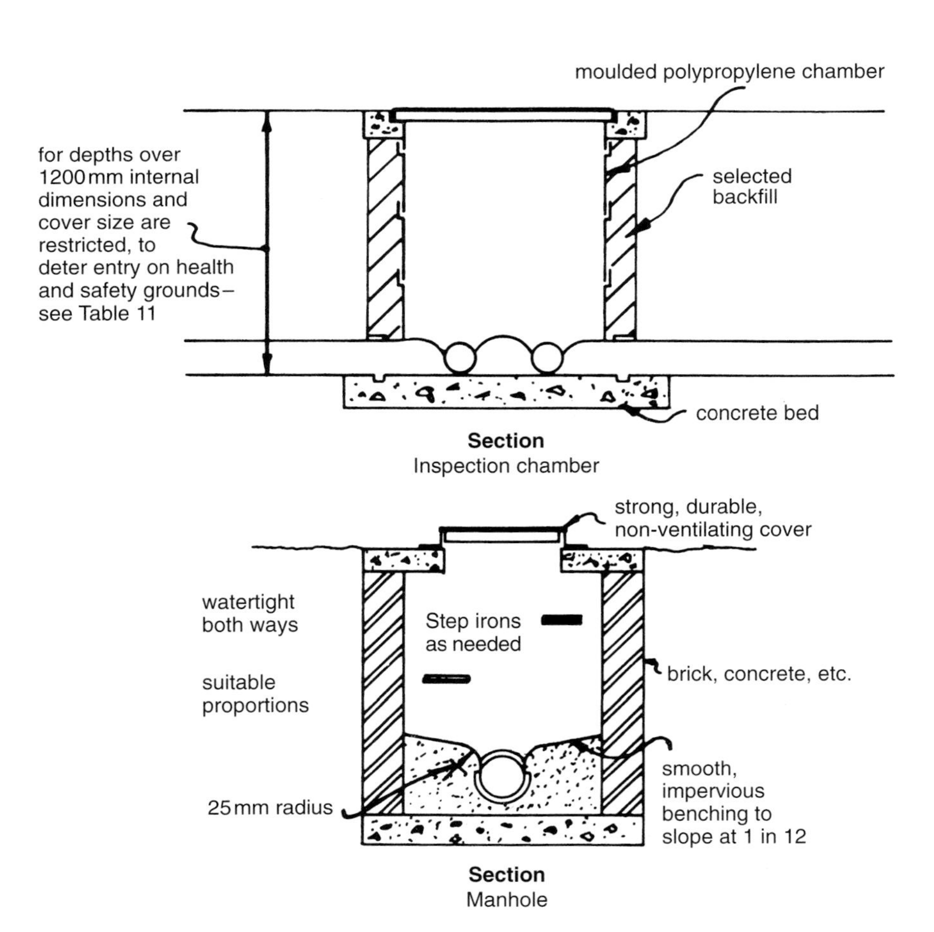

Fig. 13.13 Access points.

each type of access point. Where a large number of branches enter an inspection chamber or manhole the sizes given in Tables 11 and 12 may need to be increased. It is usual to allow 300mm for each branch connection (thus a 1200 mm long manhole could cater for up to four branch connections on each side). The Tables are set out below.

13.3.22 Access points – siting and spacing

Access points should be provided:

- at or near the head of any drain run
- at any change of direction or gradient
- at a junction, unless each drain run can be rodded separately from another access point
- at a change of pipe size, unless this occurs at a junction where each drain run can be rodded separately from another access point
- at regular intervals on long drain runs.

The spacing of access points will depend on the type of access used. Table 13 to section 2 of AD HI gives details of the maximum distances that should be allowed for drains up to 300 mm in diameter and is set out below.

AD H1, section 2

Table 11 Minimum dimensions for access fittings and chambers.

Type		Depth to invert from cover level (m)	Internal sizes		Cover sizes	
			Length × width (mm × mm)	**Circular (mm)**	**Length × width (mm × mm)**	**Circular (mm)**
Rodding eye			As drain but min 100			Same size as pipework[1]
Access fitting						
small	150 diam	0.6 or less, except where situated in a chamber				
	150 × 100		150 × 100	150	150 × 100[1]	Same size as access fitting
large	225 × 100		225 × 100	225	225 × 100[1]	
Inspection chamber						
	shallow	0.6 or less	225 × 100	190[2]	–	190[1]
		1.2 or less	450 × 450	450	Min 430 × 430	430
	deep	> 1.2	450 × 450	450	max 300 × 300[3]	Access restricted to max 350[3]

Notes:
[1] The clear opening may be reduced by 20 mm in order to provide proper support for the cover and frame.
[2] Drains up to 150 mm.
[3] A larger clear opening cover may be used in conjunction with a restricted access. The size is restricted for health and safety reasons to deter entry.

AD H1, section 2

Table 12 Minimum dimensions for manholes.

Type	Size of largest pipe (DN)	Min internal dimensions[1] Rectangular length and width	Circular diameter	Min clear opening size[1] Rectangular length and width	Circular diameter
Manhole < 1.5 m deep to soffit	150	750 × 675[7]	1000[7]	750 × 675[2]	na[3]
	225	1200 × 675	1200	1200 × 675[2]	
	300	1200 × 750	1200		
	> 300	1800 × (DN + 450)	The larger of 1800 or (DN + 450)		
> 1.5 m deep to soffit	225	1200 × 1000	1200	600 × 600	600
	300	1200 × 1075	1200		
	375–450	1350 × 1225	1200		
	> 450	1800 × (DN + 775)	The larger of 1800 or (DN + 775)		
Manhole shaft[4] > 3.0 m deep to soffit of pipe	Steps[5]	1050 × 800	1050	600 × 600	600
	Ladder[5]	1200 × 800	1200		
	Winch[6]	900 × 800	900	600 × 600	600

Notes:
[1] Larger sizes may be required for manholes on bends or where there are junctions.
[2] May be reduced to 600 by 600 where required by highway loading considerations, subject to a safe system of work being specified.
[3] Not applicable due to working space needed.
[4] Minimum height of chamber in shafted manhole 2 m from benching to underside of reducing slab.
[5] Min clear space between ladder or steps and the opposite face of the shaft should be approximately 900 mm.
[6] Winch only – no steps or ladders, permanent or removable.
[7] The minimum size of any manhole serving a sewer (i.e. any drain serving more than one property) should be 1200 mm × 675 mm rectangular or 1200 mm diameter.

Where an access point is provided to a sewer (i.e. serving more than one property) it should be positioned so that it is both accessible and apparent for use in emergencies. Typically, it could be positioned in a highway, public open space, unfenced front garden or shared and unfenced driveway.

13.3.23 Access points – construction

Generally, access points should:

- be constructed of suitable and durable materials
- exclude subsoil or rainwater
- be watertight under working and test conditions.

Table 14 to section 2 of AD HI is shown below and lists materials which are suitable for the construction of access points.

AD H1, section 2

Table 13 Maximum spacing of access points in metres.

From	To	Access Fitting Small	Access Fitting Large	Junction	Inspection chamber	Manhole
Start of external drain[1]		12	12	—	22	45
Rodding eye		22	22	22	45	45
Access fitting small 150 diam 150 × 100		—	—	12	22	22
large 225 × 100		—	—	45	22	45
Inspection chamber		22	45	22	45	45
Manhole		—	—	—	45	90[2]

Note
[1] Stack or ground floor appliance
[2] May be up to 200 for man-entry size drains and sewers

AD H1, section 2

Table 14 Materials for access points.

Material	British Standard
1 Inspection chambers and manholes	
Clay	
bricks and blocks	**BS 3921**
Vitrified clay	BS EN 295, BS 65
Concrete	
precast	BS 5911
in situ	BS 8110
Plastics	BS 7158
2 Rodding eyes and access fittings (excluding frames and covers)	as pipes see Table 7 ETA Certificates

Inspection chambers and manholes should fulfil the following.

- Have smooth impervious surface benching up to at least the top of the outgoing pipe to all channels and branches. The purpose of benching is to direct the flow into the main channel and to provide a safe foothold. For this reason the benching should fall towards the channel at a slope of 1 in 12 and should be rounded at the channel with a minimum radius of 25 mm (see Fig. 13.13 above).
- Be constructed so that branches up to and including 150 mm diameter discharge into the main channel at or above the horizontal diameter where half-round open

channels are used. Branches greater than 150mm diameter should be set with the soffit level with that of the main drain. Branches which make an angle of more than 45° with the channel should be formed using a three-quarter section branch bend.

- Have strong, removable, non-ventilating covers of suitable durable material (e.g. cast iron, cast or pressed steel or pre-cast concrete or plastics).
- Be fitted with step irons, ladders, etc., if over 1.0 m deep.
- Small lightweight access covers should be secured to deter unauthorised access (e.g. by children). Commonly, such covers are screwed down.
- A manhole or inspection chamber which is situated *within* a building should have an airtight cover that is mechanically fixed (e.g. screwed down with corrosion resistant bolts). This requirement does not apply if the inspection chamber or manhole gives access to part of a drain which itself has inspection fittings and these are provided with watertight covers.

13.3.24 Test for watertightness

After laying and backfilling, gravity below-ground drains and private sewers not exceeding 300 mm in diameter should be pressure tested using air or water.

For the air test, the pipe should be pressurised up to 110 mm water gauge and held for about five minutes prior to testing. Subsequently, a head loss of up to 25 mm at 100 mm water gauge is permitted in a period of seven minutes during the test.

For the water test, the section of drain to be tested should be filled with water up to a depth of 500 mm above the lowest invert and at least 100 mm above the highest invert in the test section and left to stand for about one hour to condition the pipe. Over the next 30 minutes the test pressure should be maintained by topping up the water level so that it is within 10mm of the levels given above. The leakage rate per square metre of surface area should not exceed:

- 0.15 litres for pipelines only, or
- 0.20 litres for test lengths which include pipelines and manholes, and
- 0.40 litres for tests on manholes and inspection chambers alone (i.e. no pipelines).

Using this method it is easy to check the leakage rate simply by measuring the quantity of water used to top up during the test and dividing by the surface area of the manhole or inspection chamber.

For tests on pipelines exceeding 300mm diameter, reference should be made to BS 8000: Part 14:1989 *Workmanship on building sites. Code of practice for below ground drainage*, or BS EN 1610:1998 *Construction and testing of drains and sewers. Code of practice for design and construction.*

13.3.25 Special protection for drains adjacent to or under buildings

Where drains pass under buildings or through foundations and walls there is a risk that settlement of the building may cause pipes to fracture, with consequential blockages and leakage. In the past it was common practice to require pipes (which were rigid jointed) to be encased in concrete. Since the development of flexible pipe systems it has become essential to maintain this flexibility in order that any slight settlement of the building will not cause pipe fracture.

Therefore, drain runs under buildings should be surrounded with at least 100 mm of granular or other flexible filling. On some sites unusual ground conditions may lead to excessive subsidence. To protect drain runs from fracture it may be necessary to have additional flexible joints or use other solutions such as suspended drainage especially where the pipe is adjacent to structures or where there is a change in soil conditions in the length of the pipe run. Shallow drain runs under concrete floor slabs should be protected as described in section 13.3.20 above and as shown in Fig. 13.12 where the crown of the pipe is less than 300 mm from the underside of the slab.

Where a drain is built into a structure (e.g. a wall, foundation, ground beam, inspection chamber, manhole etc.) suitable measures should be taken to prevent damage or misalignment. The following solutions are possible.

- The wall may be supported on lintels over the pipe. A clearance of 50 mm should be provided round the pipe perimeter and this gap be masked on both sides of the wall with rigid sheet material to prevent the ingress of fill or vermin. The void should be filled with a compressible sealant to prevent ingress of gas.
- A length of pipe may be built in to the wall with its joints not more than 150 mm from each face. Rocker pipes not exceeding 600 mm in length should then be connected to each end of the pipe using flexible joints (see Fig. 13.14).

Where a drain or private sewer is laid close to a load-bearing part of a building, precautions should be taken to ensure that the drain or sewer trench does not impair the stability of the building.

Where any drain or sewer trench is within 1 m of the foundation of a wall, and the bottom of the trench is lower than the wall foundation, the trench should be filled with concrete up to the level of the underside of the foundation.

Where a drain or sewer trench is 1 m or more from a wall foundation, and the trench bottom is lower than the foundation, the trench should be filled with concrete to within a vertical distance below the underside of the foundation of not more than the horizontal distance from the foundation to the trench less 150 mm (see Fig. 13.15).

Where it is necessary to adopt unusual design solutions for buried pipelines due to special ground conditions (e.g. pipes are to be laid on piles or beams, ground may prove to be unstable, there may be a high water table etc.) or where pipes are to be laid in a common trench, guidance may be found in the Department of Transport publication: *Guide to design loadings for buried rigid pipes*.

Additionally, local authorities may be able to provide information regarding subsoil conditions on many sites.

13.3.26 Special protection – drain surcharging

Under conditions of heavy rainfall, combined and rainwater sewers are designed to surcharge, whereby the water level in the manhole rises above the top of the pipe. This may also happen to some foul sewers if they receive rainwater. Therefore, on

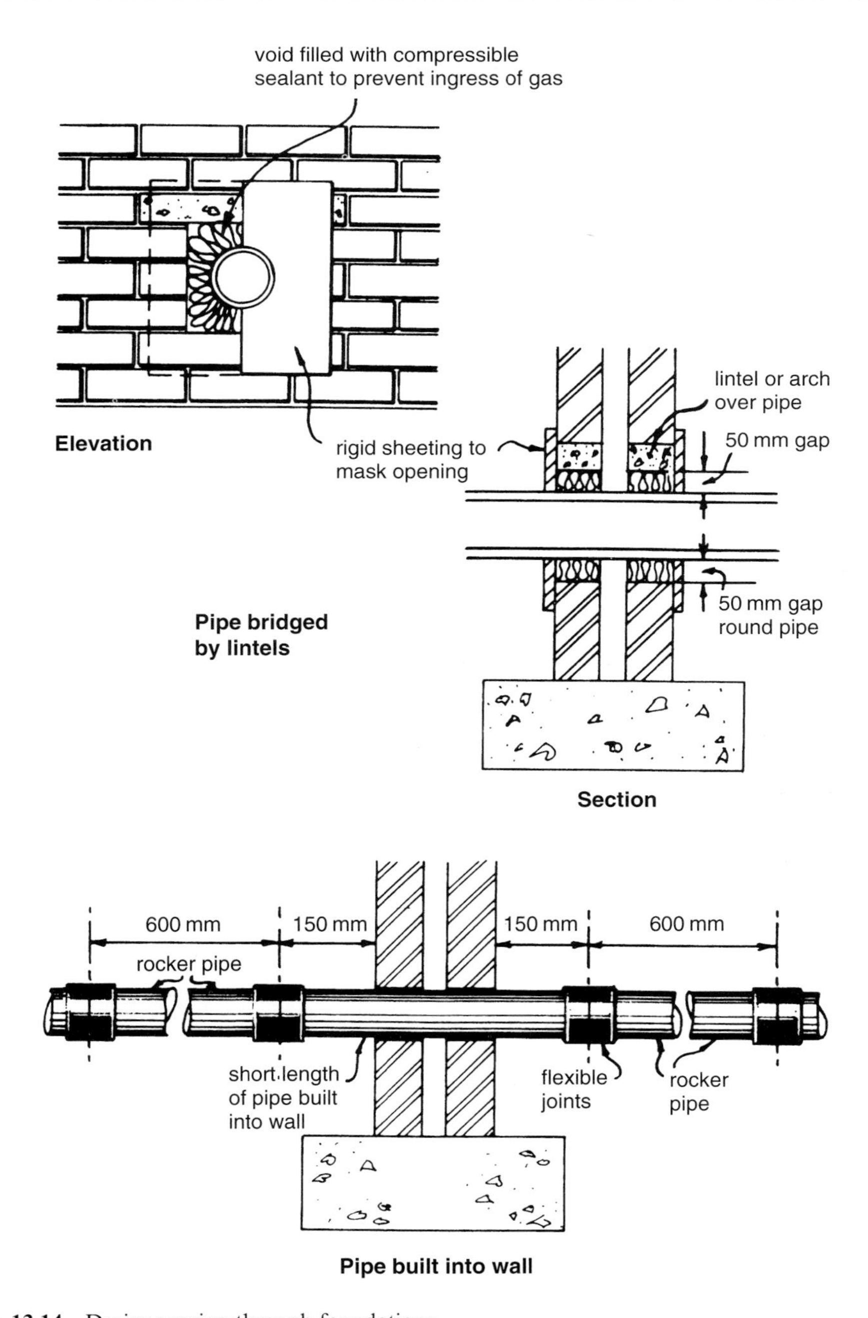

Fig. 13.14 Drains passing through foundations.

some low lying sites properties may be at increased risk of flooding if the ground level of the site (or the level of a basement) is below the level at which the drainage connects to the public sewer. The sewerage undertaker should consulted in such cases to determine the extent and frequency of the likely surcharge.

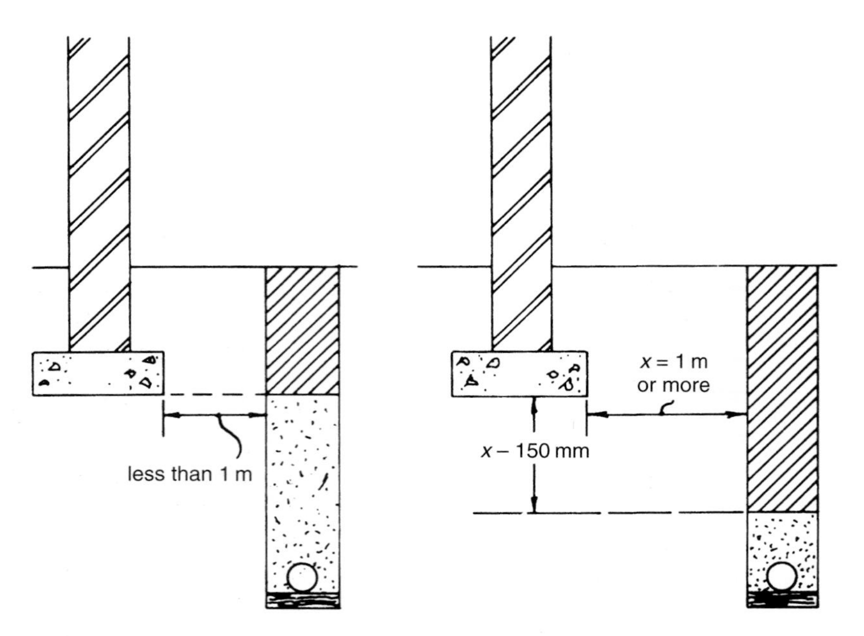

Fig. 13.15 Drain trenches.

Where a basement contains sanitary appliances and the sewerage undertaker considers that the risk of flooding due to surcharging is high, the drainage from the basement should be pumped (see section 13.3.28 below). For low risks, an anti-flooding valve should be installed on the drainage from the basement.

For low lying sites (i.e. those not containing basements) where the risk is low, protection for the building may be achieved by the provision of an external gully sited at least 75 mm below floor level in a position so that any flooding from the gully will not damage any buildings. Higher risk areas should have anti-flooding valves or pumped drainage systems (see section 13.3.28 below).

Anti-flooding valves should:

- be of the double valve type
- be suitable for foul water
- have a manual closure device
- comply with prEN 13564 *Anti flooding devices for buildings*.

Normally, a single valve should serve only one building and information about the valve should be provided on a notice inside the building. The notice should indicate the location of any manual override and include necessary maintenance information.

Some parts of the drainage system may be unaffected by surcharging. These parts should by-pass any protective measures and should discharge by gravity.

13.3.27 Special protection – rodent control

Generally, rodent infestation (especially by rats) is on the increase. Since rats use drains and sewers as effective communication routes, on previously developed sites the local authority should be consulted to ascertain if any special rodent control precautions are thought necessary.

Special precautions could include the following.

- By providing inspection chambers with screwed access covers on the pipework instead of open channels. These should only be used in inspection chambers where maintenance can be carried out from the surface without personnel entry.
- Intercepting traps may also be provided as in the past, although they do increase the incidence of blockages unless adequately maintained. Trap stoppers should be of the locking type and should be easy to remove and replace after clearing any blockage. They should always be replaced after maintenance operations and should only be used in inspection chambers where maintenance can be carried out from the surface without personnel entry.
- A number of different kinds of rodent barriers might be considered. These include enlarged sections of discharge stacks which prevent rats climbing, flexible downward facing fins in discharge stacks and one way valves in underground drainage.
- The provision of metal cages on ventilator stack terminals and fixed plastic covers or metal gratings on gullies to discourage rats from leaving the drainage system.

13.3.28 Pumping installations

Reference has been made above to the use of pumping installations where gravity drainage is impracticable or where protection is required against flooding due to surcharge in downstream sewers.

AD H1 gives details of packaged pumping systems for use both inside and outside buildings as follows.

Inside buildings

Floor mounted units available for use in basements should comply with BS EN 12050 *Wastewater lifting plants for buildings and sites – principles of construction and testing*. The pumping installation itself should be designed in accordance with BS EN 12056:2000 *Gravity drainage systems inside buildings*: Part 4: *Effluent lifting plants, layout and calculation*.

Outside buildings

Package pumping installations for use outside buildings are also available. The pumping installation should be designed in accordance with BS EN 752 *Drain and sewer systems outside buildings*: Part 6: 1998 *Pumping installations*.

Foul water drainage pumping installations should comply with the following.

- To allow for disruption in service, the effluent receiving chamber should be sized to contain 24-hour inflow.
- For domestic use the minimum daily discharge of foul drainage should be taken as 150 litres per person per day.
- For non-domestic uses the capacity of the receiving chamber should be based on the calculated daily demand of the water intake for the building (and should be assessed on a pro-rata basis where only a proportion of the foul sewage is pumped).
- For all pumped systems the controls should be arranged to optimise pump operation.

13.3.29 Workmanship

In general, workmanship should be in accordance with BS 8000 *Workmanship on building sites* Part 14: *Code of practice for below ground drainage*.

In particular, drains and sewers which are left open during construction should be covered when work is not in progress to prevent entry by rats.

A number of measures are necessary to protect drains during construction work. For example, drains can be damaged by construction traffic and heavy machinery. Barriers should be provided where necessary to keep traffic away from the line of the sewer and heavy materials should not be stored over drains or sewers. Additionally, piling works can cause damage to drains and sewers unless certain precautions are taken. This would include carrying out a survey to establish the exact location of any drain runs and connections before piling commences. Piling should not be carried out where the distance from the outside of the sewer to the outside of the pile is less than two times the diameter of the pile.

13.3.30 Altemative method of design

Additional information on the design and construction of building drainage which meets the requirements of the 2000 Regulations may be found in the relevant parts of:

- BS EN 12056: *Gravity drainage systems inside buildings*. This standard also describes the discharge unit method of calculating flows;
- BS EN 752 *Drain and sewer systems outside buildings*: Part 3:1997 *Planning*, Part 4:1997 *Hydraulic design and environmental aspects* and Part 6:1998 *Pumping installations*;
- BS EN 1610:1998 *Construction and testing of drains and sewers*;
- BS EN 1295: Part 1:1998 *Structural design of buried pipelines under various conditions of loading*;
- BS EN 1091:1997 *Vacuum sewerage systems outside buildings*; and
- BS EN 1671:1997 *Pressure sewerage systems outside buildings*.

BS EN 752 together with BS EN 1610 and BS EN 1295 contain additional information about design and construction.

13.4 Wastewater treatment systems and cesspools

Any septic tank and its form of secondary treatment, other wastewater treatment system or cesspool must be sited and constructed so that:

- it is not prejudicial to health;
- it will not contaminate any watercourse, underground water or water supply;
- it is accessible for emptying and maintenance;
- it will continue to function in the event of a power failure to a standard sufficient for the protection of health, where this is relevant (i.e. where a power supply is needed for normal operation of the system).

Furthermore, any septic tank, holding tank which is part of a wastewater treatment system or cesspool must be:

- adequately ventilated
- of adequate capacity
- constructed to be impermeable to liquids.

Since all wastewater treatment systems and cesspools rely on adequate maintenance in order to continue to operate in a safe and healthy manner the Regulations require that maintenance instructions be provided in the form of a durable notice which must be affixed in a suitable place in the building. Examples of a typical notices are given in the text below.

It should be noted that the use of non-mains foul drainage should only be considered where connection to mains drainage is not practicable and any discharge from a wastewater treatment system is likely to require consent from the Environment Agency. Contact with the Environment Agency should be made as early as possible in the design process (usually when the planning process is being initiated and before a Building Regulation application is made for non-mains drainage). This will determine whether a consent to discharge is required and what parameters apply, which in turn can have an impact on the type of system that may be installed. Further guidance may be obtained from Pollution Prevention Guideline No 4 *Disposal of sewage where no mains drainage is available*: Environment Agency 1999.

13.4.1 Wastewater treatment systems

A wastewater treatment system typically includes a septic or settlement tank which provides primary treatment to the effluent from a building. This is likely to be the most economic form of treating wastewater for one to three dwellings. The discharge from the tank can, however, still be harmful, therefore there is a need for a system of drainage which completes the treatment process after the effluent has passed through the tank, thus providing a means of secondary treatment. The term 'wastewater treatment system' can also include small sewage treatment works (see section 13.4.3 below).

In the past, the Regulations have tended to concentrate on the design and construction of the means of primary treatment but have failed to provide guidance on the ultimate means of disposal of the effluent. This is an area where considerable research and development has taken place in recent years and guidance is now given on the design and construction of drainage fields and mounds, and on constructed wetlands and reedbeds. The performance of a wastewater treatment system will depend on the capacity, siting, design and construction of both the septic or settlement tank and the drainage field or other means of secondary treatment.

13.4.2 Primary treatment systems – septic tanks and settlement tanks

Capacity

The primary treatment system should have sufficient capacity and should provide suitable conditions for the settlement, storage and partial decomposition of solid matter in the wastewater from the building. It should also be sited and constructed so as to prevent overloading of the receiving water.

For up to four users, a minimum capacity of 2.7 m^3 (2700 litres) below the level of the inlet is set for septic tanks and settlement tanks in order to reduce danger of overflowing and malfunctioning. This size should be increased by 0.18 m^3 (180 litres) for each additional user.

Siting and construction

Septic tanks should be designed and constructed to prevent leakage of contents and the ingress of subsoil water. They should also be provided with adequate ventilation, which should be kept away from buildings. Therefore, they should be kept at least 7 m from any habitable parts of buildings, preferably on a downslope.

Septic tanks must be periodically desludged and cleaned. This is usually carried out mechanically using a tanker. Because of the length of piping involved it is necessary that the cesspool or tank be sited within 30 m of a vehicular access; however, where the invert level of the tank is more than 3 m below the vehicle access level the 30 m distance will need to be reduced accordingly. Emptying and cleaning should not involve the contents being taken through a dwelling or place of work, and there should be a clear route for the hose so that the emptying and cleaning can be carried out without creating a hazard for the building's occupants. Access covers for emptying and cleaning should be sufficiently durable to resist the corrosive nature of the tank contents and should be designed to prevent unauthorised access (by being lockable or otherwise engineered to prevent personnel entry).

Tanks should also be constructed of materials which are impervious to the contents and to ground water. This would include engineering brickwork in 1:3 cement mortar at least 220 mm thick and concrete at least 150 mm thick (C/25/P mix to BS 5328), roofed with heavy concrete slabs. Prefabricated cesspools and tanks are available made of glass reinforced plastic, polyethylene or steel. These should follow the guidance in BS EN 12566 *Small wastewater treatment plants less than 50 PE*: Part 1: 2000 *Prefabricated septic tanks*. Care should be exercised over the stability of these tanks.

The inlet and outlet of the tank should be provided with access for sampling and inspection of the contents, and be designed to avoid excessive disturbance of the surface scum or settled contents by incorporating at least two chambers operating in sequence. The velocity of flow into the tank can be limited by laying the last 12 m of the incoming drain at a gradient of 1 in 50 or flatter for all pipes up to 150 mm in diameter. Alternatively, a dip pipe inlet may be provided (see Fig. 13.16) where the tank width does not exceed 1200 mm.

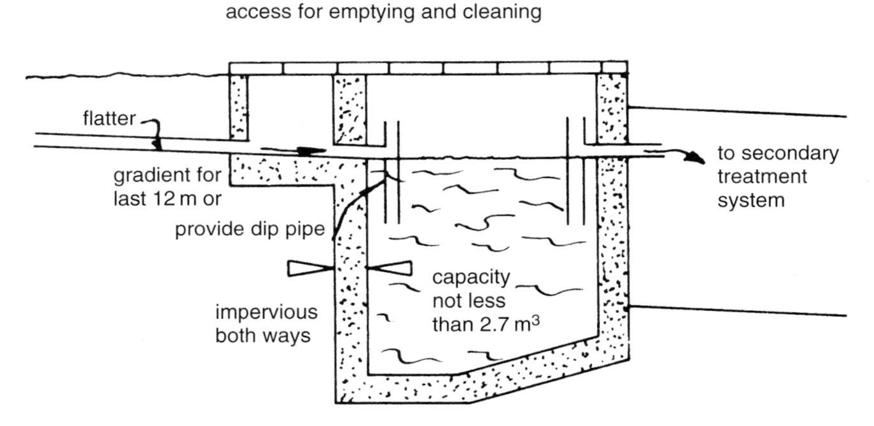

Fig. 13.16 Septic tank.

Septic tank maintenance

It is essential that building owners are kept informed about the maintenance requirements of septic tanks. Septic tanks need to be inspected monthly to make sure that they are working properly. This involves an inspection of the inlet chamber and the outlet from the tank to ensure that the effluent is flowing freely. Additionally, the effluent at the outlet should be clear.

If these conditions are not met the tank should be emptied by a licensed contractor who will make a charge. It may be more economical to take out an annual maintenance contract with a suitable contractor. Septic tanks should be emptied annually and it is usual to leave a small amount of sludge to act as an anaerobic seed.

Failure to adequately maintain the tank may result in solids being carried into the drainage field or mound. The sediments deposited may block the pores in the soil necessitating early replacement of the field or mound and in exceptional circumstances this can even render the site unsuitable for future use as a drainage field or mound.

A durable notice should be fixed within the building describing the necessary maintenance. Fig. 13 .17 below is an example of a typical notice.

Wastewater Treatment System

Details of Necessary Maintenance

Address of Property ____

Location of treatment system _____________________________________

The foul drainage system from this property discharges to a septic tank and a [*insert type of secondary treatment*]

The tank requires monthly inspections of the outlet chamber or distribution box to observe that the effluent is free-flowing and clear.

The septic tank requires emptying at least every 12 months by a licensed contractor. The [*insert type of secondary treatment*] should be [*insert details of maintenance of secondary treatment*].

The owner is legally responsible to ensure that the system does not cause pollution, a health hazard or a nuisance.

Fig. 13.17 Septic tank – typical maintenance notice.

13.4.3 Primary treatment systems – packaged treatment works

Packaged treatment works, which are engineered to treat a given hydraulic and organic load to a higher standard than septic tanks, use prefabricated components which can be installed with a minimum amount of site work. They are normally more economic than septic tanks for larger developments and can also discharge direct to a suitable watercourse. They should be considered where there are space limitations or where other options are not possible. AD H2 does not really deal with such installations in any detail since specialist knowledge is needed in their detailed design and installation. However, it does recommend that the discharge from the treatment plant should be sited at least 10 m from watercourses or any other buildings. Furthermore, since many of these systems are powered by electricity it is important that the system should be able to adequately function for up to six hours without power or have an uninterruptible power supply.

Guidance on packaged treatment works may be obtained from BS 6297: 1983 *Code of practice for design and installation of small sewage treatment works and cesspools*. Additionally, packaged treatment works should be type-tested in accordance with BS 7781 or otherwise tested by a notified body.

The guidance regarding maintenance requirements mentioned above in connection with septic tanks also applies generally to packaged treatment works; however there will be variations in maintenance needs depending on the type of plant installed. The manufacturer's instructions regarding maintenance and inspection should always be adhered to. A durable notice should be fixed within the building describing the necessary maintenance. Fig. 13.18 below is an example of a typical notice.

Wastewater Treatment System

Details of Necessary Maintenance

Address of Property ______________________________________

Location of treatment system ______________________________________

The foul drainage system from this property discharges to a packaged treatment works.

Maintenance is required [insert frequency] and should be carried out by the owner in accordance with the manufacturer's instructions.

The owner is legally responsible to ensure that the system does not cause pollution, a health hazard or a nuisance.

Fig. 13.18 Septic tank – typical maintenance notice.

13.4.3 Secondary treatment systems – drainage fields and drainage mounds

Drainage fields and drainage mounds are used to provide secondary treatment to the discharge from a septic tank or packaged treatment plant. Drainage fields normally consist of below ground irrigation pipes which allow the partially treated effluent from the septic tank to percolate into the surrounding soil. Further biological treatment takes place naturally in the aerated soil layers. They may be used in subsoils with good percolation characteristics on sites which are not prone to flooding or waterlogging at any time of the year. Drainage mounds consist of drainage fields placed above the ground surface thus providing an aerated soil layer to treat the discharge. On sites where there is a high water table or impervious ground where occasional waterlogging is possible, drainage mounds could be used.

It should be noted that drainage fields and mounds are not permitted by the Environment Agency in prescribed Zone 1 groundwater source protection zones.

Siting

Care has to be taken with siting in order to protect underground water sources and watercourses and to ensure that the system will operate effectively. Therefore, a drainage field or mound serving a wastewater treatment system or septic tank should be sited:

- not less than 10 m from any permeable drain or watercourse;
- not less than 50 m from the abstraction point of any groundwater supply;
- away from any Zone 1 groundwater protection zone;
- not less than 15 m from any building;
- far enough away from other drainage fields, mounds or soakaways so that the overall soakage capacity of the ground is not exceeded;
- on the downslope side of any groundwater sources.

The disposal area should be isolated and should not contain any access roads, driveways or paved areas. Additionally, no water supply pipes or other underground services should be located within the disposal area other than those required by the disposal system itself.

Ground conditions and percolation

Some indication of the likely percolation characteristics of a site may be gained by taking a sample and observing the nature of the subsoil. Table 13.3 below gives an indication of the likely percolation characteristics of different subsoil colours and types. Percolation characteristics should be ascertained under both summer and winter conditions. This usually takes the form of a preliminary assessment followed by a percolation test.

Table 13.3 Likely percolation characteristics of different soil types.

Likely percolation characteristics	Soil colour	Likely soil type
Well drained and well aerated	Brown, yellow or reddish	Sand, gravel, chalk, sandy loam, clay loam
Poorly drained or saturated	Grey, blue	Sandy clay, silty clay, clay
Indicative of periodic saturation	Grey or brown mottling	Sandy clay, silty clay, clay

The preliminary assessment should involve:

- consultation with the Environment Agency and the local authority to determine the possible suitability of the site;
- an assessment of the on-site natural vegetation (most plants generally grow best on well drained land, the presence of certain plants may indicate wet or boggy conditions etc.);
- a determination of the position of the standing ground water table.

The standing ground water table is determined by excavating a trial hole. This should be at least 1 m^2 in area and should be at least 2 m deep or 1.5 m below the invert of the proposed drainage field pipework. The ground water table should be at least 1m below the invert level of the proposed effluent distribution pipes in both summer and winter.

The preliminary assessment should be followed by a percolation test of the proposed disposal area. Fig. 13.19 below illustrates the three stages of the percolation test. Where deep drains are needed the 300 mm hole should be excavated at the base of a wider excavation to allow room for working. Alternatively a modified test procedure can be adopted using a 300 mm earth auger bored vertically into the ground. All debris should be removed from the hole before the test is carried out.

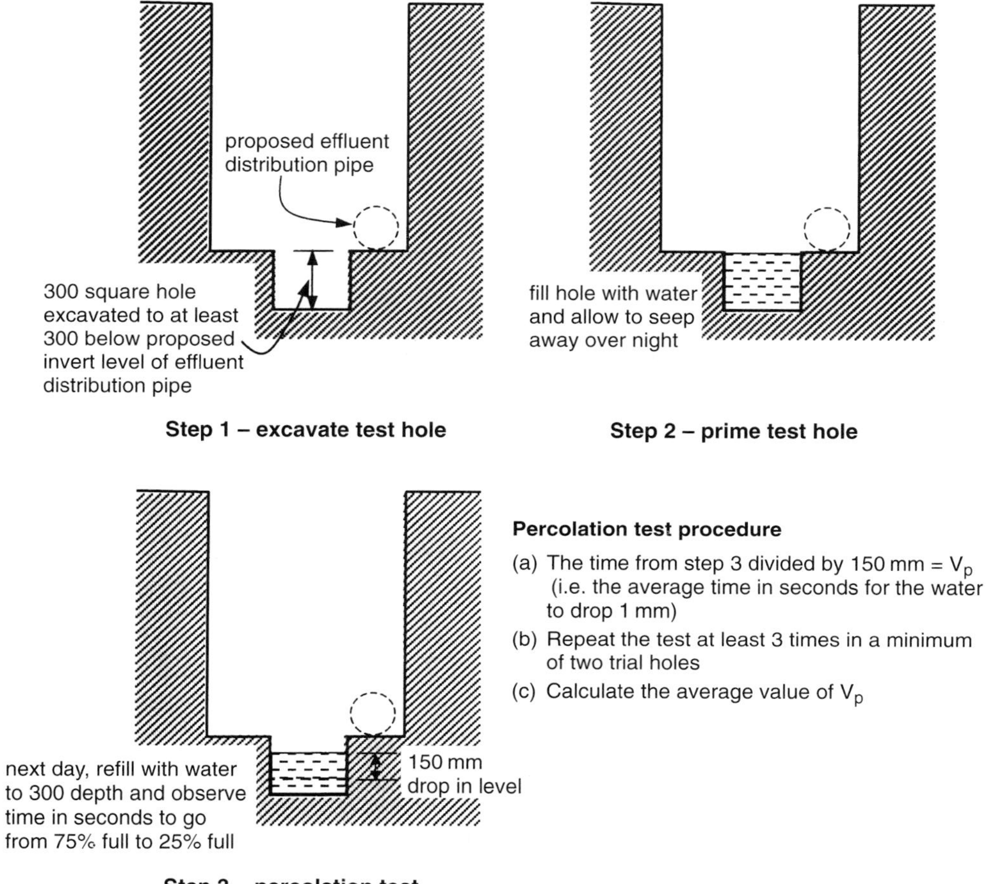

Fig. 13.19 Percolation test method.

The test should only be carried out when weather conditions are suitable (i.e. not during heavy rain, severe frost or drought).

Disposal via a drainage field should only be considered when:

- the percolation tests give values of V_p between 12 and 100 sec/mm; and
- the preliminary site assessment report is favourable; and
- the trial hole tests give acceptable results.

The values of V_p given above ensure that untreated effluent cannot percolate too rapidly into the ground water. Where V_p is outside the quoted range, treatment via a drainage field is unlikely to be successful. In these circumstances it may still be possible to use a septic tank provided that an alternative method of secondary treatment can be used to treat the effluent from the septic tank. It might then be possible to take the final discharge to a soakaway.

Design and construction

The main features of a typical drainage field are illustrated in Fig. 13.20, whilst Fig. 13.21 shows the main features of a drainage mound. Both are designed to ensure aerobic contact between the liquid effluent and the subsoil.

Pipes for drainage fields should be:

- perforated and laid in trenches of uniform gradient not exceeding 1 in 200;
- laid on a 300 mm layer of clean shingle or broken stone graded between 20 mm and 50 mm;
- laid at a minimum depth of 500 mm below the surface of the ground;
- laid in a continuous loop fed from an inspection chamber sited between the septic tank and the drainage field.

Pipe trenches for drainage fields should be:

- filled to a level of 50 mm above the pipe with the shingle or broken stone and covered with a layer of geotextile to prevent entry of silt;
- topped up with soil to ground level;
- between 300 mm and 900 mm wide;
- at least 2 m from other trenches in the same drainage field.

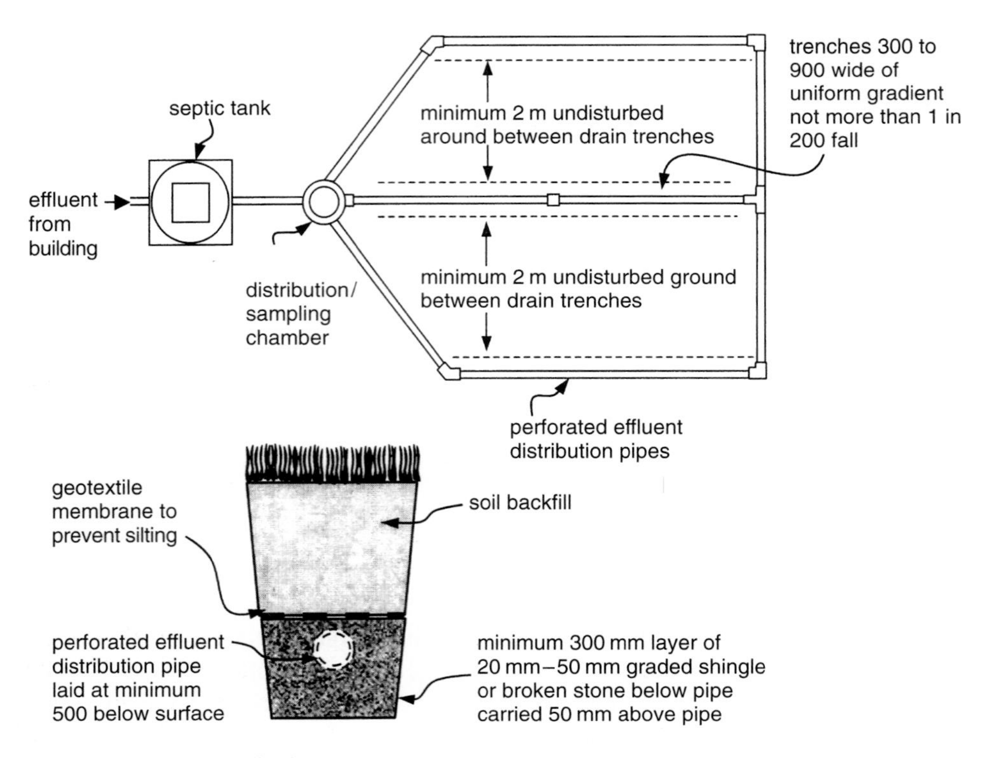

Fig. 13.20 Drainage field.

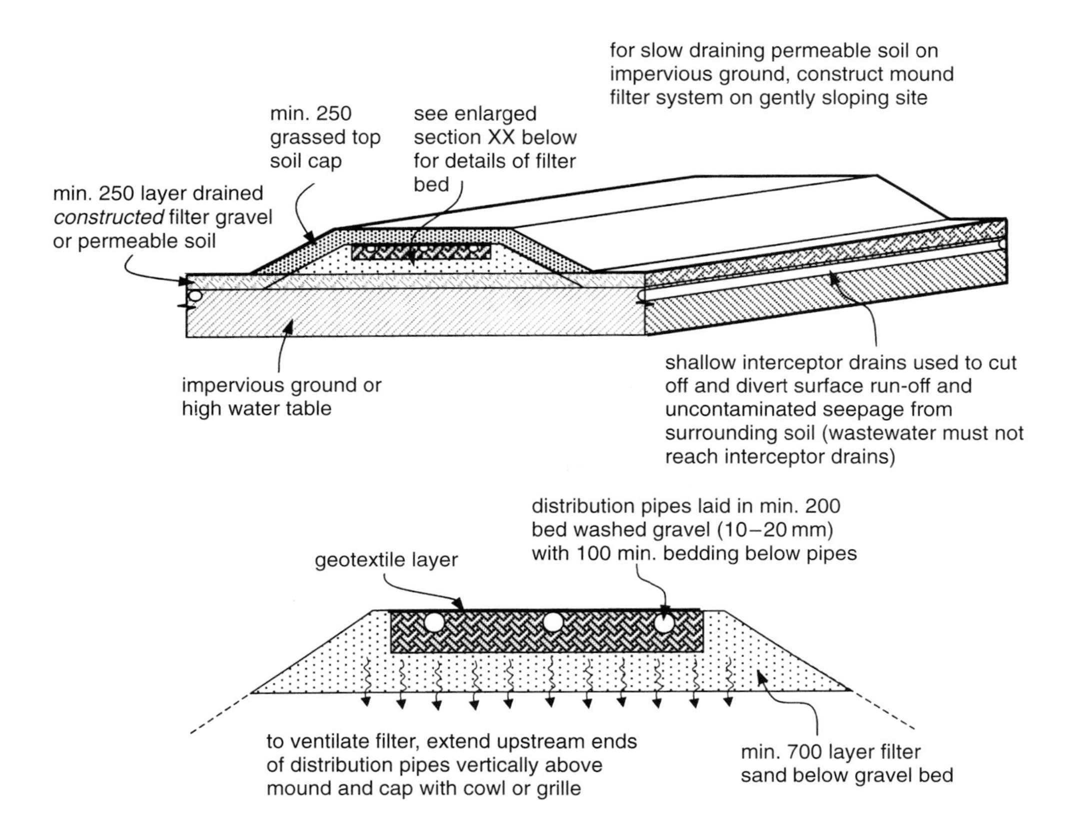

Fig. 13.21 Drainage mound details.

The area needed to be covered by the drainage field (the floor area, A_t m^2) can be calculated from the formula $\mathbf{A_t = p \times V_p \times 0.25}$
where p = number of persons served by the septic tank and V_p = the percolation value in sec/mm obtained as described above.

Drainage mounds should contain the features shown in Fig. 13.21.

13.4.4 Secondary treatment systems – constructed wetlands/reed beds

Where drainage fields or mounds are not a practical solution, it may be possible to treat septic tank effluent by means of constructed wetlands discharging to a suitable watercourse. Constructed wetlands (e.g. consisting of reed beds) are man-made systems which exploit the natural treatment capacity of certain wetland plants such as the common reed (*Phragmites communis*). The Environment Agency's consent may be required for this.

Constructed wetlands – general comments

Constructed wetland treatment systems purify wastewater by a combination of filtration, bacterial oxidation, sedimentation and chemical precipitation as the effluent moves through a gravel bed and around the rhizomes and roots of the wetland plants. In this way the biological oxygen demand (BOD) and suspended solids of the effluent are reduced, ammonia is oxidised, nitrates are reduced and a small amount of phosphorous is removed.

Plants used for reed beds and in constructed wetlands include:

- common reed (*Phragmites communis*);
- reed maces (*Typha latifolia*);
- rushes (*Juncus effusus*);
- bulrush (*Schoenoplectus lacustris*);
- members of the sedge family (*Carex*); and
- yellow flag (*Iris pseudocorus*).

In general, shaded areas (under trees or close to buildings) should not be used as the site for a constructed wetland as this will lead to poor and patchy growth. Additionally, the likely winter performance of the wetland should be taken into account during the design stage, as the lower temperatures tend to lead to poorer removal of ammonia although the other functions mentioned above are not affected.

Constructed wetlands – design

There are two principal designs for constructed wetland systems, horizontal flow and vertical flow. These can be used separately or can be combined to give superior treatment. The reed bed systems that produce good quality effluents with nitrification use vertical flow reed beds followed by a horizontal flow bed. Whether such a high level of treatment is appropriate depends on the quality and dilution of the receiving water body.

Vertical flow systems

In a vertical flow system, the top surface of the reed bed is intermittently flooded with wastewater. There are usually two or more beds provided side by side allowing a regime of rest and loading so that the surface, which might become clogged in use, can recover its permeability. The flow of wastewater passes down through layers of free-draining sand and gravel to an outlet at the bottom where it is collected by a system of drains at the base. In practice, depth is limited by available falls and construction techniques, however, a bed depth of about 1 m is typical and good results should be achieved with a single bed between 1 and 2 m deep.

In general, vertical flow systems are able to achieve more complete treatment of the effluent (particularly of ammonia) than horizontal flow systems because they can deliver much better oxygen transfer. Unfortunately, they do require more maintenance.

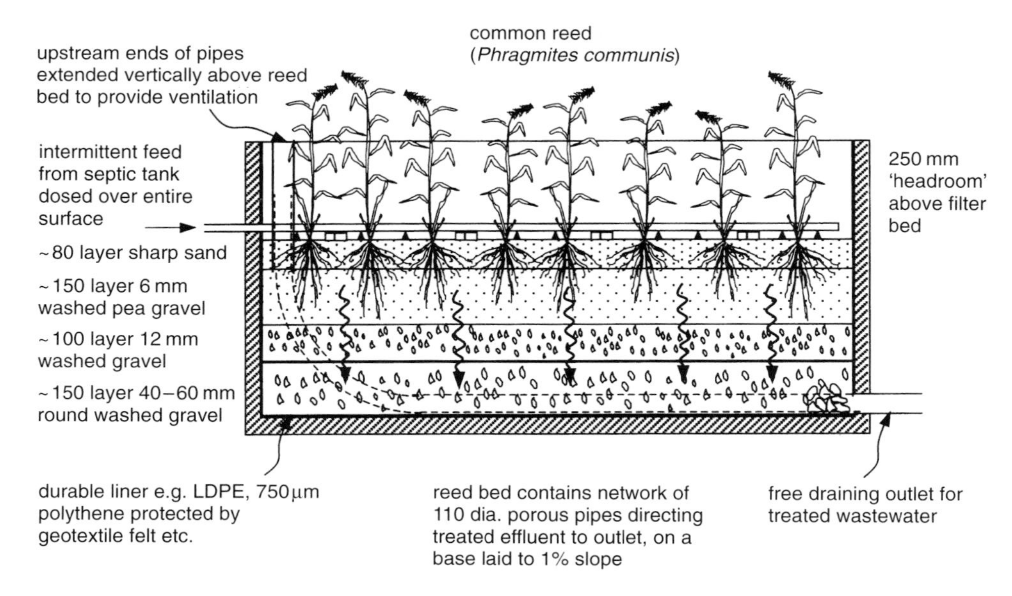

Fig. 13.22 Typical vertical flow reed bed treatment system.

A typical vertical flow reed bed treatment system is shown in Fig. 13.22.

Horizontal flow systems

In a horizontal flow system, wastewater is continuously fed in from the upstream end and passes over the full width of a gravel bed to an outlet at the downstream end. Horizontal flow systems have the disadvantage of being oxygen-limited and therefore incapable of fully treating concentrated effluents, especially those containing high levels of ammonia. They also require a relatively level site but have lower maintenance needs than vertical flow systems since only one bed is needed. A typical horizontal flow reed bed system is illustrated in Fig. 13.23.

The guidance provided in AD H2 on vertical and horizontal flow reed bed systems is extremely limited and has been enhanced in the above notes and illustrations by reference to BRE Good Building Guide 42 (GBG 42) to which reference should be made. There are many other forms of constructed wetland treatment systems available and being developed both nationally and internationally. GBG 42 contains a number of references to such systems which will normally require specialist advice.

Maintenance of constructed wetlands

It is essential that building owners are kept informed about the maintenance requirements where constructed wetlands are provided. The main maintenance

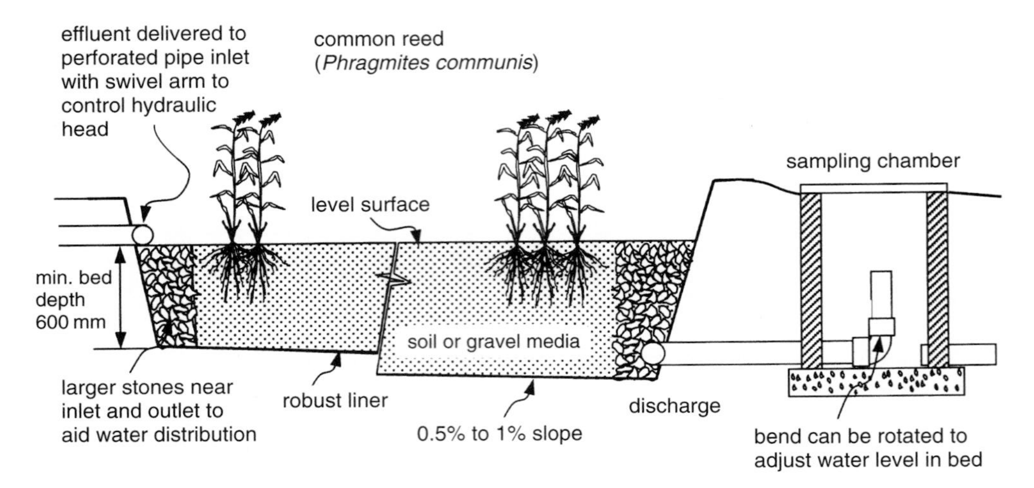

Fig. 13.23 Typical horizontal flow reed bed.

requirements are weeding, annual reed cutting and general grounds-care around the system. There will also be the need for periodic resting where multiple beds are provided in horizontal flow systems. Horizontal flow reed beds require less routine maintenance than vertical flow beds and where weeds are a problem the bed can be flooded to kill off the weeds. Full details of guidance on maintenance of reedbeds can be found in GBG 42 Part 2. A durable notice should be fixed within the building describing the necessary maintenance. Fig. 13.24 below is an example of a typical notice.

Wastewater Treatment System

Details of Necessary Maintenance

Address of Property --

Location of treatment system ------------------------------------

The foul drainage system from this property discharges to a [*insert type of primary treatment*] and a constructed wetland.

The [*insert type of primary treatment*] requires [*insert details of maintenance of the primary treatment*].

The constructed wetland system requires [*insert details of maintenance of the constructed wetland*].

Fig. 13.24 Constructed wetland – typical maintenance notice.

13.4.5 Cesspools

Where no other drainage disposal option is available it may be acceptable to provide a cesspool. Quite simply, a cesspool is a watertight underground tank provided for the storage of raw sewage. No treatment is involved.

Siting

Cesspools should be sited:

- on sloping ground away from and lower than nearby buildings;
- below, and at least 7 m from, the habitable parts of buildings;
- within 30 m of a vehicle access;
- at such levels that emptying and cleaning can be carried out without creating a hazard for the building's occupants and without the contents being taken through a dwelling or place of work.

Access for emptying and cleaning may be through a covered space which may be lockable.

Design and construction

The minimum capacity of the cesspool measured below the level of the inlet should be 18 m^3 (18,000 litres) based on two users. This capacity should be increased by 6.8 m^3 (6800 litres) for each additional user.

Additionally, cesspools should:

- have no openings except for the inlet from the drain, access for emptying and cleaning and ventilation;
- prevent leakage of the contents and ingress of subsoil water;
- be ventilated;
- be provided with access for emptying, cleaning, and inspection at the inlet.

Access covers for emptying and cleaning should be sufficiently durable to resist the corrosive nature of the tank contents and should be designed to prevent unauthorised access (by being lockable or otherwise engineered to prevent personnel entry).

Cesspools should also be constructed of materials which are impervious to the contents and to ground water. This would include engineering brickwork in 1:3 cement mortar at least 220 mm thick and concrete at least 150 mm thick (C/25/P mix to BS 5328), roofed with heavy concrete slabs. Prefabricated cesspools are available made of glass reinforced plastics, polyethylene or steel. These should follow the guidance in BS EN 12566 *Small wastewater treatment plants less than 50 PE:* Part 1: 2000 *Prefabricated septic tanks*. Care should be exercised over the stability of these tanks. Fig. 13.25 illustrates a typical cesspool designed for two users.

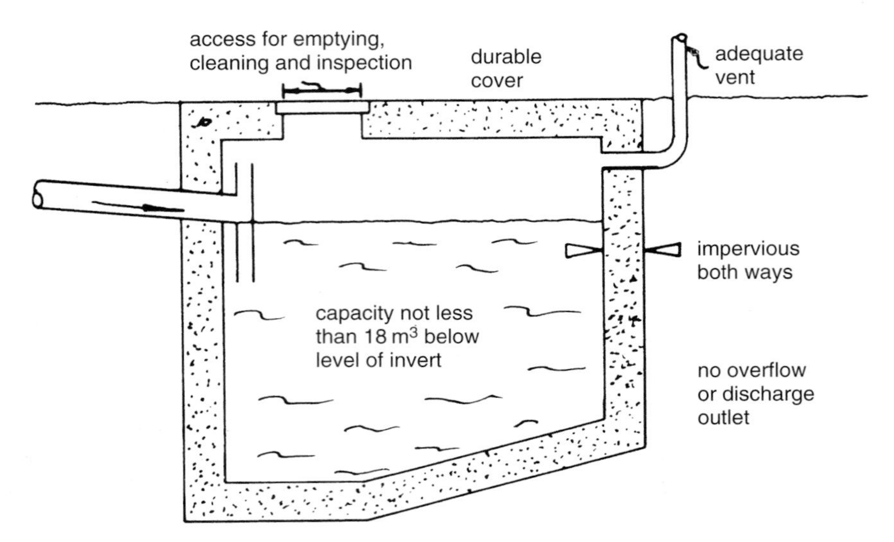

Fig. 13.25 Cesspools.

Maintenance of cesspools

It is essential that building owners are kept informed about the maintenance requirements where cesspools are provided. Cesspools should be inspected every two weeks for overflow and emptied as necessary. A typical emptying frequency is once per month and this can be estimated by assuming a filling rate of 150 litres per person per day. Cesspools which do not fill within the expected period may be leaking and should be checked out. A durable notice should be fixed within the building describing the necessary maintenance. Fig. 13.26 below is an example of a typical notice.

Cesspool foul drainage system

Details of Necessary Maintenance

Address of Property ____

Location of treatment system ____________________________________

The foul drainage system from this property is served by a cesspool

The system should be emptied approximately every *[insert design emptying frequency*] by a licensed contractor and inspected fortnightly for overflow.

The owner is legally responsible to ensure that the system does not cause pollution, a health hazard or a nuisance.

Fig. 13.26 Cesspool – typical maintenance notice.

13.5 Greywater and rainwater storage tanks

AD H2 concludes with limited guidance information on tanks for storage of greywater or rainwater for reuse within the building. The guidance does not apply to water butts used for storing rainwater for use in gardens.

Reclaimed water systems aid water conservation by reducing the amount of mains supply water used in houses and commercial buildings. Clearly there is a potential for reclaimed water systems to contaminate potable mains water supplies by inadvertent cross connection or backflow. Therefore, it is essential that water installations comply with the Water Supply (Water Fittings) Regulations 1999 in England and Wales, the Water Byelaws 1999 in Scotland or the Water Regulations in Northern Ireland.

Greywater should only be used for irrigation purposes and even then, certain precautions should be taken to prevent possible health problems occurring. Section 9 of the Water Regulations Advisory Scheme leaflet 09-02-04 *Reclaimed Water Systems. Information about installing, modifying or maintaining reclaimed water systems* gives general tips on reclaimed water uses and treatment. It also includes a great deal more information than can be found in AD H2, including a definition of greywater. This means water originating from the mains potable water supply that has been used for bathing or washing, washing dishes or laundering clothes.

Therefore, greywater and rainwater tanks should be:

- ventilated and prevent ingress of subsoil water or leakage of the contents;
- fitted with an anti-backflow device on any overflow connected to a drain or sewer to prevent contamination should a surcharge occur in the drain or sewer;
- provided with access for emptying and cleaning.

Access covers for emptying and cleaning should be sufficiently durable to resist the corrosive nature of the tank contents and should be designed to prevent unauthorised access (by being lockable or otherwise engineered to prevent personnel entry).

13.5.1 Alternative approach

Requirement H2 of Schedule 1 to the Building Regulations 2000 may also be met by complying with the relevant recommendations of BS 6297: 1983 *Code of practice for the design and installation of small sewage treatment works and cesspools*. These are – sections one, two, four and the appendices and clauses 6 – 11 of section three.

13.6 Rainwater drainage

Paragraph H3 of Schedule 1 to the Building Regulations 2000 requires that:

- any system carrying rainwater from the roof of a building is adequate; and
- paved areas around the building are so constructed as to be adequately drained.

It should be noted that only the following paved areas are covered by the Regulations:

- those which provide access for disabled people in accordance with requirement M2 (see Chapter 17 below);
- those which provide access to or from a place used for the storage of solid waste (see requirement H6(2) below); and
- those which give access to the building where this is intended to be used in common by the occupiers of one or more other buildings.

Rainwater from the roof of the building and any relevant paved areas must be taken to one of the following, listed in order of priority:

- an adequate soakaway or some other adequate infiltration system; or
- a watercourse; or
- a sewer.

Movement to a lower level in the order of priority may only be on the grounds of reasonable practicability since it is the purpose of the Regulation to encourage drainage connections to other than surface water sewers where this is technically feasible. This is an attempt to lessen the effect of flash flooding occurring in times of exceptionally heavy rainfall. This requirement does not apply to the gathering of rainwater for re-use.

The requirements of Paragraph H3 will be met if the following are fulfilled.

- Rainwater from paved areas and roofs is carried away from the relevant surface by a drainage system or some other appropriate means.
- Any rainwater drainage system:
  (a) conveys the flow of rainwater to a suitable outfall (soakaway, watercourse, surface water or combined sewer);
  (b) reduces to a minimum the risk of leakage or blockage;
  (c) is accessible for clearing blockages.
- Rainwater soaking into the ground (to a soakaway etc.) is sufficiently distributed so as not to damage the foundations of the building or any adjacent structure.

The emphasis in H3 on infiltration systems to dispose of rainwater means that it is essential to ensure adequate distribution of rainwater where it will not harm the structure of the building. Rainwater or surface water should never be discharged to a cesspool or septic tank.

13.6.1 Gutters and rainwater pipes

A rainwater drainage system should be capable of carrying the anticipated flow at any point in the system. The flow will depend on the area of roofs to be drained and on the intensity of the rainfall. For eaves gutters, a design rainfall intensity of 0.021 litres/second/m^2 (i.e. 75 mm of rainfall in any one hour) should be assumed in design

AD H3, Section 1

Diagram 1 Rainfall intensities for design of gutters and rainfall pipes (litres per second per square metre)

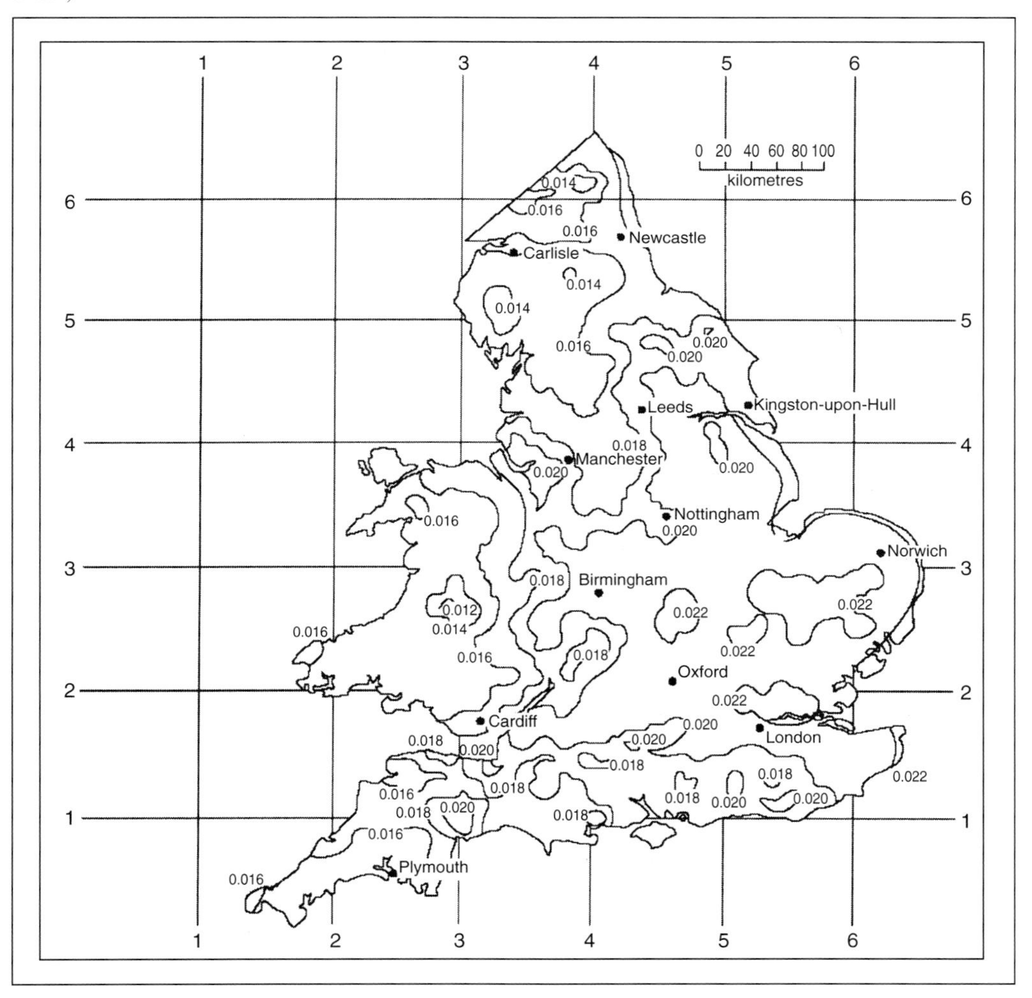

calculations. For valley gutters, parapet gutters, siphonic systems and other rain gathering systems the rainfall intensity should be obtained from Diagram 1 of AD H3 which is reproduced above.

For some roof designs incorporating valley gutters, parapet gutters or drainage from flat roofs it is possible that intense rainfall could cause overtopping of the construction resulting in water entering the building causing damage, wetting of insulation etc. The design of such systems should be carried out in accordance with BS EN 12056 *Gravity drainage systems inside buildings*.

The ultimate capacity of gutters and rainwater pipes depends on their length, shape, size and gradient and on the number, disposition and design of outlets. AD

AD H3, section 1

Table 2 Gutter sizes and outlet sizes.

Max effective roof area (m^2)	Gutter size (mm dia)	Outlet size (mm dia)	Flow capacity (litres/sec)
6.0	—	—	—
18.0	75	50	0.38
37.0	100	63	0.78
53.0	115	63	1.11
65.0	125	75	1.37
103.0	150	89	2.16

Note
Refers to nominal half round eaves gutters laid level with outlet at one end sharp edged.
Round edged outlets allow smaller downpipe sizes.

H3 contains design data for half-round gutters up to 150 mm in diameter. They are assumed to be laid level and to have a sharp-edged outlet at one end only. Table 2 to section 1 of AD H3 is reproduced above and gives gutter and outlet sizes for the drainage of different roof areas for lengths of gutter up to 50 times the water depth. The gutter capacity should be reduced for greater lengths.

The maximum roof areas given in Table 2 are the largest effective areas which should be drained into the gutters given in the table. The effective area of a roof will depend on whether the surface is flat or pitched. Table 1 to section 1 of AD H3 shows how the effective area may be calculated for different roof pitches.

AD H3, section 1

Table 1 Calculation of area drained.

Type of surface	Effective design area (m^2)
1 flat roof	plan area of relevant portion
2 pitched roof at 30° pitched roof at 45° pitched roof at 60°	plan area of portion × 1.29 plan area of portion × 1.50 plan area of portion × 1.87
3 pitched roof over 70° or any wall	elevational area × 0.5

Gutters should also be fitted so that any overflow caused by abnormal rainfall will be discharged clear of the building. Additional outlets may be necessary on flat roofs, valley gutters and parapet gutters to avoid over-topping.

Where it is not possible to comply with the conditions assumed in Table 2, further guidance is given in AD H3.

- Where an end outlet is not practicable the gutter should be sized to take the larger of the roof areas draining into it.
- If two end outlets are provided they may be 100 times the depth of flow apart.
- It may be possible to reduce pipe and gutter sizes if:
  (a) the gutter is laid to fall towards the nearest outlet; or
  (b) a different shaped gutter is used with a larger capacity than the half round gutter; or
  (c) a rounded outlet is used.

In these cases reference should be made to the following parts of BS EN 12056:

- Part 3: *Roof drainage layout and calculation*, clauses 3 to 7
- Annex A and National Annexes
- Part 5: *Installation, testing instructions for operation and maintenance and use*, clauses 3, 4, 6 & 11.

Rainwater pipes should comply with the following rules.

- Discharge should be to a drain, gully, other gutter or surface which is drained.
- Any discharge into a combined system of drainage should be through a trap (e.g. into a trapped gully).
- Rainwater pipes should not be smaller than the size of the gutter outlet. Where more than one gutter serves a rainwater pipe the pipe should have an area at least as large as the combined areas of the gutter outlets and be large enough to take the flow from the whole contributing area.
- Discharge from a rainwater pipe onto a lower roof or paved area should be via a pipe shoe to divert water away from the building.
- Where a single downpipe serves a roof with an effective area greater than $25\,m^2$ and discharges onto a lower roof, a distributor pipe should be fitted to the shoe to ensure that the width of flow at the receiving gutter is great enough to prevent overtopping of the gutter.

13.6.2 Siphonic roof drainage systems

Using siphonic action to accelerate the flow of water from the gutters to the below ground drainage system enables small-diameter pipes to achieve high rates of discharge. This permits a reduction in the number of downpipes that need to be provided when compared with traditional systems of roof drainage, and this is particularly useful in large single-storey commercial buildings with restricted numbers of columns.

For the siphonic action to start, the pipework must be airtight. This requires special rainwater outlets with baffle plates, correctly dimensioned pipes and fully sealed joints. Additionally, care should be taken that other trades do not connect their internal drainage pipes into a siphonic system, and thus break the vacuum.

The following considerations should also be taken into account.

- Possible surcharging in the downstream drainage system as this can cause reductions in flow rates in downpipes.
- The time taken to prime the siphonic action may be excessive where long gutters are specified. In this case, overflow arrangements to prevent gutters from overtopping, should be provided.

More information on the design of siphonic systems of roof drainage may be obtained from Report SR 463 *Performance of syphonic drainage systems for roof gutters* published by Hydraulics Research Ltd. Reference should also be made to BS EN 12056: Part 3.

13.6.3 Eaves drop systems

In modern buildings it is normal for rainwater from roofs to be collected by a system of guttering and downpipes to be transmitted to a system of below-ground drainage. In fact, the requirement of H3 for adequacy or rainwater systems means that an eaves drop system (where rainwater is allowed to fall from the roof freely to the ground) can be a perfectly acceptable solution provided that the following design considerations are taken into account.

- The fabric of the building should be protected against ingress of water caused by splashing against the external walls.
- The entry of water into doorways and windows should be prevented.
- Persons should be protected from falling water in doorways etc.
- Splashback caused by water hitting the ground should be prevented from affecting people and the fabric of the building (e.g. by providing a gravel layer or angled concrete apron to deflect water away).
- Foundations should be protected from concentrated discharges which occur at valleys, valley gutters or from excessive flows caused by large roofs (where the area of roof per unit length of eaves is high).

13.6.4 Rainwater recovery systems

In order to conserve water supplies, it is possible to collect rainwater for re-use within the building provided that the following considerations are taken into account.

- Storage tanks should follow the guidance given in section 13.5 above.
- Pipework, valves and washouts used for recovered water should be clearly identified on marker plates in accordance with the recommendations the Water Regulations Advisory Scheme leaflet 09-02-04 *Reclaimed Water Systems. Information about installing, modifying or maintaining reclaimed water systems* where further guidance on the use of rainwater recovery systems will be also be found.

13.6.5 Materials

Materials used should be adequately strong and durable. Additionally:

- Gutters should have watertight joints under working conditions.
- Downpipes placed inside a building should be capable of withstanding the test for airtightness described in section 13.3.13 above.
- Gutters and rainwater pipes should be adequately supported with no restraint on thermal movement.
- Pipes and gutters of different metals should be separated by non-metallic material to prevent electrolytic corrosion.
- Siphonic roof drainage pipework should be designed to resist negative pressures.

13.6.6 Alternative method of design

The requirements of the 2000 Regulations for rainwater drainage can also be met by following the relevant recommendations of BS EN 12056 *Gravity drainage systems inside buildings*. These are:

- In Part 3 *Rainwater drainage, layout and calculation*, clauses 3 to 7.
- Annex A and National Annexes.
- In Part 5 *Installation and testing, instructions for operation, maintenance and use*, clauses 3, 4, 6, & 11.

13.6.7 Drainage of paved areas

Section 2 of AD H3 contains information on the design of rainwater drainage systems for paved areas around buildings and small car parks up to 4000 m^2. For the design of systems serving larger catchment areas the guidance in BS EN 752 *Drain and sewer systems outside buildings* Part 4: 1998 *Hydraulic design and environmental aspects* should be followed. Rainfall intensities of 0.014 litres/sec/m^2 (i.e. 50 mm of rainfall in any one hour) are assumed for normal situations. More accurate local figures can be obtained from Diagram 2 of Section 2 to AD H3 which is reproduced below. In very high risk areas where ponding could lead to flooding of buildings, drainage of paved areas should be designed in accordance with BS EN 752: Part 4.

AD H3 describes three methods for draining paved areas:

- Allow pavings to drain freely onto adjacent pervious surfaces
- Use pervious paving
- Use impervious paving discharging to gullies or channels connected to a drainage system.

13.6.8 Design of freedraining surfaces

Surface water should not be allowed to soak into ground where the conditions are not suitable; however, paved areas do not always have to be served by underground

AD H3, Section 2

Diagram 2 Rainfall intensities for design of drainage from paved areas and underground rainwater drainage (litres per second per square metre)

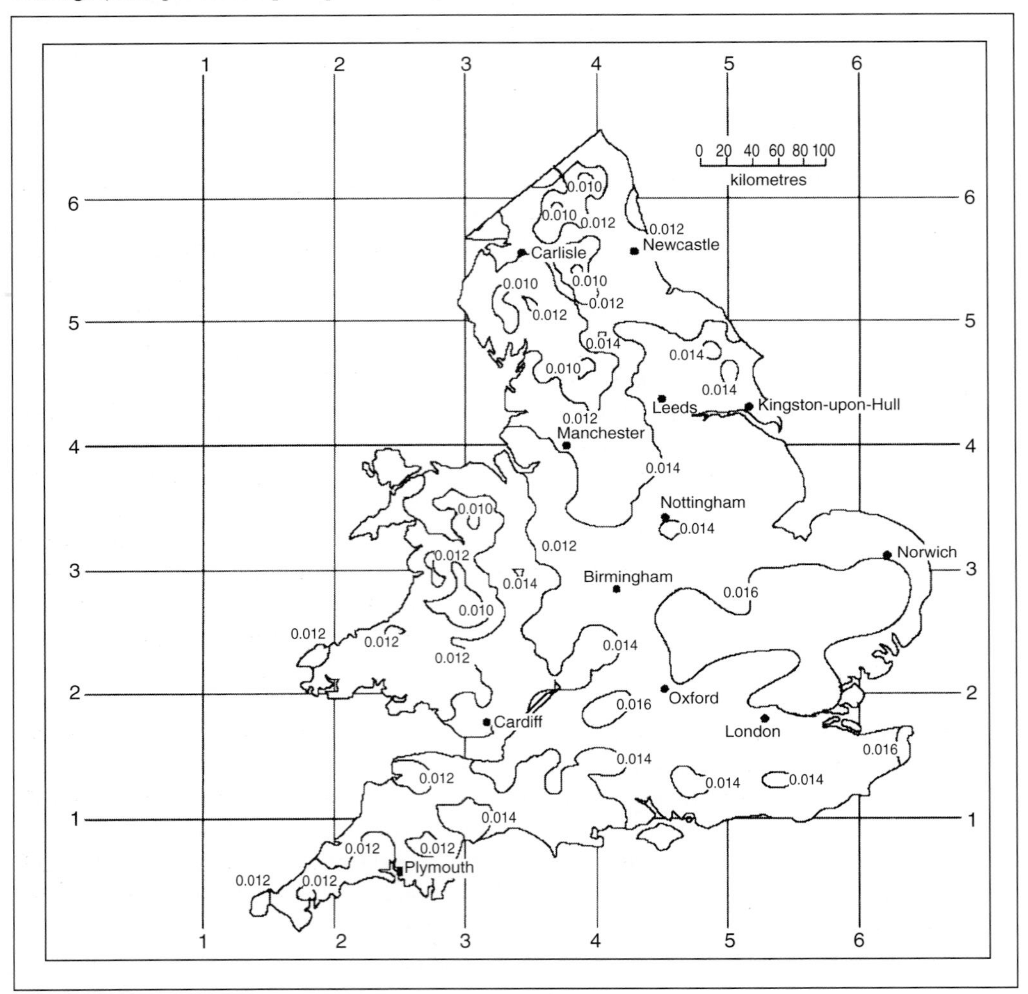

drainage systems. It is acceptable for paths, driveways and other narrow areas of paving to be freedraining to pervious areas (e.g. grassland) if the following conditions are met.

- Water should be directed away from buildings where foundations could be damaged. This can be achieved by suitable surface gradients (e.g. where ground levels would cause water to collect along the wall of a building a reverse gradient could be created at least 500 mm wide to divert water away).
- Impervious surfaces should have a cross fall of at least 1 in 60 to permit rapid draining. The fall across a path should not exceed 1 in 40.

- Where paving drains onto adjacent ground it should be finished flush with, or above the level of the surrounding ground to permit the water to run off.
- The soakage capacity of the ground should not be overloaded and where the adjacent ground is not sufficiently permeable to take the flow it may be necessary to provide filter drains (see section 13.7.4 below).

13.6.9 Pervious paving

As an alternative, and where it is not possible to drain large paved areas to adjacent pervious surfaces, it may be possible to construct pervious paving to deal with surface drainage. Pervious paving is made up of a porous or permeable surface material placed onto a granular layer which acts as a storage reservoir, retaining peak water flows until soakage into the underlying subsoil takes place. The storage layer should be designed on a similar basis to the design of the storage volume in a soakaway (see section 13.7.4 below).

On steeply sloping surfaces it will be necessary to check that the water level can rise sufficiently in the storage reservoir to enable its full capacity to be used. It will also be necessary to check that water is not inadvertently accumulating around the building foundations.

Pervious paving, on flat or sloping sites, may even be used where infiltration drainage is not possible (see section 13.7.4). In this case an impermeable barrier is placed below the storage layer to act as a detention tank or pond prior to discharge of the stored water to a drainage system (see section 13.6.10 below).

Pervious paving should not be used:

- where excessive amounts of sediment are present since these can enter the pavement and block the pores;
- in oil contaminated areas or where run-off may be contaminated with pollutants.

More information on the design of pervious paving can be found on pages 64 to 66 of CIRIA report C522: *Sustainable urban drainage systems – design manual for England and Wales.*

13.6.10 Paving connected to drainage system

Where it is not possible for the paving to be freedraining or for pervious paving to be used, impervious paving should be used in conjunction with gullies or channels connected to a drainage system. Gullies should comply with the following guidance.

- Be provided as necessary at low points to ensure that ponding does not occur.
- Be provided at intermediate positions so that individual gullies are not overloaded and channels do not have excessive depths of flow.
- Have their gratings set about 5 mm below the surrounding paving to allow for settlement of the paving.

Since it is possible that drainage from pavings may encourage silt and grit to enter the drainage system this should be intercepted by providing suitably sized gully pots or catchpits.

13.6.11 Alternative method of design

The requirements of the 2000 Regulations for drainage of pavings can also be met by following the relevant recommendations of BS EN 752 *Drain and sewer systems outside buildings*. These are:

- in Part 4, *Hydraulic design and environmental considerations*, clause 11.
- National Annexes ND and NE.

13.7 Rainwater drainage below ground

13.7.1 Connections and outlets

Section 3 of AD H3 deals specifically with drainage systems carrying only rainwater. Where practicable, surface water drainage should discharge to a soakaway or other infiltration system. Discharge to a watercourse is the next best option but the consent of the Environment Agency may be required and they may put a limit on the rate of discharge, although this can be attenuated by the use of detention ponds or basins. Where these forms of outlet are not practicable, discharge should be made to a suitable sewer.

Combined systems (those carrying both foul and rainwater) are permitted by some drainage authorities where allowance is made for the additional capacity. Where a combined system does not have sufficient capacity, rainwater will need to be taken via a separate system to its own outfall. Even where a sewer is operated as a combined system and has sufficient capacity it may still be necessary to provide separate systems of drainage to the building in accordance with the provisions of Requirement H5 (see below section 13.9.2). Surface water drainage connected to a combined system should have traps on all inlets.

Pumped systems of surface water drainage may be needed where there is a tendency to surcharging or gravity connections are impracticable (see also section 13.3.28 above).

The design information contained below is suitable for the drainage of small impervious catchment areas up to 2 hectares with an assumed design rainfall intensity of 0.014 litres/sec/m^2 for normal situations. Rainfall intensity may also be obtained from Diagram 2 of AD H3 illustrated in section 13.6.7 above. Where it is intended to drain larger areas than 2 hectares or where low levels of surface flooding could cause flooding of buildings, reference should be made to BS EN 742: Part 4.

With the exception of pipe gradients and sizes, the recommendations given above for below ground foul drainage (materials, bedding and backfilling, clearance of blockages, workmanship and testing and inspection) apply equally to rainwater drainage below ground.

13.7.2 Pipe sizes and gradients

Drains should be laid to falls and should be large enough to carry the expected flow. The rate of flow will depend on the area of the surfaces (including paved or other hard surfaces) being drained. The capacity will depend on the diameter and gradient of the pipes. The minimum permitted diameter of any rainwater drain is 75 mm. Surface water sewers (i.e. drains serving more than one building) should have a minimum diameter of 100 mm. Diagram 3 to Section 3 of AD H3 is reproduced below and gives discharge capacities for rainwater drains running full where it will be seen that the capacity increases proportionately with the pipe diameter and gradient.

AD H3, Section 3

Diagram 3 Discharge capacities of rainwater drains running full.

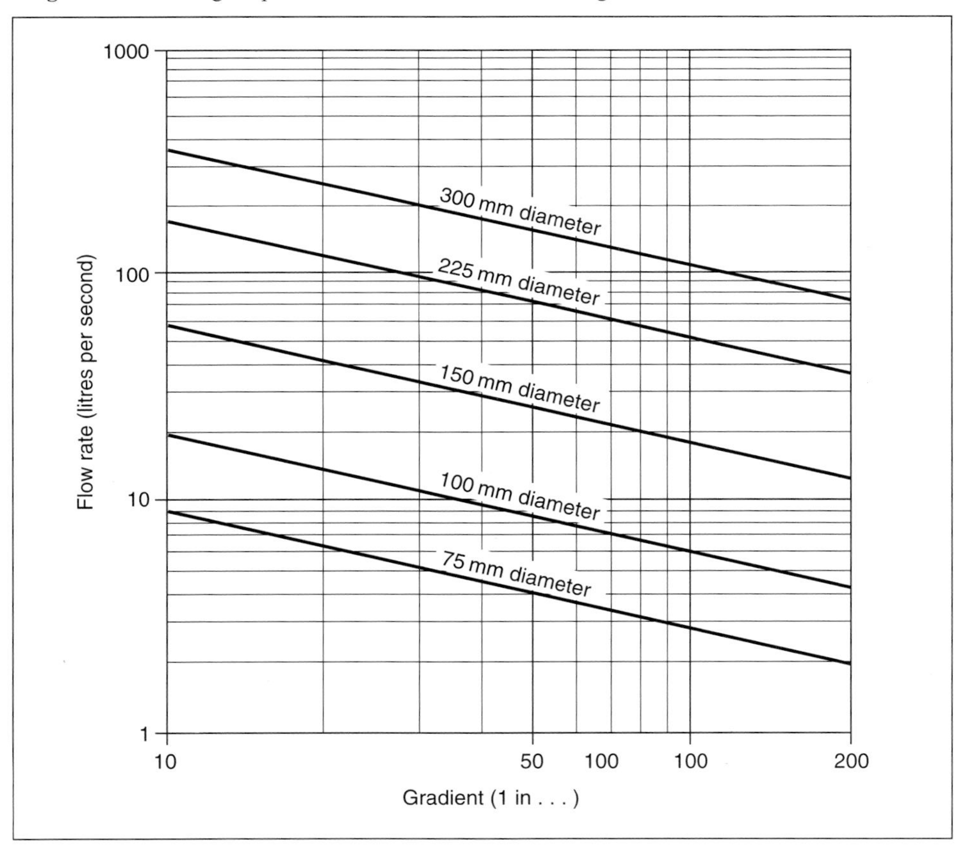

In general, the minimum permitted gradient of a pipe is related to its diameter as shown in Table 13.4 below.

Table 13.4 Pipe sizes and minimum gradients.

Pipe diameter mm	Minimum gradient
75	1 in 100
100	1 in 100
150	1 in 150
225	1 in 225
Over 225	See BS EN 752: Part 4

13.7.3 Contaminated runoff

It is an offence under section 85 (*offences of polluting controlled waters*) of the Water Resources Act 1991 to discharge any polluting or noxious material into coastal or underground water, or a watercourse. Since most surface water sewers discharge to watercourses, separate drainage systems should be provided where materials are stored or used which could cause pollution. The drainage system should include:

- an appropriate form of separator (see oil separators below); or
- an appropriate treatment system; or
- discharge of the flow into a system suitable for receiving polluted effluent.

Certain areas, such as petrol filling stations and car parks suffer from leakage or spillage of oil and the surface water runoff from these areas can find its way via the drainage system to a watercourse where pollution can occur. Since it is an offence under section 111 (*restrictions on the use of public sewers*) of the Water Industry Act 1991 to discharge petrol into any drain connected to a public sewer, oil separators should be provided in risk situations.

There are, of course, other controls over the storage of petrol, due to its combustible nature. Premises used for keeping petrol must be licensed under the Petroleum (Consolidation) Act 1928. A licence can be granted with or without conditions. Guidance on the storage of oil may be obtained from the Health and Safety Executive.

Oil separators

The type of oil separator that should be provided will depend on the risk of contamination represented by the site.

For comparatively low risk areas, such as paved areas around buildings and car parks, a bypass separator should be provided with a nominal size (NSB) of 0.0018 times the contributing area and a silt storage volume in litres equal to 100 × NSB. Bypass separators treat all flows generated by rainfall rates of up to 5 mm/hr, thus accounting for 99% of all rainfall events. Flows above this rate are allowed to bypass the separator.

For fuel storage and other high risk areas, full retention separators should be provided with nominal size (NS) equal to 0.018 times the contributing area and a silt storage volume in litres equal to 100 × NS. Full retention separators treat the full flow that can be delivered by the drainage system which is normally equivalent to the flow generated by a rainfall intensity of 50 mm/hr.

Separators should:

- be Class 1 when discharging to infiltration devices or surface water sewers (i.e. designed to achieve a concentration of less than 5mg/litre of oil under standard test conditions);
- be leaktight and adequately ventilated;
- have inlet arrangements that avoid directing the inflow to the surface of the water already in the separator;
- comply with the requirements of the Environment Agency and prEN 858 *Installations for separation of light liquids (e.g. petrol or oil)*;
- comply with the requirements of the licensing authority where the Petroleum Act applies;
- be regularly maintained to ensure continued effectiveness. It is normal for routine inspections to be carried out every six months including the completion of a log which details the inspection date, depth of oil and any cleaning undertaken; and
- be provided with sufficient access points to allow for inspection and cleaning of all internal chambers.

More information on the provision of oil separators may be found in the Environment Agency publication – Pollution Prevention Guidelines 3 (PPG 3) *Use and design of oil separators in surface water drainage systems.*

13.7.4 Infiltration drainage systems

Infiltration drainage systems are designed to return rainwater from roofs and pavings to the ground in the vicinity of the building, without involving connection to sewers or watercourses. They include such devices as soakaways, swales, infiltration basins, filter drains and detention ponds (but see the comments on these below). AD H3 gives a very brief summary of the various infiltration devices which are available; however the information provided is too brief to be of any real use for the designer. The notes which follow have been enhanced using the various documents referred to in the text. These reference sources are essential for anyone seriously interested in infiltration drainage systems.

It is not always possible to provide infiltration drainage to a building. For example, infiltration devices should not be provided in the following situations.

- Within 5 m of a road or building.
- In areas of unstable land. (Annex 1 of Planning Policy Guidance Note 14 warns against the use of infiltration systems in areas subject to landslip).
- In ground with a high water table (i.e. where the water table reaches the base of the device at any time of the year.
- Where ground water source or resource might be polluted by the presence of contamination in the runoff.

- At such a distance from drainage fields, drainage mounds or other soakaways so that the overall soakage capacity of the ground would be exceeded and the effectiveness of any drainage field would be impaired.

Soakaways

Soakaways should be designed to store the immediate surface water runoff and allow for its efficient infiltration into the surrounding soil. Stored water must be discharged sufficiently quickly to provide the necessary capacity to receive runoff from a subsequent rainfall event. The time taken for discharge depends upon the soakaway shape and size, and the infiltration characteristics of the surrounding soil.

Soakaways serving catchment areas of less than $100\,m^2$ are usually built as square or circular pits filled with rubble or lined with dry-jointed masonry or pre-cast perforated concrete ring units surrounded by suitable granular backfill. For drained areas above $100\,m^2$, soakaways can be lined pits or of trench type and usually a depth of 3 to 4 m is adequate if ground conditions allow. Trench soakaways are cheaper to dig with readily available excavating equipment.

Although the design of soakaways should be carried out by considering storms of different durations over a ten year period in order to determine the maximum storage volume, for small soakaways serving $25\,m^2$ or less, a design rainfall of 10 mm in five minutes can be taken to represent the worst case. For soakaways serving larger areas reference should be made to BRE Digest 365 *Soakaway design* or BS EN 752: Part 4. Where the percolation characteristics of the ground are marginal it may still be possible to use soakaways in conjunction with overflow drains.

The percolation test described in AD H2 (see Fig. 13.19 above) may be carried out to determine the capacity of the soil to receive infiltration. The value of V_p from the percolation test may be used in the equation below to determine the soil infiltration rate:

$$f = 10^{-3}/2V_p$$

where f = the soil infiltration rate

Therefore, assuming a value of V_p of 20,

$$f = 1/1000 \times 2 \times 20 = 0.000025\,m/sec$$

The storage volume of the soakaway should be able, during storm conditions, to accommodate the difference between the inflow volume and the outflow volume.

The inflow volume is simply calculated by considering the design rainfall depth during a storm multiplied by the drainage area. Therefore, if a rainfall depth of 10mm is considered over an area of $25\,m^2$ in a five minute period:

$$\text{the inflow volume} = 0.01 \times 25 = 0.25\,m^3$$

The outflow volume (O) is calculated from the equation:

$$O = a_{s50} \times f \times D$$

Where:
a_{s50} is the area of the side of the storage volume when filled to 50% of its effective depth, and D is the duration of the storm in minutes.

Using the figures from the example given above and assuming a soakaway 2m deep and 2m x 1m in area with the inlet 1m below ground:

$$O = 4 \times 0.000025 \times 5 \times 60 = 0.03\,\text{m}^3$$

Therefore the difference between the inflow volume and the outflow volume equals the storage volume $= 0.25 - 0.03 = 0.22\,\text{m}^3$

The actual volume of soakaway below inlet $= 1 \times 1 \times 2 = 2\,\text{m}^3$ which is more than adequate.

Swales

Swales are simply grass-lined channels with shallow side-slopes used to carry rainwater from a site. They can also control the flow and quality of surface runoff and allow a certain amount of the flow to infiltrate into the ground. To increase the infiltration and detention capacity of swales they can be provided with low check dams across their width. To prevent overtopping during wet spells, it is possible to provide an overflow at one end discharging into another form of infiltration device or watercourse. They can be used to treat runoff from small residential developments, parking areas and roads.

Infiltration basins

These are dry grass-lined basins for storage of surface runoff that are free from water under dry weather flow conditions. They can be designed to manage water quantity and quality and are used to encourage surface water infiltration into the ground.

Filter drains

Otherwise known as french drains, filter drains consist of geotextile-lined trenches filled with gravel, sometimes containing perforated pipes to assist drainage. They are designed so that most of the flow enters the filter drain directly from the runoff or is discharged into it through other drains from where it infiltrates into the ground.

Detention ponds

The term 'detention pond' appears to be a mistake in the AD since the reference material given for this section in AD H3 (*Sustainable urban drainage systems – a design manual for England and Wales* published by CIRIA) makes reference to 'detention basins' and 'retention ponds' but not 'detention ponds'.

According to this reference source '*detention basins are vegetated depressions. They are formed below the surrounding ground, and are dry except during and immediately following storm events. Detention basins only provide flood storage to attenuate flows. Extending the detention times improves water quality by permitting the settlement of course silts*'.

On the other hand, '*retention ponds are permanently wet ponds with rooted wetland and aquatic vegetation – mainly around the edge. The retention time of several days provides better settlement conditions than offered by extended detention ponds and provides a degree of biological treatment*'. The description given in AD H3 could apply to either or both of the above.

13.7.5 Alternative method of design

Requirement H3 can also be met by following the relevant recommendations of BS EN 752 *Drain and sewer systems outside buildings*.

These are:

- in Part 4, *Hydraulic design and environmental considerations*, clauses 3 to 12;
- National Annexes NA, NB and ND to NI.

Additionally, detailed information about design and construction can be found in:

- BS EN 1295, *Structural design of buried pipelines under various conditions of loading:* Part 1: 1998 *General requirements*; and
- BS EN 1610: 1998 *Construction and testing of drains and sewers.*

13.8 Building over existing sewers

13.8.1 Introduction

Control of building works over or near existing sewers has long been subject to control in England and Wales. Until the coming into effect of the first amendment to the 2000 Regulations on 1 April 2002 this control was exercised by local authorities through the medium of section 18 of the Building Act 1984. The first amendment introduced a new Building Regulation requirement H4, which replaced section 18 and can be administered by both local authorities and approved inspectors, although the substance of section 18 has been little altered by the change. What has altered is the substantial amount of guidance provided by Approved Document H4.

13.8.2 Interpretation

The following terms apply in AD H4:

DISPOSAL MAIN – Any pipe, tunnel or conduit used for the conveyance of effluent to or from a sewage disposal works, which is not a public sewer.

MAP OF SEWERS – Any records kept by a sewerage undertaker under section 199 of the Water Industry Act 1991.

13.8.3 Building over sewers

H4 requires that where it is intended to:

- erect a building; or
- extend a building; or
- carry out works of underpinning to a building

near to or over a drain, sewer or disposal main, then the work must be carried out so that it is not detrimental to the building or extension or to the continued maintenance of the drain, sewer or disposal main.

H4 is limited to work carried out:

- near to or over a drain, sewer or disposal main which is shown on any map of sewers; or
- which will result in interference with the use of, or obstruction of any person's access to, any drain, sewer or disposal main shown on any map of sewers.

In order to meet the requirements of H4 it is necessary to ensure the following.

(1) That the work of building, extending or underpinning:
  - is expedited so as not to overload or otherwise damage the drain, sewer or disposal main both during construction and after it is completed; and
  - will not prevent reasonable access to any manhole or inspection chamber situated on the drain, sewer or disposal main.

(2) That where the drain, sewer or disposal main needs to be replaced:
  - a satisfactory diversionary route can be provided; or
  - the building or extension will not unduly obstruct the replacement work if the current alignment is maintained.

(3) That if the drain, sewer or disposal main fails, the risk of damage to the building will not be excessive. To assess the risk of damage to the building it is necessary to consider:
  - the nature of the ground;
  - the location, construction and condition of the drain, sewer or disposal main;
  - the nature, volume and pressure of the flow in the drain, sewer or disposal main; and
  - the design and construction of the building's foundations.

13.8.4 Application

The provisions of H4 apply where it is intended to erect, extend or underpin a building that is situated over, or within 3 m of the centreline of, an existing drain, sewer or disposal main shown on the sewer records of the sewerage undertaker, even if the sewer is not a public sewer.

The public have access to copies of sewer record maps during normal office hours, these being held by both sewerage undertakers and local authorities.

13.8.5 Consultation

When it is proposed to carry out any work to which H4 applies, the developer should always consult the owner of the drain or sewer (for public sewers this would be the sewerage undertaker) unless, of course, the developer is also the owner. In the case of public sewers the sewerage undertaker should be able to provide useful information regarding the age, location, condition and depth to invert of the sewer. They may also be able to arrange an inspection and if a public sewer needs to be repaired or replaced, they will carry out this work. The sewerage undertaker should also be consulted where it is proposed to build or extend over a sewer that is later intended for adoption.

In order to ensure compliance with H4 it will be necessary to apply to the relevant building control body (local authority or approved inspector) so that the works can be properly controlled. This will involve the carrying out of further consultations as follows.

- If using the local authority you must deposit full plans. This enables the local authority to carry out its duties under Regulation 14A of the Building Regulations 2000 to consult the sewerage undertaker as soon as practicable after the plans have been deposited. The local authority is not permitted to pass the plans or issue a completion certificate until the consultation has taken place (the sewerage undertaker has up to 15 days to reply) and it must have regard to the views expressed by the sewerage undertaker.
- If using an approved inspector, he must consult the sewerage undertaker where an initial notice or amendment notice is to be given (or has been given). The consultation must take place at the following stages:
  (a) before or as soon as is practicable after giving an initial notice or an amendment notice;
  (b) before giving a plans certificate (whether or not this is combined with an initial notice); and
  (c) before giving a final certificate.

  Additionally, he must allow the sewerage undertaker up to 15 working days to comment, and have regard to the views it expresses, before giving a plans certificate or final certificate to the local authority.

13.8.6 Building near drains or sewers in risk situations

Unless special measures are taken, buildings should not be constructed or extended over or within 3 m of any of the following:

- drains or sewers in poor condition (pipes which are cracked, fractured, misaligned or more than 5% deformed);

- drains or sewers constructed from brick or other masonry;
- rising mains (except those used only to drain the building);

since failure of the drain or sewer would expose the building to a high level of risk.

Additionally, certain soil types (fine sands, fine silty sands, saturated silts and peat) are easily eroded by groundwater leaking into drains or sewers. Therefore, failure of a drain or sewer could result in erosion of soil from around the foundations thereby exposing the building to undue risk. Where such soils are present, buildings should not be constructed or extended over or within 3 m of any drain or sewer to which H4 applies unless special measures are taken in the design and construction of the foundations to mitigate the effect of drain or sewer failure. Special measures are not needed if the invert of the drain or sewer is:

- above the level of the foundations; and
- above the level of the groundwater; and
- no deeper than 1 m.

13.8.7 Access for maintenance

Fig 13.27 below gives details of the precautions that should be taken to ensure that sewers remain accessible when buildings are constructed over or within 3m of them. The following main points should be observed.

- Do not construct a building or extension over a manhole, inspection chamber or access fitting on a sewer (i.e. a drain serving more than one property).
- Locate access points to sewers where they are accessible and apparent for use in emergency. Where this provision is already met by the existing sewer, do not construct a building or extension which would remove this provision unless a satisfactory alternative on the line of the sewer can be agreed with the sewer owner.
- Ensure that a satisfactory diversionary route is available at least 3 m from the building to allow the drain or sewer to be reconstructed without affecting the building. Where existing drains or sewers more than 1.5 m deep have access for mechanical excavators, ensure that the diversionary route also has such access.
- Unless the sewer owner agrees, the length of drain or sewer under a building should not exceed 6 m in length.
- Do not build over or near an existing sewer more than 3 m deep or more than 225 mm diameter without the sewer owner's permission.

13.8.8 Protection of drains and sewers

Approved Document H4 contains details of protection which should be provided both during construction of the building over the drain or sewer, and subsequently to prevent damage by settlement of the building. Details of other protection

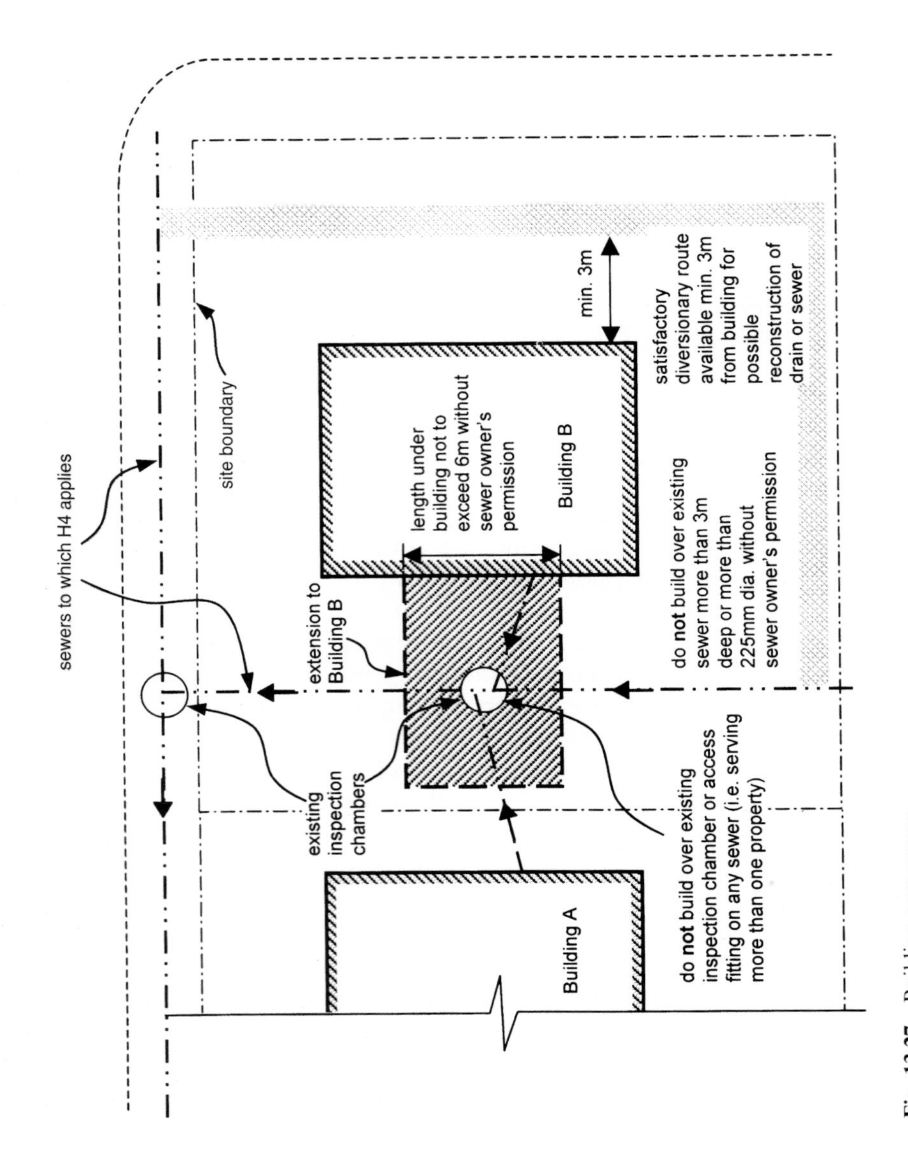

Fig. 13.27 Building over sewers.

measures for below ground drainage also apply to drains or sewers covered by the following notes. They may be found in sections 13.3.20 (pipes); 13.3.25 (drains); 13.3.26 (surcharging); and 13.3.27 (rodent control).

Protection during construction

During construction activities drains and sewers should be protected from damage by:

- providing barriers to keep construction traffic and heavy machinery away from the line of the sewer;
- not storing heavy materials over drains or sewers.

Piling works present a special risk and care should be taken to avoid damage to drains and sewers in the vicinity of such activities. The following precautions should be taken:

- a survey should be carried out to establish the position of the drain or sewer;
- where piling will take place within 1 m of a drain or sewer trial holes should be excavated to establish its exact position and the location of any connections;
- piling should not be carried out where the distance from the outside of the pile to the outside of the drain or sewer is less than twice the pile diameter.

Protection from settlement

Drains or sewers passing under buildings should comply with the following guidance.

- Provide at least 100 mm of granular or other flexible filling round the pipe.
- Where excessive subsidence is possible provide additional flexible joints or adopt other solutions (e.g. suspended drainage).
- Provide special protection where the crown of the pipe is within 300 mm of the underside of the slab (see section 13.3.20 above).
- For drains or sewers less than 2 m deep, increase the depth of the foundations in the vicinity of the drain or sewer so that it may pass through the wall.
- For drains or sewers greater than 2 m deep, design the foundations as a lintel spanning the drain or sewer. The 'lintel' should extend at least 1.5 m on either side of the pipe and should be designed so that no loads are transmitted to the drain or sewer.
- Where the drain or sewer passes through a wall or foundation follow the guidance given in section 13.3.25 above and shown in Fig. 13.14.
- Trenches for drains and sewers should only be excavated below the level of the building foundations if the precautions described in section 13.3.25 and illustrated in Fig.13.15 are taken.

13.9 Separate systems of drainage

13.9.1 Introduction

Control over the provision of separate systems of drainage was first introduced into the Building Regulations with the coming into force of the first amendment to the 2000 Regulations on 1 April 2002. The provisions are aimed at helping to minimise the volume of rainwater which enters the public foul sewer system since this can lead to overloading of the capacity of sewers and treatment works, and can cause flooding. At the date of this edition, this requirement sits uncomfortably beside six local Acts of Parliament which cover broadly the same area of control and are listed in Appendix A. A consultative document published by the Office of the Deputy Prime Minister in July 2002 aims to repeal these six Acts and will probably come into force some time in 2003.

13.9.2 Requirement H5

Any system for discharging water to a sewer which is provided to take rainwater from roofs or from paved areas around buildings covered by the requirements of H3 (see section 13.6), must be separate from that provided for the conveyance of foul water from the building.

For H5 to apply:

- the drainage system must be provided in connection with the erection or extension of a building; and
- it must be reasonably practicable for the system to discharge directly or indirectly to a sewer for the separate conveyance of surface water.

Additionally, the sewer must be:

- shown on a map of sewers (see definition in section 13.8.2 above); or
- under construction either by the sewerage undertaker or by some other person (although in this case the sewerage undertaker must have agreed in advance to adopt the drain or sewer in accordance with section 104 of the Water Industry Act 1991).

13.9.3 Meeting the requirement

The requirements of H5 can be met in either of two ways:

- by connecting to separate public sewers which are already in existence; or
- by providing separate drainage systems on the site of the building which will later be connected to separate public sewers which are under construction at the time of the building works.

13.9.4 Provision where separate sewer systems already exist

Where the sewerage undertaker has provided separate sewer systems, the owner or occupier of a building has a right to connect to the public sewers (see section 106 of the Water Industry Act 1991) provided that the following restrictions are observed.

- The surface water drainage from the building must be connected to the appropriate public surface water sewer.
- The foul water drainage from the building must be connected to the appropriate public foul water sewer.
- The way in which the connection is made must not be prejudicial to the public sewer system.
- 21 days notice must be given to the sewerage undertaker of the intention to make the connection.

It is normal for the sewerage undertaker to carry out the work of making the connection and recover its reasonable costs from the developer (see section 107 of the Water Industry Act 1991). Alternatively the developer may be permitted to carry out the work under the supervision of the sewerage undertaker.

13.9.5 Provision where separate sewer systems are proposed

Separate sewer systems should still be provided to drain the building even if only a combined system exists at the time of the building works, provided that separate public sewers are under construction by the sewerage undertaker, or by some other person for later adoption by the sewerage undertaker. Depending on the respective programmes for the building works and the public sewer construction, it may be necessary initially to connect the separated site drainage to the existing combined sewer. Later reconnection to the separate sewer systems can be made when these are completed thus minimising disruption to the building occupiers.

13.9.6 Dealing with contaminated surface water

It should be noted that the necessity to connect to a separate surface water sewer would only apply if the surface water was uncontaminated. Drainage from areas where materials are stored could contaminate runoff and lead to pollution if discharged to a surface water sewer. The alternative of discharging such contaminated water to a foul sewer needs to be discussed with the sewerage undertaker (see section 106 of the Water Industry Act 1991 and notes above) whose consent is required. It will also be necessary to consult the sewerage undertaker when connecting such contaminated runoff via a new foul sewer to an existing combined sewer, if it is intended that this will eventually be reconnected to a foul sewer that is proposed or under construction.

13.10 Solid waste storage

13.10.1 Introduction

The efficacy of the refuse storage system is dependent on its capacity and ease of collection by the waste collection authority. Under section 46 (*Receptacles for household waste*) and section 47 (*Receptacles for commercial or industrial waste*) of the Environmental Protection Act 1990, the waste collection authority has powers to specify the type and number of receptacles which should be provided and the position where the waste should be placed for collection. Therefore it is important that consultations take place with the waste collection authority to establish its specific requirements regarding the storage and collection of waste.

The opportunity has been taken in AD H6 to give general recommendations regarding the separate storage of waste for recycling. This is interesting since the Building Regulations do not cover the recycling of household or other waste. However, there are moves afoot to amend sections 46 and 47 of the Environmental Protection Act 1990 to allow for separate storage, and of course, there are a number of national initiatives on recycling and waste reduction which the AD is attempting to support. From a legal standpoint it is unlikely that the recommendations can, in fact, be enforced.

13.10.2 Requirement H6

Buildings are required to have:

- adequate means of storing solid waste;
- adequate means of access for the users of the building to the place of storage; and
- adequate means of access from the place of storage to:
  - (a) a collection point where one has been specified by the waste collection authority under section 46 (for household waste) or section 47 (for commercial waste) of the Environmental Protection Act 1990; or
  - (b) to a street (in the case of no collection point being specified).

The requirements of paragraph H6 may be met by providing solid waste storage facilities which are:

- large enough, bearing in mind the requirements of the waste collection authority for the number and size of receptacles (this relates to the quantity of refuse generated and the frequency of removal, see sections 46 and 47 of the Environmental Protection Act 1990);
- designed and sited so as not to present a health risk; and
- sited so as to be accessible for filling by people in the building and for removing to the access point specified by the waste collection authority.

13.10.3 Domestic buildings – storage capacity

Assuming weekly collection, dwellinghouses, flats and maisonettes up to four storeys high should have, or have access to, a location large enough to accommodate at least two movable, individual or communal containers, which meet the requirements of the waste collection authority.

The location should cater for separated waste (i.e. one container taking waste for recycling and another taking all other waste). The combined capacity of the two containers should not be less than 0.25 m^3 per dwelling (or such other capacity as is agreed with the waste collection authority). If the waste collection authority does not provide weekly collections then larger capacity containers or more individual containers will need to be provided.

The size of the location will depend on whether this is based on the provision of communal or individual storage. Where individual storage is provided for each dwelling this should be an area with dimensions of at least 1.2 m × 1.2 m. The waste collection authority should be consulted regarding space requirements for communal storage areas.

Dwellings in buildings above the fourth storey may either:

- share a container fed by a chute for non-recyclable waste, plus be provided with separate storage for waste which is to be recycled; or
- be provided with storage compounds or rooms for both types of waste if suitable management arrangements can be assured for conveying the waste to the place of storage.

For large blocks, recyclable waste can also be dealt with by providing 'Residents Only' recycling centres (places where residents can bring their own recyclable waste for storage in large containers, such as bottle banks).

13.10.4 Siting of waste containers and storage areas

Waste containers should comply with the following rules with regard to siting.

- For new buildings it should be possible to take a container to a collection point without taking it through a building. (It is permissible to pass through a porch, garage, carport or other covered space.) Buildings should not be extended or converted in such a way as to remove such an access facility where it already exists.
- Waste containers and chutes should not be sited more than 25 m from the waste collection point specified by the waste collection authority.
- Householders should not be required to carry refuse more than 30 m to a storage area for a waste container or chute (excluding any vertical distance).
- For waste containers with capacities up to 250 litres:
  (a) steps should be avoided wherever possible on the route between the container store and the waste collection point and where unavoidable they should be restricted to three in number;

(b) ideally, slopes should not be greater than 1 in 12 but where this is unavoidable they should be of restricted length and not in a series.

- For waste containers with capacities greater than 250 litres the storage area should be located so that steps are avoided altogether.
- The waste collection authority should be consulted to ensure that the collection point can be accessed by its normal size of waste collection vehicle.
- External locations for waste containers should be sited in shade or under shelter away from windows or ventilators.
- Waste storage areas should not obstruct pedestrian or vehicle access routes to buildings.

13.10.5 Design of waste containers and storage areas

Enclosures, compounds or storage rooms

Enclosures, compounds and storage rooms should comply with the following.

- Be designed to allow room for filling and emptying.
- Have a clear space of 150 mm provided between and around containers.
- Be permanently ventilated at top and bottom.
- Have paved impervious floors.
- Be secure to prevent access by vermin (unless, in the case of compounds, the refuse is stored in secure containers with close fitting lids).
- If enclosing communal containers, have:
  (a) clear headroom of 2 m;
  (b) provision for washing down and draining the floor into a drainage system designed to receive polluted effluent;
  (c) gullies incorporating traps which maintain their seals even after prolonged periods of disuse.
- If enclosing individual containers, be sufficiently high to allow the lid to be opened for filling.
- Where storage rooms are provided these should contain separate rooms for recyclable and non-recyclable waste.

A waste storage facility which is located in a publicly accessible area or in an open area around a building (such as a front garden) should be provided with an enclosure or shelter.

Refuse chutes

For high rise domestic developments where refuse chutes are provided, AD H6 recommends that they should be at least 450 mm in diameter and constructed with:

- smooth, non-absorbent inner surfaces
- close fitting access doors at each storey containing a dwelling
- ventilation at top and bottom.

Alternatively, refuse chutes may be designed in accordance with the relevant clauses in BS 5906: 1980 *Code of practice for storage and on-site treatment of solid waste from buildings*. Figure 13.28 is based on the recommendations of BS 5906: 1980 and illustrates a typical refuse chute installation.

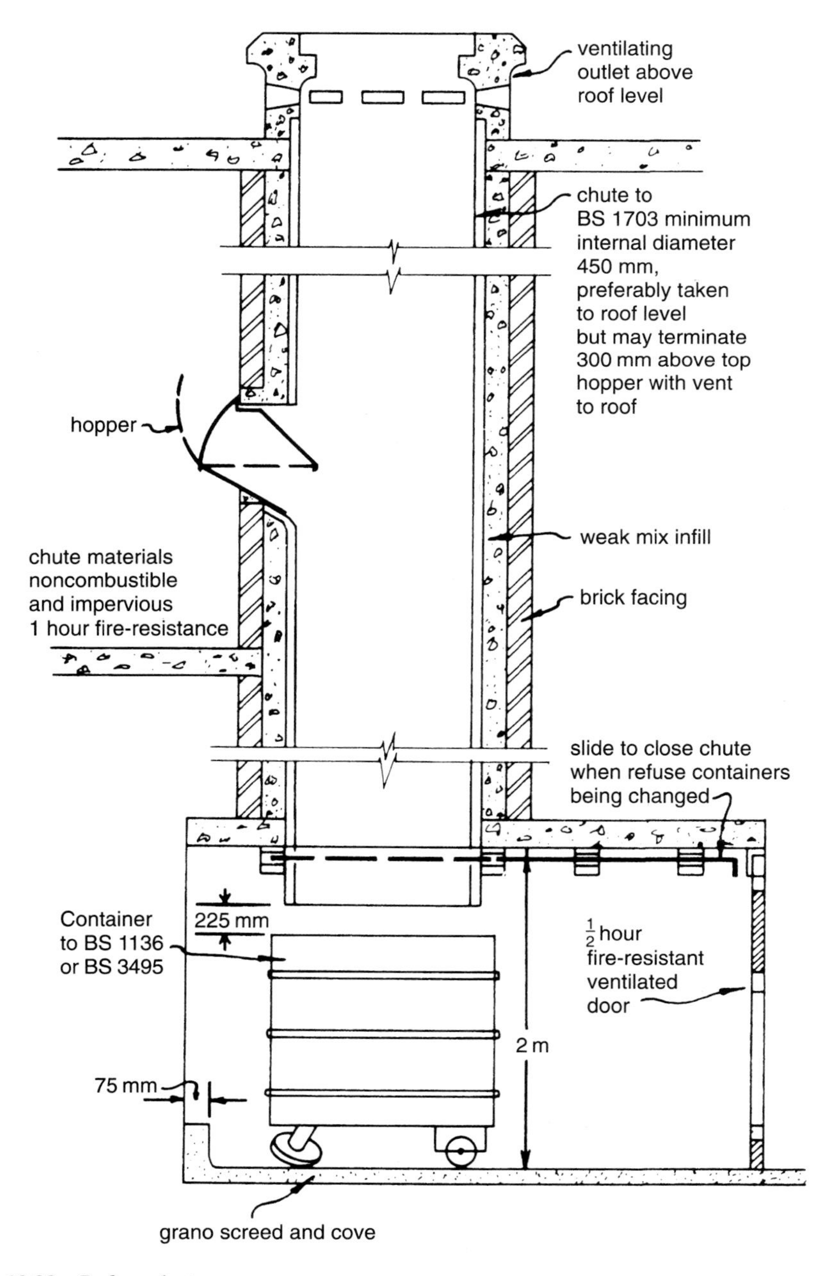

Fig. 13.28 Refuse chutes.

13.10.6 Non-domestic buildings

In the development of non-domestic buildings special problems may arise. It is therefore essential to consult the refuse collection authority for their requirements with regard to the following.

- The storage capacity required for the volume and nature of the waste produced. (The collection authority will be able to give guidance as to the size and type of container they will accept and the frequency of collection).
- Storage method. This may include details of any proposed on-site treatment and should be related to the future layout of the development and the building density.
- Location of storage and treatment areas and collection points, including access for vehicles and operatives.
- Measures to ensure adequate hygiene in storage and treatment areas.
- Measures to prevent fire risks.
- Segregation of waste for recycling.

The following recommendations should also be considered.

- Rooms and compounds provided for the open storage of waste should be secure to prevent access by vermin (unless in the case of compounds, the refuse is stored in secure containers with close fitting lids).
- Waste storage areas should:
  (a) have an impervious floor and provision for washing down and draining the floor into a drainage system designed to receive polluted effluent;
  (b) have gullies incorporating traps which maintain their seals even after prolonged periods of disuse;
  (c) be marked to show their use and signs should be provided to indicate their location.

13.10.7 Alternative approach

As an alternative to the recommendations listed above for waste disposal, it is permissible to use BS 5906: 1980 especially clauses 3 to 10, 12 to 15 and Appendix A. However, BS 5906 does not contain information on recycling. It is currently being revised and the new edition is expected to contain such information.

14 Combustion appliances and fuel storage systems

14.1 Introduction

Part J of Schedule 1 to the Building Regulations 2000 (as amended) is concerned with the safe installation and use of combustion appliances in buildings. In 2002 Part J was extended to include the protection of liquid fuel storage systems (oil and liquefied petroleum gas) from fire in neighbouring buildings and protection against pollution resulting from spills from oil storage tanks.

It is limited to fixed appliances burning solid fuel, oil or gas and to incinerators. (It is assumed that this means incinerators that burn solid fuel, oil or gas since no further guidance is given.) This excludes all electric heating appliances and small portable heaters such as paraffin stoves.

In order that it may function safely a combustion appliance needs an adequate supply of combustion air and it must be capable of discharging the products of combustion to outside air. This must be achieved without allowing noxious fumes to enter the building and without causing damage by heat or fire to the fabric of the building.

14.1.1 The Clean Air Acts 1956–1993

As a direct result of the great London smog of 1952 and after the Report of the Committee on Air Pollution (Cmnd. 9322, November 1954) the Clean Air Act 1956 was passed to give effect to some of the Committee's recommendations. The 1956 Act, amended in 1968, was consolidated in the Clean Air Act 1993. Its main provisions may be summarised briefly, and should be borne in mind in considering the effect of Part J of Schedule 1 to the 2000 Regulations.

The Act makes it an offence to allow the emission of *dark smoke* from a chimney, but certain special defences are allowed, e.g. unavoidable failure of a furnace. DARK SMOKE is defined as smoke which appears to be as dark as, or darker than, shade 2 on the Ringelmann Chart. Regulations made under the Act amplify its provisions in relation to industrial and other buildings, and it should be noted that the prohibition on dark smoke applies to all buildings, railway engines and ships. However, its chief effect is on industrial and commercial premises.

House chimneys rarely emit dark smoke, but local authorities may, by order confirmed by the Secretary of State for the Environment, declare *smoke control areas*. In a smoke control area the emission of smoke from chimneys constitutes an offence, although it is a defence to prove that the emission of smoke was not caused by the use of any fuel other than an authorised fuel. Regulations prescribe the following authorised fuels: anthracite; briquetted fuels carbonised in the process of manufacture; coke; electricity; low temperature carbonisation fuels; low volatile steam coals; and fluidised char binderless briquettes.

The 1956 Act, as amended, provides for the payment of grants by local authorities, and Exchequer contributions, towards the cost of any necessary adaptation or conversion of fireplaces to smokeless forms of heating in private dwellings in smoke control areas.

14.1.2 Combustion appliances and fuel storage systems

Although Part J applies to any combustion installation and liquid fuel storage system within the limits on application, the guidance in Approved Document J deals mainly with domestic installations, such as those which comprise space and water heating systems and cookers and their flues, and their attendant oil and LPG fuel storage systems.

Therefore, the guidance is concerned only with combustion installations having the following power ratings:

- solid fuel installations of up to 50 kW rated output;
- gas installations of up to 70 kW net (77.7 kW gross) rated input; and
- oil installations of up to 45 kW rated heat output;

and with fuel storage installations with the following capacities:

- heating oil storage installations with capacities up to 3500 litres; and
- liquefied petroleum gas (LPG) storage installations with capacities up to 1.1 tonne.

It should be noted, however, that no upper size limit is specified on the application of paragraph J5 of Part J of Schedule 1 to the Building Regulations 2000 (which deals with the protection of liquid fuel storage systems).

There are no specific references to incinerators or to the installation of appliances with a higher rating than those given above in AD J even though these are covered by the requirements of Part J and it is evident that specialist guidance will usually be needed (since these will almost invariably be installed under the supervision of a heating engineer). Some larger installations can be shown to comply by adopting the relevant recommendations of:

- CIBSE Design Guide Volume B, and
- practice standards produced by the British Standards Institute and the Institution of Gas Engineers.

14.2 Interpretation

Compared with the previous edition of AD J in which only one definition appeared, in an attempt to provide clarity, the current edition contains no fewer than 39!

APPLIANCE COMPARTMENT means an enclosure specifically constructed or adapted to accommodate one or more gas or oil-fired appliances.

BALANCED COMPARTMENT refers to a method of installing an open-flued appliance into a compartment so that it is sealed from the remainder of the building and whose ventilation is so arranged in conjunction with the appliance flue as to achieve a balanced flue effect (see below for definitions of BALANCED FLUED APPLIANCE and OPEN FLUED APPLIANCE).

BALANCED FLUED APPLIANCE means a combustion appliance that draws its combustion air from a point immediately adjacent to the point where it discharges its combustion products. The inlet and outlet are so arranged as to substantially balance any wind effects. Balanced flues can run vertically, but they usually discharge horizontally through the external wall on which the appliance is situated.

BOUNDARY. This definition should not be confused with the definition of relevant boundary given in AD B (see Chapter 7) and applies solely for the purposes of AD J. What is referred to here is the boundary of the land or buildings belonging to and under the control of the building owner. Some sections of AD J relate to the distance that outlets from chimneys and flues can be from the boundary. In these cases the measurements may be taken up to the centreline of adjacent routes or waterways. Other sections refer to the distance that oil and LPG storage tanks must be from the boundary. In these cases the measurements may only be taken to the physical boundaries of the site.

BUILDING CONTROL BODY may be either the local authority or an approved inspector. These are fully described in Chapters 3 and 4.

CAPACITY of an oil tank means its nominal volume as stated by the manufacturer. This is usually about 97% of the volume of liquid required to totally fill it.

CHIMNEY includes a wall or walls enclosing one or more flues (see Fig. 14.1). (The chimney for a gas appliance may be referred to as the flue in the gas industry.)

COMBUSTION APPLIANCE means an apparatus which burns fuel to generate energy for space heating, water heating, cooking etc. (e.g. boilers, warm air heaters, water heaters, fires, stoves and cookers), but not including fuel delivery or heat distribution systems.

DECORATIVE FUEL EFFECT (DFE) FIRES are described in BS 5871: Part 3. These are gas-fired imitations, which can be substituted for the solid fuel appliances in open fires. Where suitable, they can also be used in flueboxes designed for gas appliances only. Common designs include beds of artificial coals shaped to fit into a fireplace recess or baskets of artificial logs for use in larger fireplaces or under canopies (see Fig. 14.2(a)).

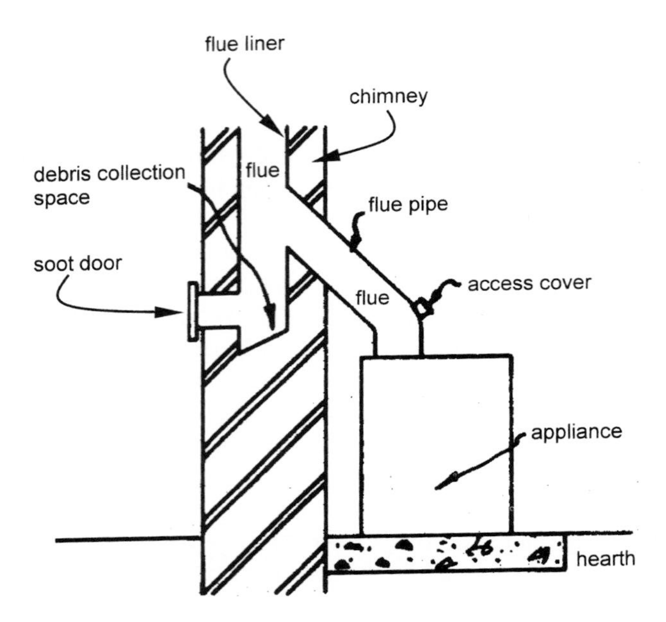

Fig. 14.1 Interpretation – chimneys and flues.

DESIGNATION SYSTEM for the performance characteristics of a chimney or its components means designations which are referred to throughout AD J and details of the designation system are contained in BS EN 1443:1999. This allows the performance characteristics to be expressed by means of a code such as *Chimney EN 1443 T400 P1 S W 1 R22 C50*. In this example the various parts of the code stand for:

Chimney EN 1443 – the number of corresponding European standard
T400 – the temperature class (i.e. a product with a normal working temperature of 400°C, see Table 1 of EN 1443)
P1 – the pressure class N or P or H (negative pressure, positive pressure or high positive pressure)
S – sootfire resistance class S or O (S – chimney has sootfire resistance, O – no sootfire resistance)
W – resistance to condensate class (W – chimneys operating under wet conditions, D – dry conditions)
1 – corrosion resistance class (1 – gas fuel, 2 – oils with sulphur content less than 0.2% and natural wood, 3 – oils with sulphur content over 0.2 % and solid mineral fuels and peat)
R22 – the thermal resistance
C50 – distance to combustible material (i.e. 50 mm)

Sometimes the designation is shortened, e.g. clay ceramic flue liners with the designation EN 1457 T600 N2 S D 3 can be described as Class A1N2.

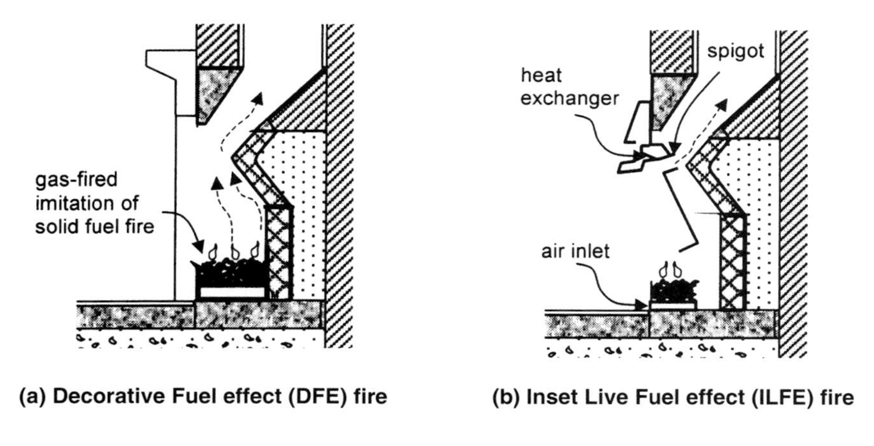

(a) Decorative Fuel effect (DFE) fire

(b) Inset Live Fuel effect (ILFE) fire

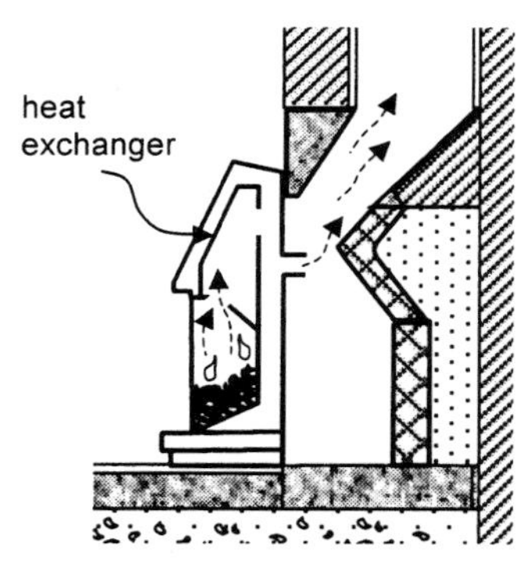

(c) Radiant convector gas fire

Fig. 14.2 Interpretation – gas fires.

DRAUGHT BREAK is an opening into any part of the flue serving an open flued appliance, formed by a factory-made component. This can allow dilution air to be drawn into a flue or can be used to lessen the effects of down-draught on combustion in the appliance.

DRAUGHT DIVERTER is a form of draught break which allows the appliance to operate without interference from down-draughts that may occur in adverse wind conditions and excessive draught.

DRAUGHT STABILISER is a factory made counter-balanced flap device usually mounted in the fluepipe or chimney, but sometimes located on the appliance, which admits air to the flue from the same space as the combustion air. It is designed to prevent excessive variations in the draught.

FACTORY-MADE METAL CHIMNEYS are prefabricated chimneys that are usually manufactured as sets of components for assembly on site. Commonly available types range from single-walled metal chimneys suitable for some gas appliances to chimneys with insulation sandwiched between an inner liner and an outer metal wall designed for oil or solid fuel use. They are also known as system chimneys.

FANNED DRAUGHT INSTALLATION. Sometimes known as forced draught appliances, these incorporate a fan to enable the proper discharge of the flue gases. The fan may be separately installed in the flue or may be an integral part of the combustion appliance. Fans can be installed in most oil-fired and many gas-fired boilers and may either extract flue gases from the combustion chamber or may cause the flue gases to be displaced from the combustion chamber if the fan is supplying it with air for combustion. Depending on the location of the fan, flues in fanned draught installations can run horizontally or vertically and can be at higher or lower pressures than their surroundings.

FIREPLACE RECESS means a structural opening formed in a wall or chimney breast, from which a chimney leads and including a hearth at its base. For closed appliances such as stoves, cookers or boilers a simple structural opening may be suitable. For accommodating open fires it will usually be necessary to form a gather to reduce the recess to the size of the flue. Fireplace recesses are often lined with firebacks to accommodate inset open fires and lining components and decorative treatments may be fitted around openings to reduce the opening area. It is this finished fireplace opening area which determines the size of flue required for an open fire in such a recess (see Fig. 14.14 in section 14.7.1).

FIRE WALL is a means of shielding a fuel tank from the heat of a fire. For LPG tanks, a fire wall will lengthen the path that has to be travelled by gas accidentally leaking from the tank or fittings. This will allow it more time to disperse safely, before reaching a hazard such as other potential ignition sources, an opening in a building or a boundary.

FLUE means a passage conveying the products of combustion to the external air (see Fig. 14.1).

FLUEBLOCK CHIMNEY means a chimney constructed from a set of factory-made components. Flueblock chimneys may be made from precast concrete, clay or other masonry units, designed for assembly on site to provide a complete chimney with the performance appropriate for the intended appliance. Two types of common systems are available:

- for use solely with gas burning appliances, and
- for solid fuel burning appliances (sometimes called chimney block systems).

FLUE BOX means a factory-made unit (usually of metal), which is designed to accommodate a gas burning appliance in conjunction with a factory-made chimney. (See also PREFABRICATED APPLIANCE CHAMBER below.)

FLUELESS APPLIANCE means an appliance, which is designed to be used without being connected to a flue. Its products of combustion mix with the surrounding room air and are eventually ventilated from the room to the outside. Examples include gas cookers, gas instantaneous water heaters and some types of gas space heaters.

FLUE LINER is the wall of the chimney that is in contact with the products of combustion. This could be a concrete flue liner, the inner liner of a factory-made chimney system or a flexible liner fitted into an existing chimney (see Fig. 14.1).

FLUE OUTLET is that part of the combustion installation where the products of combustion are discharged from the flue to the outside air, such as the top of a chimney pot or flue terminal.

FLUEPIPE (see Fig. 14.1) means a pipe that connects a combustion appliance to a flue in a chimney (sometimes called a CONNECTING FLUEPIPE). It may be either single-walled (bare or insulated) or double-walled. (The term FLUEPIPE is also used to describe the tubular components from which some factory made chimneys for gas and oil appliances are made or from which plastic flue systems are made.)

HEARTH is a base on which a combustion appliance is placed (see Fig. 14.1). It provides safe isolation between the appliance, and people, combustible parts of the building fabric and soft furnishings. The exposed surface of the hearth usually extends beyond the appliance and provides a region which can be kept clear of anything at risk of fire. The hearth may be constructed of:

- thin insulating board;
- a substantial thickness of material such as concrete; or
- some intermediate form of construction depending on the weight and downward heat emission characteristics of the appliance(s) upon it.

For solid fuel open fires the substantial thickness of material necessary may be provided by a constructional hearth (often as part of the building structure, floor slab etc.), on which may be placed a decorative superimposed hearth to provide the clear surface.

HEAT INPUT RATE. For a gas appliance, this is the maximum rate of energy flow that could be provided by the prevailing rate of fuel flow into the appliance if the fuel were to be burned with full oxidation. It is calculated as the rate of fuel flow to the appliance multiplied by either the fuel's gross or net calorific value. The gross calorific value takes account of the latent heat due to the condensation of water in combustion products and allows this to be included in the heat obtained from the fuel (such as in a gas condensing boiler). It is thus a larger figure than the net heat input rate. Either heat input rating can be used for any given appliance; however, it is now usual to express the rating of a gas appliance as a net heat input rate (kW (net)).

INDEPENDENTLY CERTIFIED means that the product conforms to a product certification scheme that has been approved by an independent certification body. Such schemes certify compliance with the requirements of a recognised document that is appropriate to the purpose for which the material is to be used. Materials which are not so certified may still conform to a relevant standard. Many certification bodies, which approve such schemes, are accredited by UKAS.

INSET LIVE FUEL EFFECT (ILFE) FIRES are described in BS 5871: Part 2. These gas fires stand fully or partially within a fireplace recess or suitable fluebox and give the impression of an open fire. The appliance covers the full height of the fireplace opening so that air only enters through purpose designed openings and the flue gases only discharge through the spigot (see Fig. 14.2(b)).

INSTALLATION INSTRUCTIONS means the manufacturer's instructions, to enable installers to correctly install and test appliances and flues and to commission them into service.

NATURAL DRAUGHT flue – this is the traditional concept for a flue whereby the draught which takes the flue gases up the flue to outside air relies on the difference between the temperature of the gases within the flue and the temperature of the ambient air. Draught increases with the height of the flue and a satisfactory natural draught requires an essentially vertical run of flue. This concept can be contrasted with balanced flue appliances which are designed to discharge directly through the wall adjacent to the appliance (see above).

NON-COMBUSTIBLE means capable of being classed as non-combustible:

- when subjected to the non-combustibility test of BS 476, Part 4: 1970 (1984) *Non-combustibility test for materials*; and
- any material which when tested to BS 476, Part 11: 1982 (1988) *Method for Assessing the Heat Emission from Building Materials* does not flame nor cause any rise in temperature on either the centre (specimen) or furnace thermocouples.

(See also Chapter 7, Fire, section 7.18.5.)

NOTIFIED BODY for the purposes of the Gas Appliances (Safety) Regulations (1995) means a body that:

- is approved by the Secretary of State for Trade and Industry as being competent to carry out the required Attestation procedures for gas appliances and whose name and identification number has been notified by him/her to the Commission of the European Community and to other member states in accordance with the Gas Appliances (Safety) Regulations 1995;
- has been similarly approved for the purposes of the Gas Appliances Directive by another member state and whose name and identification number has been notified to the Commission and to other member states pursuant to the Gas Appliances Directive.

OPEN FLUED APPLIANCE is, for example, the traditional open fire or stove that draws its combustion air from the room or space in which it is installed and which requires a flue to discharge its products of combustion to the outside air (see Fig. 14.3).

PREFABRICATED APPLIANCE CHAMBER means a set of factory-made pre-cast concrete components designed to provide a FIREPLACE RECESS (see above). It is normal for the chamber to be positioned against a wall and it may be designed to support a chimney. The chamber and chimney can be enclosed to create a false chimney breast. (See also FLUE BOX above.)

RADIANT CONVECTOR GAS FIRES, CONVECTOR HEATERS AND FIRE/BACK BOILERS, are described in BS 5871: Part 1. These stand in front of a closure plate which is fitted to the fireplace opening of a fireplace recess or suitable fluebox. The appliance covers the full height of the fireplace opening so that air only enters through purpose designed openings and the flue gases only discharge through the flue spigot (see Fig. 14.2(c)).

RATED HEAT INPUT (or rated input) for a gas appliance means the maximum heat input rate at which it can be operated. This will be declared on the appliance data plate and for gas appliances it is now usual to express this rating as a net value (kW (net)) although the gross value (kW (gross)) was used until recently. (For details of net and gross values see HEAT INPUT RATE above.)

RATED HEAT OUTPUT for an oil appliance is the maximum declared energy output rate (kW) as declared on the appliance data plate. For a solid fuel appliance, this is the maximum manufacturers' declared energy output rate (kW) for the appliance. This may be different for different fuels.

ROOM-SEALED APPLIANCE means an appliance whose combustion system is sealed from the room in which the appliance is situated. The appliance obtains combustion air either from a ventilated uninhabited space within the building or directly from the open air outside the building. The products of combustion will be vented directly to open air outside the building (see Fig. 14.3).

THROAT means a narrowing part of the flue between a fireplace recess and its chimney. Throats can be formed from prefabricated components or can be built in brickwork by a process of corbelling.

14.3 Rules for measurement

When measuring the size of a duct or flue (to establish the area, diameter etc. for the purposes of the approved document guidance) the dimensions should be taken at right angles to the direction of gas flow. Minimum requirements for flue sizes are given in the text below and where offset components are used they should not reduce the flue area to less than the quoted figures.

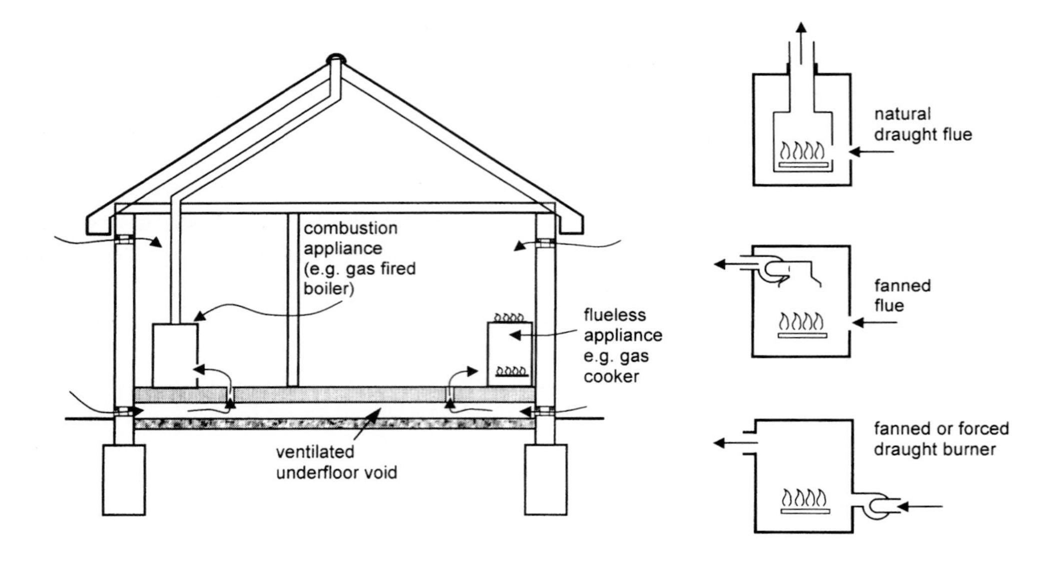

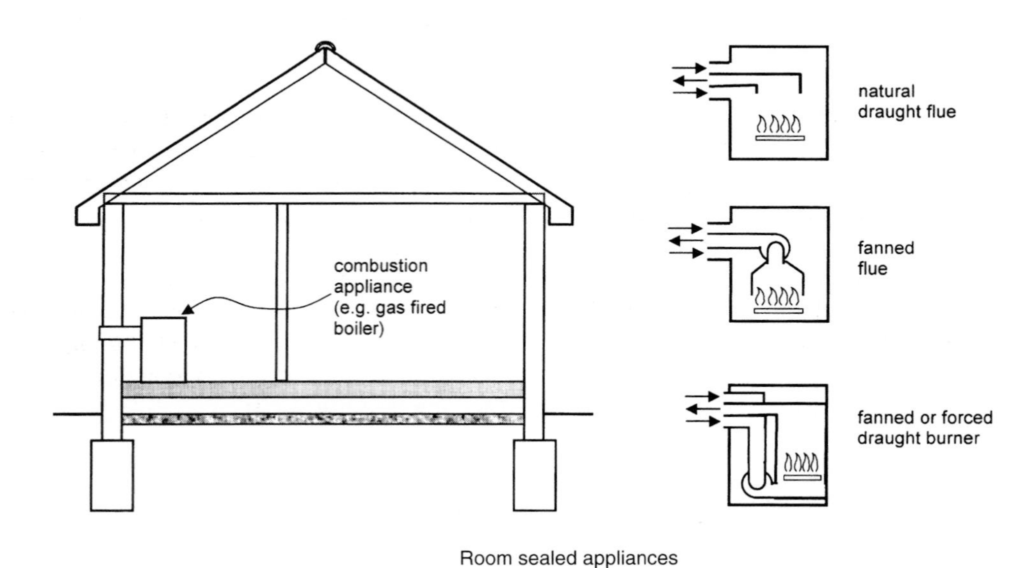

Fig. 14.3 Types of combustion installation.

14.4 Checking the condition of combustion installations before use

Combustion installations must be designed and constructed in accordance with the requirements of Part J. Additionally, before being used they should be inspected and tested to prove that they are in fact in compliance. This applies not only to new build but also to repairs, refurbishment and re-use of existing flues (see section 14.8.3 below for details of this). Responsibility for such proof of compliance rests with the person carrying out the work. This could be a specialist firm working directly for the client, a developer, a main contractor or even a sub-contractor working for the main contractor.

Proving compliance can involve a number of steps, which should be documented in the form of a report drawn up by a 'specialist' firm for the main contractor, client or developer.

To be acceptable to the Building Control Body, the specialist firm would probably need to be a registered member of one of the following organisations although there is no clear guidance in the AD to this effect:

- The Council for Registered Gas Installers (CORGI) contactable at www.corgi-gas.com;
- The Heating Equipment Testing and Approval Scheme (HETAS) contactable at www.hetas.co.uk;
- The Oil Firing Technical Association for the Petroleum Industry (OFTEC) contactable at www.oftec.org;
- The National Association of Chimney Sweeps (NACS) contactable at www.chimneyworks.co.uk;
- The National Association of Chimney Engineers (NACE) contactable at www.nace.org.uk.

The Building Control Body can ask for this report as a way of proving compliance. An example of a completed checklist for such a report is given in Appendix A to AD J, which is reproduced below. The report will need to show that materials and components have been used which are suitable for the intended application and that flues have passed appropriate tests.

In addition to checking a combustion installation for compliance, where a material change of use takes place in a building (e.g. conversion to flats, see Chapter 2) the fire resistance of the existing chimney walls should be checked and improved as necessary. This can be done by applying additional layers of non-combustible material to the existing chimney walls.

14.5 Air supply

Paragraph J1 of Part J requires that combustion appliances be installed so that an adequate supply of air is provided for combustion of the fuel, to prevent overheating and for efficient operation of the flue.

14.5.1 Air supply – general provisions

Where combustion appliances are installed in a building, Paragraph J1 can be met if provisions are made to enable the admission of sufficient air for:

AD J, Appendix A

Hearths, Fireplaces, flues and chimneys

This checklist can help you to ensure hearths, fireplaces, flues and chimneys are satisfactory. If you have been directly engaged, copies should also be offered to the client and to the Building Control Body to show what you have done to comply with the requirements of Part J. If you are a sub-contractor, a copy should be offered to the main contractor.

1. **Building address, where work has been carried out** . . . .
. . . .
. . . .

2. **Identification of hearth, fireplace, chimney or flue**	*Example: Fireplace in lounge*	*Example: Gas fire in rear addition bedroom*	*Example: Small boiler room*
3. **Firing capability: solid fuel/gas/oil/all**	*All*	*Gas only*	*Oil only*
4. **Intended type of appliance. State type or make. If open fire give finished fireplace opening dimensions**	*Open fire 480 W × 560 H (mm)*	*Radiant/convector fire 6 kW input*	*Oil fired boiler 18 kW output (pressure jet)*
5. **Ventilation provision for the appliance: State type and area of permanently open air vents**	*2 through wall ventilators each 10,000 mm^2 (100 cm^2)*	*Not fitted*	*Vents to outside: Top 9,900 mm^2 Bottom 19,800 mm^2*
6. **Chimney or flue construction**			
(a) **State the type or make and whether new or existing**	*New. Brick with clay liners*	*Existing masonry*	*S.S. prefab to BS 4543-2*
(b) **Internal flue size (and equivalent height, where calculated – natural draught gas appliances only)**	*200 mm Ø*	*125 mm Ø (H_e = 3.3 m)*	*127 mm Ø*
(c) **If clay or concrete flue liners used confirm they are correctly jointed with socket end uppermost and state jointing materials used**	*Sockets uppermost Jointed by fire cement*	*Not applicable*	*Not applicable*
(d) **If an existing chimney has been refurbished with a new liner, type or make of liner fitted**	*Not applicable to BS 715*	*Flexible metal liner*	*Not applicable*
(e) **Details of flue outlet terminal and diagram reference**			
Outlet Detail:	*Smith Ltd Louvred pot 200 mm Ø*	*125 mm Ø GC1 terminal*	*Maker's recommended terminal*
Complies with:	*As Diagram 2.2 AD J*	*As BS 5440-1: 2000 Figure C.1*	*As Diagram 4.2 AD J*
(f) **Number and angle of bends**	*2 × 45°*	*2 × 45°*	*1 × 90° Tee*
(g) **Provision for cleaning and recommended frequency**	*Sweep annually via fireplace opening*	*Annual service by CORGI engineer*	*Sweep annually via base of Tee and via appliance*
7. **Hearth. Form of construction. New or existing?**	*New. Tiles on concrete floor. 125 mm thick. As Diagram 2.9 AD J*	*Existing hearth for solid fuel fire, with fender*	*New. Solid floor Min 125 mm concrete above DPM. As Diagram 4.3 AD J*
8. **Inspection and testing after completion** **Tests carried out by:**	*Inspected and tested by J Smith, Smith Building Co*	*Tested by J Smith, CORGI Reg no. 12345*	*Tested by J Smith, The Oil Heating Co*
Tests (Appx E in AD J 2002 ed) and results			
Flue Inspection — **visual**	*Not possible, bends*	*Not possible, bends*	*Checked to Section 10, BS 7566: Part 3: 1992 – OK*
sweeping	*OK*	*Not applicable*	
coring ball	*OK*	*Not applicable*	
smoke	*OK*	*Not applicable*	*OK*
Appliance (where included) spillage	*Not included*	*OK*	*OK*

I/We the undersigned confirm that the above details are correct. In my opinion, these works comply with the relevant requirements in Part J of Schedule 1 to the Building Regulations.

Print name and title . . . . Profession . . . .

Capacity . . . (e.g. 'Proprietor of Smith's Flues', Authorising Engineer for Brown plc) . . . . Tel no . . . .

Address . . . . Postcode . . . .

Signed . . . . Date . . . .

Registered membership of . . . (e.g. CORGI, OFTEC, HETAS, NACE, NACS) . . . .

- proper combustion of the fuel;
- proper operation of any flues, (or for flueless appliances, safe dispersal of the products of combustion to the outside air); and
- cooling control systems and/or to make sure that appliance casings do not become too hot to touch, where this is deemed necessary.

AD J gives a range of air vent sizes, which vary with the type of fuel being burned. These are described below and should be read with the following general notes.

- The figures given are for single combustion appliances only, and will need to be increased if more than one appliance is installed in a room (e.g. where a kitchen contains an open-flued boiler and a flueless appliance such as a cooker).
- Where an open-flued appliance is installed in a room it will receive a small amount of combustion air from infiltration through the building fabric. Depending on the type of appliance installed this will need to be supplemented by permanently open air vents.
- Where an open-flued appliance is installed in an appliance compartment:
  (a) all the air necessary for combustion and proper operation of the flue must be supplied through adequately sized permanent vents which may be situated in an outside or internal wall;
  (b) where cooling air is needed, the compartment should be large enough to allow air to circulate via high and low level vents;
  (c) the appliance and ventilation system manufacturer's instructions should be followed where appliances are to be installed within balanced compartments since special provisions will be necessary.
- Ventilation direct to outside air should be provided for other rooms or spaces within the building where:
  (a) a room-sealed appliance takes its combustion air from another space (e.g. a roof void);
  (b) a flue has a permanent opening to another space (e.g. where it feeds a secondary flue in the roof void).

In the case of ventilation via roof voids, the ventilation provisions contained in Approved Document F (see Chapter 11) would normally be adequate, where the combustion installation serves a dwelling.

In other cases, the room or space from which the combustion air is obtained should have air vent openings direct to outside air of at least the same size as the internal openings serving the appliance, although air vents for flueless appliances should always open direct to outside air (i.e. not through an adjoining space). Figure 14.4 below gives examples of locations for permanent air vent openings.

Where permanently open air vents are called for (see guidance related to specific fuels below) they should be:

- non-adjustable;
- appropriately sized to admit the correct amount of air taking into account their free area (or equivalent free area, see Fig. 14.5) and any obstructions such as

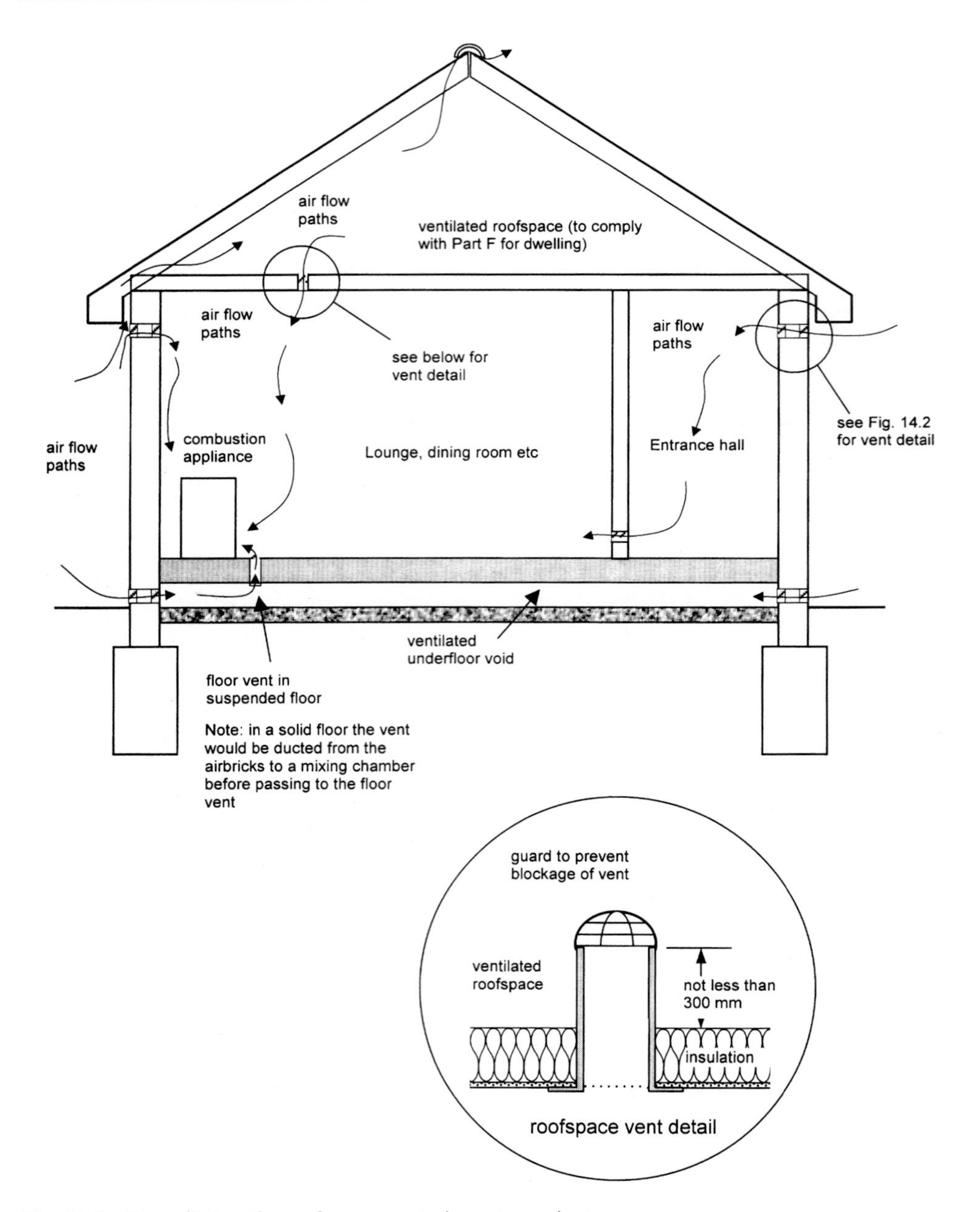

Fig. 14.4 Example locations of permanent air vent openings.

grilles and anti-vermin mesh (which should have aperture dimensions no smaller than 5 mm);

- positioned so that they are unlikely to become blocked;
- located so that occupants are not provoked to seal them against cold draughts and noise (e.g. place vents close to appliances; draw air from hallways or other intermediate spaces; place air vents next to ceilings to ensure good mixing of

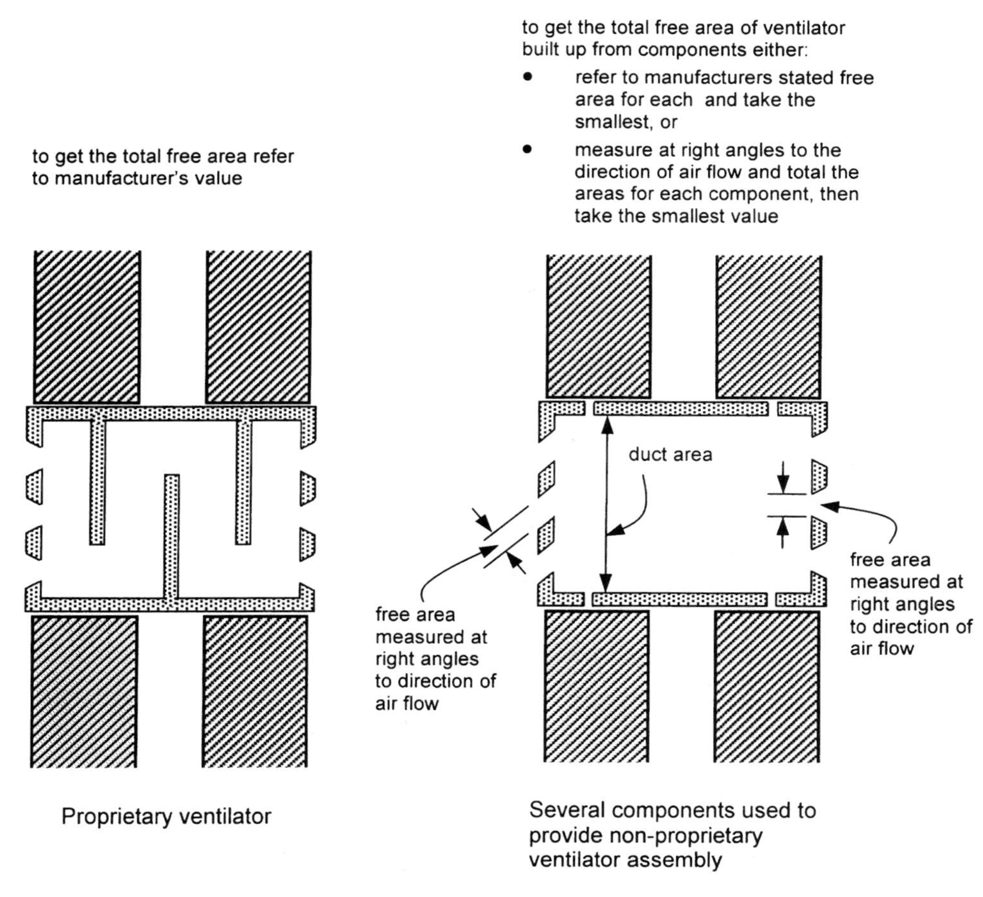

Fig. 14.5 Free area of ventilators.

incoming cold air; install noise attenuated ventilators to cut down on unwanted external noise; place ventilators outside fireplace recesses and beyond the hearths of open fires where dust or ash might be disturbed by draughts).

Permanently open air vents should not be installed in fire-resisting walls. Although external walls are excluded from this provision, this exclusion will not apply to parts of external walls shielding LPG tanks. Additionally, vents should not be sited in fireplace recesses unless expert advice has been sought.

Ventilation via permanently open vents can be provided in a number of ways. Where proprietary components or assemblies are used the manufacturer will usually be able to give a value for the free area (or equivalent free area). Where this is not available (e.g. some air bricks, grilles or louvres) it will be necessary to calculate the free area as shown in Fig. 14.5 by aggregating the individual apertures and taking the smallest free area in the assembly (or the overall duct area if this is less).

Where buildings have air tight membranes in their floors (such as radon or landfill gas barriers, see Chapter 8) ventilation ducts or vents should be installed so as not to compromise the effectiveness of the membrane.

14.5.2 Air supply – compliance with other parts of the Regulations

By referring to Chapter 11 (which deals with Part F, Ventilation of the 2000 Regulations), it will be apparent that buildings need to be ventilated not only to provide combustion air for fuel burning appliances (where these are present) but also to provide general ventilation for health reasons. A possible conflict might occur between the two forms of ventilation since the guidance in Approved Document F allows the background ventilation to be adjustable, whereas Part J ventilation must be permanently open. Previous editions of Part J have been unsuccessful in addressing this conflict, however the 2002 edition gives the following guidance.

Where rooms or spaces contain open-flued appliances:

- permanently open vents provided for combustion appliances are acceptable to replace some or all of the Part F adjustable background ventilation (depending on location and amount of opening area);
- adjustable Part F vents can only be used for Part J combustion air ventilation if the are fixed permanently open.

Where rooms or spaces contain flueless appliances (see definition in section 14.2 above) it may be necessary to provide permanent, adjustable and rapid ventilation (e.g. an openable window) to comply with Parts F and J. For such appliances:

- permanent and adjustable ventilation provisions for Part J and Part F compliance can be used as described for open-flued appliances above;
- rapid ventilation provided by opening elements for Part F compliance can also be accepted for Part J compliance provided that the minimum opening areas are achieved.

Where mechanical extract ventilation is provided for Part F compliance, dangerous conditions can be created where open-flued appliances are also present due to the spillage of flue gases (even where the fans and appliances are in different rooms). Approved Document F (see Chapter 11 section 11.3.5 above) contains recommendations designed to avoid flue gas spillage, for different types of combustion appliances. They should be read in conjunction with the following.

- Specialist advice may be needed for commercial and industrial installations, regarding the possible need for the interlocking of gas heaters and any mechanical ventilation systems.
- Suitable spillage tests for gas appliances may be provided by appliance manufacturers in their installation instructions. Alternatively, the procedure given in BS 5440: *Installation and maintenance of flues and ventilation for gas appliances of rated input not exceeding 70 Kw net (1st, 2nd and 3rd family gases)*, Part 1: 2000 *Specification for installation and maintenance of flues* can also be used. The object

of the test is to check for spillage when appliances are subjected to the greatest possible depressurisation. To cater for this the following should be taken into account.

(a) All external doors, windows and other adjustable ventilators to outside should be closed.
(b) Several tests may be necessary to demonstrate the safe operation of the appliance with reasonable certainty since various combinations of fans in operation and open internal doors will be possible, and the specific combination causing the greatest depressurisation at the appliance will depend on the circumstances in each case. One test should of course be carried out with the door leading into the room of installation closed and all fans in that room switched on.
(c) The effect of ceiling fans should be taken into account during the tests.
(d) It is important to consider all fans which might be in use and not only the obvious ones, such as those on view in kitchens. Others include fans installed in domestic appliances such as tumble dryers, fans fitted to other open-flued combustion appliances, and fans installed to draw radon gas from the ground below a building (see also *BRE Good Building Guide GBG 25* and Chapter 8 section 8.5.5 above).

14.5.3 Air supply – specific provisions relating to appliances burning solid fuel

In addition to the general recommendations given above, appliances burning solid fuel with rated outputs up to 50 kW should have permanently open air vents at least as great as the sizes shown in Table 2.1 to Approved Document J, which is reproduced below. It should be noted that where an appliance is installed that is capable of burning a range of different solid fuels the ventilation requirements should suit the fuel that produces the greatest heat output.

Manufacturers' installation instructions should be followed where these vary from the Table recommendations (e.g. they may specify even larger areas of permanently open air vents or, in the case of a cooker, omit to specify a rated output).

14.5.4 Air supply – specific provisions relating to appliances burning gas

Appliances burning gas with rated inputs up to 70 kW (net) should comply with the general recommendations given above. Additionally, the amount of permanently open air vents which should be provided will vary with the type of appliance (and whether it is room-sealed, open-flued or flueless). The various combinations are discussed below.

Flued decorative fuel effect (DFE) fires

These are defined in section 14.2 above. Any room or space containing a DFE fire should have ventilation provided in accordance with Table 14.1 below.

AD J Section 2

Table 2.1 Air supply to solid fuel appliances.

Type of appliance	Type and amount of ventilation (1)	
Open appliance, such as an open fire with no throat, e.g. a fire under a canopy as in Diagram 2.7	Permanently open air vent(s) with a total free area of at least 50% of the cross sectional area of the flue	
Open appliance, such as an open fire with a throat as in Diagram 2.6 and 2.13	Permanently open air vent(s) with a total free area of at least 50% of the throat opening area (2)	
Other appliance, such as a stove, cooker or boiler, with a flue draught stabiliser	Permanently open air vent(s) as below: (3)	Total free area
	First 5 kW of appliance rated output	300 mm^2/kW
	Balance of rated output	850 mm^2/kW
Other appliance, such as a stove, cooker or boiler, with no flue draught stabiliser	A permanent air entry opening or openings with a total free area of at least 550 mm^2 per kW of appliance rated output above 5 kW	

Notes:
1. Divide the area given in mm^2 by 100 to find the corresponding area in cm^2
2. For simple open fires (see Fig. 14.15) the requirement can be met with room ventilation areas as follows:

Nominal fire size (fireplace opening size)	500 mm	450 mm	400 mm	350 mm
Total free area of permanently open air vents	20,500 mm^2	18,500 mm^2	16,500 mm^2	14,500 mm^2

3. Example: an appliance with a flue draught stabiliser and a rated output of 7 kW would require a free area of: $[5 \times 300] + [2 \times 850] = 3200\ mm^2$

Table 14.1 Supply of combustion air to flued decorative fuel effect fires.

Type of appliance	Type of ventilation
1 DFE fire in a fireplace recess with a throat	Air vent free area of at least 10,000 mm^2 (100 cm^2)
2 DFE fire in a fireplace with no throat (e.g. under a canopy)	Air vent free area sized as for a solid fuel fire (see Table 2.1 to AD J section 2 above)
3 DFE fire with rating not exceeding 7 kW (net)	Permanently open air vents not necessary for appliances certified by a Notified Body (see section 14.2 above) as having a flue gas clearance rate (without spilling) not exceeding 70 m^3/hour

Flued appliances other than DFE fires

These include inset live fuel effect fires (ILFE), radiant convector gas fires, convector heaters and fire/back boilers (all as defined in section 14.2 above). All these combustion appliances come in both room-sealed and open-flued variants. Table 14.2 gives the free areas of permanently open vents for these appliances.

Table 14.2 Supply of combustion air to gas appliance installations (other than DFE fires or flueless appliances).

Location/type of appliance	Amount/type of ventilation
1 Appliance in a room or space:	Ventilation direct to outside air
(a) Open-flued	Permanently open vents of at least 500 mm^2 per kW (net) of rated input over 7 kW (net)
(b) Room-sealed	No vents needed
2 Appliance in appliance compartment:	Ventilation via adjoining room or space
(a) Open-flued	From adjoining room or space to outside air: permanently open vents of at least 500 mm^2 per kW (net) of rated input over 7 kW (net) Between adjoining room or space and appliance compartment, permanently open vents: at high level – 1000 mm^2 per kW input (net) at low level –2000 mm^2 per kW input (net)
(b) Room-sealed	Between adjoining space and appliance compartment, permanently open vents: both high and low levels – 1000 mm^2 per kW input (net)
3 Appliance in appliance compartment:	Ventilation direct to outside air:
(a) Open-flued	Permanently open vents: at high level – 500 mm^2 per kW input (net) at low level – 1000 mm^2 per kW input (net)
(b) Room-sealed	Permanently open vents: both high and low levels – 500 mm^2 per kW input (net)

Example calculation
An open-flued boiler with a rated input of 20 kW (net) is installed in a boiler room (appliance compartment, row 3 above) ventilated directly to the outside. The design of the boiler is such that it requires cooling air in these circumstances.
The cooling air will need to be exhausted via a high level vent.
From above, the area of ventilation needed = 20 kW × 500 mm^2/kW = 10,000 mm^2.
A low level vent will need to be provided to allow cooling air to enter, as well as admitting the air needed for combustion and the safe operation of the flue.
From above the area of ventilation needed = 20 kW × 1000 mm^2/kW = 20,000 mm^2.
These ventilation areas can be converted to cm^2 by dividing the results given above in mm^2 by 100.
The calculated areas are the free areas of the vents (or equivalent free areas for proprietary ventilators) described in section 14.5.1 above.

Air supply to flueless gas appliances

Flueless appliances are designed to be used without being connected to a flue. The products of combustion mix with the surrounding room air and are eventually ventilated from the room to the outside. Examples include gas cookers, gas instantaneous water heaters and some types of gas space heaters.

It will usually be necessary to comply with Part F (see Chapter 11) regarding background, rapid and extract ventilation as well as the permanent ventilation provisions in Part J.

Table 14.3 below gives details of the amounts of permanent ventilation that

Table 14.3 Ventilation of flueless gas appliances.

Type of flueless appliance	Maximum rated heat input of appliance	Volume of room, space or internal space[1] (m^3)	Free area of permanently open air vents (mm^2)
1 Cooker, oven, hob, grill or combination of these	Not applicable	Up to 5 5 to 10 over 10	10,000 5000[3] permanently open vent not needed
2 Instantaneous water heater	11 kW (net)	5 to 10 10 to 20 over 20	10,000 5000 permanently open vent not needed
3 Space heater[2]:			
(a) not in an internal space[4]	0.045 kW (net) per m^3 volume of room or space	All room sizes	10,000 **plus** 5500 per kW (net) greater than 2.7 kW
(b) in an internal space[5]	0.09 kW (net) per m^3 volume of room or space	All room sizes	10,000 **plus** 2750 per kW (net) greater than 5.4 kW

Notes:
[1] In this Table 'internal space' means e.g. a hallway or landing (a space which communicates with several other rooms spaces).
[2] For LPG fired space heaters which conform to BS EN 449: 1997 *Specification for dedicated liquified petroleum gas appliances. Domestic flueless space heaters (including diffusive catalytic combustion heaters)*, follow the guidance in BS 5440: Part 2: 2000.
[3] If the room or space has a door direct to outside air, no permanently open air vent is needed.
[4] A space heater is to be installed in a lounge measuring $5 \times 4 \times 2.5 = 50\ m^3$. Its maximum rated input should not exceed $50 \times 0.045 = 2.25$ kW (net).
[5] A space heater installed in a hallway to provide background heating has a rated input of 6 kW (net). It will need to be provided with $10{,}000 + 2750 \times (6 - 5.4) = 11{,}650\ mm^2$ of permanently open ventilation.

should be provided in different circumstances. These are in addition to any openable elements or extract ventilation needed to comply with Part F. The Table should be read in conjunction with the general provisions listed above. Where a gas point is installed in a room which is intended to be used with a flueless appliance, the room should have the ventilation provisions required for the intended appliance. The ventilation provision should be calculated on the basis that the largest rated appliance consistent with the Table 14.3 recommendations will be installed in the room.

A flueless instantaneous water heater should never be installed in a room or space which has a volume of less than $5m^3$.

14.5.5 Air supply – specific provisions relating to appliances burning oil

The guidance in Approved Document J is relevant to combustion appliances burning oils meeting the BS 2869: 1998 (or equivalent) specifications for Class C2 (Kerosene), and Class D (Gas oil). Appliances burning oil with rated outputs up to 45 kW should comply with the general recommendations given above. In rooms such as bathrooms and bedrooms, where there is an increased risk of carbon monoxide poisoning, open-flued oil-fired combustion appliances should not be installed. Instead, room-sealed appliances could be installed.

Table 14.4 below gives details of the amounts of permanent ventilation that should be provided in different circumstances. Manufacturers' installation instructions should be followed where these require greater areas or permanent ventilation than shown in the table recommendations.

Table 14.4 Supply of combustion air to oil-fired appliance installations.

Location/type of appliance	Amount/type of ventilation
1 Appliance in a room or space:	Ventilation direct to outside air
(c) Open-flued	Permanently open vents of at least 550 mm^2 per kW of rated output over 5 kW[1]
(d) Room-sealed	No vents needed
2 Appliance in appliance compartment:	Ventilation via adjoining room or space
(c) Open-flued	From adjoining room or space to outside air: permanently open vents of at least 550 mm^2 per kW of rated output over 5 kW Between adjoining room or space and appliance compartment, permanently open vents: at high level – 1100 mm^2 per kW output at low level – 1650 mm^2 per kW output
(d) Room-sealed	Between adjoining space and appliance compartment, permanently open vents: both high and low levels – 1100 mm^2 per kW output
3 Appliance in appliance compartment:	Ventilation direct to outside air:
(c) Open-flued	Permanently open vents: at high level – 550 mm^2 per kW output at low level – 1100 mm^2 per kW output
(d) Room-sealed	Permanently open vents: both high and low levels – 550 mm^2 per kW output

Notes:

[1] Increase the area of permanent ventilation by a further 550 mm^2 per kW output if appliance fitted with draught break.

Example calculation

An open-flued boiler with a rated output of 15 kW is installed in a cupboard (appliance compartment, row 2 above) ventilated via an adjacent room. Since the boiler output exceeds 5 kW, permanent ventilation openings will be needed in the adjacent room in addition to the vents between the cupboard and the room designed to provide combustion and cooling air.

Area of permanent vents to outside air needed in adjacent room = (15 kW − 5 kW) × 550 mm^2/kW = 5500 mm^2.

The cooling air will need to be exhausted via a high level vent.

From above, the area of ventilation needed = 15 kW × 1100 mm^2/kW = 16,500 mm^2.

A low level vent will need to be provided to allow cooling air to enter, as well as admitting the air needed for combustion and the safe operation of the flue.

From above the area of ventilation needed = 15 kW × 1650 mm^2/kW = 24,750 mm^2

These ventilation areas can be converted to cm2 by dividing the results given above in mm^2 by 100.

The calculated areas are the free areas of the vents (or equivalent free areas for proprietary ventilators) described in section 14.5.1 above.

14.6 Discharge of products of combustion

Paragraph J2 of Part J requires that combustion appliances have adequate provision for the discharge of the products of combustion to the outside air.

In general, this means that the combustion installation must enable normal operation of the appliances without a hazard to health being created by the products of combustion.

Apart from flueless appliances, this is achieved by providing each combustion appliance with a suitable flue which discharges to outside air.

14.6.1 Provision of flues and chimneys

The guidance in AD J is based on the provision of a separate flue for each appliance. Although every solid fuel appliance should be connected to its own flue, for oil and gas-fired installations it is possible to connect more than one appliance to a single flue by following the alternative guidance in:

- BS 5410: *Code of practice for oil firing,* Part 1: 1977 *Installations up to 44 kW output capacity for space heating and hot water supply purposes*, AMD 3637, for oil-fired installations; and
- BS 5440: *Installation and maintenance of flues and ventilation for gas appliances of rated input not exceeding 70 Kw net (1st, 2nd and 3rd family gases),* Part 1: 2000 *Specification for installation and maintenance of flues*, for gas-fired installations.

A chimney is defined above as a structure consisting of a wall or walls enclosing one or more flues and the old practice of building chimneys with unlined flues became obsolete many years ago.

Interestingly, AD J now contains provisions for the repair and testing of flues in existing chimneys when these are brought back into use or re-use. In many cases this will involve relining the old flues.

14.6.2 Chimneys, flues and flue pipes – general provisions

AD J gives a range of flue sizes, and a number of specific recommendations, which vary with the type of fuel being burnt. These are described below and should be read with the following general notes.

Liners to masonry chimneys

Liners to masonry chimneys should be installed in accordance with manufacturer's instructions and the following notes:

- the flue should be formed with appropriate components, keeping joints to a minimum and avoiding cutting;
- matching factory-made components should be used for bends and offsets;
- liners should be built into the chimney with sockets or rebate ends uppermost to

keep moisture and other condensates in the flue (this also prevents condensate from running out of the joints where it might adversely affect any caulking material);

- joints between liners should be sealed with fire cement, refractory mortar or installed in accordance with manufacturer's instructions;
- the space between the liners and the masonry should be filled with weak mortar or insulating concrete, with mixes such as:
  - (a) 1 : 20 ordinary Portland cement : suitable lightweight expanded clay aggregate (minimally wetted);
  - (b) 1 : 6 ordinary Portland cement : vermiculite;
  - (c) 1 : 10 ordinary Portland cement : perlite.

The following liners for masonry chimneys are suitable for all fuels.

- Liners as described in BS EN 1443: 1999 with performance at least equal to the designation T450 N2 S D 3 (see section 14.2 above), such as:
  - (a) clay flue liners with rebated or socketed joints which meet the requirements for Class A1 N2 or Class A1 N1 to BS EN 1457: 1999
  - (b) concrete flue liners independently certified as meeting the requirements for the classification Type A1, Type A2, Type B1 or Type B2 as described in prEN 1857(e18) January 2001; or
  - (c) other products that are independently certified as meeting the BS EN 1443 criteria.
- Imperforate clay pipes with sockets for jointing to comply with BS 65: 1991 (1995) Specification for vitrified clay pipes, fittings, joints and ducts, also flexible mechanical joints for use solely with surface water pipes and fittings, AMD 8622.

Construction of flueblock chimneys

For all fuels, flueblock chimneys should:

- be constructed using factory-made components suitable for the intended application;
- be installed in accordance with manufacturer's instructions;
- have joints sealed in accordance with the flueblock manufacturer's instructions;
- have bends and offsets formed only with matching factory-made components;
- be constructed using:
  - (a) liners as described in BS EN 1443: 1999 with performance at least equal to the designation T450 N2 S D 3 (see section 14.2 above), such as:
    - (i) clay flueblocks meeting the requirements for the class FB1 N2 to BS EN 1806: 2000 *Chimneys. Clay/ceramic flue blocks for single wall chimneys. Requirements and test methods*; or
    - (ii) other products that are independently certified as meeting the BS EN 1443 criteria;
  - (b) blocks lined as for masonry chimneys above (and independently certified as suitable for the purpose).

Condensates in flues

The products of combustion of all fuels contain considerable quantities of water vapour which can condense on the inside of the flue and cause damage if not satisfactorily controlled. Some modern appliances (called condensing appliances) are designed to extract the latent heat of vaporisation from the condensate to improve efficiency.

In the case of chimneys that do not serve condensing appliances, satisfactory control of condensation can be achieved by insulating flues so that flue gases do not condense under normal conditions of operation.

For chimneys serving condensing appliances, satisfactory control of condensation can be achieved by:

- using lining components that are impervious to condensates and suitably resistant to corrosion;
- making appropriate drainage provisions and avoiding ledges, crevices etc;
- providing for the disposal of condensate from condensing appliances.

Plastic fluepipe systems

Under certain circumstances it is possible to use plastic fluepipe systems. For example, with condensing boiler installations, the fluepipes should be:

- supplied by or specified by the appliance manufacturer, and
- approved by a Notified Body or independently certified as being suitable for purpose.

Design of flues serving natural draught open-flued appliances

Flue systems serving natural draught open-flued appliances rely on the stack effect (the natural buoyancy of hot air) to ensure that the products of combustion are successfully discharged to outside air. Therefore, the system should offer as little resistance as possible to the passage of the flue gases and it should be possible to inspect and sweep the whole flue after the appliance (if any) has been installed. This can be done by minimising changes in direction and avoiding long horizontal sections of flue. The following design factors illustrated in Fig. 14.6 below should be considered:

- build flues so that they are as straight and vertical as possible;
- restrict horizontal sections of flue to connections to appliances with rear outlets and do not allow these to be longer than 150 mm;
- if bends have to be included in the flue make sure that they are angled at not greater than 45° to the vertical;
- have no more than four 45° bends between the appliance outlet and the flue outlet, and not more than two of these between:
  (a) adjacent cleaning access points, or
  (b) a cleaning access point and the flue outlet.

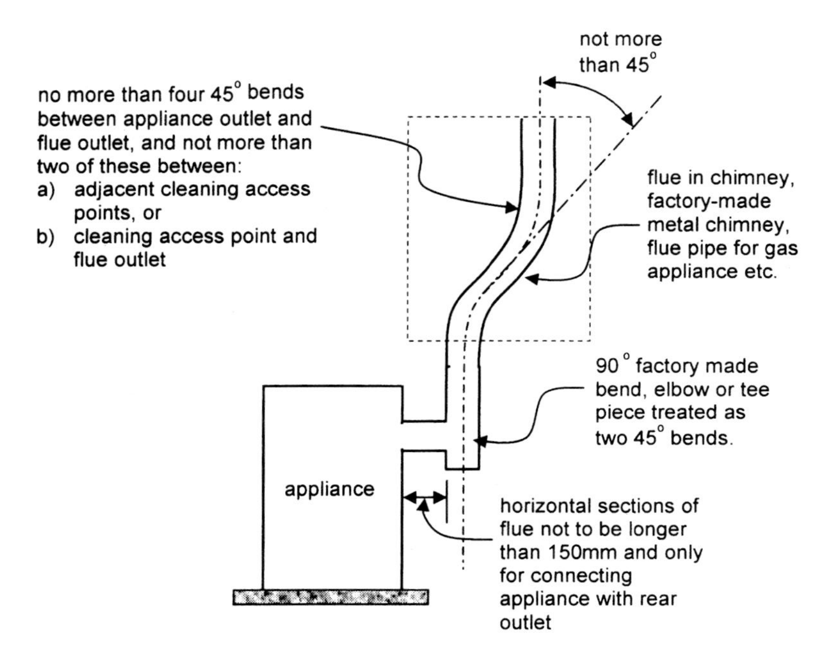

Fig. 14.6 Configuration of natural draught flues serving open-flued appliances.

It should be noted that a 90° factory-made bend, elbow or tee piece should be treated as two 45° bends.

Openings into flues for inspection and cleaning

No flue should communicate with more than one room or internal space in a building except for the purposes of:

- inspection or cleaning, or
- the fitting of a draught diverter, draught stabiliser, draught break or explosion door.

Openings for inspection and cleaning should:

- be formed using purpose factory-made components which are compatible with the flue system;
- have an access cover with the same level of gas-tightness as the flue system and an equal thermal insulation level;
- allow easy passage of the sweeping brush.

Covers should be non-combustible unless fitted to a combustible fluepipe (e.g. a plastic fluepipe). A chimney which cannot be cleaned through an appliance should be provided with:

(a) suitably sized openings for cleaning provided at a sufficient number of locations in the chimney (see Fig. 14.7); and
(b) a debris collecting space with access for emptying.

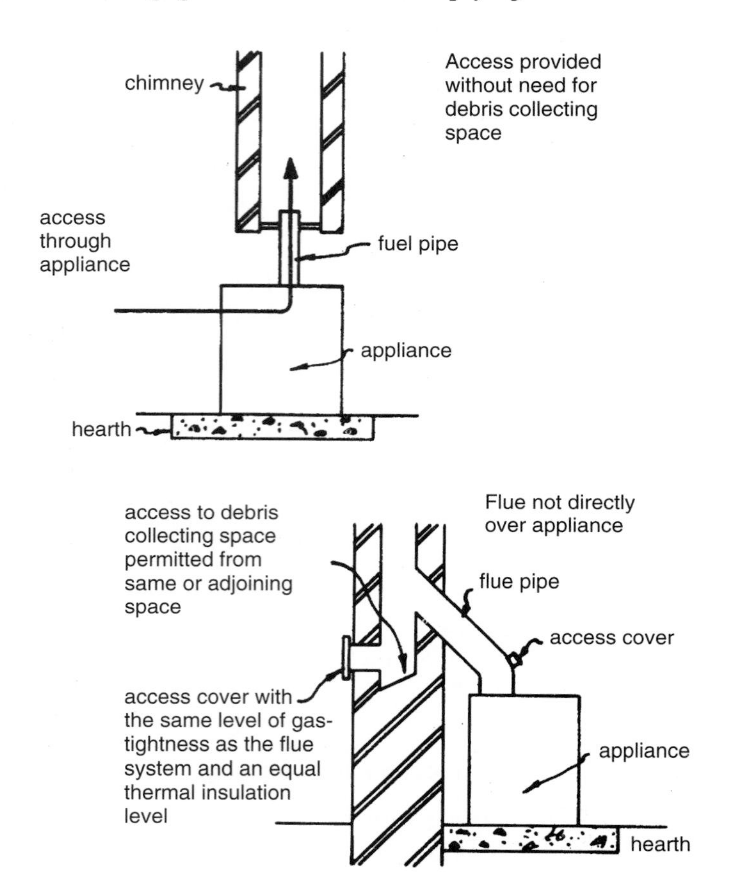

Fig. 14.7 Access for inspection and cleaning.

For appliances burning gas, the debris collection space should be provided at the base of a flue unless it is a factory-made metal chimney with a flue box, is lined or is constructed of flue blocks. To achieve this provide a space which:

- has a volume of at least 12 litres and a depth of at least 250 mm below the point where flue gases discharge into the chimney;
- is readily accessible for clearing debris, (e.g. by removing the appliance).

For radiant convector gas fires, convector heaters, fire/back boilers and inset live fuel effect fires, provide a clearance of at least 50 mm between the end of the appliance flue outlet and any surface.

Connecting fluepipes

A connecting fluepipe is used to connect a combustion appliance to a flue in a chimney. Suitable components for constructing connecting fluepipes include:

- cast iron, in accordance with BS 41: 1973 (1988) *Specification for cast iron spigot and socket flue or smoke pipes and fittings*;
- mild steel complying with BS 1449: *Steel plate, sheet and strip. Carbon and carbon manganese plate, sheet and strip*. Part 1: 1991 *General specifications*, with a flue wall thickness of at least 3 mm;
- stainless steel pipes at least 1 mm thick as described in BS EN 10088: *Stainless Steels:* Part 1: 1995. *List of stainless steels*, grades 1.4401, 1.4404, 1.4432 or 1.4436;
- vitreous enamelled steel complying with BS 6999: 1989 *Specification for vitreous enamelled low carbon steel flue pipes, other components and accessories for solid fuel burning appliances with a maximum rated output of 45 kW*;
- other fluepipes which have been independently certified to have the necessary performance designation for suitable use with the intended appliance.

Where spigot and socket fluepipes are used the sockets should be placed uppermost to contain moisture and other condensates in the flue. In order to achieve gas-tight joints it is usual either to use proprietary jointing accessories or, where appropriate, to pack the joints with noncombustible rope and fire cement.

Inspection and testing of flues

As explained in section 14.4 above, combustion installations should be inspected and tested before use to prove that they are in compliance with the Building Regulations. For flues this entails carrying out checks to show that they are:

- free from obstructions;
- satisfactorily gas-tight; and
- constructed of materials and components of the correct size to suit the intended application.

If building work also includes installation of the combustion appliance, tests can be carried out which involve firing up the appliance. These should include:

- tests for gas-tightness of the flue and at joints between the flue and the combustion appliance outlet;
- spillage tests as part of the commissioning process to check compliance with:
  - (a) Part J2 and Part L (see Chapter 16); and
  - (b) The Gas Safety (Installation and Use) Regulations 1998 (see Chapter 5).

It will normally be necessary for a suitably qualified person to prepare a report showing that the above considerations have been taken into account. An example checklist for such a report is given in section 14.4 above). For more information on methods of checking compliance with Regulation J2 see the section on *repair and re-use of existing flues* in section 14.8.3 below.

Dry lining around fireplace openings

It is usual to finish off around a fireplace opening with decorative treatment, such as a fireplace surround, masonry cladding or dry lining. Care must be taken to avoid the creation of any gaps that could allow flue gases to escape from the fireplace opening into the void behind the decorative treatment, by applying a suitable sealant at the junction. The sealant should be able to remain in place despite any relative movement between the decorative treatment and the fireplace recess.

14.6.3 Access to combustion appliances for maintenance purposes

Combustion appliances should be provided with a permanent means of access for maintenance purposes. The means of access should suit the location of the appliance, e.g. where an appliance is installed in a roof space it would probably be necessary to provide an access walkway.

14.6.4 Factory-made metal chimneys

These are defined in section 14.2 above and consist of prefabricated chimneys that are usually manufactured as sets of components for assembly on site.

Specification of factory-made metal chimneys

Where it is proposed to use component systems they should be independently certified as complying with:

- BS 4543 *Factory-made insulated chimneys*,
  (a) Part 1: 1990 (1996) *Methods of test*, (this part was withdrawn in April 2000 and partially replaced by BS EN 1859:2000);
  (b) Part 2: 1990 (1996) *Specification for chimneys with stainless steel flue linings for use with solid fuel fired appliances*; and
  (c) Part 3: 1990 (1996) *Specification for chimneys with stainless steel flue lining for use with oil-fired appliances*.

Such component systems should be installed in accordance with the relevant recommendations of:

- BS 7566: *Installation of factory-made chimneys to BS 4543 for domestic appliances*,
  (a) Part 1: 1992 (1998) *Method of specifying installation design information*;
  (b) Part 2: 1992 (1998) *Specification for installation design*;
  (c) Part 3: 1992 (1998) *Specification for site installation*;
  (d) Part 4: 1992 (1998) *Recommendations for installation design and installation*.

Gas and oil-fired appliances with flue gas temperatures not exceeding 250°C using twin wall component systems (oil-fired) or single wall component systems (gas-fired) should comply with BS 715: 1993. They should be installed in accordance with BS 5440 *Installation and maintenance of flues and ventilation for gas appliances of rated input not exceeding 70 Kw net (1st, 2nd and 3rd family gases)*, Part 1: 2000 *Specification for installation and maintenance of flues.*

Apart from the above, any other chimney system can be used provided that it is independently certified as being suitable for its purpose and it is installed in accordance with BS 7566 or BS 5440 as appropriate.

Using factory-made metal chimneys – general precautions

Where a factory-made metal chimney passes through a fire compartment wall or floor it must comply with the requirements of Part B (see Chapter 7) regarding the integrity of the compartmentation. Additionally, it may meet the requirements if:

- it is surrounded in non-combustible material having at least half the fire resistance required for the compartment wall or floor; or
- it has an appropriate level of fire resistance.

No combustible material should be placed nearer to the outer surface of the chimney than the distance (X) derived from the test procedures specified in BS 4543 *Factory-made insulated chimneys*, Part 1: 1990 (1996) *Methods of test*. The distance X will be the flue manufacturer's declared minimum distance as derived from the test. Additionally, a chimney passing through a cupboard, storage space or roof space may be separated from that space by a guard of suitable imperforate material provided that no combustible material is enclosed within the guard, and the distance between the inside of the guard and the outside of the chimney is not less than the distance (X) specified above.

Although it is not permissible to take a connecting fluepipe serving a solid fuel appliance (or an oil-fired appliance with flue gas temperatures which exceed 250°C) through a roof space, partition, internal wall or floor, it is permissible to do this for a factory-made metal chimney if the following precautions are taken:

- when passing through a wall, provide sleeves so that thermal movement is allowed to take place without damaging the flue or the building;
- do not conceal joints between chimney sections within ceiling joist spaces or walls, since this might prevent the flue from being checked for gas-tightness.

The chimney should be designed and constructed so that the appliance can be changed at a later date without the need for the chimney to be dismantled.

Like any other flue, factory-made metal chimneys should be guarded if they present a burn hazard which is not immediately apparent to people or they are so sited as to be at risk of damage.

14.6.5 Flues – specific provisions relating to appliances burning solid fuel

In addition to the general recommendations given above, appliances burning solid fuel with rated outputs up to 50 kW should comply with the following specific recommendations.

Size of flues for appliances burning solid fuel

Fluepipes and flues should:

- be at least the size shown in Table 2.2 to Approved Document J, which is reproduced below;
- have the same diameter or equivalent cross sectional area as that of the appliance flue outlet;
- not be smaller than the size recommended by the appliance manufacturer; and
- where a multifuel appliance is installed, be sized to accommodate burning the fuel that requires the largest flue.

Table 2.2 should used in conjunction with the following notes.

- A fireplace with an opening larger than 500 mm × 550 mm should be provided with a flue with a cross sectional area equal to 15% of the total face area of the fireplace opening.
- Specialist advice should be sought when proposing to construct flues having an area of:
  (a) more than 15% of the total face area of the fireplace openings; or
  (b) more than 120,000 mm^2 (0.12 m^2).
- Where a fireplace is exposed on two or more sides the area of the opening should be calculated from the formula:
  Fireplace opening area (mm^2) = L × H, where
  L = total horizontal length of fireplace opening in mm, and
  H = height of fireplace opening in mm
  For examples, see Fig. 14.8 below.

Outlets of flues for appliances burning solid fuel

Flue outlets should be located above the roof of the building where, whatever the wind conditions, the products of combustion can discharge freely and will not present a fire hazard.

Where wind exposure, surrounding tall buildings, high trees or high ground could have adverse effects on flue draught, these chimney heights and/or separations may need to be increased.

The outlet of any flue should be at least:

- 1 m above the highest point of contact between the flue and the roof, for roofs pitched at less than 10° (i.e. flat roofs);

AD J Section 2

Table 2.2 Size of flues in chimneys.

Installation (1)	Minimum flue size
Fireplace with an opening of up to 500 mm × 550 mm	200 mm diameter or rectangular/square flues having the same cross sectional area and a minimum dimension not less than 175 mm
Fireplace with an opening in excess of 500 mm × 550 mm or a fireplace exposed on two or more sides	See Paragraph 2.7. If rectangular/square flues are used the minimum dimension should not be less than 200 mm
Closed appliance of up to 20 kW rated output which: (a) burns smokeless or low volatiles fuel (2); or (b) is an appliance which meets the requirements of the Clean Air Act when burning an appropriate bituminous coal (3)	125 mm diameter or rectangular/square flues having the same cross sectional area and a minimum dimension not less than 100 mm for straight flues or 125 mm for flues with bends or offsets
Other closed appliance of up to 30 kW rated output burning any fuel	150 mm diameter or rectangular/square flues having the same cross sectional area and a minimum dimension not less than 125 mm
Closed appliance of above 30 kW and up to 50 kW rated output burning any fuel	175 mm diameter or rectangular/square flues having the same cross sectional area and a minimum dimension not less than 150 mm

Notes:
1. Closed appliances include cookers, stoves, room heaters and boilers.
2. Fuels such as bituminous coal, untreated wood or compressed paper are not smokeless or low volatiles fuels.
3. These appliances are known as 'exempted fireplaces'.

- 2.3 m measured horizontally from the roof surface for roofs pitched at 10° or more;
- 1 m above the top of any openable part of a dormer window, rooflight, or similar opening which is in a roof or external wall and is not more than 2.3 m horizontally from the top of the flue; and
- 600 mm above the top of any part of an adjoining building which is not more than 2.3 m horizontally from the top of the flue.

Additionally, if the flue passes through the roof within 2.3 m of the ridge and both slopes are at 10° or more to the horizontal, the top of the flue should be not less than 600 mm above the ridge (see Fig. 14.9).

Where flues discharge on or close to roofs with readily ignitable surfaces (e.g. roofs covered in thatch or shingles), the clearances shown in Fig. 14.9 might be insufficient to avoid a fire hazard. In such cases the outlet of any flue should be:

- at least 2.3 m measured horizontally from, and at least 1.8 m above, the roof surface;

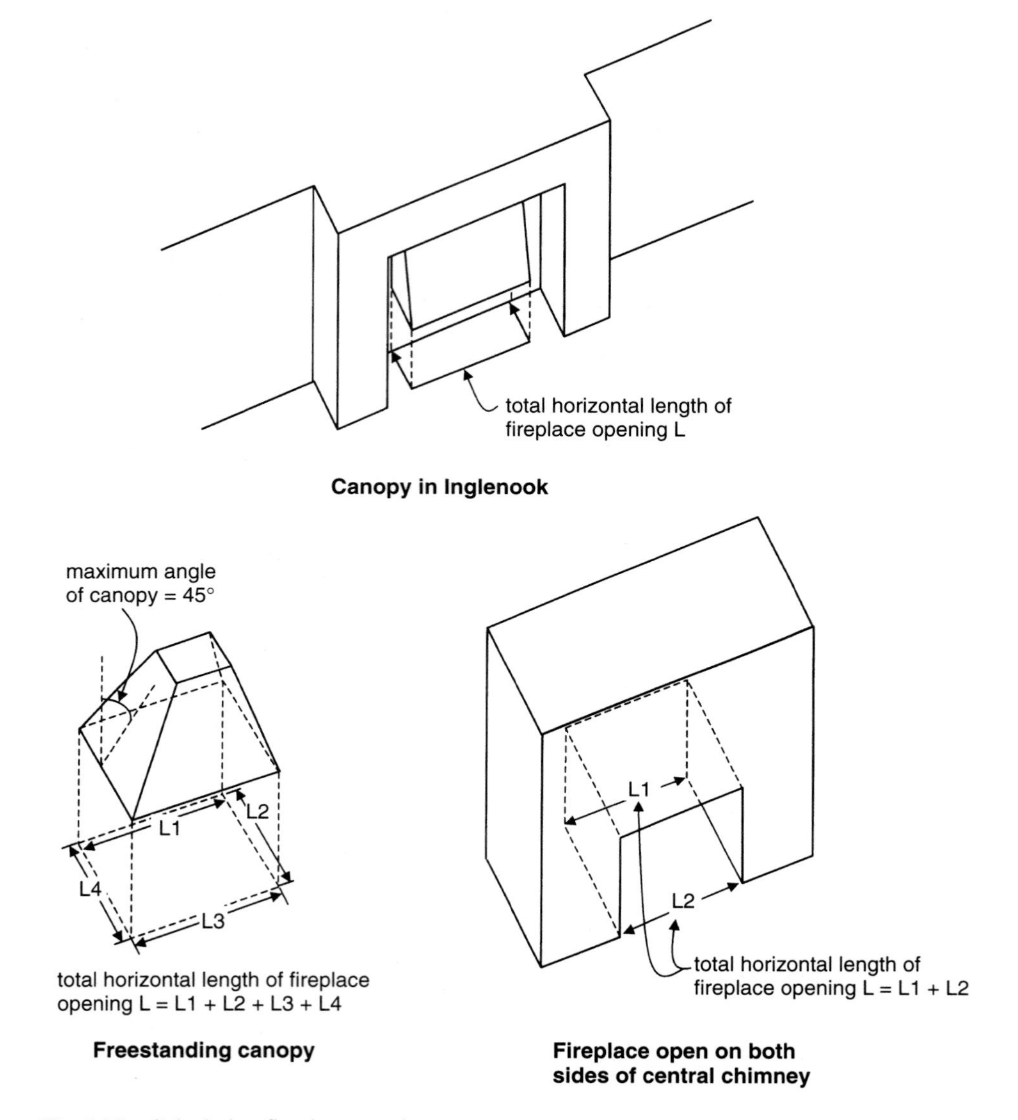

Fig. 14.8 Calculating fireplace opening.

- at least 600 mm above the ridge and at least 1800 mm vertically above the roof surface.

Fig. 14.10 gives details of these clearances.

Heights of flues for appliances burning solid fuel

The height of flue needed to ensure that sufficient draught is provided to clear the products of combustion, will depend on:

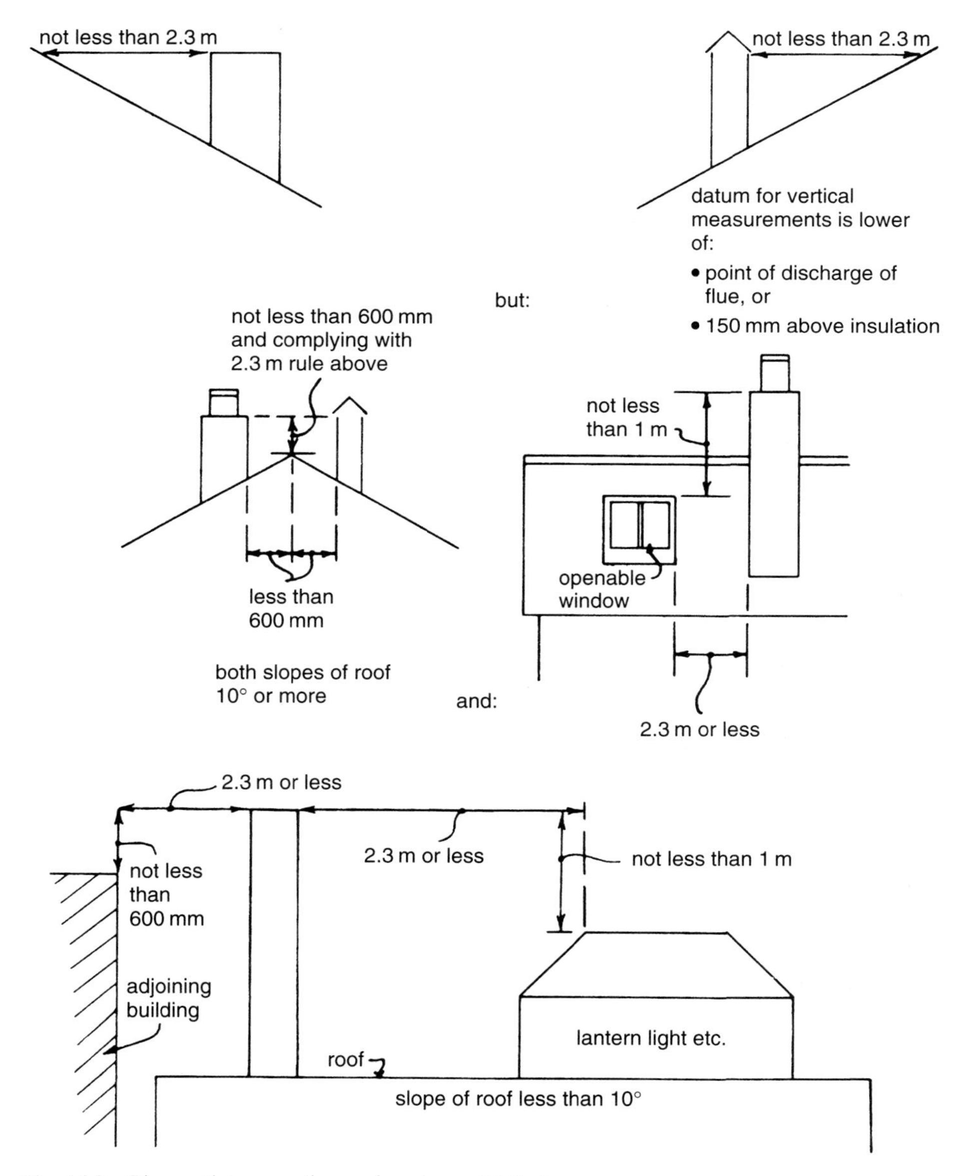

Fig. 14.9 Flue outlets – appliances burning solid fuel.

- the type of the appliance;
- the height of the building;
- the type of flue and the number of bends in it; and
- a careful assessment of local wind patterns.

It is possible that a flue height of 4.5 m could be satisfactory if the guidance on the position of flue outlets in sections 14.6.2 and 14.6.5 is followed.

Alternatively, the calculation procedure shown in BS 5854:1980 (1996): *Specifi-*

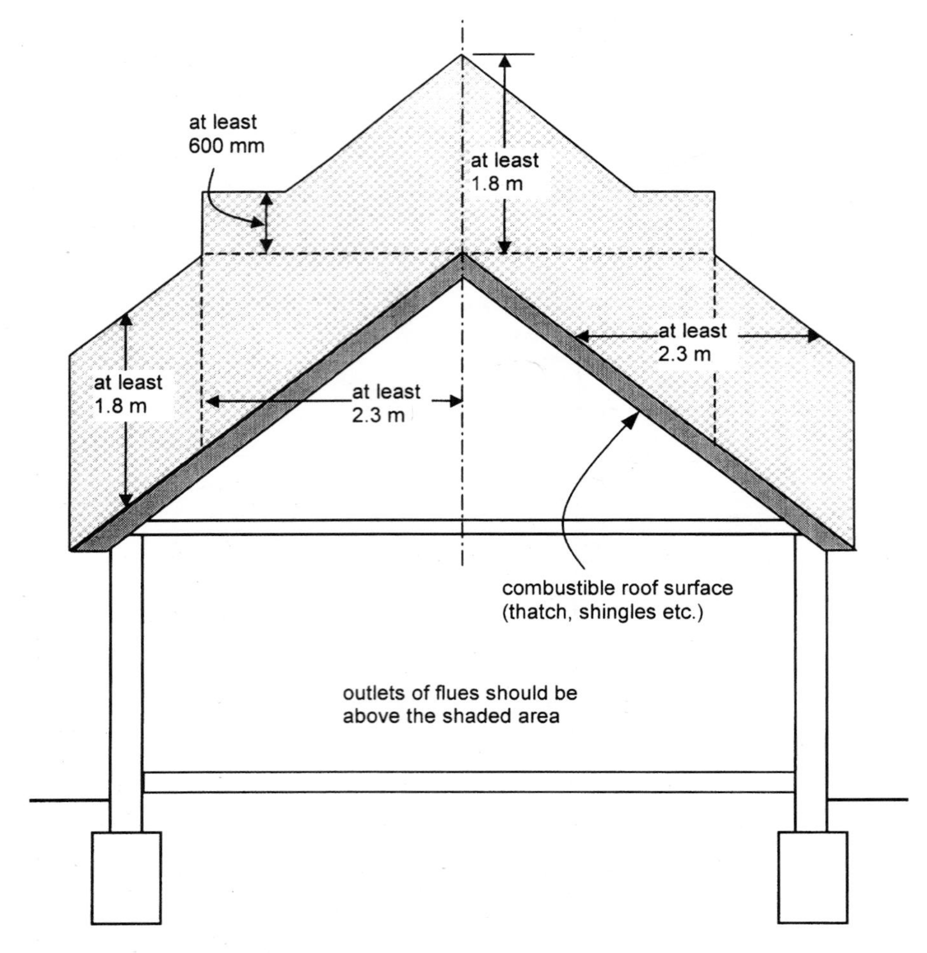

Fig. 14.10 Flue outlets from solid fuel appliances above easily ignitable roof surfaces.

cation for installation in domestic premises of gas-fired ducted-air heaters of rated input not exceeding 60 kW, AMD 8130, can be used to decide whether or not a chimney design will provide sufficient draught.

When serving an open fire, the flue height is measured vertically from the highest point at which air can enter the fireplace (e.g. the top of the fireplace opening or, for a fire under a canopy, the bottom of the canopy) to the level at which the flue discharges into the outside air.

14.6.6 Flues – specific provisions relating to appliances burning gas

In addition to the general recommendations given above, appliances burning gas with rated inputs up to 70 kW (net) should comply with the following specific recommendations.

Additionally, for new appliances of known type, satisfactory provision for chimneys and flues will be achieved by:

- using factory-made components which have been independently certified (when tested to an appropriate BS EN European chimney standard) as complying with the performance requirements corresponding to the designations given in Table 3.2 from AD J (reproduced below); and
- installing these components in accordance with the appliance and component manufacturer's installation instructions and the guidance on location and shielding of fluepipes, connecting fluepipe components, masonry and flueblock chimneys and factory-made metal chimneys given in the text below.

AD J Section 3

Table 3.2 Minimum performance designations for chimney and fluepipe components for use with new gas appliances.

Appliance type		Minimum designation (See Notes)
Boiler: open-flued	natural draught	T250 N2 O D 1
	fanned draught	T250 P2 O D 1
	condensing	T250 P2 O W 1
Boiler: room-sealed	natural draught	T250 N2 O D 1
	fanned draught	T250 P2 O D 1
Gas fire – Radiant/convector, ILFE or DFE		T300 N2 O D 1
Air heater	natural draught	T250 N2 O D 1
	fanned draught	T200 P2 O D 1
	SE – duct	T450 N2 O D 1

Notes:
1. The designation of chimney products is described in section 14.2 above. The BS EN for the product will specify its full designation and marking requirements.
2. These are default designations. Where appliance manufacturer's installation instructions specify a higher designation, this should be complied with.

Connecting DFE, ILFE and other gas appliances to flues

The building provisions needed to safely accommodate each of the main categories of gas appliance will differ for each type. Provided that the safety of the installation can be assured, it is permissible to install gas fires into fireplaces with flues designed for solid fuel appliances. Certain types of gas fire may also be installed in fireplaces with flues designed specifically for gas appliances. Reference to the Gas Appliances (Safety) Regulations 1995 will show that it is a requirement for particular combinations of appliance, flue box (if required) and flue to be selected from those stated in the manufacturer's instructions as having been shown to be safe by a Notified Body (see section 14.2 above).

Size of natural draught flues for open-flued appliances burning gas

Flues for gas-fired appliances should:

- be at least the size shown in Table 3.1 to Approved Document J (see below), where the builder is responsible for providing (or refurbishing) the flue but is not responsible for supplying the appliance;
- for a connecting fluepipe, have the same diameter or equivalent cross sectional area as that of the appliance flue outlet (the chimney flue should also have the same cross sectional area as the appliance flue outlet);
- be sized in accordance with the appliance manufacturer's installation instructions for appliances that are CE marked as complying with the Gas Appliances (Safety) Regulations 1995.

AD J Section 3

Table 3.1 Size of flues for gas fired appliances.

Intended installation	Minimum flue size	
Radiant/Convector gas fire	New flue: Circular Rectangular	 125 mm diameter 16,500 mm^2 cross sectional area with a minimum dimension of 90 mm
	Existing flue: Circular Rectangular	 125 mm diameter 12,000 mm^2 cross sectional area with a minimum dimension of 63 mm
ILFE fire or DFE fire within a fireplace opening up to 500 mm × 550 mm	Circular or Rectangular	Minimum flue dimension of 175 mm
DFE fire installed in a fireplace with an opening in excess of 500 mm × 550 mm	Calculate in accordance with Fig. 14.8 and the notes referring to similar sized openings for appliances burning solid fuel in section 14.6.5	

Outlets of flues for appliances burning gas

Flue outlets should be located externally so as to permit dispersal of the products of combustion and, for balanced flues, the intake of air. Suitable positions for outlets for both balanced and open-flued appliances are shown in Diagram 3.4. Minimum separation distances of outlets from various elements of buildings are given in the *Table to Diagram 3.4*. Both Diagram 3.4 and its accompanying Table are reproduced below. It should be noted that Diagram 3.4 and its Table are substantially the same as Figure C.1 and Table C.1 from BS 5440: *Installation and maintenance of flues and ventilation for gas appliances of rated input not exceeding 7 kW net (1st, 2nd and 3rd family gases):* Part 1: 2000 *Specification for installation and maintenance of flues*. The version of Diagram 3.4 shown in this chapter has been corrected for typographical errors found in the original in AD J.

Where a flue passes through a roof and the outlet is near to a roof window it should not be sited nearer to the window than:

- for a flat roof – 600 mm
- for a pitched roof – 600 mm if sited alongside or above, and 2000 mm if below the window.

AD J Section 3

Diagram 3.4 Location of outlets from flues serving gas appliances.

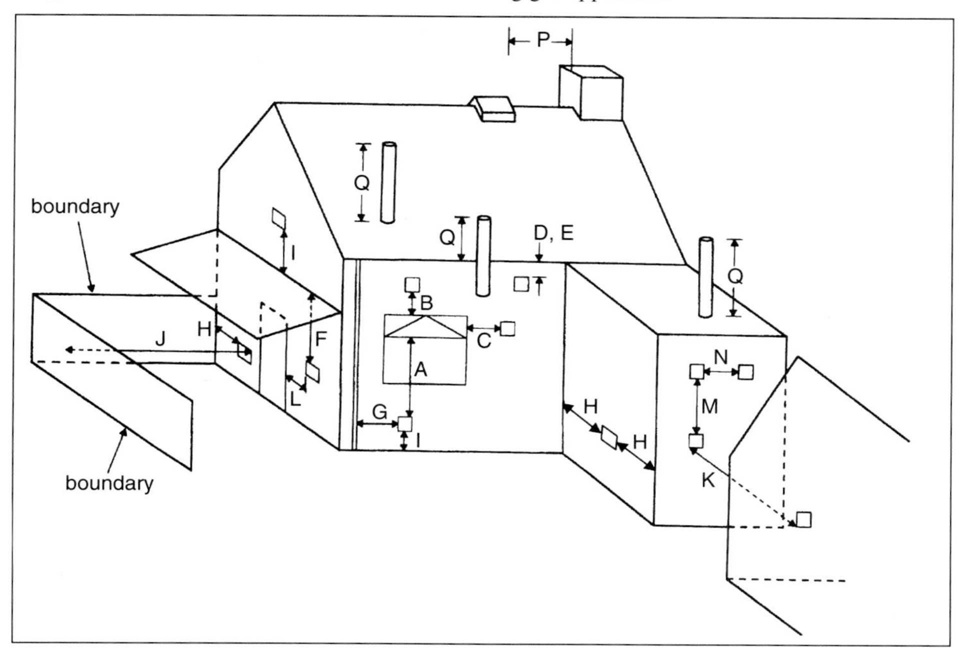

Where they are at significant risk of blockage flue outlets should be suitably protected as follows.

- For flues serving natural draught open-flued appliances consideration should be given to the following:
  - (a) Protect with a suitable outlet terminal if the flue is less than 170 mm diameter. Suitable terminals are specified in BS 715: 1993 *Specification for metal flue pipes, fittings, terminals and accessories for gas-fired appliances with a rated input not exceeding 60 kW*, AMD 8413 and BS 1289: 1986 *Flue blocks and masonry terminals for gas appliances,* Part 1: 1986 *Specification for precast concrete flue blocks and terminals.*
  - (b) For flues of over 170 mm diameter local conditions should be taken into account when assessing the risk of blockage. This could include the risk of blockage from nesting squirrels or jackdaws in areas where these creatures are prevalent. Protective cages designed for use with solid fuel appliances with mesh sizes between 6 mm and 25 mm could be used provided that the total free area of the outlet openings in the cage is at least twice the cross sectional area of the flue.
- Any flue outlet should also be protected with a guard to prevent it being damaged and to protect people who might otherwise come in contact with it.
- Flue outlets in vulnerable positions (e.g. in reach of a balcony, veranda, window

AD J Section 3

Table to Diagram 3.4 Location of outlets from flues serving gas appliances.

Minimum separation distances for terminals in mm					
Location		**Balanced flue**		**Open flue**	
		Natural draught	**Fanned draught**	**Natural draught**	**Fanned draught**
A	Below an opening (1)	Appliance rated heat input (net)	300	(3)	300
		0–7 kW 300 > 7–14 kW 600 > 14–32 kW 1500 > 32 kW 2000			
B	Above an opening (1)	0–32 kW 300 > 32 kW 600	300	(3)	300
C	Horizontally to an opening (1)	0–7 kW 300 > 7–14 kW 400 > 14 kW 600	300	(3)	300
D	Below gutters, soil pipes or drain pipes	300	75	(3)	75
E	Below eaves	300	200	(3)	200
F	Below balcony or car port roof	600	200	(3)	200
G	From a vertical drain pipe or soil pipe	300	150 (4)	(3)	150
H	From an internal or external corner or to a boundary alongside the terminal (2)	600	300	(3)	200
I	Above ground, roof or balcony level	300	300	(3)	300
J	From a surface or a boundary facing the terminal (2)	600	600	(3)	600
K	From a terminal facing the terminal	600	1200	(3)	1200
L	From an opening in the car port into the building	1200	1200	(3)	1200
M	Vertically from a terminal on the same wall	1200	1500	(3)	1500
N	Horizontally from a terminal on the same wall	300	300	(3)	300
P	From a structure on the roof	Not applicable	Not applicable	1500 mm if a ridge terminal. For any other terminal, as given in BS 5440-1:2000	N/A
Q	Above the highest point of intersection with the roof	Not applicable	Site in accordance with manufacturer's instructions	Site in accordance with BS 5440-1:2000	150

Notes:

1. An opening here means an openable element, such as an openable window, or a fixed opening such as an air vent. However, in addition, the outlet should not be nearer than 150 mm (fanned draught) or 300 mm (natural draught) to an opening into the building fabric formed for the purpose of accommodating a built in element, such as a window frame.
2. Boundary as defined in section 4.2. Smaller separations to the boundary may be acceptable for appliances that have been shown to operate safely with such separations from surfaces adjacent to or opposite the flue outlet.
3. Should not be used.
4. This dimension may be reduced to 75 mm for appliances of up to 5 kW input (net).

or from the ground) should be designed so that the entry of any matter which might restrict the flue is prevented.

Heights of natural draught flues for open-flued appliances burning gas

The height of flue needed to ensure that sufficient draught is provided to clear the products of combustion, will depend on:

- the type of the appliance;
- the height of the building;
- the type of flue and the number of bends in it; and
- a careful assessment of local wind patterns.

The requirement for sufficient flue height can be met:

- for appliances that are CE marked as complying with the Gas Appliances (Safety) Regulations 1995, by following the appliance manufacturer's installation instructions;
- for older appliances that are not CE marked (but have manufacturer's installation instructions) by following:
  - (a) the guidance in BS 5871: *Specification for installation of gas fires, convector heaters, fire/back boilers and decorative fuel effect gas appliances:* Part 3: 2001 *Decorative fuel effect gas appliances of heat input not exceeding 20 kW (2nd and 3rd family gases)*, AMD 7033, for decorative fuel effect fires; or
  - (b) the calculation procedure BS 5440: Part 1: 2000, for appliances other than decorative fuel effect fires.

Components for connecting fluepipes for appliances burning gas

Suitable components for constructing connecting fluepipes include:

- any components given in section 14.6.2 'Connecting fluepipes' above;
- sheet metal fluepipes complying with BS 715: 1993 *Specification for metal flue pipes, fittings, terminals and accessories for gas-fired appliances with a rated input not exceeding 60 kW*, AMD 8413; or
- fibre cement pipes as described in BS 7435 *Fibre cement flue pipes, fittings and terminals,* Part 1: 1991 (1998) *Specification for light quality fibre cement flue pipes, fittings and terminals:* Part 2 1991 *Specifications for heavy quality cement flue pipes, fittings and terminals*; or
- any other material or component that has been independently certified as suitable for this purpose.

Construction of flueblock chimneys for installations burning gas

In addition to the general recommendations for flueblock chimneys (see section 14.6.2 'Construction of flueblock chimneys' above), for gas fired installations, flueblock chimneys:

- should be supported and restrained in accordance with manufacturer's instructions where they are not intended to be bonded into surrounding masonry;
- may be constructed from factory-made flueblock systems (consisting of, for example, straight blocks, lintel blocks, offset blocks, transfer blocks, recess units and jointing materials) complying with:
  - (a) BS 1289: *Flue blocks and masonry terminals for gas appliances* Part 1: 1986 *Specification for Precast Concrete Flue Blocks and Terminals*, or
  - (b) BS EN 1806: 2000 (with a performance class of at least FB4 N2).

14.6.7 Flues – specific provisions relating to appliances burning oil

In addition to the general recommendations given above, appliances burning oil with rated outputs up to 45 kW should comply with the following specific recommendations.

Size of flues for appliances burning oil

Unlike previous editions of AD J, no specific minimum flue sizes are given in the 2002 edition. This could reflect the different types of oil-fired appliances that are available (and their different modes of operation, e.g. pressure jet or vaporising) since they are likely to have different discharge velocities. Therefore, in general, flues should be sized to suit the appliance being installed, so that adequate discharge velocities are achieved to prevent flow reversal problems and excessive flow resistances are avoided.

Oil-fired appliances can be connected to discharge:

- via a connecting fluepipe to a flue in a chimney or flueblock chimney; or
- to a balanced flue (or a flue designed to discharge through or adjacent to a wall).

In the first case the connecting fluepipe should be the same size as the appliance flue outlet and any flue in a chimney should have the same cross sectional area as the appliance flue outlet. If the make and model of the appliance is known when the chimney (or flueblock chimney) is being designed and built then the flue can be made the same size as the appliance flue outlet. Otherwise, the flue can be made large enough to allow the later insertion of a suitable flexible flue liner matching the appliance to be installed.

In all cases, the flue size should always be provided to suit the appliance manufacturers' installation instructions.

Outlets of flues for appliances burning oil

Flue outlets should be located externally so as to permit:

- dispersal of the products of combustion;
- correct operation of a natural draught flue; and
- for balanced flues, the intake of air.

Suitable positions for outlets for both balanced and open-flued appliances are shown in Diagram 4.2. Minimum separation distances of outlets from various elements of buildings are given in the Table to Diagram 4.2. Both Diagram 4.2 and its accompanying Table are reproduced below.

AD J Section 4

Diagram 4.2 Location of outlets from flues serving oil-fired appliances.

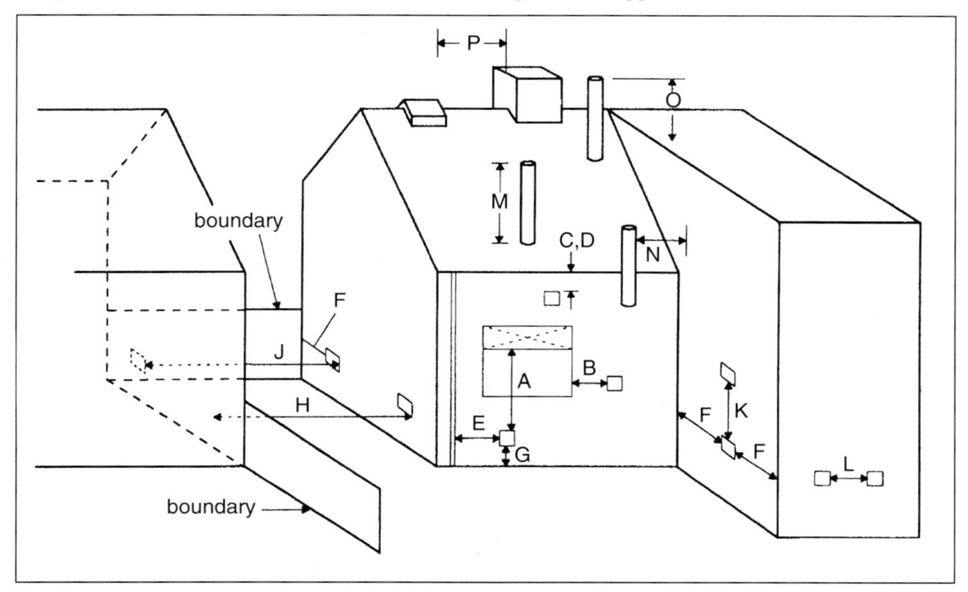

The minimum separation distances given in Table to Diagram 4.2 may need to be increased where:

- local factors such as wind patterns, might disrupt operation of the flue; or
- the height of a natural draught flue may be insufficient to disperse the products of combustion from an open-flued appliance.

Flue outlets should be protected with guards to prevent them being damaged and to protect people who might otherwise come in contact with them. Flue outlets in vulnerable positions (e.g. in reach of a balcony, veranda, window or from the ground) should be designed so that the entry of any matter which might restrict the flue is prevented.

Appliances burning oil – the effect of flue gas temperatures on the design of chimneys and fluepipes

In order to provide a satisfactory chimney or fluepipe for an oil-fired appliance it is necessary to establish whether the flue gas temperature is above or below 250°C as measured by a suitable method (e.g. as in *OFTEC Standards A 100 or A 101*).

AD J Section 4

Table to Diagram 4.2 Location of outlets from flues serving oil-fired appliances.

Minimum separation distances for terminals in mm			
Location of outlet (1)		**Appliance with pressure jet burner**	**Appliance with vaporising burner**
A	Below an opening (2, 3)	600	should not be used
B	Horizontally to an opening (2, 3)	600	should not be used
C	Below a plastic/painted gutter, drainage pipe or eaves if combustible material protected (4)	75	should not be used
D	Below a balcony or a plastic/painted gutter, drainage pipe or eaves without protection to combustible material	600	should not be used
E	From vertical sanitary pipework	300	should not be used
F	From an external or internal corner or from a surface or boundary alongside the terminal	300	should not be used
G	Above ground or balcony level	300	should not be used
H	From a surface or boundary facing the terminal	600	should not be used
J	From a terminal facing the terminal	1200	should not be used
K	Vertically from a terminal on the same wall	1500	should not be used
L	Horizontally from a terminal on the same wall	750	should not be used
M	Above the highest point of an intersection with the roof	600 (6)	1000 (5)
N	From a vertical structure to the side of the terminal	750 (6)	2300
O	Above a vertical structure which is less than 750 mm (pressure jet burner) or 2300 mm (vaporising burner) horizontally from the side of the terminal	600 (6)	1000 (5)
P	From a ridge terminal to a vertical structure on the roof	1500	should not be used

Notes:

1. Terminals should only be positioned on walls where appliances have been approved for such configurations when tested in accordance with BS EN 303-1:1999 or OFTEC standards OFS A100 or OFS A101.
2. An opening means an openable element, such as an openable window, or a permanent opening such as a permanently open air vent.
3. Notwithstanding the dimensions above, a terminal should be at least 300 mm from combustible material, e.g. a window frame.
4. A way of providing protection of combustible material would be to fit a heat shield at least 750 mm wide.
5. Where a terminal is used with a vaporising burner, the terminal should be at least 2300 mm horizontally from the roof.
6. Outlets for vertical balanced flues in locations M, N and O should be in accordance with manufacturer's instructions.

As a guide to establishing the likely flue gas temperatures for a particular appliance the following notes may prove useful.

- Since flue gas temperatures depend on the appliance type and its age, older and re-used appliances will probably have flue gas temperatures in excess of 250°C.
- Modern appliances bearing the CE Mark (showing compliance with the *Boiler (Efficiency) Regulations 1993*) will usually have flue gas temperatures which do not exceed 250°C.
- Manufacturers of appliances should be able to supply information on flue gas temperatures for individual appliances, in their installation instructions.
- The Oil Firing Technical Association for the Petroleum Industry (OFTEC) can be contacted to obtain information on individual appliances. They may be contacted at Century House, 100 High Street, Banstead, Surrey, SM7 2NN.

Where information is unobtainable, the flue gas temperature for a particular appliance should be assumed to be greater than 250°C.

Design of flues for oil-fired appliances with flue gas temperatures exceeding 250°C

Where the flue gas temperature exceeds 250°C, a satisfactory design for the flue could be achieved by discharging the appliance:

- via a connecting fluepipe, masonry or flueblock chimney designed for use with solid fuel;
- to a suitable factory-made metal chimney designed for flue gas temperatures exceeding 250°C (see section 14.7.3 below);
- via other products which have been independently certified as being suitable for this purpose.

Design of flues for oil-fired appliances with flue gas temperatures not exceeding 250°C

Where the flue gas temperature does not exceed 250°C, a satisfactory design for the flue could be achieved:

- for any appliances (new or existing), by:
  - (a) following the guidance for installations where the flue gas temperature exceeds 250°C listed above; and
  - (b) the provisions for location and shielding of fluepipes and connecting fluepipe components, outlined in section 14.7.3 below; or
- for new appliances of known type, by:
  - (a) using factory-made components which have been independently certified (when tested to an appropriate BS EN European chimney standard) as complying with the performance requirements corresponding to the designations given in Table 4.1 from AD J (reproduced below); and

AD J Section 4

Table 4.1 Minimum performance designations for chimneys and fluepipe components for use with new oil-fired appliances with flue gas temperature less than 250°C.

Appliance type	Minimum designation (See Notes)	
Boiler – pressure jet (including combination)	T160 P2 O D 1	Class C2 oil
Boiler – condensing	T160 P2 O D 2	Class D oil
Cooker – pressure jet		
Cooker – vaporising burner	T160 N2 O D 1	Class C2 oil
Room heater – vaporising burner	T160 N2 O D 2	Class D oil

Notes:
3. The designation of chimney products is described in section 14.2 above. The BS EN for the product will specify its full designation and marking requirements.
4. These are default designations. Where appliance manufacturer's installation instructions specify a higher designation, this should be complied with.

(b) installing these components in accordance with the appliance and component manufacturer's installation instructions and the guidance on location and shielding of fluepipes, connecting fluepipe components, flueblock chimneys and factory-made metal chimneys given in section 14.7.3 below.

Components for connecting fluepipes for appliances burning oil

Suitable components for constructing connecting fluepipes include:

- any components given in section 14.6.2 'Connecting fluepipes' above;
- any components given in section 14.6.5 'Connecting fluepipes' above; or
- any other material or component that has been independently certified as suitable for this purpose.

Construction of flueblock chimneys for installations burning oil

In addition to the general recommendations for flueblock chimneys (see section 14.6.2 'Construction of flueblock chimneys' above), for oil fired installations, flueblock chimneys:

- should be supported and restrained in accordance with manufacturer's instructions where they are not intended to be bonded into surrounding masonry;
- may be constructed from factory-made flueblock systems (consisting of, for example, straight blocks, lintel blocks, offset blocks, transfer blocks, recess units and jointing materials) complying with:
  (a) BS 1289: *Flue blocks and masonry terminals for gas appliances* Part 1: 1986 *Specification for precast concrete flue blocks and terminals*, or
  (b) BS EN 1806: 2000 (with a performance at least equal to the Table 4.1 of AD J Section 4 above, designations for oil-fired appliances).

14.7 Protection of building against fire and heat

Paragraph J3 of Part J requires that the construction of fireplaces and chimneys and the installation of combustion appliances and fluepipes must be carried out so as to reduce to a reasonable level the risk of people suffering burns or the building catching fire in consequence of their use.

In general, this means that the combustion installation must enable normal operation of the appliances without danger being caused through damage to the fabric of the building by heat or fire. Additionally, the combustion installation should undergo inspection and testing to ensure that it is suitable for its intended purpose.

Therefore, there are implications for the design and construction of hearths, fireplaces, chimneys and flues to ensure that they are:

- of sufficient size
- constructed of suitable materials
- suitably isolated from any adjacent combustible materials.

14.7.1 Protection of the building against fire and heat – appliances burning solid fuel

Hearths for appliances burning solid fuel – general provisions

In most cases constructional hearths should be provided where an appliance burning solid fuel is to be installed. However, where it can be independently certified that the appliance cannot cause the temperature of the hearth to exceed 100°C (and the appliance is not designed to stand in an appliance recess), it is possible to provide a hearth made of non-combustible board/sheet material or tiles at least 12 mm thick.

Constructional hearths should be constructed of solid non-combustible material at least 125 mm thick (including the thickness of any non-combustible floor under the hearth).

Constructional hearths built in connection with a fireplace recess should:

- extend within the recess to the back and jambs of the recess;
- project at least 500 mm in front of the jamb; and
- extend outside the recess to at least 150 mm beyond each side of the opening.

If not built in connection with a fireplace recess, the plan dimensions of the hearth should be such as to accommodate a square of at least 840 mm. See Fig. 14.11 for details.

Hearths for appliances burning solid fuel – proximity of combustible materials

Hearths are provided to prevent combustion appliances setting fire to the building fabric and furnishings and to limit the risk of people being accidentally burnt.

Minimum plan dimensions

with fireplace recess

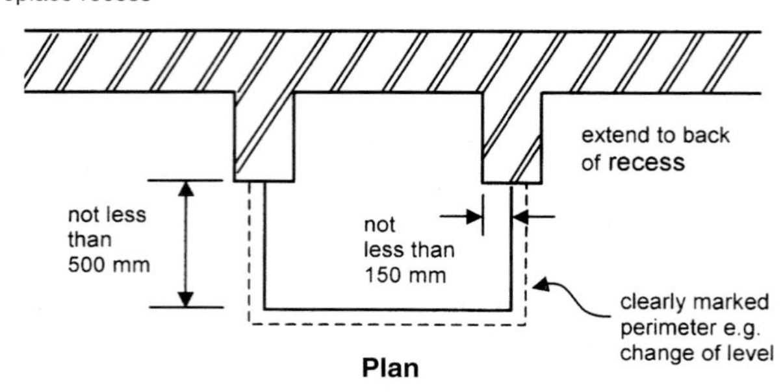

without fireplace recess

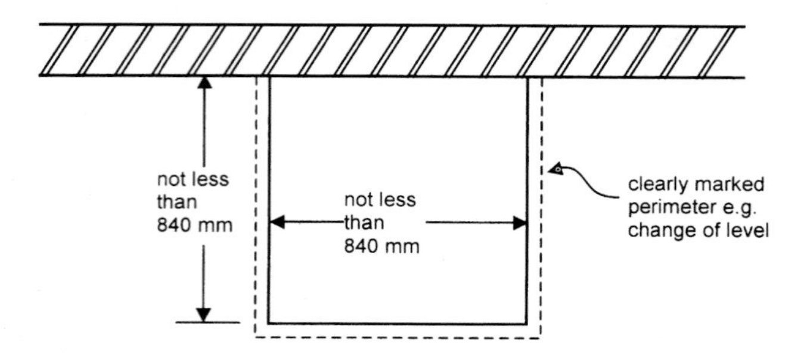

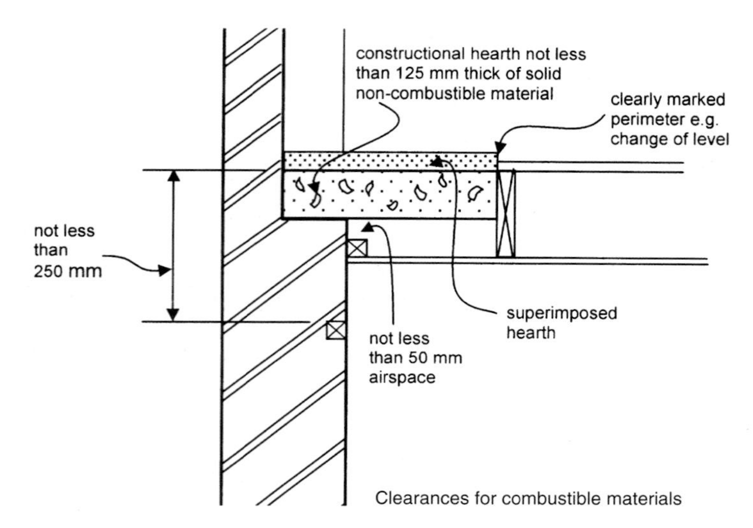

Fig. 14.11 Hearths for appliances burning solid fuel.

Therefore, they should be separated from adjacent combustible materials and should be satisfactorily delineated from surrounding floor finishes (carpets etc.) as follows.

- Combustible material should not be placed under a constructional hearth for a solid fuel appliance within a vertical distance of 250 mm from the upper surface of the hearth, unless there is an airspace of at least 50 mm between the combustible material and the underside of the hearth (see Fig. 14.11).
- Where a superimposed hearth has been placed onto a constructional hearth, combustible material placed on or beside the constructional hearth should not extend under the superimposed hearth by more than 25 mm or closer to the appliance than 150 mm (see Fig. 14.12).
- Ensure that the hearth (superimposed or constructional) is suitably delineated to discourage combustible floor finishes from being laid too close to the appliance, by marking the edges or providing a change of level.
- Position the appliance on the hearth such that combustible material cannot be laid closer to the base of the appliance than:
  - (a) *at the front*, 300 mm if the appliance is an open fire or stove which can, when opened, be operated as an open fire, or 225 mm in any other case;
  - (b) *at the back and sides*, 150 mm or in accordance with the recommendations below which relate to distance from hearth to walls.

If any part of the back or sides of the appliance lies within 150 mm horizontally of the wall, then the wall should be of solid non-combustible construction at least 75 mm thick from floor level to a level of 300 mm above the top of the appliance and 1200 mm above the hearth.

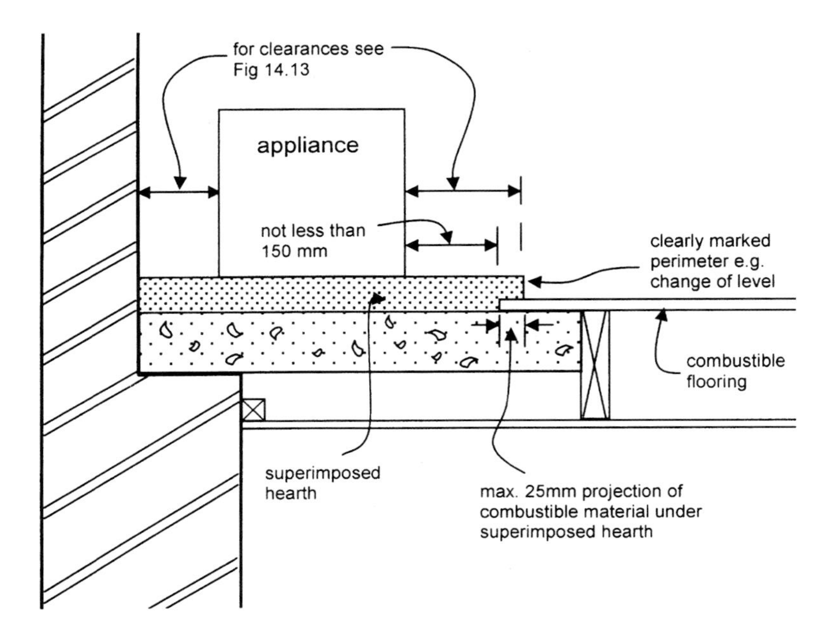

Fig. 14.12 Appliances burning solid fuel – superimposed hearths.

If, however, any part of the back or sides of the appliance lies within 50 mm of the wall, then the wall should be of solid non-combustible construction at least 200 mm thick from floor level to a level of 300 mm above the top of the appliance and 1200 mm above the hearth (see Fig. 14.13). Where the hearth itself is at least 150 mm from an adjacent wall there is no requirement for protection of the wall. It should be noted that these thicknesses of solid non-combustible material can be substituted by thinner material if the same overall level of protection can be achieved.

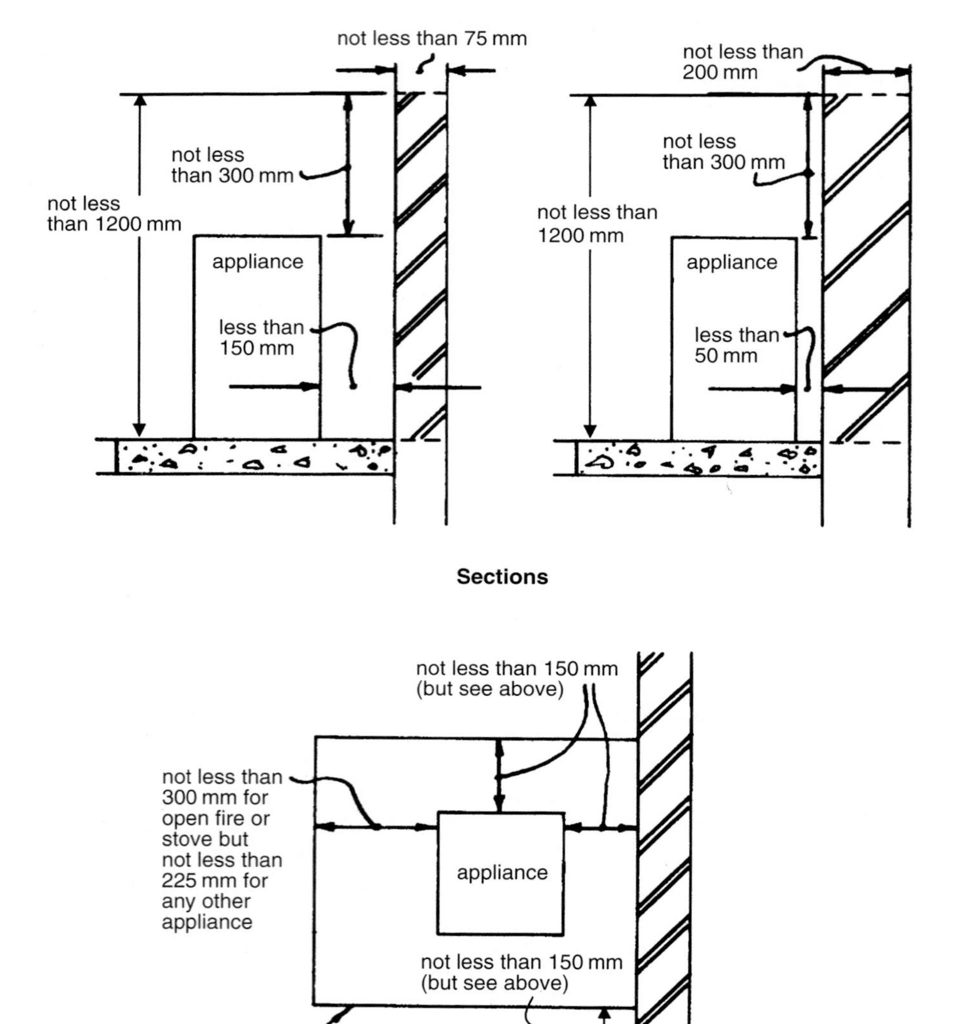

Fig. 14.13 Positioning of appliances burning solid fuel.

Fireplace recesses for appliances burning solid fuel – general provisions

Fireplace recesses need to be constructed so that they protect the building from the risk of fire. Traditionally, fireplace recesses have been constructed in masonry or concrete and guidance on these forms is still available in AD J. It is also possible to construct prefabricated factory-made appliance chambers and the 2002 edition of AD J contains guidance on these for the first time.

Fireplace recesses constructed in masonry or concrete should have a jamb on each side at least 200 mm thick, a solid back wall at least 200 mm thick, or a cavity wall back with each leaf at least 100 mm thick. These thicknesses are to run the full height of the recess. However, if a fire-place recess is in an external wall the back may be a solid wall of not less than 100 mm thickness. Similarly, if part of a wall acts as the back of two recesses on opposite sides of the wall, it may be a solid wall not less than 100 mm thick. It is assumed that this latter exemption does not apply to a wall separating buildings or dwellings within a building since the requirements for chimney walls (see below) specify a minimum thickness of 200mm in these circumstances (see Fig. 14.14).

Fireplace recesses constructed from prefabricated factory-made appliance chambers use components made from insulating concrete with a density between 1200 kg/m^2 and 1700 kg/m^2. AD J lays down the minimum thicknesses for the various components as follows:

- base – 50 mm
- side sections (forming walls on each side if the chamber) – 75 mm
- back section (forming the rear of the chamber) – 100 mm
- top slab, lintel or gather (forming the top of the chamber) – 100 mm.

These components should be supplied as sets which must be assembled and jointed in accordance with the manufacturer's instructions.

Fireplace recesses for appliances burning solid fuel – linings

In most cases a fireplace recess will need to be lined so that it is able to provide an acceptable setting for the installation of an appliance (such as an inset open fire). There are a great many lining components on the market and one example of a traditional fireplace lining and throat forming lintel is shown in Fig. 14.15. In the example, the tapered gather is formed by a throat forming lintel. Gathers are needed to ensure the proper working of the flue and can be formed in other ways, such as by:

- using a combined prefabricated lintel and gather unit built into the fireplace recess; or
- traditional corbelling of the masonry; or
- using a prefabricated appliance chamber incorporating the gather; or
- using a suitable canopy (see Fig. 14.8 above).

Recesses built in masonry

Fireplace recesses generally

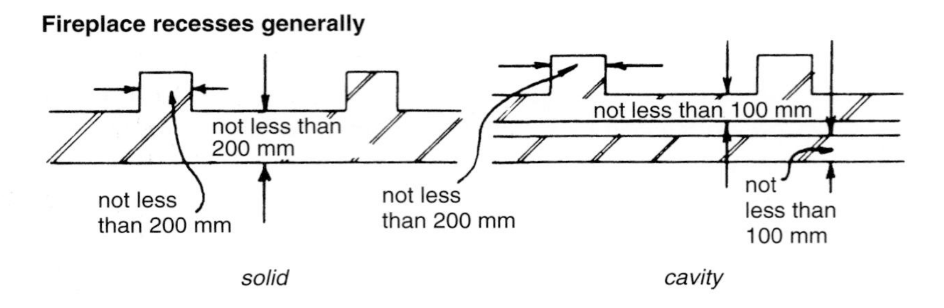

Recesses in external wall

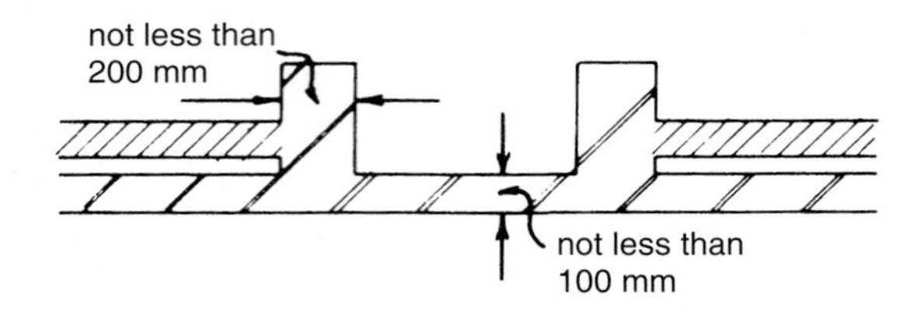

Back-to-back recesses

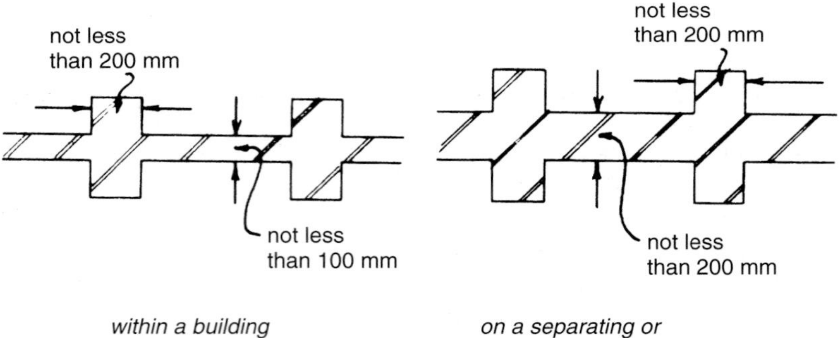

Fig. 14.14 Fireplace recesses – appliances burning solid fuel.

Fireplace recesses for appliances burning solid fuel – proximity of combustible material

Combustible materials should not be placed where they could be ignited by heat dissipating through the walls of the fireplace recess. This means that combustible material should be at least:

- 200 mm from the *inside* surface of the fireplace recess; or
- 40 mm from the *outside* surface (although this does not apply to floorboards, skirting boards, dado and picture rails, mantelshelves and architraves).

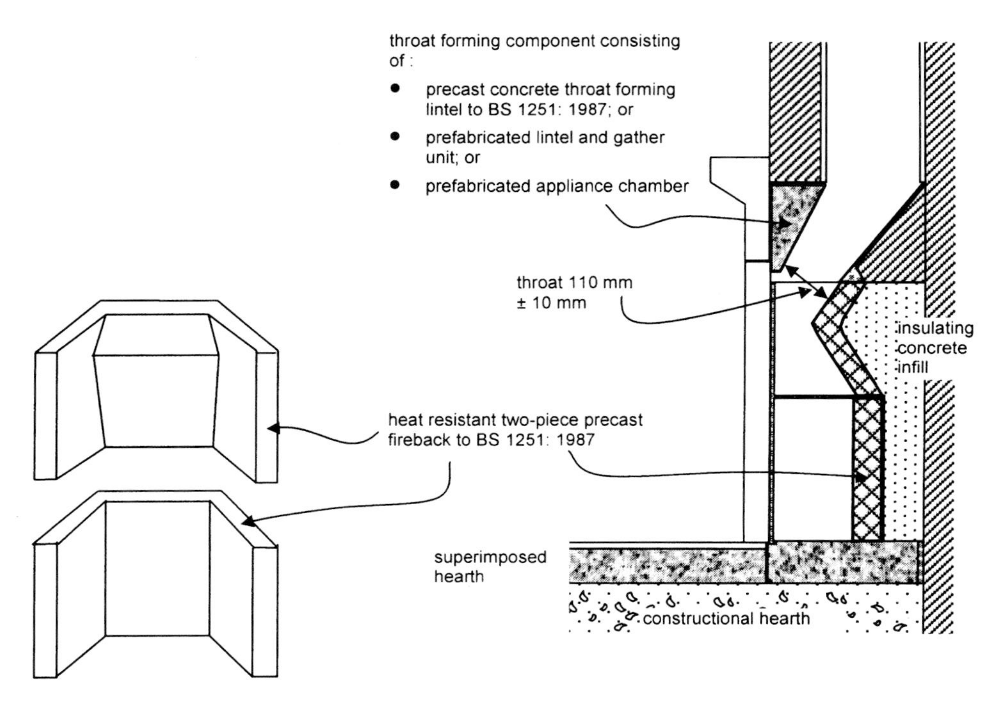

Fig. 14.15 Lining and fireplace components to open fireplace burning solid fuel.

Masonry and flueblock chimneys for appliances burning solid fuel

A masonry chimney provides structural support to suitably caulked flue liners and flueblocks, and fire protection to the building, by being constructed of bricks, medium weight concrete blocks or stone set in suitable mortar. The thickness of the masonry will vary with the type of fuel. Therefore, for appliances burning solid fuel, if a chimney is built of masonry and is lined as described above, any flue in that chimney (including a flue composed of flueblocks) should be:

- surrounded and separated from any other flue in that chimney by at least 100 mm thickness of solid masonry material, excluding the thickness of any flue lining material;
- separated by at least 200 mm of solid masonry material from another compartment of the same building, another building or another dwelling;
- separated by at least 100 mm of solid masonry material from the outside air (see Fig. 14.16).

Masonry and flueblock chimneys for appliances burning solid fuel – proximity of combustible material

Combustible materials should not be placed where they could be ignited by heat dissipating through the walls of flues. This means that combustible material should be at least:

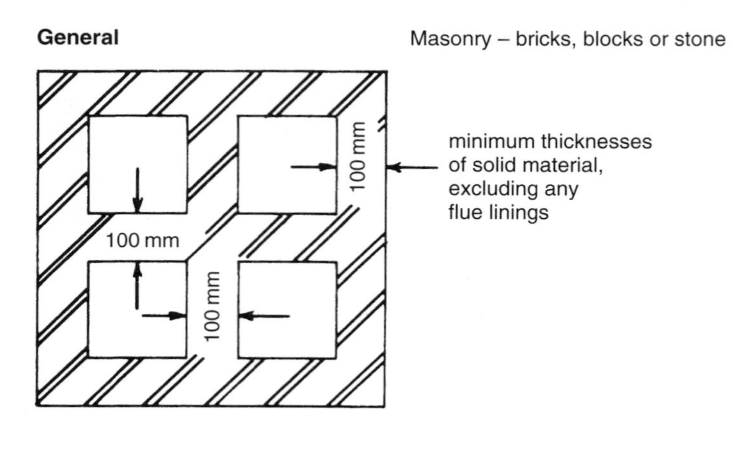

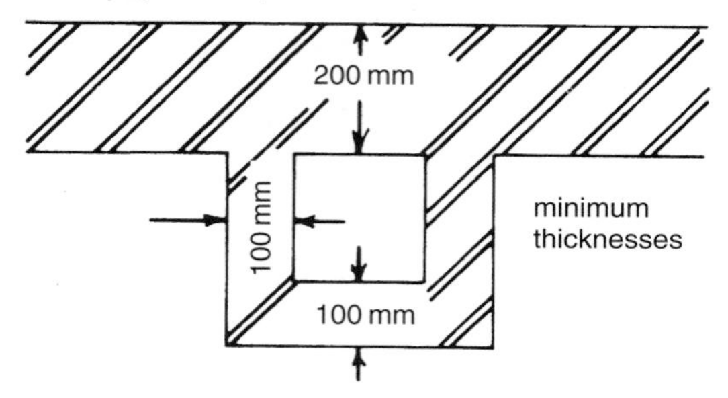

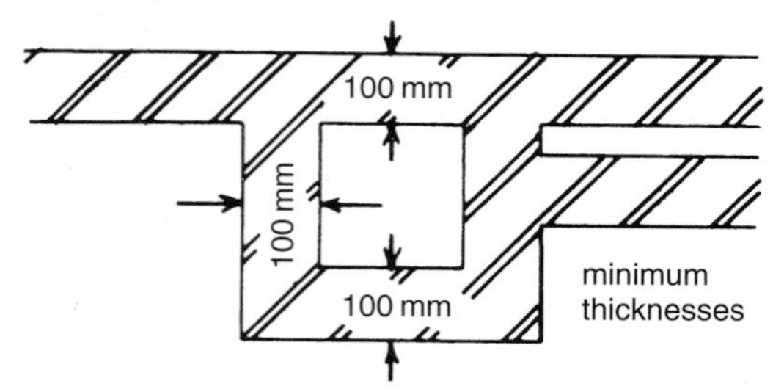

Fig. 14.16 Masonry and flueblock chimneys – wall thicknesses.

- 200 mm from the *inside* surface of a flue; or
- 40 mm from the *outside* surface of the chimney (although this does not apply to floorboards, skirting boards, dado and picture rails, mantelshelves and architraves).

No metal fastening in contact with combustible material should be placed within 50 mm of the inside of the flue (see Fig. 14.17).

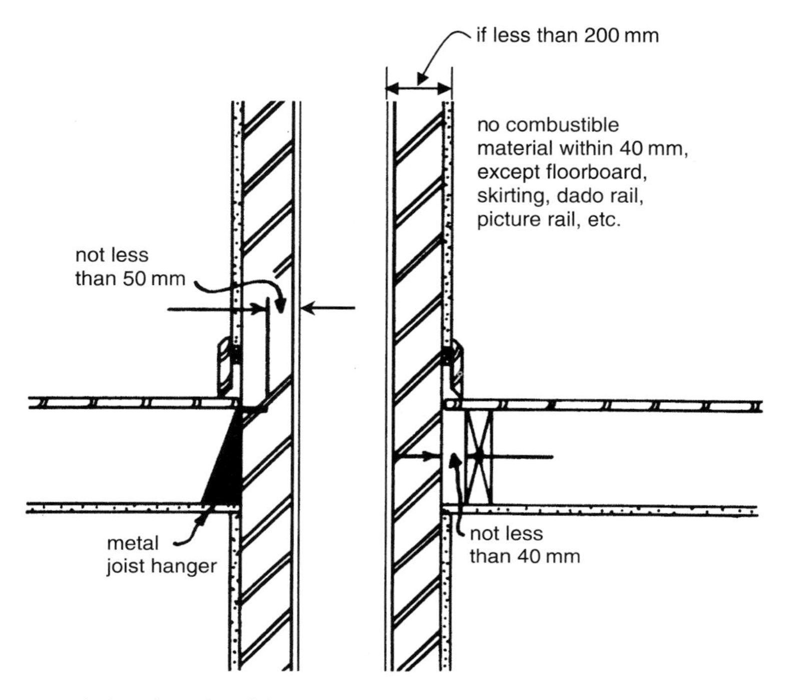

Fig. 14.17 Proximity of combustible materials – appliances burning solid fuel.

Connecting fluepipes serving appliances burning solid fuel

Connecting fluepipes serving appliances burning solid fuel should not pass through any roof space, partition, internal wall or floor (although they are allowed to pass through the wall of a chimney and a floor supporting a chimney!) and should only be used to connect an appliance to a chimney (see Fig. 14.18). The obvious intention is that connecting fluepipes should be as short as possible and should only be used to connect an appliance to a proper chimney. They should also be guarded if:

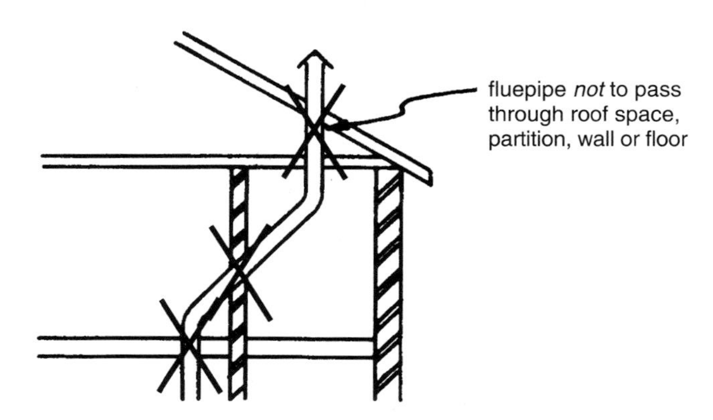

Fig. 14.18 Connecting the fluepipes for appliances burning solid fuel.

- the burn hazard they present is not immediately apparent to people; or
- they are so sited as to be at risk of damage.

A horizontal connection is permitted to connect a back outlet appliance to a chimney but this should not exceed 150 mm in length (see Fig. 14.6).

Placing and shielding of connecting fluepipes for appliances burning solid fuel

A connecting fluepipe should be located so that it does not risk igniting combustible material. Horizontal and sloping runs should be limited in length and where the connecting fluepipe is adjacent to combustible material, the following examples show how the risk could be minimised.

- By allowing sufficient separation, e.g. setting a minimum distance from any combustible material forming part of the wall or partition of at least:
  - (a) three times the external diameter of the pipe; or
  - (b) where the pipe is insulated (at least 12 mm of insulation with thermal conductivity not more than 0.065 W/mK) at least 0.75 times the outside diameter of the insulated pipe.
- By a combination of separation and shielding so that the pipe is at least 1.5 times its external diameter from the combustible material, and a shield of non-combustible material is placed so that there is an airspace of at least 12 mm between the shield and the combustible material. The non-combustible shield may either:
  - (a) extend past the fluepipe by at least 1.5 times its external diameter; or
  - (b) be positioned to ensure that the minimum distance from the fluepipe to any combustible material is at least three times the pipe diameter (see Fig. 14.19).
- By using a factory-made metal chimney and following the guidance in section 14.6.4 above.

14.7.2 Protection of the building against fire and heat – appliances burning gas

Hearths for appliances burning gas

The decision as to the type of hearth to provide (or indeed whether one is needed at all) will depend on the type of appliance installed. For example, a hearth is not required:

- if the appliance is installed so that no part of any flame or incandescent material is less than 225 mm above the floor; or
- the manufacturer's instructions state that a hearth is not required.

For a back boiler behind a gas fire, the hearth should be constructed of solid non-combustible material:

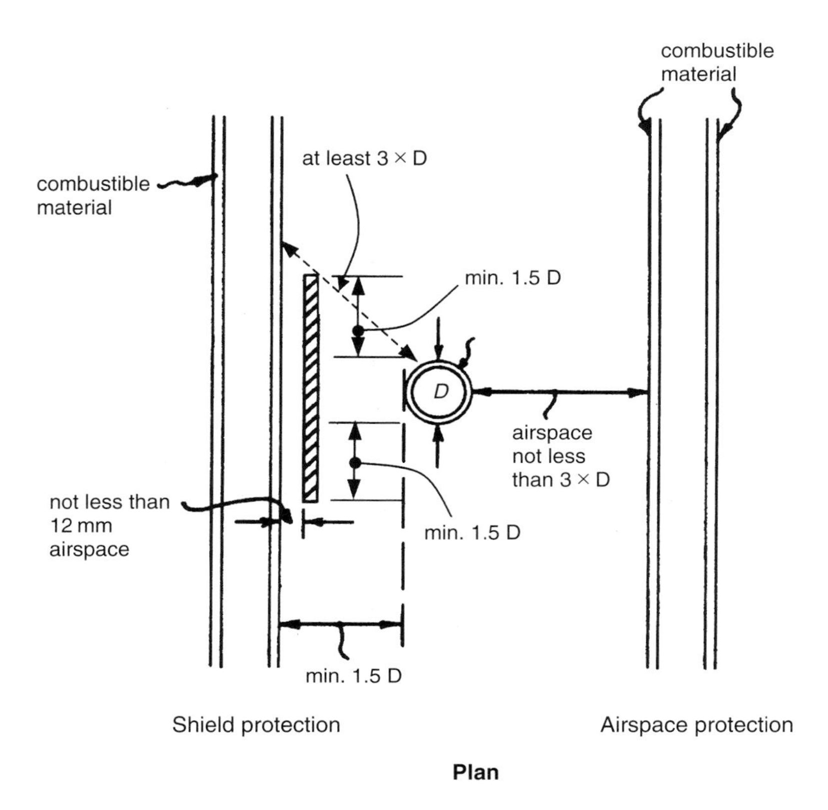

Fig. 14.19 Protection of combustible materials next to uninsulated fluepipes for appliances burning solid fuel.

- at least 125 mm thick (i.e. as for solid fuel appliances); or
- at least 25 mm thick on 25 mm non-combustible supports.

It should extend at least 150 mm beyond the back and sides of the back boiler and extend forward in front as required for the type of fire fitted (see Fig. 14.20).

The hearth for a decorative fuel effect fire (DFE) or inset live fuel effect fire (ILFE) should consist of a top layer of non-combustible, non-friable material at least 12 mm thick. The extent of any projections will depend on the type of appliance and whether it is freestanding or situated in a fireplace recess (see Fig. 14.21). The edges of the hearth should be designed to be apparent to building occupiers so as to discourage the laying of combustible floor finishes too close to the appliance. This could be achieved by providing a change in levels.

Shielding of appliances burning gas

Gas-fired appliances should be positioned so that the possibility of accidental contact is minimised, and separated from combustible materials as shown in Fig. 14.22 by a non-combustible surface such as:

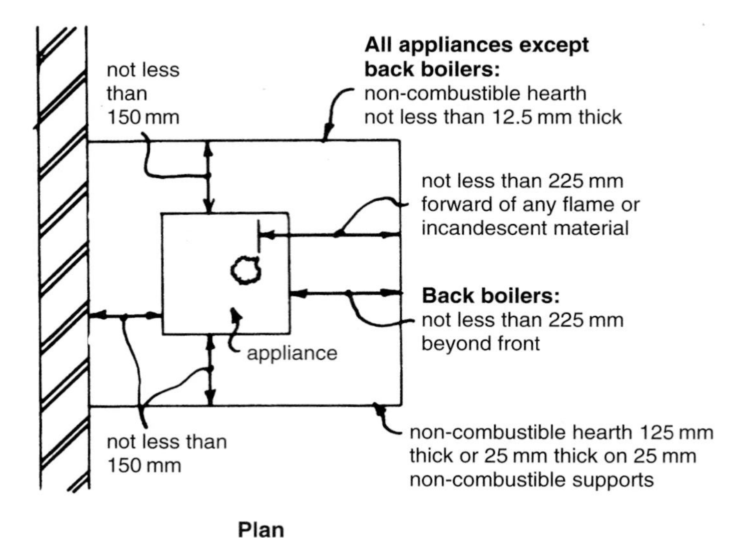

Unless:

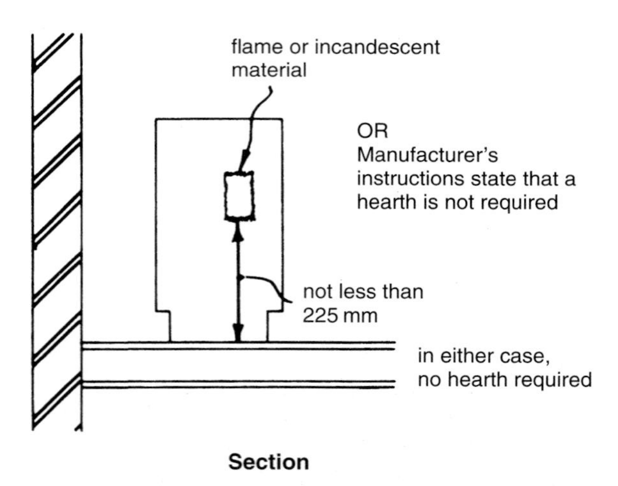

Fig. 14.20 Hearths for appliances burning gas.

- a shield of non-combustible material at least 25 mm thick, or
- at least 75 mm airspace.

Alternatively, for appliances that are CE marked as compliant with the Gas Appliances (Safety) Regulations 1995 the manufacturer's instructions regarding shielding should be followed.

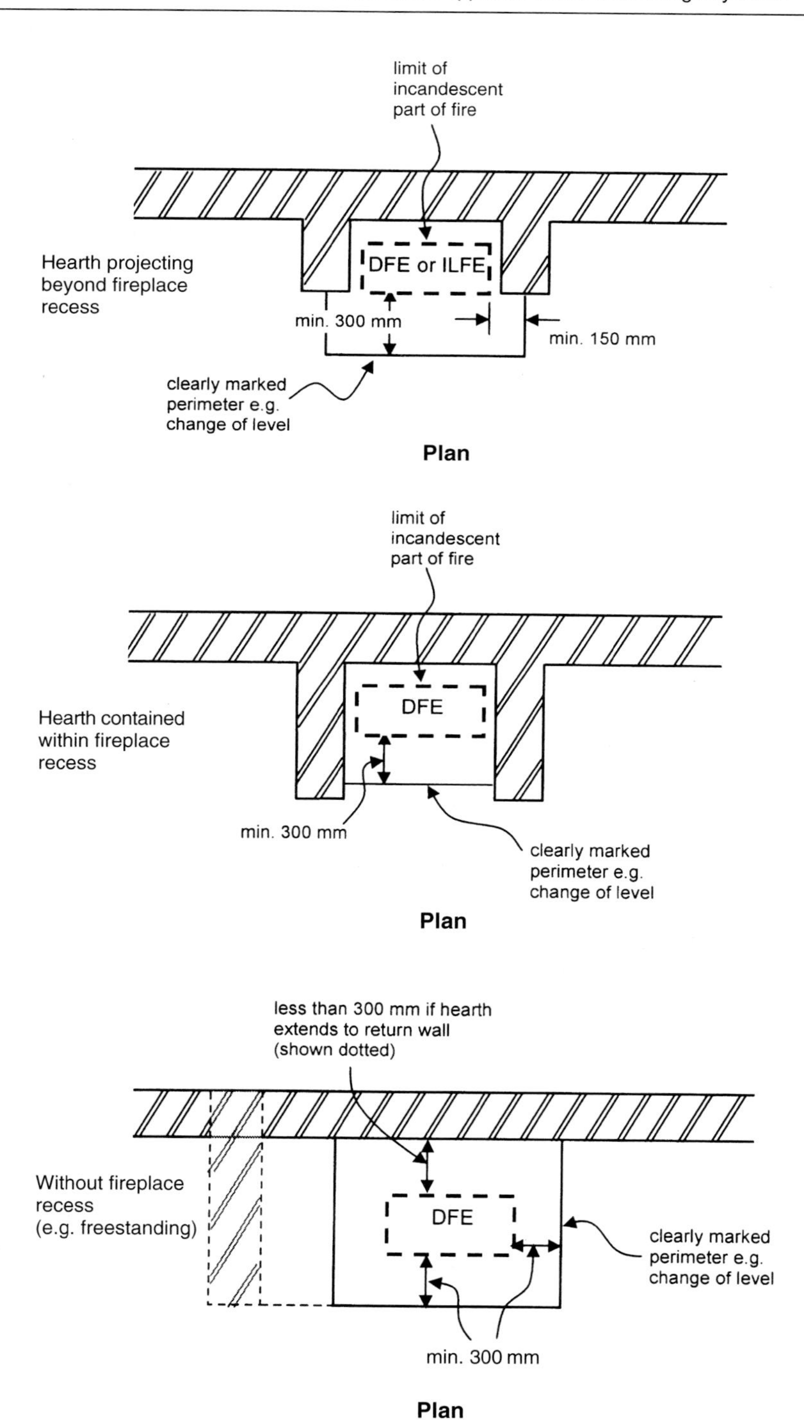

Fig. 14.21 Hearths for decorative fuel effect (DFE) and inset live fuel effect (ILFE) fires burning gas.

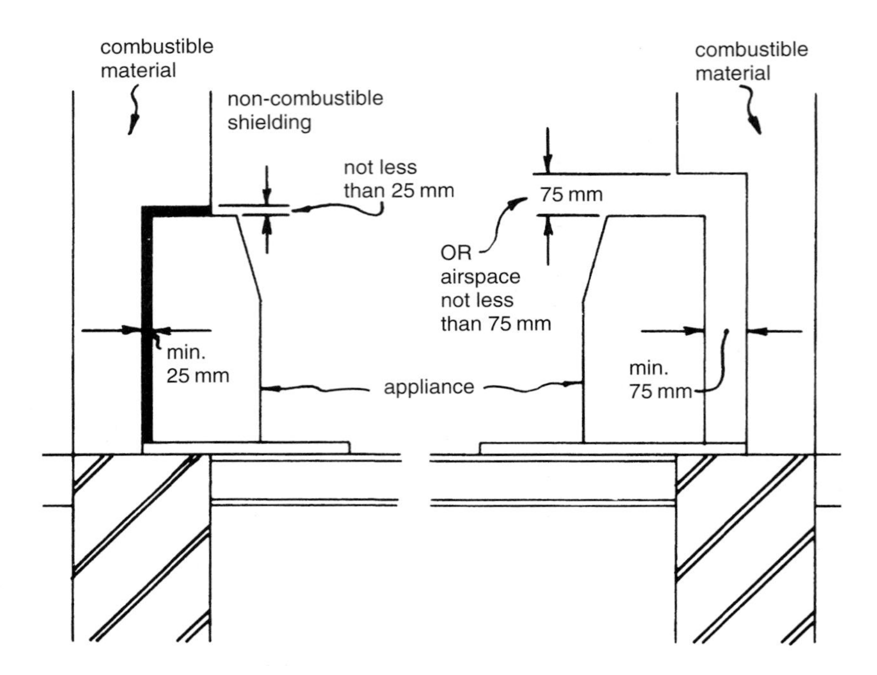

Fig. 14.22 Shielding of appliances burning gas.

Chimneys for appliances burning gas

Where gas appliances are served by masonry chimneys there should be at least 25 mm of masonry between the flues and any combustible material. Similarly, where a flueblock chimney serves a gas appliance the flueblock walls should be at least 25 mm thick. Where a chimney penetrates a fire compartment wall or floor it must also comply with the fire separation requirements of Part B (see Chapter 7).

Placing and shielding of flues for appliances burning gas

Connecting flues and factory-made chimneys complying with BS 715: 1993 *Specification for metal flue pipes, fittings, terminals and accessories for gas-fired appliances with a rated input not exceeding 60 kW*, serving appliances burning gas should be placed as shown in Fig. 14.23 to ensure the following.

- Every part of the flue is at least 25 mm from any combustible material. The distance is measured from the outer surface of the flue wall, or the outer surface of the inner wall for multi-walled products.
- Where the flue passes through a roof, floor or wall formed of combustible materials (other than a compartment roof, floor or wall), it is enclosed in a sleeve of non-combustible material and there is at least 25 mm airspace between the flue and the sleeve. The airspace could be wholly or partially filled with non-combustible insulating material.

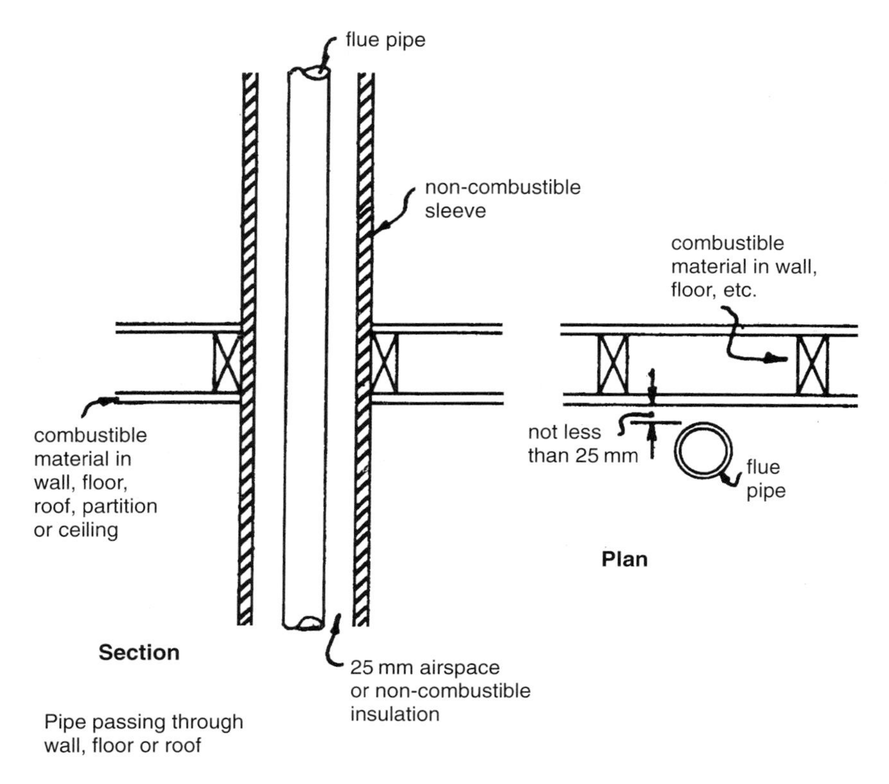

Fig. 14.23 Placing and shielding of flues from combustible materials – appliances burning gas.

Factory-made chimneys complying with BS 4543: 1996 *Factory-made insulated chimneys* should also be separated from combustible materials. This is dealt with in section 14.6.4 above. Factory-made chimneys and connecting fluepipes should also be guarded if:

- the burn hazard they present is not immediately apparent to people; or
- they are so sited as to be at risk of damage.

14.7.3 Protection of the building against fire and heat – appliances burning oil

Hearths for appliances burning oil

The basic function of a hearth is to stop the building from catching fire. Where oil is the fuel it is customary to place a non-combustible tray on top of the hearth to collect any spilled fuel although this is not a health and safety provision.

The decision as to the type of hearth to provide will depend on the temperature reached by the floor below the hearth as a result of the operation of the appliance as follows.

- For hearth temperatures which are unlikely to exceed 100°C (as shown by a suitable test procedure such as that contained in *OFTEC Oil-Fired Appliance Standards OFS A 100 and OFS A 101*) it is usual to provide a rigid, imperforate, non-absorbent, non-combustible sheet, (e.g. a steel tray) which can be part of the appliance. No other special measures are necessary.
- Where hearth temperatures could exceed 100°C, the guidance on hearths for appliances burning solid fuel should be followed (see section 14.7.1 above).

The hearth surface surrounding the appliance should be kept clear of any combustible material by maintaining the following minimum clearances:

- at the back and sides – 150 mm (clear space or distance to a suitably heat resistant wall);
- at the front –
  (a) 150 mm; or
  (b) 225 mm if the appliance provides space heating by means of visible flames or radiating elements.

The edges of the hearth should be designed to be apparent to building occupiers so as to discourage the laying of combustible floor finishes too close to the appliance. This could be achieved by providing a change in levels.

Shielding of appliances burning oil

Combustible materials adjacent to oil-fired appliances may only need special protection if they are subjected to temperatures which exceed 100°C. In these cases they should be separated from combustible materials by a non-combustible surface such as:

- a shield of non-combustible material at least 25 mm thick; or
- at least 75 mm airspace.

Alternatively, shielding would not usually be needed for appliances that are independently certified as having surface temperatures of not more than 100°C during normal operation.

Chimneys and flues for appliances burning oil

Where oil-fired appliances are served by masonry chimneys there should be at least 25 mm of masonry between the flues and any combustible material. Similarly, where a flueblock chimney serves an oil-fired appliance the flueblock walls should be at least 25 mm thick.

Where a chimney penetrates a fire compartment wall or floor it must also comply with the fire separation requirements of Part B (see Chapter 7).

Flues which are likely to serve appliances burning Class D oil (gas oil) should be made of materials which are resistant to acids of sulphur.

Placing and shielding of flues for appliances burning oil

Where flue gas temperatures are unlikely to exceed 250°C, the building fabric may be protected from heat dissipation from connecting flues and factory-made chimneys complying with BS 715: 1993 serving appliances burning oil by making sure of the following.

- Every part of the flue is at least 25 mm from any combustible material. The distance is measured from the outer surface of the flue wall, or the outer surface of the inner wall for multi-walled products.
- Where the flue passes through a roof, floor or wall formed of combustible materials (other than a compartment roof, floor or wall), it is enclosed in a sleeve of non-combustible material and there is at least 25 mm airspace between the flue and the sleeve. The airspace could be wholly or partially filled with non-combustible insulating material.

Factory-made chimneys complying with BS 4543: 1996 *Factory-made insulated chimneys* should also be separated from combustible materials. This is dealt with in section 14.6.4 above. Factory-made chimneys and connecting fluepipes should also be guarded if:

- the burn hazard they present is not immediately apparent to people (e.g. when they cross intermediate floors which are not visible from the appliance); or
- they are so sited as to be at risk of damage.

Where a flue assembly for a room-sealed appliance passes through a combustible wall it should be:

- surrounded by insulating material with a thickness of at least 50 mm; and
- provided with a minimum clearance of 50 mm, from any combustible wall cladding to the edge of the flue outlet.

14.8 Repair and re-use of existing flues

14.8.1 Introduction

Until the coming into force of the current edition of Part J on 1 April 2002 it was possible to open up disused flues in buildings undergoing refurbishment for the purposes of connecting new appliances without having to comply with the Building Regulations. In many cases this led to the installation of appliances burning inappropriate fuels for the old flue and, of course, the use of flues which were in a poor state of repair. Inevitably this resulted in danger to the health and safety of building occupants from fire and noxious fumes.

14.8.2 Repairs to flues

Under Regulations 2 and 3 of the Building Regulations 2000, building work to which Part J imposes requirements, constitutes work to a controlled service or fitting. This means that the following types of work must be notified to a Building Control Body (local authority or approved inspector) and will need to comply with Part J.

- Renovation, refurbishment or repair work involving the provision of a new or replacement flue liner (e.g. the insertion of new linings such as rigid or flexible prefabricated components; or a cast in situ liner that significantly alters the flue's internal dimensions);
- Proposals to bring a flue in an existing chimney back into use or to re-use a flue with a different type or rating of appliance.

The Building Control Body will need to be assured that any altered flues comply with the Regulations. This will involve the inspection and testing of the altered flue by a competent person and is described more fully below.

14.8.3 Re-lining and repairing existing flues

General provisions

In many older buildings with flues designed for use with open fires, flues are likely to be larger than normal and unlined (or they may originally have been parged with lime mortar which has subsequently deteriorated). Since oversize flues can be unsafe, lining may be necessary in order to reduce the flue area to suit the intended appliance.

It is usually possible to refurbish old or defective flues by relining them using materials and components designed to suit the appliance and/or fuel which it is intended to burn. Before being relined, flues should be swept to remove deposits.

Of course, it may be the case that a chimney has been relined in the past using a metal lining system to suit a particular appliance. Where a decision is taken to replace the appliance, the metal liner should also be replaced unless it is in good condition or it can be shown that it has recently been installed. Details of the lining materials and components that are suitable for the various appliances and fuels are considered below.

Chimneys can be relined using independently certified flexible metal flue liners which have been specifically designed to suit the particular fuel being burnt. Such flue liners should only be used for relining purposes and should not be used as the primary liner of a new chimney.

Relining flues serving appliances burning solid fuel

Liners for existing flues serving appliances burning solid fuel should have an independently certified performance at least equal to that corresponding to the BS EN 1443 designation of T450 N2 S D 3 i.e.:

T450 – a product with a normal working temperature of 450°C, see Table 1 of EN 1443
N2 – the pressure class (N = negative pressure)
S – chimney has sootfire resistance
D – resistance to condensate class (D – dry conditions)
3 – corrosion resistance class (3 – Oils with sulphur content over 0.2% and solid mineral fuels and peat).

Examples of liners which comply with this include the following, provided that they are independently certified as being suitable for appliances burning solid fuel:

- factory-made flue lining systems (e.g. double skin flexible stainless steel);
- cast in situ flue relining systems (materials and installation procedures must be independently certified as suitable); and
- any other systems meeting the BS EN 1443 designation and independent certification criteria.

Relining flues serving appliances burning gas

Liners for existing flues serving appliances burning gas can be:

- any of the liners described in section 14.6.2 above as being suitable for all fuels;
- any of the liners described in the section on relining flues serving appliances burning solid fuel immediately above;
- flexible stainless steel liners independently certified as complying with BS 715: 1993 and installed in accordance with BS 5440: Part 1: 2000;
- any other lining systems independently certified as being fit for purpose.

When installing flexible metal flue liners the following points should be noted:

- install the liner in one complete length (without joints in the chimney);
- leave the space between the liner and the chimney empty (other than for sealing at top and bottom) unless the manufacturer's instructions say otherwise;
- double skin flexible flue liners should be installed in accordance with manufacturer's instructions.

Relining flues serving appliances burning oil

Liners for existing flues serving appliances burning oil, with flue gas temperatures which are expected to exceed 250°C can be:

- any of the liners described in section 14.6.2 above as being suitable for all fuels;
- any of the liners described in the section on relining flues serving appliances burning solid fuel immediately above;
- flexible stainless steel liners independently certified as complying with BS 715: 1993 and installed in accordance with BS 5440: Part 1: 2000;
- any other lining systems independently certified as being fit for purpose.

Liners for existing flues serving appliances burning oil, with flue gas temperatures which are unlikely to exceed 250°C can be:

- any of the liners described above for flue gas temperatures exceeding 250°C;
- any other lining systems independently certified as being fit for purpose;
- for new appliances of known type, flue lining systems which have been independently certified as complying with the performance requirements corresponding to the designations given in Table 4.1 from AD J (see section 14.6.7 above).

When installing flexible metal flue liners the following points should be noted:

- install the liner in one complete length (without joints in the chimney);
- leave the space between the liner and the chimney empty (other than for sealing at top and bottom) unless the manufacturer's instructions say otherwise;
- double skin flexible flue liners should be installed in accordance with manufacturer's instructions.

14.8.4 Checking and testing refurbished and repaired flues

Reference has been made in section 14.4 above regarding the need to inspect and test combustion installations before they are first used. This requirement also applies to flues which have been refurbished or repaired in line with the guidance given in this section. The tests, inspections and report should be done by a specialist firm and copies should be supplied to the main contractor, client or developer, and the Building Control Body.

Guidance on methods of checking and testing for compliance with Regulation J2 (*Discharge of products of combustion*) is given in Appendix E of AD J. The guidance applies to new, and existing re-used or relined natural draught flues intended for open flued appliances.

The described procedures cover flues in chimneys, connecting fluepipes, and flue gas passages in appliances, to ensure that these are acceptably gastight and free of obstruction. Furthermore, when an appliance is commissioned to check for compliance with Part L (see Chapter 16) and as required by the Gas Safety (Installation and Use) Regulations 1998, it will also be necessary to carry out appliance performance tests (including flue spillage tests to check for compliance with J2).

It is not necessary to wait for all the building work to be complete before carrying out tests on flues etc. In fact, the most appropriate time to carry out a test may be before the application of plaster finishes or drylining to the structure of a chimney, because at such times possible smoke leakage will be unobscured by surface finishes.

14.9 Test methods

14.9.1 Tests on existing flues

Flues in existing chimneys may suffer from the following defects:

- obstruction caused by bird nests, soot, tar and debris resulting from deterioration of the structure (e.g. pieces of brickwork and chimney pot, and decaying flue lining materials);
- leakage of flue gases as a result of holes or cracks appearing in the structure and linings, particularly at joints;
- decay of the exposed part of the chimney above the roof.

The following methods can be used to determine the state of repair of a flue before it is brought back into re-use.

- Sweeping – to clean the flue and show that it is substantially free from obstructions. It will also enable better visual inspection and testing of the flue. Some deposits (such as tar from burning wood) may be especially hard to dislodge and should be removed. Examine the debris that comes down the chimney when sweeping, to see if it contains excessive quantities of flue lining or brick. These are signs that further repairs may be necessary.
- Visual inspection of the accessible parts – to identify the following.
  (a) Deterioration in the structure, connections or linings which could affect the gastightness and safe performance of the flue and its associated combustion appliance. Examine the exterior of the chimney (including the part in the roofspace) and the interior of the flue. Look for the presence of smoke or tar stains on the exterior of a chimney/breast. These are signs of leaks and could indicate possible damage.
  (b) Modifications made whilst the flue was no longer in service. Look out for the presence of ventilator terminals. These could be incompatible when the flue is used with the intended appliance.
  (c) Correct specification and size of the lining for the proposed new application.
  (d) Freedom from restriction in the flue. A visual inspection may be sufficient where the full length of the flue can be seen. In other cases it may be better to carry out a coring ball test.
- Smoke testing to check the operation and gastightness of the flue.

14.9.2 Tests on new masonry and flueblock chimneys

Flues in new masonry and flueblock chimneys should be checked to demonstrate that they have been correctly constructed, are free of restrictions and are acceptably gas tight. The following defects may occur.

- Incorrect installation of flue liners. Checks should be made during construction that liners are installed with sockets facing upwards and joints are sealed so that moisture and condensate will be contained in the chimney.
- Obstructions (particularly at bends), caused by:
  (a) debris left during construction;
  (b) excess mortar falling into the flue;
  (c) jointing material extruded from between liners and flueblocks.

Before bringing a new flue into use its condition should be checked as follows:

- carry out a visual inspection of the accessible parts to check that the lining, liners or flueblocks are correctly specified and of the correct size for the proposed application;
- carry out a coring ball test, or sweep the flue to remove flexible debris if a visual examination cannot confirm that the flue is free from restrictions;
- carry out a smoke test to confirm the correct operation and gastightness of the flue.

14.9.3 Tests on new factory-made metal chimneys

Newly completed factory-made metal chimneys should follow the checklist for visual inspection given in section 10 of BS7566 *Installation of factory-made chimneys to BS 4543 for Domestic appliances:* Part 3:1992 (1998) *Specification for site installation*. This covers the following:

- chimney route – to be in accordance with installation design;
- openings for testing, cleaning and maintenance – must be accessible;
- components, joints, connections, locking bands, etc. – to be secure;
- combustible material – to be at least the manufacturer's specified distance *X* mm from the chimney (see section 14.6.4 above);
- fire-stops, fire-stop spacers and ceiling supports – to be in position;
- weatherproofing where the chimney penetrates the roof – use only cover flashing and sealing material specified by either the installation designer or the chimney manufacturer;
- enclosure – to be clear of all rubbish and extraneous matter;
- aerials, etc. – not to be attached to the chimney structure.

Additional checks or particular variants may be included in manufacturer's installation instructions. A smoke test should be carried out after the inspection.

14.9.4 Tests on relined flues

Apply the same tests for freedom from restrictions and gastightness as for newly built flues. A flue for a gas appliance which has been relined with a flexible metal liner (see section 14.8.3 above) may be assumed to be unobstructed and acceptably gastight. (The use of a coring ball or inappropriate sweeps brushes can seriously damage a flexible metal flue liner.)

14.9.5 Testing appliances

Where the building work involves the complete installation of combustion appliance and flue system, the entire system should be tested for gastightness in addition to testing the flue separately as above. Appropriate spillage test procedures for different fuels are given in:

- BS 5440: Part 1: 2000 – gas appliances;
- BS 5410 Part 1: 1997 – oil fired appliances;
- BS 6461: *Installation of chimneys and flues for domestic appliances burning solid fuel (including wood and peat):* Part 1: 1984 (1998) *Code of practice for masonry chimneys and flue pipes* – solid fuel appliances.

14.9.6 Test procedures for flues

Coring ball test

In this test a heavy ball, (about 25 mm less in diameter than the flue) is lowered on a rope from the flue outlet to the bottom of the flue. On encountering any obstruction the flue should be unblocked and the test repeated. It should be carried out before smoke testing.

The test can be used for:

- proving the minimum diameter of circular flues;
- checking for obstructions in square flues (but it will not detect obstructions in the corners unless a purpose made coring ball or plate for rectangular flues is used).

The test is not applicable to fluepipes and, as explained above, should not be used with flexible metal flue liners.

Smoke testing

Smoke tests are carried out in order to check that flue gases can rise freely through the flue and to identify any faults that would cause the flue gases to escape into the building (e.g. incorrectly sealed joints or damage to the flue).

Two types of smoke test are described in Appendix E of AD J. Smoke Test 1 is the most stringent and should be used to test flues serving appliances burning solid fuel or oil, since it tests the gastightness of the whole flue. It can be used for flues serving appliances burning gas if there is any doubt about the state of the flue. Smoke Test 2 is used exclusively for appliances burning gas and does not involve sealing the flue.

These tests are in addition to any spillage test carried out when the appliance is commissioned. Where an approved flue or relining system is installed it is possible that other tests could be a requirement of the installation procedure.

Smoke Test 1 is carried out as follows:

- close all doors and windows in the room served by the flue;
- warm the flue to establish a draught (e.g. with a blow lamp or electric heater);
- place and ignite a suitable number of flue testing smoke pellets at the base of the flue, (e.g. in the fireplace recess or in the appliance, where fitted);
- when smoke starts to form, seal off the base of the flue or fireplace opening, or close the appliance if fitted, so that smoke can only enter the flue (e.g. close off the recess opening with a board or plate, sealed at the edges or, where an appliance is fitted, close its doors, ashpit covers and vents);

- establish that there is a free flow of smoke from the flue outlet or terminal;
- seal the top of the flue;
- check the entire length of the flue to establish that there is no significant leakage (see also notes below on checking flues for leakage);
- allow the test to continue for at least 5 minutes;
- remove the closures at the top and bottom.

Smoke Test 2 is carried out as follows:

- close all doors and windows in the room served by the flue;
- warm the flue to establish a draught (e.g. with a blow lamp or electric heater);
- place and ignite a suitable flue testing smoke pellet at the base of the flue, (e.g. in the fireplace recess or in the intended position of the appliance);
- partially close off the opening between the recess and the room with a board to leave an air entry gap of about 25 mm at the bottom;
- establish that smoke is issuing freely from the flue outlet or terminal and not to spilling back into the room;
- check the entire length of the flue to establish that there is no significant leakage inside or outside the building.

Smoke Tests 1 and 2 conform to the recommendations for testing given in BS 6461: Part 1: 1984 (1998) and BS 5440: Part 1: 2000.

Further guidance on smoke testing of flues

The following notes should read in conjunction with the procedures for smoke testing referred to above.

(1) Flues should be warmed for a minimum of 10 minutes in order to establish a draught. Large or cold flues may take longer than this.
(2) Where an appliance is fitted it should not be under fire at the time of carrying out the test.
(3) When being tested, smoke should only emerge from the flue under test. Smoke issuing from any other flue would indicate leakage between the flues.
(4) If leaks do occur during the test they may be quite remote from the area of the fault. This may make it difficult to pinpoint the exact location. For example, where a chimney is located on a gable wall smoke can track down the verge and emerge from under the bargeboard overhang, or it may issue from window reveals where it has penetrated into the cavity of the wall
(5) Smoke pellets create a significantly higher pressure than that required in the product standards for natural draught chimneys and for flues having a gas tightness designation of N1 (negative pressure chimney tested at a pressure of 40 Pa). BS EN 1443 permits flues to this designation to have a leakage rate of up to 2 litres/second per m^2 of flue wall area. Therefore, when assessing the extent of any smoke leakage seen during the test the following points should be considered:

- wisps of smoke seen on the outside of the chimney or near joints between chimney sections are not necessarily indicative of a fault;
- evidence of forceful plumes, or large volumes of smoke could indicate a major fault (e.g. an incorrectly made connection or joint, or a damaged section of chimney).

Therefore, evidence of smoke leakage during a test may not necessarily indicate failure and it can be a matter of expert judgement as to whether this is, in fact, the case. Where leakage is established it will be necessary to conduct an investigation and carry out any remedial action before repeating the smoke test.

14.10 Provision of information

It has been shown in the preceding parts of this chapter that safe installation of a combustion appliance depends on making sure that the appliance is compatible with the hearth, fireplace, flue or chimney to which it is connected. It is vital therefore, that anyone carrying out work to such an installation is aware of its performance capabilities so that an appliance is not connected to an inappropriate flue etc.

Accordingly, paragraph J4 of Part J requires that where a hearth, fireplace, flue or chimney is provided or extended, a durable notice containing information on the performance capabilities of the installation must be affixed in a suitable place in the building.

14.10.1 Notice plates for hearths, fireplaces, chimneys and flues

A robust and indelibly marked notice plate containing information essential to the correct application and use of the combustion installation should be permanently posted up in an unobtrusive but obvious position in the building where any of the following are being provided or extended:

- a hearth and fireplace (including a flue box); and/or
- a flue or chimney (including the provision of a flue as part of the refurbishment work).

A suitable location in the building for the notice plate might be next to the:

- electricity consumer unit; or
- chimney or hearth described on the plate; or
- water supply stop-cock.

The notice plate should convey the following information:

- location of the hearth, fireplace (or flue box) or the location of the start of the flue;
- category of flue and the generic types of appliances that the flue can safely accommodate;

- type and size of the flue (or liner for relined flues) and the name of the manufacturer;
- date of installation.

An example of a typical notice plate is given in Diagram 1.9 from Section 1 of AD J, which is reproduced below. Where a chimney product has had its performance characteristics assessed in accordance with a European Standard (EN) and it is supplied or marked with a designation as described in section 14.2 above, this designation may be included on the notice plate at the option the installer (see Diagram 1.9).

AD J Section 1

Diagram 1.9 Example notice plate for hearths and flues.

Essential information	**IMPORTANT SAFETY INFORMATION** **This label must not be removed or covered**	
	Property address	*20 Main Street* *New Town*
	The hearth and chimney installed in the	*lounge*
	are suitable for	*decorative fuel effect gas fire*
	Chimney liner	*double skin stainless steel flexible, 200 mm diameter*
	Suitable for condensing appliance	*no*
	Installed on	*date*
Optional additional information	Other information (optional) .... *e.g. installer's name, product trade names, installation and maintenance advice, European chimney product designation, warnings on performance limitations of imitation elements e.g. false hearths*	*Designation of stainless steel liner stated by manufacturer to be T450 N2 S D 3*

14.11 Alternative means of compliance

As has been shown in Chapter 2, it is possible to achieve the levels of performance described in the Approved Documents (and which are needed to satisfy the requirements of the Building Regulations) by using other sources of guidance, such as British and European Standards. AD J lists a number of alternative guidance sources according to the fuel being burnt. These are outlined below.

Alternative sources of guidance for appliances burning solid fuel

For appliances burning solid fuel the requirements of the Regulations will be met if the relevant recommendations of the following publications are adopted:

- BS 6461: *Installation of chimneys and flues for domestic appliances burning solid*

fuel (including wood and peat). Code of practice for masonry chimneys and flue pipes. Part 1: 1984 (1998); and
- BS 7566: *Installation of factory-made chimneys to BS 4543 for domestic appliances* Parts 1 to 4: 1992 (1998); and
- BS 8303: *Installation of domestic heating and cooking appliances burning solid mineral fuels.* Parts 1 to 3: 1994.

Alternative sources of guidance for appliances burning gas

For appliances burning gas the requirements of the Regulations will be met if the relevant recommendations of the following publications are adopted:

- BS 5440: *Installation and maintenance of flues and ventilation for gas appliances of rated input not exceeding 70 kW net (1st, 2nd and 3rd family gases)*, Part 1: 2000 *Specification for installation and maintenance of flues;* Part 2: 2000 *Specification for installation and maintenance of ventilation for gas appliances.*
- BS 5546: 2000 *Specification for installation of hot water supplies for domestic purposes, using gas-fired appliances of rated input not exceeding 70 kW.*
- BS 5864: 1989 *Specification for installation in domestic premises of gas-fired ducted-air heaters of rated input not exceeding 60 kW.*
- BS 5871: *Specification for installation of gas fires, convector heaters, fire/back boilers and decorative fuel effect gas appliances,* Part 1: 2001 *Gas fires, convector heaters and fire/back boilers and heating stoves (1st, 2nd and 3rd family gases);* Part 2: 2001 *Inset live fuel effect gas fires of heat input not exceeding 15 kW and fire/back boilers (2nd and 3rd family gases);* Part 3: 2001 *Decorative fuel effect gas appliances of heat input not exceeding 20 kW (2nd and 3rd family gases).*
- BS 6172: 1990 *Specification for installation of domestic gas cooking appliances (1st, 2nd and 3rd family gases).*
- BS 6173: 2001 *Specification for installation of gas-fired catering appliances for use in all types of catering establishments (2nd and 3rd family gases).*
- BS 6798: 2000 *Specification for installation of gas-fired boilers of rated input not exceeding 70 kW net.*

Alternative sources of guidance for appliances burning oil

For appliances burning oil the requirements of the Regulations will be met if the relevant recommendations of the following publication are adopted:

- BS 5410: *Code of practice for oil firing,* Part 1: 1997 *Installations up to 45 kW output capacity for space heating and hot water supply purposes.*

14.12 Protection of liquid fuel storage systems

Under the requirements of Paragraph J5 of Part J, liquid fuel storage systems and the pipes connecting them to combustion appliances must be constructed and separated from buildings and the boundary of the premises so as to reduce to a reasonable level the risk of the fuel igniting in the event of fire in adjacent buildings or premises.

This requirement is limited in application to the following:

- fixed oil storage tanks with capacities in excess of 90 litres and their connecting pipes; and
- fixed liquefied petroleum gas (LPG) storage installations with capacities in excess of 150 litres and associated connecting pipes

provided that these are located outside the building, serving fixed combustion appliances (including incinerators) inside the building.

14.12.1 Limitations on the Approved Document guidance

It will be noticed that there are no upper limits set on the size of the liquid fuel storage systems described above and that the requirements apply to all building types and all types of tanks (even those that are buried). The guidance in AD J, however is restricted to the following types of installation:

- oil storage systems with above-ground or semi-buried tanks of up to 3500 litres capacity, used exclusively for heating oil burning:
  (a) Class C2 oil (kerosene); or
  (b) Class D oil (gas oil);
- LPG storage systems of up to 1.1 tonne capacity comprising one tank standing in the open air, or LPG storage systems consisting of sets of cylinders, under the conditions described below.

For oil storage tanks with capacities exceeding 3500 litres, AD J recommends that advice should be sought from the Fire Authority regarding suitable fire precautions. Further guidance may also be obtained from BS 5410: Part 1: 1997 for oil storage systems, and from Code of Practice 1: *Bulk LPG storage at fixed installations* Part 1: 1998 *Design, installation and operation of vessels located above ground*, published by the LP Gas Association for LPG storage systems.

14.12.2 Protection of oil storage tanks against fire

Oil and LPG fuel storage installations (including the pipework connecting them to combustion appliances in the buildings they serve) should be located and constructed so that they are reasonably protected from fires which may occur in buildings or beyond boundaries. To this end Table 5.1 from Section 5 of AD J (which is reproduced below) contains guidance on measures that can be taken to protect the tank from a fire occurring within the building. Additionally the guidance offered would help to reduce the risk of fuel storage fires igniting buildings although this is thought (in BS 5410) to be unlikely to happen.

In addition to the measures described in Table 5.1, the following issues should also be addressed.

AD J Section 5

Table 5.1 Fire protection for oil storage tanks.

Location of tank	Protection usually satisfactory
Within a building	Locate tanks in a place of special fire hazard which should be directly ventilated to outside. Without prejudice to the need for compliance with all the requirements in Schedule 1, the need to comply with Part B should particularly be taken into account
Less than 1800 mm from any part of a building	(a) Make building walls imperforate (1) within 1800 mm of tanks with at least 30 minutes fire resistance (2) to internal fire and construct eaves within 1800 mm of tanks and extending 300 mm beyond each side of tanks with at least 30 minutes fire resistance to external fire and with non-combustible cladding; or (b) Provide a fire wall (3) between the tank and any part of the building within 1800 mm of the tank and construct eaves as in (a) above. The fire wall should extend at least 300 mm higher and wider than the affected parts of the tank
Less than 760 mm from a boundary	Provide a fire wall between the tank and the boundary or a boundary wall having at least 30 minutes fire resistance to fire on either side. The fire wall or the boundary wall should extend at least 300 mm higher and wider than the top and sides of the tank
At least 1800 mm from the building and at least 760 mm from a boundary	No further provisions necessary

Notes:
1. Excluding small openings such as air bricks etc.
2. Fire resistance in terms of insulation, integrity and stability.
3. Fire walls are imperforate non-combustible walls or screens, such as masonry walls or steel screens.

- To prevent the storage installation becoming overgrown by weeds, ensure that the ground under the tank is hard surfaced with 100 mm concrete or paving slabs at least 42 mm thick extending at least 300 mm beyond the edge of the tank (or the face of the external skin of the tank if it is of the integrally bunded type).
- Ensure that the recommended fire walls are structurally stable and do not pose a danger to people coming near them. Since the Building Regulations do not apply to freestanding walls of this description, other guidance should be sought. The former DTLR has published a guide to the construction of garden walls entitled *Your Garden Walls. Better Safe than Sorry*. This is obtainable from DTLR Free Literature, PO Box 236, Wetherby, West Yorkshire, LS23 7NB. Guidance will be found in this publication relating wall thickness to height.

A way to protect the oil supply pipework is to install it so that it is resistant to the effects of fire and to fit a proprietary fire valve system. More information on this can be gained by reading sections 8.2 and 8.3 of BS 5410: Part 1: 1997.

14.12.3 Protection of LPG storage installations against fire

Because of the potential hazards presented by liquefied petroleum gas (LPG) installations, they are controlled by legislation enforced by the Health and Safety Executive or its agents. There is a considerable amount of legislation dealing with the safe use of LPG in different branches of industry and compliance is usually demonstrated by constructing an LPG storage installation in accordance with an appropriate industry Code of Practice, prepared in consultation with the HSE. Therefore, the amount of building work that is needed in order for an installation to comply will depend on:

- the capacity of the installation;
- whether or not tanks are installed above or below ground; and
- the nature of the premises served.

However, by following the guidance in Approved Document J as outlined in this section, and the relevant guidance in Approved Document B (see Chapter 7), no further building work will normally be necessary to comply with other legislation if the tank stands in the open air and the installation has a capacity not exceeding 1.1 tonne.

LPG tank location

Since liquefied petroleum gas is heavier than air the LPG tank should be installed outdoors and not within an open pit or bund which would trap the vapour and not allow it to disperse. In fact, safe dispersal is essential should accidental venting or leakage of the tank occur and this necessitates adequate separation of the tank from buildings, boundaries and any fixed sources of ignition to reduce the risk of fire spread. Safe separation distances from buildings, boundaries and fire walls are shown in Fig. 14.24 and Fig. 14.25 below. It also follows that drains, gullies and cellar hatches should be protected from gas entry if they are within the separation distance shown in the Figures.

Fire walls

Fire walls can be freestanding walls which separate the tank from the building, the boundary or a fixed source of ignition. They can also be part of the building itself or a boundary wall belonging to the property. All these options are illustrated in Fig 14.24 and Fig. 14.25 together with their respective separation distances.

Acceptable fire walls should:

- be imperforate and constructed of masonry, concrete or similar materials;
- have fire resistance in terms of stability, integrity and insulation (see Chapter 7 for details of these concepts) of at least:
  - (a) 30 minutes if a freestanding or boundary wall
  - (b) 60 minutes if part of a building

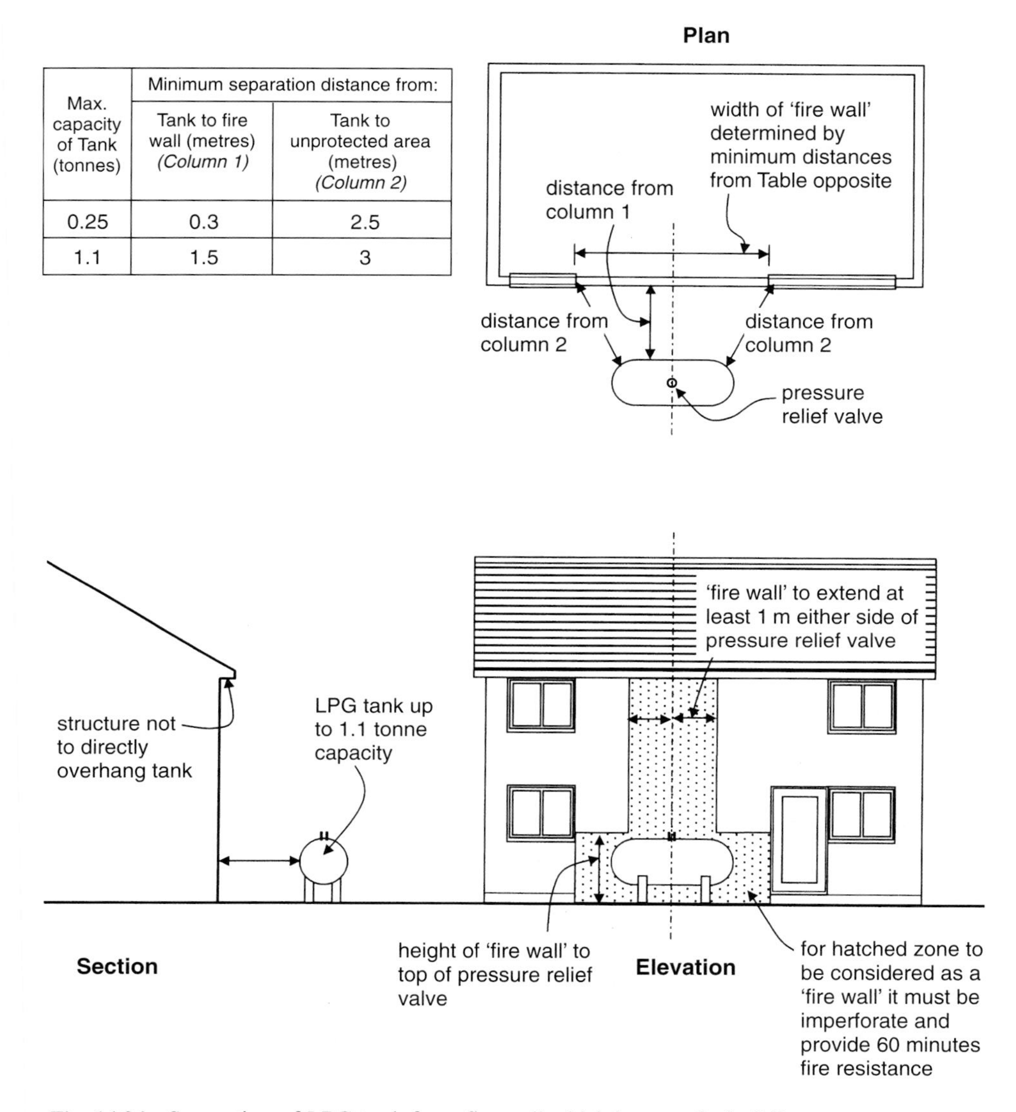

Max. capacity of Tank (tonnes)	Minimum separation distance from:	
	Tank to fire wall (metres) *(Column 1)*	Tank to unprotected area (metres) *(Column 2)*
0.25	0.3	2.5
1.1	1.5	3

Fig. 14.24 Separation of LPG tank from fire wall which is part of a building.

- only be built on one side of a tank (to assist ventilation and dispersal of the gas in the event of a leak);
- be at least as high as the pressure relief valve on the tank;
- extend horizontally so that the separation shown in Figs. 14.24 and 14.25 is maintained.

Sets of cylinders – location and support

LPG storage installations that consist of sets of cylinders should be adequately supported and sufficiently far enough away from openings into the building to not

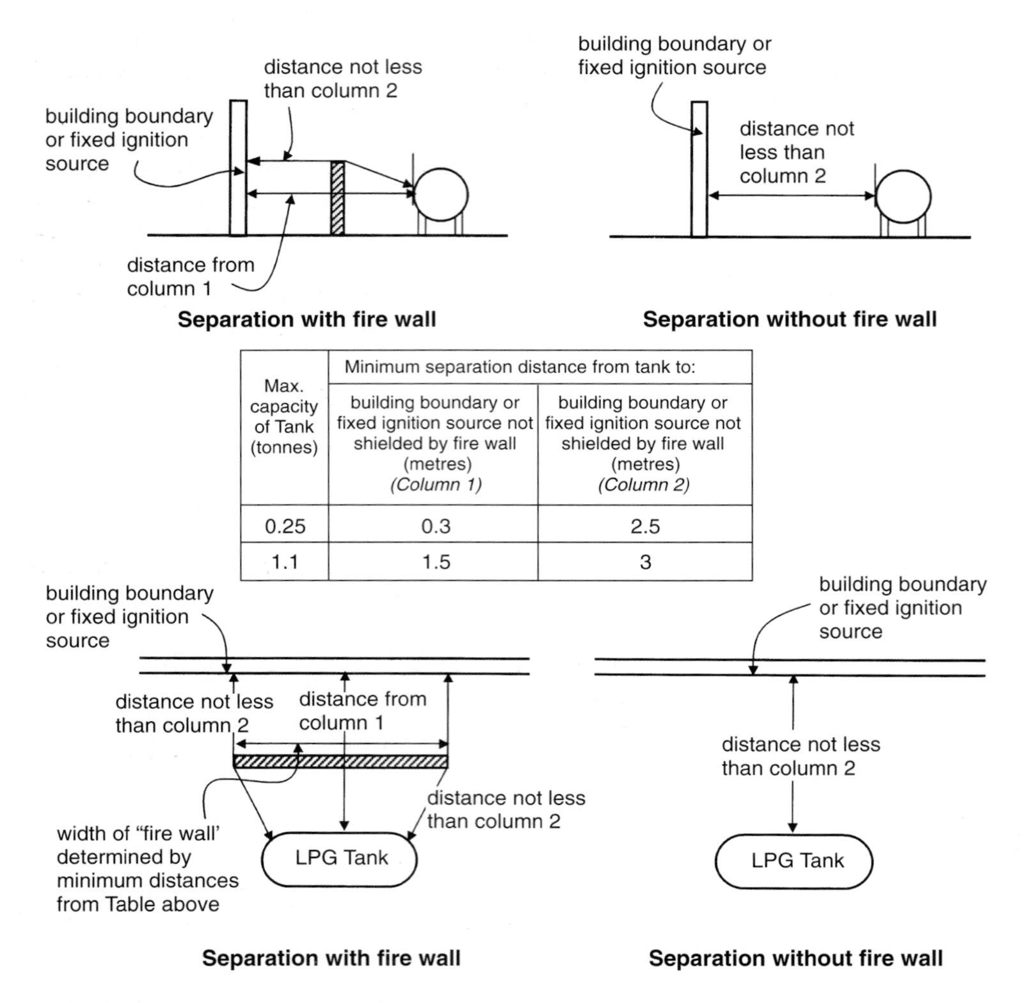

Max. capacity of Tank (tonnes)	Minimum separation distance from tank to:	
	building boundary or fixed ignition source not shielded by fire wall (metres) *(Column 1)*	building boundary or fixed ignition source not shielded by fire wall (metres) *(Column 2)*
0.25	0.3	2.5
1.1	1.5	3

Fig. 14.25 Separation of LPG tank from building boundary or fixed source of ignition with and without fire wall.

present a risk to the health and safety of the occupants. In terms of support they should:

- stand upright;
- be readily accessible;
- be reasonably protected from physical damage; and
- not obstruct exit routes from the building.

It is usual to secure the cylinders by means of chains or straps against a wall outside the building at ground level, in a well-ventilated position. They should be placed on a firm, level base (e.g. concrete at least 50 mm thick or paving slabs bedded in mortar).

In terms of separation distances, these should be sufficient to prevent the entry of gas into the building should venting or a leak occur. Examples of acceptable separation distances include the following from AD J:

- from openings through the wall of the building (doors, windows, airbricks etc.) and from heat sources such as flue terminals or tumble-dryer vents – at least 1 m horizontally and 300 mm vertically measured from the nearest cylinder valve;
- from drains without traps, unsealed gullies, cellar hatches and similar entries at ground level – at least 2 m horizontally measured from the nearest cylinder valve, unless an intervening wall is provided at least 250 mm high.

Fig. 14.26 below shows how sets of cylinders should be located in relation to any doors, windows or other openings into the building.

14.13 Protection against pollution

Under the requirements of Paragraph J6 of Part J, oil storage tanks and the pipes connecting them to combustion appliances, must:

- be constructed and protected so as to reduce to a reasonable level the risk of oil escaping and causing pollution; and
- have a durable notice affixed in a prominent position that contains information on how to respond to an escape of oil so that the risk of pollution is reduced to a reasonable level.

This requirement is limited in application to the following:

- fixed oil storage tanks with capacities of 3500 litres or less and their connecting pipes which are:
  (a) located outside the building; and
  (b) serve fixed combustion appliances (including incinerators) inside a building which is used wholly or mainly as a private dwelling;
- above-ground storage systems (i.e. it does not apply to buried systems).

In addition to paragraph J6 of Part J of the Building regulations, there are controls exercised over storage tanks in England under the provisions of the Control of Pollution (Oil Storage) (England) Regulations 2001 (SI 2001/2954). These regulations apply to a wide range of oil storage installations but they do not apply to any premises which are wholly or mainly used as private dwellings where the storage capacity of the oil installation is less than 3500 litres. These regulations are policed by the Environment Agency in the event of any oil pollution.

14.13.1 Provisions to prevent oil pollution

As has been stated above, Paragraph J6 applies to oil storage tanks with capacities of 3500 litres or less serving combustion appliances in buildings used

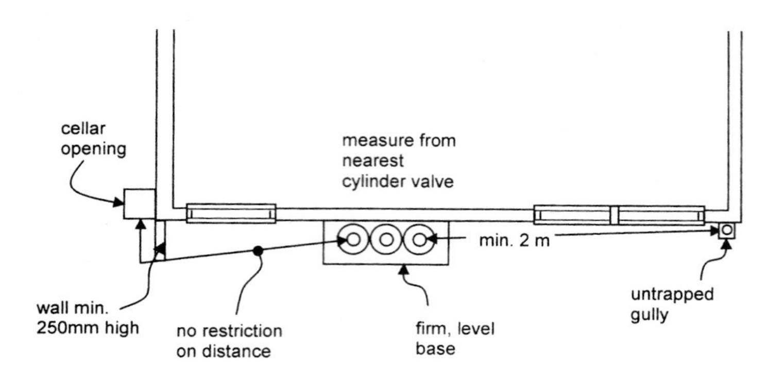

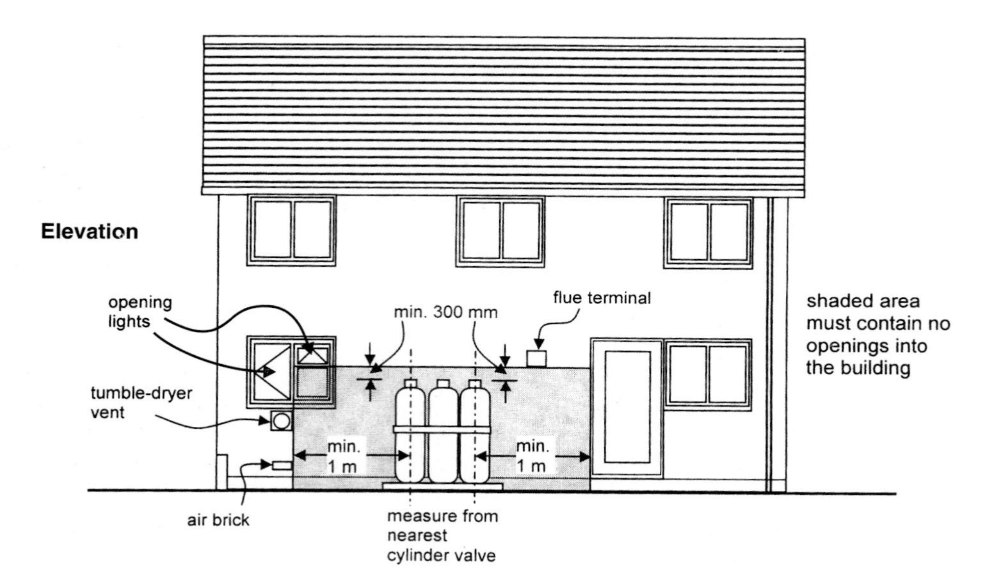

Fig. 14.26 Location and separation of LPG cylinders.

wholly or mainly as private dwellings. In these circumstances, secondary containment (the provision of bunds etc. see below) should be provided where there is considered to be a significant risk of oil pollution, i.e. if the oil storage installation:

- has a total capacity exceeding 2500 litres; or
- is closer than 10 m from inland freshwaters (see definition below) or coastal waters; or
- is located where an open drain or loose fitting manhole cover could be affected by spillage; or
- is closer than 50 m from sources of drinkable water (e.g. wells, bore-holes or springs); or

- is situated where oil spillage from the installation could reach inland freshwaters or coastal waters by running across hard ground; or
- is located where tank vent pipe outlets are not visible from the intended filling point (possibly resulting in accidental overfilling of the tank).

INLAND FRESHWATERS mean streams, rivers, reservoirs and lakes, and the ditches and ground drainage (including perforated drainage pipes) that run into them.

Secondary containment, when considered necessary, can be provided by means of:

- integrally bunded prefabricated tanks; or
- bunds constructed from masonry or concrete.

General advice on the construction of masonry or concrete bunds can be obtained from the Environment Agency in their publication – Pollution Prevention Guidelines PPG2 – *Above Ground Oil Storage Tanks*. They can also supply specific advice for bunds constructed of masonry or concrete in their publications – *Masonry Bunds for Oil Storage Tanks and Concrete Bunds for Oil Storage Tanks*. All these guidance documents are available free of charge from DMS Mailing House, Fax No. 0151 604 1222.

When constructing bunds the following points should also be taken into account.

- They should have a capacity of at least 110% of the largest tank they contain, whether they are part of a prefabricated tank system or constructed on site.
- Where the walls of a chamber or building enclosing a tank act as a bund for the purposes of this section, any door through such walls should be above bund level.
- Where the bund has a structural role as part of a building, specialist advice should be sought.

Labelling of oil storage installations

A label should be placed on an oil storage installation in a prominent position, to advise on what to do in the event of an oil spill. It should contain the Environment Agency's Emergency Hotline number 0800 80 70 60.

Useful addresses

Reference is made throughout this chapter to a number of organisations concerned with various aspects of the design, construction and use of combustion installations and who were involved in the development of the 2002 edition of Approved Document J, either directly or as consultees. Contact details are given in the list below.

ACE (Amalgamated Chimney Engineers): White Acre, Metheringham Fen, Lincoln LN4 3AL
Tel 01526 32 30 09 Fax01526 32 31 81
BFCMA (British Flue and Chimney Manufacturers Association): Henley Road, Medmenham, Marlow, Bucks SL7 2ER
Tel 01491 57 86 74 Fax 01491 57 50 24
info@feta.co.uk www.feta.co.uk
BRE (Building Research Establishment Ltd.): Bucknalls Lane, Garston, Watford, Hertfordshire WD25 9XX
Tel 01923 66 4000 Fax 01923 66 4010
enquiries@bre.co.uk www.bre.co.uk
BSI (British Standards Institution): 389 Chiswick High Road, London W4 4AL
Tel 020 8996 9000 Fax 020 8996 7400
www.bsi-global.com
CIBSE (Chartered Institution of Building Services Engineers): 222 Balham High Road, London SW12 9BS
Tel 020 8675 5211 Fax 020 8675 5449
www.cibse.org
CORGI (The Council for Registered Gas Installers): 1, Elmwood, Chineham Business Park, Crockford Lane, Basingstoke, Hampshire RG24 8WG
Tel 01256 37 22 00 Fax 01256 70 81 44
www.corgi-gas.com
Environment Agency: Rio House, Waterside Drive, Aztec West, Almondsbury, Bristol BS32 4UD
Tel 0845 9333111 Fax 01454 624 409
www.environment-agency.gov.uk
(Publication enquiries to: Tel 01454 624 411 Fax 01454 624 014)
Environment Agency Emergency Hotline 0800 80 70 60
HETAS (Heating Equipment Testing and Approval Scheme): PO Box 37, Bishops Cleeve, Gloucestershire, GL52 4TB
Tel 01242 673257 Fax 01242 673463
www.hetas.co.uk
HSE (Health and Safety Executive): Rose Court, 2 Southwark Bridge, London SE1 9HS
Tel 020 7717 6000 Fax 020 7717 6717
www.hse.gov.uk
Gas safety advice line 0800 300 363
IGasE (Institution of Gas Engineers): 21 Portland Place, London W1B 1PY
Tel 020 7636 6603 Fax 020 7636 6602
www.igaseng.com
LP Gas Association: Pavilion 16, Headlands Business Park, Salisbury Road, Ringwood, Hampshire BH24 3PB
Tel 01425 461612 Fax 01425 471131
www.lpga.co.uk
NACE (National Association of Chimney Engineers): PO Box 5666, Belper, Derbyshire, DE56 0YX

Tel 01773 599095 Fax 01773 599195
www.nace.org.uk
NACS (National Association of Chimney Sweeps): Unit 15, Emerald Way, Stone Business Park, Stone, Staffordshire, ST15 0SR
Tel 01785 811732 Fax 01785 811712
nacs@chimneyworks.co.uk
www.chimneyworks.co.uk
NFA (National Fireplace Association): 6th Floor, McLaren Building, 35 Dale End, Birmingham B4 7LN
Tel 0121 200 13 10 Fax 0121 200 13 06
www.nationalfireplaceassociation.org.uk
OFTEC (Oil Firing Technical Association for the Petroleum Industry): Century House, 100 High Street, Banstead, Surrey, SM7 2NN.
Tel 01737 37 33 11 Fax 01737 37 35 53
enquiries@oftec.org www.oftec.org
SFA (Solid Fuel Association): 7 Swanwick Court, Alfreton, Derbyshire, DE55 7AS
Tel 0800 600 000 Fax 01773 834 351
sfa@solidfuel.co.uk www.solidfuel.co.uk

15 Protection from falling, collision and impact

15.1 Introduction

The control of stairways, ramps and guards has always formed an important part of Building Regulations since stairways represent, in many cases, the only way out of a building in the event of fire and there is a need to make buildings accessible to disabled people. Additionally, provisions are necessary to protect people from the risk of falling when they use exposed areas such as landings, balconies and accessible roofs.

The 1998 edition of Approved Document K introduced new provisions governing:

- the safe use of vehicle loading bays,
- measures to reduce the risk of collisions with open windows, skylights and ventilators; and
- measures designed to reduce the risk of injury when using various types of sliding or powered doors and gates,

mainly in order to ensure compliance with the Workplace (Health, Safety and Welfare) Regulations 1992.

Therefore, compliance with the revised regulations in Part K prevents action being taken against the occupier under the Workplace Regulations when the building is eventually in use. This involves considering aspects of design which will affect the way a building is used and applies only to workplaces. Although dwellings are excluded from the changes, in mixed use developments the requirements for the non-domestic part of the use would apply to any shared parts of the building (such as common access staircases and corridors). With flats the situation may be less clear for although certain sections of Part K do not apply to dwellings, it is still necessary for people such as cleaners, wardens and caretakers to work in the common parts. Therefore, the requirements of the Workplace Regulations may still apply even though the Building Regulations do not. Additionally, it is now necessary to provide safe access to areas used exclusively for maintenance purposes.

15.2 Stairways, ramps and ladders

Stairways, ramps and ladders which form part of the building must be designed, constructed and installed so that people may move safely between levels, in or about the building. Regulation K1 applies to all areas of a building which need to be accessed, including those used only for maintenance.

It should be noted that compliance with Regulation K1 prevents action being taken against the occupier of a building under regulation 17 of the Workplace (Health, Safety and Welfare) Regulations 1992 when the building is eventually in use. (Regulation 17 relates specifically to permanent stairs, ladders and ramps on routes used by people in places of work and includes access to areas used for maintenance.)

15.3 Application

The provisions contained in AD K1 regarding stairs, ramps and ladders only apply if there is a change in level of:

- more than 600 mm in dwellings; or
- two or more risers in other buildings (the difference in levels in this case will depend on the recommended height for the risers), or 380 mm where there is no stair.

Since stairs, etc. must provide safe access for people, an acceptable level of safety in a building will be dictated by the circumstances. Therefore, the standard in a dwelling will be lower than that recommended for a public building because there are likely to be less people in the dwelling and they will be familiar with the stairs. Similarly, the standard of access to maintenance areas will be lower than that recommended for normal use to reflect the greater care expected of those gaining access.

Outside stairways and ramps, (e.g. entrance steps), are covered by the regulations if they form part of the building (obviously, the proximity of the steps to the building and the way they are associated with it will dictate whether or not they are part of the building, in most cases). Therefore, steps in paths leading to the building would not be covered by Part K. These access routes may, of course, need to comply with other parts of the regulations if:

- they form part of a means of escape in case of fire (see Approved Document B: Fire safety);
- they are intended for use by disabled people (see Approved Document M: Access and facilities for disabled people).

In general, normal access routes in assembly buildings (e.g. sports stadia, theatres, cinemas etc.), should follow the guidance in AD K. Where special consideration needs to be given to guarding spectator areas or there are steps in gangways serving these areas then it may be preferable to follow more specialised guidance contained in the following:

- BS 5588 *Fire Precautions in the Design, Construction and Use of Buildings*, Part 6:1991 *Code of practice for places of assembly* – for new assembly buildings.

- *Guide to fire precautions in existing places of entertainment and like premises*, Home Office 1990 – for existing assembly buildings.
- *Guide to Safety at Sports Grounds*, The Stationery Office 1997 – for stands at sports grounds.

15.4 Interpretation

A number of definitions are given which apply generally throughout AD K:

CONTAINMENT – A barrier to prevent people from falling from a floor to the storey below.

FLIGHT – The section of a ramp or stair running between landings with a continuous slope or series of steps.

RAMP – A slope which is steeper than 1 in 20 intended to enable pedestrians or wheelchair users to get from one floor level to another.

A number of definitions are given which are specific to stairways and ladders:

STAIR – Steps and landings designed to enable pedestrians to get to different levels.

PRIVATE STAIRS – Stairs in or serving only one dwelling.

INSTITUTIONAL AND ASSEMBLY STAIRS – Stairs serving places where a substantial number of people gather.

OTHER STAIRS – Stairs serving all other buildings apart from those referred to above.

GOING – The distance measured in plan across the tread less any overlap with the next tread above (see Fig. 15.1).

RISE – The vertical distance between the top surfaces of two consecutive treads (see Fig. 15.1).

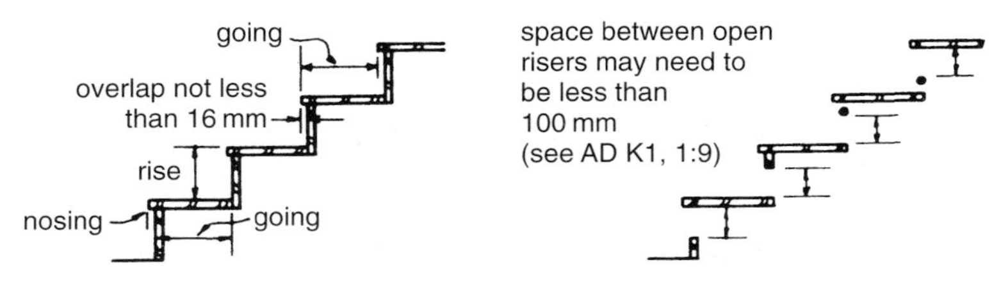

Fig. 15.1 Rise and going.

PITCH AND PITCH LINE – These terms are illustrated in AD K1 but are not defined; however

PITCH LINE – May be defined as a notional line connecting the nosings of all treads in a flight, including the nosing of the landing or ramp at the top of the flight. The line is taken so as to form the greatest possible angle with the horizontal, subject to the special recommendations for tapered treads (see below).

PITCH – May be defined as the angle between the pitch line and the horizontal (see Fig. 15.2).

ALTERNATING TREAD STAIRS – Stairs with paddle shaped treads where the wider portion alternates from one side to the other on each consecutive tread.

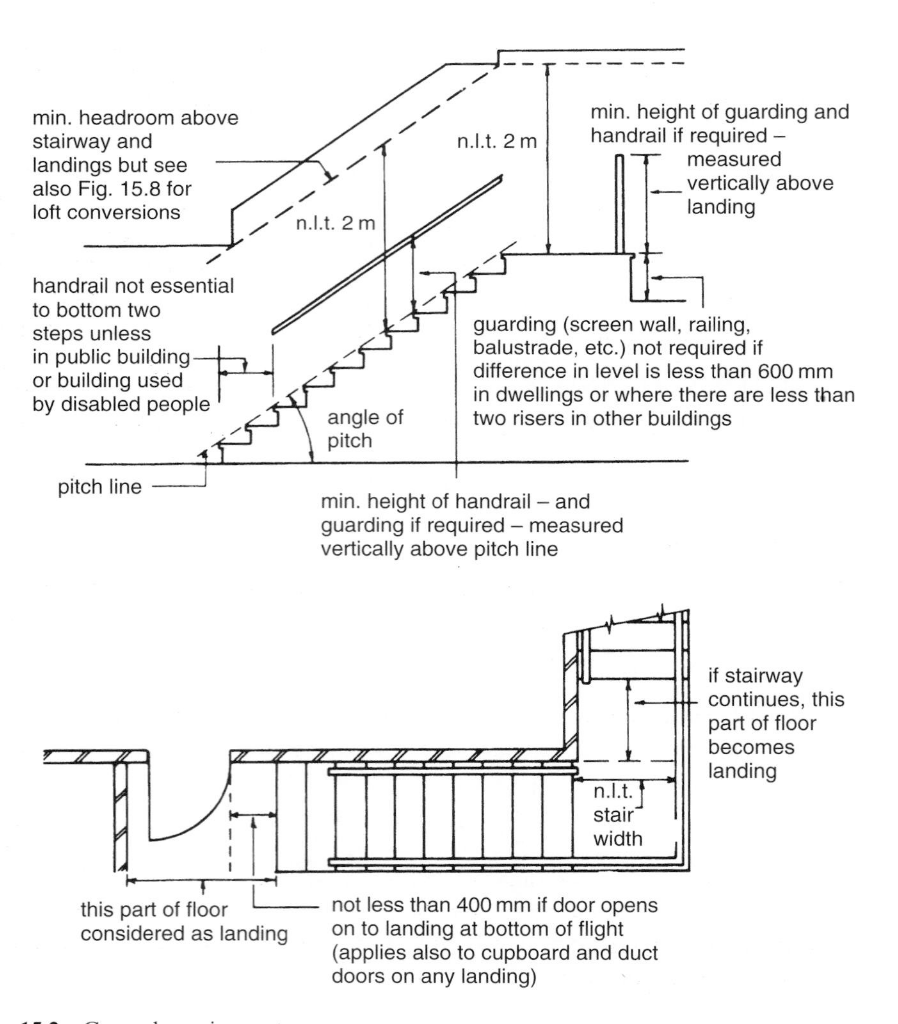

Fig. 15.2 General requirements.

HELICAL STAIRS – Stairs which form a helix round a central void.

SPIRAL STAIRS – Stairs which form a helix round a central column.

LADDER – A series of rungs or narrow treads used as a means of access from one level to another which is normally used by a person facing the ladder.

TAPERED TREAD – A tread with a nosing which is not parallel to the nosing of the tread or landing above it.

15.5 General recommendations for stairways and ramps

15.5.1 Landings

As a general rule a landing should be provided at the top and bottom of every flight or ramp. Where a stairway or ramp is continuous, part of the floor of the building may count as a landing. The going and width of the landing should not be less than the width of the flight or ramp.

Landings should be level and free from permanent obstructions. (This would allow, for example, the placing of a temporary barrier such as a child's safety gate between a landing and a flight.) A door is permitted to swing across a landing at the bottom of a flight or ramp but only if it leaves an area 400 mm wide across the full width of the flight or ramp. This rule also applies to cupboard and duct doors, but in this case they are permitted to open over any landing (including a landing at the top of a flight). Approved Document B contains details of restrictions on the use of cupboards situated in means of escape staircases.

A landing of firm ground or paving at the top or bottom of an external flight or ramp may slope at a gradient of not more than 1 in 20 (see Fig. 15.2).

15.5.2 Handrails

Stairs and ramps should have a handrail on at least one side. Where the width of the stair or ramp is 1m or more, then a handrail should be fixed on both sides.

Handrails are not required:

- beside the bottom two steps in a stairway unless it is in a public building or is intended for use by disabled people (see Approved Document M and Chapter 17).
- where the rise of a ramp is 600 mm or less.

Handrails should provide firm support and grip and be fixed at a height of between 900 mm and 1 m vertically above the pitch line or floor. They can also form guarding where the heights can be matched. Further details regarding handrails for disabled people can be found in Chapter 17 (see also Fig. 15.2).

15.5.3 Headroom

Clear headroom of 2 m should be provided over the whole width of any stairway, ramp or landing. There are reduced dimensions for headroom over stairs in loft conversions (see section 15.6.5, stairs to loft conversions).

Headroom is measured vertically from the pitch line, or where there is no pitch line, from the top surface of any ramp, floor or landing (see Fig. 15.2).

15.5.4 Width

AD K contains no recommendations for minimum stair or ramp widths; however there may be width recommendations in other Approved Documents. Reference should be made to AD B: Fire safety (Chapter 7) and AD M: Access and facilities for disabled people (Chapter 17), where minimum width guidance can be found.

15.6 Stairway recommendations

15.6.1 Rules applying to all stairways

- In any stairway there should not be more than thirty-six rises in consecutive flights, unless there is a change in the direction of travel of at least 30° (see Fig. 15.3).
- For any step the sum of twice its rise plus its going ($2R + G$) should not be more than 700 mm nor less than 550 mm. This rule is subject to variation at tapered steps, for which there are special rules.
- The rise of any step should generally be constant throughout its length and all steps in a flight should have the same rise and going.
- Open risers are permitted in a stairway but for safety the treads should overlap each other by at least 16 mm.
- Each tread in a stairway should be level.

Tapered treads should comply with the following rules:

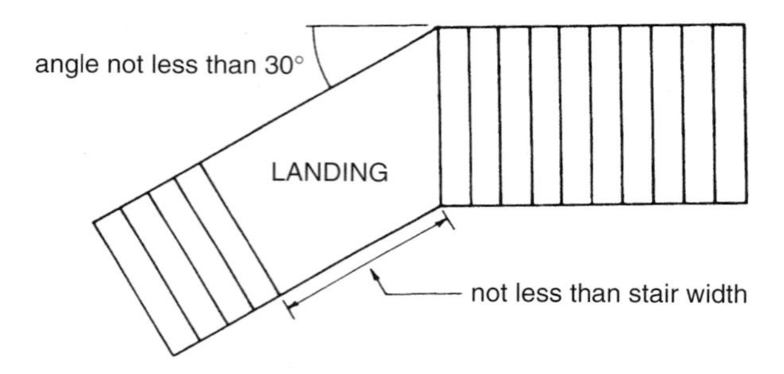

Fig. 15.3 Length of flights.

- The minimum going at any part of a tread within the width of a stairway should not be less than 50 mm.
- The going should be measured:
  - (a) if the stairway is less than 1m wide, at the centre point of the length or deemed length of a tread; and
  - (b) if the stairway is 1 m or more wide, at points 270 mm from each end of the length or deemed length of a tread. (When referring to a set of consecutive tapered treads of different lengths, the term 'deemed length' means the length of the shortest tread. This term is not used in AD K1.) (See Fig. 15.4.)
- All consecutive tapered treads in a flight should have the same taper.
- Where stairs contain straight and tapered treads the goings of the tapered treads should not be less than those of the straight flight.
- In order to prevent small children from becoming trapped between the treads of open riser staircases there should be no opening in a riser of such size as to allow the passage of a sphere of 100 mm diameter. This rule applies to all stairways which are likely to be used by children under the age of five years. (See Fig. 15.1.)

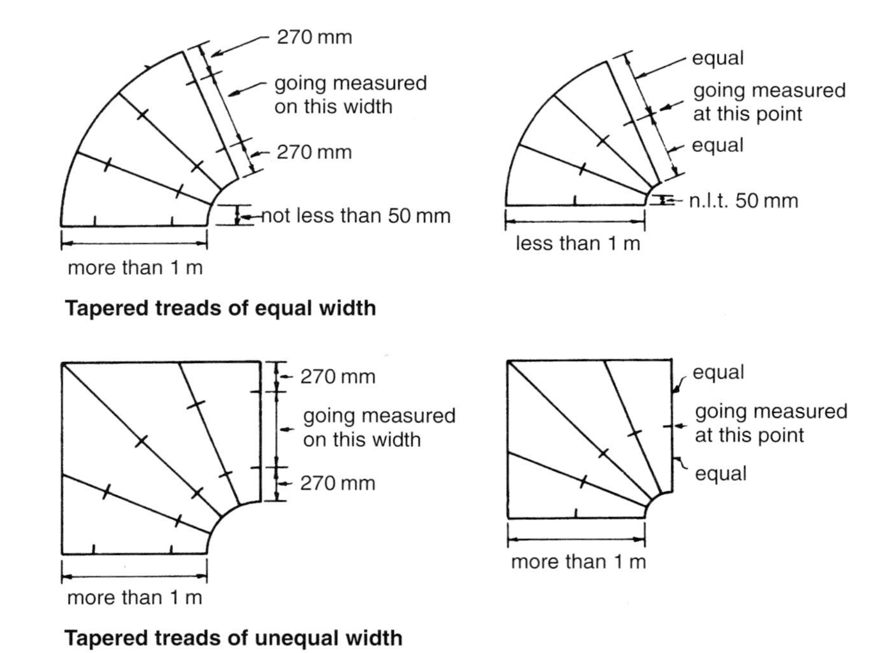

Fig. 15.4 Tapered treads.

15.6.2 Rules applying to private stairways

There are a number of recommendations in AD K which control the steepness, rise and going of private stairs as follows.

- The height of any rise should not be more than 220 mm.
- The going of any step should generally not be less than 220 mm (but see the rules relating to tapered treads above).
- The pitch should not be more than 42°. This means that it is not possible to combine a maximum rise with a minimum going.

Figure 15.5 illustrates the practical limits for rise and going. This figure also incorporates the $(2R + G)$ relationship mentioned above.

The rules governing rise and going for private stairs will also satisfy the recommendations contained in AD M for access for disabled people.

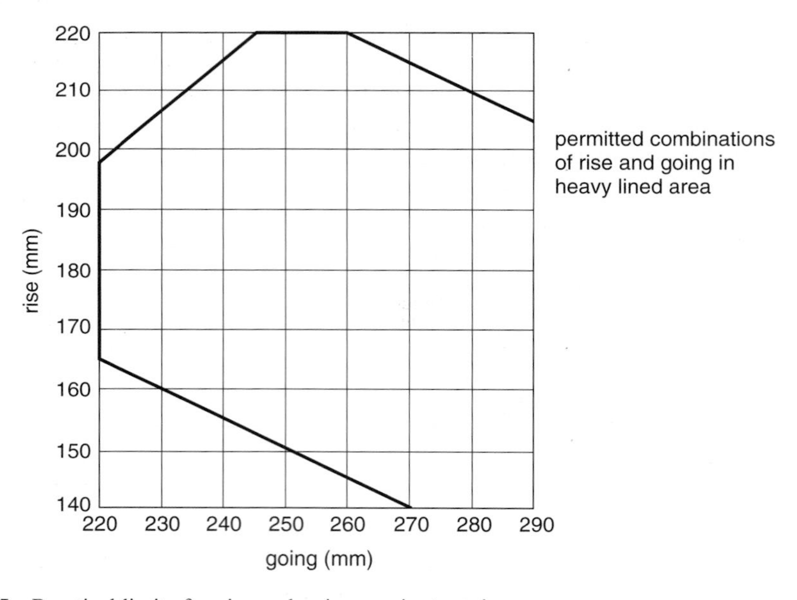

Fig. 15.5 Practical limits for rise and going – private stairways.

15.6.3 Rules applying to institutional and assembly stairs

Institutional buildings are usually occupied by young children, old people or people with physical or mental disabilities. It is necessary, therefore, that staircases should be of slacker pitch than in other types of buildings and further recommendations regarding rise and going may be found in AD M. Additionally, they may need to comply with width recommendations in AD B or AD M. Assembly buildings contain large numbers of people and similar considerations will apply.

- The height of any rise should not be more than 180 mm.
- Subject to the rules governing tapered treads, the going of any step should generally not be less than 280 mm. This may be reduced to 250 mm if the floor area of the building served by the stairway is less than 100 m^2.

Figure 15.6 illustrates the practical limits for rise and going. This figure also incorporates the $(2R + G)$ relationship mentioned above.

- In order to maintain sightlines for spectators in assembly buildings, gangways are permitted to be pitched at up to 35°.
- There should not be more than sixteen risers in a single flight where a stairway serves an area used for assembly purposes. (See also Rules applying to other stairs, below.)
- The width of a stairway in a public building which is wider than 1800 mm should be sub-divided with handrails or other suitable means so that the sub-divisions do not exceed 1800 mm in width.

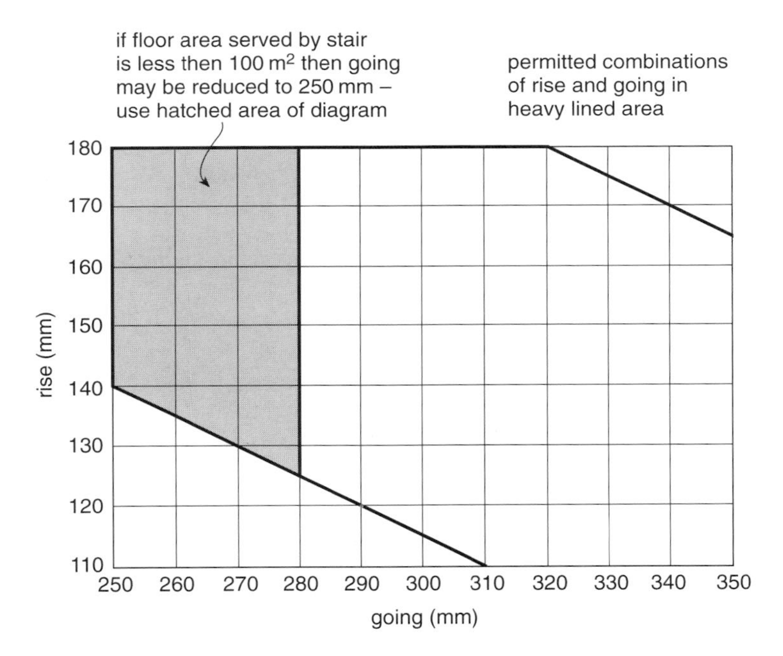

Fig 15.6 Practical limits for rise and going – institutional and assembly buildings.

15.6.4 Rules applying to other stairs

These recommendations apply to all stairs other than those in private dwellings, institutional or assembly buildings. Many of the buildings covered by these recommendations will also need to be accessible to disabled people and should, therefore, conform to the guidance given in AD M.

- The height of any rise should not be more than 190 mm.
- The going of any step should generally not be less than 250 mm (but see the rules relating to tapered treads above).

Figure 15.7 illustrates the practical limits for rise and going. This figure also incorporates the $(2R + G)$ relationship mentioned above.

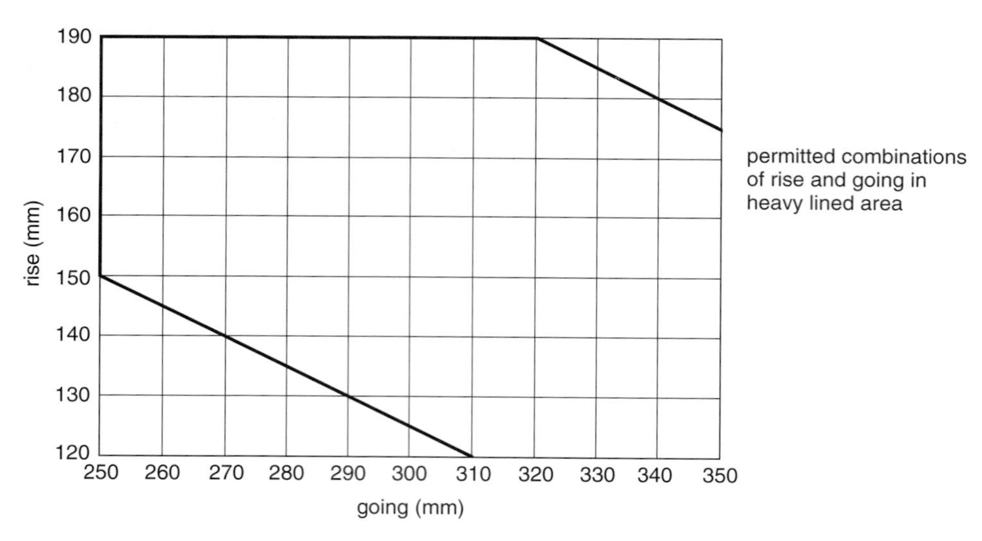

Fig. 15.7 Practical limits for rise and going – other buildings.

- There should not be more than sixteen risers in a single flight where a stairway serves an area used for shop purposes.

15.6.5 Stairs to loft conversions

When carrying out loft conversions it is often extremely difficult to fit a conventional staircase without substantial alteration to the existing structure. This often results in serious loss of the space which the conversion is intended to provide. Approved Document K contains a number of alternative recommendations which are intended to assist in the better use of space where a conventional staircase would prove difficult to install.

Headroom

Where there is insufficient height to achieve the recommended 2 m headroom over a stairway, 1.9 m at the centre of the stair is acceptable reducing to 1.8 m at the side (see Fig. 15.8).

Ladders

Fixed ladders may be installed for access in a loft conversion where there is insufficient room to install a conventional staircase without alteration to the existing space if they conform to the following:

- there should be fixed handrails on both sides;
- the ladder should only serve one habitable room;
- retractable ladders are not acceptable for means of escape in case of fire.

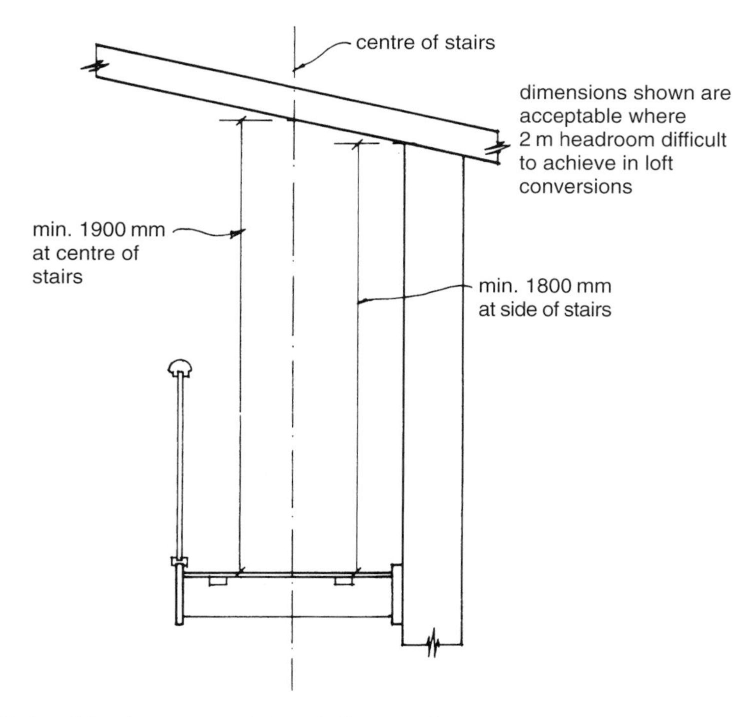

Fig. 15.8 Reduced headroom over stairs – loft conversions.

The definition of habitable room is not included in AD K and it varies from one Approved Document to another. The only safe conclusion which can be drawn is that bathrooms or WCs are not habitable rooms. Therefore, it would seem that a ladder could be used to access one habitable room and a bathroom or WC, but see Alternating tread stairs below.

It would also seem to be the case that retractable ladders can be used where the roof of a bungalow is converted to provide a two-storey house. These houses do not need to have means of escape to a stairway since escape from a suitable first floor window is acceptable in AD B. (See AD B, Section 1, 1.1 and Chapter 7 above).

15.6.6 Spiral and helical stairs

Generally stairs designed in accordance with BS 5395 *Stairs, Ladders and Walkways*, Part 2: 1984 *Code of practice for the design of helical and spiral stairs* will satisfy the recommendations of AD K. It is permissible to provide stairs with lesser goings for conversion work if space is limited, but the stair should only serve one habitable room (and, perhaps a bathroom or WC). The degree of variation from the norm is, however, not specified.

15.6.7 Alternating tread stairs

When first introduced a few years ago, these stairs caused quite a controversy since the treads are not of uniform width and the staircase is steeper than a conventional flight. They rely on a certain degree of familiarity on the part of the user since it is necessary to start the ascent or descent on the correct foot or disaster will ensue!

Alternating tread stairs should only be installed for loft conversion work where insufficient space is available to accommodate a conventional staircase and they should:

- be in one or more straight flights;
- provide access to only one habitable room plus bathroom or WC provided it is not the only WC in the dwelling;
- be fitted with handrails on both sides;
- contain treads which have slip-resistant surfaces;
- have uniform steps with parallel nosings;
- have a minimum going of 220 mm and a maximum rise of 220 mm when measured over the wider part of the tread;
- conform to the recommendations regarding maximum gap sizes for open riser stairs.

A typical design for an alternating tread stair is shown in Fig. 15.9.

15.7 Ramp recommendations

In addition to the general recommendations outlined above, section 2 of AD K contains the following particular guidance for ramps:

- the slope should not be greater than 1 in 12;
- there should be no permanent obstructions placed across any ramp.

15.8 Guarding of stairways, ramps and landings

Guarding should generally be provided at the sides of every flight, ramp or landing. However, guarding need not be provided in dwellings where there is a drop of 600 mm or less, or where there are fewer than two steps in other buildings.

The rules which apply to prevent small children from being trapped in open riser staircases also apply to guarding; that is, there should be no opening of such size as to allow the passage of a sphere of 100 mm diameter. This relates to the guarding on any staircase except that which is in a building which is unlikely to be used by children under the age of five years. It may be difficult in certain circumstances to decide where this exemption should apply. For example, some public houses have a policy of not admitting young children whereas others actually encourage them (accompanied by their parents, of course!). Guarding should also be designed so

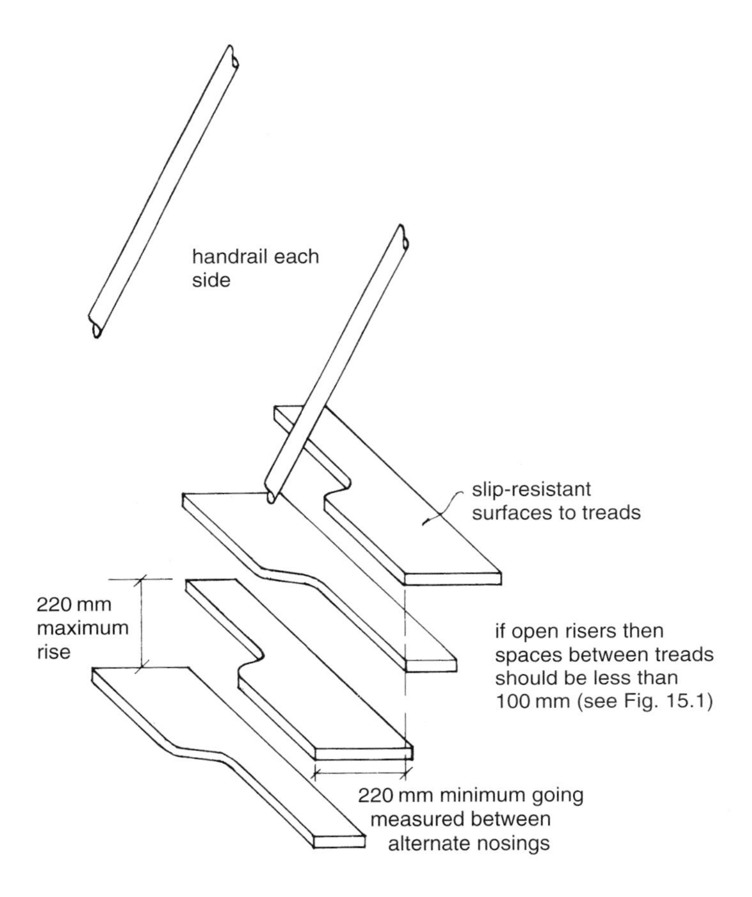

Fig. 15.9 Alternating tread stairs.

that it cannot easily be climbed by small children. (This might preclude the use of horizontal 'ranch' style balustrading, for example.)

Certain minimum heights for the guarding to flights, ramps and landings are given in AD K2/3, Section 3, Diagram 11. These are illustrated in Fig. 15.10 below and it should be noted that the guarding should be strong enough to resist at least the horizontal forces given in BS 6399: Part 1:1996.

15.9 Access to maintenance areas

It is important to establish how frequently access will be required to areas needing maintenance. If this is likely to be more than once per month then permanent stairs or ladders (such as those suggested in AD K1 for private stairs in dwellings) may need to be provided. Alternatively, the requirement may be satisified by following the guidance in BS 5395: Part 3: 1985 *Code of Practice for the Design of Industrial*

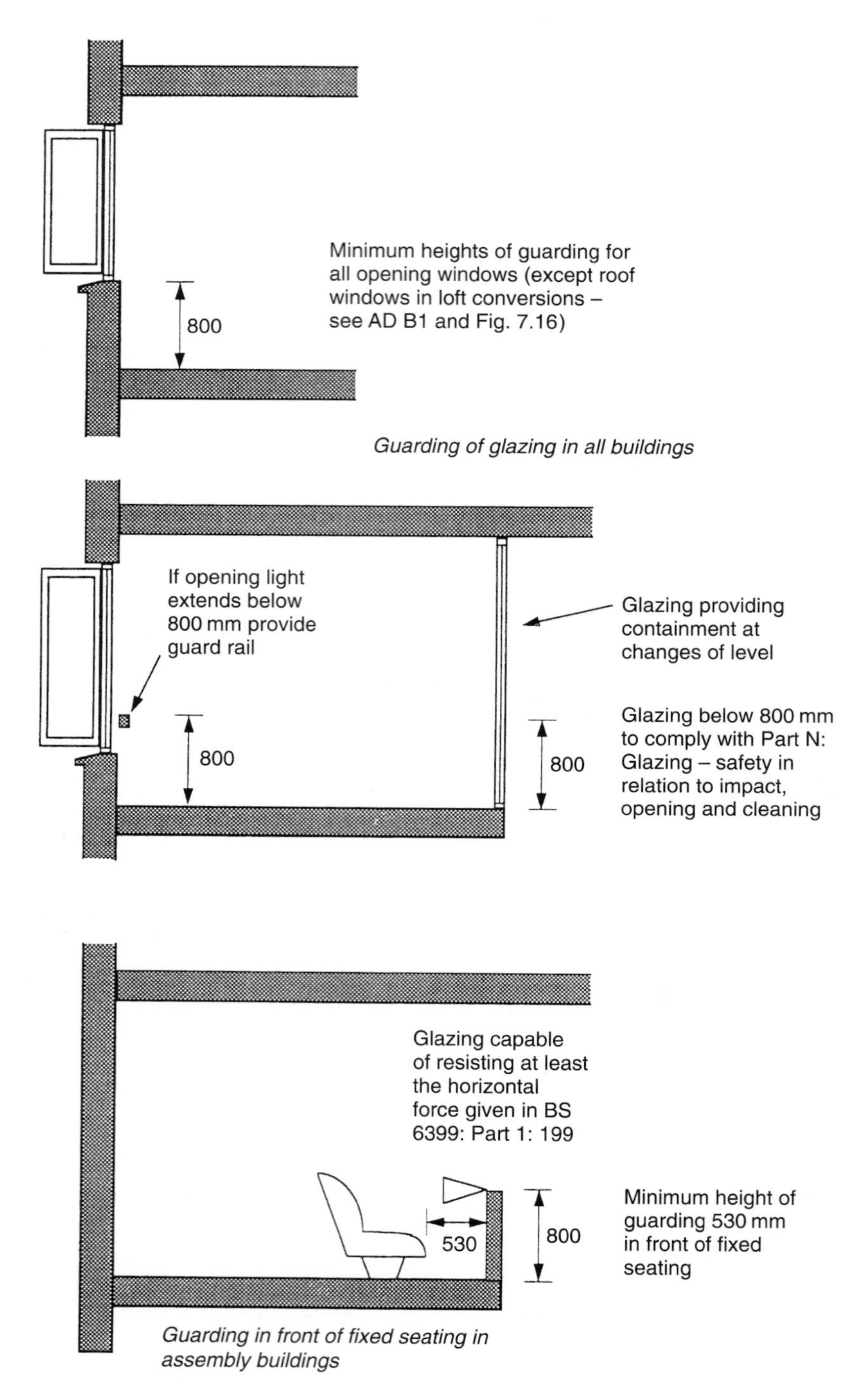

Fig. 15.10 Minimum height of guarding in all buildings.

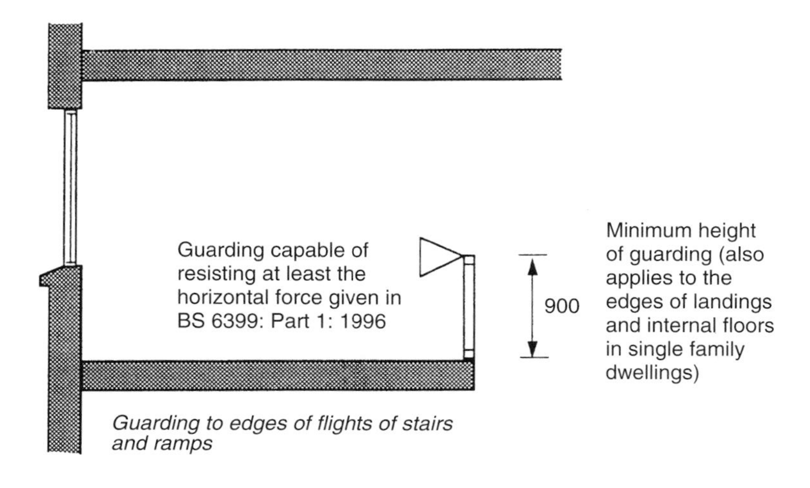

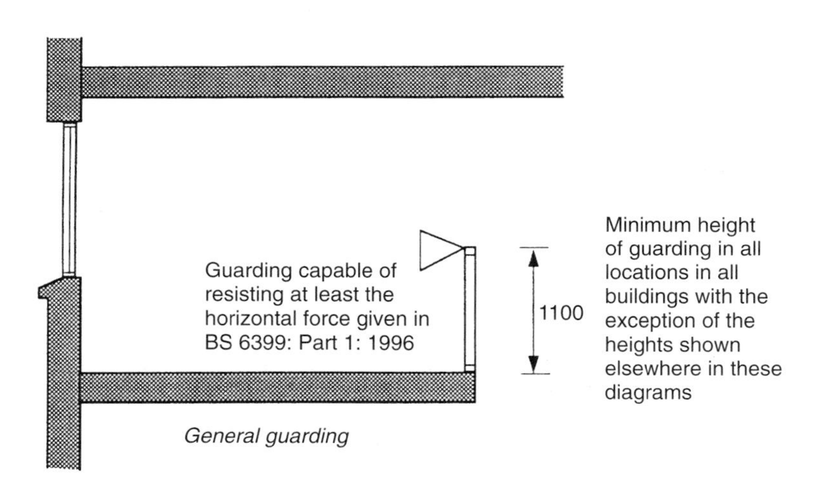

Fig. 15.10 continued.

Type Stairs, Permanent Ladders and Walkways. Less frequent access to maintenance areas may be achieved using portable ladders.

15.10 Alternative approach to stairway design

It is permissible to use other sources of guidance when designing stairs.

- BS 5395 *Stairs, ladders and walkways*, Part 1: 1977 *Code of practice for the design of straight stairs* contains recommendations which will meet the steepness requirements of AD K.

- Wood stairs designed in accordance with BS 585, Part 1: 1989 will offer reasonable safety to users.
- Stairs, ladders or walkways in industrial buildings should follow the recommendations of BS 5395, Part 3: 1985 *Code of practice for the design of industrial stairs, permanent ladders and walkways* or BS 4211: 1987 *Specification for ladders for permanent access to chimneys, other high structures, silos and bins.*

15.11 Protection from falling

The following areas of the building must be provided with barriers to protect people in or about the building from falling.

- Stairways and ramps which form part of the building.
- Floors, balconies, and any roof to which people normally have access.
- Light wells, basements or similar sunken areas which are connected to the building.

Regulation K2 applies to all areas of a building which need to be accessed, including those used only for maintenance.

It should be noted that compliance with Regulation K2 prevents action being taken against the occupier of a building under Regulation 13 of the Workplace (Health, Safety and Welfare) Regulations 1992 when the building is eventually in use. (Regulation 13 relates to requirements designed to protect people from the risk of falling a distance likely to cause personal injury.)

The recommendations contained in AD K2 regarding the provision of pedestrian barriers to provide protection from falling only apply if there is a change in level of:

- more than 600 mm in dwellings; or
- more than the height of two risers in other buildings (the difference in levels in this case will depend on the recommended height for the risers), or 380 mm where there is no stair.

As in the case of stairs (see section 15.3) circumstances will usually dictate what constitutes an acceptable level of safety for the guarding in a building. Therefore, the standard in a dwelling may be lower than that recommended for a public building because there are likely to be less people in the dwelling and they will be familiar with its layout, etc. Similarly, the standard of access to maintenance areas will be lower than that recommended for normal use to reflect the greater care expected of those gaining access.

The provisions for guarding contained in AD K1 are extended by Approved Document K2/3 to cover the edges of any part of:

- a balcony, gallery, roof (including rooflights or other openings), floor (including the edge below any window openings) or other place to which people have access;
- any light well, basement or sunken area adjoining a building;
- a vehicle park (but not, of course, on any vehicle access ramps).

Guarding is not needed where it would obstruct normal use (for example, at the edges of loading bays).

15.11.1 Guarding recommendations

Where guarding is provided to meet the requirements of AD K2, then it should:

- have a minimum height as shown in Fig. 15.10;
- be capable of resisting the horizontal force given in BS 6399: Part 1: 1996;
- have any glazed part designed in accordance with the recommendations of AD N; Glazing – safety in relation to impact, opening and cleaning (see Chapter 18);
- have no opening of such size as to permit the passage of a 100 mm diameter sphere if it is in a building which is likely to be used by children under 5 years of age.

Guarding can consist of a wall, balustrade, parapet or similar barrier but should be designed so that it cannot be easily climbed by small children (i.e. horizontal rails should be avoided) if they are likely to be present in the building.

15.11.2 Guarding recommendations for maintenance areas

It is important to establish how frequently access will be required to areas needing maintenance. If this is likely to be more than once per month then guarding provisions such as those suggested for dwellings (see Fig. 15.10) will be satisfactory. For less frequent access it may only be necessary to provide temporary guarding and in many circumstances the posting of warning notices may be sufficient to satisfy the requirement. Building Regulations, of course, do not cover the design and installation of temporary guarding or warning signs although they are covered by the Construction (Design and Management) Regulations 1994. (For information on signs see the Health and Safety (Signs and Signals) Regulations 1996.)

15.12 Vehicle barriers and loading bays

Vehicle ramps and any levels in a building to which vehicles have access are required to have barriers in order to protect people in or about the building. Additionally, vehicle loading bays must either be constructed in such a way as to protect people in them from collision with vehicles or they must contain features which achieve the same end.

15.12.1 Vehicle barriers

If the perimeter of any roof, ramp or floor to which vehicles have access, forms part of a building, it should have barriers to protect it, provided that it is level with or above any adjacent floor, ground or vehicular route.

Vehicle barriers can be formed by walls, parapets, balustrading or similar obstructions and should be at least the heights shown in Fig. 15.11. Barriers should be capable of resisting the horizontal forces as set out in BS 6399 *Loading for buildings*, Part 1: 1984 *Code of practice for dead and imposed loads*.

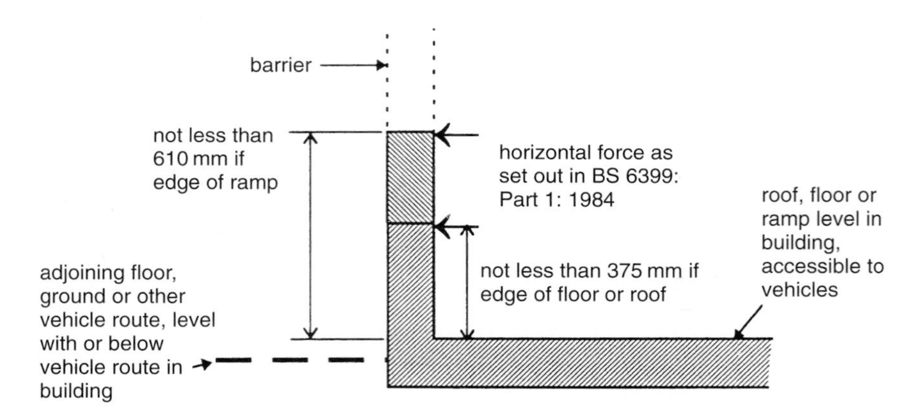

Fig. 15.11 Vehicle barriers – any building.

15.12.2 Loading bays

Loading bays for less than three vehicles should be provided with:

- one exit point (e.g. steps) from the lower level, ideally near to the centre of the rear wall; or
- a refuge into which people can go if in danger of being struck or crushed by a vehicle.

Larger loading bays (i.e. for three or more vehicles) should be provided with:

- one exit point at each side, or
- a refuge.

See Fig. 15.12 for details.

15.13 Protection from collision with open windows, skylights or ventilators

Regulation K4 requires that provision be made to prevent people who are moving in or about a building from colliding with open windows, skylights or ventilators. This requirement does not apply to dwellings but it does apply to all parts of other buildings including, in a limited form, to areas of the building used exclusively for maintenance purposes.

Again, compliance with this regulation prevents action being taken against the

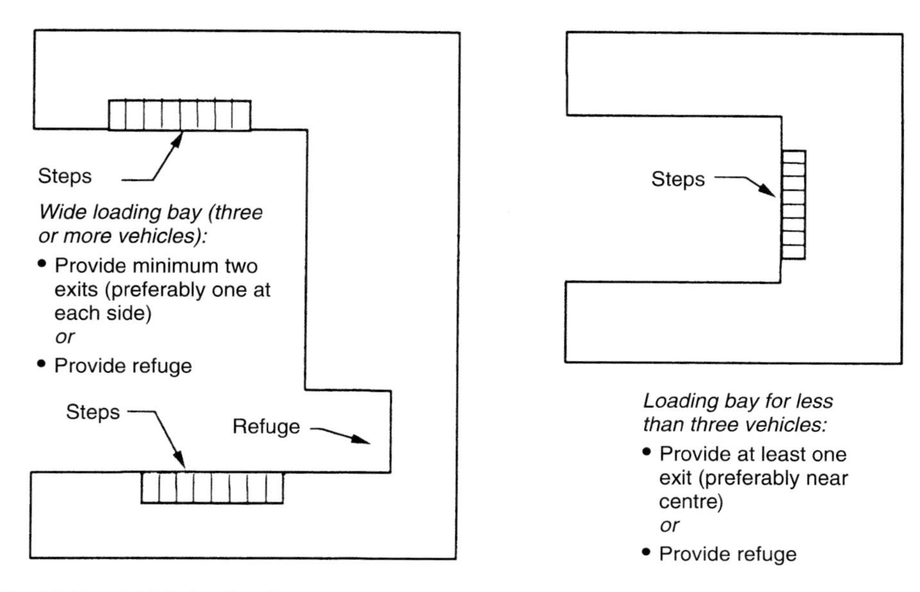

Fig. 15.12 Vehicle loading bays.

occupier of a building under Regulation 15(2) of the Workplace (Health, Safety and Welfare) Regulations 1992 (Regulation 15(2) also relates to projecting windows, skylights and ventilators) when the building is eventually in use.

Since it is desirable to leave windows, etc. open for ventilation purposes, this should be possible without causing danger to people who might collide with them. Generally, this can be achieved by:

- Providing windows, skylights and ventilators so that projecting parts are kept away from people moving in and about the building; or
- Providing features which guide people away from these projections.

In certain special cases (e.g. in spaces used only for maintenance purposes) it is reasonable to expect the exercise of greater care by those gaining access, therefore less demanding provisions than those for normal access could satisfy the regulation.

15.13.1 Avoiding projecting parts

AD K4 makes it clear that the requirements of Regulation K4 do not apply to the opening parts of windows, skylights or ventilators:

- which are at least 2 m above ground or floor level; or
- which do not project more than about 100 mm internally or externally into spaces in or about the building where people are likely to be present.

Since the siting of opening lights above 2 m may not always be possible, an acceptable solution might be to fit projecting lights with restraint straps designed to restrict the projection to about 100 mm. The restraint should only be removed for maintenance and cleaning purposes. An ordinary casement stay would probably not be considered suitable for this purpose. Any such solution should, of course, be checked for acceptability with the building control authority (local authority or approved inspector).

Alternative solutions given in AD K4 also include:

(a) Marking the projection with a feature such as a barrier or rail about 1100 mm high to prevent people walking into it; or
(b) Guiding people away from the projection by providing ground or floor surfaces with strong tactile differences or suitable landscaping features.

These solutions are shown in Fig. 15.13 below.

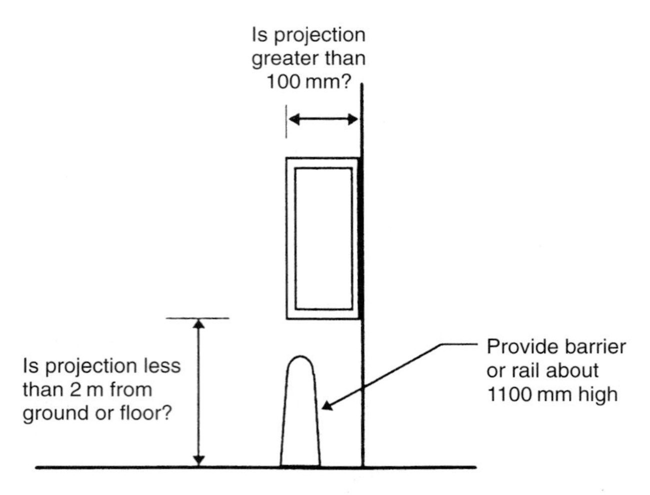

Fig. 15.13(a) Marking projections – barriers.

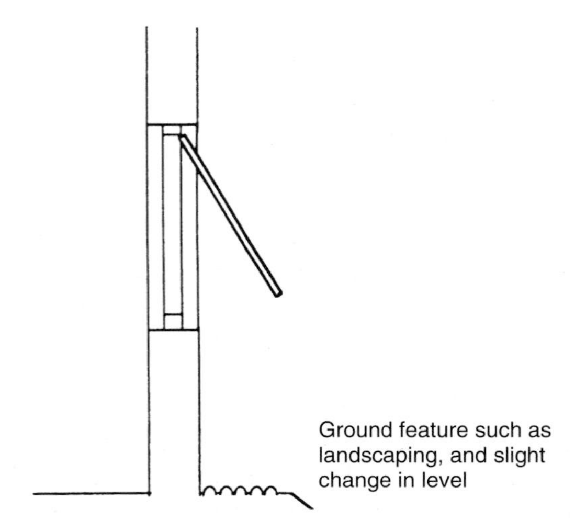

Fig. 15.13(b) Marking projections – ground features.

15.13.2 Maintenance areas

Where such areas are used infrequently, projecting parts of windows, skylights and ventilators should be made easier to see by being clearly marked, in order to satisfy Regulation K4.

It should be noted that there are other provisions in the Building Regulations which relate to safety on common circulation routes in buildings. In particular, reference should be made to Approved Document B – Fire safety (see section 7.9.3 for guidance on the clear widths of escape routes), and Approved Document M – Access and facilities for disabled people (see section 17.5.5 below for guidance on the avoidance of hazards on circulation routes).

15.14 Safe use of doors

In order that they may be safely used, provisions must be made to prevent doors or gates:

- which slide or open upward, from falling onto any person; and
- which are powered, from trapping anyone.

Furthermore, powered doors and gates must be openable in the event of a power failure and all swing doors and gates are required to be designed to allow a clear view of the space on either side.

Regulation K5 does not apply to dwellings, or to any door or gate which is part of a lift. Compliance with this regulation prevents action being taken against the occupier of a building under Regulation 18 (requirements for doors and gates) of the Workplace (Health, Safety and Welfare) Regulations 1992 when the building is eventually in use.

15.14.1 Door and gate safety features

Typically, the following safety features will satisfy the requirements of Regulation K5 by preventing the opening and closing of doors and gates from presenting a safety hazard:

- Vision panels should be provided in doors and gates on traffic routes and in those which can be pushed open from either side (i.e. double swing doors) unless they can be seen over (e.g. lower than about 900 mm for a person in a wheelchair). For example, this means providing such doors and gates with a glazed panel giving a zone of visibility between 900 mm and 1500 mm from floor level (see also Fig. 17.3 in section 17.6 below).
- Sliding doors and gates should be prevented from leaving the end of the suspension track by a stop or other suitable means. Additionally, if the suspension system fails or the rollers leave the track they should be prevented from falling by means of a retaining rail.

- Upward opening doors and gates should be designed to prevent them from falling and causing injury.
- If doors and gates are power operated the following safety features are examples of those which could be incorporated:
  (a) devices to prevent injury to people from being struck or trapped (e.g. doors with pressure sensitive edges)
  (b) readily accessible and identifiable stop switches
  (c) manual or automatic opening provisions in the event of power failure if this is necessary for health and safety reasons.

Reference should also be made to other provisions in the Building Regulations which relate to the design and use of doors in buildings. In particular, see Approved Document B – Fire safety (see section 7.22.3 above for guidance on the provision of doors on escape routes) and Approved Document M – Access and facilities for disabled people (see sections 17.6, 17.6.1 and 17.6.2 below for guidance on the design of internal and external doors).

16 Conservation of fuel and power

J.R. Waters

16.1 Introduction

Requirements for the conservation of fuel and power are dealt with by Part L of Schedule 1, and there are two Approved Documents providing practical guidance for meeting the requirements. The first, Approved Document L1 (AD L1), is applicable to dwellings, and the second, Approved Document L2 (AD L2), is applicable to all buildings other than dwellings. The two documents are separate publications and are substantially different. However, there are some common elements between them and these areas of commonality are listed in Table 16.1. All other material differs between the two documents, even though there is some similarity in the wording in places. It is therefore important to maintain the dis-

Table 16.1 Material common to L1 and L2.

Section	Paragraph in L1	Paragraph in L2	Comment
Use of Guidance			Identical except for 'Mixed Use Development' in L2
Technical risk	0.5	0.9	
Thermal conductivity and transmittance	0.6 to 0.9	0.10 to 0.13	
U-value reference tables	0.10	0.14	
Calculation of U-values	0.11 to 0.13	0.15 to 0.17	
Roof window	0.14	0.18	
Basis for calculating areas	0.15	0.19	
Air permeability	0.16	0.20	
Limiting thermal bridging at junctions and around openings	1.30 to 1.32	1.90 to 1.11	Identical except for some additional references in L2
Appendix A: Tables of U-values	The whole Appendix	The whole Appendix	
Appendix B: Calculating U-values	The whole Appendix	The whole Appendix	
Appendix C: U-values of ground floors	The whole Appendix	The whole Appendix	
Appendix D: Determining U-values for glazing	The whole Appendix	The whole Appendix	The principle is the same but the examples differ

tinction between the Regulations and Requirements of L1 and L2 when applying them to a specific building project. Consequently, in this chapter parts L1 and L2 are treated separately. However, to avoid undue repetition, the common material in 'Use of Guidance' and in 'Introduction to the Provisions' (L1 paragraphs 0.5 to 0.16; L2 paragraphs 0.9 to 0.20) is considered first, then L1 and L2 are described separately. The common material in Appendices A, B, C and D is presented next, followed by the remaining Appendices of L1 and L2.

16.1.1 Use of guidance

Although this section is common to both Approved Documents, there are two points of note. The first point arises from the fact that the thermal performance of a building is the result of a complex interaction between many different systems (building fabric, services systems, fuel supply, availability of ambient energy, solar gains, building usage, etc.). It is therefore possible that novel design solutions can be found which cannot easily be assessed for compliance by the guidance and methods contained in the Approved Documents, but which nevertheless can be shown to satisfy the requirements for the conservation of fuel and power by some other means. In any case, it is not legally mandatory to use the Approved Documents, provided that it can be demonstrated, to the satisfaction of a Building Control Body, that the legal requirement of Part L of Schedule 1 to conserve fuel and power has been met.

The second point of note concerns the statement that Building Regulations do *not* require anything to be done *except* for the purpose of securing reasonable standards of health and safety. Parts L1 and L2 (along with Part M and paragraphs H2 and J6) are excluded from this statement because the requirement to conserve fuel and power is *in addition* to health and safety, and not a substitute for it.

16.1.2 General definitions applicable to L1 and L2

Both L1 and L2 begin with a reminder that designing to minimise energy consumption may carry the risk of technical problems in other areas. High levels of thermal insulation, and careful attention to draught proofing can cause problems due to:

- Interstitial condensation
- Surface condensation in roof spaces
- Inadequate ventilation for occupants
- Inadequate ventilation and air supply to combustion systems and flues.

Other potential problems are these.

- Rain penetration causing, among other things, damage to thermal insulation. This is often associated with failure of flat roof coverings due to the high levels of thermal stress induced by high levels of thermal insulation within the roof construction.
- Sound transmission, due to the fact that materials which provide good thermal

insulation are normally very poor at providing insulation against the transmission of sound.

Guidance on the avoidance of these related technical risks must be found elsewhere, and a number of sources of information are suggested:

- BRE Report No. 262: *Thermal Insulation: avoiding risks*, 2002 edition [1]
- Approved Document F, *Ventilation*
- Approved Document J, *Combustion appliances and fuel storage systems*
- Approved Document E, *Resistance to the passage of sound*
- *Limiting thermal bridging and air leakage: Robust construction details for dwellings and similar buildings* [2].

Both L1 and L2 also contain a series of definitions, as follows, that apply throughout the documents.

Thermal conductivity (the λ-value)

This is the rate at which heat will pass through unit area of a material when there is unit temperature gradient across the material. It is usually expressed in watts per square metre for a temperature gradient of one degree Kelvin per metre, and is given the symbol λ (Greek lambda). The units of λ are thus $\frac{W}{m^2} / \frac{K}{m}$, which simplifies to $\frac{W}{m.K}$.

Thermal transmittance (the U-value)

This is the rate at which heat will pass through unit area of a material (or a construction made up of several materials) when there is unit temperature difference between the environments on the opposite sides of the material. It is usually expressed in watts per square metre for a temperature difference of one degree Kelvin, and is given the symbol U. The units of U are thus $\frac{W}{m^2.K}$.

If measured test results for λ and/or U-values are available, these should be used. Manufacturers are often able to provide such information for the λ-values of their materials, but where these are not available, values may be obtained from data published in Appendix A of the Approved Documents or in any other authoritative publication (e.g. BS EN 12524 [3] or CIBSE Guide, section A3 [4]). The measurement standards that should be followed are BS EN 12664 [5], BS EN 12667 [6] and BS EN 12939 [7]. Manufacturers sometimes also provide test results for the U-values of various construction elements that incorporate their materials, the relevant measurement standards being BS EN ISO 8990 [8] or, for windows and doors, BS EN ISO 12567-1 [9]. More usually, the U-values supplied by manufacturers are obtained by calculation from λ-values. Where calculated values are used, care must be taken to ensure that proper allowance has been made for thermal bridging effects. These are most likely to arise when the construction includes joists, structural or other types of framing, or any material or component that breaks through an insulation layer. The bridging effect can only be ignored when the bridged material and the bridging material have sufficiently similar thermal properties (i.e. if the difference in their thermal resistances is less than 0.1 $\frac{m^2K}{W}$). This will normally apply to mortar joints in brickwork, but not necessarily to mortar joints in lightweight blockwork. Calculation procedures for U-values are specified in:

- BE EN ISO 13789 [10] and BRE/CRC [11] for calculation methods and conventions
- BS EN ISO 6946 [12] for walls and roofs
- BS EN ISO 13370 [13] for ground floors
- BS EN ISO 10077-1 [14] or prEN ISO 10077-2 [15] for windows and doors
- BS EN ISO 13370 [13] or BCA/NHBC Approved Document [16] for basements
- BS EN ISO 10211-1 [17] and BS EN ISO 10211-2 [18] for thermal bridges.

Appendix B of AD L1 and L2 provides a simplified method, based on BS EN ISO 6946 [12], which is suitable for the calculation of the U-values of most wall and roof constructions. Section 16.5 gives example calculations.

Exposed element

This means an element exposed to the outside air, and includes:

- A suspended floor over a ventilated or unventilated void
- An element exposed to the outside air indirectly via an unheated space
- An element in a floor or basement that is in contact with the ground.

It should be noted that an element exposed to the outside air indirectly via an unheated space was previously known as a 'semi-exposed' element. The calculation of the U-value of such an element must now use the method given in SAP 2001 [19]. It should also be noted that a wall separating a dwelling from any other premises that are heated to the same temperature does not require thermal insulation.

Roof window

This is defined as a window in the plane of a pitched roof. Within AD L1 and L2 such a window may be considered as a rooflight.

Basis for calculating areas

When evaluating areas, measurements should be taken on the internal faces of external elements. The areas of projecting bays must be included. Roof areas should be measured in the same plane as the roof insulation, and floor areas should include non-useable space (e.g. stairwells and builders' ducts).

Air permeability

This is a measure of the leakiness of the building fabric to unwanted internal to external air exchange. It is defined as the average volume of air, per unit area, which passes through the building envelope, when there is an internal to external pressure difference. It is normally expressed in cubic metres per hour, per square metre of building envelope area, at a pressure difference of 50 Pa. Within AD L1 and L2, the envelope area is taken to be the total area of walls, floor and roof separating the interior volume from the external environment. Air permeability does not include deliberate leakage paths between the inside and the outside, such as flues, ventilation ducts, air bricks, etc. These are sealed up during any measurement procedure, so that only cracks, gaps at joints and similar leakage paths are included. The lower the air permeability the better, as this shows that the building fabric is more airtight.

Standard assessment procedure (SAP)

This is the UK Government's procedure for rating the energy cost performance of dwellings [19], and is defined in section 16.9. Separately and independently of Part L1, Building Regulation 16 requires that whenever a new dwelling is created (either by building work or by a material change of use), the energy rating of that dwelling must be calculated by means of the standard assessment procedure. The result of the SAP calculation cannot be used by itself to demonstrate compliance with Part L1. However, the SAP calculation can be continued to find the carbon index of the dwelling. If the carbon index is above a specified minimum, compliance has been demonstrated (see section 16.2.3).

16.1.3 Testing

Building Regulation 18 has been extended. Previously it gave local authorities the power to test drains and private sewers for compliance with Part H. Now, Regulation 18 allows a local authority to make such tests as may be necessary:

- for compliance with Building Regulation 7, which specifies that all building work should be carried out with proper materials and in a workmanlike manner; and also
- for compliance with any of the applicable requirements of Schedule 1.

The inclusion of Schedule 1 within Regulation 7 confers very wide powers on a local authority to require testing. Much of that testing could be done beforehand on components, using approved test procedures, and the results provided in the form of appropriate certificates. However, there are several areas where testing can only be carried out and be effective when the building is complete, and two of these are of particular significance to Part L:

- airtightness testing of buildings to test for compliance with the air permeability criterion; and
- testing for continuity of insulation and the avoidance of thermal bridging.

In the case of airtightness testing, the Approved Documents point to only one type of test, the fan pressurisation test, and the criterion for compliance is written specifically in terms of the result which this test provides. Indeed, for large buildings above 1000 m^2 floor area, Approved Document L2 appears to offer no alternative to the test as a means of demonstrating compliance.

For the testing of the continuity of insulation, there is a similar lack of flexibility. Unless an authoritative certificate can be provided stating that the design details and building techniques are appropriate, the Approved Documents recommend satisfactory results from a test of the whole of the visible external envelope using infrared thermography.

16.2 The conservation of fuel and power in dwellings

Part L1 of Schedule 1 to the 2000 Regulations (as amended) is concerned with the conservation of fuel and power in dwellings. This is supported by Approved Document L1 which covers the following topics.

- Compliance by the elemental method
- Compliance by the target U-value method
- Compliance by the carbon index method
- Limiting thermal bridging and air leakage
- Space heating controls and hot water systems
- Commissioning, operating and maintenance of heating and hot water systems
- Insulation of pipes and ducts
- External lighting
- Conservatories.

The purpose of Part L1 is to minimise the environmental impact of dwellings with respect to the depletion of fuel reserves and the degradation of the atmosphere by carbon dioxide emissions.

16.2.1 The legal requirement for the conservation of fuel and power in dwellings

The regulation requires that a dwelling should be so designed and constructed as to make reasonable provision for the conservation of fuel and power, and that this shall be achieved by attention to all of the following four points.

- Limiting heat loss through the fabric of the building, from hot water pipes, from air ducts used for space heating and from hot water vessels.
- Ensuring that the space heating and hot water systems which are provided are energy efficient.
- Ensuring that internal and external lighting systems are designed to use energy efficiently.
- Ensuring that, by the provision of sufficient information, building occupiers are able to operate and maintain the heating and hot water services efficiently, so as to use no more energy than is reasonable in the circumstances.

In addition, under Regulation 16, all new dwellings (whether created by new construction or by change of use) are required to be given a SAP Energy Rating. This must be supplied to the local authority or approved inspector.

16.2.2 AD L1 Section 0 – general guidance

Approved Document L1 gives general and specific guidance on the energy efficient measures which will satisfy the requirements. These include not only means to limit heat loss from the building and its heating and hot water services, but also means to limit the energy consumed by the heating, hot water and lighting equipment.

Heat losses through walls, roofs, floors, windows and doors, etc. must be kept within acceptable limits by means of suitable levels of thermal insulation. Also, where it is possible and appropriate, any beneficial gains from solar heat and more efficient heating systems should be taken into account.

Heat may also be lost by unplanned air leakage through the fabric of a building, particularly around doors, windows, and junctions between elements or components. This requires attention not only to draught stripping of doors and windows, but also consideration of potential air leakage paths through all other elements of the building.

Heat losses from hot water pipes, hot air ducts, hot water vessels, and all their connections must also be controlled by the application of sufficient insulation. If, however, a particular component contributes to the space heating in an efficient manner, the insulation may be omitted.

The energy efficiency of the systems that provide space heating and hot water has two aspects. Firstly, the heat generating device (i.e. the boiler) must extract heat from the fuel in an efficient manner. This means that it must have a sufficiently high combustion efficiency. Secondly, the heat must be delivered and distributed efficiently. This means that the equipment must have controls for space and water temperatures and for the timing of its operation. Furthermore, the systems must have been appropriately commissioned and be capable of being operated in an efficient manner by the user.

The requirements for lighting systems apply both to internal lighting and to external lighting fixed to the building. All lighting systems should, where appropriate, be fitted with energy efficient lamps. In addition, external lighting systems should have a control system, either fully automatic or combined manual/automatic, which acts in such a way as to conserve energy.

The energy efficiency of heating and hot water systems can be affected by the manner in which they are operated. Incorrect usage may result in poor energy performance from an otherwise energy efficient system. Consequently, there is a recommendation to provide easily understandable information that will enable the occupier to operate and maintain these systems so that they do not waste energy. This information should also include the results of performance tests on the systems.

16.2.3 AD L1 Section 1 – design and construction

Specific guidance

For specific guidance, Approved Document L1 offers three methods for demonstrating that reasonable provision has been made for limiting heat loss through the building fabric.

The elemental method has the advantages that it involves a minimum of calculation effort, and is appropriate for alterations and extensions as well as for new construction. However, it allows less flexibility in the design of the dwelling than other methods and can only be used with certain heating systems:

- systems based on an efficient gas or oil boiler
- community heating with CHP
- systems based on biogas or biomass fuel.

The elemental method *cannot* be used for dwellings using any other heating system (such as direct electric heating).

The target U-value method can only be used for complete dwellings. However, it can be used with any heating system. Its advantage is that it allows adjustment of the areas of windows, doors and rooflights by taking into account insulation levels, heating system efficiency and the possibility of solar gain.

The carbon index method can also be used for complete dwellings only and with any heating system. It is intended to allow substantial flexibility in design, and is likely to be the most suitable method if the design includes unconventional or novel features. In practice it is based on the calculation of the SAP rating, which is then converted to the carbon index. The requirement is that the carbon index for a dwelling (or for each dwelling in a block of flats or converted building) should be not less than 8.0. Values below 8.0 fail the requirement whereas values of 8.0 and above satisfy it.

The Approved Document provides a 'Summary Guide', which is a comprehensive check-list to help the designer choose the most suitable of the above three methods. This check-list is shown in flow chart form in Fig. 16.1a. The summary guide also indicates the additional checks which should be made by builders; these are shown in Fig. 16.1b.

The elemental method for dwellings

In the elemental method, compliance is demonstrated by specifying:

- Maximum U-values for walls, floors and roofs
- Maximum area-weighted U-values for windows, doors and rooflights
- Maximum combined area of windows, doors and rooflights
- Minimum boiler efficiencies for the heating system boiler (the SEDBUK rating, or Seasonal Efficiency of a Domestic Boiler in the UK).

The maximum U-values allowed by the elemental method are shown in Table 16.2, and are illustrated in Fig. 16.2.

If an element is exposed to the outside through an unheated space (e.g. garage, atrium, etc.) the unheated space can be ignored, provided it is assumed that the element itself is exposed to outside air, i.e. the element is assigned its normal U-value. Alternatively, the unheated space may be included in the U-value calculation for the element, as demonstrated in section 16.5.1. The slightly higher per-

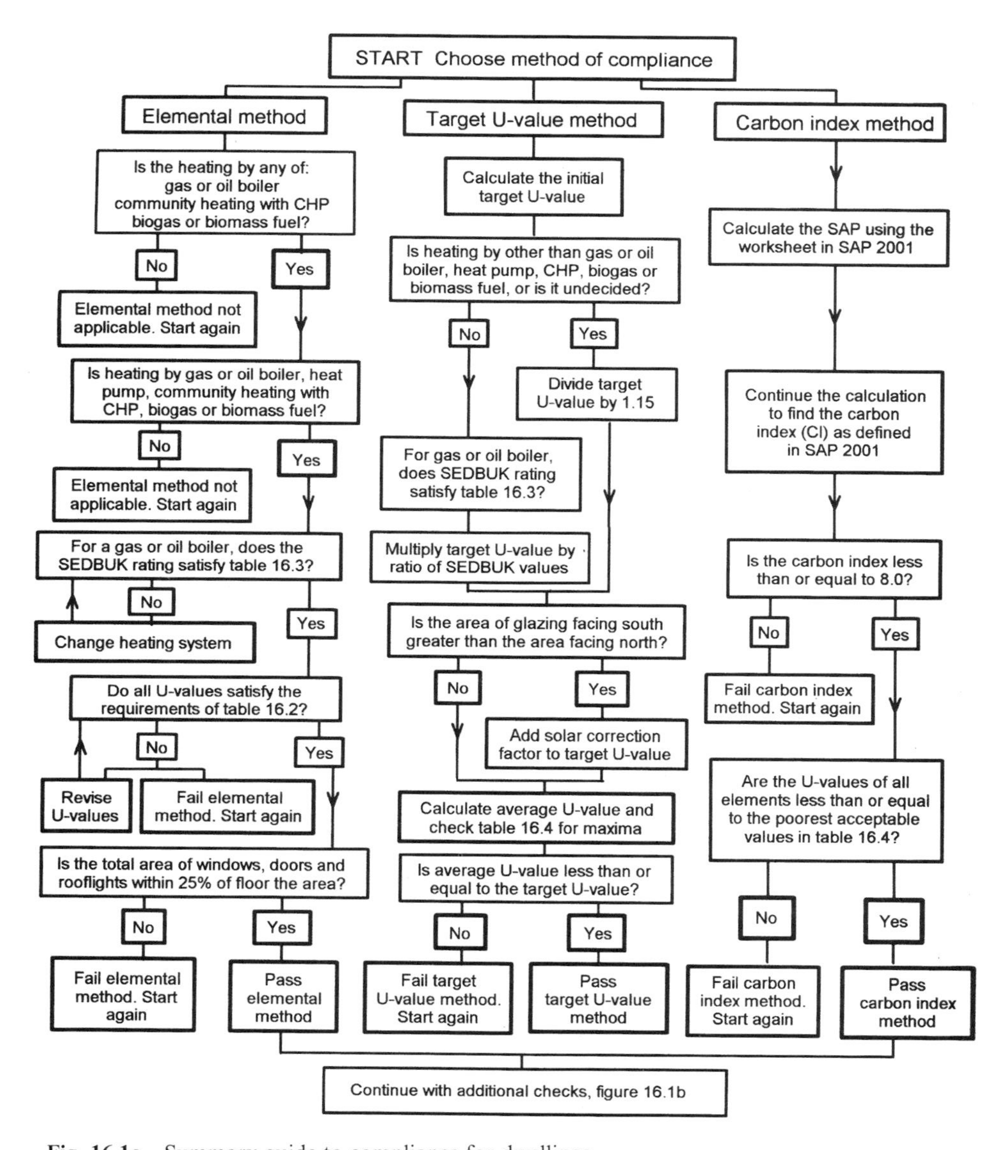

Fig. 16.1a Summary guide to compliance for dwellings.

mitted U-value for openings in metal frames is intended to allow for the fact that, because the frames themselves are usually more slender than wood or UPVC and therefore have a higher proportion of glazing, there is on average additional solar gain to offset the heat load of the dwelling.

The area weighted U-value of the windows, doors and rooflights, U_{WDR}

The maximum values in Table 16.2 refer to the area weighted average of all the windows, doors and rooflights in the dwelling; this can be calculated from the formula at the bottom of page 16.11.

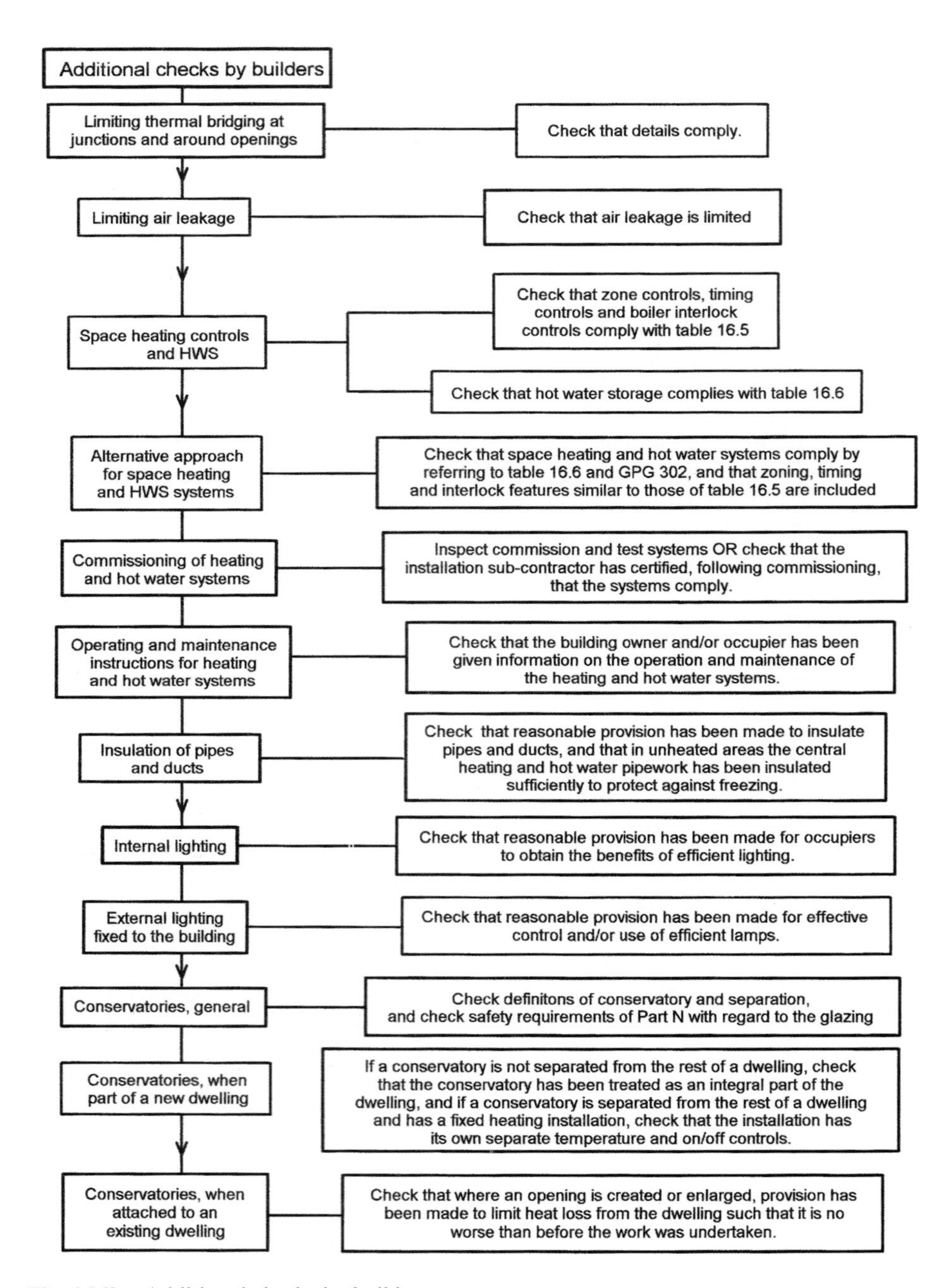

Fig. 16.1b Additional checks by builders.

Table 16.2 Maximum U-values for the elemental method.

Exposed element	Maximum U-value, W/m^2K	Comment
Pitched roof, insulation between rafters	0.20	Any part of a roof having a pitch of 70° or more can be considered as a wall.
Pitched roof with integral insulation	0.25	
Pitched roof, insulation between joists	0.16	
Flat roof	0.25	Roof of pitch 10° or less
Walls, including basement walls	0.35	
Floors, including ground floors and basement floors	0.25	
Windows, doors and rooflights, glazing in metal frames, area-weighted average	2.20	Rooflights include roof windows. The higher U-value for metal frames allows for extra solar gain due to greater glazed proportion
Windows, doors and rooflights, glazing in wood or PVC frames, area-weighted average	2.00	

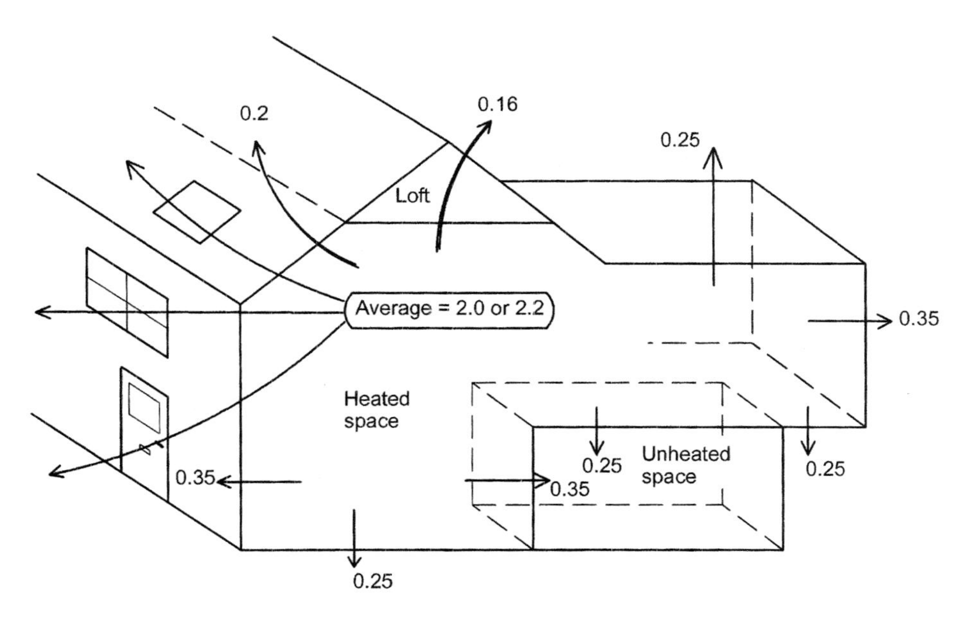

Fig. 16.2 Maximum U-values allowed by the elemental method for dwellings.

$$\text{Average U-value, } U_{WDR} = \frac{A_W U_W + A_D U_D + A_R U_R}{A_W + A_D + A_R}$$

where A_W, A_D and A_R are, respectively, the areas of windows, doors and rooflights
U_W, U_D and U_R are their U-values.

Maximum areas of windows, doors and rooflights

The elemental method allows a maximum combined area, called the *standard area provision*, which is equal to 25% of the total floor area. If the average U-value, U_{WDR} of the windows, doors and rooflights, does not exceed the value given in Table 16.2, *and also* their combined total area does not exceed 25% of the total floor area, then the requirement is met.

Minimum boiler efficiencies

The minimum boiler efficiencies allowed by the elemental method are shown in Table 16.3. There is no provision to use lower values. The SEDBUK efficiency (also called the SAP seasonal efficiency) of a large number of boilers can be found on the internet (www.SEDBUK.com). If the SEDBUK of the proposed boiler is not available, it is permissible to use the appropriate seasonal efficiency value as given in SAP 2001.

Table 16.3 Minimum boiler seasonal efficiencies (SEDBUK) for the elemental method.

Central heating system fuel	SEDBUK efficiency, %
Mains natural gas	78
LPG	80
Oil	85
Oil-fired combination (as calculated using the SAP-98 method)	82

Adjustments allowed by the elemental method

When applying the elemental method to a complete dwelling, the only adjustment available concerns the standard area provision for windows, doors and rooflights. If the U-value of any or all of the windows, doors and rooflights is greater than normal, due say to lower performance glazing, then the area-weighted average U-value may exceed the value given in Table 16.2, and may therefore lead to a failure to meet the requirement. In such a case, the total area may be reduced to compensate. The over-riding principle is that the actual heat loss through the windows, doors and rooflights should be no more than if the maximum average U-value and the standard area provision had been met. If A_F is the total floor area, the maximum allowable heat loss through windows, doors and rooflights may be found from:

- For metal frames: maximum heat loss $= 2.2 \times 0.25A_F = 0.55A_F$
- For all other frames: maximum heat loss $= 2.0 \times 0.25A_F = 0.50A_F$

The actual heat loss is given by:

$$\text{Actual heat loss} = A_W U_W + A_D U_D + A_R U_R$$

Combining these equations, the test for whether or not the areas satisfy the requirement is:

- For metal frames: $A_W U_W + A_D U_D + A_R U_R \leq 0.55A_F$
- For all other frames: $A_W U_W + A_D U_D + A_R U_R \leq 0.50A_F$

When the total area of openings is less than the standard area provision, there is an additional heat loss through the extra area of wall or roof. This is not included in the above test, but the resulting error is small, and in most cases can be ignored. Examples of how the adjustment may be made are given in section 16.7. However, within the elemental method, these formulae *cannot* be used to *increase* the areas above the standard area provision. If this is desired, a different approach must be used, such as the target U-value or the carbon index method.

The elemental method applied to extensions to dwellings

Unlike other methods, the elemental method can be applied to an extension to a dwelling. If the target U-value or the carbon index method is used, the extended dwelling must be considered as a whole, and the method applied to the complete newly enlarged structure. However, when applied to extensions, the elemental method does allow some flexibility. The ways in which the requirements can be satisfied are:

- Adhere to the maximum U-values given in Table 16.2 for all elements in the extension, including the area-weighted average U-value of the windows, doors and rooflights. Alternatively, use U-values which would give the same total rate of heat loss from the extension if the U-values in Table 16.2 had been used.
- *In addition*, ensure that *either*:
  (1) the area of openings (i.e. windows plus doors plus rooflights) in the extension does not exceed the area of any original windows or doors which no longer exist or are no longer exposed because of the extension, plus 25% of the floor area of the extension, *or*
  (2) the total area of openings in the new enlarged dwelling does not exceed the total area of openings in the existing dwelling prior to the extension, *or*
  (3) the area of openings in the enlarged dwelling does not exceed 25% of the total floor area of the enlarged dwelling.

In order to interpret condition (1) above, first take the area of the original windows and/or doors which no longer exist or are no longer exposed because of the extension. Then take 25% of the floor area of the extension. Add these two areas together. The resulting total is the maximum permissible area of openings in the extension.

A further relaxation is allowed in the case of a small extension where the additional heated space has a maximum floor area of about 6 m^2. In such cases, the construction details of the extension will be acceptable if they have an energy performance as least as good as those of the existing dwelling. This relaxation would typically apply to a porch or small kitchen extension.

The target U-value method for dwellings

In order to demonstrate compliance using the target U-value method, it is necessary to calculate and compare two U-values. The first, U_T, is a target U-value. This is a theoretical index which is calculated from an initial basic formula, and which is then adjusted to allow for any design features that affect the energy consumption of the dwelling. The second, U_{AVG}, is the area-weighted average U-value of the actual dwelling, taking into account all exposed surfaces including walls, floors, roofs, windows, doors, rooflights and all elements adjacent to unheated spaces. The two values are then compared. The target U-value procedure can be arranged in several stages:

- Calculate an initial target U-value, U_1, using the standard formula
- Adjust U_1 for the efficiency of the proposed boiler, and obtain U_2
- For dwellings with metal window frames, allow for additional solar heat gain by adjusting U_2 to U_3
- For dwellings with sufficient south facing openings, allow for additional solar heat gains by adjusting U_3 to the target U-value, U_T
- Calculate the average U-value, U_{AVG}, of the proposed dwelling, using the actual U-values and areas of the elements.

The target U-value, U_T, is the overall maximum U-value of the dwelling, and as such, replaces the maximum U-values specified in Table 16.2 for the elemental method. Provided the average U-value, U_{AVG}, is less than (or at worst equal to) U_T, the requirement for compliance is satisfied.

The calculation procedure and the relevant formulae may be arranged as follows.

(1) Find U_1, the initial target U-value, from:

$$U_1 = \left[0.35 - 0.19\frac{A_R}{A_T} - 0.10\frac{A_{GF}}{A_T} + 0.413\frac{A_F}{A_T}\right]$$

where U_1 is the initial target U-value
A_R is the area of exposed roof
A_{GF} is the ground floor area
A_F is the total floor area, including all storeys
A_T is the total area of all exposed elements.

(2) Adjust U_1 according to the boiler efficiency ratio, f_e, found from:

$$f_e = \frac{\text{Proposed boiler SEDBUK(\%)}}{\text{Reference boiler SEDBUK(\%)}}$$

where the reference boiler SEDBUK is the minimum value as given in Table 16.3. A convenient source of SEDBUK values may be found on the internet web site www.SEDBUK.com. The adjusted target U-value, U_2, is then:

$$U_2 = F_e\, U_1$$

If there is no suitable data for the boiler, either because there is no SEDBUK, no seasonal efficiency value from SAP 2001, or if the heating system for the dwelling is not known, then the target U-value must be made more demanding by dividing by 1.15 to compensate for a possible higher carbon emission rate. In such cases, the adjusted target U-value, U_2, is obtained from:

$$U_2 = \frac{U_1}{1.15}$$

(3) Allow for additional solar gain when the dwelling has metal windows. This is because the framing ratio (the fraction of the area of the window opening taken up by the frame material) is usually less for metal frames than for other frame materials. In the elemental method the allowance is given by means of the slightly higher maximum U-value for metal frames. In the target U-value method, if the windows have metal frames (including thermally broken frames), the extra solar energy is included by multiplying U_2 by a factor of 1.03 to obtain U_3. Thus:

For metal window frames: $U_3 = 1.03\ U_2$
For all other frames: $U_3 = U_2$

(4) Allow for additional solar gain when the area of glazed openings on the southern elevation exceeds that on the northern elevation. In this context, the area of the glazed openings includes the frames, and the definitions of the two elevations are:
Southern elevation: facing within an arc which is 30° either side of South
Northern elevation: facing within an arc which is 30° either side of North.
The solar adjustment factor, ΔS, is found from:

$$\Delta S = 0.04 \left[\frac{A_S - A_N}{A_{TG}} \right]$$

where A_S is the area of glazing facing south
A_N is the area of glazing facing north
A_{TG} is the total area of all glazed openings in the dwelling.

If A_S is less than A_N then ΔS is taken as zero.

(5) Convert U_3 to the final target U-value, U_T, using:

$$U_T = U_3 + \Delta S$$

(6) Find U_{AVG} from:

$$U_{AVG} = \frac{\Sigma AU}{A_T}$$

where ΣAU is the summation of area times U-value for all exposed elements
A_T is the total area of all exposed elements.

Finally, U_{AVG} and U_T are compared. If U_{AVG} is less than or equal to U_T then the requirement is satisfied, that is:

$$U_{AVG} \leq U_T$$

If the calculation results in a fail, i.e. U_{AVG} is greater than U_T, then the situation can be rectified either by taking action to reduce the U-values of exposed elements, or by reducing the area of openings. It may also be possible to increase U_T by selecting a boiler with a higher efficiency, or by increasing the proportion of glazing on the southern elevation. Care must be taken to repeat the calculation of U_T from the beginning, otherwise it is possible to introduce an error in the allowances for solar gain. Examples of the target U-value method are given in section 16.8.

The carbon index method for dwellings

There are two indices used for measuring the amount of carbon dioxide put into the atmosphere by a dwelling. The *carbon factor* (CF) is the carbon dioxide emission in kilograms per year per m^2 of floor area. As this measures the actual quantity of carbon emitted, it follows that *decreasing* its value is good. The *carbon index* (CI) is calculated from the carbon factor, converting it from a linear to a logarithmic scale and also inverting it. The conversion formula was chosen to provide a more convenient scale for measuring the energy efficiency of a dwelling, in which *increasing* values represent higher energy efficiency and are therefore good. The formulae which connect these indices are:

$$CF = \frac{CO_2}{A_F + 45.0}$$

$$CI = 17.7 - 9.0\log_{10}(CF)$$

where CO_2 is the carbon dioxide emission rate in kilograms per year
A_F is the total floor area.

The procedure for determining the carbon index is:

- Follow the standard assessment procedure for the energy rating of dwellings, SAP 2001 edition [19], using the worksheets provided, to obtain the SAP rating of the dwelling
- Continue following the worksheet calculation to obtain the carbon factor and the carbon index.

The requirement is satisfied if:

$$CI \geq 8.0$$

As with the target U-value method, there is no restriction on the area or U-value of any individual element in the dwelling, thus providing substantial flexibility in design. Nevertheless the same considerations apply with respect to the provision of daylight and to the risk of cold internal surfaces and surface condensation. Thus, even if the minimum carbon index requirement of 8.0 is met, the constraints on the area of glazing and the poorest acceptable U-values given below still apply.

Constraints applicable to all three methods of demonstrating compliance

Area of glazing

All three methods of demonstrating compliance allow the area of windows, doors and rooflights to be reduced. However, consideration must also be given to the need to provide adequate daylight. As a general rule, the minimum standard for daylight provision occurs when the total area of openings is 17% of the total floor area. Anything less than this would be inadequate. Further information is available in BS 8206: Part 2 [20].

Poorest acceptable U-values

The flexibility built in to the target U-value and the carbon index methods allows the U values of some parts of the roof, walls or floor of a dwelling to be worse (i.e. higher) than the values given in Table 16.2, provided that this poorer performance is compensated for elsewhere. However, if the U-value of any part of a roof, wall or floor is too high, there is an increased risk of both surface and interstitial condensation. Consequently, regardless of the results of the target U-value or carbon index calculations, AD L1 gives values for poorest (i.e. maximum) acceptable U-values, as shown in Table 16.4.

Table 16.4 Poorest (i.e. maximum) acceptable U-values.

Element	Maximum acceptable U-value, W/m^2K
Part of roof	0.35
Part of exposed wall or floor	0.70

It should be noted that the values given in Table 16.4 would normally only be applied to parts of a roof, wall or floor. If they were applied to the whole of an element, it is very unlikely that compliance would be achieved, whatever compensatory measures had been taken elsewhere.

Additional requirements for fabric thermal performance

As overall standards of thermal insulation improve, so heat losses via thermal bridges and unwanted air leakage become more significant and therefore more important. Consequently there is a requirement to take action to minimise these effects.

Thermal bridging

Thermal bridges are most often due to:

- Gaps in insulation layers within the fabric
- Structural elements, especially lintels and frames (in timber, concrete, etc.)
- Joints between elements
- Joints around windows and doors.

Because of this, there is a requirement to construct in such a way as to avoid significant thermal bridges. This can be done in one of two ways.

- Demonstrating, by calculation, that the thermal performance is satisfactory; although there are several methods, none are simple (see, for example, [17], [18] and [55]).
- Adopting robust construction details that have an authoritative recommendation, such as [2].

Air leakage

Air leakage is the unwanted passage of air through the building fabric, and must be distinguished from the planned ventilation which is necessary for health and safety. Air leakage is most often due to:

- Poor joints within or between elements of the external fabric
- The use of construction elements which are inherently porous to air movement
- Gaps around service pipes, ducts and flues where they penetrate the external fabric
- Ill-fitting windows, doors and rooflights.

The solution is to provide a continuous barrier to air movement around the habitable space (including separating walls and the edges of intermediate floors), penetrated only in those places where it is intentional. As with thermal bridges, this can be demonstrated by adopting published robust construction details. Alternatively, the dwelling can be pressure tested according to the procedure given in CIBSE TM 23 [21], the standard for acceptability being:

$$AP_{50} \leq 10\ m^3/h/m^2$$

where AP_{50} is the air permeability in cubic metres per hour per square metre of external surface area at an applied pressure difference of 50 pascals.

Heating and hot water systems for dwellings

The requirements can be considered under four headings:

- Controls for space heating
- The provision of hot water by means of systems which incorporate hot water storage

- Insulation of pipes and ducts
- Commissioning, operating and maintenance instructions.

Controls for space heating

The guidance given in AD L1 is most suitable for systems in which heat is distributed from a central heat source. For such systems, it is necessary to consider the inclusion of zone controls, timing controls and boiler interlock controls. Although there are a large number of possible system types, Table 16.5 gives guidance for the most usual situations. No guidance is given on stand-alone heaters, whether supplied by solid fuel, gas or electricity.

Table 16.5 Controls for space heating.

Temperature zone controls	Timing zone controls	Boiler interlock controls
Control temperatures independently in areas with different heating needs, such as separate living and sleeping areas	In most dwellings with two temperature zones, one timing zone is sufficient to control both temperature zones	For gas and oil fired hot water central heating systems, the boiler must switch off when no heat is required, regardless of the type and location of thermostats
For large dwellings, ensure that no temperature zone exceeds $150m^2$ floor area. If so, subdivide into smaller zones	For large dwellings, ensure that no timing zone exceeds $150m^2$ floor area. If so, subdivide into smaller zones	In systems controlled by thermostats, the boiler must only switch on when the thermostats call for space heating and/or hot water and/or heat for a storage vessel
For small dwellings such as single-storey open-plan flats and bed-sitters, one temperature zone may be sufficient	For small dwellings such as single-storey open-plan flats and bed-sitters, one timing zone may be sufficient	Where thermostatic radiator valves are used, a room thermostat must also be provided to switch off the boiler when there is no demand for heat or hot water
Provide temperature control by room thermostats *and/or* thermostatic radiator valves *or* any suitable device for sensing temperature. Temperature sensors must be linked to suitable control elements	Provide separate timing control for space heating and water heating (except for combination boilers and solid fuel appliances)	

The provision of hot water systems in dwellings

Although there are several acceptable ways of providing hot water, AD L1 provides guidance only for systems which include integral or separate hot water storage. Possible ways of satisfying the requirement, together with the relevant standards, are given in Table 16.6.

Table 16.6 Standards for hot water systems.

Requirement	Standard
Meet the insulation standards of the most appropriate of these listed standards	BS 1566 [22] BS 699 [23] BS 3198 [24] BS 7206 [25]
or	
In ordinary cases, use insulating vessels with a factory-applied coating of polyurethane foam to meet this stated specification	Minimum thickness of insulation, 35mm Maximum density of insulation, 30kg/m^3
and	
Avoid excessive boiler firing and primary circuit losses, and enable efficient operation. For indirectly heated hot water storage systems, size the heat exchanger according to the most appropriate of these standards, and feed it by a pumped primary system	BS 1566 [22] BS 3198 [24] BS 7206 [25]
or	
For primary storage systems meet the requirements of the WMA standard for thermal stores.	Performance specification for thermal stores Waterheater Manufacturers Association, 1999 [26]

Alternative approach for space heating and the controls for hot water systems

Good practice guide GPG 302 [27] and BS 5864 [28] are both relevant. If an installation follows the recommendations of either of these documents, and also includes the zoning, timing and interlock features described above, then it should meet the requirement.

Commissioning, operating and maintenance instructions for heating and hot water systems

When completed, heating and hot water systems should be inspected and commissioned to ensure that they:

- Operate correctly and to specification
- Comply with health and safety requirements
- Operate efficiently with respect to the need to conserve fuel and power
- Achieve compliance with the requirements of AD L1.

In this context, commissioning means moving from static completion to full operation, and includes:

- Setting the system to work
- Regulation (i.e. repetitive testing and adjustment) to achieve correct performance
- Calibration, setting up and testing of associated automatic control systems
- Recording of the system settings and performance test results that have been accepted as satisfactory.

Responsibility for achieving compliance with AD L1 lies with the person carrying out the work. Such a person may be a:

- Developer or contractor who has directly carried out the work
- Sub-contractor engaged by a developer or contractor
- Specialist organisation directly engaged by a private client.

Written certification that commissioning has been successfully carried out and that compliance with Part L1 (b) and (d) has been achieved must be provided and made available to both the client and the building control body. The person providing the certificate should normally have a recognised qualification, and should be the person responsible for achieving compliance, i.e. the developer, contractor, sub-contractor or specialist. If the person providing the certificate does not have a relevant qualification, or if a suitably qualified certifier is not available, a written declaration of successful commissioning must still be obtained and made available to both the client and the building control body.

Information on the operation and maintenance of the heating and hot water services must be provided and given to the building owner and/or occupier. Operating and maintenance instructions must:

- Be provided for each new dwelling
- Be provided for an existing dwelling when the systems are substantially altered
- Be in an accessible format
- Be directly related to the system or systems in the dwelling
- Explain to householders how to operate the systems, and what routine maintenance is advisable, so that the systems perform efficiently in terms of the conservation of fuel and power.

Insulation of pipes and ducts

The insulation of pipes and ducts is required in order to conserve heat and thus maintain the temperature of water or air being supplied to the heat emitters of a heating system. Insulation in hot water services systems is also required to prevent excessive losses between useful draw-off points. Guidance on methods of achieving the requirements is given in Table 16.7.

Lighting systems for dwellings

There is a requirement to apply energy conservation principles to both internal and external lighting by providing systems with appropriate lamps and sufficient controls. However, although the regulation itself appears to demand that *all* lamps are 'energy efficient', the guidance allows some flexibility. Furthermore, the requirement for sufficient controls does *not* apply to internal lighting, only to external lighting.

Care must be taken to interpret the term 'luminous efficacy' correctly. When referring to dwellings in AD L1, it refers to the amount of light given out by the lamp. The definition, therefore, is:

Table 16.7 Insulation of pipes and ducts.

Type of pipe or duct	Insulation requirement
Space heating pipe-work located outside the insulation layer(s) of the building fabric	Wrap the pipe with insulation which has maximum thermal conductivity of 0.035W/mK at 40°C, to a minimum thickness equal to the outside diameter of the pipe or duct, up to a maximum thickness of 40 mm
Hot pipes connected to hot water storage vessels, including the vent pipe and the primary flow and return to the heat exchanger	As above for space heating pipe-work, for at least 1 metre from the point of connection to the storage vessel (or to the point where the pipes become concealed)
Pipes and warm air ducts	Provide insulation in accordance with the recommendations of BS 5422 [29]

$$\text{Luminous efficacy} = \frac{\text{Total lumens emitted by lamp}}{\text{Total circuit-watts consumed by lamp}}$$

Note that no account is taken of the fitting or lampshade in which the lamp is placed.

Internal lighting

The guidance recommends the following approach:

> At a minimum number of locations, provide fixed lighting comprising
> - *either* basic lighting outlets
> - *or* complete luminaires
>
> that will *only* take lamps of luminous efficacy greater than 40 lumens per circuit-watt.

The minimum number of locations must include those that are expected to have most use, and can be calculated from the number of rooms in the dwelling by the formula:

$$\text{Minimum number of locations} = \frac{\text{Number of rooms}}{3}$$

the result being rounded up to the nearest whole number. Thus, for three rooms the minimum number of locations is one, whereas for four rooms the calculation gives $1\frac{1}{3}$, which is rounded up to two. When counting the rooms in the house, note that:

- Hall, stairs and landing count as one room, even if they contain more than one light fitting
- An integral conservatory in a new dwelling is included in the count, but in other cases a conservatory is excluded
- Garages lofts and outhouses are excluded.

Note also that in the definition of luminous efficacy, circuit-watt includes the electrical power consumed by the lamps plus all associated control gear and power factor equipment. Typically, it may be expected that the 40 lumens per circuit-watt minimum would be achieved by fluorescent tubes and compact fluorescent lamps, *but not* by GLS tungsten lamps with bayonet caps or Edison screw bases.

Further guidance on internal lighting is given in General Information Leaflet GIL 20, *Low energy domestic lighting* [30].

External lighting

The requirement is in respect of systems fixed to the building, and includes porches but not lighting in garages and car ports. The guidance suggests two alternative solutions:

(1) Provide controls which automatically extinguish the system when there is enough daylight *and* when the the system is not required at night, *or*
(2) Install sockets that will *only* take lamps of luminous efficacy greater than 40 lumens per circuit-watt. As for internal lighting, this means that typically fluorescent tubes and compact fluorescent lamps (*but not* GLS tungsten lamps with bayonet caps or Edison screw bases) will be acceptable.

Conservatories

A conservatory is defined as a space that has not less than three-quarters of the area of its roof and not less than one-half of the area of its external walls made of translucent material. It is also necessary to define 'separation'. In this context, separation between a dwelling and a conservatory means:

- Separating walls and floors insulated to at least the same degree as the exposed walls and floors of the dwelling
- Separating windows and doors with the same U-value and draught-stripping provisions as the exposed windows and doors elsewhere in the dwelling.

A conservatory may be attached to and constructed as part of a new dwelling, or it may be attached to an existing dwelling. The requirements of these two cases differ slightly.

A conservatory attached to and built as part of a new dwelling

There are three possibilities, one where there is no separation between the conservatory and the dwelling, and the other two where there is separation:

- No separation – the conservatory should be treated as an integral part of the dwelling
- Separation, unheated – energy savings can be achieved if the conservatory is not heated
- Separation, heated – if fixed heating installations are proposed, they should have their own separate temperature and on/off controls.

A conservatory attached to an existing dwelling

Reasonable provision must be made to limit heat loss from the dwelling. Ways of doing this depend on whether or not the opening to the conservatory is newly created or enlarged:

- If the opening is *not* to be enlarged, retain the existing separation
- If the opening *is* to be newly created or enlarged, provide separation the same as or equivalent to windows and doors having the same maximum average U-value given in Table 16.2.

16.2.4 AD L1 Section 2 – work on existing dwellings

As well as extensions and conservatories, the regulations also cover certain other categories of work on existing dwellings. There are four such categories:

- Replacement of controlled services or fittings
- Material alterations
- Material change of use
- Historic buildings.

Replacement of controlled services or fittings

A controlled service or fitting is defined in Regulation 2(1) of the Building Regulations (2000 as amended 2001) as:

> '. . . a service or fitting in relation to which Part G, H, J or L of Schedule 1 imposes a requirement'.

In this context, whether or not replacement of controlled services or fittings falls within the definition of building work (and is therefore subject to the requirements of Part L) is stated in Regulation 3(1), as qualified in Regulation 3(1A), as follows:

> 'The provision or extension of a controlled service or fitting
> (a) in or in connection with an existing dwelling, *and*
> (b) being a service or fitting in relation to which Part L1, but not Parts G, H, or J, of Schedule 1 imposes a requirement shall only be building work where that work consists of the provision of any of the following':

- Window
- Rooflight
- Roof window
- Door, glazed to more than 50% of its total area, including its frame, measured internally
- Space heating or hot water service boiler
- Hot water vessel.

Part L1 applies to replacement work on controlled services or fittings when:

- Replacing old with new identical equipment
- Replacing old with new but different equipment
- The work is solely in connection with controlled services or includes work on them.

Ways of satisfying the requirements of Part L1 may depend on the circumstances of the particular case. Specific guidance is as follows.

Windows, doors and rooflights

The requirement does *not* apply to repair work on *parts* of these elements such as:

- Replacing broken glass
- Replacing sealed double-glazing units
- Replacing rotten framing members.

However, where these elements are to be replaced rather than repaired, it becomes necessary to provide new draught-proofed ones with either:

- An average U-value not exceeding the appropriate value in Table 16.2, *or*
- A centre-pane U-value not exceeding $1.2\,W/m^2K$.

The replacement work should comply with the requirements of Parts L and N. Furthermore, after completion of the work, the building should not have a worse level of compliance with other applicable parts of Schedule 1, including Parts B, F and J.

Heating boilers

If the dwelling has a floor area greater than $50\,m^2$, the new boiler should satisfy the same requirements as for a new dwelling. This includes:

- For ordinary gas or oil boilers, adherence to the minimum SEDBUK ratings given in Table 16.3
- For back boilers, adherence to a minimum SEDBUK rating that is 3 percentage points less than the appropriate value in Table 16.3
- For solid fuel boilers, provision of a boiler having an efficiency not less than that recommended for its type in the HETAS [31] certification scheme.

Hot water vessels

When replacing hot water vessels, new equipment should be provided which would satisfy the requirements for a new dwelling.

Boiler and hot water controls

To ensure that replacement boilers (except solid fuel boilers) and hot water vessels achieve a reasonable and acceptable seasonal efficiency, it may be necessary to replace or provide:

- The time switch or programmer
- The room thermostat
- The hot water vessel thermostat
- A boiler interlock
- Fully pumped circulation.

Advice and information on how this can be done is given in Good Practice Guide GPG 302, section 3 [27].

Alternative approach using the carbon index

As an alternative to the procedures of the above four paragraphs, it may be acceptable to follow the guidance in Good Practice Guide GPG 155 [32], provided that an equivalent improvement in the carbon index of the dwelling is achieved.

Commissioning, and operating and maintenance instructions

If heating and/or hot water systems are altered or replaced, it is reasonable to expect that commissioning would be carried out, and operating and maintenance instructions provided, as if for a new dwelling.

Material alterations

It is necessary to define the terms 'Material alteration' and 'Relevant requirement'.

Material alteration (Regulation 3(2))

'An alteration is material for the purposes of these regulations if the work, or any part of it, would at any stage result –

(a) in a building or controlled service or fitting not complying with a relevant requirement where previously it did, or

(b) in a building or controlled service or fitting which before the work commenced did not comply with a relevant requirement, being more unsatisfactory in relation to such a requirement.'

Relevant requirement (Regulation 3(3))

As used in Regulation 3(2):

'... relevant requirement means any of the following applicable requirements of Schedule 1, namely –
Part A (structure)
Paragraph B1 (means of warning and escape)
Paragraph B3 (internal fire spread – structure)
Paragraph B4 (external fire spread)
Paragraph B5 (access and facilities for fire service)
Part M (access and facilities for disabled people).'

The consequence of these definitions is that an alteration is only considered material if it affects the existing building with respect to any one or more of Parts A, B1, B3,

B4, B5 or M. However, as soon as it is established that an alteration is material, then account should be taken of:

- All the relevant requirements of Schedule 1, including Parts L1, F and J
- Insulation of roofs, floors and walls
- Sealing measures
- Controlled services and fittings (as above).

Ways of satisfying the requirements of Part L1 may depend on the circumstances of the particular case. Specific guidance is as follows.

Roof insulation

If the material alteration includes substantial replacement of any of the major elements of a roof structure, insulate to the U-value standard of a new dwelling.

Floor insulation

If the structure of a ground floor or exposed floor is to be substantially replaced or re-boarded, and if the room is heated, insulate to the U-value standard of a new dwelling.

Wall insulation

Provide a reasonable thickness of insulation when substantially replacing:

- Complete exposed walls
- The external rendering or cladding of an exposed wall
- The internal surface finishes of an exposed wall
- The internal surfaces of separating walls.

Sealing measures

When carrying out any of the above work on roofs, floors or walls, include reasonable sealing measures to improve airtightness, but remember to take into account the requirements of Parts F and J.

Controlled services or fittings

These are dealt with above.

Material changes of use

A material change of use is defined differently from a material alteration. According to Regulation 5:

> '... for the purposes of these Regulations, there is a material change of use where there is a change in the purpose for which or the circumstances in which a building is used, so that after the change –
> (a) the building is used as a dwelling, where previously it was not;
> (b) the building contains a flat, where previously it did not;

(c) the building is used as an hotel or a boarding house, where previously it was not;
(d) the building is used as an institution, where previously it was not;
(e) the building is used as a public building, where previously it was not;
(f) the building is not a building described in Classes I to VI in Schedule 2, where previously it was;
(g) the building, which contains at least one dwelling, contains a greater or lesser number of dwellings than it did previously.'

Regulation 6 includes a list of all those parts of Schedule 1 that apply when works comprising a change of use are undertaken. The list includes Part L1. It also includes two other parts of relevance to the conservation of fuel and power and to which particular attention should also be paid: Part F (ventilation) and Part J (combustion appliances). As far as Part L1 is concerned, when undertaking material changes of use, account should be taken of:

- Accessible lofts
- Insulation of roofs, floors and walls
- Sealing measures
- Lighting
- Controlled services and fittings (as above).

Clearly, if the whole of a building is subject to a material change of use, then the whole building should comply with all the relevant parts of Schedule 1 that are listed in Regulation 6. If the change of use applies to only part of a building, then in general only that part must comply. Ways of satisfying the requirements of Part L1 may depend on the circumstances of the particular case, and are mostly the same as for a material alteration. Specific guidance is as follows.

Accessible lofts

Where the existing insulation in accessible lofts is worse than 0.35 W/m^2K, replace or add extra insulation to upgrade the U-value to a maximum of 0.25 W/m^2K.

Roof insulation

If the material alteration includes substantial replacement of any of the major elements of a roof structure, insulate to the U-value standard of a new dwelling.

Floor insulation

If the structure of a ground floor or exposed floor is to be substantially replaced or re-boarded, and if the room is heated, insulate to the U-value standard of a new dwelling.

Wall insulation

Provide a reasonable thickness of insulation when substantially replacing:

- Complete exposed walls
- The external rendering or cladding of an exposed wall

- The internal surface finishes of an exposed wall
- The internal surfaces of separating walls.

Sealing measures

When carrying out any of the above work on roofs, floors or walls, include reasonable sealing measures to improve airtightness, but remember to take into account the requirements of Parts F and J.

Lighting

Provide lighting in accordance with requirements for a new dwelling.

Controlled services or fittings

These are dealt with above.

Historic buildings

Historic buildings include:

- Listed buildings
- Buildings situated in conservation areas
- Buildings of architectural and historical interest and which are referred to as a material consideration in a local authority development plan
- Buildings of architectural and historical interest within national parks, areas of outstanding natural beauty, and world heritage sites.

Any work on an historic building must balance the need to improve energy efficiency against the following factors:

- The need to avoid prejudicing the character of the historic building
- The danger of increasing the risk of long-term deterioration of the building fabric
- The danger of increasing the risk of long-term deterioration of the building's fittings
- The extent to which energy conservation measures are a practical possibility.

Advice on achieving the correct balance should be sought from the conservation officer of the local authority. Advice from other published sources, e.g. PPG15 [33], BS 7913 [34] and SPAB Information sheet 4 [35], may also be appropriate, particularly regarding:

- Restoration of the historic character of a building that had been the subject of inappropriate alteration, such as the replacement of windows, doors or rooflights
- Rebuilding of a former historic building, which may have been damaged or destroyed due to some mishap (such as a fire), or in-filling a gap in a terrace
- Providing a means for the fabric of an historic building to 'breathe' so that moisture movement may be controlled and the potential for long-term decay problems reduced.

16.3 The conservation of fuel and power in buildings other than dwellings

Part L2 of Schedule 1 to the 2000 Regulations (as amended) is concerned with the conservation of fuel and power in buildings other then dwellings. This is supported by Approved Document L2 which includes the following topics:

- Compliance by the elemental method
- Compliance by the whole-building method
- Compliance by the carbon emissions calculation method
- Limiting thermal bridging and air leakage
- Avoiding solar overheating
- Heating systems
- Controls for space heating and hot water systems
- Insulation of pipes, ducts and vessels
- Inspection and commissioning of the building services systems
- Lighting
- Air conditioning and mechanical ventilation (ACMV).

The purpose of Part L2 is to minimise the environmental impact of buildings with respect to the depletion of fuel reserves and the degradation of the atmosphere by carbon dioxide emissions.

16.3.1 The legal requirement for the conservation of fuel and power in buildings other than dwellings

The regulation requires that a building (other than a dwelling) should be so designed and constructed as to make reasonable provision for the conservation of fuel and power, and that this shall be achieved by attention to all of the following eight points:

- Limiting heat losses and gains through the fabric of the building
- Limiting the heat loss from hot water pipes and hot air ducts used for space heating, and limiting the heat loss from hot water vessels and hot water service pipes
- Ensuring that the space heating and hot water systems which are provided are energy efficient
- Limiting exposure to solar overheating
- Where a floor area greater than 200 m^2 is served by an ACMV system, ensuring that no more energy needs to be used than is reasonable in the circumstances
- Limiting the heat gains to chilled water and refrigerant vessels and pipes and air ducts that serve air conditioning systems
- Providing lighting systems which use energy efficiently
- Ensuring that, by the provision of sufficient information with the relevant services, the building can be operated and maintained in such a manner as to use no more energy than is reasonable in the circumstances.

Approved Document L2 gives general and specific guidance on the energy efficient measures which will satisfy the requirements.

16.3.2 AD L2 Section 0 – general guidance

The general guidance for Part L2 is broadly similar to that given for Part L1. However, because of the large range of building types in the non-domestic sector, the guidance is broader in scope, and there are some differences in the detail.

Heat losses through walls, roofs, floors, windows and doors, etc. must be kept within acceptable limits by means of suitable levels of thermal insulation. Also, where it is possible and appropriate, any beneficial gains from solar heat and more efficient heating systems should be taken into account. However, it is also expected that, where necessary, provision be made to limit heat gains in summer.

Heat may also be lost by unplanned air leakage through the fabric of a building, particularly around doors, windows and junctions between elements or components. This requires attention not only to draught stripping of doors and windows, but also consideration of potential air leakage paths through all other elements of the building.

Heat losses from hot water pipes, hot air ducts, hot water vessels, and all their connections must also be controlled by the application of sufficient insulation. If, however, a particular component contributes to the space heating in an efficient manner, the insulation may be omitted.

The energy efficiency of the systems which provide space heating and hot water has two aspects. Firstly, the equipment itself must generate heat from fuel in an efficient manner, and secondly the equipment must have controls for space and water temperatures and for the timing of its operation. Furthermore, the systems must have been appropriately commissioned and be capable of being operated in an efficient manner by the user.

In order to limit solar overheating, a combination of techniques may be used. These include passive measures such as limiting the area of unshaded glazing and designing the external fabric so that it limits and delays heat penetration, plus active measures such as night ventilation.

Conservation of energy in mechanically ventilated and air conditioned buildings requires that the energy demand is limited and that the energy which is needed is supplied efficiently. The energy demand includes heating, cooling, air circulation, water and refrigerants. The energy supply must include efficient plant, controls for timing, temperature and flow, and energy consumption metering. All services must have been appropriately commissioned.

Heat gains to chilled water and refrigerant vessels, and to the ducts and pipes which may be carrying chilled air or fluids, must be limited by suitable thicknesses of insulation, with vapour barriers to prevent the insulation becoming damp.

There are four requirements for lighting systems:

(1) Systems must use energy efficient lamps and luminaires.
(2) The switching must be manual or automatic, or a combination of manual and automatic.

(3) The lighting systems must have their own energy consumption metering.
(4) The complete lighting system must have been appropriately commissioned.

The energy efficiency of heating and hot water systems can be affected by the manner in which they are operated. Incorrect usage may result in poor energy performance from an otherwise energy efficient system. Consequently, there is a requirement to provide easily understandable information that will enable the occupier to operate and maintain the building and its building services so that they do not waste energy. This information should also include the results of performance tests.

Carbon and carbon dioxide indices

In order to be consistent with the general objective of Part L, which is to reduce pollution of the atmosphere (especially by carbon dioxide) caused by energy consumption in buildings, the fuel consumed by a building is most conveniently expressed in terms of the amount of carbon that has been generated. In AD L2, therefore, performance targets are expressed in kilograms of carbon, and not in terms of an energy unit such as gigajoules or megawatt-hours. Occasionally, there may be a preference for expressing performance in terms of gaseous carbon dioxide rather than solid carbon. The conversion factor from one to the other is the ratio of the molecular weights, i.e. 44/12. Thus, 6 kg per square metre of carbon per year is equivalent to $6 \times (44/12) = 22$ kg of carbon dioxide per square metre per year.

Special cases

Exemptions

There are two types of exemption from the requirements of L2: when there are low levels of heating and/or when there are low levels of use.

Low levels of heating

A building with a heat requirement which does not exceed 25 W/m^2 may be regarded as having a low heat requirement, and therefore does not require measures to limit heat loss through the fabric. In such a case, the fabric insulation is likely to be chosen for operational reasons. Typical examples are:

- A warehouse where general heat is provided as a protection against frost or condensation damage to goods, and where higher temperatures are only necessary for local work stations
- A cold store, where the major concern may be to avoid heat gain rather than heat loss through the building fabric.

Low levels of use

A building which, because of its function, is used for a smaller number of hours than is normal, may where appropriate be fitted with heating and lighting systems to a lower standard than that required by AD L2. However, the building must adhere to the fabric insulation standards of AD L2. A typical example is:

- A building used solely for worship at set times.

Historic buildings

Historic buildings are considered in section 16.3.6.

Buildings constructed from sub-assemblies

A new building, however it is constructed, must normally comply with all the requirements of Schedule 1. However, with respect to the external fabric of buildings constructed from sub-assemblies, reasonable provisions for the conservation of fuel and power may vary according to the circumstances of the particular case. Examples where this may apply are:

- A building created from the external fabric sub-assemblies of an existing building by dismantling, transporting and re-erecting on the same site; this would normally be considered to meet the requirements
- A building which uses external fabric assemblies manufactured before 31 December 2001, or obtained from other premises; this would normally be considered to meet the requirements if the fabric thermal resistance, or the predicted annual energy consumption, satisfies the relevant requirements of Approved Document L, 1995 edition.

Enclosed heating and cooling links (which may also be made from sub-assemblies) should be insulated and made airtight to the same standards as the buildings themselves. In temporary accommodation, the requirements that must be satisfied by the heating and lighting may depend on the particular case. The normal expectation would be:

- *For heating and hot water systems* provide on/off time and temperature controls as specified in section 16.3.3 (page 16.46) under 'Space heating controls' and 'Hot water systems and their controls'
- *For general and display lighting* follow the guidance in section 16.3.3 (page 16.50) under 'General lighting efficacy in all other building types' and 'Display lighting for all buildings'.

Mixed use development

Where part of a building is used as a dwelling and the rest of it is used for other purposes, the requirements for dwellings must also be taken into consideration.

16.3.3 AD L2 – Section 1 design

There are a number of general considerations which should be applied to a building design project. These include:

- Preparing designs for the building and its services which are appropriate to the need to achieve energy efficiency
- Providing information so that the performance of the building in use may be assessed
- Making provisions in the design of the building services installations which facilitate inspection and commissioning.

Reference may be made to the CIBSE Guide on energy efficiency in buildings [36].

It is permissible and also perhaps preferable, if a building is large and complex, to consider the building in parts and apply the principles and requirements for energy conservation to each part. Also, where a building has alternative building services (e.g. dual fuel boilers, combined heat and power with standby boilers, etc.), the building should meet the requirements of AD L2 in all possible operating modes.

Specific guidance

For specific guidance, Approved Document L2 offers three methods for demonstrating that reasonable provision has been made for limiting heat loss through the building fabric:

The elemental method has the advantage that it does not require complex calculations. Each aspect of the building is considered individually, and a minimum level of performance must be achieved in each of the elements. Some flexibility is provided by allowing trade-off between different elements of the construction, and between insulation standards and heating system performance.
The whole-building method considers the performance of the whole building. It is applicable to office buildings, but similar alternative methods are available for schools (see DfEE Building Bulletin 87 [47]) and hospitals (see NHS Estates Guides [48]).
The carbon emissions calculation method considers the performance of the whole building and can be applied to any building type. To comply, the annual carbon emissions from the proposed building should be no greater than those from an equivalent notional building that satisfies the criteria of the elemental method. The carbon emissions from both the proposed building and the notional building must be estimated by means of an appropriate calculation tool.

Although there is no obligation to adopt any particular method or solution for satisfying the requirements, the Approved Document provides a check-list, in the form of a 'Summary Guide', that helps the designer choose the most suitable of the above three methods. This summary guide is shown in flow chart form in Figs 16.3a, 16.3b and 16.3c.

The elemental method for buildings other than dwellings

In the elemental method, compliance is demonstrated by specifying:

- Maximum U-values for walls, floors and roofs and rooflights
- Maximum area-weighted U-values for windows, personnel doors and rooflights

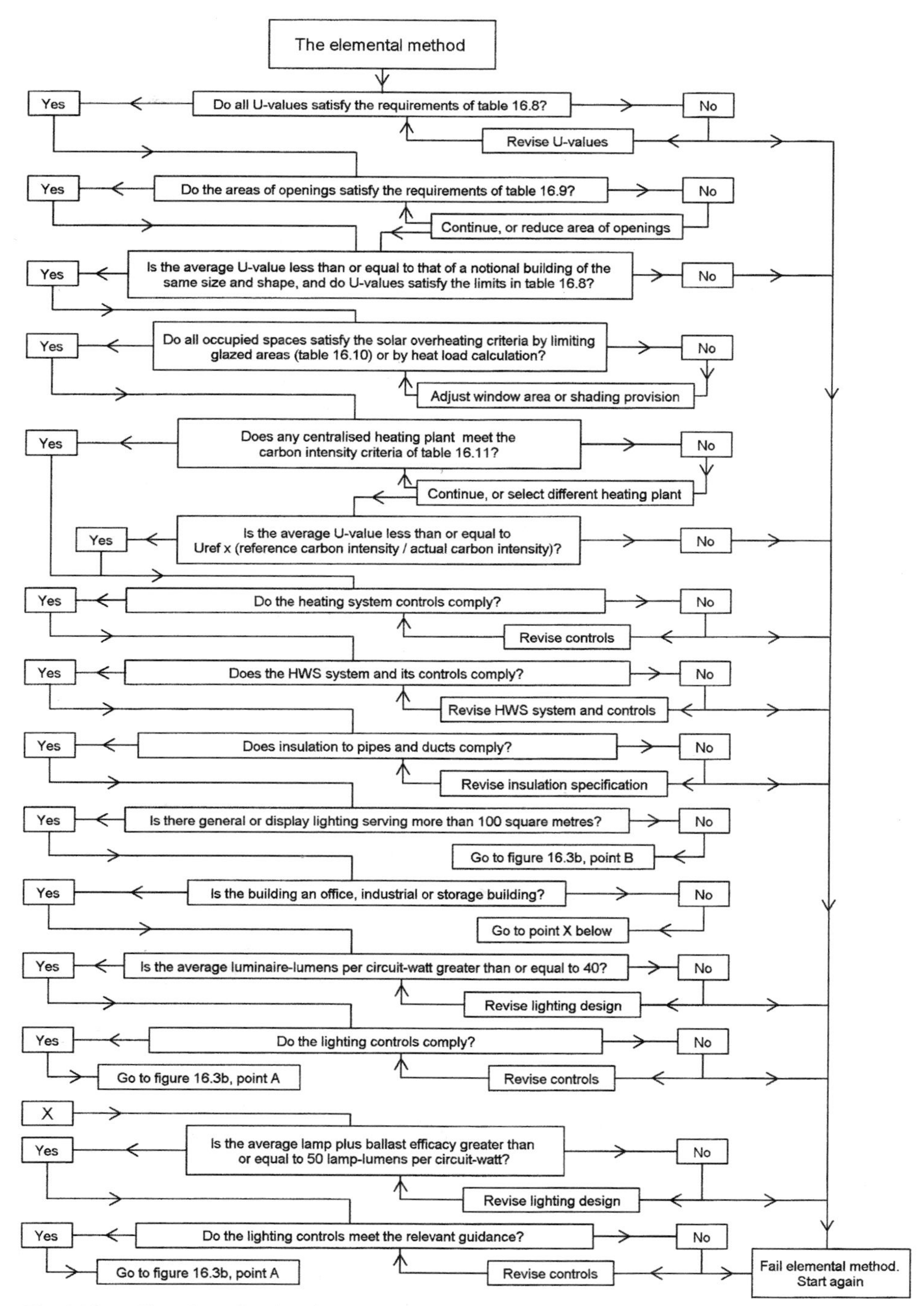

Fig. 16.3a Flowchart for the elemental method, Part L2.

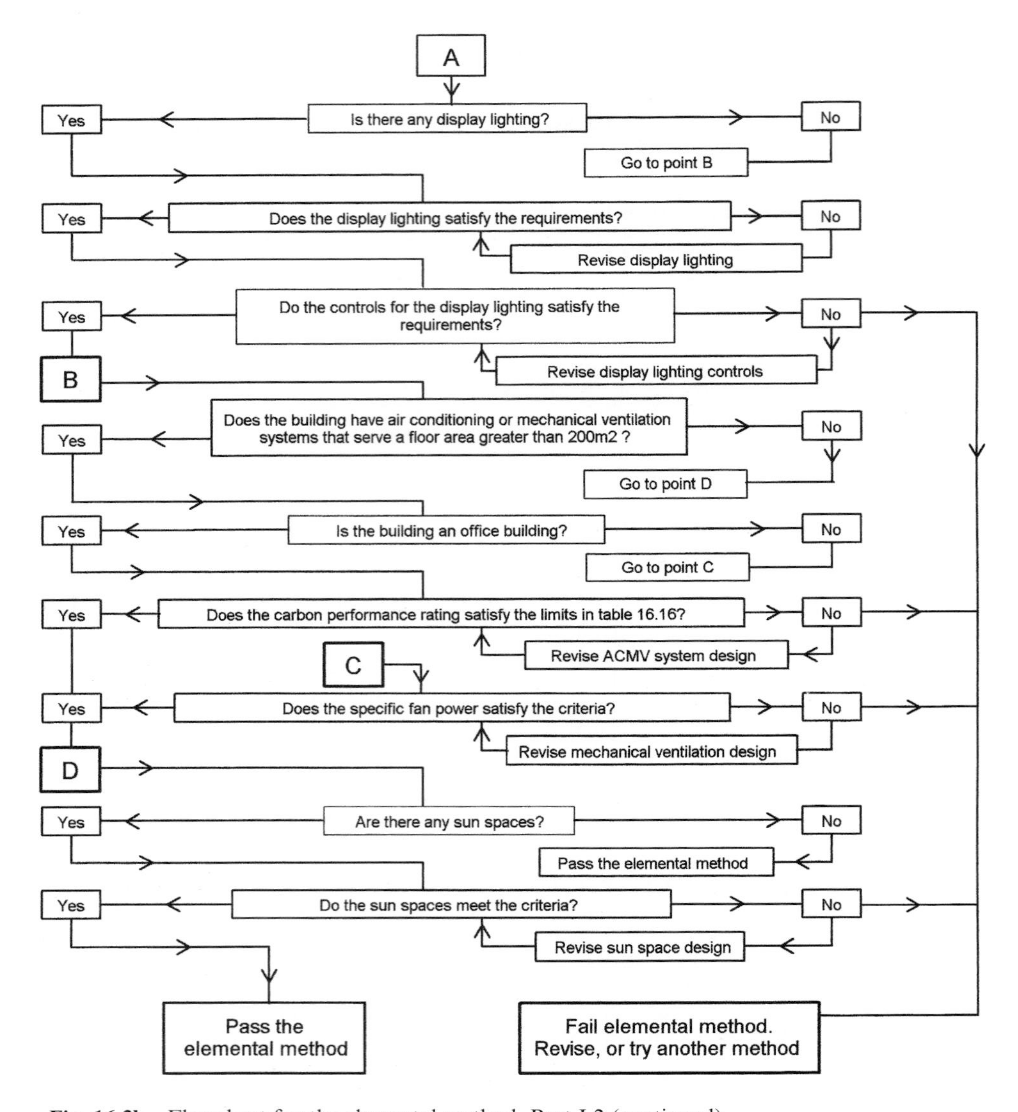

Fig. 16.3b Flowchart for the elemental method, Part L2 (continued).

- Maximum combined area of windows and doors (as a percentage of exposed wall area)
- Maximum area of rooflights (as a percentage of roof area)
- Minimum performance standards for all energy consuming building services
- Design details to meet air leakage standards and to avoid solar overheating.

With regard to windows within part L2 the following should be noted:

- Display windows, shop entrance doors and similar glazing are exempt from the maximum U-value standard of Table 16.8, but not from the maximum area standards of Table 16.9

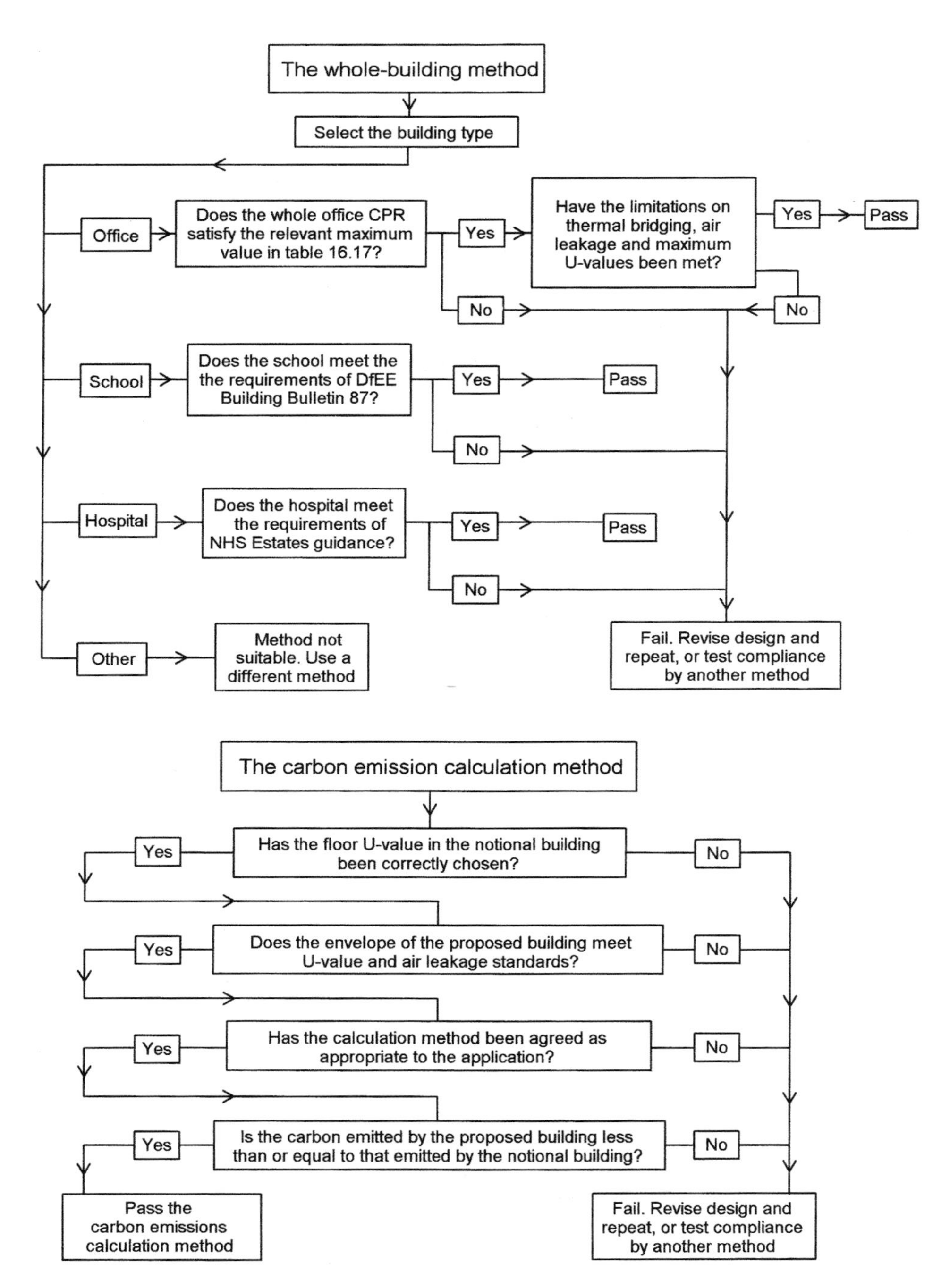

Fig. 16.3c Flowcharts for the whole-building method and the carbon emissions calculation method.

Table 16.8 Maximum U-values for the elemental method.

Exposed element	Maximum U-value, W/m^2K	Comment
Pitched roof, insulation between rafters	0.20	Any part of a roof having a pitch of 70° or more can be considered as a wall
Pitched roof with integral insulation	0.25	
Pitched roof, insulation between joists	0.16	
Flat roof	0.25	Roof of pitch 10° or less
Walls, including basement walls	0.35	
Floors, including ground floors and basement floors	0.25	
Windows, personnel doors and roof windows, glazing in metal frames, area-weighted average for the whole building	2.20	A roof window may be considered as a rooflight. The higher U-value for metal frames allows for extra solar gain due to greater glazed proportion
Windows, personnel doors and roof windows, glazing in wood or PVC frames, area-weighted average for the whole building	2.00	
Rooflights	2.20	Applies only to the unit, excluding any upstand which must be insulated
Vehicle access and similar large doors	0.70	

Table 16.9 Maximum areas of windows and rooflights.

Building type	Maximum window and door area as % of exposed wall area	Maximum rooflight area as % of area of roof
Residential buildings where people temporarily or permanently reside	30	20
Places of assembly, offices and shops	40	20
Industrial and storage buildings	15	20
Vehicle access doors and display windows and similar glazing	As required	

- The terms rooflight and roof window are synonymous
- For the purposes of calculating maximum areas, dormer windows in a roof may be included in the rooflight area.

Maximum U-values are shown in Table 16.8.

With the exception of vehicle access doors, Table 16.8 is the same as Table 16.2 for dwellings, and so the illustration of these U-values in Fig. 16.3 is still applicable. Also, the same comments apply with respect to unheated spaces and metal frame windows.

With regard to area and average U-value, windows and doors are treated separately from rooflights. The area-weighted U-value of the windows and doors can be calculated from the formula:

$$\text{Average U} - \text{value,} \quad U_{WD} = \frac{A_W U_W + A_D U_D}{A_W + A_D}$$

where A_W and A_D are respectively the areas of windows and doors
U_W and U_D are their U-values.

The maximum areas of windows, doors and rooflights depend on the building type, and are given in Table 16.9.

Adjustments allowed by the elemental method

The elemental method allows the U-values of the construction elements and the areas of windows, doors and rooflights to be varied from the values given in Tables 16.8 and 16.9, provided that the rate of heat loss from the proposed building is no worse than that from a notional building of the same size and shape that does meet the maximum U-values and maximum areas in these tables. There are, however, four constraints on the allowable variations:

(1) If the U-value of any part of a roof, wall or floor in the proposed building is greater than the value given in Table 16.8, it must nevertheless not exceed the following absolute maxima:

- Any part of a roof: maximum acceptable U-value = 0.35 W/m^{2}K
- Any part of an exposed wall or floor: maximum acceptable U-value = 0.70 W/m^{2}K

(2) If in the proposed building the floor has no added insulation and its U-value is better (i.e. lower) than the value given in Table 16.8, then this lower value must be used in the calculation for the notional building.
(3) If the areas of any of the openings in the proposed building are less than the maxima in Table 16.9, then the average U-value of the roof, wall or floor must not exceed the relevant value in Table 16.8, by more than 0.02 W/m^{2}K.
(4) No more than half of the allowable rooflight area can be used to increase the area of windows and doors. The increase is not calculated on a simple area basis, but by taking into account the difference in U-values between the roof and the wall. The method is given in Appendix E to AD L2, and is demonstrated in the example in Section 16.10.

Additional requirements for fabric thermal performance

As overall standards of thermal insulation improve, so heat losses via thermal bridges and unwanted air leakage become more significant and therefore more important. Consequently there is a requirement to take action to minimise these effects.

Thermal bridging
Thermal bridges are most often due to:

- Gaps in insulation layers within the fabric
- Structural elements, especially lintels and frames (in timber, concrete, etc.)
- Joints between elements
- Joints around windows and doors.

Because of this, there is a requirement to construct in such a way as to avoid significant thermal bridges. This can be done most easily by adopting robust construction details that have an authoritative recommendation. Although the detailing given in the 'robust construction details' document [2] is intended mainly for dwellings, it is also relevant to other buildings of similar construction and with similar internal environmental conditions. The detailing may not be satisfactory for a building with an unusual temperature or humidity regime. Alternatively, it may be possible to demonstrate by calculation that thermal bridging is not a problem and that the thermal performance is satisfactory. Although there are several methods of doing this, none are simple. The method described in IP 17/01 [55] is one of the least complex.

Air leakage
Air leakage is the unwanted passage of air through the building fabric, and must be distinguished from the planned ventilation that is necessary for health and safety. Air leakage is most often due to:

- Poor joints within or between elements of the external fabric
- The use of construction elements which are inherently porous to air movement
- Gaps around service pipes, ducts and flues where they penetrate the external fabric
- Ill-fitting windows, doors and rooflights and junctions around them.

These problems can be alleviated by providing sealing measures such as a continuous barrier to air movement in contact with the insulation layer over the whole thermal envelope, including separating walls and the edges of intermediate floors, and penetrated only in those places where it is intentional. For domestic type constructions, some suitable design details and installation practice are available in the robust details publication. For metal cladding and roofing systems, design guidance is given in the MCRMA Technical Report No. 14 [37]. The sealing of gaps around service penetrations and the draught-proofing of external doors and windows must also be attended to.

For buildings of any size, the rules for showing compliance with the requirements of AD L2 for resistance to air leakage are as follows:

- Carry out an air leakage test in accordance with CIBSE TM 23 [21] to show that:

$$AP_{50} \leq 10 \ m^3/h/m^2$$

where AP_{50} is the air permeability in cubic metres per hour per square metre of external surface area at an applied pressure difference of 50 pascals.

For buildings of less than 1000 m² gross floor area, the rules for showing compliance with the requirements of AD L2 for resistance to air leakage are:

- *Either* carry out air leakage tests and satisfy the requirements as above for a building of any size
- *Or* provide certificates or declarations that the design details, construction techniques and the execution of the work are such that reasonable conformity with the air leakage requirement can be expected.

Avoidance of solar overheating

Solar overheating can manifest itself in two ways. In naturally ventilated spaces it can lead to uncomfortably or even unacceptably high internal temperatures. In spaces with mechanical ventilation or cooling it can lead to excessive cooling capacity, and if the cooling capacity is insufficient, to high internal temperatures as well. AD L2 requires that both these problems should be addressed. The general objectives are therefore that buildings should be constructed so that:

- For occupied and naturally ventilated spaces, there should be no overheating when they are subjected to a 'moderate level' of internal heat gain
- For occupied spaces with mechanical ventilation or cooling, there should be no need for excessive cooling plant capacity in order to maintain the desired conditions.

The methods by which these objectives may be achieved could include some or all of:

- Specification of an appropriate type and quantity of glazing
- The incorporation in the design of passive measures such as shading
- The use of night ventilation in conjunction with exposed thermal capacity.

Three alternative possible criteria are suggested for demonstrating compliance.

(1) For spaces with glazing facing one orientation only, limit the area of the glazed opening. The maximum area of glazed opening depends on the orientation and is expressed as a percentage of the area (measured internally) of the wall or roof element in which it occurs. Table 16.10 gives details.

Table 16.10 Maximum allowable areas of glazed opening.

Orientation of glazed opening	Maximum allowable area of opening %
North	50
North-east, north-west, south	40
East, south-east, west, south-west	32
Horizontal	12

(2) For conditions of summer sunshine, show that the solar heat load per unit floor area, averaged between the hours of 07.30 and 17.30, does not exceed 25 W/m^2. For this purpose, the condition of summer sunshine is defined as:

> 'The solar irradiances for the month of July at the location of the building that were not exceeded on more than 2.5% of occasions during the period 1976 to 1995.'

Strictly speaking, the solar irradiances vary with location, especially latitude. However, for the purposes of demonstrating compliance, Appendix H AD L2 (see section 16.13) simplifies matters by giving a calculation procedure based on a single table that may be used for any location. This does not introduce a significant error over the range of latitudes to which Part L2 applies.

(3) Show by detailed calculation procedures (e.g. CIBSE Guide A, Chapter 5 [38]), that in the absence of mechanical cooling or mechanical ventilation the space will not overheat when subjected to an internal gain of 10 W/m^2. However, no criterion is provided to indicate when a space has become overheated.

Heating systems

Clearly, despite all the requirements in Schedule 1, the installation of combustion appliances and combustion systems (e.g. boiler plant and direct-fired gas heaters) must not infringe the necessary supply of air for health and for combustion. Reference must be made to:

- Part F for the provision of adequate ventilation for health, *and*
- Part J for the provision of adequate air for combustion appliances.

The carbon intensity of heating plant

The requirement is that heating plant should be reasonably efficient. The method of demonstrating compliance is to show that the amount of carbon generated by the heating plant is within specified limits at both maximum output and part load. The method applies to:

- Heating plant serving hot water and steam heating systems
- Electric heating
- Heat pumps (irrespective of the form of heat distribution).

The carbon intensity of the heat generating equipment must be calculated at maximum heat output of the heating system, *and* at 30% of this maximum, and the results must not exceed the values given in Table 16.11.

Table 16.11 Maximum allowable carbon intensities for heating systems.

Fuel	Maximum carbon intensities, kgC/kWh	
	At maximum heat output of the heating system	**At 30% of maximum heat output of the heating system**
Natural gas	0.068	0.065
Other fuels	0.091	0.088

The carbon intensity of the heating system depends on three factors:

- The rated output of each individual element of heat raising plant, R
- The gross thermal efficiency of each element of heat raising plant, measured in kWh of heat per kWh of delivered fuel, η_t
- The carbon emission factor of the fuel supplying each element of heat raising plant, measured in kg of carbon emitted per kWh of delivered fuel consumed, C_f.

The efficiency of consuming fuel in any equipment that supplies heat and power to the building is included in the gross thermal efficiency, the value of which depends on both the fuel and the equipment in which it is burnt. The carbon emission factor gives the total amount of carbon emitted and includes allowances for energy used in extracting, processing and delivering the fuel to the user. For example, electricity may be consumed with very high efficiency by equipment within the building, but the generation of electricity from the primary fuel such as oil has a relatively low efficiency, and so losses have already occurred before the electricity itself is consumed. The same is true for other fuels. Values for the gross thermal efficiency must be obtained for each particular combination of equipment and fuel. However, values for C_f, the carbon emission factor, depend only on the fuel. A change in the production, manufacture or method of delivery of a fuel could cause the value of C_f to change, but any such change is unlikely to occur suddenly, and for the purposes of AD L2, C_f may be taken to be a constant for each fuel. The generally agreed values for C_f are shown in Table 16.12.

The carbon intensity, ε_c, is found by summing over all elements of heat generating plant:

$$\varepsilon_c = \frac{1}{\Sigma R}\Sigma\left(\frac{RC_f}{\eta_t}\right)$$

This equation is applicable to boilers, heat pump systems and electrical heating, but not combined heat and power (CHP) systems. In order to use the equation, data for the gross thermal efficiency for equipment must be obtained from the manufacturers or suppliers. For most practical cases, the efficiency may be taken as the efficiency at

Table 16.12 Carbon emission factors, C_f.

Delivered fuel	Carbon emission factor kgC/kWh	Comments
Natural gas	0.053	
LPG	0.068	
Biogas	0	
Oil	0.074	This value applies to all grades of fuel oil
Coal	0.086	
Biomass	0	
Electricity	0.113	Average for grid-supplied electricity, 2000–2005. For on-site generation by photovoltaics or wind power, use carbon emissions method. 'Green tarifs' etc. not appropriate
Waste heat	0	Includes waste heat from industrial processes, and from power stations of more than 10 MW electrical output and power efficiency better than 35%

full load. However, where appropriate, a part load efficiency based on certified data supplied by the manufacturer may be used instead.

The carbon intensity of combined heat and power systems

A combined heat and power (CHP) system can create benefits because the on-site generation of electricity reduces the carbon emissions that would have occurred from the power stations feeding the national grid. This is taken into account by calculating the carbon intensity for CHP, ε_{chp}, from the equation:

$$\varepsilon_{chp} = \frac{C_f}{\eta_t} - \frac{C_{displaced}}{HPR}$$

where:

- η_t is the gross thermal efficiency of the CHP engine, measured in kWh of useful heat per kWh of fuel burned. Note that useful heat must be taken as the net amount supplied to the building after subtraction of any excess heat that the CHP system 'dumps' to the environment. For any given installation, it may be difficult to estimate the proportions of 'useful' and 'dumped' heat. It may therefore be necessary to demonstrate that η_t has been correctly evaluated, and AD L2 suggests the CHPQA certification scheme [39] as a means for doing this.
- C_f is the carbon emission factor of the fuel burned by the CHP engine, in kg of carbon emitted per kWh of delivered fuel consumed.
- $C_{displaced}$ is the carbon emission factor of the grid-supplied electricity which has been displaced by the CHP. Its value is that for the generating capacity which would otherwise be built if the CHP had not been provided, and is taken as 0.123 kgC/kWh (the value for new-generation gas-fired stations).

- HPR is the heat to power ratio of the CHP engine, in kWh of useful heat produced per kWh of electrical output. It is equivalent to the ratio of the thermal efficiency to the power efficiency of the CHP unit.

As for other heating systems, the carbon intensity for the CHP must be calculated for 100% and 30% of heating system output. If the CHP system operates alongside other heat generation equipment, then the carbon intensity for the complete system is found from:

$$\varepsilon_c = \frac{1}{\Sigma R}\Sigma\left(\frac{RC_f}{\eta_t} + R_{chp}\varepsilon_{chp}\right)$$

where R_{chp} is the rated output of the heat raising elements supplied by the CHP system.

The carbon intensity of community heating and other heating methods

The calculation of the carbon intensity of heat supplied to a building by a community heating system should take account of:

- The performance of the whole system, including the distribution circuits, all heat generating plant, and any CHP or waste heat recovery, *and*
- The carbon emission factors of all the different fuels.

The CHPQA Standard [39] provides a certification scheme for demonstrating that the thermal and power efficiencies have been estimated in a satisfactory manner.

In some buildings, especially factories, warehouses and workshops, local warm air or radiant heating systems may be acceptable and more efficient than centrally provided systems. Guidance on local systems is given in BRECSU publication GPG 303 [40].

Trade-off between construction elements and heating system efficiency

Throughout AD L2, rate of carbon emissions is taken to be an indicator of whether or not the requirement to conserve fuel and power has been met. Consequently, the elemental method allows the designer to trade off, in either direction, between the U-values of the building envelope and the carbon intensity of the heating system, provided the rate of carbon emissions is unchanged (or, although AD L2 does not specifically say, less). Compliance may be demonstrated by adjusting the area-weighted average U-value of the building fabric according to the equation:

$$U_{req} = U_{ref}\frac{\varepsilon_{ref}}{\varepsilon_{act}}$$

where U_{req} is the required maximum area-weighted average U-value of the building fabric

U_{ref} is the area-weighted average U-value of the building fabric when constructed to the elemental standards of Table 16.8

ε_{ref} is the carbon intensity of the reference heating system at an output of 30% of the installed design capacity, taken from Table 16.11 for the fuel type used in the actual heating system
ε_{act} is the carbon intensity of the actual heating system at an output of 30% of the installed design capacity.

Examples of the application of trade-off are given in section 16.10.

Space heating controls

The building should be provided with controls that maintain the required temperature of each functional area *only during the period when it is occupied*. This means that the building and its heating system must have sufficient and appropriate zone, timing and temperature control devices. Additional controls may be provided to allow:

- Heating during extended unusual occupation hours
- Heating to prevent condensation or frost damage when the heating would otherwise be switched off.

AD L2 gives two ways of meeting the requirement:

- Buildings with a heating system maximum output of not more than 100 kW follow the guidance in BRECSU GPG 132 [41]
- Larger or more complex buildings follow the guidance in CIBSE Guide H [42].

Building control bodies may accept certification by a competent person that the requirements have been met.

Hot water systems and their controls

The requirement is to provide hot water safely while making efficient use of energy and thereby minimising carbon emissions. Ways of achieving this include:

- Avoidance of:
  (1) the over-sizing of hot water storage systems
  (2) low-load operation of heat raising plant
  (3) the use of grid-supplied electric water heating, except where demand is low.
- Provision of solar water heating
- Minimisation of
  (1) the length of circulation loops
  (2) the length and diameter of dead legs.

For conventional hot water storage systems, ways of satisfying the requirement would be to:

(1) Provide controls that shut off heating when the required water temperature is reached
(2) Shut off the supply of heat during periods when hot water is not required.

Guidance on ways of meeting the requirements include:

(1) In small buildings, Good Practice Guide GPG 132 [41]
(2) In larger, more complex buildings, or for non-conventional systems such as solar water heating, CIBSE Guide, Part H [42].

Building control bodies may accept certification by a competent person that the provisions meet the requirements.

Insulation of pipes and ducts

The requirements apply only to pipework, ductwork and vessels for the provision of:

- Space heating
- Space cooling, including chilled water and refrigerant pipework
- Hot water supply for normal occupancy.

The Building Regulations do not apply to pipework, ductwork and vessels for process use. The standards given in BS 5422 [29] are suitable for determining the amount of insulation to be applied to pipework, ductwork and storage vessels. In the case of storage vessels, the recommendations in BS 5422 [29] for flat surfaces should be used.

When, as a result of fluid flowing or being stored, the heat lost from pipes, ducts or vessels is always making a contribution to the conditioning of the surrounding space, insulation may not be necessary. Nevertheless, insulation may still be advisable to maintain stability and control of the fluid temperatures.

Lighting efficiency standards

AD L2 states that lighting systems should be reasonably efficient and, where appropriate, make effective use of daylight. Beyond this statement there is no further mention of daylight, and hence no guidance as to what level of daylight provision might be considered effective and acceptable. AD L2 is, however, concerned with conserving the energy used by electric lighting systems. The efficiency and the control of lighting systems are considered separately, and the guidance varies according to building type.

General lighting efficacy in office, industrial and storage buildings

It should first be noted that the definition of luminous efficacy used here differs from that used for domestic lighting in section 16.2.3. Thus:

$$\text{Luminous efficacy} = \frac{\text{Total lumens emitted by lamp and luminaire combined}}{\text{Total circuit-watts consumed by lamp}}$$

With this definition in mind, the electric lighting system should be provided with reasonably efficient lamp/luminaire combinations. The requirement can be met by a luminous efficacy, averaged over the whole building, of not less than 40 luminaire-lumens per circuit-watt.

When interpreting this criterion, it should be noted that:

- The figure of 40 refers to the number of lumens emitted by the lamp and luminaire combination, and not just the lamp.
- The circuit-watts includes all the power consumed in the lighting circuits, including the lamps, their associated control gear and power factor correction equipment.
- The effect of the above is to allow flexibility in the choice of lamp, luminaire and control gear. For example, a low efficiency luminaire (chosen perhaps for reasons of appearance or low glare) can be compensated by a lamp of high efficacy and/or more efficient control gear.
- A maximum of 500 W of installed lighting may be excluded from this guidance and hence from the calculations. This is to allow flexibility for the use of feature lighting, etc.

The initial luminaire efficacy, η_{lum}, in luminaire-lumens per circuit-watt may be calculated by means of the equation:

$$\eta_{lum} = \frac{1}{P}\sum\left(\frac{LOR \times \phi_{lamp}}{C_L}\right)$$

where LOR is the light output ratio of the luminaire. The LOR is defined as the ratio of the total light output of the luminaire under stated practical conditions to the total light output of the bare lamp or lamps used in the luminaire under reference conditions

ϕ_{lamp} is the sum of the average initial (100 hour) lumen output of all the lamps in the luminaire

P is the total circuit-watts for all luminaires

C_L is a control factor which takes account of controls which reduce the output of the luminaire when electric light is not required. The control factor, C_L, is an empirical parameter which allows for the energy savings which are expected to accrue on average from lighting control strategies. It is based in part on observations of occupant behaviour, and cannot be derived from first principles. Some values are given in Table 16.13.

General lighting efficacy in all other building types

In other building types, it may be appropriate to use luminaires for which photometric data is not available (i.e. the value of the LOR is not known), and/or luminaires which take low power less efficient lamps. For these cases, the criterion is based on the light output of the lamps themselves, without the need to allow for the performance of the luminaire. That is to say, the definition of luminous efficacy corresponds to that used for domestic lighting. The requirement is that the installed lighting capacity has an initial (100 hour) lamp plus ballast efficacy of not less than 50 lamp-lumens per circuit-watt. This would be taken as achieved if at least 95% of the installed lighting capacity used lamps as described in Table 16.14.

Table 16.13 Luminaire control factors.

Control function	C_L
The luminaire is in a daylit space* and its light output is controlled by: *either* a photoelectric switching or dimming control, with or without manual override *or* local manual switching**	0.80
The luminaire is in a space that is likely to be unoccupied for a significant proportion of working hours and where a sensor switches off the luminaires in the absence of occupants but switching on is done manually	0.80
Both of the above circumstances combined	0.75
None of the above	1.00

Definitions
* *Daylit space:* Any space within 6 m of a window wall provided that the glazed area is at least 20% of the internal area of the window wall or a roof-lit space with a glazing area at least 10% of the floor area. The glazing must have a light transmittance at normal incidence (i.e. to light at an angle perpendicular to its surface) of at least 70%, or to a lower value if the glazed area is increased in proportion to its area (provided that the maximum areas to limit heat loss and to avoid solar overheating are not exceeded).
** *Local manual switch:* A switch whose distance from the luminaire it controls is not more than 8 m on plan, or three times the height of the luminaire above the floor if this is greater *and* is operated by deliberate action of the occupants (rocker switch, push button, pull cord, etc.) *or* is operated by remote control (infrared transmitter, sonic or ultrasonic devices, telephone handset controls, etc.).

Table 16.14 Light sources which meet the criterion for general lighting.

Light source	Types and ratings
High pressure sodium	All types and ratings
Metal halide	All types and ratings
Induction lighting	All types and ratings
Tubular fluorescent	38 mm diameter (T12) linear fluorescent lamps 2400 mm in length
Tubular fluorescent, with high efficiency control gear*	26 mm diameter (T8) lamps, and 16 mm diameter (T5) lamps rated above 11 W
Compact fluorescent	All ratings above 11 W
Other	Any type and rating with an efficacy greater than 50 lamp-lumens per circuit-watt

* *High efficiency control gear:* This means low loss or high frequency control gear that has a power consumption, including the starter component, not exceeding a specified value. The specified value depends on the nominal lamp rating and may be obtained from Table 16.15.

Table 16.15 Maximum power consumption of high efficiency control gear.

Nominal lamp rating, watts	Maximum power consumption, watts
Less than or equal to 15	6
Greater than 15, not more than 50	8
Greater than 50, not more than 70	9
Greater than 70, not more than 100	12
Greater than 100	15

Display lighting for all buildings
Within AD L2, display lighting is defined as:

- Lighting intended to highlight displays of exhibits or merchandise
- Lighting used in spaces for public entertainment (e.g. dance halls, auditoria, conference halls and cinemas).

The special requirements of display lighting may make it necessary to accept lower standards of energy performance than for general lighting. Nevertheless, it is still necessary for display lighting to be energy efficient, and AD L2 offers two ways of demonstrating compliance:

- *either* ensure that the installed capacity of the display lighting has an initial (100 hour) efficacy of not less than 15 lamp-lumens per circuit-watt
- *or* ensure that at least 95% of the installed display lighting capacity in circuit-watts is made up of lamps and fittings that have circuit efficacies no worse than the following:
  (a) High pressure sodium — all types and fittings
  (b) Metal halide — all types and fittings
  (c) Tungsten halogen — all types and fittings
  (d) Compact and tubular fluorescent — all types and fittings.

The circuit-watts should include the power consumed by transformers and ballasts.

Emergency escape lighting and specialist process lighting
Lighting for these purposes is not subject to the requirements of Part L. These types of lighting are defined as:

- *Emergency escape lighting:* that part of emergency lighting that provides illumination for the safety of people leaving an area or attempting to terminate a dangerous process before leaving an area.
- *Specialist process lighting:* lighting that is intended to illuminate specialist tasks within a space rather than the space itself.

Specialist process lighting could include:

- Theatre spotlights
- Projection equipment
- Lighting in TV and photographic studios
- Medical lighting in operating theatres and doctors' and dentists' surgeries
- Illuminated signs
- Coloured or stroboscopic lighting
- Integral lighting for art objects such as sculptures
- Decorative fountains and chandeliers.

Lighting controls

The aim of lighting controls should be to encourage the maximum use of daylight and to avoid unnecessary use of lighting when spaces are not occupied. This should not, however, create a situation in which the operation of an automatically switched lighting system endangers the passage of building occupants. Guidance on lighting controls is given in BRE IP 2/99 [43].

Controls in offices and storage buildings

The requirement can be met by providing local switches in easily accessible positions within each working area or at boundaries between working areas and circulation routes. In the context of AD L2, a switch is taken to include a dimmer switch, and switching includes dimming. However, as a general rule, dimming should be achieved by reducing rather than by diverting the energy supply. Local switches could include:

- A switch that can be operated by the deliberate action of an occupant either manually or by remote control, *and/or*
- Automatic switching systems that switch off the lighting when a sensor senses the absence of occupants.

Local switches include:

- Rocker switches, push buttons and pull cords operated manually, *and/or*
- Infrared, sonic and ultrasonic transmitters, and telephone handset controls, operated remotely.

Local switching can be supplemented by other controls such as timers and photo-electric devices.

The positioning of local switches is important. The distance on plan from any local switch to the luminaire it controls should not normally be more than 8 m, or three times the height of the luminaire above the floor if this is greater.

Controls in buildings other than offices and storage buildings

Again, the designer is urged to maximise the beneficial use of daylight without being given any specific criterion for daylight provision. The guidance is limited to a requirement to use, as appropriate, strategies to ensure that artificial lighting is switched off whenever daylight levels within a building are sufficiently high. The lighting controls should be one or more of:

- Local switching (see above)
- Time switching (e.g. in major occupational areas which have clear timetables of occupation)
- Photo-electric switching.

Controls for display lighting

Connect display lighting in dedicated circuits that can be switched off when not required. For example, in a retail store, timers could be used that switch off display lighting outside store hours, except for displays intended to be viewed from outside through display windows.

Air conditioning and mechanical ventilation (ACMV)

Within the context of AD L2, the acronym ACMV includes mechanical ventilation systems with no cooling facility as well as mechanical ventilation systems with full air conditioning. Definitions are provided for the terms:

- mechanical ventilation
- air conditioning
- treated areas
- process requirements.

Mechanical ventilation: This is a system that uses fans to supply outdoor air and/or extract indoor air to meet ventilation requirements. The system may be extensive and may include components such as air handling units, air filtration and heat reclamation. The system does *not* provide active cooling from refrigeration equipment. The definition does *not* apply to naturally ventilated buildings which use individual extract fans (mounted in either a wall or a window) to improve the ventilation of a small number of rooms.
Air conditioning: This is any system where refrigeration is included to provide cooling for the comfort of the building's occupants. The cooling function can be provided from stand-alone refrigeration equipment in the cooled space, or from centralised or partly centralised equipment, or from systems that combine cooling with mechanical ventilation.
Treated area is the floor area, measured between the internal faces of the surrounding walls, of the spaces that are served by a mechanical ventilation or air conditioning system. Treated area does *not* apply to spaces (plant rooms, service ducts, lift wells, etc.) which are *not* served by a mechanical ventilation or air conditioning system; such spaces should be excluded from the treated area total.
Process requirements: In an office building, process requirements include any significant area within which an activity takes place that is not typical of an ordinary commercial office. The performance of the mechanical ventilation and air conditioning systems for such areas are determined by the process requirements, and these areas, together with the plant capacity (or proportion of plant capacity) associated with them, should be excluded from calculations. In office buildings, activities and areas that may be considered to fall within the definition of process requirements include:

- staff restaurants and kitchens
- large dedicated conference rooms
- sports facilities
- dedicated computer or communications rooms.

The general requirement for ACMV is that buildings should be designed and constructed such that:

- The form and fabric of the building do not create a need for excessive installed ACMV capacity. Glazing type, glazing ratios, and solar shading are important in limiting cooling requirements
- Fans, pumps and refrigeration equipment are reasonably efficient and are not over-sized, so that the capacity for demand and standby is no more than necessary
- Facilities for the management, control and monitoring of the operation of equipment and systems are provided.

The method of demonstrating compliance depends on the building type. For buildings which are used as offices, or are of a similar type and usage to an office development, the carbon performance rating (CPR) method [44] can be used. For all other building types the only criterion is the specific fan power (SFP) of the mechanical systems.

The CPR method for office buildings with ACMV
The CPR method can be used if there are no innovative building or building services provisions. Otherwise, the carbon emissions calculation method, or some other acceptable alternative, must be used. The formulae and method of calculation of the CPR are given in Appendix G of AD L2 (see section 16.12). Compliance is demonstrated if the CPR does not exceed the relevant maximum value given in Table 16.16. If part of a building is served by a new air conditioning system and part served by a new mechanical ventilation system, the two parts should be considered separately and the relevant CPR must be met in each part.

Table 16.16 Maximum allowable carbon performance ratings.

	Maximum CPR, kgC/m²/year	
System type	**New building**	**Existing building**
Air conditioning	10.30	11.35
Mechanical ventilation	6.50	7.35

Where substantial alteration is made to an existing ACMV system, compliance would be achieved if:

- *Either* the CPR is reduced by at least 10% by the work
- *Or* the new CPR does not exceed the relevant value in Table 16.16.

Where there is only replacement of existing equipment, the criterion is the product of the installed capacity per unit area, PD (or PR), and the control management factor, FD (or FR). (See Appendix G of AD L2 for the full definitions of PD, etc.). The product PD $\times$ FD (or PR $\times$ FR) should:

- *Either* be reduced by at least 10%
- *Or* meet a level of performance equivalent to the component benchmarks given in CIBSE TM 22 [45].

Methods for other buildings with ACMV

Other mechanically ventilated buildings may be assessed for compliance by means of the specific fan power (SFP). This is the sum of all the circuit-watts used to drive all the supply and extract ventilation fans, including switchgear, inverters, etc., divided by the design ventilation rate, in litres per second, of the building. The SFP may be used regardless of whether or not the air supply is heated or cooled. For typical spaces ventilated for human occupancy, compliance is demonstrated if:

- for ACMV systems in new buildings, the SFP is no greater than 2.0 W/litre/second
- for new ACMV in refurbished buildings, or where an existing ACMV system is being substantially altered, the SFP is no greater than 3.0 W/litre/second.

For spaces where higher ventilation rates are required because of, say, specialist processes or higher than normal external pollution levels, higher values of SFP may be appropriate.

It is also important that mechanical ventilation systems are reasonably efficient at part load. This could be demonstrated by providing efficient variable flow control systems such as variable speed drives or variable pitch axial flow fans. Detailed guidance is given in BRESCU GIR 41 [46].

The whole-building method

This method is separate and independent of the elemental method and therefore allows much more design flexibility. To show compliance it must be shown that:

- *Either* the total carbon emissions
- *Or* the primary energy consumption

for the complete building are reasonable for the purposes of conserving fuel and power. Three building types are considered in AD L2, though for all three types the details and calculation procedures are given not in AD L2 but in other publications.

Office buildings

Office buildings can be treated by means of the whole-office carbon performance rating method. In principle this is the same as the CPR method for assessing the ACMV systems of office buildings (described above, page 16.53), and which is part of the elemental method. However, when used as part of the elemental method, the CPR calculation deals only with the building's ACMV systems, whereas in the whole-building method, the CPR calculation is expanded to include lighting and space heating. Full details are to be found in BRE Digest No. 457 [44]. There are several assumptions implicit in using the method.

- The proposed design of the building and its energy consuming systems is capable of creating internal environmental conditions which are normal and acceptable for the occupants and the functional requirements of the building. Otherwise, it

would be possible to meet the relevant criteria by deliberately undersizing the installed equipment.
- The design of the ACMV, space heating and lighting systems is conventional and does not require the use of novel equipment. This means that the heating system is likely to be supplied from a conventional boiler, and that cooling will be provided by conventional refrigeration plant.
- The CPR method applies only to the energy consuming systems themselves. Therefore, while it provides the flexibility for adjustment between the energies consumed by the ACMV, the heating system and the lighting system, the method is not intended to provide an excuse for poor fabric design.
- The CPR calculation is intended to be a relatively simple method for assessing a building and its systems for compliance. If certain features of the design are likely to make the calculation unduly complex, one of the alternative methods should be used.

The question of whether or not particular equipment may be considered conventional and within the scope of the method must be carefully considered. In some cases, it depends on the detail:

- *Heat recovery* can be included if the effect is to reduce the size of the heat generator. It can also be included, though not so easily, if the effect is to reduce the hours of use of the heat generator.
- *Thermal storage* can be included for ice thermal storage which is part of a cooling/refrigeration system. For thermal stores whose purpose is to even out the peaks in demand on heat generators, the CPR method is not recommended.
- *Space heating using heat pumps* is possible but not advised. The designer could not use the factors given as part of the method, and would have to make his own calculation of the carbon emissions.
- *Combined heat and power* is also possible but not advised. The calculation of the carbon emissions could be excessively complex.
- *Renewable energy* is possible in some cases. The recommended technique is to calculate the carbon emissions assuming all the energy is supplied from conventional sources, and then to calculate the saving due to the renewable energy source. The amount saved is then subtracted from the initial result.

The calculation procedure is described in section 16.12. When the CPR value of an office building has been found, it can be considered to comply if it meets *all* the following:

- A whole-office CPR which is no worse than the relevant maximum in Table 16.17
- The requirements for avoiding thermal bridging at junctions and around openings (as described previously)
- The requirements for meeting air leakage standards
- The upper limits for U-values, i.e for parts of a roof $U \leq 0.35\,W/m^2K$, and for parts of an exposed wall or floor $U \leq 0.70\,W/m^2K$.

Table 16.17 Maximum whole-office CPR.

	Maximum allowable CPR, kgC/m^2/year	
Office building type	**New office**	**Refurbished office**
Naturally ventilated	7.1	7.8
Mechanically ventilated	10.0	11.0
Air-conditioned	18.5	20.4

Schools
For schools, compliance may be demonstrated by showing that the building conforms with DfEE Building Bulletin 87 [47].

Hospitals
For hospitals, compliance may be demonstrated by showing that the building conforms with the NHS Estates Guide [48].

Carbon emissions calculation method

The carbon emissions method allows considerable flexibility in the design of a building. The design may take advantage of any energy conservation measure and may take into account heat gains due to solar radiation and internal heat sources. To show compliance using the carbon emissions method, four conditions must be met:

- The calculated annual carbon emissions of the proposed building should be no greater than from a notional building of the same size and shape which has been designed to comply with the elemental method
- For the notional building, the U-value of the floor must be taken as *either* 0.25 W/m^{2}K *or* the U-value of the floor in the proposed building, whichever is the lower
- For the proposed building, the poorest acceptable U-values for parts of the roof and walls are 0.35 W/m^{2}K for the roof and 0.70 W/m^{2}K for the walls
- The fabric of the proposed building must be constructed to provide resistance to air leakage to at least the same standard as required by the elemental method.

A critical feature of the carbon emissions method is the procedure which is used to carry out the calculation. In general, simple methods which do not require the aid of a computing device are very unlikely to be sufficient, and so in nearly all cases the calculations must be performed by computer using either in-house or commercially purchased software. The reliability of the method being used is therefore a major consideration, and the Approved Document requires that the method is acceptable without specifying those methods which would be acceptable. Tests of acceptability are:

- That the method has been approved by a relevant authority responsible for issuing professional guidance

- The organisation responsible for carrying out the calculations is using a method which satisfies their own in-house quality assurance procedures
- Either of the above can be demonstrated by submitting with the calculations a completed copy of Appendix B of CIBSE AM11 [49]; this is a checklist for choosing BEEM software, and should show that the software which has been used is appropriate for the purpose.

Other matters – conservatories, atria, sun-spaces, etc.

Definitions

Sun-space (including conservatory and atrium): A sun-space is a building or part of a building having not less than three-quarters of the area of its roof and not less than half the area of its external walls (if any) made of translucent material.

Separation: Separation between a building and a sun-space means:

- The separating walls and floors are insulated to at least the same degree as the exposed walls and floors of the building
- The separating windows and doors have U-values and draught proofing to at least the same standard as the windows and doors elsewhere in the building.

A sun-space attached to and built as part of a new building

Where there is *no separation* between the sun-space and the building, the sun-space should be treated as an integral part of the building.

Where there is *separation* between the sun-space and the building, energy savings can be achieved if the sun-space is neither heated nor mechanically cooled. If fixed heating or mechanical cooling is installed, they should have their own separate temperature and on/off controls.

A sun-space attached to an existing building

When attaching a sun-space to an existing building, reasonable provision should be made to limit heat loss from, or summer solar heat gain to, the building. Ways of meeting this requirement are:

- If the opening is not to be enlarged, retain the existing separation, *or*
- If an opening has to be enlarged or newly created as a material alteration, provide separation as or equivalent to windows and doors having the average U-value given in Table 16.8 for the elemental method.

16.3.4 AD L2 Section 2 – construction

The persons or organisations who construct and assemble the building have to satisfy the building control body of a number of matters, two of which are dealt with in this section:

- Certain aspects of the building fabric
- Inspection and commissioning of the building services systems.

Building fabric

Continuity of insulation

Continuity of insulation is necessary in order to avoid excessive thermal bridging. The design requirements for achieving this are as described for the elemental method in section 16.3.3. In addition, the person carrying out the work has a responsibility to ensure that compliance with Part L is achieved. For a new building, that person will normally be:

- The developer who has carried out the work subject to Part L
- The contractor who has carried out the work subject to Part L
- The sub-contractor who has carried out the work subject to Part L.

If he is suitably qualified, the person responsible for achieving compliance should provide a certificate or declaration stating that the requirements of Part L2(a) have been met. Otherwise, a certificate or declaration to that effect should be obtained from a suitably qualified person. The certificate/declaration must be based on:

- *Either* a statement confirming that the design details, building techniques and manner in which the work has been carried out can be expected to achieve reasonable conformity with specifications that have been approved for compliance with Part L2
- *Or* post-completion testing using infrared thermography which shows that the fabric insulation is reasonably continuous over the whole visible envelope. Information on thermography for building surveys is given in BRE Report 176 [50].

Airtightness

Details of the requirements for minimising air infiltration through the building fabric are given under the elemental method in section 16.3.3, together with the standards that must be achieved. In addition, the person carrying out the building work must obtain certificates or declarations that confirm that the required standard of airtightness has been achieved:

- *Either* for buildings of any size, by means of air leakage tests carried out in accordance with CIBSE TM 23 [21]
- *Or* for buildings of less than 1000 m^2 gross floor area, by using appropriate design details and building techniques, and by carrying out the work in ways that should achieve reasonable conformity with the specifications that have been approved for compliance with Part L2.

Certificates and testing

Certificates/declarations such as those described in the two sections above may be accepted by building control bodies as evidence of compliance. However, it is necessary to establish, to the satisfaction of the building control body and in advance of the work, that the person who will give the certificates/declarations is suitably qualified.

Inspection and commissioning of building services systems

When describing or discussing the requirements for building services systems in Part L2, the terms 'providing' and 'making provision' include, where relevant, inspection and commissioning, the definitions of which are as follows.

Inspection of building services systems

This is defined as the establishment, at completion of installation, that the specified and approved provisions for efficient operation have been put in place.

Commissioning of building services systems

This means the advancement of these systems from the state of static completion to working order to the specifications relevant to achieving compliance with Part L2, without prejudice to the need to comply with health and safety requirements. For each system, this includes:

- Setting-to-work
- Regulation, i.e. testing and adjusting repetitively to achieve the specified performance
- Calibration, setting up and testing of the associated automatic control systems
- Recording of the system settings and the performance test results that have been accepted as satisfactory.

Responsibility for achieving compliance with the requirements of Part L lies with the person carrying out the work. For building services systems, this person may be:

- The developer who has carried out the work subject to Part L
- The main contractor who has carried out the work subject to Part L
- A sub-contractor who has carried out the work subject to Part L
- A specialist firm directly engaged by a client.

A report must be provided, either by the person responsible for achieving compliance or by a suitably qualified person, that indicates that the inspection and commissioning activities necessary to establish that the work complies with Part L have been completed to a reasonable standard. The report should include:

- A commissioning plan that shows that every system has been inspected and commissioned in an appropriate sequence

- The results of the tests that confirm that the performance is in reasonable accordance with the approved designs, including written commentaries where excursions are proposed to be accepted.

Compliance may be demonstrated by following the guidance in the CIBSE Commissioning Codes [51] and TM1 of the Commissioning Specialists Association [52].

Reports and testing

A report such as that described above may be accepted by building control bodies as evidence of compliance. However, it is necessary to establish, to the satisfaction of the building control body and in advance of the work, that the person who will provide the report is suitably qualified.

16.3.5 AD L2 Section 3 – providing information

There is an obligation to provide the owner and/or the occupier of the building with certain information. There are two aspects to this: the provision of a building log-book and the installation of energy meters.

Building log-book

The building owner and/or occupier must be provided with a log-book. The log-book should contain details of:

- The installed building services plant
- The installed building services controls
- The method of operation of the plant and controls
- Maintenance requirements
- Any other matters which collectively enable energy consumption to be monitored and controlled.

The log-book information should be provided in summary form and be suitable for use on a day-to-day basis. The log-book may refer to information contained in other documents, for example, operation and maintenance manuals, health and safety files, etc.

Log-book contents

The log-book could include the following:

(1) A description of the whole building, including its intended usage and the philosophy of the design.
(2) A description of the intended purpose of the individual building services systems.
(3) A schedule of the floor areas of each building zone, broken down according to the type of environmental service provided to that zone, i.e. air conditioning, natural ventilation, etc.

(4) The location of all relevant plant and equipment, including simple schematic diagrams.
(5) The installed capacity of the services plant, expressed as the input power and the output rating.
(6) Simple descriptions of the proper strategies for operating and controlling the energy consuming services of the building.
(7) A copy of the report that confirms that the building services equipment has been commissioned and found to be satisfactory.
(8) Inclusion, in the operating and maintenance instructions, of the provisions that enable the specified performance to be sustained during occupation.
(9) A schedule of the building's energy supply meters and sub-meters; for every meter and sub-meter, this schedule should include:
    - location
    - identification and description
    - the type of fuel being monitored
    - instructions on its use – these instructions should indicate how the energy performance of the building (or, if relevant, each separate tenancy in the building) can be calculated from the metered energy readings for comparison with published benchmarks (see also Appendix G of AD L2, and see the next section below for metering strategies).
(10) For systems serving an office floor area of more than 200 m^2, a design assessment of the building services systems' carbon emissions and the comparable performance benchmark (see Appendix G of AD L2).
(11) The measured air permeability of the building.

Installation of energy meters

The building engineering services should be provided with sufficient energy meters and sub-meters to enable owners or occupiers to measure their actual energy consumption. Sufficient instructions, including an overall metering strategy, must be provided so that owners or occupiers are able to attribute energy consumed to the end use of that energy, and to be able to compare operating performance to published benchmarks (see item 9 of section above). In order to develop a metering strategy, it is first necessary to know how the energy from each fuel will be used in the building, and to have a good estimate of the amount of each fuel that will be consumed. CIBSE TM22 [45] provides a standardised procedure for doing this, including prepared spreadsheets and a worked example on CD-ROM. The procedure is in three stages.

- Stage 1 is a quick assessment in terms of energy use per unit floor area.
- Stage 2 is an improved assessment accounting for special energy uses, occupancy and weather.
- Stage 3 is a detailed assessment of the building and all its energy systems.

Once the information on energy usage has been collected, a metering strategy can be developed. Detailed guidance on metering strategies, including worksheets and worked examples, is given in GIL 65 [53].

Reasonable provision for energy metering

Provision for energy metering is considered to be reasonable when it is possible to measure at least 90% of the estimated annual energy consumption of each fuel to be accounted for. Possible techniques by which the allocation of energy consumption to each end use can be achieved are:

- Direct metering
- Measuring the run-hours of equipment that operates at a constant known load
- Estimating the energy consumption indirectly; for example, this could be done for hot water supply by combining measurements of the amount of water supplied and the delivery temperatures of the water, together with the known efficiency of the water heater
- Estimating the energy consumption by difference; for example, by measuring the total consumption of gas and deducting the measured gas consumption for heating and hot water, the amount of gas used for some other function (e.g. catering) can be found
- Estimating non-constant small power loads by means of the procedure given in CIBSE Energy Efficiency Guide, Chapter 11 [36].

Reasonable provision of energy meters and sub-meters

Reasonable provision of meters would be to install incoming meters in every building greater than 500 m^2 gross floor area (including separate buildings on multi-building sites). This would include:

- Individual meters for direct measurement of the total electricity, gas, oil and LPG consumed by a building
- A heat meter capable of direct measurement of the total heating and/or cooling energy supplied to the building by a district heating or cooling scheme.

In the case of sub-metering, it would be reasonable to provide additional meters to directly measure or reliably estimate (see section immediately above):

- The electricity, natural gas, oil and LPG supplied to each separately tenanted area that is greater than 500 m^2
- The energy consumed by plant items with input powers greater than or equal to those listed in Table 16.18
- Any heating or cooling supplied to separately tenanted spaces: for tenancies of floor area greater than 2500 m^2, direct metering of the heating and cooling may be appropriate, but for smaller tenanted areas, the heating and cooling end uses may be proportioned on an area basis
- Any process load (see ‘Process requirements’ under ‘ACMV’ in section 16.3.3) that is discounted from the building’s energy consumption when comparing measured consumption against published benchmarks.

Table 16.18 Size of plant for which separate metering would be reasonable.

Plant item	Rated input power, kW
Boiler installations comprising one or more boilers or CHP plant feeding a common distribution circuit	50
Chiller installations comprising one or more chiller units feeding a common distribution circuit	20
Electric humidifiers	10
Motor control centres providing power to fans and pumps	10
Final electrical distribution boards	50

16.3.6 AD L2 Section 4 – work on existing buildings

Replacement of a controlled service or fitting

Attention must be paid to the definitions of 'controlled service or fitting' and 'building work'. These are the same as for dwellings and are given in section 16.2.4. The requirement to make reasonable provision applies when:

- Replacing old with new identical equipment, *or*
- Replacing old with different equipment.

It also applies:

- When the work is solely in connection with controlled services, *or*
- When the work includes work on controlled services.

Ways of meeting the requirements include the following.

Windows, doors and rooflights

When these elements are replaced:

- Provide units that meet the requirements for new buildings, *or*
- Provide units with a centre-pane U-value no worse than $1.2\,W/m^2K$.

The replacement work should comply with the requirements of both Part L2 and (unless non-glazed fittings are involved) Part N. In addition, after the work the building should not have a worse level of compliance with other relevant parts of Schedule 1, such as Parts B, F and J.

However, note that the requirement does not apply to repair work on parts of these elements, such as replacing broken glass, sealed double glazing units or rotten framing members.

Heating systems

If the heating system is substantially replaced, then provide a new heating system and new controls to the standards required for new installations (i.e. a new building). In lesser work, it is acceptable to provide insulation, zoning, timing, temperature and interlock controls.

Again, when meeting other relevant requirements of Schedule 1, particular account should be taken of Parts F and J.

Hot water systems

When substantial replacement to hot water systems, pipes and vessels takes place, provide controls and insulation as if for a new building. In lesser work, it is sufficient to provide insulation, and timing and thermostatic controls.

Lighting systems

If a complete lighting system serving more than $100\,m^2$ of floor area is to be replaced, then the new system should comply with the requirements for a new building. In the case of partial replacement, then:

- If only the complete luminaires are replaced, provide appropriate new luminaires as specified in section 16.3.3 (note however that this requirement does not apply when only components such as lamps or louvres are replaced)
- If only the control system is to be replaced, provide appropriate new controls as specified in section 16.3.3 (note however that this requirement does not apply when only components such as switches or relays are replaced).

Air conditioning or mechanical ventilation systems

For office buildings, when replacing systems which serve more than $200\,m^2$ of floor area, the carbon performance rating should be improved to the standards given in section 16.3.3.

For all other buildings, provide mechanical ventilation systems that meet the SFP requirements of section 16.3.3.

Commissioning, etc.

When carrying out any of the work described in the four sections immediately above:

- The work should be inspected and commissioned as described in section 16.3.4 'Inspection and commissioning of building service systems'.
- A building log-book should be prepared, or the existing log-book updated, with details of the replacement controlled service or fitting, as described in section 16.3.5.
- In order that the replacement controlled service or fitting can be effectively monitored (as described in section 16.3.5), the relevant part of the metering strategy should be prepared or revised as necessary, and additional metering provided where needed.

Material alterations

Attention must be paid to the definitions of 'material alteration' and 'relevant requirement'. These are the same as for dwellings and are given in section 16.2.4. When undertaking material alterations, account should be taken of:

- All the relevant requirements of Schedule 1, including Parts L2, F and J
- Insulation of roofs, floors and walls
- Sealing measures
- Controlled services and fittings.

When undertaking material alterations, reasonable provision for satisfying the requirements of Part L2 may depend on the circumstances of the particular case, and would need to take into account any historic value in the structure being altered. Specific guidance is as follows.

Roof insulation

If the material alteration includes substantial replacement of any of the major elements of a roof structure, insulate to the U-value standard required for a new building.

Floor insulation

If the structure of a ground floor is to be substantially replaced or re-boarded, and if the room is heated, insulate to the U-value standard of a new building.

Wall insulation

Provide a reasonable thickness of insulation when substantially replacing:

- Complete exposed walls
- The external rendering or cladding of an exposed wall
- The internal surface finishes of an exposed wall
- The internal surfaces of separating walls to unheated spaces.

Sealing measures

When carrying out any of the above work on roofs, floors or walls, include reasonable sealing measures to improve airtightness.

Controlled services or fittings

Follow the guidance at the beginning of section 16.3.6 above.

Material changes of use

The definition of a material change of use given in section 16.2.4 for dwellings also applies to other buildings. When undertaking material alterations, account should be taken of:

- All the relevant requirements of Schedule 1, including Parts L2, F and J
- Insulation of accessible lofts, roofs, floors and walls
- Sealing measures
- Controlled services and fittings.

When undertaking material changes of use, reasonable provision for satisfying the requirements of Part L2 may depend on the circumstances of the particular case, and would need to take into account any historic value in the structure which is subject to the material changes of use. Specific guidance is as follows.

Accessible lofts
Where the existing insulation in accessible lofts is worse than 0.35 W/m^2K, replace or add extra insulation to upgrade the U-value to a maximum of 0.25 W/m^2K.

Roof insulation
If the material change of use includes substantial replacement of any of the major elements of a roof structure, insulate to the U-value standard of a new building.

Floor insulation
If the structure of a ground floor is to be substantially replaced, and if the room is heated, insulate to the U-value standard of a new building.

Wall insulation
Provide a reasonable thickness of insulation when substantially replacing:

- Complete exposed walls
- The external rendering or cladding of an exposed wall
- The internal surface finishes of an exposed wall
- The internal surfaces of separating walls to unheated spaces.

Sealing measures
When carrying out any of the above work on roofs, floors or walls, include reasonable sealing measures to improve airtightness.

Controlled services or fittings
Follow the guidance at the beginning of section 16.3.6 above.

Historic buildings

Historic buildings include:

- Listed buildings
- Buildings situated in conservation areas
- Buildings of architectural and historical interest and which are referred to as a material consideration in a local authority development plan

- Buildings of architectural and historical interest within national parks, areas of outstanding natural beauty, and world heritage sites.

Any work on an historic building must balance the need to improve energy efficiency against the following factors:

- The need to avoid prejudicing the character of the historic building
- The danger of increasing the risk of long-term deterioration of the building fabric
- The danger of increasing the risk of long-term deterioration of the building's fittings
- The extent to which energy conservation measures are a practical possibility.

Information on the special characteristics of historic buildings and their conservation should be obtained, for example from Planning Policy document PPG15 [33] and BS 7913 [34]. Advice on achieving the correct balance should also be sought from the conservation officer of the local authority. Advice from other sources would may also be appropriate, particularly regarding:

- Restoration of the historic character of a building that has been the subject of inappropriate alteration, such as the replacement of windows, doors or rooflights
- Rebuilding of a former historic building, which may have been damaged or destroyed due to some mishap (such as a fire), or infilling a gap in a terrace
- Providing a means for the fabric of an historic building to 'breathe' so that moisture movement may be controlled and the potential for long-term decay problems reduced (see SPAB Information Sheet No. 4 [35]).

16.4 Tables of U-values

Appendix A of the Approved Documents provides look-up tables for obtaining U-values for a range of windows, doors and rooflights. For all other elements, it provides look-up tables which allow the user to find, by relatively simple calculations, the minimum thickness of insulation required to meet a specified U-value.

16.4.1 Windows, doors and rooflights

When available, manufacturers' certified U-values (by approved methods of measurement or calculation) should be used. If these are not available, values for single, double and triple glazing may be taken from Tables 16.19, 16.20 and 16.21, modified where necessary for metal frames according to Table 16.22. Low emissivity (low-E) coatings are of two main types, 'hard' and 'soft'. If the exact value of the emissivity, ε_n, is not known, then for hard coatings or where the type of coating is unknown use the data for $\varepsilon_n = 0.2$, and for soft coatings use the data for $\varepsilon_n = 0.1$. For doors that are half-glazed, the U-value is the average of the non-glazed door and the appropriate U-value for the glazing. For windows and rooflights with metal frames where

Table 16.19 Single glazing U-values for windows, rooflights and doors.

Single glazing description	W/m^2K
Windows in wood or PVC-U frames	4.8
Rooflights in dwellings in wood or PVC-U frames	5.1
Rooflights in buildings other than dwellings in wood or PVC-U frames	4.8
Windows in metal frames (4 mm thermal break)	5.7
Solid wooden door	3.0

Table 16.20 Double glazing U-values for windows and rooflights, W/m^2K.

Double glazing description	Gap between panes			Adjustment for rooflights in dwellings
	6 mm	12 mm	16 mm or more	
Wood or PVC-U frames				
Air filled	3.1	2.8	2.7	
Low-E, $\varepsilon_n = 0.2$	2.7	2.3	2.1	
Low-E, $\varepsilon_n = 0.15$	2.7	2.2	2.0	
Low-E, $\varepsilon_n = 0.1$	2.6	2.1	1.9	For dwellings only,
Low-E, $\varepsilon_n = 0.05$	2.6	2.0	1.8	add 0.2 for all wood
Argon filled	2.9	2.7	2.6	and PVC-U frames
Low-E, $\varepsilon_n = 0.2$, argon filled	2.5	2.1	2.0	
Low-E, $\varepsilon_n = 0.1$, argon filled	2.3	1.9	1.8	
Low-E, $\varepsilon_n = 0.05$, argon filled	2.3	1.8	1.7	
Metal frames, 4 mm thermal break				
Air filled	3.7	3.4	3.3	
Low-E, $\varepsilon_n = 0.2$	3.3	2.8	2.6	
Low-E, $\varepsilon_n = 0.1$	3.2	2.6	2.5	
Low-E, $\varepsilon_n = 0.05$	3.1	2.5	2.3	For metal frames, see
Argon filled	3.5	3.3	3.2	Table 16.22
Low-E, $\varepsilon_n = 0.2$, argon filled	3.1	2.6	2.5	
Low-E, $\varepsilon_n = 0.1$, argon filled	2.9	2.4	2.3	
Low-E, $\varepsilon_n = 0.05$, argon filled	2.8	2.3	2.1	

the thermal break differs from 4 mm, the corrections in Table 16.22 should be applied. Note that if corrections for thermal break *and* rooflight are applicable, both should be made.

Minimum specifications for windows

Inspection of the U-values in Tables 16.20 and 16.21 reveals the required design specification for a window to meet the maximum U-values of 2.0 or 2.2 W/m^2K given in Tables 16.2 and 16.8. For double glazing with a 6 mm air gap in a wood or UPVC frame, it is not possible to keep within the maximum U-value for any of the listed types of glass. With a 12 mm air gap there are three possibilities, and with a 16 mm air gap there are six. For double glazing in metal frames there is only one possibility, and that will pass only if it includes a 4 mm (or better) thermal break. Extracting these cases from Table 16.20, the double glazed window designs which can be expected to be satisfactory are therefore:

Table 16.21 Triple glazing U-values for windows and rooflights, W/m^2K.

Triple glazing description	Gap between panes 6 mm	12 mm	16 mm or more	Adjustment for rooflights in dwellings
Wood or PVC-U frames				
Air filled	2.4	2.1	2.0	
Low-E, $\varepsilon_n = 0.2$	2.1	1.7	1.6	
Low-E, $\varepsilon_n = 0.1$	2.0	1.6	1.5	For dwellings only, add 0.2 for all wood and PVC-U frames
Low-E, $\varepsilon_n = 0.05$	1.9	1.5	1.4	
Argon filled	2.2	2.0	1.9	
Low-E, $\varepsilon_n = 0.2$, argon filled	1.9	1.6	1.5	
Low-E, $\varepsilon_n = 0.1$, argon filled	1.8	1.4	1.3	
Low-E, $\varepsilon_n = 0.05$, argon filled	1.7	1.4	1.3	
Metal frames, 4 mm thermal break				
Air filled	2.9	2.6	2.5	
Low-E, $\varepsilon_n = 0.2$	2.6	2.2	2.0	
Low-E, $\varepsilon_n = 0.1$	2.5	2.0	1.9	
Low-E, $\varepsilon_n = 0.05$	2.4	1.9	1.8	For metal frames, see Table 16.22
Argon filled	2.8	2.5	2.4	
Low-E, $\varepsilon_n = 0.2$, argon filled	2.4	2.0	1.9	
Low-E, $\varepsilon_n = 0.1$, argon filled	2.2	1.9	1.8	
Low-E, $\varepsilon_n = 0.05$, argon filled	2.2	1.8	1.7	

Table 16.22 Corrections for metal frames with various thermal breaks.

Thermal break mm	Correction to U-value, W/m^2K: Window, or rooflight in buildings other than dwellings	Rooflight in dwellings
0 (no break)	+0.3	+0.7
4	0	+0.3
8	−0.1	+0.2
12	−0.2	+0.1
16	−0.2	+0.1

- In wood or UPVC frames:
  (a) Air filled low-E, $\varepsilon_n = 0.15$, minimum 16 mm air gap
  (b) Air filled low-E, $\varepsilon_n = 0.10$, minimum 16 mm air gap
  (c) Air filled low-E, $\varepsilon_n = 0.05$, minimum 12 mm air gap
  (d) Argon filled low-E, $\varepsilon_n = 0.20$, minimum 16 mm air gap
  (e) Argon filled low-E, $\varepsilon_n = 0.10$, minimum 12 mm air gap
  (f) Argon filled low-E, $\varepsilon_n = 0.05$, minimum 12 mm air gap
- In a metal frame with 4 mm thermal break:
  (g) Argon filled low-E, $\varepsilon_n = 0.05$, minimum 16 mm air gap.

For triple glazing, even though there is a much higher number of glass/frame combinations which satisfy the U-value standard, a significant number do not, especially metal frames with a small air gap.

16.4.2 Roofs, walls and floors

The tables in AD L1 and AD L2 do not give U-values for complete constructions. Rather, they enable the calculation of the minimum thickness of the insulation layer that is necessary to achieve a desired U-value. However, the minimum thickness is applicable only when the insulation layer is perfectly continuous. In practice, U-values may vary because of:

- Air gaps in the insulation
- Mechanical fasteners penetrating the insulation layer
- Precipitation on inverted roofs.

The effect of each of these factors is to add a correction factor to the U-value. Thus, if one or more of these factors is present in a construction, it is necessary to make an appropriate adjustment to the U-value. However, for the purposes of the 'look-up' tables of Appendix A, these correction factors are combined into a single correction factor, ΔU, for which values are given in Table 16.23. If this correction factor is applicable to a particular construction, it is then necessary to select the thickness of insulation to give a lower U-value than that which is desired. Then, when the correction is added on, the net result is the desired U-value. The procedure is therefore first to select the *desired* U-value and decide on the position and method of fixing of the insulation layer, then, using Table 16.23, to obtain ΔU, the correction to the U-value. The *design* U-value is found from the *desired* U-value using:

$$U_{design} = U_{desired} - \Delta U$$

Table 16.23 The ΔU correction term for U-values.

Roofs	**ΔU, W/m^{2}K**
Insulation fixed with nails or screws	0.02
Insulation between joists or rafters	0.01
Insulation between and over joists or rafters	0
Walls	
Timber frame where the insulation partly fills the space between the studs	0.04
Timber frame where the insulation fully fills the space between the studs	0.01
Internal insulation fixed with nails or screws which penetrate the insulation	0.02
External insulation with metal fixings which penetrate the insulation	0.02
Insulated cavity wall with cavity greater than 75 mm and tied with steel vertical-twist ties	0.02
Insulated cavity wall with cavity less than or equal to 75 mm tied with ties other than steel vertical-twist ties	0
Floors	
Suspended timber floor with insulation between joists	0.04
Floor insulation fixed with nails or screws	0.02

The tables are then used to find the thickness of insulation which will meet the *design* U-value. If this thickness of insulation is used, the desired U-value will be achieved, because:

$$U_{desired} = U_{design} + \Delta U$$

Once the desired U-value has been chosen, and the ΔU correction term (if applicable) has been found, the design U-value may be calculated and the tables used to find the *base* thickness of insulation, defined as the smallest thickness of insulation required to meet a specified U-value *without* allowances for other components in the structure.

Of course other elements in the construction also provide some thermal resistance, thus allowing a reduction to be made on the base thickness. The true minimum is therefore:

$$\text{Minimum thickness of insulation} = \text{base thickness} - \text{allowable reductions}$$

The calculated minimum thickness is unlikely to correspond exactly to any of the thicknesses in which a particular material is supplied, and so the nearest thickness above the minimum must be specified. Nor is it certain that the minimum thickness will be reasonably practical in terms of its installation and fixing. If the result appears impractical, a redesign of the element may be necessary.

Determining the thickness of insulation for roofs

Figure 16.4 illustrates three common roof types, and Table 16.24 gives the base thickness of insulation for these roofs.

Because the base thickness may not be the minimum, the reductions in heat flow due to other elements of the construction must be considered, and further corrections may be made to reduce the base thickness to give the minimum thickness. The amounts by which the base thickness may be reduced are given in Table 16.25.

Example calculations for roofs

Example 16.4.1 Pitched roof with insulation between the joists
Figure 16.5a shows insulation laid between the ceiling joists of a pitched roof which is covered with 19 mm roof tiles. The maximum U-value allowed by the elemental method is, from Table 16.2, 0.16 W/m^{2}K, and it is required to determine the necessary thickness of insulation. First, from Table 16.23, there is a ΔU correction of 0.01 W/m^{2}K. The design (or 'look-up') U-value is therefore:

Insulation between joists $\qquad U_{design} = 0.16 - 0.01 = 0.15\ W/m^2K$

Next, use Table 16.24 to find the base thickness of insulation, and Table 16.25 to determine allowable reductions to the base thickness:

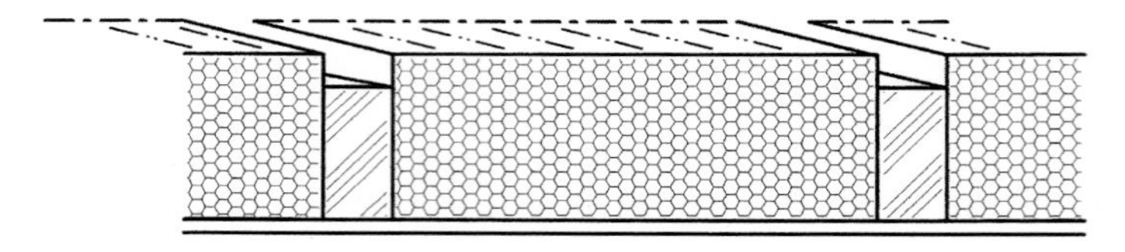

(a) Insulation laid between ceiling joists or rafters

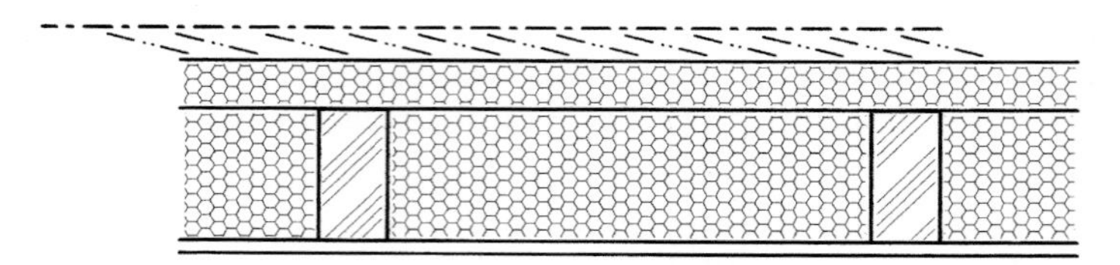

(b) Insulation laid between and over joists or rafters

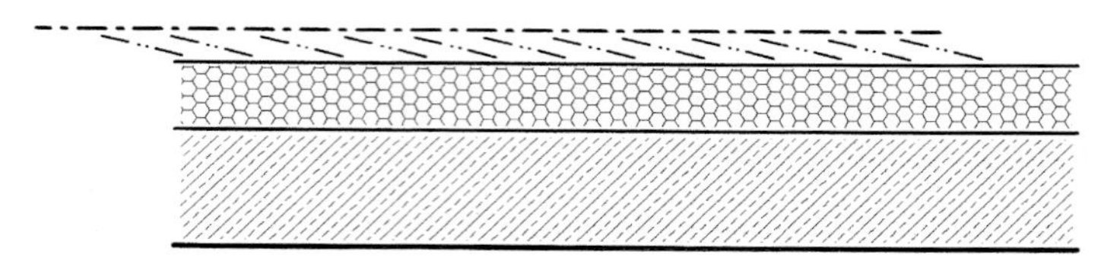

(c) Continuous layer of insulation over a structural base

Fig. 16.4 Three common roof types.

From Table 16.24	Base thickness = 557 mm		
From Table 16.25	Reduction for	19 mm roof tiles	= 1 mm
		Pitched roofspace	= 6 mm
		10 mm plasterboard	= 2 mm
	Total reduction		= 9 mm

Minimum insulation thickness = base thickness − total reduction = 557 − 9
= 548 mm

In this case, the ceiling joists create a thermal bridge that is not protected by the insulation layer, with the result that a very large thickness of insulation is required. As the joists are likely to be about 100 mm in depth, this thickness of insulation laid between them is not a practicable design solution.

Example 16.4.2 Pitched roof with insulation between the rafters
Figure 16.5b shows a pitched roof with the insulation laid between the rafters. The

Table 16.24 Base thickness of insulation for roofs.

Design U-value W/m²K	0.020	0.025	0.030	0.035	0.040	0.045	0.050
	Thermal conductivity of the insulation material, W/mK						
	Base thickness of insulation layer, mm						
Insulation laid between ceiling joists or rafters							
0.15	371	464	557	649	742	835	928
0.20	180	224	269	314	359	404	449
0.25	118	148	178	207	237	266	296
0.30	92	110	132	154	176	198	220
0.35	77	91	105	122	140	157	175
0.40	67	78	90	101	116	130	145
Insulation laid between and over joists or rafters							
0.15	161	188	217	247	277	307	338
0.20	128	147	167	188	210	232	255
0.25	108	122	137	153	170	187	205
0.30	92	105	117	130	143	157	172
0.35	77	91	103	113	124	136	148
0.40	67	78	90	101	110	120	130
Continuous layer of insulation							
0.15	131	163	196	228	261	294	326
0.20	97	122	146	170	194	219	243
0.25	77	97	116	135	154	174	193
0.30	64	80	96	112	128	144	160
0.35	54	68	82	95	109	122	136
0.40	47	59	71	83	94	106	118

Table 16.25 Reduction in base thickness of insulation for roof components.

Concrete slab density kg/m³	0.020	0.025	0.030	0.035	0.040	0.045	0.050
	Thermal conductivity of the insulation material, W/mK						
	Reduction (mm) in the base thickness of the insulation, for each 100 mm thickness of the concrete slab						
600	10	13	15	18	20	23	25
800	7	9	11	13	14	16	18
1100	5	6	8	9	10	11	13
1300	4	5	6	7	8	9	10
1700	2	2	3	3	4	4	5
2100	1	2	2	2	3	3	3
Various materials and components	**Reduction in the base thickness of the insulation mm**						
10 mm plasterboard	1	2	2	2	3	3	3
13 mm plasterboard	2	2	2	3	3	4	4
13 mm sarking board	2	2	3	3	4	4	5
12 mm calcium silicate liner board	1	2	2	2	3	3	4
Roof space (pitched)	4	5	6	7	8	9	10
Roof space (flat)	3	4	5	6	6	7	8
19 mm roof tiles	0	1	1	1	1	1	1
19 mm asphalt (or 3 layers of felt)	1	1	1	1	2	2	2
50 mm screed	2	3	4	4	5	5	6

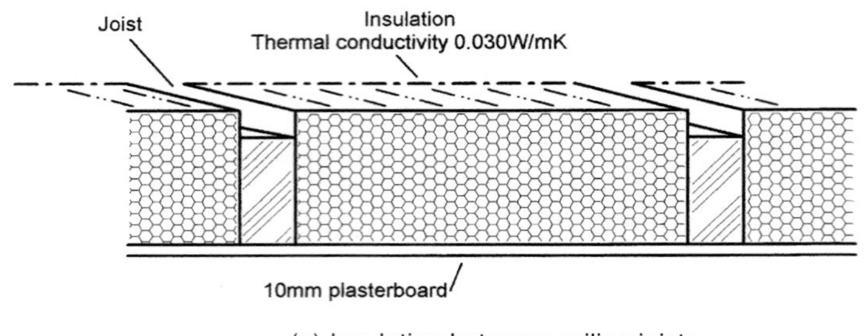

(a) Insulation between ceiling joists

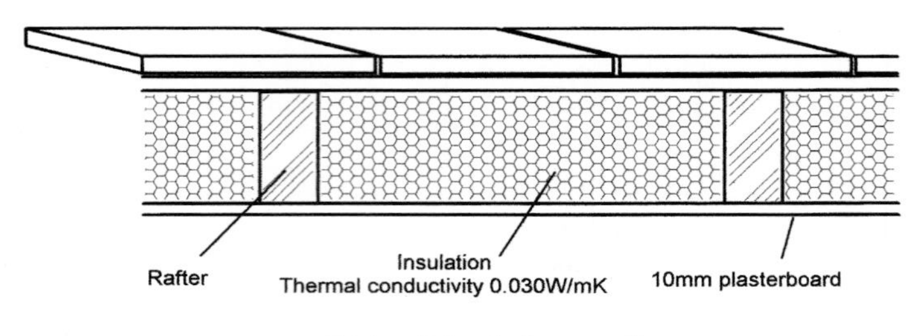

(b) Insulation between rafters

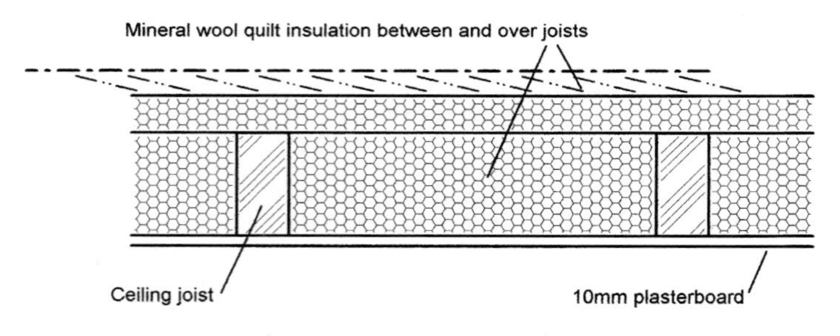

(c) Insulation between and over ceiling joists

Fig. 16.5 Examples 16.4.1, 16.4.2 and 16.4.3.

maximum U-value allowed by the elemental method is, from Table 16.2, $0.20\,W/m^2K$, and it is required to determine the necessary thickness of insulation. First, from Table 16.23, there is a ΔU correction of $0.01\,W/m^2K$. The design (or 'look-up') U-value is therefore:

$$\text{Insulation between rafters} \qquad U_{design} = 0.20 - 0.01 = 0.19\ W/m^2K$$

Next, use Table 16.24 to find the base thickness of insulation. This requires an interpolation:

At U = 0.19, thickness at λ = 0.030 is 557 − (0.04/0.05) × (557 − 269)
= 327 mm

Thus, from Table 16.24 Base thickness = 327 mm

Table 16.25 is now used to determine allowable reductions to the base thickness:

From Table 16.25	Reduction for	19 mm roof tiles	= 1 mm
		10 mm plasterboard	= 2 mm
	Total reduction		= 3 mm

Minimum insulation thickness = base thickness − total reduction = 327 − 3
= 324 mm

As in the previous example, the bridging effect of the timber has produced a requirement for an insulation layer much thicker than depth of the rafters. Unless the rafters themselves are also over 300 mm deep, there will be insufficient space between the plasterboard and the tiling for the insulation to be fitted.

Example 16.4.3 Pitched roof with insulation between and over ceiling joists
Figure 16.5c shows a pitched roof with insulation between and over ceiling joists. The maximum U-value allowed by the elemental method is, from Table 16.2, 0.16 W/m^2K, and it is required to determine the necessary thickness of insulation. The insulation material is mineral wool quilt of thermal conductivity (Table 16.34, section 16.4.3) of 0.042 W/mK. First, from Table 16.23, the ΔU correction is zero. The design (or 'look-up') U-value is therefore:

Insulation between and over joists $U_{design} = 0.16 - 0 = 0.16\ W/m^2K$

Next, use Table 16.24 to find the base thickness of insulation. This requires a double interpolation:

At U = 0.15, thickness at λ = 0.042 is 277 + (0.002/0.005) × (307 − 277)
= 289 mm

At U = 0.20, thickness at λ = 0.042 is 210 + (0.002/0.005) × (232 − 210)
= 219 mm

At U = 0.16, thickness at λ = 0.042 is 289 − (0.01/0.05) × (289 − 219)
= 275 mm

Thus, from Table 16.24 Base thickness = 275 mm

Table 16.25 is now used to determine allowable reductions to the base thickness:

From Table 16.25	Reduction for	19 mm roof tiles	= 1 mm
		Pitched roofspace	= 8 mm
		10 mm plasterboard	= 3 mm
	Total reduction		= 12 mm

Minimum insulation thickness = base thickness − total reduction = 275 − 12
= 263 mm

In this case, the thermal bridge created by the ceiling joists is protected by part of the insulation. The resulting thickness of insulation is much less and more practicable than the first two examples, but is still considerable. If access to the roof space is required, then the problem of providing walkways over the insulation without compressing it or penetrating it with fixings will have to be solved.

Example 16.4.4 Flat roof

Figure 16.6 shows a concrete deck flat roof. The maximum U-value allowed by the elemental method is, from Table 16.2, 0.25 W/m^{2}K, and it is required to determine the necessary thickness of insulation. There are no ΔU corrections, and so:

Continuous layer of insulation $\quad U_{design} = 0.25 - 0 = 0.25$ W/m^{2}K

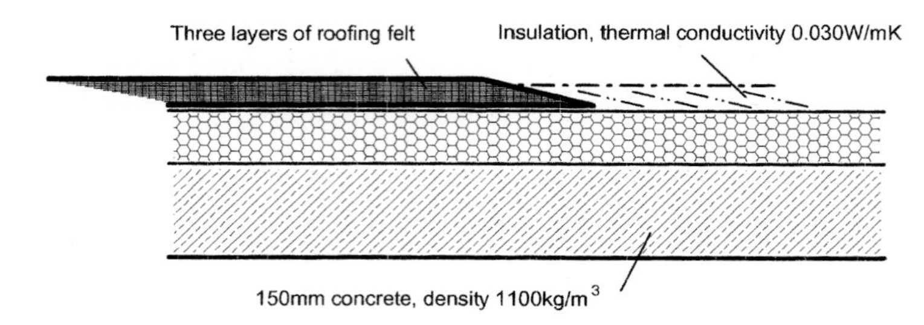

Fig. 16.6 Example 16.4.4, concrete deck roof.

Referring to Table 16.24 for a thermal conductivity of 0.030 W/mK, the base thickness of insulation is:

From Table 16.24 $\quad$ Base thickness = 116 mm

There are allowable reductions for the roofing felt and for the concrete deck itself. For 150 mm of concrete of density 1100 kg/m^3, the reduction is $1.5 \times 8 = 12$ mm. Thus:

From Table 16.25	Reduction for	3 layers felt	= 1 mm
		150 mm concrete deck	= 12 mm
	Total reduction		= 13 mm

Minimum insulation thickness = base thickness − total reduction = 116 − 13
= 103 mm

A solution using this thickness of insulation is feasible but has a number of possible drawbacks. The insulation material itself is unlikely to have any significant compressive strength (otherwise it would not have a thermal conductivity as low as 0.030 W/mK), and so access to the roof deck could not be permitted. Also, because the insulation is immediately beneath the weatherproof membrane, solar radiation on the roof will generate very high surface temperatures, with the possibility of degradation of the membrane due to thermal movement or chemical action.

Determining the thickness of insulation for walls

The procedure is the same as for roofs. Select the desired U-value, decide on the position of the insulation layer and method of fixing, and use Table 16.23 to obtain ΔU, the correction to the U-value. The *design* U-value is found from the *desired* U-value using:

$$U_{design} = U_{desired} - \Delta U$$

Then use Table 16.26 to find the base thickness of insulation, followed by Table 16.27 for any allowable reduction in the base thickness, and hence obtain the minimum thickness.

Table 16.26 Base thickness of insulation for walls.

Design U-value W/m²K	Thermal conductivity of the insulation material, W/mK						
	0.020	0.025	0.030	0.035	0.040	0.045	0.050
	Base thickness of insulation layer, mm						
0.20	97	121	145	169	193	217	242
0.25	77	96	115	134	153	172	192
0.30	63	79	95	111	127	142	158
0.35	54	67	81	94	107	121	134
0.40	47	58	70	82	93	105	117
0.45	41	51	62	72	82	92	103

For timber framed walls, where the timber frame contains its own integral insulation, there is an additional and often significant reduction in the base thickness of the separate insulation layer. However, this reduction depends on the proportion of the area of the frame that is timber. The Approved Document gives a table for one common case when the proportion of timber is 15% of the wall area, corresponding to 38 mm wide studs at 600 mm centres, with additional timbers at junctions and around openings. This table is given here as Table 16.28. For other area proportions, or frames with insulation material of different thermal conductivity, it is necessary to use an acceptable calculation procedure, such as that given in Appendix B of the Approved Documents (see section 16.5), to determine either the U-value or the necessary minimum thickness of an insulation layer.

Example calculations for walls

Example 16.4.5 Masonry cavity wall

Figure 16.7 shows a masonry cavity wall. The maximum U-value allowed by the elemental method is, from Table 16.2, 0.35 W/m²K, and it is required to determine the necessary thickness of expanded polystyrene board (EPS) insulation, of thermal conductivity 0.040 W/mK. From Table 16.23, there are no ΔU reductions, and so:

$$U_{design} = 0.35 - 0 = 0.35\ W/m^2K$$

Table 16.27 Reduction in base thickness of insulation for wall components.

	Thermal conductivity of the insulation material, W/mK						
	0.020	0.025	0.030	0.035	0.040	0.045	0.050
Concrete blockwork density kg/m^3	**Reduction (mm) in the base thickness of the insulation, for each 100 mm thickness of the concrete blockwork**						
Inner leaf							
600	9	11	13	15	17	20	22
800	7	9	10	12	14	16	17
1000	5	6	8	9	10	11	13
1200	4	5	6	7	8	9	10
1400	3	4	5	6	7	8	8
1600	3	3	4	5	6	6	7
1800	2	2	3	3	4	4	4
2000	2	2	2	3	3	3	4
2400	1	1	2	2	2	2	3
Outer leaf or single leaf wall							
600	8	11	13	15	17	19	21
800	7	9	10	12	14	15	17
1000	5	6	7	8	10	11	12
1200	4	5	6	7	8	9	10
1400	3	4	5	6	6	7	8
1600	3	3	4	5	5	6	7
1800	2	2	3	3	3	4	4
2000	1	2	2	3	3	3	4
2400	1	1	2	2	2	2	3
Various materials and components	**Reduction in the base thickness of the insulation mm**						
Cavity, 25 mm or more	4	5	5	6	7	8	9
Outer leaf brickwork	3	3	4	5	5	6	6
13 mm plaster	1	1	1	1	1	1	1
13 mm lightweight plaster	2	2	2	3	3	4	4
9.5 mm plasterboard	1	2	2	2	3	3	3
12.5 mm plasterboard	2	2	2	3	3	4	4
Airspace behind plasterboard drylining	2	3	4	4	5	5	6
9 mm sheathing ply	1	2	2	2	3	3	3
20 mm cement render	1	1	1	1	2	2	2
13 mm tile hanging	0	0	0	1	1	1	1

Table 16.28 Reduction in base thickness of insulation for insulated timber frame walls.

	Thermal conductivity of the insulation material, W/mK						
	0.020	0.025	0.030	0.035	0.040	0.045	0.050
Thermal conductivity of insulation within frame W/mK	**Reduction (mm) in the base thickness of insulation, for each 100 mm thickness of the timber frame**						
0.035	39	49	59	69	79	89	99
0.040	36	45	55	64	73	82	91

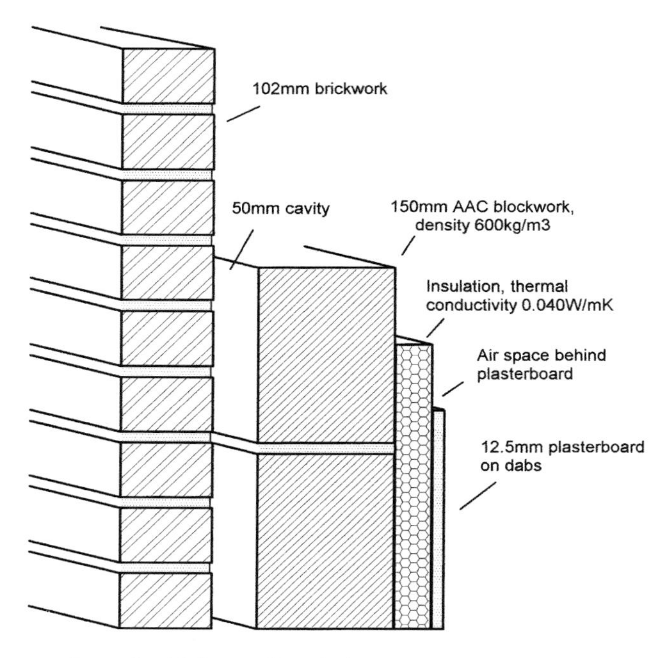

Fig. 16.7 Masonry cavity wall with internal insulation.

Now use Tables 16.26 and 16.27 to find the base thickness and the allowable reductions. Note that the reduction for 150 mm of 600 kg/m^3 blockwork is $1.5 \times 17 = 25.5$ mm. As this is a reduction, it should be rounded to 25 mm.

From Table 16.26	Base thickness = 107 mm		
From Table 16.27	Reduction for	brickwork outer leaf	= 5 mm
		cavity	= 7 mm
		concrete blockwork	= 25 mm
		air space behind plasterboard	= 5 mm
		plasterboard	= 3 mm
	Total reduction		= 45 mm

Minimum insulation thickness = base thickness − total reduction = 107 − 45
= 61 mm

The next available thickness of EPS board above the 62 mm minimum is likely to be 75 mm. This should have sufficient rigidity to be suitable for fixing, as shown in Fig. 16.7.

Example 16.4.6 Masonry wall with cavity fill

Figure 16.8 shows a masonry wall with the cavity completely filled with polyurethane foam insulation of thermal conductivity 0.040 W/mK. The brickwork and blockwork are tied with stainless steel vertical-twist ties. The maximum U-value allowed by the elemental method is, from Table 16.2, 0.35 W/m^{2}K, and it is required

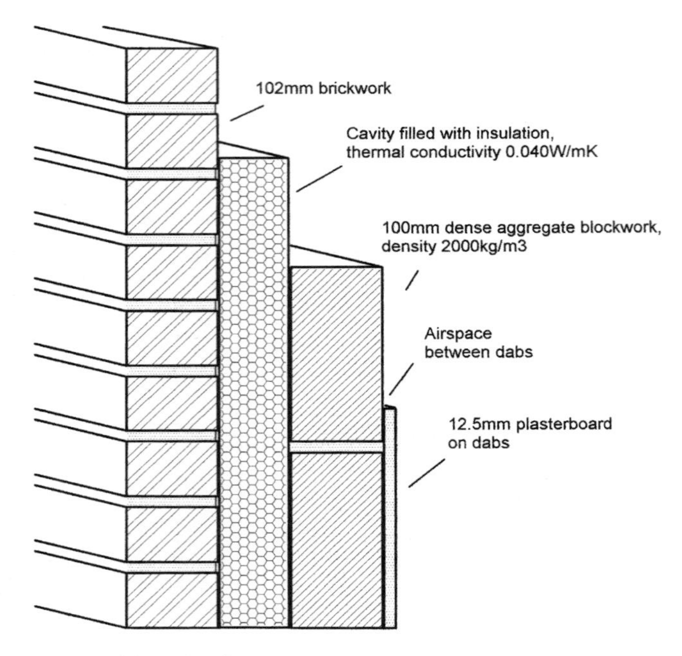

Fig. 16.8 Masonry wall with cavity fill.

to determine the necessary thickness of insulation. From Table 16.23, there is a ΔU reduction of 0.02 W/m^{2}K for the wall ties, and so:

$$U_{design} = 0.35 - 0.02 = 0.33 \text{ W/m}^2\text{K}$$

Next, use Table 16.26 to find the base thickness of insulation. This requires an interpolation:

At U = 0.33, thickness at $\lambda = 0.040$ is $127 - (0.03/0.05) \times (127 - 107)$
$= 115$ mm

Thus, from Table 16.26 Base thickness = 115 mm

Now use Table 16.27 to obtain allowable reductions:

From Table 16.27	Reduction for	brickwork outer leaf	=	5 mm
		concrete blockwork	=	3 mm
		air space behind plasterboard	=	5 mm
		plasterboard	=	3 mm
	Total reduction		=	16 mm

Minimum insulation thickness = base thickness − total reduction = 115 − 16
= 99 mm

This shows that the wall will have to be constructed with a 100 mm cavity. If this is not acceptable, an alternative solution would be to use concrete blocks with a lower density and/or a greater thickness. The calculation would have to be repeated to see if the reduction in the minimum insulation thickness is sufficient.

Example 16.4.7 Masonry cavity wall with partial cavity fill
Figure 16.9 shows a masonry wall with the cavity partially filled with insulation. The brickwork and blockwork are tied with stainless steel vertical-twist ties. The maximum U-value allowed by the elemental method is, from Table 16.2, 0.35 W/m²K, and it is required to determine the necessary thickness of insulation. From Table 16.23, there is a ΔU reduction of 0.02 W/m²K for the wall ties, and so:

$$U_{design} = 0.35 - 0.02 = 0.33 \text{ W/m}^2\text{K}$$

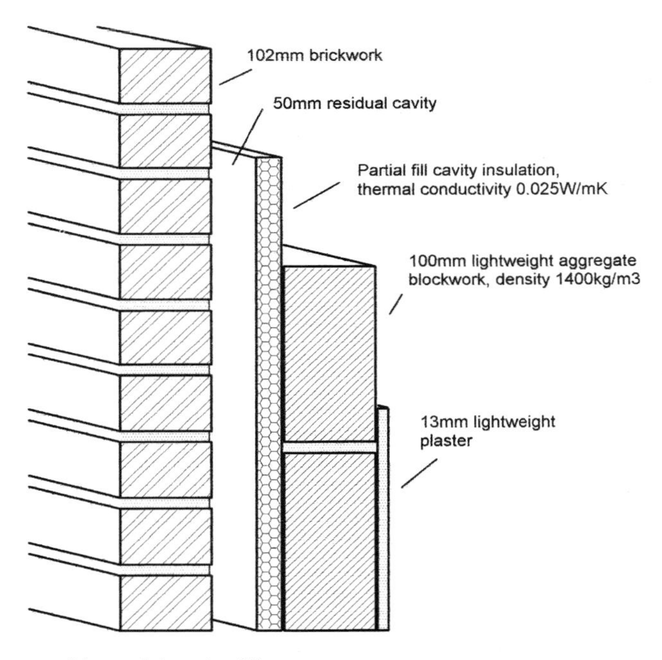

Fig. 16.9 Masonry with partial cavity fill.

Next, use Table 16.26 to find the base thickness of insulation. This requires an interpolation:

At $U = 0.33$, thickness at $\lambda = 0.025$ is $79 - (0.03/0.05) \times (79 - 67)$
$= 72$ mm

Thus, from Table 16.26 Base thickness = 72 mm

Now use Table 16.27 to obtain allowable reductions:

From Table 16.27	Reduction for	brickwork outer leaf	=	3 mm
		cavity	=	5 mm

	concrete blockwork	= 4 mm
	lightweight plaster	= 2 mm
Total reduction		= 14 mm

Minimum insulation thickness = base thickness − total reduction = 72 − 14
= 58 mm

The minimum insulation thickness is much less than in the previous example. Nevertheless, if a 50 mm residual cavity is to be preserved, the overall cavity width will have to be of the order of 100 mm.

Example 16.4.8 Timber frame wall

Figure 16.10 shows a timber frame wall. The maximum U-value allowed by the elemental method is, from Table 16.2, 0.35 W/m^{2}K. The 90 mm timber frame has insulation that fully fills the space between the studs. It is required to determine the necessary thickness of a continuous insulation layer of EPS board that is to be fixed between the sheathing ply and the timber frame. From Table 16.23, there is a ΔU reduction of 0.01 for a timber frame where the insulation fully fills the space between the studs. However, the insulation layer which is being added is continuous, and so the ΔU reduction may be taken as zero.

$$U_{design} = 0.35 - 0 = 0.35\ \mathrm{W/m^2K}$$

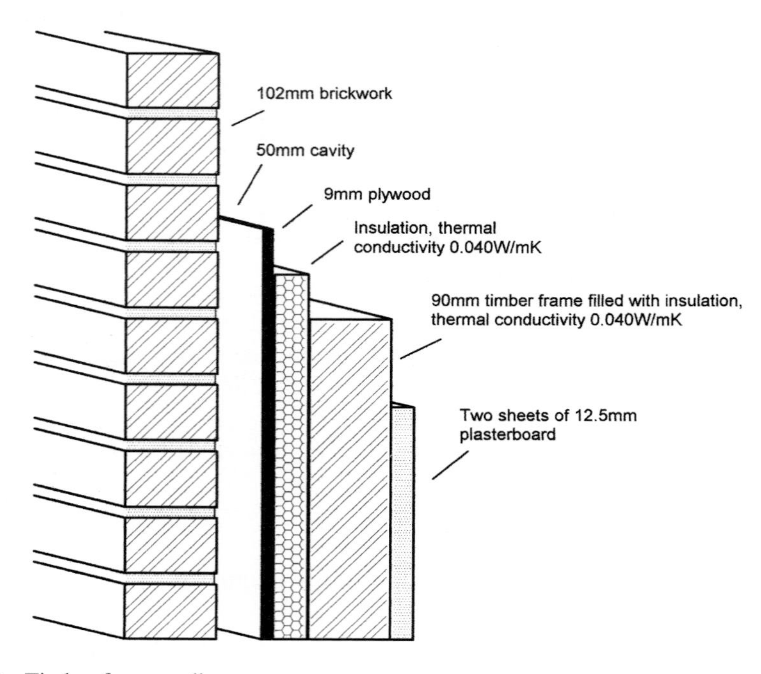

Fig. 16.10 Timber frame wall.

Now use Table 16.26 to find the base thickness:

From Table 16.26 Base thickness = 107 mm

Now use Tables 16.27 and 16.28 to obtain allowable reductions:

From Table 16.27	Reduction for	brickwork outer leaf	=	5 mm
		cavity	=	7 mm
		sheathing ply	=	3 mm
		2 sheets plasterboard	=	6 mm
	Reductions from Table 4.9		=	21 mm
From Table 16.28	Reduction for timber frame ($73 \times 90/100$)		=	66 mm
	Total reduction		=	87 mm

Minimum insulation thickness = base thickness − total reduction = 107 − 87
= 20 mm

This thickness of insulation should provide a suitable solution.

Determining the thickness of insulation for floors

Ground floors must be considered separately from upper floors. In the case of ground floors, the U-value depends not only on the insulation but also on the size and shape of the floor. Both size and shape are taken into account by means of the ratio P/A, where P is the perimeter length in metres of the whole of the ground floor, and A is the total area in square metres of the ground floor. Further, AD L1 does not give any allowable reductions for components in the floor structure, and so the 'base' thickness of insulation is also the minimum thickness. Data is given for the three principal types of floor, i.e. solid floors in contact with the ground (Fig. 16.11 and Table 16.29), suspended timber ground floors (Fig. 16.12 and Table 16.30) and suspended concrete beam and block floors (Fig. 16.13 and Table 16.31).

Solid floor in contact with the ground

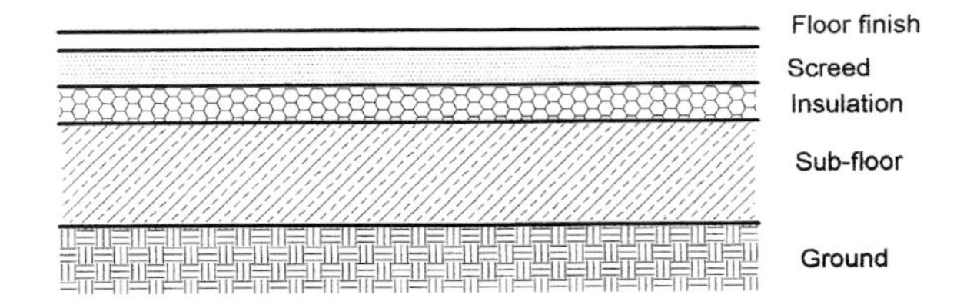

Fig. 16.11 Solid floor in contact with the ground.

Table 16.29 Thickness of insulation for solid floors in contact with the ground.

Design U-value W/m^2K	P/A ratio m/m^2	Thermal conductivity of the insulation material, W/mK						
		0.020	0.025	0.030	0.035	0.040	0.045	0.050
		Base (minimum) thickness of insulation layer, mm						
0.020	1.00	81	101	121	142	162	182	202
	0.90	80	100	120	140	160	180	200
	0.80	78	98	118	137	157	177	196
	0.70	77	96	115	134	153	173	192
	0.60	74	93	112	130	149	167	186
	0.50	71	89	107	125	143	160	178
	0.40	67	84	100	117	134	150	167
	0.30	60	74	89	104	119	134	149
	0.20	46	57	69	80	92	103	115
0.25	1.00	61	76	91	107	122	137	152
	0.90	60	75	90	105	120	135	150
	0.80	58	73	88	102	117	132	146
	0.70	57	71	85	99	113	128	142
	0.60	54	68	82	95	109	122	136
	0.50	51	64	77	90	103	115	128
	0.40	47	59	70	82	94	105	117
	0.30	40	49	59	69	79	89	99
	0.20	26	32	39	45	52	58	65
0.30	1.00	48	60	71	83	95	107	119
	0.90	47	58	70	81	93	105	116
	0.80	45	56	68	79	90	102	113
	0.70	43	54	65	76	87	98	108
	0.60	41	51	62	72	82	92	103
	0.50	38	47	57	66	76	85	95
	0.40	33	42	50	59	67	75	84
	0.30	26	33	39	46	53	59	66
	0.20	13	16	19	22	25	28	32

Suspended timber ground floor

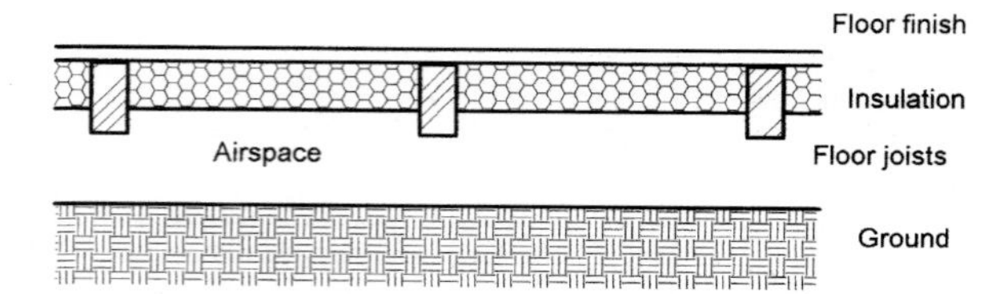

Fig. 16.12 Suspended timber ground floor.

Table 16.30 Thickness of insulation for suspended timber ground floors.

Design U-value W/m^2K	P/A ratio m/m^2	Thermal conductivity of the insulation material, W/mK						
		0.020	**0.025**	**0.030**	**0.035**	**0.040**	**0.045**	**0.050**
		Base (minimum) thickness of insulation layer, mm						
0.020	1.00	127	145	164	182	200	218	236
	0.90	125	144	162	180	198	216	234
	0.80	123	142	160	178	195	213	230
	0.70	121	139	157	175	192	209	226
	0.60	118	136	153	171	188	204	221
	0.50	114	131	148	165	181	198	214
	0.40	109	125	141	157	173	188	204
	0.30	99	115	129	144	159	173	187
	0.20	82	95	107	120	132	144	156
0.25	1.00	93	107	121	135	149	162	176
	0.90	92	106	119	133	146	160	173
	0.80	90	104	117	131	144	157	170
	0.70	88	101	114	127	140	153	166
	0.60	85	98	111	123	136	148	161
	0.50	81	93	106	118	130	142	154
	0.40	75	87	99	110	121	132	143
	0.30	66	77	87	97	107	117	127
	0.20	49	57	65	73	81	88	96
0.30	1.00	71	82	93	104	114	125	135
	0.90	70	80	91	102	112	122	133
	0.80	68	78	89	99	109	119	129
	0.70	66	76	86	96	106	116	126
	0.60	63	73	82	92	102	111	120
	0.50	59	68	78	87	96	104	113
	0.40	53	62	70	79	87	95	103
	0.30	45	52	59	66	73	80	87
	0.20	28	33	38	42	47	51	56

Suspended concrete ground floor

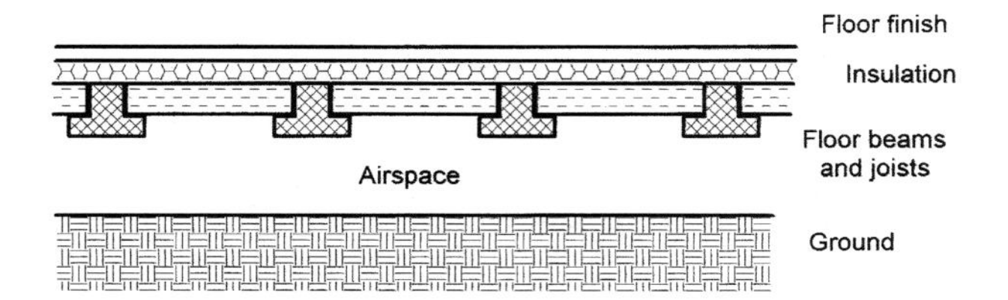

Fig. 16.13 Suspended concrete ground floor.

Table 16.31 Thickness of insulation for suspended concrete beam and block ground floors.

Design U-value W/m²K	P/A ratio m/m²	Thermal conductivity of the insulation material, W/mK: 0.020	0.025	0.030	0.035	0.040	0.045	0.050
		Base (minimum) thickness of insulation layer, mm						
0.20	1.00	82	103	123	144	164	185	205
	0.90	81	101	122	142	162	183	203
	0.80	80	100	120	140	160	180	200
	0.70	79	99	118	138	158	177	197
	0.60	77	96	116	135	154	173	193
	0.50	75	93	112	131	150	168	187
	0.40	71	89	107	125	143	161	178
	0.30	66	82	99	115	132	148	165
	0.20	56	69	83	97	111	125	139
0.25	1.00	62	78	93	109	124	140	155
	0.90	61	76	92	107	122	138	153
	0.80	60	75	90	105	120	135	150
	0.70	59	74	88	103	118	132	147
	0.60	57	71	86	100	114	128	143
	0.50	55	68	82	96	110	123	137
	0.40	51	64	77	90	103	116	128
	0.30	46	57	69	80	92	103	115
	0.20	36	45	54	62	71	80	89
0.30	1.00	49	61	73	85	97	110	122
	0.90	48	60	72	84	96	108	120
	0.80	47	59	70	82	94	105	117
	0.70	45	57	68	80	91	102	114
	0.60	44	55	66	77	88	98	109
	0.50	41	52	62	72	83	93	104
	0.40	38	48	57	67	76	86	95
	0.30	33	41	49	57	65	73	81
	0.20	22	28	33	39	44	50	56

Determining the thickness of insulation for upper floors

For upper floors, i.e. floors above an external space, Table 16.32 applies, together with the allowable reductions in Table 16.33. Note that in Table 16.32 it is assumed that the proportion by area of structural timber in the timber floor construction is 12%, corresponding to 48 mm wide timber joists at 400 mm centres.

Table 16.32 Thickness of insulation for upper floors of timber construction.

	Design U-value W/m²K	Thermal conductivity of the insulation material, W/mK: 0.020	0.025	0.030	0.035	0.040	0.045	0.050
		Base thickness of insulation layer, mm						
Timber construction	0.20	167	211	256	298	341	383	426
	0.25	109	136	163	193	225	253	281
	0.30	80	100	120	140	160	184	208
Concrete construction	0.20	95	119	142	166	190	214	237
	0.25	75	94	112	131	150	169	187
	0.30	62	77	92	108	123	139	154

Table 16.33 Reduction in base thickness for upper floor components.

	Thermal conductivity of the insulation material, W/mK						
	0.020	**0.025**	**0.030**	**0.035**	**0.040**	**0.045**	**0.050**
Component	**Reduction in the base thickness of the insulation, mm**						
10 mm plasterboard	1	2	2	2	3	3	3
19 mm timber flooring	3	3	4	5	5	6	7
50 mm screed	2	3	4	4	5	5	6

The U-value for other proportions of timber must be calculated using the procedures in section 16.5.

Example calculations for floors

Example 16.4.9 Solid floor in contact with the ground
Figure 16.14 shows a solid floor. The maximum U-value allowed by the elemental method, from Tables 16.2 and 16.8, is $0.25\,W/m^2K$, and it is desired to find the necessary thickness of insulation of thermal conductivity 0.025 W/mK. Table 16.29 must be used, and this requires the perimeter to area ratio to be calculated:

$$\text{Floor perimeter} = 6+2+4+4+10+6 = 32\,m$$
$$\text{Floor area} = (6 \times 6) + (4 \times 4) = 52\,m^2$$
$$\text{Perimeter to area ratio, P/A: } \frac{P}{A} = \frac{32}{52} = 0.615 \cong 0.6$$

From Table 16.29, the necessary minimum thickness of insulation is 68 mm.

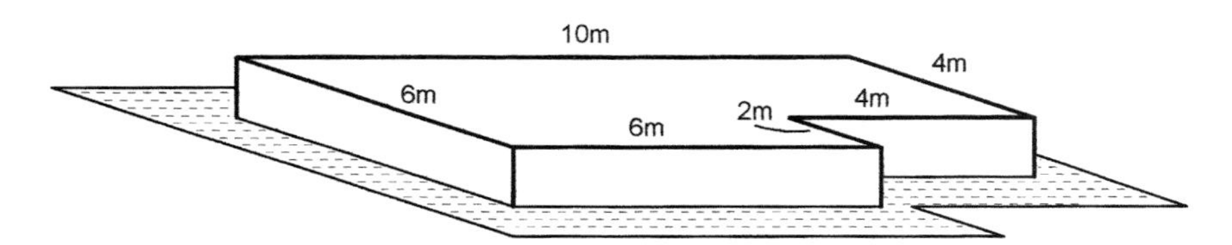

Fig. 16.14 Solid floor.

Example 16.4.10 Suspended timber ground floor
If the floor shown in Fig. 16.14 was a suspended timber floor of the same perimeter shape and dimensions, the perimeter to area ratio would still be 0.6, but it would be necessary to use Table 16.30. To achieve the same U-value of $0.25\,W/m^2K$, and using insulation of thermal conductivity 0.025 W/mK, the necessary minimum thickness of insulation placed between the joists is 98 mm.

Example 16.4.11 Upper floor, timber construction
Figure 16.15 shows an upper floor in timber construction. The joists are 45 mm wide and are at 400 mm centres. The maximum U-value allowed by the elemental method is, from Tables 16.2 and 16.8, $0.25\,W/m^2K$, and it is desired to find the necessary

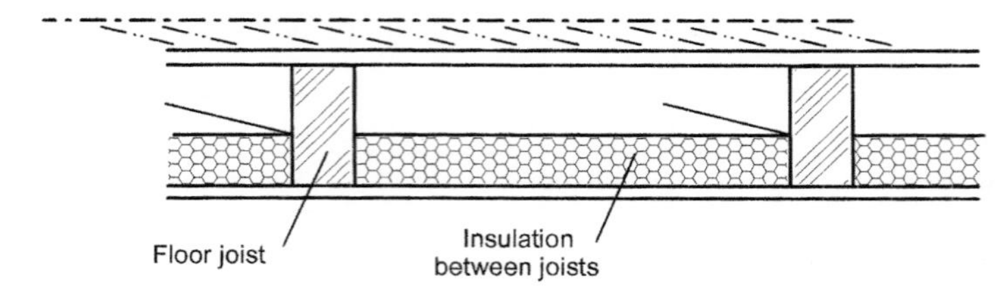

Fig. 16.15 Upper floor, timber construction.

thickness of insulation of thermal conductivity 0.030W/mK. The proportion by area of structural timber in the floor is (45/400) × 100 = 11.25%. As this is less than 12%, Table 16.32 will very slightly overestimate the insulation thickness and therefore can still be used. If the proportion of structural timber had been greater than 12%, Table 16.32 would be inappropriate and a suitable calculation method [17, 18] would have to be used instead. Using Tables 16.32 and 16.33:

From Table 16.32	Base thickness = 163 mm	
From Table 16.33	Reduction for 10 mm plasterboard	= 2 mm
	19 mm floorboards	= 4 mm
	Total reductions	= 6 mm

Minimum insulation thickness = base thickness − total reduction = 163 − 6
= 157 mm

Floor joists are typically not less than 175 mm deep, and so this thickness of insulation can be accommodated.

Example 16.4.12 Upper floor, concrete

Figure 16.16 shows an upper floor in concrete construction. The required U-value is 0.25 W/m^{2}K, and insulation of thermal conductivity 0.030 W/mK is to be used.

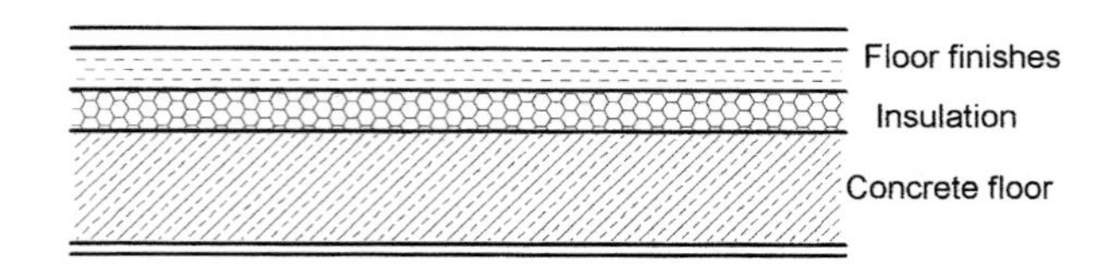

Fig. 16.16 Upper floor, concrete.

From Table 16.32	Base thickness = 112 mm	
From Table 16.33	Reduction for 50 mm screed	= 4 mm
	Total reduction	= 4 mm

Minimum insulation thickness = base thickness − total reduction = 112 − 4
= 108 mm

The stability and compressive strength of this thickness of insulation immediately beneath the floor finishes must be considered.

16.4.3 Thermal conductivity and density of building materials

Table 16.34 lists the thermal conductivities and densities of some common building materials.

Table 16.34 Thermal conductivity and density of common building materials.

	Density Kg/m^3	Thermal conductivity W/mK
Walls		
Brickwork (outer leaf)	1700	0.77
Brickwork (inner leaf)	1700	0.56
Lightweight aggregate concrete block	1400	0.57
Autoclaved aerated concrete block	600	0.18
Concrete, medium density (inner leaf)	1800	1.13
	2000	1.33
	2200	1.59
Concrete, high density	2400	1.93
Reinforced concrete, 1% steel	2300	2.30
Reinforced concrete, 2% steel	2400	2.50
Mortar, protected	1750	0.88
Mortar, exposed	1750	0.94
Gypsum	600	0.18
	900	0.30
	1200	0.43
Gypsum plasterboard	900	0.25
Sandstone	2600	2.30
Limestone, soft	1800	1.10
Limestone, hard	2200	1.70
Fibreboard	400	0.10
Plasterboard	900	0.25
Tiles, ceramic	2300	1.30
Timber, softwood	500	0.13
Timber, hardwood	700	0.18
Timber, plywood and chipboard	500	0.13
Wall ties, stainless steel	7900	17.00
Surface finishes		
External rendering	1300	0.57
Plaster, dense	1300	0.57
Plaster, lightweight	600	0.18
Roofs		
Aerated concrete slab	500	0.16
Asphalt	2100	0.70
Felt/bitumen layers	1100	0.23
Screed	1200	0.41
Stone chippings	2000	2.00
Tiles, clay	2000	1.00
Tiles, concrete	2100	1.50
Wood wool slab	500	0.10

Contd

Table 16.34 *Contd*

	Density Kg/m^3	Thermal conductivity W/mK
Floors		
Cast concrete	2000	1.35
Metal tray, steel	7800	50.00
Screed	1200	0.41
Timber, softwood	500	0.13
Timber, hardwood	700	0.18
Timber, plywood and chipboard	500	0.13
Insulation		
Expanded polystyrene (EPS) board	15	0.040
Mineral wool quilt	12	0.042
Mineral wool batt	25	0.038
Phenolic foam board	30	0.025
Polyurethane board	30	0.025

16.5 The calculation of U-values for walls

Appendix B of the Approved Documents deals with the calculation of U-values from the thickness and thermal conductivity of the materials that make up a construction element. The methods described in Appendix B are not comprehensive but are adequate for many typical constructions. More sophisticated methods are necessary if the construction is complex, particularly if there is significant thermal bridging or if there is two or three dimensional heat flow at corners or reveals.

16.5.1 Background theory

The lower the U-value of a construction element, the more significant is the effect of thermal bridging on the calculation of the U-value. Consequently it is usually necessary to include thermal bridging in the calculation method. The theory is based on the calculation of thermal resistances. For a single layer of material, the thermal resistance R is given by:

$$R = \frac{d}{\lambda}$$

where d is the thickness of the layer in metres, and λ is the thermal conductivity. The combined resistance of several materials depends on whether the heat flows through them sequentially, i.e. in series or in parallel, as shown in Figure 16.17.

Resistances in series

For three materials in series (Fig. 16.17a), the total combined resistance R_T is given by:

$$R_T = R_1 + R_2 + R_3$$

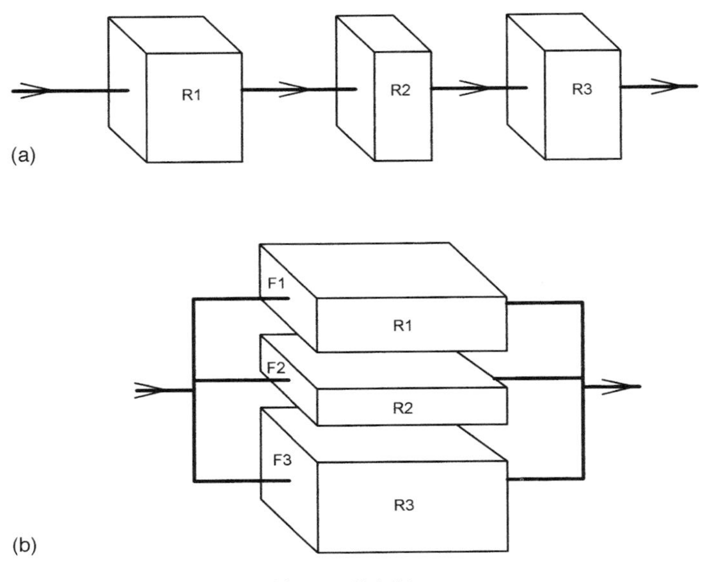

Fig. 16.17 Resistances in series (a) and in parallel (b).

Resistances in parallel

For three materials in parallel (Fig. 16.17b), the total combined resistance R_T is given by:

$$\frac{1}{R_T} = \frac{F_1}{R_1} + \frac{F_2}{R_2} + \frac{F_3}{R_3}$$

where F_1, F_2 and F_3 are the cross-sectional areas of each material expressed as a fraction of the total.

Once the total resistance of a structure has been found, the U-value is its reciprocal:

$$U = \frac{1}{R_T}$$

Construction elements with materials in series and in parallel

Many practical construction elements consist of several layers through which heat passes in series, with some components embedded within them through which the heat passes in parallel (the thermal bridges). The total thermal resistance can be calculated in several ways, and the Approved Documents present one of the simplest and most direct. The method is suitable when the bridging material is timber or mortar, or some other material which is thermally similar. It is not suitable when the bridging material is metal, nor is it suitable for ground floors and basements. The method calculates an upper resistance limit for the construction element, R_{upper}, and a lower resistance limit, R_{lower}, and then finds R_T by taking the average:

$$R_T = \frac{1}{2}\left(R_{upper} + R_{lower}\right)$$

The upper resistance limit is found by taking each possible heat flow path separately and calculating its resistance as the sum of the resistances of each component taken in series. When all paths have been calculated, their resistances are combined in parallel, taking account of their relative areas to give R_{upper}. The lower resistance limit is found by first converting each thermal bridge in the construction into a single equivalent resistance by combining the elements of the bridge in parallel. When all bridges have been converted, the construction should consist of a single heat flow path whose resistance is found by summing its components in series to give R_{lower}. The example calculations which follow should clarify the method.

Corrections to calculated U-values

The ΔU correction factors described in Appendix A and tabulated in Table 16.23 can be treated more rigorously when calculating U-values (see Appendix D of BS EN ISO 6946 [12] for details). The procedure is:

Corrected U-value $U_c = U + \Delta U$

$\Delta U = \Delta U_g + \Delta U_f + \Delta U_r$

ΔU_g = correction for air gaps

ΔU_f = correction for mechanical fasteners

ΔU_r = correction for inverted roofs

Correction for air gaps

$$\Delta U_g = \Delta U^{11} \times \left(\frac{R_1}{R_T}\right)^2$$

where R_1 is thermal resistance of layer containing the gaps
R_T is total thermal resistance of the whole component
ΔU^{11} is obtained from Table 16.35.

Correction for mechanical fasteners

$$\Delta U_f = \alpha \lambda_f n_f A_f$$

Table 16.35 Correction factors for air gaps.

Level	ΔU^{11}, W/m²K	Type of air gap
0	0.00	Insulation installed in such a way that no air circulation is possible on the warm side of the insulation. No air gaps penetrating the entire insulation layer
1	0.01	Insulation installed in such a way that no air circulation is possible on the warm side of the insulation. Air gaps may penetrate the insulation layer
2	0.04	Air circulation is possible on the warm side of the insulation. Air gaps may penetrate the insulation layer

where λ_f is thermal conductivity of the fastener
n_f is number of fasteners per square metre
A_f is cross-sectional area of one fastener
α is obtained from Table 16.36.

Table 16.36 Corrections for mechanical fasteners.

Type of fastener	**α, m^{-1}**
Wall tie between masonry leaves	6
Roof fixing	5

Reproduced with permission from BS EN ISO 6946

Corrections for fasteners must *not* be applied when:

- The wall ties are across an empty cavity
- The wall ties are between a masonry leaf and timber studs
- The thermal conductivity of the fastener, or part of it, is less than 1 $Wm^{-1}K$.

Correction for inverted roofs
The correction for an inverted roof is not yet available.

The U-value via an unheated space

The precise calculation of the heat flow through a building element, and then via an unheated space to the outside, requires complex procedures. These can be found in BS EN ISO 13789 [10]. However, for the purposes of Part L, a simpler procedure in which the unheated space is assumed to behave like an additional homogeneous plane layer is usually adequate. With this assumption, the extra thermal resistance of an unheated space may be included in the calculation of the U-value of an element using the formula:

$$U = \frac{1}{R_o + R_{extra}}$$

where U is the U-value of the element including the effect of the unheated space
R_0 is the thermal resistance of the element as if exposed directly to the outside
R_{extra} is the extra thermal resistance due to the unheated space.

This formula is acceptable provided R_{extra} is small compared to R_0, say if $R_{extra} < 0.3R_0$. Values of R_{extra} are given in SAP 2001 [19] for some typical unheated spaces attached to dwellings, including:

- Single and double garages in various configurations
- Stairwells
- Access corridors
- Conservatories
- Roof spaces adjacent to a room in a roof.

These values, which are shown in Tables 16.37 and 16.38, can also be applied to similar situations in other buildings. For other unheated spaces it may be possible to calculate R_{extra} from:

$$R_{extra} = 0.09 + 0.4(A_{INT}/A_{EXT})$$

where A_{INT} is the total area of the elements separating the internal heated space from the unheated space, and A_{EXT} is the total area of the elements separating the unheated space from the outside. However, if this formula yields a result for R_{extra} greater than $0.5\,W/m^2K$, it should not be used.

Table 16.37 Extra thermal resistance due to unheated spaces – garages.

Garage type and description	R_{extra} m^2K/W	
	Garage inside insulation layer of building	**Garage outside insulation layer of building**
Single, fully integral, sharing side wall, end wall and floor with building	0.68	0.33
Single, fully integral, sharing side wall and floor with building	0.54	0.25
Single, partially integral, projecting forward, sharing part of side wall, part of floor and end wall with building	0.56	0.26
Single, adjacent, sharing side wall only with building	0.23	0.09
Double, fully integral, sharing side wall, end wall and floor with building	0.59	0.28
Double, half integral, sharing side wall, half of end wall and half of floor with building	0.34	0.17
Double, partially integral, projecting forward, sharing part of side wall, part of floor and end wall with building	0.28	0.13
Double, adjacent, sharing side wall only with building	0.13	0.05

Adapted with permission from SAP 2001 [19]

16.5.2 Example calculations

The calculation method is most conveniently explained by means of examples.

Example 16.5.1 Cavity wall

Figure 16.18 shows a cavity wall consisting of external brickwork, cavity, lightweight blockwork, mineral wool insulation within a timber sub-frame, and internal plasterboard. The blockwork and the mineral wool are the main providers of thermal insulation in this construction, and both suffer from thermal bridging. The blockwork is bridged by the mortar joints, and the mineral wool is bridged by the timber frame. In each case, the proportion of the area bridged is:

Table 16.38 Extra thermal resistance due to unheated spaces – various.

Type and description of unheated space	R_{extra} m^2K/W
Stairwell between heated space and external (exposed) wall	0.82
Stairwell between heated space and internal (not exposed) wall	0.90
Access corridor between heated space and external (exposed) wall, with another corridor above *or* below	0.31
Access corridor between heated space and external (exposed) wall, with another corridor above *and* below	0.28
Access corridor between heated space and internal (not exposed) wall, with another corridor above *or* below	0.43
Conservatory, double glazed, sharing one wall with heated space	0.06
Conservatory, double glazed, sharing two walls with heated space, i.e. in the angle between two walls	0.14
Conservatory, double glazed, sharing three walls with heated space, i.e. recessed	0.25
Conservatory, single glazed, sharing any number of walls with heated space	0.10
Loft space, between the roof covering and the wall of a heated room formed within a pitched roof above an insulated ceiling, for heat flow horizontally through the wall of the heated room	0.50
Loft space, between the roof covering and the wall of a heated room formed within a pitched roof above an insulated ceiling, for heat flow vertically from/ through the ceiling of the room below	0.50

Adapted with permission from SAP 2001 [19]

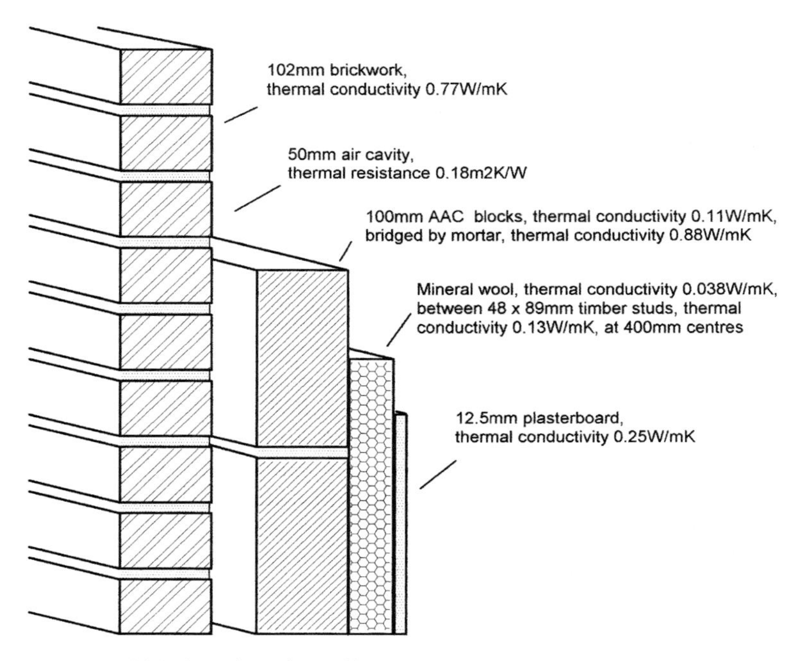

Fig. 16.18 Brick and blockwork cavity wall.

Table 16.39 Thermal data for cavity wall.

Material	Thickness mm	Thermal conductivity W/mK	Thermal resistance m^2K/W
External surface	—	—	0.040
Outer brickwork	102	0.77	—
Cavity, unvented	—	—	0.180
AAC blocks	100	0.11	0.909
Mortar	100	0.88	0.114
Mineral wool insulation	89	0.038	2.342
Timber battens	89	0.13	0.685
Plasterboard	12.5	0.25	0.050
Internal surface	—	—	0.130

Blockwork 93% of area mortar joints 7% of area
Mineral wool 88% of area timber battens 12% of area.

Table 16.39 gives the thermal data for the wall.

The upper resistance limit, R_{upper}

Each possible heat flow path through the wall is considered separately, and in this case it can be seen that there are four such paths, as shown in Fig. 16.19. The resistance of each path is calculated on the basis that the materials are in series, and then the four paths are combined on the basis that they are in parallel. The first part of the calculation is illustrated in Table 16.40. The four paths are then combined in parallel to find R_{upper}:

$$\frac{1}{R_{upper}} = \frac{F_1}{R_1} + \frac{F_2}{R_2} + \frac{F_3}{R_3} + \frac{F_4}{R_4} = \frac{0.818}{3.783} + \frac{0.062}{2.988} + \frac{0.112}{2.126} + \frac{0.008}{1.331} = 0.2957$$

$$R_{upper} = 3.382\,m^2K/W$$

Table 16.40 Calculation of the upper resistance limit, cavity wall.

	Thermal resistance, m^2K/W			
	Path 1	Path 2	Path 3	Path 4
External surface resistance	0.040	0.040	0.040	0.040
Resistance of brickwork	0.132	0.132	0.132	0.132
Resistance of cavity	0.180	0.180	0.180	0.180
Resistance of AAC blocks	0.909	—	0.909	—
Resistance of mortar	—	0.114	—	0.114
Resistance of mineral wool	2.342	2.342	—	—
Resistance of timber	—	—	0.685	0.685
Resistance of plasterboard	0.050	0.050	0.050	0.050
Internal surface resistance	0.130	0.130	0.130	0.130
Total thermal resistance of path	3.783	2.988	2.126	1.331
Fractional area of path	93% × 88% = 0.818	7% × 88% = 0.062	93% × 12% = 0.112	7% × 12% = 0.008

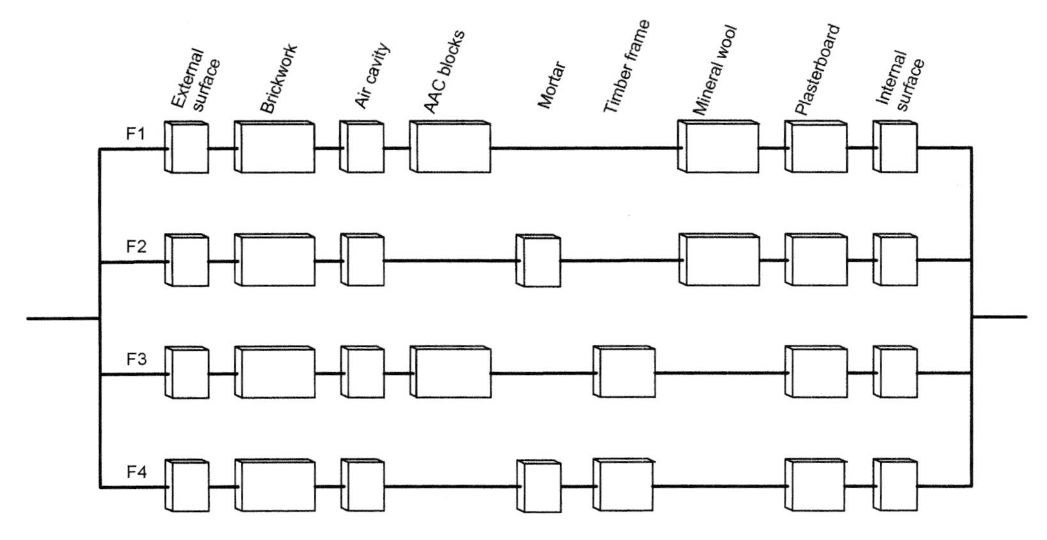

Fig. 16.19 Brick and blockwork cavity wall – upper resistance limit.

The lower resistance limit, R_{lower}

Each thermal bridge in the construction element is first converted to a single combined resistance, as shown in Fig. 16.20. Using these combined resistances, the construction can then be considered as a single heat flow path with all components in series. Thus in the present example, the AAC blocks and the mortar form one thermal bridge, and their combined resistance, R_{bm}, is found from:

$$\frac{1}{R_{bm}} = \frac{F_{blocks}}{R_{blocks}} + \frac{F_{mortar}}{R_{mortar}} = \frac{0.93}{0.909} + \frac{0.07}{0.114} = 1.637$$

$$R_{bm} = 0.611\,m^2K/W$$

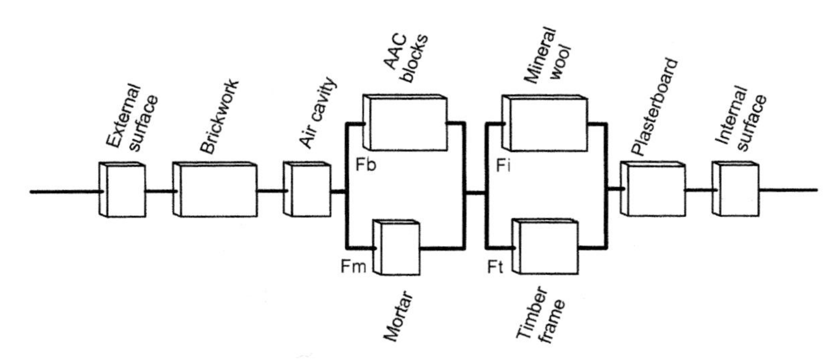

Fig. 16.20 Brick and blockwork cavity wall – lower resistance limit.

The mineral wool insulation in its timber frame form another thermal bridge, and their combined resistance, R_{it}, is found from:

$$\frac{1}{R_{it}} = \frac{F_{insulation}}{R_{insulation}} + \frac{F_{timber}}{R_{timber}} = \frac{0.88}{2.342} + \frac{0.12}{0.685} = 0.5509$$

$$R_{it} = 1.815\,m^2K/W$$

The combined resistances may now be used with the other resistances in the chain to find the lower resistance limit, as shown in Table 16.41.

Table 16.41 Calculation of the lower resistance limit, cavity wall.

	Thermal resistances, m^2K/W Thermal bridges		
	Components	**Combined**	
External surface resistance	—	—	0.040
Resistance of brickwork	—	—	0.132
Resistance of cavity	—	—	0.180
Resistance of AAC blocks (93%)	0.909	0.611	0.611
Resistance of mortar (7%)	0.114		
Resistance of mineral wool (88%)	2.342	1.815	1.815
Resistance of timber (12%)	0.685		
Resistance of plasterboard	—	—	0.050
Internal surface resistance	—	—	0.130
Total thermal resistance, R_{lower}			2.958

Note that R_{upper} is an overestimate of the true resistance, whereas R_{lower} is an underestimate. The average of these is very close to the true value. Hence, the total resistance of the wall is found from:

$$R_T = \frac{1}{2}\left(R_{upper} + R_{lower}\right) = \frac{1}{2}(3.382 + 2.958) = 3.170\ m^2K/W$$

and the U-value is

$$U = \frac{1}{R_T} = \frac{1}{3.170} = 0.315\ W/m^2K$$

Corrections to the U-value for air gaps and mechanical fixings

If there are small air gaps or mechanical fixings (such as wall ties) penetrating the insulation layer, it may be necessary to add a correction, ΔU_g , to the U-value. The correction is required if ΔU_g is 3% or more of the uncorrected U-value, but may be ignored if it is less than 3%. The correction is calculated from:

$$\Delta U_g = \Delta U^{11} \times \left(\frac{R_1}{R_T}\right)^2$$

In this wall, the fixing of the mineral wool insulation in its timber sub-frame is such that there is no air movement on the warm side, but there are some air gaps penetrating the insulation layer. As the air gaps are in the mineral wool and timber sub-frame, $R_l = R_{it} = 1.815$. Referring to Table 16.35, the correction for air gaps is level 1, and so $\Delta U^{11} = 0.01$. With $R_T = 3.170$, the correction is thus:

$$\Delta U_g = 0.01 \times \left(\frac{1.815}{3.170}\right)^2 = 0.003 \text{ W/m}^2\text{K}$$

As this is less than 3% of U, it may be ignored. The final U-value is rounded to two decimal places:

$$U = 0.32 \text{ W/m}^2\text{K}$$

Example 16.5.2 Timber framed wall

Figure 16.21 shows a timber framed wall consisting of an outer layer of brickwork, a clear ventilated cavity, 10 mm plywood, 38 × 140 mm timber stud framing with 140 mm mineral wool quilt insulation placed between the studs, and two sheets of 12.5 mm plasterboard with an integral vapour check. The timber studs account for 15% of the area, corresponding to 38 mm studs at 600 mm centres, with allowances for horizontal noggins and additional framing at junctions and around openings. The thermal data for this wall is given in Table 16.42.

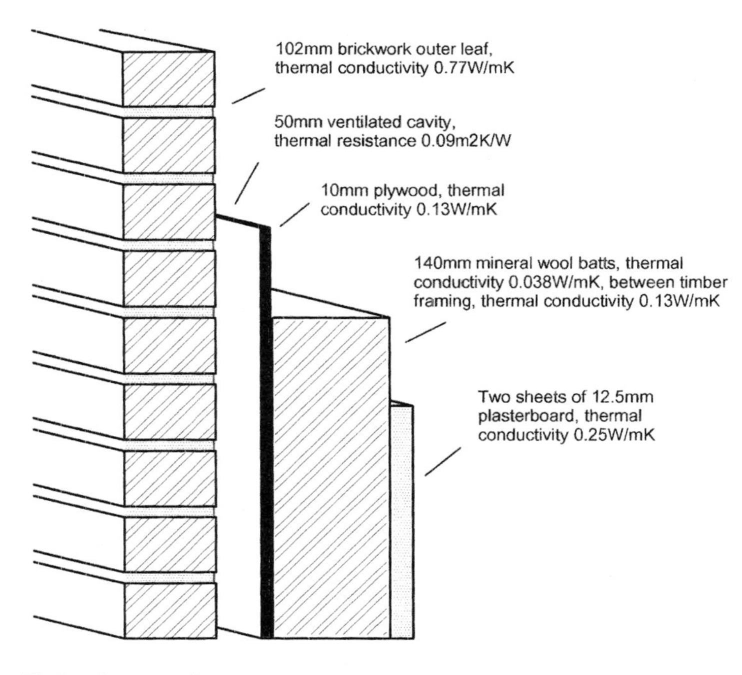

Fig. 16.21 Timber frame wall.

Table 16.42 Thermal data for timber frame wall.

Material	Thickness mm	Thermal conductivity W/mK	Thermal resistance m^2K/W
External surface	—	—	0.040
Outer brickwork	102	0.77	0.132
Cavity, vented	—	—	0.090
Plywood	10	0.13	0.077
Mineral wool quilt insulation	140	0.038	3.684
Timber framing	140	0.13	1.077
Plasterboard	25	0.25	0.100
Internal surface	—	—	0.130

The upper resistance limit, R_{upper}

Each possible heat flow path through the wall is considered separately, and in this case it can be seen that there are two such paths. This is illustrated in Fig. 16.22. The resistance of each path is calculated on the basis that the materials are in series, and then the two paths are combined on the basis that they are in parallel. The first part of the calculation is illustrated in Table 16.43. The two paths are now combined in parallel to find R_{upper}:

$$\frac{1}{R_{upper}} = \frac{F_1}{R_1} + \frac{F_2}{R_2} = \frac{0.85}{4.253} + \frac{0.15}{1.646} = 0.291$$

$$R_{upper} = 3.437\,m^2K/W$$

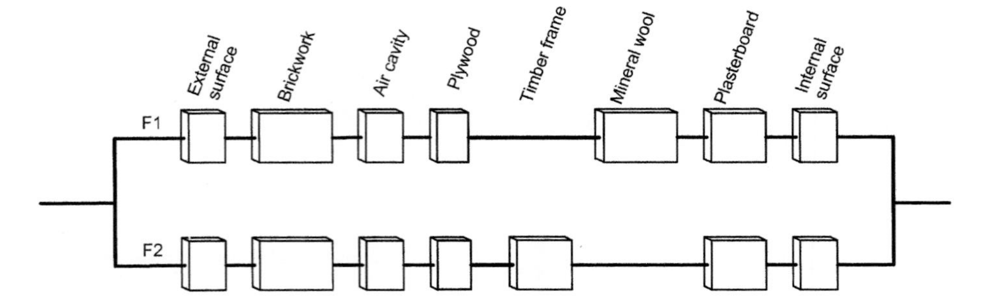

Fig. 16.22 Upper resistance limit – timber frame wall.

The lower resistance limit, R_{lower}

Each thermal bridge in the construction element is first converted to a single combined resistance. There is only one bridge in this case, formed by the mineral wool in its timber frame, as shown in Fig. 16.23. The combined resistance of the thermal bridge, R_{it} is found from:

$$\frac{1}{R_{it}} = \frac{F_{insulation}}{R_{insulation}} + \frac{F_{timber}}{R_{timber}} = \frac{0.85}{3.684} + \frac{0.15}{1.077} = 0.370$$

$$R_{it} = 2.703\,m^2K/W$$

Table 16.43 Calculation of the upper resistance limit, timber frame wall.

	Thermal resistance, m^2K/W Path 1	Path 2
External surface resistance	0.040	0.040
Resistance of brickwork	0.132	0.132
Resistance of cavity	0.090	0.090
Resistance of plywood	0.077	0.077
Resistance of mineral wool quilt	3.684	—
Resistance of timber	—	1.077
Resistance of plasterboard	0.100	0.100
Internal surface resistance	0.130	0.130
Total thermal resistance of path	4.253	2.988
Fractional area of path	85% = 0.85	15% = 0.15

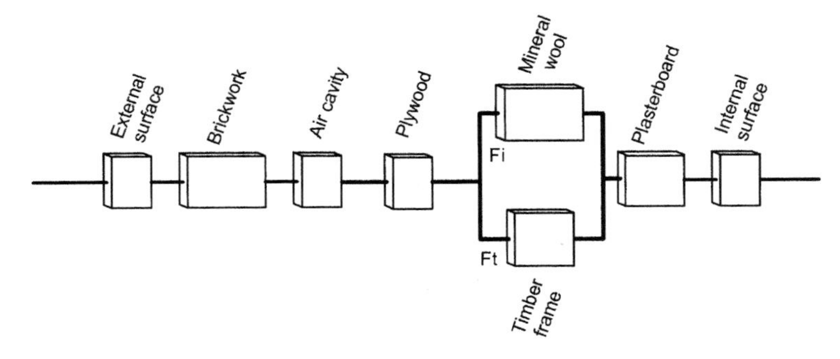

Fig. 16.23 Lower resistance limit – timber frame wall.

This combined resistance may now be used to find the lower resistance limit, as shown in Table 16.44. Note again that R_{upper} is an overestimate of the true resistance, whereas R_{lower} is an underestimate. The average of these is very close to the true value. Hence, the total resistance of the wall is found from

$$R_T = \frac{1}{2}\left(R_{upper} + R_{lower}\right) = \frac{1}{2}(3.437 + 3.272) = 3.354 \text{ m}^2\text{K/W}$$

and the U-value is:

$$U = \frac{1}{R_T} = \frac{1}{3.354} = 0.298 \text{ W/m}^2\text{K}$$

Corrections to the U-value for air gaps and mechanical fixings

As in the previous example, there is the possibility of a correction for small air gaps. Again it can be assumed that the fixing of the mineral wool insulation in its timber sub-frame is such that there is no air movement on the warm side, but there are some air gaps penetrating the insulation layer. As the air gaps are in the mineral wool and timber sub-frame, $R_l = R_{it} = 2.703$. Referring to Table 16.35,

Table 16.44 Calculation of the lower resistance limit, timber frame wall.

	Thermal resistances, m^2K/W Thermal bridges		
	Components	Combined	
External surface resistance	—	—	0.040
Resistance of brickwork	—	—	0.132
Resistance of cavity	—	—	0.180
Resistance of plywood	—	—	0.077
Resistance of mineral wool (85%)	3.684	2.703	2.703
Resistance of timber (15%)	1.077		
Resistance of plasterboard	—	—	0.100
Internal surface resistance	—	—	0.130
Total thermal resistance, R_{lower}			3.272

the correction for air gaps is level 1, and so $\Delta U^{11} = 0.01$. With $R_T = 3.354$, the correction is thus:

$$\Delta U_g = 0.01 \times \left(\frac{2.703}{3.354}\right)^2 = 0.006\ W/m^2K$$

As this is less than 3% of U, it may be ignored. The final U-value is rounded to two decimal places, and so the result is

$$U = 0.30\ W/m^2K$$

16.6 The calculation of U-values for ground floors

16.6.1 Introduction

The accurate calculation of the U-value of ground floors is difficult and requires the rigorous procedures given in BS EN ISO 13370 [13] or in CIBSE Guide, section A3 (1999 edition) [4]. However, the full rigour of these methods may not be necessary, and AD L provides a simple approach that is adequate for most of the common constructions and ground conditions to be found in the UK. The method is based on precalculated tabulated values. There are several points to be noted.

- For solid ground floors, if the perimeter to area ratio is less than $0.12\ m/m^2$, the U-value will normally be $0.25\ W/m^2K$ or less without the need for insulation.
- For suspended ground floors, if the perimeter to area ratio is less than $0.09\ m/m^2$, the U-value will normally be $0.25\ W/m^2K$ or less without the need for insulation.
- For ground floors the U-value depends on the type of soil beneath the building; clay soil is the most typical in the UK and this is assumed to be the case in the following tables.
- Where the soil is neither clay nor silt, the U-value must be calculated in accordance with BS EN ISO 13370 [13].

As the U-value of a ground floor depends on the ratio of the perimeter to the area, the following rules for calculating this ratio must be observed.

- Floor dimensions should be measured between the finished internal faces of the external elements of the building and must include any projecting bays.
- For semi-detached houses, terraced houses, blocks of flats and similar structures, the floor dimensions can be either those of the individual unit or the whole building.
- When considering extensions to existing buildings, the floor dimensions may be taken as those of the complete building including the extension.
- Unheated spaces outside the insulated fabric (e.g. attached garages and porches) are excluded from the perimeter and area calculation, but the length of common wall between them must be included in the perimeter.

In addition to meeting U-value requirements, it is also important that the floor design should prevent excessive thermal bridging at the floor edge. This is to reduce the risk of condensation and mould growth.

16.6.2 Solid ground floors

For solid ground floors with all-over insulation, the U-values in Table 16.45 apply.

Table 16.45 U-values for solid ground floors.

Perimeter to area ratio m/m²	Thermal resistance of all-over insulation, m²K/W					
	0	0.5	1	1.5	2	2.5
	U-value of solid ground floor, W/m²K					
0.05	0.13	0.11	0.10	0.09	0.08	0.08
0.10	0.22	0.18	0.16	0.14	0.13	0.12
0.15	0.30	0.24	0.21	0.18	0.17	0.15
0.20	0.37	0.29	0.25	0.22	0.19	0.18
0.25	0.44	0.34	0.28	0.24	0.22	0.19
0.30	0.49	0.38	0.31	0.27	0.23	0.21
0.35	0.55	0.41	0.34	0.29	0.25	0.22
0.40	0.60	0.44	0.36	0.30	0.26	0.23
0.45	0.65	0.47	0.38	0.32	0.27	0.23
0.50	0.70	0.50	0.40	0.33	0.28	0.24
0.55	0.74	0.52	0.41	0.34	0.28	0.25
0.60	0.78	0.55	0.43	0.35	0.29	0.25
0.65	0.82	0.57	0.44	0.35	0.30	0.26
0.70	0.86	0.59	0.45	0.36	0.30	0.26
0.75	0.89	0.61	0.46	0.37	0.31	0.27
0.80	0.93	0.62	0.47	0.37	0.32	0.27
0.85	0.96	0.64	0.47	0.38	0.32	0.28
0.90	0.99	0.65	0.48	0.39	0.32	0.28
0.95	1.02	0.66	0.49	0.39	0.33	0.28
1.00	1.05	0.68	0.50	0.40	0.33	0.28

Solid ground floors with edge insulation

It is often more practical to insulate a floor by means of edge insulation. The edge insulation may be laid horizontally or vertically, as shown in Fig. 16.24. Where horizontal or vertical edge insulation is used *instead of* all-over insulation, a correction factor is subtracted from the U-value for an *uninsulated* solid ground floor. The correction factor is a combination of the perimeter to area ratio and an edge insulation factor, Ψ. Thus:

$$U = U_0 - \frac{P}{A}\Psi$$

where U_0 is the value for an uninsulated floor taken from Table 16.45 (the column for zero thermal resistance), and Ψ is obtained from Table 16.46.

Ground floors with both all-over insulation and edge insulation

There are no tables for floors with both types of insulation. For such cases, the calculation method of BS EN ISO 13370 [13] must be used.

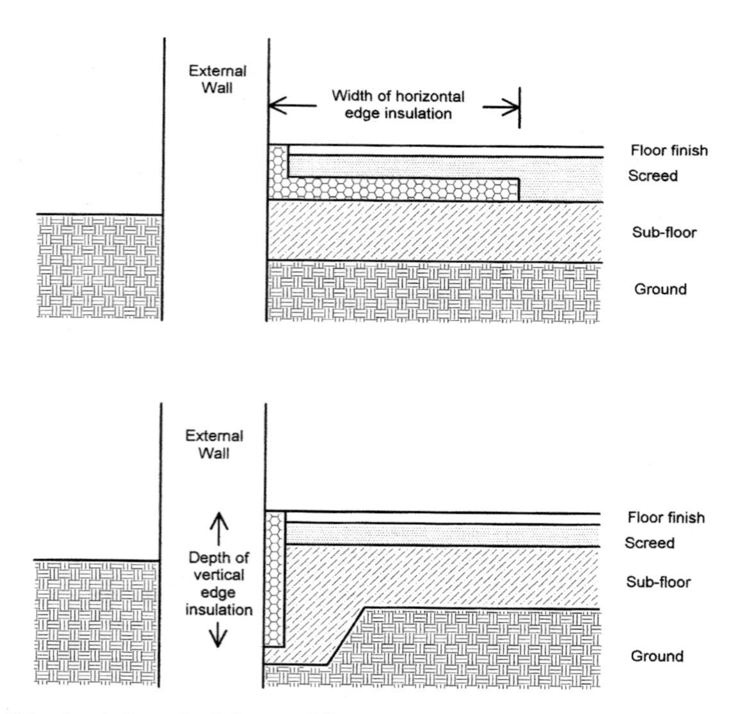

Fig. 16.24 Edge insulation of solid ground floors.

Table 16.46 Edge insulation factors for solid ground floors.

Width of horizontal insulation m	Thermal resistance of insulation, m^2K/W 0.5	1.0	1.5	2.0
	Edge insulation factor Ψ, W/mK			
0.50	0.13	0.18	0.21	0.22
1.00	0.20	0.27	0.32	0.34
1.50	0.23	0.33	0.39	0.42
Depth of vertical insulation m				
0.25	0.13	0.18	0.21	0.22
0.50	0.20	0.27	0.32	0.34
0.75	0.23	0.33	0.39	0.42
1.00	0.26	0.37	0.43	0.48

16.6.3 Suspended floors

Uninsulated suspended ground floors

The U-values for uninsulated suspended ground floors are given in Table 16.47. These values can be used when:

- The floor deck is not more than 500 mm above the external ground level
- The wall surrounding the underfloor space is uninsulated.

Table 16.47 U-values for uninsulated suspended ground floors.

Perimeter to area ratio m/m^2	Ventilation opening area per unit area of underfloor space $0.0015\,m^2/m$	$0.0030\,m^2/m$
	U-value of suspended ground floor, W/m^2K	
0.05	0.15	0.15
0.10	0.25	0.26
0.15	0.33	0.35
0.20	0.40	0.42
0.25	0.46	0.48
0.30	0.51	0.53
0.35	0.55	0.58
0.40	0.59	0.62
0.45	0.63	0.66
0.50	0.66	0.70
0.55	0.69	0.73
0.60	0.72	0.76
0.65	0.75	0.79
0.70	0.77	0.81
0.75	0.80	0.84
0.80	0.82	0.86
0.85	0.84	0.88
0.90	0.86	0.90
0.95	0.88	0.92
1.00	0.89	0.93

The U-values depend on the amount of ventilation which is provided to the underfloor space. This is expressed as the area in square metres of ventilation opening per unit length in metres of floor perimeter, and the table provides data for two typical values.

Insulated suspended floors

The U-value of an insulated suspended floor is calculated from the parameters U_o, R_f and U_f where:

U_o is the U-value of the equivalent uninsulated floor taken from Table 16.47
U_f is the U-value of the floor deck, including allowances for thermal bridging, and calculated according to the methods recommended in BS EN ISO 6946 [12], or by a numerical modelling method
R_f is the thermal resistance of the floor deck itself.

The procedure is first to obtain U_f and then to find R_f from the formula:

$$R_f = \frac{1}{U_f} - 0.17 - 0.17$$

The two values of 0.17 are the surface resistances. The U-value of the floor is then found from:

$$U = \frac{1}{[(1/U_o) - 0.2 + R_f]}$$

16.6.4 Example calculations

Example 16.6.1 Solid ground floor over clay sub-soil
Figure 16.25 illustrates the floor plan of a detached house. The sub-soil is clay, and the floor is uninsulated. First, calculate the perimeter to area ratio:

$P = 2 \times (10.2 + 6.7) = 33.8\,\text{m}$
$A = 10.2 \times 6.7 - 3.5 \times 2.7 = 58.89\,\text{m}^2$
$P/A = 33.8/58.89 = 0.57\,\text{m/m}^2$

The U-value is found from Table 16.45, using the column for zero thermal resistance. The Approved Document recommends that the row nearest to the actual P/A value is used; in this case the nearest row to 0.57 is P/A = 0.55. It is not necessary to interpolate between rows because the change in U-value between rows is not large enough to warrant the extra work. The U-value of this floor is therefore:

$$U = 0.74\,\text{W/m}^2\text{K}$$

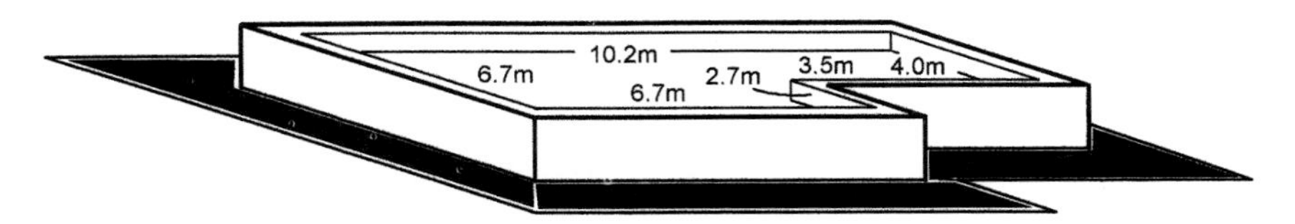

Fig. 16.25 Solid ground floor.

Example 16.6.2 Solid ground floor over clay sub-soil with all-over insulation
The floor in Fig. 16.25 is now provided with all-over insulation between the screed and the structural floor. The insulation layer is 75 mm thick and has a thermal conductivity of 0.040 W/mK. The resistance of the insulation layer is:

$$R_{ins} = 0.075/0.040 = 1.875\,m^2K/W$$

In Table 16.45 we again use the row for $P/A = 0.55$, but this time we must interpolate between the columns for $R_{ins} = 1.5$ and $R_{ins} = 2.0$. This interpolation is necessary because the change in U-value between columns is more significant than the change between rows. Thus:

$$\text{At } R_{ins} = 1.875, U = 0.34 - \left(\frac{1.875 - 1.5}{2.0 - 1.5}\right) \times (0.34 - 0.28) = 0.295 \text{ W/m}^2\text{K}$$

Example 16.6.3 Solid ground floor over clay sub-soil with vertical edge insulation
The floor in Fig. 16.25 is provided with vertical edge insulation instead of all-over insulation. The insulation is to a depth of 750 mm, and the insulation is 75 mm thick with a thermal conductivity of 0.040 W/mK. The resistance of the insulation layer is:

$$R_{ins} = 0.075/0.040 = 1.875\,m^2K/W$$

From Example 16.6.1, the perimeter to area ratio of this floor is 0.57, and its uninsulated U-value is 0.74 W/m^{2}K. We require the edge insulation factor from Table 16.46, and it is necessary to interpolate between the columns for $R_{ins} = 1.5$ and $R_{ins} = 2.0$. Hence:

$$\Psi = 0.39 + \left(\frac{1.875 - 1.5}{2.0 - 1.5}\right) \times (0.42 - 0.39) = 0.413 \text{ W/mK}$$

The U-value of the floor is now:

$$U = 0.74 - 0.57 \times 0.413 = 0.50\,W/m^2K$$

Example 16.6.4 Uninsulated, suspended ground floor
Assume that the floor in Fig. 16.25 is an uninsulated suspended timber ground floor. The floor deck is less than 500 mm above external ground level, and the under-floor ventilation openings amount to approximately 0.0015 m^2 per metre of floor peri-

meter. The perimeter to area ratio is 0.57, and taking 0.55 as the nearest value in Table 16.47, the U-value is:

$$U = 0.69\,W/m^2K$$

Example 16.6.5 Insulated, suspended ground floor
Let the floor in Example 16.6.4 be insulated, with insulation fitted between the floor joists. The U-value of this floor deck may be calculated in the same way as the U-value of a wall, using the method in section 5. Assuming the result of this calculation is a U-value of 0.45 W/m^{2}K, the U-value of the floor is found as follows:

$$R_f = \frac{1}{U_f} - 0.17 - 0.17 = 2.22 - 0.34 = 1.88\ m^2K/W$$

The uninsulated U-value, from Example 16.6.4, is $U_o = 0.69$, and so:

$$U = \frac{1}{[(1/U_o) - 0.2 + R_f]} = \frac{1}{[1.45 - 0.2 + 1.88]} = 0.32\ W/m^2K$$

16.7 Compensation calculations for glazing

16.7.1 Introduction

The elemental method of both AD L1 and AD L2 allows any or all of the windows, doors or rooflights to have U-values which are greater than the standard maximum values, provided that compensating measures are taken. The compensation may be either a lower U-value in one element to offset a higher U-value in another, or it could be a reduction in the area of one or all of the elements. The guiding principle is that the total heat loss through windows, doors and rooflights should not exceed that which would occur when the building has the maximum allowable area of openings with maximum allowable U-values. However, in the case of dwellings, this flexibility cannot be used within the elemental method to increase the area of openings above the standard area provision of 25% of the total floor area. On the other hand, for buildings other than dwellings, the elemental method does not have this constraint. The examples and the procedure which follow here have wider applicability than those in the Approved Documents.

16.7.2 Example calculations

Example 16.7.1 Detached dwelling
A detached dwelling has a total floor area of 130 m^2. There are two solid wood doors and a pair of glazed patio doors. There is also a rooflight. The standard area provision for the windows, doors and rooflights taken together is 25% of the floor area, i.e. $0.25 \times 130 = 32.5\,m^2$. It is intended to use the whole of this allowance, and so the proposed areas for windows, doors and rooflights in the dwelling are:

- Windows $A_W = 23.7\,m^2$
- External doors, wood (2 No.) $A_D = 3.8\,m^2$
- External doors, glazed (2 No.) $A_D = 3.8\,m^2$
- Rooflights $A_R = 1.2\,m^2$
- Total area $A_W + A_D + A_R = 32.5\,m^2$

The solid wooden doors have a U-value of 3.0 W/m^2K, and the glazed patio doors have a U-value of 2.7 W/m^2K. The rooflight has a U-value (including adjustment) of 1.9 W/m^2K. The windows have UPVC frames, and so the maximum average U-value for windows, doors and rooflights taken together is 2.0 W/m^2K.

Adjusting the window U-value

To achieve compliance, we must select glazing with an appropriate U-value. As the doors have U-values greater than 2.0, it is likely that the windows will need to have U-values that are less than 2.0. To find the required U-value for the windows, we first calculate the maximum and actual heat losses:

$$\text{Maximum heat loss} = 2.0 \times (0.25A_F) = 0.5A_F = 0.5 \times 130 = 65\,W/K$$
$$\text{Actual heat loss} = A_W U_W + A_D U_D + A_R U_R$$
$$\text{Actual heat loss} = 23.7\,U_W + 3.8 \times 3.0 + 3.8 \times 2.7 + 1.2 \times 1.9$$
$$= 23.7\,U_W + 23.94\,W/K$$

Equating these two heat losses gives the maximum U-value for the windows:

$$23.7\,U_W + 23.94 = 65$$
$$U_W = 1.73\,W/m^2K$$

Hence the maximum permissible U-value for the window is 1.73 W/m^2K. Referring to Table 16.20 it can be seen that this can be achieved with 16 mm low-E ($\varepsilon_n = 0.05$) argon filled double glazing ($U = 1.7\,W/m^2K$), or several different types of triple glazing.

Adjusting the window area

Alternatively, if argon filled units are unavailable, it may be possible to use 16 mm low-E ($\varepsilon_n = 0.05$) air filled double glazing, with a U-value of 1.8 W/m^2K. In this case the window area must be reduced, and the calculation to find the maximum permissible window area is:

$$\text{Actual heat loss} = A_W U_W + A_D U_D + A_R U_R$$
$$\text{Actual heat loss} = 1.8\,A_W + 3.8 \times 3.0 + 3.8 \times 2.7 + 1.2 \times 1.9$$
$$= 1.8\,A_W + 23.94\,W/K$$

Equating these two heat losses gives the maximum area of the windows:

$$1.8\,A_W + 23.94 = 65$$
$$A_W = 22.8\,m^2$$

Therefore, provided the window area is reduced from 23.7 m^2 to 22.8 m^2, the requirement will be satisfied with double glazed units of U = 1.8 W/m^2K.

Example 16.7.2 Semi-detached dwelling
A semi-detached dwelling has a total floor area of 90 m^2. There are two insulated wood doors, each of area 1.9 m^2 and U-value 1.0 W/m^2K. There is no rooflight. For the windows it is intended to use 12 mm air gap air filled double glazing units in wooden frames, with a U-value (Table 16.20) of 2.8 W/m^2K. The standard area provision for the windows, doors and rooflights taken together is 25% of the floor area, i.e. $0.25 \times 90 = 22.5\,m^2$, and it is intended to install as much of this allowance as possible. The maximum window area is therefore $22.5 - (2 \times 1.9) = 18.7\,m^2$. We first check the average U-value of the windows, doors and rooflights in the dwelling:

$$\text{Average U} - \text{value, } U_{WDR} = \frac{A_W U_W + A_D U_D + A_R U_R}{A_W + A_D + A_R}$$

$$U_{WDR} = \frac{18.7 \times 2.8 + 3.8 \times 1.0 + 0}{22.5} = \frac{56.16}{22.5} = 2.5\ W/m^2K$$

This is above the value given in Table 16.2 for wood frames, and so the window area must be reduced, and the calculation to find the maximum permissible window area is:

$$\text{Maximum heat loss} = 2.0 \times (0.25A_F) = 0.5A_F = 0.5 \times 90 = 45\,W/K$$
$$\text{Actual heat loss} = A_W U_W + A_D U_D + A_R U_R$$
$$\text{Actual heat loss} = 2.8\,A_W + 3.8 \times 1.0 + 0 = 2.8\,A_W + 3.8\ W/K$$

Equating these two heat losses gives the maximum area of the windows:

$$2.8\,A_W + 3.8 = 45$$
$$A_W = 14.7\,m^2$$

Therefore, provided the window area is reduced from 18.7 m^2 to 14.7 m^2, the requirement will be satisfied with double glazed units of U = 2.8 W/m^2K.

If instead of being reduced the window area is maintained at 18.7 m^2, it may be possible to achieve compliance by selecting a double glazed unit with a better thermal performance. The calculation to find the necessary U-value follows the same procedure as in the first part of the detached house example above. Alternatively, one can simply select a different double glazing unit and then check for compliance. If, say, a 12 mm air gap low-E ($\varepsilon_n = 0.15$) air filled double glazing, U = 2.2 W/m^2K, is installed in the windows, it is necessary to recalculate the average U-value:

$$\text{Average U} - \text{value, } U_{WDR} = \frac{A_W U_W + A_D U_D + A_R U_R}{A_W + A_D + A_R}$$

$$U_{WDR} = \frac{18.7 \times 2.2 + 3.8 \times 1.0 + 0}{22.5} = \frac{44.94}{22.5} = 2.0\ W/m^2K$$

This satisfies the relevant maximum average U-value requirement in Table 16.2, and therefore complies.

Example 16.7.3 An office building

An office building has a total external exposed wall area of 1875 m^2. It is proposed to fit metal framed windows to a total area of 732 m^2, and the windows will be fitted with 16 mm air gap low-E ($\varepsilon_n = 0.05$) argon filled double glazing, $U = 2.1\ W/m^2K$ (Table 16.20). The building is to have metal part-glazed personnel doors of U-value 3.3 W/m^2K to a total area of 15.9 m^2. There are no rooflights. Check for compliance by the elemental method. The requirements are:

- from Table 16.8, that the average U-value of the windows and doors does not exceed the figure for metal window frames, i.e. $U = 2.2\ W/m^2K$, and
- from Table 16.9, that the total area of windows and doors does not exceed 40% of the exposed wall area.

The total area of openings is $732 + 15.9 = 747.9\ m^2$. Thus:

$$\text{Openings as a percentage of exposed wall area} = \frac{747.9 \times 100}{1875} = 39.9\%$$

$$\text{Average U-value,} \quad U_{WD} = \frac{732 \times 2.1 + 15.9 \times 3.3}{747.9} = 2.13\ W/m^2K$$

Taken together, these results satisfy the requirements of the elemental method for openings.

Example 16.7.4 An office building

For the office building described above, it is found that the proposed 16 mm air gap low-E argon filled double glazing units are unavailable, and that is necessary to use 12 mm air gap low-E ($\varepsilon_n = 0.05$) air filled double glazing units instead. These have a U-value of $U = 2.5\ W/m^2K$ (Table 16.20). As both the windows and the doors have U-values greater than the maximum of 2.2 W/m^2K, the average U-value must also be too high. Therefore compensatory measures are required, and in this case it is proposed to reduce the window area. The means that there will be a corresponding increase in the wall area, and hence an increase in the wall heat loss. Unlike dwellings, this extra heat loss may not be small enough to ignore, and so it is better to include it in the calculation. The calculation may proceed as follows.

Total area of walls, windows and doors	=	A_T
Maximum area of windows and doors	=	A_M
Corresponding external wall area	=	$(A_T - A_M)$
Maximum permissible heat loss	=	$0.35(A_T - A_M) + 2.2A_M$
Actual window area/U-value	=	A_W/U_W
Actual door area/U-value	=	A_D/U_D
Actual heat loss	=	$0.35(A_T - A_W - A_D) + U_WA_W + U_DA_D$

Equating:

$$0.35(A_T - A_M) + 2.2A_M = 0.35(A_T - A_W - A_D) + U_W A_W + U_D A_D$$

and simplifying:

$$1.85A_M = A_W(U_W - 0.35) + A_D(U_D - 0.35)$$

Applying this formula to determine A_W for this office:

$$\text{Maximum area of openings, } A_M = \frac{40}{100} \times 1875 = 750 \text{ m}^2$$

$$U_W = 2.5 \text{ W/m}^2\text{K} \quad A_D = 15.9 \text{ m}^2 \quad U_D = 3.3 \text{ W/m}^2\text{K}$$

$$1.85 \times 750 = A_W(2.5 - 0.35) + 15.9(3.3 - 0.35)$$

$$A_W = 624 \text{ m}^2$$

If the extra heat loss through the external wall had been ignored, the result would have been 639 m^2, an error of about 2.4%. If the metal window frames were replaced by wood or UPVC, the formula would be:

$$1.65A_M = A_W(U_W - 0.35) + A_D(U_D - 0.35)$$

16.8 Target U-value examples

The target U-value method can only be applied to a complete dwelling. The following examples illustrate the application of the procedure given in section 16.2.

16.8.1 Semi-detached dwelling

Figure 16.26 is a plan view of the ground floor and first floor of a semi-detached dwelling. The dwelling lies approximately on a north-south axis. The windows have wood frames, with 11.8 m^2 on the southern face, 9.2 m^2 on the northern face, and 1.2 m^2 to the side. It is proposed to fit a mains gas fired boiler with a SEDBUK rating of 81%. The areas and U-values of the elements of the dwelling are given in Table 16.48.

The total area of openings is 26.00 m^2 and the total floor area is 94.08 m^2. The openings are therefore 27.6% of the floor area. Therefore the dwelling does not meet the requirements of the elemental method, but can be assessed using the target U-value method.

Step 1 The initial target U-value

$$U_1 = \left[0.35 - 0.19\frac{47.04}{191.08} - 0.10\frac{47.04}{191.08} + 0.413\frac{94.08}{191.08}\right] = 0.482 \text{ W/m}^2\text{K}$$

Step 2 Adjust for boiler efficiency

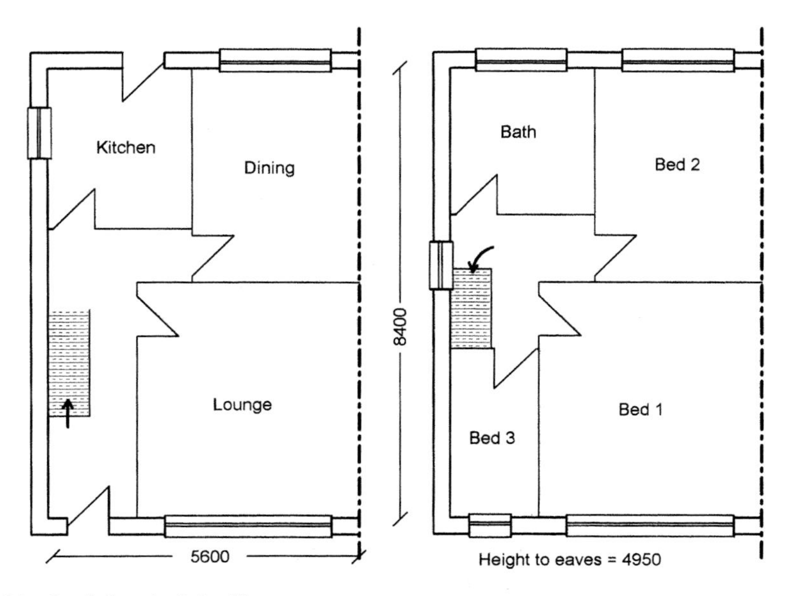

Fig. 16.26 Semi-detached dwelling.

Table 16.48 Semi-detached dwelling.

Element	Area m^2	U-value W/m^2K	Heat loss per degree W/K
Wall	71.00	0.35	24.85
Roof	47.04	0.20	9.41
Ground floor	47.04	0.25	11.76
Windows	22.20	1.90	42.18
Doors (2 No.)	3.80	3.00	11.40
Totals	A_T = 191.08		ΣAU = 99.60

SEDBUK of proposed boiler = 81%
SEDBUK of reference boiler = 78% (from Table 16.3)

$$f_e = \frac{81}{78} = 1.04$$

$$U_2 = 1.04 \times 0.482 = 0.501 \text{ W/m}^2\text{K}$$

Step 3 Allowance for additional solar gain due to window frame material

The window frames are wood, and so there is no change:

$$U_3 = U_2 = 0.501 \text{ W/m}^2\text{K}$$

Step 4 Allowance for additional solar gain due to orientation of windows

South facing area	A_S	$= 11.8\,m^2$
North facing area	A_N	$= 9.2\,m^2$
Total window area	A_{TG}	$= 22.2\,m^2$

Solar adjustment factor $\quad \Delta S = 0.04\left[\frac{11.8 - 9.2}{22.2}\right] = 0.005$

Step 5 Convert to final target U-value

$$U_T = 0.501 + 0.005 = 0.506\ W/m^2K$$

Step 6 Find U_{AVG}

$$U_{AVG} = \frac{\Sigma AU}{A_T} = \frac{99.6}{191.08} = 0.521\ W/m^2K$$

For compliance, U_{AVG} must be less than or equal to U_T. However in this case U_{AVG} is greater than U_T, and so the dwelling does not comply by the target U-value method. There are several ways, applied either singly or in combination, in which the design of the dwelling could be altered to improve the possibility of compliance. These may either increase the target U-value, or reduce U_{AVG}, or affect both. The following are examples.

(1) *Raise boiler efficiency – and hence raise U_T*
A boiler with a higher SEDBUK rating, say 85%, could be used. The boiler efficiency factor is then 85/78 = 1.09, and this would alter U_T as follows:

$U_1 = 0.482$
$U_2 = 1.09 \times 0.482 = 0.525$
$U_3 = U_2 = 0.525$
$U_T = 0.525 + 0.005 = 0.530\ W/m^2K$

This has resulted in U_{AVG} being less than the target U-value, and so the dwelling now complies.

(2) *Reduce the U-values of some of the elements – and hence reduce U_{AVG}*
This may be attempted by trial and error by choosing lower U-values and repeating the calculation of U_{AVG}. Alternatively, the required reduction can be estimated by setting U_{AVG} equal to U_T (which would satisfy the target U-value requirement) and calculating a new value for ΣAU.

$U_{AVG} = U_T = 0.506$
Reduced $\Sigma AU = A_T U_{AVG} = 191.08 \times 0.506 = 96.69$
Required reduction $= 99.60 - 96.69 = 2.91$

This reduction may be obtained from any one of the exposed elements, or from a combination of several. If only one element is altered, the effect on its U-value would be:

Walls	Reduction = 2.91/71.00 = 0.04
	New U-value = 0.35 − 0.04 = 0.31 W/m^2K
Roof	Reduction = 2.91/47.04 = 0.06
	New U-value = 0.20 − 0.06 = 0.14 W/m^2K
Ground floor	Reduction = 2.91/47.04 = 0.06
	New U-value = 0.25 − 0.06 = 0.19 W/m^2K
Windows	Reduction = 2.91/22.20 = 0.13
	New U-value = 1.90 − 0.13 = 1.77 W/m^2K
Doors	Reduction = 2.91/3.80 = 0.77
	New U-value = 3.00 − 0.77 = 2.23 W/m^2K

Most of these reduced U-values may be difficult to achieve, with the exception of the doors. An insulated door construction providing a U-value of 2.23 W/m^2K or less is feasible, and would be sufficient to ensure compliance. Otherwise, if the doors cannot be altered, it may be necessary to reduce the U-value of more than one element.

(3) *Reduce the window area – and hence reduce U_{AVG}*
From (2) above, the required reduction in ΣAU is 2.91. However, if the window area is reduced, the wall area is increased by the same amount. Consequently the gain from reducing the window area is offset by a corresponding increase in wall area. If δA is the reduction in window area, then we may write the following equation:

$$2.91 = 1.90 \times \delta A - 0.35 \times \delta A = 1.55 \times \delta A$$
$$\delta A = 2.91/1.55 = 1.88\,m^2$$

Therefore:

New window area = 22.20 – 1.88 = 20.32 m^2
New wall area = 71.00 + 1.88 = 72.88 m^2

In this case, adopting a window area of 20.32 m^2 or less would be sufficient to reduce U_{AVG} to 0.506 W/m^2K, and hence ensure compliance.

16.8.2 Detached dwelling

Figures 16.27a and 16.27b are plan views of the ground and first floors of a detached dwelling. The dwelling lies approximately on a north-south axis. The windows have metal frames, with 24.8 m^2 on the southern face, 11.4 m^2 on the northern face, and 1.8 m^2 to the side. It is proposed to fit a mains gas fired boiler with a SEDBUK rating of 76%. The areas and U-values of the elements of the dwelling are given in Table 16.49.

The total area of openings is 43.70 m^2, and the total floor area is 156.00 m^2. The openings are therefore 28.0% of the floor area. The dwelling is therefore below the requirements of the elemental method, but can be assessed using the target U-value method.

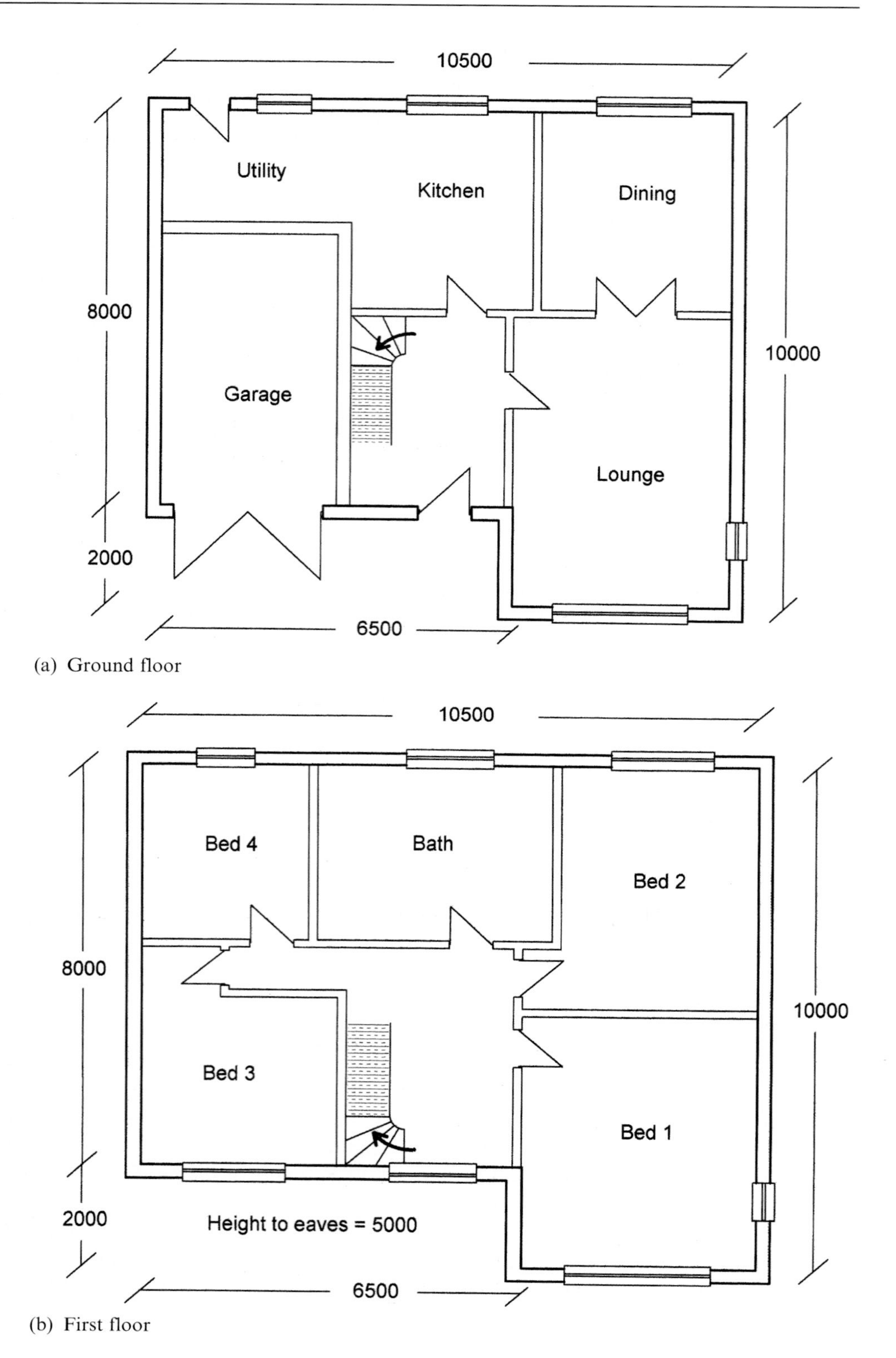

(a) Ground floor

(b) First floor

Fig. 16.27 Detached dwelling.

Table 16.49 Detached dwelling.

Element	Area m^2	U-value W/m^2K	Heat loss per degree W/K
Wall	133.80	0.35	46.83
Roof	92.00	0.20	18.40
Ground floor	64.00	0.25	16.00
First floor to garage	28.00	0.15	4.20
Windows	38.00	2.30	87.40
Doors (3 No.)	5.70	1.70	9.69
Totals	A_T = 361.50		ΣAU = 182.52

Step 1 The initial target U-value

$$U_1 = \left[0.35 - 0.19\frac{92.00}{361.50} - 0.10\frac{64.00}{361.50} + 0.413\frac{156.00}{361.50}\right] = 0.462\ W/m^2K$$

Step 2 Adjust for boiler efficiency

SEDBUK of proposed boiler = 76%
SEDBUK of reference boiler = 78%

$$f_e = \frac{76}{78} = 0.974$$

$$U_2 = 0.974 \times 0.462 = 0.450\ W/m^2K$$

Step 3 Allowance for additional solar gain due to window frame material

The window frames are metal, and so:

$$U_3 = 1.03\,U_2 = 0.464\ W/m^2K$$

Step 4 Allowance for additional solar gain due to orientation of windows

South facing area $A_S = 24.8\ m^2$
North facing area $A_N = 11.4\ m^2$
Total window area $A_{TG} = 38.0\ m^2$
Solar adjustment factor $\Delta S = 0.04\left[\frac{24.8 - 11.4}{38.0}\right] = 0.014$

Step 5 Convert to final target U-value

$$U_T = 0.464 + 0.014 = 0.478\ W/m^2K$$

Step 6 Find U_{AVG}

$$U_{AVG} = \frac{\Sigma AU}{A_T} = \frac{182.52}{361.50} = 0.505\ W/m^2K$$

For compliance, U_{AVG} must be less than or equal to U_T. However in this case U_{AVG} is greater than U_T, and so the dwelling does not comply by the target U-value method. There are several ways, applied either singly or in combination, in which the design of the dwelling could be altered to improve the possibility of compliance. These may either increase the target U-value, or reduce U_{AVG}, or affect both. The following are examples.

(1) *Raise boiler efficiency – and hence raise U_T*
A boiler with a higher SEDBUK rating, say 81%, could be used. The boiler efficiency factor is then 81/78 = 1.04, and this would alter U_T as follows:

$$U_1 = 0.462$$
$$U_2 = 1.04 \times 0.462 = 0.480$$
$$U_3 = 1.03U_2 = 0.494$$
$$U_T = 0.494 + 0.014 = 0.508 \text{ W/m}^2\text{K}$$

This has resulted in U_{AVG} being less than the target U-value, and so the dwelling now complies.

(2) *Reduce the U-values of some of the elements – and hence reduce U_{AVG}*
This may be attempted by trial and error by choosing lower U-values and repeating the calculation of U_{AVG}. Alternatively, the required reduction can be estimated by setting U_{AVG} equal to U_T (which would satisfy the target U-value requirement) and calculating a new value for ΣAU.

$$U_{AVG} = U_T = 0.476$$
$$\text{Reduced } \Sigma AU = A_T U_{AVG} = 361.5 \times 0.476 = 172.07$$
$$\text{Required reduction} = 182.52 - 172.07 = 10.45$$

This reduction may be obtained from any one of the exposed elements, or from a combination of several. If only one element is altered, the effect on its U-value would be:

Walls	Reduction = 10.45/133.80 = 0.08
	New U-value = 0.35 – 0.08 = 0.27 W/m²K
Roof	Reduction = 10.45/92.00 = 0.11
	New U-value = 0.20 – 0.11 = 0.09 W/m²K
Ground floor	Reduction = 10.45/64.00 = 0.16
	New U-value = 0.25 – 0.16 = 0.09 W/m²K
Windows	Reduction = 10.45/38.00 = 0.275
	New U-value = 2.30 – 0.28 = 2.02 W/m²K
Doors	Reduction = 10.45/5.7 = 1.83
	New U-value = 1.70 – 1.83 negative, therefore not possible

Most of these reduced U-values may be difficult to achieve. In the case of the windows, it would be possible to achieve a U-value of 2.0 or less by specifying triple glazing. In order to retain double glazing it would be necessary to change to non-metal frames. This would reduce U_T because the factor of 1.03 for metal frames would no longer apply. Nevertheless, in this particular case, the adoption of non-metal frames may be the most practical option. Otherwise it would be necessary to take a combination of measures.

(3) *Reduce the window area – and hence reduce U_{AVG}*
From (2) above, the required reduction in ΣAU is 10.45. However, if the window area is reduced, the wall area is increased by the same amount. Consequently the gain from reducing the window area is offset by a corresponding increase in wall area. If δA is the reduction in window area, then we may write the following equation:

$$10.45 = 2.30 \times \delta A - 0.35 \times \delta A = 2.65 \times \delta A$$
$$\delta A = 10.45/2.65 = 3.94\,m^2$$

Therefore:

$$\text{New window area} = 38.00 - 3.94 = 34.06\,m^2$$
$$\text{New wall area} = 133.80 + 3.94 = 137.74\,m^2$$

In this case, adopting a window area of $34.06\,m^2$ or less would be sufficient to reduce U_{AVG} to $0.476\,W/m^2K$, and hence ensure compliance.

16.9 SAP ratings and the carbon index

16.9.1 SAP

SAP is the UK government's standard assessment procedure for rating the energy cost performance of dwellings. The SAP rating is a scale from 1 to 120. A rating of SAP = 1 is the worst possible performance (i.e. the highest energy cost) and SAP = 120 is best possible energy performance (i.e. the lowest energy cost). The scale is arbitrary, and the minimum value at which the performance of a dwelling is deemed satisfactory is a matter of opinion as expressed through government policy. The 2000 edition of AD L1 does not specify a minimum acceptable SAP, but nevertheless Building Regulations require that it should be calculated for every new dwelling, and the result made public. Details of the background to the SAP worksheets for calculating its value are given in the Government's Standard Assessment Procedure for Energy Rating of Dwellings, SAP 2001 [19].

In principle, the procedure calculates the annual energy consumed by space heating and by water heating, taking into account all relevant losses and offset by certain possible gains. This gives the energy consumption in GJ/year. The energy consumption is then converted to a cost by means of fuel price factors, to give an overall energy cost factor (ECF), in monetary units per unit floor area. This is then converted to the SAP rating according to the equation:

$$\text{SAP rating} = 97 - 100 \log_{10} (\text{ECF})$$

Clearly, the lower the energy cost, the higher the SAP rating.

16.9.2 Carbon factor and carbon index

One of the most important objectives of the regulation to conserve fuel and power is to control and reduce the amount of carbon dioxide released into the atmosphere by building services systems. The carbon factor and the carbon index are measures of this. The procedure is first to calculate the annual energy consumption in GJ/year. This is the same as the SAP calculation and the same worksheet is used. However, instead of being converted to a cost, the energy consumption is converted, by means of emission factors, to the total carbon dioxide, in kg/year, emitted into the atmosphere by the dwelling. The carbon emissions are then converted to carbon factor, which in turn is converted to the carbon index, by the equations:

$$\text{Carbon factor, CF} = \frac{CO_2}{(A_T + 45)}$$

$$\text{Carbon index, CI} = 17.7 - 9.0 \log_{10} (\text{CF})$$

where CO_2 is the carbon dioxide emissions in kg/year
A_T is the total floor area of the dwelling in m^2.

It can be seen from the equations that as the CO_2 emissions fall, the carbon index rises. The carbon index is therefore used as the criterion for determining compliance under the carbon index method. The scale of values goes from 0 to 10, and the minimum value is 8.0.

Table 16.50 Comparisons of SAP rating and carbon index.

Description	Total floor area, m^2	SAP rating	Carbon index
Two bed mid-terrace house Conventional natural gas boiler, $\varepsilon_n = 78\%$	54.6	100	8.0
Three bed semi-detached house Condensing natural gas boiler, $\varepsilon_n = 88\%$	80.0	101	8.0
Three bed semi-detached house Condensing LPG gas boiler, $\varepsilon_n = 88\%$	80.0	63	7.1
Four bed mid-storey flat Conventional natural gas boiler, $\varepsilon_n = 78\%$	90.0	107	8.5
Four bed detached house Condensing natural gas boiler, $\varepsilon_n = 90\%$	100.0	101	8.0
Two bed bungalow Condensing natural gas boiler, $\varepsilon_n = 91\%$	56.7	100	8.0

16.9.3 Relationship between SAP and CI

The connection between SAP and CI is not consistent because one depends on fuel prices whereas the other depends on CO_2 emissions. Fuel prices are affected by market forces and taxation policy, while emissions are a fundamental property of the fuel. Nevertheless there is an approximate correspondence, and the examples in AD L1 Appendix F illustrate this. The results of these examples are given in Table 16.50, in which ε_n is the SEDBUK rating of the boiler. It may be noted that in all those examples which satisfy the requirement of the carbon index method with CI = 8 or more, the SAP rating is at least 100.

16.10 Example of trade-off calculations

The elemental method for buildings other than dwellings provides some design flexibility by means of trade-offs. There are two possible methods of trading off:

- Trade-off between construction elements by varying U-values and areas
- Trade-off between heating system efficiency and fabric performance, i.e. the areas and U-values of construction elements.

The principle that governs these trade-offs is that the actual building should be no worse than a notional building of the same size and shape which satisfies the requirements of the elemental method without employing trade-off.

16.10.1 Residential and conference centre

A new building is planned as a residential and conference centre. The building will be three-storey, but part of the ground floor will be two-storey in height with vehicle unloading bay doors. The building will be heated by two identical natural gas fired boilers of efficiency 78% and of combined rated heat output of 120 kW. The main characteristics of the building are:

Dimensions: rectangular, 40 m × 15 m on plan, 10.5 m high, flat roof
Windows: fitted with 12 mm air gap low-E (ε_n = 0.05) argon filled double glazing units in metal frames
linear run of 75 m on ground and first floors
linear run of 90 m on second floor
total linear run 240 m
height from sill to head 1.5 m
total area 240 × 1.5 = 360 m^2
Personnel doors: 2 double @ 3.8 m^2, 3 single @ 1.9 m^2
total area = 13.3 m^2
Vehicle door 1 roller door @ 25 m^2

From this we may calculate the remaining areas:

Total area of windows and personnel doors is thus 373.3 m^2
Total perimeter wall area $= (40 + 40 + 15 + 15) \times 10.5$ $= 1155.0\,m^2$
Area of exposed wall $= 1155 - 373.3 - 25$ $= 756.7\,m^2$
Area of roof $= 40 \times 15$ $= 600.0\,m^2$
Area of ground floor $= 40 \times 15$ $= 600.0\,m^2$

The area of windows and personnel doors as a percentage of the perimeter wall area is:

$$\frac{373.3}{1155} \times 100 = 32.3\%$$

This exceeds the allowance of 30% given in Table 16.9, and so either the window/door area must be reduced or compensating measures must be taken. This can be attempted in several ways. For example, some of the U-values for the proposed building may be lower than the maximum values specified by the elemental method, and up to half of the allowable rooflight area can be converted into an increase in window area. Consider first the U-values, as in Table 16.51.

Table 16.51 Selected U-values for proposed building.

Element	Elemental method Max U, W/m^2K	Selected U W/m^2K
Roof	0.25	0.25
Walls – extra insulation added	0.35	0.30
Windows – U-value determined by selected window and glazing design	2.20	2.30
Personnel door – same design as windows	2.20	2.30
Vehicle door – insulated to elemental standard	0.70	0.70
Ground floor – insulated with 75 mm EPS	0.25	0.20

In attempting to allow for the fact that both the window area and the window U-value exceed the requirements of the elemental method, the designers have decided to add insulation to the walls and the floor. The U-value of the floor may be determined from Table 16.45. The perimeter to area ratio is $110/600 = 0.183\,m/m^2$, which is nearest to 0.20 in the table, and which gives an uninsulated U-value of 0.37 W/m^2K. The expanded polystyrene insulation has a thermal conductivity of 0.040 W/mK, and so the thermal resistance of the extra insulation is $0.075/0.040 = 1.875\,m^2K/W$. Interpolating between the 1.5 and 2.0 columns of the table gives a U-value of 0.20 W/m^2K.

The allowable area for rooflights is 20% of the roof area, but as there are no rooflights in the proposed building, half of this allowance (i.e. 10% of the roof area) can be used to increase the allowable window area. However, this cannot be done by simply adding 10% of the roof area to the allowance for windows. This is because the maximum allowable U-values of the roof and the walls are different. This difficulty can be overcome by including rooflights in the notional building, as shown below.

The rate of heat loss from the proposed building is then compared with a notional building of the same size and shape which satisfies the elemental method. Step 1 is to prepare a heat loss table of the proposed building, incorporating the actual U-values of the proposed construction elements, as shown in Table 16.52.

Table 16.52 Proposed building heat loss table.

Element	**Area** m^2	**U-value** W/m^2K	**Heat loss** W/K
Roof	600.0	0.25	150.00
Exposed walls	756.7	0.30	227.01
Windows	360.0	2.30	828.00
Personnel doors	13.3	2.30	30.59
Vehicle loading bay doors	25.0	0.70	17.50
Ground floor	600.0	0.20	120.00
Totals	2355.0		1373.10
Average U-value, U_{AVG} = 1373.10/2355.0 = 0.583			

The notional building must have the same dimensions as the proposed building, and the combined area of windows and personnel doors must account for a maximum of 30% of the perimeter wall area. However, half of the allowable rooflight area (i.e. 10% of the roof area) can be included in the notional building to help cater for the increased window area in the proposed building. The relevant areas for the notional building are thus:

Total perimeter wall area = $(40 + 40 + 15 + 15) \times 10.5$	$= 1155.0\,m^2$
Area of windows and personnel doors = 0.30×1155	$= 346.5\,m^2$
Area of exposed wall = $1155 - 346.5 - 25$	$= 783.5\,m^2$
Area of personnel doors	$= 13.3\,m^2$
Area of vehicle doors	$= 25.0\,m^2$
Area of windows = $346.5 - 13.3$	$= 333.2\,m^2$
Area of rooflights = 0.10×600	$= 60.0\,m^2$
Area of roof = $600 - 60$	$= 540.0\,m^2$
Area of ground floor = 40×15	$= 600.0\,m^2$

Combining these with the maximum U-values allowed by the elemental method yields the heat loss table (Table 16.53). Note that the U-value of the ground floor in the notional building is 0.25 (the elemental standard) and not 0.20. This is because the value of 0.20 in the proposed building was achieved with added insulation. If it had been 0.20 *without* added insulation then it would have to be 0.20 in the notional building as well.

The results show that the proposed building has a lower heat loss rate than the notional building, and so the proposed building complies with regard to U-values and areas.

Table 16.53 Notional building heat loss table.

Element	Area m^2	U-value W/m^2K	Heat loss W/K
Roof	540.0	0.25	135.00
Rooflights	60.0	2.20	132.00
Exposed walls	783.5	0.35	274.23
Windows	333.2	2.20	733.04
Personnel doors	13.3	2.20	29.26
Vehicle loading bay doors	25.0	0.70	17.50
Ground floor	600.0	0.25	150.00
Totals	2355.0		1471.03
Average U-value, U_{AVG} = 1471.03/2355.0 = 0.625			

It is now necessary to consider the efficiency of the heating system. This can be done by calculating the carbon intensity of the heating system at 100% load and at 30% load, as follows:

At 100% output:

$$\text{Carbon intensity}, \varepsilon_c = \frac{1}{\Sigma R}\Sigma\left(\frac{RC_f}{\eta_t}\right) = \frac{1}{120}\left(\frac{60 \times 0.053}{0.78} + \frac{60 \times 0.053}{0.78}\right)$$

$$\varepsilon_c = 0.0679\,\text{kg/kWh}$$

At 30% output, with only the lead boiler operating:

$$\text{Carbon intensity}, \varepsilon_c = \frac{1}{\Sigma R}\Sigma\left(\frac{RC_f}{\eta_t}\right) = \frac{1}{36}\left(\frac{36 \times 0.053}{0.78}\right)$$

$$\varepsilon_c = 0.0679\,\text{kg/kWh}$$

Table 16.11 specifies maxima of 0.068 kg/kWh at 100% output and 0.065 kg/kWh at 30% output. The results show that the heating system is satisfactory at 100% output, but fails at 30% output. There are two ways of addressing this problem:

- Choose a more efficient lead boiler
- Check to see if boiler efficiency can be traded-off against fabric performance.

Considering the first of these options, assume that a more efficient lead boiler is available, say a condensing boiler, of efficiency 85%. The calculation at 30% load is now:

$$\text{Carbon intensity}, \varepsilon_c = \frac{1}{\Sigma R}\Sigma\left(\frac{RC_f}{\eta_t}\right) = \frac{1}{36}\left(\frac{36 \times 0.053}{0.85}\right)$$

$$\varepsilon_c = 0.0623\,\text{kg/kWh}$$

This is below the maximum of 0.065 kg/kWh and therefore the heating system complies.

Before considering taking action to implement the second option, it should be noted that the proposed building has a lower average U-value than the notional building. This lower value may already be sufficient. Thus:

$$\text{Required average U-value, } U_{req} = U_{ref}\frac{\varepsilon_{ref}}{\varepsilon_{act}} = 0.625 \times \frac{0.0650}{0.0679}$$

$$U_{req} = 0.598\,\text{W/m}^2\text{K}$$

From Table 16.52, the proposed building has an average U-value of 0.583 W/m^{2}K, and as this is less than U_{req} the heating system is satisfactory without changing either the fabric or the heating system.

16.11 Methods of meeting the lighting standards

Paragraphs 1.41 to 1.59 of AD L2 set out the lighting efficiency standards that are required within the elemental method. The standards have two aims:

- To ensure adequate efficiency of the lamps and luminaires in converting electricity to useful light
- To provide sufficient controls to ensure that the lighting is switched on only when it is required, i.e. when there is insufficient daylight and when the space is occupied.

In any lighting design, both of these aims must be kept in mind, but for clarity of explanation, it is convenient to consider them separately.

16.11.1 Lamp and luminaire efficiency

The efficiency of the lighting system is measured by its *luminous efficacy*. The basic requirement is that the initial luminous efficacy of the lamp and luminaire, averaged over the whole building, should be not less than 40 luminaire-lumens per circuit-watt. This criterion must be applied to office, industrial and storage buildings, but it can also be applied to any other type of building. Note that:

- Luminaire-lumens means the useful light emitted into the space by the lamp-luminaire combination; this is less than the light emitted by the lamp because some of the light is trapped and absorbed within the luminaire
- Circuit-watt means all the electrical power consumed by the lighting circuit, and includes the power consumed by ballasts, starters, control gear, etc.
- There is no restriction on the total amount of light which may be provided; only the efficiency is controlled

- There is no restriction on the way in which the luminaires distribute light within the space
- Up to 500 W of any form of lighting is exempt, and not included in any calculations
- Display lighting, emergency lighting and specialist local lighting are considered separately from the 40 luminaire-lumens per circuit-watt requirement.

There are two methods of meeting the 40 luminaire-lumens per circuit-watt requirement:

(1) Choose lamps and luminaires which are known to have a performance which is sufficient to meet the criterion.
(2) Calculate the initial luminous efficacy of the lamp and luminaire combination from photometric and electrical data.

Method 1 requires all luminaires in the building to have a light output ratio (LOR) of 0.6 or above *and* all luminaires to be fitted with lamps of any of the types given in Table 16.54. Note that in Appendix F of AD L2 this table refers to non-daylit areas of these buildings. However, as there is no separate guidance, it must be presumed that the table applies to daylit areas as well.

Table 16.54 Lamps for offices, industrial and storage buildings.

Lamp type	Lamp rating
High pressure sodium	All ratings above 70 W
Metal halide	All ratings above 70 W
Tubular fluorescent	All 26 mm diameter (T8) lamps and all 16 mm diameter (T5) lamps rated above 11 W, provided that all these lamps have low-loss or high frequency control gear
Compact fluorescent	All ratings above 26 W

Method 2 is used when either the luminaires or the lamps do not meet the requirements of method 1. In this case, it is necessary to calculate the initial luminaire efficacy, η_{lum}, using the formula given previously. This will be illustrated by example 16.11.1 below.

For all buildings other than offices, industrial units and storage buildings, it is occasionally impossible to meet the 40 luminaire-lumens per circuit-watt criterion because it may be necessary to use luminaires of unknown light output ratio, or to use less efficient lamps. In such circumstances, neither method 1 nor method 2 can be applied, and the alternative criterion of 50 lamp-lumens per circuit-watt must be used instead. The efficacy is 50 rather than 40 in order to compensate for the fact that, by specifying lamp-lumens instead of luminaire-lumens, the performance of the luminaires is being ignored. There are two methods of meeting the criterion:

- Calculate the average lamp-lumen efficacy from the lamp data
- Ensure that at least 95% of the installed lighting capacity uses lamps whose circuit efficacies are no worse than those given in Table 16.14.

These two methods are illustrated in Examples 16.11.2 and 16.11.3 below.

16.11.2 Lighting controls

The main text of AD L2 gives guidance on the selection of controls. In addition, when calculating η_{lum} it is necessary to know the control strategy in order to select a value for the luminaire control factor.

16.11.3 Example calculations

Example 16.11.1 Calculation of the average luminaire efficacy, η_{lum}
A building consists of a manufacturing unit, a storage area and offices. The offices are two storey, and the ceiling height in the other areas corresponds to the full height of the offices. The relevant data for the lighting design is:

(1) Manufacturing unit
- Usage – 7-day shift patterns
- Non-daylit – electric light only
- Controls – timed switching according to manufacturing shift patterns

(2) Storage area
- Usage – occasional use
- Non-daylit – electric lighting only
- Controls – manual switch on, automatic switch off by local absence detection

(3) Offices
- Usage – normally daytime only
- Daylit – window area 30% of office external wall area, glazed with clear low-E double glazing units; furthest luminaire less than 6 m from window
- Controls – local infrared switches

(4) Entrance foyer, corridors and toilets
- Usage – as for office
- Non-daylit – electric lighting only
- Controls – automatic on and off by occupancy sensing.

Comparing the specification for the lighting controls with the requirements of the approved document (see section 16.3.3 'Lighting controls', page 16.51) shows that they meet the requirements. Note, however, that if the offices had been glazed with tinted glass having a normal light transmittance 40%, the effective window area would be reduced to 30 × 40/70 = 17.1%. As this is less than 20%, the offices would not have been considered to be daylit.

The efficiency of the lamps and luminaires can be checked by calculating the overall luminaire efficacy, and comparing it with the required minimum of 40 luminaire-lumens per circuit-watt. Table 16.55 illustrates a convenient format for collecting the data and carrying out the calculation. The values for luminaire control factors are derived from Table 16.13 according to the control strategy of each area. The table shows that the total light output from the luminaries is 321820 lumens for a total power consumption of 6956 W. This yields an overall efficacy of η_{lum} = 46 luminaire-lumens per circuit-watt. As this exceeds the minimum value of 40, the lighting design satisfies the requirements of AD L2. In addition and independently of the calculation, the building can have a further 500 W of installed lighting.

Although it was not necessary in this case, it is useful to calculate the efficacy of each lamp and the overall luminaire efficacy of each luminaire and lamp combi-

Table 16.55 Lamp schedules for manufacturing and office building.

	Lamp schedule				
	1	**2**	**3**	**4**	**Totals**
Position	Production area	Offices	Storage	Circulation, toilets and foyer	
Description	250 W high bay metal halide	4 lamp 12 W fluorescent with aluminium Cat 2 louvres and high frequency control gear	58 W fluorescent with aluminium louvres and mains frequency control gear	24 W compact fluorescent mains frequency downlights	
Number, N	16	16	10	24	—
Circuit power per lamp, W	271	18 × 4 = 72	70	32	—
Light output Φ per lamp, lm	17000	1150 × 4 = 4600	4600	1800	—
Efficacy per lamp, lm/W	63	64	66	56	—
LOR of luminaire	0.80	0.57	0.60	0.40	—
Luminaire control factor, C_L	1.00	0.80	0.80	1.00	—
Total luminaire output, lm N.Φ.LOR/C_L	217600	52440	34500	17280	**321820**
Total circuit power, W	4336	1152	700	768	**6956**
Overall luminaire efficacy, lm/W	50	46	49	23	**46**

nation. These results are included in Table 16.55. It can be seen that all the selected lamps have efficacies above 50 lm/W, indicating that they are all of high efficiency. However, because of losses in the luminaires, the overall luminaire efficacy of the lighting for Schedule 4 (the circulation areas, toilets and foyer) is only 23 lm/W, well below the 40 lm/W criterion. If instead of being 46 the overall result had been less than 40, then the low result for lamp Schedule 4 in Table 16.55 indicates that this is where corrective action would have to be taken.

Example 16.11.2 Calculation of the average lamp efficacy

A new restaurant consists of a reception/bar area, a dining area, kitchens and circulation/toilet areas. It is desired to use decor lighting appropriate to the building's function, and so, in the bar and dining areas, it is proposed to use a combination of concealed perimeter lighting and local lighting over tables. Controls for the reception and dining areas will be by local switching from behind the bar, and lighting to all other areas will be by local switching. The building is neither an office, an industrial unit nor a storage building, and so compliance will be attempted by calculating the average lamp efficacy. Table 16.56 gives details of the lamp schedule for the whole building, and calculates the required results.

The average lamp efficacy is 45.0 lumens per circuit-watt. This is below the criterion of 50, and so the lighting scheme does not meet the requirements of AD L2. Inspection of the results for each lamp type shows that the very low efficacy for the tungsten lamps over the dining tables is responsible for the poor overall result. One

Table 16.56 Lamp schedules for restaurant.

	Lamp schedule				
	1	2	3	4	Totals
Position	Over dining tables	Concealed perimeter and bar lighting	Toilets and circulation	Kitchens	
Description	60 W tungsten	32 W T8 fluorescent high frequency ballast	18 W compact fluorescent mains frequency ballast	50 W T8 fluorescent high frequency ballast	
Number	30	22	8	8	—
Circuit power per lamp, W	60	36	23	56	—
Light output per lamp, lm	710	3300	1200	5200	—
Total circuit power, W	1800	792	184	448	**3224**
Total lamp light output, lm	21300	72600	9600	41600	**145100**
Efficacy per lamp, lm/W	11.8	91.7	52.2	92.9	**45.0**

solution would be to replace the tungsten lamps with compact fluorescent lamps of similar light output. The nearest would be 11 W rated lamps with a circuit power of 14 W and a light output of 690 lumens. Recalculating the table with these lamps over the dining tables raises the overall average lamp efficacy from 45.0 to 78.4 lumens per circuit-watt, thus complying with AD L2.

Example 16.11.3 Calculation of installed circuit power
A new sports hall consists of a sports area for badminton courts, changing room/toilet facilities, an entrance/reception area, and an office. The lighting schedule is:

- Sports playing area: 8 No. 100 W high pressure sodium downlighters
- Changing room/toilets: 12 No. 15 W compact fluorescent lamps
- Entrance/reception area: 10 No. ceiling recessed 50 W tungsten halogen downlighters
- Office: 2 No. 85 W 38 mm diameter, 2400 mm long tubular fluorescent.

Controls for the playing area, entrance/reception area and the office will be by manual control from the office. Controls for the changing room/toilets will be by occupancy detector. Table 16.57 gives details of the lamp schedules and calculations.

Table 16.57 Lamp schedules for sports hall.

	Lamp schedule				
	1	**2**	**3**	**4**	**Totals**
Position	Sports playing area	Changing rooms and toilets	Entrance and reception areas	Office	
Description	100 W high pressure sodium	18 W compact fluorescent	50 W tungsten halogen	85 W T12 linear fluorescent	
Number	8	12	10	2	—
Circuit power per lamp, W	115	23	50	103	—
Total circuit power, W	920	276	500	206	**1902**

Only the entrance/reception area is fitted with lamps which are not listed in Table 16.14. The percentage of circuit-watts consumed by these lamps is (500/1902) × 100 = 26.3%, and so only 73.7% of the installed circuit-watts is from the list of approved light sources. It is therefore necessary to choose different lamps for the entrance/reception area.

Note that the concession allowing an additional uncontrolled 500 circuit-watts of installed lighting applies only when using the 40 luminaire-lumens per circuit-watt criterion. It cannot be applied to the 500 W of tungsten halogen lamps proposed for the entrance and reception areas.

16.12 CPR calculations – methods for office buildings

The carbon performance rating (CPR) calculation method is a simple technique, derived from three sources, for assessing the amount of carbon emitted into the atmosphere in units of $kgC/m^2/year$. It is used within the elemental method for assessing the contribution to carbon emissions arising from the operation of mechanical ventilation and air conditioning systems. It is also used in the whole-building method for offices in order to assess the total emissions due to mechanical ventilation, air conditioning, heating and lighting.

16.12.1 Origins of the CPR method for office buildings

The three principal sources relevant to the CPR method are these.

- Energy Consumption Guide 19 *Energy use in offices* (ECON19) [54]. This provides benchmarks for energy consumption which have been derived from surveys of operational office buildings.
- *Energy assessment and reporting methodology (EARM): office assessment method* [45]. This provides a technique for estimating operational energy consumption, and comparing actual performance with ECON 19 benchmarks.
- CIBSE Guide *Energy efficiency in buildings* [36]. This provides a means of comparing services design with benchmarks of installed load and energy use.

The EARM has now been extended to include:

- Banks and agencies assessment method
- Hotels assessment method
- Mixed-use buildings assessment method.

16.12.2 The carbon performance rating for mechanical ventilation, $CPR_{(MV)}$

The value of $CPR_{(MV)}$ is found from four components.

- PD The total installed capacity of the fans which provide mechanical ventilation. This is expressed as the sum of the input kW ratings per square metre of the floor area of the treated space, kW/m^2.
- HD The typical annual equivalent hours of full load operation, assumed to be 3700 hours per year.
- CD The carbon emission factor, in kgC/kWh, for the fuel used to power the fans. This is nearly always electricity, for which CD = 0.113.
- FD A plant operating efficiency factor, which depends on provisions made to improve annual efficiency or on measures which reduce annual hours of use.

The equation for $CPR_{(MV)}$ is:

$$CPR_{(MV)} = PD \times HD \times CD \times FD$$

With HD = 3700 hours, and CD = 0.113, the equation for electrically powered fans is:

$$CPR_{(MV)} = 418 \times PD \times FD$$

16.12.3 The carbon performance rating for air conditioning and mechanical ventilation, $CPR_{(ACMV)}$

The value of $CPR_{(ACMV)}$ is found from PD, HD, CD and FD, as defined above, plus four more components:

- PR The total installed capacity of the plant that provides cooling or refrigeration. This is expressed as the sum of the input kW ratings per square metre of the floor area of the treated space, kW/m^2.
- HR The typical annual equivalent hours of full load operation of the cooling or refrigeration plant, assumed to be 1000 hours per year.
- CR The carbon emission factor, in kgC/kWh, for the fuel used to power the cooling or refrigeration plant. This is most often electricity (CD = 0.113) or natural gas (CD = 0.053).
- FR A plant control and management factor, which depends on provisions made to improve annual efficiency or on measures which reduce annual hours of use.

The equation for $CPR_{(ACMV)}$ is:

$$CPR_{(ACMV)} = (PD \times HD \times CD \times FD) + (PR \times HR \times CR \times FR)$$

Values for PD, PR, CD and CR

Values for PD and PR are determined by the ratings of the installed equipment. Values for CD and CR are taken from the carbon emission factors given in Table 16.12.

Values for FD

Values for FD depend on a combination of plant management features and monitoring and reporting features, and must be obtained from Table 16.58. In this table, values are selected from the most relevant column. If it is appropriate to select more than one value from that column, then the final value of FD is the product of the selected values.

Plant management features for FD (Table 16.58)
The plant management features in Table 16.58 should be interpreted as follows.

Operation in mixed mode with natural ventilation
If there are sufficient openable windows to provide the required internal environment by means of natural ventilation when external conditions permit, then this will

Table 16.58 The factor FD.

Plant management features	Monitoring and reporting features		
	Provision of energy metering of plant and/or metering of plant hours run, and/or monitoring of internal temperature in zones, plus the ability to draw attention to out-of-range values	**Provision of energy metering of plant and/ or metering of plant hours run, and/or monitoring of internal temperature in zones**	**No monitoring provided**
Operation in mixed mode with natural ventilation	0.85	0.90	0.95
Controls which restrict the hours of operation of distribution system	0.90	0.93	0.95
Efficient means of controlling air flow rate	0.75	0.85	0.95

only qualify as mixed mode operation if the perimeter zone exceeds 80% of the treated floor area. In addition, systems with cooling or refrigeration must have interlock controls to inhibit the air conditioning supply in zones where windows are open.

Controls which restrict the hours of operation of distribution system
This applies to controls capable of limiting plant operation to occupancy hours, with operation outside occupancy hours allowed only as necessary for efficient use of the system in order to:

- Control condensation, *or*
- Allow optimum start/stop control, *or*
- Allow the adoption of a night cooling strategy.

Efficient means of controlling air flow rate
This applies when a reduction in the air flow rate can be achieved with an efficient reduction in the input power to the fans. The reduced power may be achieved by means of motors with a variable speed control or by variable pitch fan blades. It does *not* apply to damper, throttle or inlet guide vane controls.

Values for FR

Values for FR also depend on a combination of plant management features and monitoring and reporting features, and must be obtained from Table 16.59. Values are selected from the most relevant column, and if it is appropriate to select more

Table 16.59 The factor FR.

Plant management features	Monitoring and reporting features		
	Provision of energy metering of plant and/or metering of plant hours run, and/or monitoring of internal temperature in zones, plus the ability to draw attention to out-of-range values	Provision of energy metering of plant and/ or metering of plant hours run, and/or monitoring of internal temperature in zones	No monitoring provided
Free cooling from cooling tower	0.90	0.93	0.95
Variation of fresh air using economy cycle or mixed mode operation	0.85	0.90	0.95
Controls to restrict hours of operation	0.85	0.90	0.95
Controls to prevent simultaneous heating and cooling in the same zone	0.90	0.93	0.95
Efficient control of plant capacity, including modular plant	0.90	0.93	0.95
Partial ice thermal storage	1.80	1.86	1.90
Full ice thermal storage	0.90	0.93	0.95

than one value from that column, then the final value of FR is the product of the selected values.

Plant management features for FR (Table 16.59)

Free cooling from cooling tower
This applies to systems that, when conditions allow, permit cooling to be obtained without the operation of the refrigeration equipment. Examples are the 'strainer cycle' and the 'thermosyphon'.

Variation of fresh air using economy cycle or mixed mode operation
This refers to systems that incorporate an economy cycle in which the mix of fresh and recirculated air is controlled by dampers, or to mixed mode operation as defined for Table 16.58.

Controls to restrict hours of operation
This applies to controls capable of limiting plant operation to occupancy hours, with operation outside occupancy hours allowed only as necessary for efficient use of the system in order to:

- Control condensation, *or*
- Allow optimum start/stop control, *or*
- Allow the adoption of a strategy to pre-cool the building overnight using outside air.

Controls to prevent simultaneous heating and cooling in the same zone
These are controls that include an interlock or dead band capable of precluding simultaneous heating and cooling in the same zone.

Efficient control of plant capacity, including modular plant
This refers to refrigerant plant capacity which is controlled on-line in such a way as to reduce input power in proportion to cooling demand while maintaining good part load efficiencies. Examples include modular plant with sequence controls, and variable speed compressors. It does *not* include hot gas bypass control.

Partial ice thermal storage
The chiller is intended to operate continuously, charging the ice store overnight and supplementing its output during occupancy.

Full ice thermal storage
The chiller only operates to recharge the thermal store overnight and outside occupancy hours.

Selecting and substituting the relevant factors in the equation gives the carbon performance rating (CPR) in kgC/m^2/year. This result is then compared with the maximum allowable CPR in Table 16.16. The process is demonstrated in Examples 16.12.1, 16.12.2 and 16.12.3 below.

16.12.4 The carbon performance rating and the whole-building method, $CPR_{(HLAC)}$

The whole-building CPR takes account of heating, lighting and air conditioning. In principle it is calculated in the same way as the CPR for air conditioning and mechanical ventilation, but with three additional sets of data, two for the heating system and one for the lighting. The equation for $CPR_{(HLAC)}$ is:

$$\begin{aligned} CPR_{(HLAC)} = {} & (PD \times HD \times CD \times FD) + (PR \times HR \times CR \times FR) \\ & + (PB \times HB \times CB \times FB) + (PH \times HH \times CH \times FH) \\ & + (PL \times HL \times CL \times FL) \end{aligned}$$

As before, the terms PD, HD, CD and FD refer to the mechanical ventilation system, and the terms PR, HR, CR and FR refer to the refrigeration plant. The additional terms are defined as follows:

- PB The total installed capacity of the heat raising plant, kW/m^2
- HB The typical annual equivalent hours of full load operation
- CB The carbon emission factor, in kgC/kWh, for the fuel used in the heat raising plant
- FB A plant management factor for the heat raising equipment.

- PH The total installed capacity of the heat distribution system, kW/m^2
- HH The typical annual equivalent hours of full load operation
- CH The carbon emission factor, in kgC/kWh, for the fuel used by the heat distribution system
- FH A plant management factor for heat distribution equipment

- PL The total installed power of the lighting system, kW/m^2
- HL The typical annual equivalent hours of full load operation
- CL The carbon emission factor, in kgC/kWh, for the fuel used by the lighting system. This is nearly always electricity for which CL = 0.113
- FL A plant management factor for the lighting system.

Values for PD, HD, CD, FD, PR, HR, CR and FR are as before. Values of PB, PH and PL are obtained from the ratings of the installed equipment, and the carbon emission factors CB, CH and CL are obtained from Table 16.12. For HB, HH and HL, typical annual equivalent hours of full load operation are:

HB = 1250 hours per annum
HH = 2500 hours per annum
HL = 2000 hours per annum

Values for the remaining parameters, FB, FH and FL are given in Tables 16.60, 16.61 and 16.62 respectively.

The three time and occupancy factors, TOF, are obtained from the following formulae:

$$TOF(A) = 1 - 0.15(TC + 2 \times OC - TC \times OC)$$
$$TOF(B) = 1 - 0.10(TC + 2 \times OC - TC \times OC)$$
$$TOF(C) = 1 - 0.05(TC + 2 \times OC - TC \times OC)$$

where TC is the fraction of building fitted with time-based control
OC is the fraction of building fitted with occupancy control, i.e. either with sufficient zonal switching to allow occupants to control lighting in their own workspace, or with occupancy-sensing control.

The daylight management factor, DMF, is obtained from:

$$DMF = 1.3 - 0.3 \times DA \times (DC + 1)$$

Table 16.60 The factor FB.

Plant management features	Monitoring and reporting features		
	Provision of energy metering of plant and/or metering of plant hours run, and/or monitoring of internal temperature in zones, plus the ability to draw attention to out-of-range values	Provision of energy metering of plant and/or metering of plant hours run, and/or monitoring of internal temperature in zones	No monitoring provided
Controls which restrict the hours of operation of boiler plant	0.75	0.80	0.85
Efficient capacity control of boiler plant	0.75	0.80	0.85
Controls to prevent simultaneous heating and cooling in the same zone	0.90	0.93	0.95

Adapted with permission from BRE Digest 457 [44]

Table 16.61 The factor FH.

Plant management features	Monitoring and reporting features		
	Provision of energy metering of plant and/or metering of plant hours run, and/or monitoring of internal temperature in zones, plus the ability to draw attention to out-of-range values	Provision of energy metering of plant and/or metering of plant hours run, and/or monitoring of internal temperature in zones	No monitoring provided
Controls which restrict the hours of operation of heating system	0.85	0.90	0.95
Efficient capacity control of distribution system	0.90	0.93	0.95

Adapted with permission from BRE Digest 457 [44]

where DC is the fraction of building fitted with daylight control
DA is the fraction of building which falls within the definition of a daylit space, i.e. any space within 6 m of a window wall provided that the glazed area is at least 20% of the internal area of the window wall, or a roof-lit space with a glazing area at least 10% of the floor area.

Table 16.62 The factor FL.

	Monitoring and reporting features		
Plant management features	**Provision of energy metering of lighting system and/or metering of hours of use, plus the ability to draw attention to out-of-range values**	**Provision of energy metering of lighting system and/or metering of hours of use**	**No monitoring provided**
Time-based lighting control and occupancy based lighting control	TOF(A)	TOF(B)	TOF(C)
Daylight-based lighting control and ability to utilise daylight	Daylight management factor, DMF		

Adapted with permission from BRE Digest 457 [44]

As with FD and FR, the final value of FB, FH or FL is found by multiplying together all applicable factors in the relevant column.

16.12.5 Example calculations

Example 16.12.1 Mechanical ventilation
In a new mechanically ventilated office building the total floor area treated by the ventilation system is 4200 m^2. The input power rating of the fans will be 80 kW. The fans will be driven by variable speed motors, with the speed being controlled from CO_2 sensors in the exhaust ducts. The controller will include a time function to limit operation to occupancy hours, but there is no separate metering of the ventilation plant. The required data is therefore:

$PD = 80/4200 = 0.019\ kW/m^2$
$HD = 3700$ hours
$CD = 0.113$ kgC/kWh for electricity
FD As there is no monitoring, FD must be chosen from the third column of Table 16.58. Two of the parameters in this column are relevant: the parameter for controls which restrict the hours of operation, and the parameter for efficient means of controlling air flow. Hence:

$$FD = 0.95 \times 0.95 = 0.9025$$

The CPR for the mechanical ventilation system in this building is thus:

$$CPR_{(MV)} = PD \times HD \times CD \times FD = 0.019 \times 3700 \times 0.113 \times 0.9025$$
$$CPR_{(MV)} = 7.17\ kgC/m^2/year$$

This must be compared with the maximum allowable CPR from Table 16.16. As the maximum for a new building is 6.5, this mechanical ventilation system fails the requirement. The result could be improved by providing separate metering of the power supplied to the fans and their control gear, and monitoring the run of plant hours. This means that the two parameters for FD would be chosen from column 2 of Table 16.58, giving:

$$FD = 0.93 \times 0.85 = 0.7905$$

The revised CPR is then:

$$CPR_{(MV)} = 0.019 \times 3700 \times 0.113 \times 0.7905 = 6.28 \text{ kgC/m}^2\text{/year}$$

This is now less than the target of 6.5, and so the mechanical ventilation system complies with the requirements of AD L2.

Example 16.12.2 Air conditioning

An air conditioning system is to be installed in a new air conditioned office building. The building does not have openable windows. Cooling will be provided by a pair of electrically powered speed-controlled compressors. Metering will be provided to measure the energy consumption of the refrigerant compressors, and the energy consumption of the fans used to distribute the cooled air will also be metered. Timing controls will also be provided to restrict the operation of the refrigeration and air distribution system to occupancy hours. The system data is:

Total treated floor area (TFA) = 3600 m^2
Total rated input power of compressors = 160 kW
Total rated input power of air distribution fans = 40 kW

The required data for the air distribution system is:

PD = 40/3600 = 0.011 kW/m^2
HD = 3700 hours
CD = 0.113 kgC/kWh for electricity
FD Both the compressors and the fans will be metered and so the components of FD are chosen from column 2 of Table 16.58 However, only the factor for restriction of the hours of operation applies, and so:

$$FD = 0.93$$

The required data for the refrigeration system is:

PR = 160/3600 = 0.044 kW/m^2
HR = 1000 hours
CR = 0.113 kgC/kWh for electricity
FR Both the compressors and the fans will be metered and so the components of FR are chosen from column 2 of Table 16.59. Two of the parameters in

this column are relevant: the parameter for controls which restrict the hours of operation, and the parameter for efficient control of plant capacity. Hence:

$$FD = 0.90 \times 0.93 = 0.837$$

The CPR for the air conditioning system in this building is thus:

$$CPR_{(ACMV)} = (PD \times HD \times CD \times FD) + (PR \times HR \times CR \times FR)$$
$$CPR_{(ACMV)} = (0.011 \times 3700 \times 0.113 \times 0.93) + (0.044 \times 1000 \times 0.113 \times 0.837)$$
$$CPR_{(ACMV)} = 4.28 + 4.16 = 8.44\ kgC/m^2/year$$

This is less than the target value of 10.3 $kgC/m^2/year$ for a new air conditioned building, and so the proposed design complies with the requirements of AD L2.

Example 16.12.3 Air conditioning

One area of a new office building is to be used as an archive area for the storage, retrieval and examination of ancient documents, many of which are of a delicate nature. The building will have a centralised air conditioning system, but the archive area will be treated as a separately controlled zone. Cooling will be provided by three electrically powered speed-controlled compressors. Metering will be provided to measure the energy consumption of the refrigerant compressors, and the energy consumption of the fans used to distribute the cooled air will also be metered. Timing controls will also be provided to restrict the operation of the refrigeration and air distribution system to occupancy hours in the office areas. The archive area requires continuous operation to maintain constant environmental conditions. The system data is:

Total treated floor area (TFA) = 2800 m^2
Total rated input power of compressors = 180 kW
Total rated input power of air distribution fans = 36 kW

The archive area can be considered as a significant process load, and so its floor area and associated plant capacity can be excluded from the calculation. The data for the archive area is:

Treated floor area (TFA) = 250 m^2
Estimated required input power of compressors = 22 kW
Estimated required input power of distribution fans = 6 kW

Subtracting the archive area data from that for the whole building gives the relevant data for the CPR calculation:

Relevant treated floor area (TFA) = 2800 − 250 = 2550 m^2
Relevant rated input power of compressors = 180 − 22 = 158 kW
Relevant rated input power of air distribution fans = 36 − 6 = 30 kW

The data for the air distribution system is therefore:

PD = 30/2550 = 0.012 kW/m^2
HD = 3700 hours
CD = 0.113 kgC/kWh for electricity
FD Both the compressors and the fans will be metered and so the components of FD are chosen from column 2 of Table 16.58. Two of the parameters in this column are relevant: the parameter for controls which restrict the hours of operation, and the parameter for efficient means of controlling air flow. Hence:
FD = 0.93 × 0.85 = 0.7905

The required data for the refrigeration system is:

PR = 158/2550 = 0.062 kW/m^2
HR = 1000 hours
CR = 0.113 kgC/kWh for electricity
FR Both the compressors and the fans will be metered and so the components of FR are chosen from column 2 of Table 16.59. Two of the parameters in this column are relevant: the parameter for controls which restrict the hours of operation, and the parameter for efficient control of plant capacity. Hence:

FD = 0.90 × 0.93 = 0.837

The CPR for the air conditioning system in this building is thus:

$$CPR_{(ACMV)} = (PD \times HD \times CD \times FD) + (PR \times HR \times CR \times FR)$$
$$CPR_{(ACMV)} = (0.012 \times 3700 \times 0.113 \times 0.7905) + (0.062 \times 1000 \times 0.113 \times 0.837)$$
$$CPR_{(ACMV)} = 3.97 + 5.86 = 9.83 \text{ kgC/m}^2\text{/year}$$

This is less than the target value of 10.3 kgC/m^2/year for a new air conditioned building, and so the proposed design complies with the requirements of AD L2.

Example 16.12.4 Whole-building CPR for a new air conditioned building
A new 10 storey office building is a conventional rectangular shape, 45 m × 15 m on plan, and 35 m in height. The office areas are arranged on either side of corridors which run centrally along the major axis of the building, so that the inner wall of all offices is within 6 m of the window wall. The total gross floor area is 6750 m^2, and the treated occupied area is 90% of this total. Glazing is restricted to the two main facades and is 40% of the area of these facades. The U-values are 0.3 W/m^{2}K for the external walls, an average of 2.0 W/m^{2}K for the glazing, 0.25 W/m^{2}K for the roof, and 0.2 W/m^{2}K for the ground floor.

Heating and cooling is provided via ceiling mounted cassettes, with hot water and chilled water supplied from a central plant room. Ventilation air is provided separately via a ducted system. The ventilation fans are fitted with variable speed drives which are controlled between a low (night-time) setting and maximum according to

CO_2 sensors in the exhaust ducts. Heating and hot water is provided by three identical gas fired boilers, and cooling is provided by a pair of chiller sets.

The lighting system is conventional, providing 500 lux in all areas, with daylight control in the office spaces (approximately 80% of the occupied area), and with time control in other areas. The total installed power of the lighting system is 110 kW. All equipment, space temperatures, etc. are controlled by a BMS system, which provides full monitoring including the ability to flag out-of-range values. However, the circulating pumps for hot water are not separately monitored from their parent equipment. There is no specific control to restrict hours of operation, as out-of-hours working is expected in parts of the building. The relevant details are:

Heating and hot water:	3 identical gas-fired boilers, each rated at 60 kW
Hot water distribution pumps:	total rated input power 20 kW
Lighting:	total installed load 110 kW
Ventilation:	total installed fan power 40 kW
Refrigeration:	2 identical chiller sets, each rated at 150 kW, including heat rejection fans, with an additional 20 kW for pumps supplying chilled water to the cassettes.

The CPR calculation proceeds as follows:

The treated floor area is 90% of 6750, i.e. 6075 m^2.

The boiler system

Total boiler input power = 3 × 60 = 180 kW
PB = 180/6075 = 0.0296 kW/m^2
HB = 1250 hours
CB = 0.053 kgC/kWh for gas
As full monitoring is installed, values are chosen from the first column. As there are three sequenced boilers there is efficient control of boiler capacity, and the cassettes are wired to prevent simultaneous heating and cooling. The two relevant factors are 0.75 and 0.90. Therefore:
FB = 0.75 × 0.90 = 0.675

The heat distribution system

PH = 20/6075 = 0.0033 kW/m^2
HH = 3700 hours
CH = 0.113 kgC/kWh for electricity
The pumps are not directly monitored, and so
FH = 0.95

The lighting system

PL = 110/6075 = 0.0181 kW/m^2
HL = 3000 hours
CL = 0.113 kgC/kWh for electricity

As there is full monitoring via the BMS, the equation for time and occupancy based control is TOF(A). The office areas, which account for 80% of floor area, are all occupancy controlled, and so OC = 0.80. The remaining 20% of floor area is timed, and so TC = 0.20. Thus:

$TOF(A) = 1 - 0.15(0.2 + 2 \times 0.8 - 0.2 \times 0.8) = 0.754$

The 80% of the floor area which is office space is all within 6 m of a window wall that is glazed to 40% of its area. These areas are therefore within the definition of being daylit, and they are also fitted with daylight control. Hence DA = 0.8 and DC = 0.8, and so:

$DMF = 1.3 - 0.3 \times 0.8(0.8 + 1) = 0.868$

$FL = 0.754 \times 0.868 = 0.654$

The air distribution system

The total installed fan power is 40 kW

$PD = 40/6075 = 0.0066\ kW/m^2$

HD = 3700 hours

CD = 0.113 kgC/kWh for electricity

There is no contribution from natural ventilation and the controls do not restrict hours of operation. On the other hand, the variable speed drives provide efficient control of air flow rate, and so:

FD = 0.75

The refrigeration system

$PR = (300 + 20)/6075 = 0.0527\ kW/m^2$

HR = 1000 hours

CR = 0.113 kgC/kWh for electricity

The controls prevent simultaneous heating and cooling in the same zone, and provide efficient control of plant capacity, and so:

$FR = 0.9 \times 0.9 = 0.81$

The CPR calculation can be arranged most conveniently in tabular form as in Table 16.63. The result of CPR = 13.54 kgC/m^2/year is well within the target of 18.5 kgC/m^2/year for a new air-conditioned office, and so the building complies. The final column of the calculation table is useful in identifying the most significant contributions to the result. In this case, the refrigeration and lighting systems were the

Table 16.63 Whole-building CPR calculation.

		P kW/m^2	H hours	C kgC/kWh	F	$P \times H \times C \times F$ kgC/m^2/yr
Boiler system	B	0.0296	1250	0.053	0.675	1.324
Heat distribution system	H	0.0033	3700	0.113	0.950	1.311
Lighting system	L	0.0181	3000	0.113	0.654	4.013
Cool/vent dist. system	D	0.0066	3700	0.113	0.750	2.070
Refrigeration system	R	0.0527	1000	0.113	0.810	4.824
					CPR =	13.542

largest. In the case of a building which failed the target, the final column would help to identify those systems which need attention in order to improve the result.

16.13 Solar overheating calculations

Appendix H of AD L2 describes a calculation method suitable for the second of the three criteria for assessing solar overheating: by specifying a maximum solar heat load per unit floor area of 25 W/m^2. In order to carry out the calculation, the building should be divided into zones and the calculation carried out for each zone. The result for every zone must be no greater than the maximum of 25 W/m^2.

16.13.1 Definitions

Perimeter zones

A perimeter zone is the space between a window wall (or walls) and a boundary drawn up to a maximum of 6 m from the wall (or walls). All windows, part windows, rooflights and part rooflights within a perimeter zone must be included in the calculations.

Interior zones

An interior zone is the space between the internal boundaries of perimeter zones and non-window walls, or between the internal boundaries of two or more perimeter zones. All rooflights and part rooflights within an interior zone must be included in the calculations.

Parameters and equations

The parameters used are as follows:

Q_{slw} the solar load from windows, per unit of the zone floor area (W/m^2)
Q_{slr} the solar load from rooflights, per unit of the zone floor area (W/m^2)

It is necessary to define Q_{slw} and Q_{slr} separately because they require different equations for their evaluation.

A_z the floor area of a zone (m^2)
A_g the area of glazed opening in the window wall(s) of a perimeter zone (m^2)
A_r the total area of rooflight(s) in a perimeter or interior zone (m^2)
q_{sw} the solar load due to a window, selected from Table 16.64 (W/m^2 of glazing)
q_{sr} the solar load due to a rooflight or similar horizontal opening (W/m^2 of the area of the rooflight or opening)
f_c the correction factor for the glazing/blind combination, selected from Table 16.65

Table 16.64 Average solar load, windows.

Orientation	q_{sw} **W/m²**
N	125
NE/NW	160
E/W	205
SE/SW	198
S	156

Table 16.65 Correction factor for glazing/blind combinations.

Glazing/blind combination (from inside to outside)	**Correction factor** f_c
Blind/clear/clear	0.95
Blind/clear/reflecting	0.62
Blind/clear/absorbing	0.66
Blind/low-e/clear	0.92
Blind/low-e/reflecting	0.60
Blind/low-e/absorbing	0.62
Clear/blind/clear	0.69
Clear/blind/reflecting	0.47
Clear/blind/absorbing	0.50
Clear/clear/blind/clear	0.56
Clear/clear/blind/reflecting	0.37
Clear/clear/blind/absorbing	0.39
Clear/clear/blind	0.57
Clear/clear/clear/blind	0.47

f_{rw} the framing ratio for windows (default value $f_{rw} = 0.1$ for vertical windows)
f_{rr} the framing ratio for rooflights (default value $f_{rr} = 0.3$ for horizontal rooflights)

The equations for the solar loads are then:

Windows $$Q_{slw} = \frac{1}{A_z} \Sigma A_g q_{sw} f_c (1 - f_{rw})$$

Rooflights $$Q_{slr} = \frac{1}{A_z} A_r q_{sr} f_c (1 - f_{rr})$$

16.13.2 Sources of data for the parameters

Values of q_{sw} and q_{sr}

The solar loads, q_{sw} and q_{sr}, are averages between 07.30 in the morning and 17.30 in the evening. A single value of q_{sr} is suitable for all horizontal surfaces (i.e. rooflights):

$$q_{sr} = 327\,W/m^2$$

For vertical surfaces (i.e. windows) the value of q_{sw} depends on orientation and must be obtained from Table 16.64.

Values of f_c

The correction factor for the glazing/blind combination, f_c, may either be selected from Table 16.65, or obtained from the appropriate shading coefficients.

The shading coefficients for various glazing and shading device combinations are often provided by manufacturers. For the purposes of Appendix H they are defined as:

S_c The ratio of the instantaneous solar heat gain through the glazing and its shading device at normal incidence to the instantaneous solar heat gain through clear unshaded 4 mm glass.
S_{cf} As S_c, but for glazing with a fixed shading device.
S_{ctot} As S_c, but for glazing with fixed and moveable shading devices.

There are three equations which relate these shading coefficients to the correction f_c:

- For fixed shading, including units with absorbing or reflecting glass:

$$f_c = \frac{S_c}{0.7}$$

- For moveable shading:

$$f_c = \frac{1}{2}\left(1 + \frac{S_c}{0.7}\right)$$

- For a combination of fixed and moveable shading:

$$f_c = \frac{S_{cf} + S_{ctot}}{1.4}$$

These equations maybe used to find f_c, or they may be inverted in order to determine the shading coefficient, and hence the glazing/shading device combination, which is required to meet the 25 W/m^2 criterion.

Values of f_{rw} and f_{rr}

If there is no information on the framing ratio of a window, or it is not possible to evaluate it from the dimensions, the default values should be used.

16.13.3 Example calculation

Part of the top floor of a college building is used as a studio for student art classes. The room is 14 m x 7 m on plan, with a floor to ceiling height of 3 m. One of the long walls is an external wall facing south-west with four windows, each 2.5 m wide by 1.5 m high. The window frames have a framing ratio of 20% of the window area. Daylight along the long internal wall is provided by a line of six horizontal rooflights, each 1 m square, with a framing ratio of 25%. The centre line of the rooflights is 1 m from the internal wall, as shown in Figure 16.28. Both the windows and the rooflights are double glazed with clear glass and an internal blind. The room can be divided into two zones. A perimeter zone can be formed by the external wall, the two end walls, and a line drawn 6 m from and parallel to the external wall. The remaining space between the perimeter zone and the inner long wall is an interior zone. Exactly half of the rooflight area falls within the perimeter zone, and the other half within the interior zone. Each zone must be considered separately.

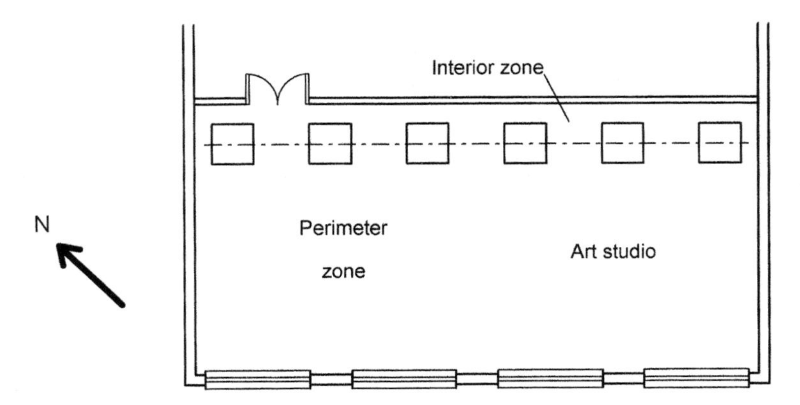

Fig. 16.28 Solar gain to a college art studio.

Perimeter zone

Area of zone $A_z = 14 \times 6 = 84\,m^2$

Area of windows $A_g = 4 \times 2.5 \times 1.5 = 15\,m^2$

Solar load, south-west window $q_{sw} = 198\,W/m^2$

Glazing/blind correction factor $f_c = 0.95$

Framing ratio for windows $f_{rw} = 0.20$

Solar load from windows $Q_{slw} = \frac{1}{84}(15 \times 198 \times 0.95 \times [1 - 0.20]) = \frac{2257.2}{84}$

$Q_{slw} = 26.9\ W/m^2$

Area of rooflights $A_r = \frac{1}{2} \times 6 \times 1 \times 1 = 3\,m^2$

Solar load, rooflights $q_{sr} = 327\,W/m^2$

Glazing/blind correction factor $f_c = 0.95$

Framing ratio for rooflights $f_{rr} = 0.25$

Solar load from rooflights $Q_{slr} = \frac{1}{84}(3 \times 327 \times 0.95 \times [1 - 0.25]) = \frac{699.0}{84}$

$Q_{slr} = 8.3\ W/m^2$

Total solar load for perimeter zone $= 26.9 + 8.3 = 35.2\,W/m^2$

Interior zone

Area of zone $A_z = 14 \times 1 = 14\,m^2$

Area of rooflights $A_r = \frac{1}{2} \times 6 \times 1 \times 1 = 3\,m^2$

Solar load, rooflights $q_{sr} = 327\,W/m^2$

Glazing/blind correction factor $f_c = 0.95$

Framing ratio for rooflights $f_{rr} = 0.25$

Solar load from rooflights $$Q_{slr} = \frac{1}{14}(3 \times 327 \times 0.95 \times [1 - 0.25]) = \frac{699.0}{14}$$

$$Q_{slr} = 49.9\ W/m^2$$

Total solar load for interior zone $= 49.9\,W/m^2$

The results for both zones exceed the limiting value of $25\,W/m^2$, and so the window and rooflight design must be reconsidered. There are several possible courses of action, but in view of the room's function as an art studio, it may be inappropriate to use tinted or reflecting glass. One possible solution would be to specify, for both windows and rooflights, a double glazing system using clear glass, but with the blind fitted between the panes, for which f_c is 0.69. This, however, would not give a sufficient reduction to the result for the interior zone. Therefore, in addition, the line of rooflights could be moved an extra 250 mm away from the inner wall towards the window wall, so that three-quarters of their area is in the perimeter zone, and only one quarter in the interior zone. The revised calculations are as follows.

Perimeter zone

Area of zone $A_z = 14 \times 6 = 84\,m^2$

Area of windows $A_g = 4 \times 2.5 \times 1.5 = 15\,m^2$

Solar load, south-west window $q_{sw} = 198\,W/m^2$

Glazing/blind correction factor $f_c = 0.69$

Framing ratio for windows $f_{rw} = 0.20$

Area of rooflights $A_r = \frac{3}{4} \times 6 \times 1 \times 1 = 4.5\,m^2$

Solar load, rooflights $q_{sr} = 327\,W/m^2$

Glazing/blind correction factor $f_c = 0.69$

Framing ratio for rooflights $f_{rr} = 0.25$

Solar load from windows $$Q_{slw} = \frac{1}{84}(15 \times 198 \times 0.69 \times [1 - 0.20]) = \frac{1639.4}{84}$$

$$Q_{slw} = 19.5\ W/m^2$$

Solar load from rooflights $$Q_{slr} = \frac{1}{84}(4.5 \times 327 \times 0.69 \times [1 - 0.25]) = \frac{761.5}{84}$$

$$Q_{slr} = 9.1\ W/m^2$$

Total solar load for perimeter zone $= 19.5 + 9.1 = 28.6\,W/m^2$

Interior zone

Area of zone $A_z = 14 \times 1 = 14\,m^2$

Area of rooflights $A_r = \frac{1}{4} \times 6 \times 1 \times 1 = 1.5\,m^2$

Solar load, rooflights $q_{sr} = 327\,W/m^2$
Glazing/blind correction factor $f_c = 0.69$
Framing ratio for rooflights $f_{rr} = 0.25$

Solar load from rooflights $Q_{slr} = \frac{1}{14}(1.5 \times 327 \times 0.69 \times [1 - 0.25]) = \frac{253.8}{14}$

$Q_{slr} = 18.1\ W/m^2$
Total solar load for interior zone $= 18.1\,W/m^2$

The interior zone is now well within the limiting value, but the result for the perimeter zone is still too high. This suggests a variation on the above solution. If the same clear-blind-clear double glazing units are used, but the rooflights are left in their original position with their centre line 1 m from the inner wall, it is possible to manipulate the calculation by drawing the boundary of the perimeter zone 5 m from the window wall instead of the maximum permissible 6 m. The interior zone then becomes 14 m × 2 m and the rooflights fall entirely within it. The calculation becomes as follows.

Perimeter zone
Area of zone $A_z = 14 \times 5 = 70\,m^2$
Area of windows $A_g = 4 \times 2.5 \times 1.5 = 15\,m^2$
Solar load, south-west window $q_{sw} = 198\,W/m^2$
Glazing/blind correction factor $f_c = 0.69$
Framing ratio for windows $f_{rw} = 0.20$

Solar load from windows $Q_{slw} = \frac{1}{70}(15 \times 198 \times 0.69 \times [1 - 0.20]) = \frac{1639.4}{70}$

$Q_{slw} = 23.4\ W/m^2$
Solar load from rooflights $Q_{slr} = 0\,W/m^2$
Total solar load for perimeter zone $= 23.4 + 0 = 23.4\,W/m^2$

Interior zone
Area of zone $A_z = 14 \times 2 = 28\,m^2$
Area of rooflights $A_r = 6 \times 1 \times 1 = 6\,m^2$
Solar load, rooflights $q_{sr} = 327\,W/m^2$
Glazing/blind correction factor $f_c = 0.69$
Framing ratio for rooflights $f_{rr} = 0.25$

Solar load from rooflights $Q_{slr} = \frac{1}{28}(6 \times 327 \times 0.69 \times [1 - 0.25]) = \frac{1015.3}{28}$

$Q_{slr} = 36.2\ W/m^2$
Total solar load for interior zone $= 36.2\,W/m^2$

The perimeter zone is now just within the limiting value, but the interior zone is too high. However, it can be seen that the interior zone can be brought within target by reducing the number of rooflights. The required reduction in the solar load is approximately two thirds, and so four rooflights instead of six should have the desired result. Thus for the interior zone:

Solar load from rooflights $$Q_{slr} = \frac{1}{28}(4 \times 327 \times 0.69 \times [1 - 0.25]) = \frac{676.9}{28}$$

$$Q_{slr} = 24.2 \text{ W/m}^2$$

Both zones are now acceptable.

References

BRE Building Research Establishment, Watford, Herts WD25 9XX

BRECSU Building Research Energy Conservation & Support Unit, Watford, Herts WD25 9XX

CRC Construction Research Communications Ltd, Watford, Herts WD25 9XX

CIBSE Chartered Institution of Building Services Engineers, 222 Balham High Road, London SW12 9BS

1. *Thermal insulation: Avoiding risks.* BR 262, BRE/CRC, 2002.
2. *Limiting thermal bridging and air leakage: Robust construction details for dwellings and similar buildings.* The Stationery Office, 2001.
3. *Building materials and products – Hygrothermal properties – Tabulated design values.* BS EN 12524: 2000.
4. *GuideA: Environmental design, Section A3: Thermal properties of building structures.* CIBSE, 1999.
5. *Thermal performance of building materials and products – Determination of thermal resistance by means of guarded hot plate and heat flow meter methods – Dry and moist products of low and medium thermal resistance.* BS EN 12664: 2001.
6. *Thermal performance of building materials and products – Determination of thermal resistance by means of guarded hot plate and heat flow meter methods – Products of high and medium thermal resistance.* BS EN 12667: 2001.
7. *Thermal performance of building materials and products – Determination of thermal resistance by means of guarded hot plate and heat flow meter methods – Thick products of high and medium thermal resistance.* BS EN 12939: 2001.
8. *Thermal insulation – Determination of steady-state thermal transmission properties – Calibrated and guarded hot box.* BS EN ISO 8990: 1996.
9. *Thermal performance of windows and doors – Determination of thermal transmittance by hot box method – Part 1: Complete windows and doors.* BS EN ISO 12567-1: 2000.
10. *Thermal performance of buildings – Transmission loss coefficient – Calculation method.* BS EN ISO 13789: 1999.
11. *Conventions for the calculation of U-values.* BRE/CRC, 2002.
12. *Building components and building elements – Thermal resistance and thermal transmittance – Calculation method.* BS EN ISO 6946: 1997.
13. *Thermal performance of buildings – Heat transfer via the ground – Calculation methods.* BS EN ISO 13370: 1998.

14. *Thermal performance of windows, doors and shutters – Calculation of thermal transmittance – Part 1: Simplified methods*. BS EN ISO 10077-1: 2000.
15. *Thermal performance of windows, doors and shutters – Calculation of thermal transmittance – Part 2: Numerical method for frames*. prEN ISO 10077-2.
16. *Basements for dwellings*. Approved Document, ISBN 0-7210-1508-5, British Cement Association and National House Building Council, 1997.
17. *Thermal bridges in building construction – Calculation of heat flows and surface temperatures – Part 1: General methods*. BS EN ISO 10211-1: 1996.
18. *Thermal bridges in building construction – Calculation of heat flows and surface temperatures – Part 2: Linear thermal bridges*. BS EN ISO 10211-2: 2001.
19. *SAP: The Government's Standard Assessment Procedure for energy rating of dwellings*. BRECSU, 2001. (The most up-to-date edition is available at www.bre.co.uk)
20. *Lighting for buildings: Code of practice for daylighting*. BS 8206: 1992; Part 2
21. *Testing buildings for air leakage*. TM23, CIBSE, 2000.
22. *Copper indirect cylinders for domestic purposes – Specification for double feed indirect cylinders*. BS 1566-1: 1984.
23. *Specification for copper direct cylinders for domestic purposes*. BS 699: 1884.
24. *Specification for copper hot water storage combination units for domestic purposes*. BS 3198: 1981.
25. *Specification for unvented hot water storage units and packages*. BS 7206: 1990.
26. *Performance specification for thermal stores*. Waterheater Manufacturers Association, 1999.
27. *Controls for domestic central heating and hot water systems*. GPG 302, BRECSU, 2001.
28. *Specification for installation in domestic premises of gas-fired ducted air heaters of rated output not exceeding 60 kW*. BS 5864: 1989.
29. *Method for specifying thermal insulating materials for pipes, tanks, vessels, ductwork and equipment operating within the temperature range* $-40°C$ *to* $+700°C$. BS 5422: 2001.
30. *Low energy domestic lighting*. GIL 20, BRECSU, 1995.
31. HETAS Limited, PO Box 37, Cheltenham GL52 9TB.
32. *Energy efficient refurbishment of existing housing*. GPG 155, BRECSU.
33. *Planning and the historic environment*. Planning Policy Guidance PPG15, DoE/DNH, September 1994. (In Wales, refer to *Planning guidance Wales planning policy first revision* 1999, and *Planning and historic environment: Historic buildings and conservation areas*, Welsh Office Circular 61/96.)
34. *The principles of the conservation of historic buildings*. BS 7913: 1998.
35. Information Sheet 4: *The need for old buildings to breathe*. Society for the Protection of Ancient Buildings, 1986.
36. *Energy efficiency in buildings*. CIBSE, 1999.
37. Technical Note 14: *Guidance for the design of metal cladding and roofing to comply with Approved Document L*. Metal Cladding and Roofing Manufacturers Association, 2002.
38. Guide A: *Environmental design*, section A5: *Thermal response and plant sizing*. CIBSE, 1999.
39. CHPQA Standard: *Quality assurance for combined heat and power, issue 1*. DETR, November 2000.
40. *The designer's guide to energy-efficient buildings for industry*. GPG 303, BRECSU, 2000.
41. *Heating controls in small commercial and multi-residential buildings*. GPG 132, BRECSU, 2001.
42. Guide H: *Building control systems*. CIBSE, 2000.

43. *Photoelectric control of lighting: Design, set-up and installation issues*. IP 2, BRE/CRC, 1999.
44. *The Carbon Performance Rating for offices*. Digest 457, BRE/CRC, 2001.
45. *Energy assessment and reporting methodology: Office assessment method*. TM22, CIBSE, 1999.
46. Variable flow control. GIR 41, BRECSU, 1996.
47. *Guidance for environmental design in schools*. Building Bulletin 87, The Stationery Office, 1997.
48. *Achieving energy efficiency in new hospitals*. NHS Estates, The Stationery Office, 1994.
49. *Building energy and environmental modelling*. AM11, CIBSE, 1998.
50. *A practical guide to infra-red thermography for building surveys*. BR 176, BRE/CRC, 1991.
51. Commissioning Code A: *Air distribution systems* (1996), Commissioning Code B: *Boiler plant* (1975), Commissioning Code C: *Automatic controls* (2000), Commissioning Code R: *Refrigerating systems* (1991), Commissioning Code W: *Water distribution systems* (1994). CIBSE, 1975–2000.
52. Technical memorandum 1: *Standard specification for the commissioning of mechanical engineering services installations for buildings*. Commissioning Specialists Association, 1999.
53. *Sub-metering new build non-domestic buildings: A guide to help designers meet Part L of the Building Regulations*. GIL 65, BRECSU, 2001.
54. Energy consumption guide 19 (ECON 19): *Energy use in offices*. DETR, 1998.
55. *Assessing the effects of thermal bridging at junctions and around openings*. IP 17/01, BRE/CRC, 2001.

Other sources of information

Avoiding or minimising the use of air conditioning. GIR 31, BRECSU, 1995.
Metal cladding: Assessing thermal performance. IP 5/98, BRE/CRC, 1998.
U-value calculation procedure for light steel frame walls. BRE/CRC, 2002.
Solar shading of buildings. BR 364, BRE/CRC, 1999.
Central heating specifications (CHeSS). GIL 59, BRECSU, 2000
Guide to good practice for assessing glazing frame U-values. Centre for Window and Cladding Technology, 1998.
Guide to good practice for assessing heat transfer and condensation risk for a curtain wall. Centre for Window and Cladding Technology, 1998.
Guide for assessment of the thermal performance of aluminium curtain wall framing. Council for Aluminium in Building, 1996.

17 Access and facilities for disabled people

17.1 Introduction

The law concerning access for disabled people to buildings has a relatively short history. The first provisions were contained in the Chronically Sick and Disabled Persons Act 1970. These provisions were mostly advisory and were only applied if it was reasonably practicable to do so. There were no enforcement powers contained in the Act and it proved to be rather ineffective. It was clear that some form of legislation with 'teeth' was required.

Therefore, it was considered that the Building Regulations were the most suitable medium for any future legislation. This resulted in the fourth amendment to the 1976 Regulations which introduced Part T, *Facilities for disabled people,* in August 1985. Since then the former Part T has been recast in the current format as Part M (supported by Approved Document M) and has been the subject of three further revisions to extend its scope and coverage.

The most recent amendments to Part M (and Approved Document M) mean that requirements M1 to M3 apply to new dwellings as well as to other types of buildings. The new regulations came into force on 25 October 1999.

When dealing with access requirements for disabled people reference should also be made to the Disability Discrimination Act 1995 (see Chapter 5, section 5.20).

17.2 Interpretation

DISABLED PEOPLE – This is a narrowly defined term in regulation M1 which applies to those people with:

- a physical impairment that limits their ability to walk or makes them dependent on a wheelchair for mobility, or
- impaired sight or hearing.

A number of other terms are defined in Approved Document M, that apply throughout the document as follows:

ACCESS – Approach or entry.

ACCESSIBLE – Access is facilitated for disabled people to buildings or parts of buildings.

BUILDING – In addition to dwellings (see section 17.11 below) the rules in AD M apply to the following buildings:

- shops, offices, factories and warehouses
- schools, other educational establishments and student residential accommodation in traditional halls of residence
- institutions
- premises which admit the public whether on immediate payment, subscription, fee or otherwise.

The rules apply to the whole building and equally to any parts of the building that comprise separate individual premises.

SUITABLE – Means of access and facilities that are designed for the use of disabled people.

PRINCIPAL ENTRANCE STOREY – The storey of a building that contains the main entrance or entrances.

Sometimes it may be necessary to provide an alternative, accessible entrance into the building (see below). In this case the storey containing the alternative entrance would be the principal entrance storey.

17.3 Application

Part M of Schedule 1 to the 2000 Regulations applies to the following:

(a) New buildings (including dwellings) or buildings which have been demolished to leave only the external walls standing.

In buildings, other than dwellings, which have been substantially reconstructed, it may prove impractical to provide an accessible entrance suitable for disabled people. Where this is the case AD M still recommends that the other requirements of Part M be applied.

(b) extensions to existing buildings (but not to dwellings, since extensions to these are excluded from Part M) if the extension has a ground storey.

The extension must comply with the requirements of Part M but there is no obligation to bring the existing building up to the standards. On the other hand the extension must not adversely affect the existing building with regard to the provision of Part M requirements which may already exist. Where the extension is accessed through the existing building then it would be unreasonable to expect a higher standard of access than that provided in the existing building. However, if the extension is capable of being independently approached and entered from the boundary of the site, then it should be treated as if it were a new building.

(c) those external features which are needed to provide access from the edge of the site to the building and from any car parking within the site.

Part M does not apply to the following.

- Material alterations – although facilities which existed before the alterations were carried out must not be made worse as a consequence of the alterations. For example, existing sanitary conveniences provided for disabled people may be moved to another, equally accessible, location in the building, but they may not be removed.
- Any part of a building used solely for inspection, maintenance or repair of the building or its services or fittings.
- *Any* extension to an existing dwelling, and an extension to any other building which does not have a ground storey (e.g. where a new floor is added to an existing building).
- Additionally, Part M4 – Audience or spectator seating – does not apply to dwellings.

17.4 Buildings other than dwellings

17.4.1 The main provisions

Reasonable provision must be made in buildings for:

- Disabled people to gain access to and use their facilities.
- Sanitary conveniences suitable for disabled people in any building where sanitary conveniences are provided.
- A reasonable number of wheelchair spaces, where the building contains audience or spectator seating.

Therefore, in buildings other than dwellings, it should be reasonably safe and convenient for disabled people to:

- Approach the principal entrance (or other entrances permitted by AD M) to the building from the edge of the site or from car parking within the site.
- Gain access into the building.
- Gain access within the building.
- Use the facilities which are provided in the building (including sanitary conveniences and accommodation for disabled people within audience or spectator seating).
- Enjoy the use of aids to communication if they have impaired hearing or sight in auditoria, meeting rooms, reception areas and ticket offices.

It should be noted that the provision of access and facilities for disabled people in buildings is not only for those who work there, but also for visitors to the building.

17.5 Means of access

17.5.1 The approach to the building

Disabled people should be able to reach the principal entrance into the building or any other entrances that are provided (see Access to the building, below).

Access should be provided from the entrance into the site curtilage or from any car parking that is provided for disabled people within the building site.

Disabled people should also be able to get from one building to another on the site.

The following recommendations are given in AD M regarding approach to the building.

- It should be level where possible and certainly no steeper than 1 in 20 since disabled people have difficulty negotiating changes in level and people with impaired vision may be unaware of abrupt level changes.
- It should have a surface width of at least 1200 mm in order to provide adequate space for wheelchairs or helpers and for people passing in the opposite direction.
- Tactile warnings should be provided on pedestrian routes for people with impaired vision where the route crosses a vehicular carriageway or at the top of steps. Examples of typical tactile pavings are shown in Fig. 17.1.
- If it needs to be steeper than 1 in 20 then a proper ramped approach should be provided. (See Ramped access section 17.5.3 below and Fig. 17.2.)
- Dropped kerbs should be provided for wheelchair users at carriageway crossings.
- Many ambulant disabled people find it easier to negotiate steps than ramps, therefore, where possible easy-going steps should complement a ramped approach. (See Complementary steps section 17.5.4 below and Fig. 17.2.)
- Care should be taken to avoid hazards to people with impaired sight when they use access routes close to the building. (See Avoidance of hazards section 17.5.5 below.)

17.5.2 Access to the building

In general, buildings should be designed so that there is a convenient access into the building suitable for disabled people. This rule applies equally to visitors or staff whether they arrive on foot or in a wheelchair.

If it is necessary to provide separate entrances for visitors and customers or staff, each entrance should be suitable and accessible.

In certain cases it may not be possible to make the main entrance accessible due to space restriction and congestion or sloping ground. Sometimes, car parking spaces are provided in areas where access to the principal entrance is not possible. In these cases an additional accessible entrance may be provided, but this should also be for general use and should give suitable internal access to the principal entrance.

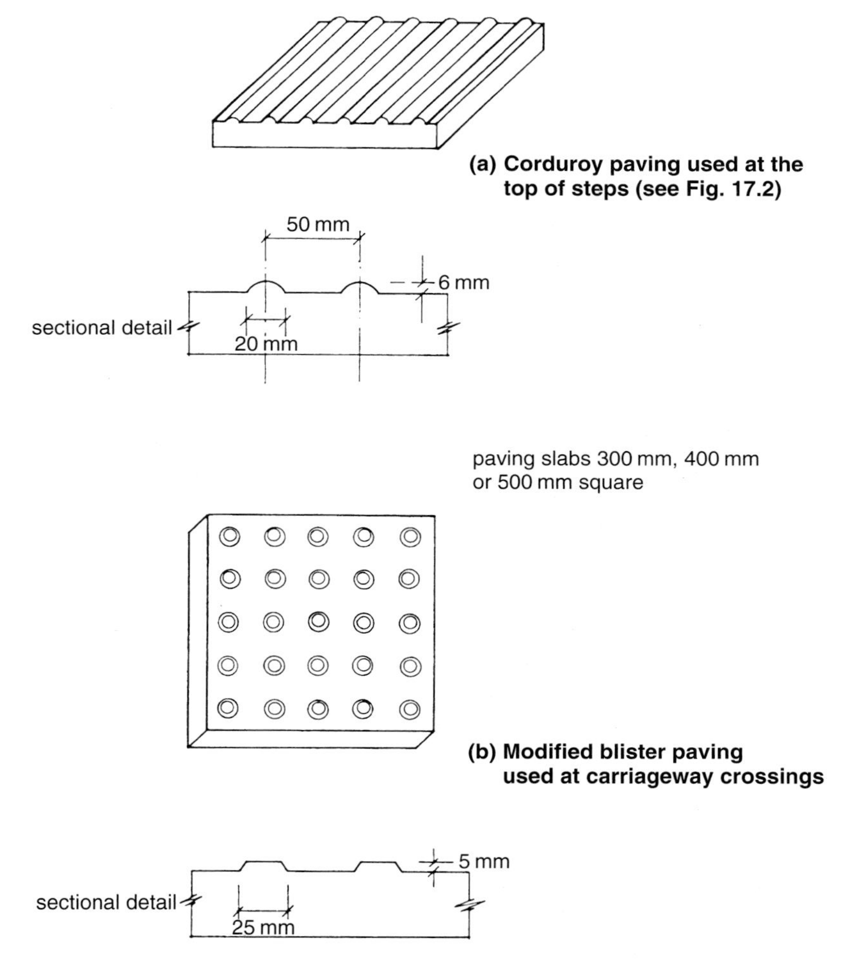

Fig. 17.1 Tactile pavings.

17.5.3 Ramped access

Steep ramps are difficult for wheelchair users to negotiate and they may add to the difficulties of an ambulant disabled person by increasing their unsteadiness in adverse weather conditions or by becoming slippery.

Equally, long ramps create difficulties since disabled people and their helpers may need to make frequent stops to regain their strength or breath, or to ease pain. Also, adequate space is needed for people passing and for the negotiation of door openings.

The recommendations for a ramped approach to the building are illustrated in Fig.17.2 and may be summarised as follows.

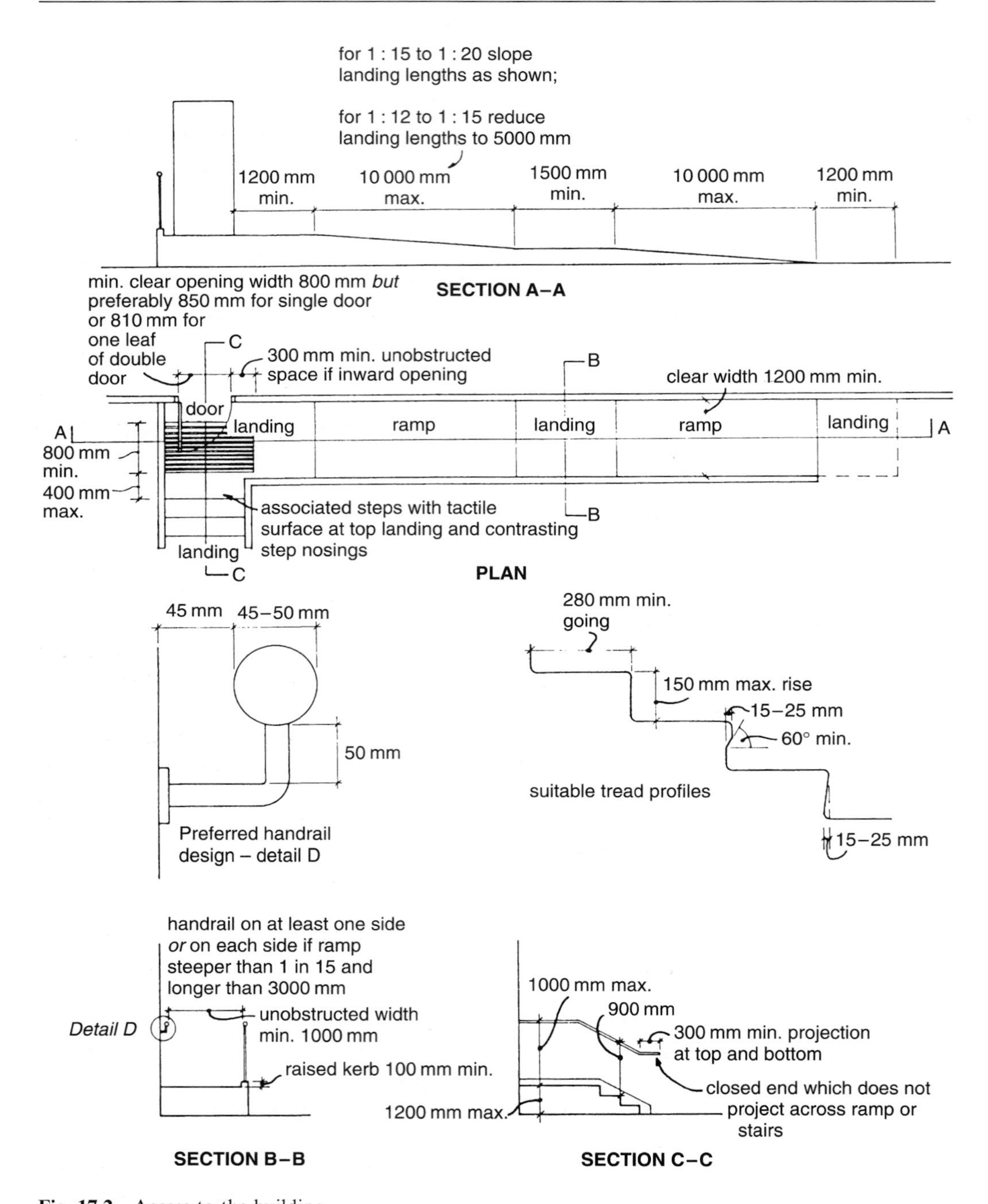

Fig. 17.2 Access to the building.

Ramps should:

- Have a non-slip surface.
- Ideally, be not steeper than 1 in 20. However, a 1 in 15 ramp is permitted if the individual flight is no longer than 10 m. Similarly a 1 in 12 ramp is allowed where the flight does not exceed 5 m in length.
- Have a landing at top and bottom at least 1200 mm long, clear of any door swing.
- Have intermediate landings at least 1500 mm long, clear of any door swing.
- Have on any open side of a flight or landing a raised kerb at least 100 mm high. This helps to avoid the risk to wheelchair users of their feet catching under or in an open balustrade.
- Have a surface width of at least 1200 mm.
- Have handrails on both sides of the flight or landing if the ramp is more than 2 m long, in order to provide support. Handrails should be designed to be easily gripped and should be well supported.

17.5.4 Complementary steps

Many of the design factors which apply to ramps also apply to steps. However, there are also a few additional factors.

Sudden changes of level marked by steps may create dangers for people with impaired sight especially at the head of a flight. Individual steps, whether on their own or in a flight, should also be apparent.

Some ambulant disabled people may have stiffness in their hip or knee joints or may need to wear calipers. Here there is a danger that they may catch their feet under nosings or treads on the staircase.

Clearly, people with physical weakness on one side or the other or who have sight impairments will need to place their feet squarely onto the treads.

Therefore, complementary steps should:

- Have a landing with a tactile surface at the top of the flight to give advance warning of the change in level. This should be the 'corduroy' type of paving shown in Fig. 17.1 and it should extend at least 400 mm beyond each side of the stairs.
- Have step nosings which are clearly visible by the use of contrasting brightness.
- Not rise more than 1200 mm between landings.
- Have landings at top and bottom of the flight and intermediate landings at least 1200 mm long, clear of any door swing.
- Have uniform rises no greater than 150 mm high and uniform goings at least 280 mm long (measured 270 mm in from the narrow edge of the flight for tapered steps).
- Have a clear width of at least 1000 mm.
- Have a suitable handrail on each side if comprising more than one riser and a suitable tread profile as illustrated in Fig. 17.2. It should be noted that open risers are not recommended.

17.5.5 Avoidance of hazards

Where a circulation route passes close to a building care should be taken to avoid projections which might be a hazard to people with sight impairments. This is most commonly caused by doors and windows which open outwards across paths. Opening lights and outward opening doors should be guarded if the path is adjacent to the building.

Alternatively, the path can be separated from the building by a slightly raised edging with a strong tactile surface such as cobbles, for example.

17.5.6 Entrance doors and lobbies

A wheelchair user will need extra space when negotiating an entrance lobby to allow for assistance and to avoid others who may be passing in the opposite direction. It should also be possible to move clear of one door when opening the next.

For this reason entrance doors and lobbies need to be built to certain minimum dimensions.

Figure 17.3 shows the design principles which should be followed.

Entrance doors should have an absolute minimum clear width of 800 mm. Ideally, the minimum clear width should be that provided by a 1000 mm single leaf external doorset or by one leaf of an 1800 mm double leaf doorset i.e. 850 mm clear or 810 mm clear – see Table 2, BS 4787: Part 1: 1980 (1985).

Whichever type of door is used it is important to allow sufficient room for the door to be opened by a person in a wheelchair. Therefore, the space into which the door opens should be unobstructed on the door handle side for at least 300 mm.

Disabled people cannot normally react quickly to avoid collisions when a door is suddenly opened. Therefore they should be able to see people who are approaching a door from the other side and should be seen themselves. This can be achieved by providing a glazed panel in the door which gives a zone of visibility between 900 mm and 1500 mm from the floor. (See Fig. 17.3.)

17.5.7 Revolving doors

Small revolving doors are not negotiable by wheelchair users and may create entry and exit difficulties for people with sight impairments or ambulatory problems. Therefore, revolving doors should always be accompanied by a conventional, accessible entrance door.

Some public buildings, supermarkets and the like are fitted with large revolving doors which may be suitable for wheelchair users. They should be capable of accommodating several people at once and should revolve very slowly. They should also be fitted with mechanisms which slow them down still further or stop them as soon as any resistance is felt.

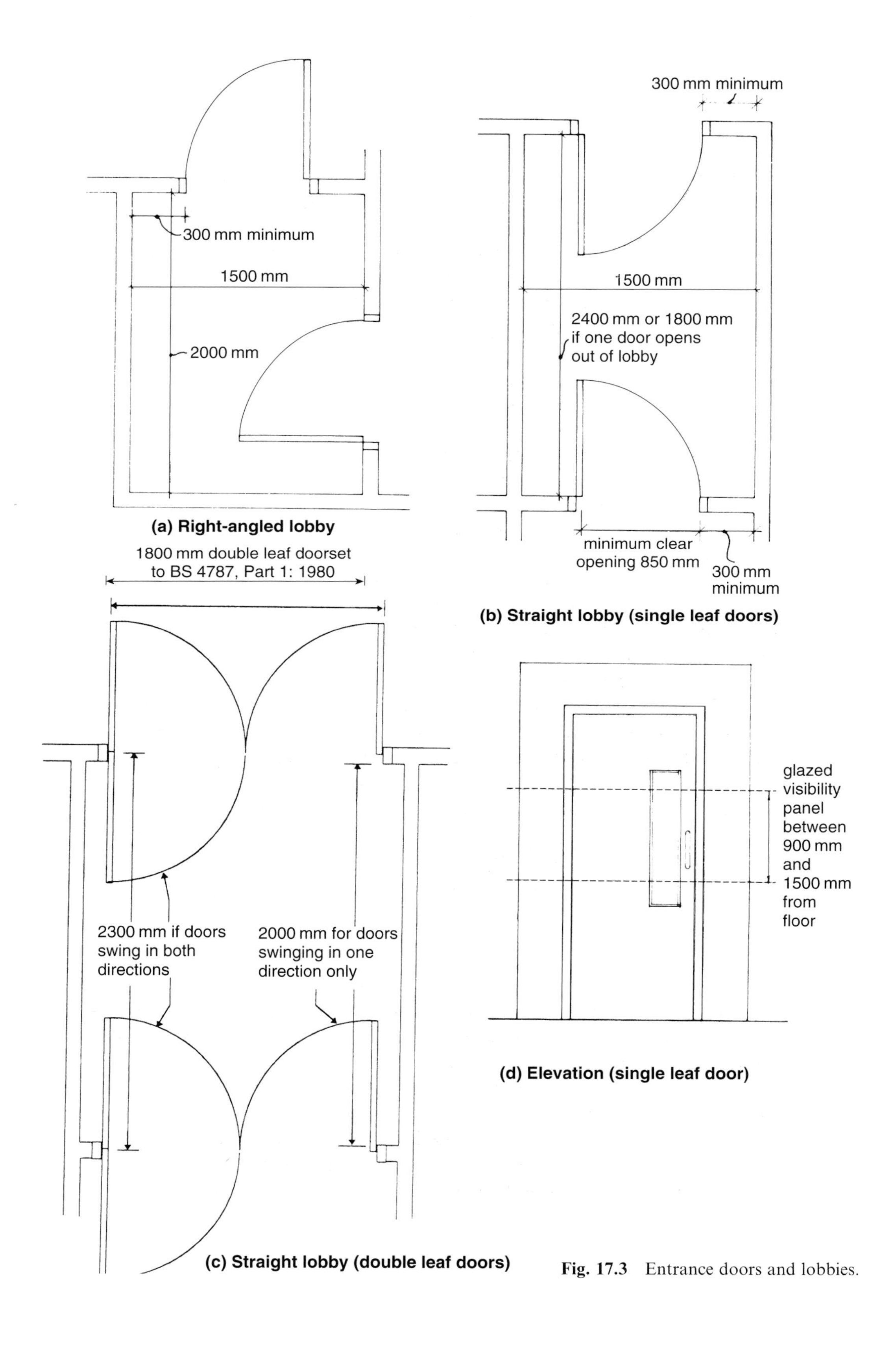

Fig. 17.3 Entrance doors and lobbies.

17.6 Access within the building

Once inside the building a disabled person must be able to reach and use the facilities provided.

Different building types contain unique facilities to which it may be reasonable to provide access for disabled people.

AD M does not attempt to provide exhaustive guidance on all the facilities that may be relevant and the choice may appear somewhat arbitrary. However, a number of aims are stated:

- People with hearing impairments should be able to play a full part in such things as conferences and committee meetings.
- Common facilities such as waiting rooms, canteens or cloakrooms should be located in an accessible storey of the building.

Specific guidance is given regarding:

- Restaurants and bars.
- Hotel and motel bedrooms.
- Changing facilities (including dressing cubicles and shower compartments).
- Aids to communication for people with impaired hearing.

These recommendations are in addition to the requirements for sanitary conveniences and audience or spectator seating.

Therefore it is necessary to provide sufficient space for wheelchair manoeuvre and convenient ways of travelling from one level to another. People with impaired hearing or sight should be catered for by providing features which enable them to find their way around the building safely and conveniently.

Corridors, passageways, internal doors and internal lobbies can present problems for disabled people unless care is taken in their design and adequate space is provided to enable a wheelchair to be manoeuvred and for other people to pass.

17.6.1 Corridors and passageways

Corridors and passageways generally should have a clear width of at least 1200 mm. However, this figure may be reduced to 1000 mm where lift access is not provided to the corridor or it is situated in an extension approached through an existing building.

With internal doors the important factor is the minimum clear opening width.

AD M quotes an absolute minimum size of 750 mm. However, the use of 900 mm single leaf internal doorsets or 1800 mm double leaf internal doorsets complying with Table 1 of BS 4787 will give clear opening sizes of 770 mm for the single leaf doorset and 820 mm for one leaf of the double doorset.

The 300 mm space requirement adjacent to the leading edge of the door mentioned above (see section 17.5.6 above) also applies to internal doors. In some circumstances, e.g. when leaving a fellow guest's hotel bedroom, it may be reasonable to assume that assistance will be on hand and therefore the 300 mm space may not be needed.

Each door across an accessible corridor should be provided with a glazed vision panel giving a zone of visibility between 900 mm and 1500 mm from the floor.

It should be noted that BS 4787 does not permit a fire resisting single leaf doorset to exceed 900 mm in overall width.

17.6.2 Internal lobbies

Internal lobbies may be smaller than principal entrance lobbies since fewer people are likely to use them at the same time. However there should still be room for a wheelchair user to move clear of one door before opening the next.

Internal lobbies should comply with the minimum dimensions shown in Fig. 17.4.

17.6.3 Vertical means of access

In the majority of buildings the most suitable form of vertical access for a wheelchair user is a lift. Lifts are expensive and take up usable space; therefore, it may be unreasonable to expect a lift to be provided in all instances especially in smaller buildings. Accordingly, lifts should be provided to serve any floor above or below the principal entrance storey where the following nett floor areas are exceeded:

- Two-storey buildings – 280 m^2 of nett floor area.
- Buildings exceeding two storeys – 200 m^2 of nett floor area.

The figures are derived by adding together the areas of all the parts of a storey which use the same entrance from the street or an indoor mall, even if they are in different parts of the same storey or are used for different purposes. Thus the figures given are for each storey above or below the principal entrance storey. In calculating the figures given above it is permissible to exclude the area of the vertical circulation, sanitary accommodation or maintenance areas in the storey. It is, of course, essential to provide means of access from the lift to the rest of the storey.

Where passenger lifts are not provided a stair suitable for ambulant disabled people should be installed. This should also be suitable for people with impaired vision. (See section 17.6.7.)

17.6.4 Lift design

Lifts for wheelchair users need to be large enough to allow access and egress, and should not be cramped internally.

Controls should be within reach and since disabled people need more time to enter and leave the lift car, suitable delay systems should be provided to lessen the risk of contact with the closing doors. For people with sensory impairments it may be necessary to provide visual and vocal floor indication.

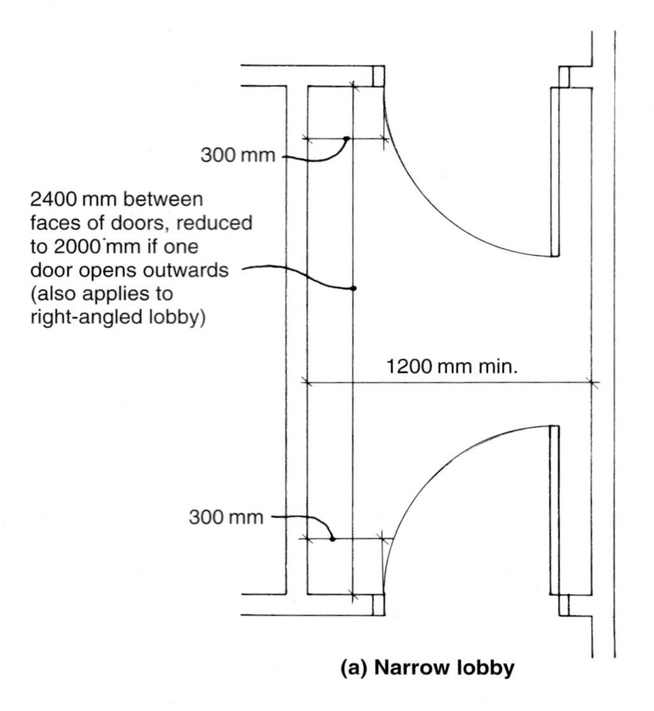

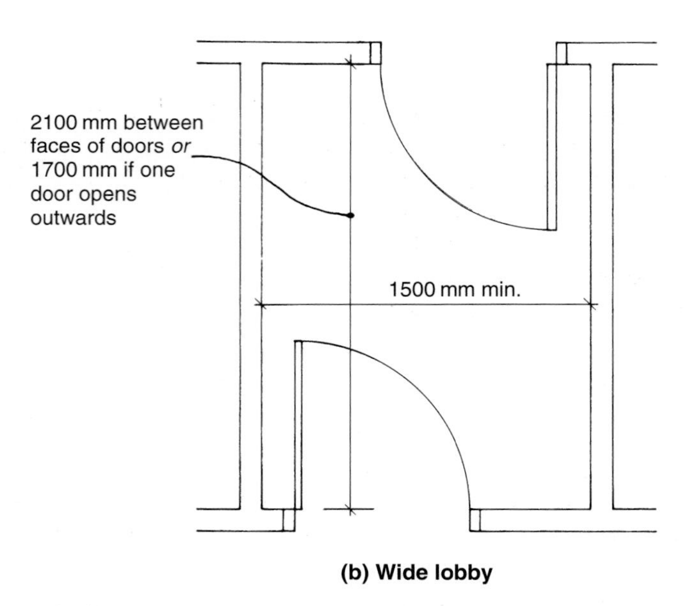

Fig. 17.4 Internal lobbies.

A suitable lift design is shown in Fig. 17.5. Its main features are:

- An unobstructed, accessible landing space at least 1500 mm square in front of the lift doors.
- A door or doors with a clear opening width of 800 mm.
- A car with minimum dimensions of 1100 mm wide by 1400 mm deep.
- Landing and car controls between 900 mm and 1400 mm from landing or car floor levels and at least 400 mm from the front wall.
- Tactile floor level indication on each landing next to the lift call button and, where the lift serves more than three floors, on or adjacent to the lift buttons in the lift car to confirm the floor selected.
- Visual and vocal floor indication where the lift serves more than three floors.
- A signalling system which gives five seconds warning that the lift is about to stop at a floor and once stopped, a minimum of five seconds before the doors begin to close after being fully open.

It is necessary to ensure that the door controls can be overridden in order that lift passengers do not get caught in the closing doors. In the past it has been common to incorporate a door edge pressure system which causes the doors to re-open when resistance is encountered. This type of system is not suitable for disabled people since it could cause them to become unbalanced. A door re-opening activator which uses a photo-eye or infra-red detector is satisfactory providing the door remains fully open for at least three seconds. BS 5655 *Lifts and service lifts* contains details of some suitable lifts systems in Parts 1, 2 and 5: 1989 and Part 7: 1983.

17.6.5 Stairlifts

Some buildings contain small areas with unique facilities such as staff rest and training rooms or small galleried libraries which are often on upper or lower floors. It is reasonable to expect access for wheelchair users to such areas but it may not be practical to provide a full passenger lift.

Storeys which have a nett floor area exceeding 100 m^2 containing unique facilities, may be accessed by a wheelchair stairlift provided it complies with BS 5776: 1996 *Specification for powered stairlifts*.

17.6.6 Platform lifts

The provision of ramps to effect level changes inside buildings can sometimes have serious planning implications since they tend to take up large areas of floor space. A platform lift which complies with BS 6440: 1983 *Code of practice for powered lifting platforms for use by disabled people* may be provided in lieu of a ramp but not at the expense of a suitable staircase.

17.6.7 Internal stairways

Where a lift is not provided in a building then a stairway suitable for ambulant disabled people or people with impaired sight should be provided.

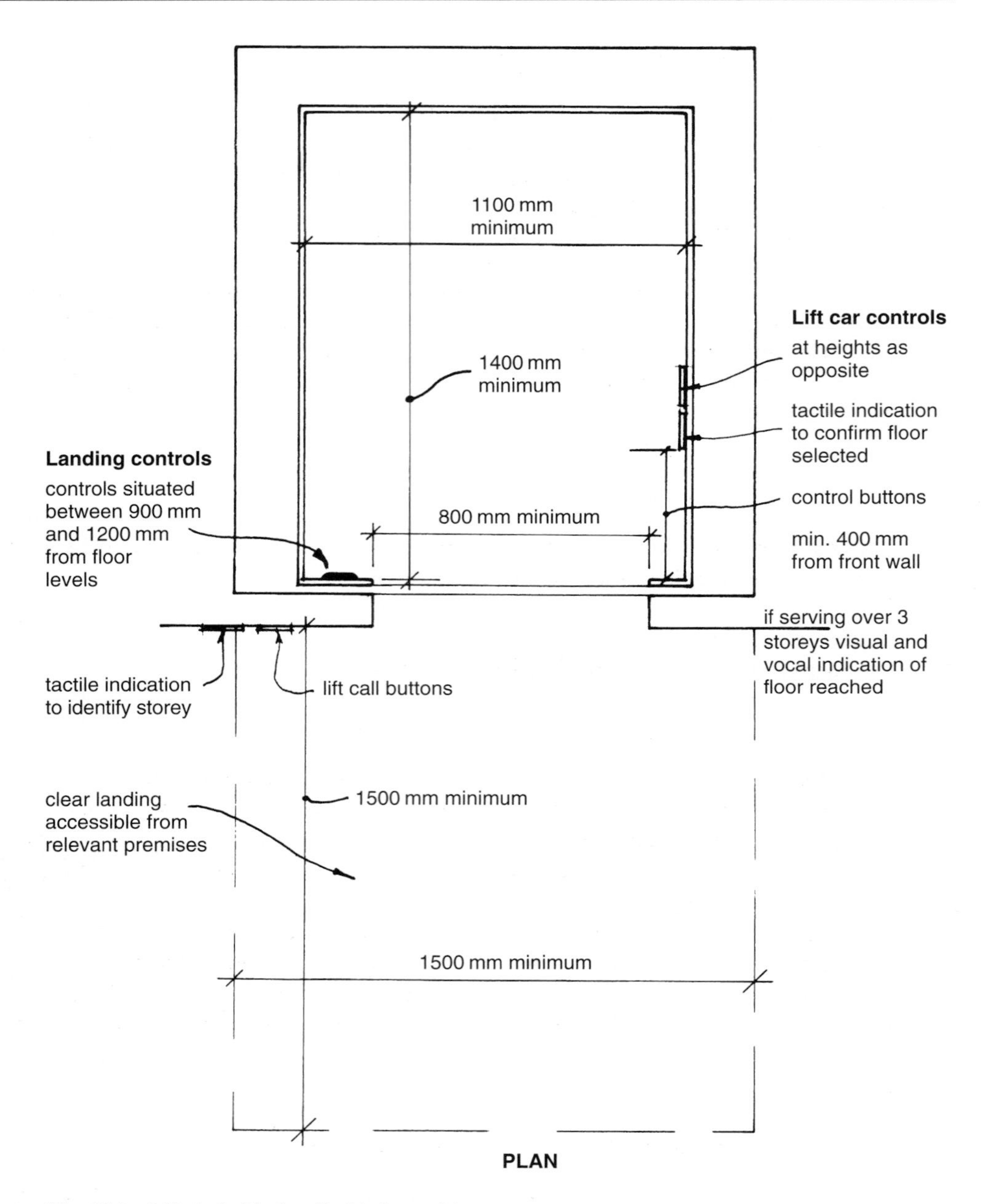

Fig. 17.5 Lift (suitable for disabled people).

With certain exceptions, the design for external stairways shown in Fig. 17.2 is also suitable for internal stairways.

Internal stairs are subject to more constraints in the form of ceiling heights and space restrictions. However, it is not considered reasonable to require the provision of tactile warnings at the start of level changes. Nevertheless, stair nosings should still be distinguishable for the benefit of people with impaired sight.

The principal variations permitted are:

- Uniform risers should not exceed 170 mm in height.
- Uniform goings should not be less than 250 mm in length (measured 270 mm in from the narrow edge of the flight for tapered steps).
- Tactile warnings of level changes are not necessary.
- The maximum rise of a flight between landings should not exceed 1800 mm.

This last figure is somewhat flexible since it depends to a large extent on site constraints such as landing levels, storey heights or the space required for the extra staircase length. If the recommendations contained in Approved Document K regarding numbers of risers are followed, then this would be acceptable in exceptional cases.

17.6.8 Internal ramps

Internal ramps should follow the recommendations which are specified for external ramps (see section 17.5.3 above and Fig. 17.2).

17.7 Use of facilities within the building

17.7.1 Access to restaurant and bar facilities

Disabled people should be able to visit and use restaurants and bars whether accompanied or alone. The choice of waiter or self-service facilities should be available and there should be suitable access to seating areas. Changes of floor levels in seating areas are permissible if kept to a reasonable scale and should be accessible by ambulent disabled people.

Therefore, the full range of services offered should be accessible, including bars and self-service counters and at least half the area where seating is provided. Sometimes, the nature of the service provided in a restaurant varies. For example, some areas may be self-service and some may be waitress service. In these cases at least half the area of each should be accessible even if they are in different storeys.

17.7.2 Access to hotel and motel guest bedrooms

It is essential that a certain proportion of guest bedrooms in hotels or motels should be suitable for wheelchair users. This means that the bedroom should be accessible and should contain sufficient space to allow wheelchair manoeuvre within it and into en-suite bathroom facilities, if provided.

Since it is usual for guests to visit each other in their bedrooms when on holiday or attending residential conferences for example, it is reasonable to limit the disabled facilities to the provision of an accessible doorway. This assumes that an able-bodied guest will open and close the door.

The following main provisions apply:

- Each guest bedroom entrance door should follow the recommendations for internal doors given in section 17.6.1 above
- One guest bedroom out of every 20 should be accessible in terms of layout, dimensions and facilities for a wheelchair user, i.e. if 21 guest bedrooms were provided then two of them would have to be suitable for a wheelchair user (see Fig. 17.6 for details).
- All other guest bedrooms should have entrance doors with a minimum clear opening width of 750 mm. It is permissible to omit the 300 mm space at the side of the door.

17.7.3 Shower and changing facilities

Many disabled people enjoy swimming and other recreational activities but find it difficult to participate due to the lack of suitable changing and showering facilities. Adequate space is required to manoeuvre the wheelchair and to transfer onto a seat. Seats, shower heads, taps, clothes hooks and mirrors all need to be mounted at suitable heights. Figure 17.7 gives details of suitable arrangements for shower and changing cubicles.

17.7.4 Communication aids

When people attend public performances or conferences it is reasonable to expect them to be able to enjoy the proceedings. People with hearing difficulties may need to receive an amplified signal some 20 dB above the level received by a person with 'normal' hearing. The amplification system should be able to suppress reverberation, and audience or other noises.

A particular problem is in glazed ticket or booking offices where it may be very difficult to hear the attendant.

There are two systems commonly in use at present – the loop induction system and the infra-red system. AD M gives a limited description of the principles of each and points out some of their disadvantages. No British Standard references are given and it is left to building owners to decide which system best suits the layout and use of their buildings. Obviously, there will be a need to consult specialist companies if it is necessary to install an amplification system.

AD M recommends that aids to communication should be provided at booking and ticket offices where a glazed screen separates customer and vendor.

Further, large reception areas, auditoria and meeting rooms where the floor area exceeds 100 m^2 are also covered.

The systems installed should incorporate features that allow the person wearing the hearing aid to receive the amplified sound without loss or distortion due to bad acoustics or extraneous noise.

It is felt that this guidance is sufficiently vague to be of little use to building owners, designers and building control authorities and will only lead to variations in interpretation and consequent delays in passing of plans and approval of work on site.

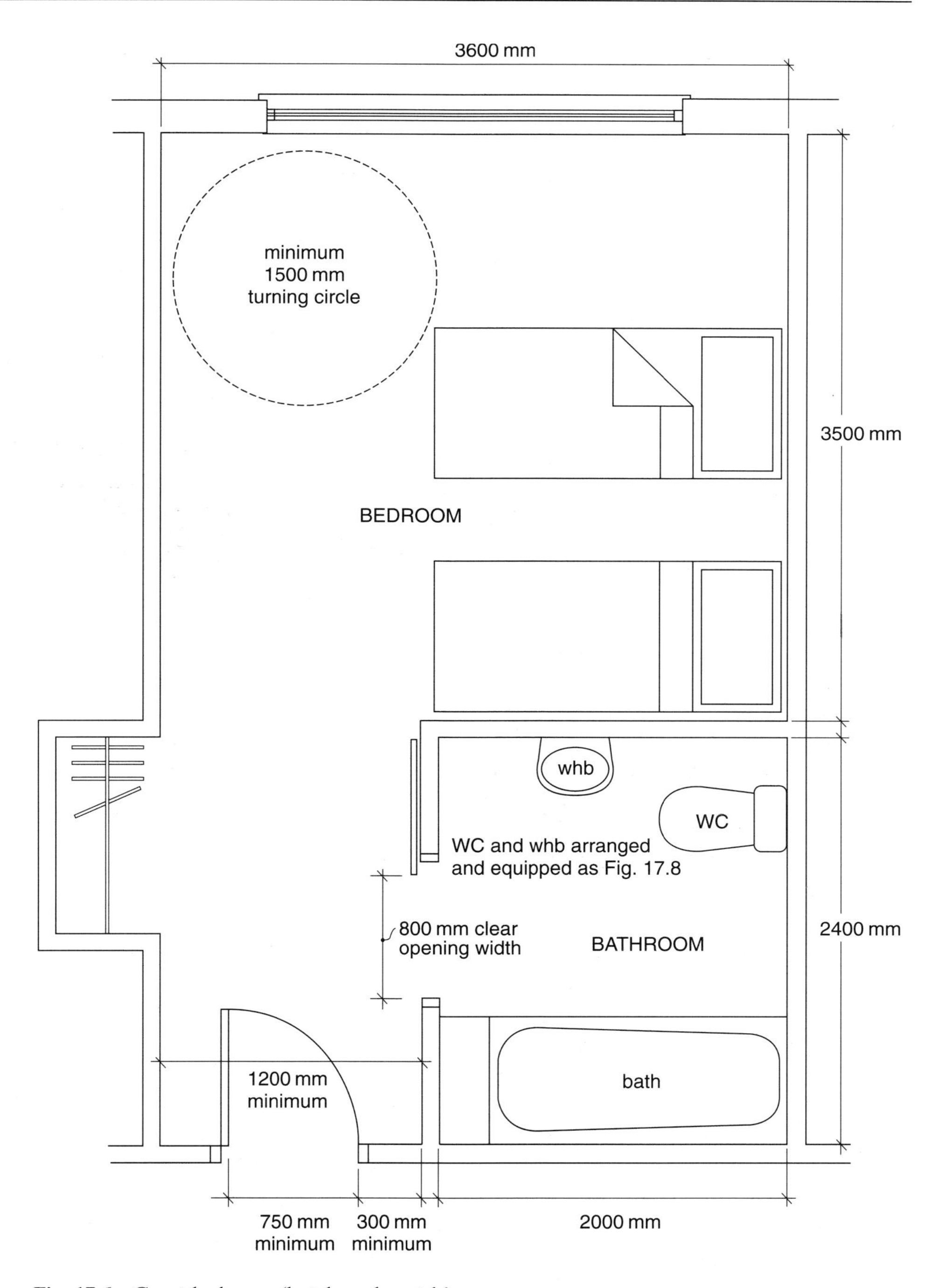

Fig. 17.6 Guest bedroom (hotels and motels).

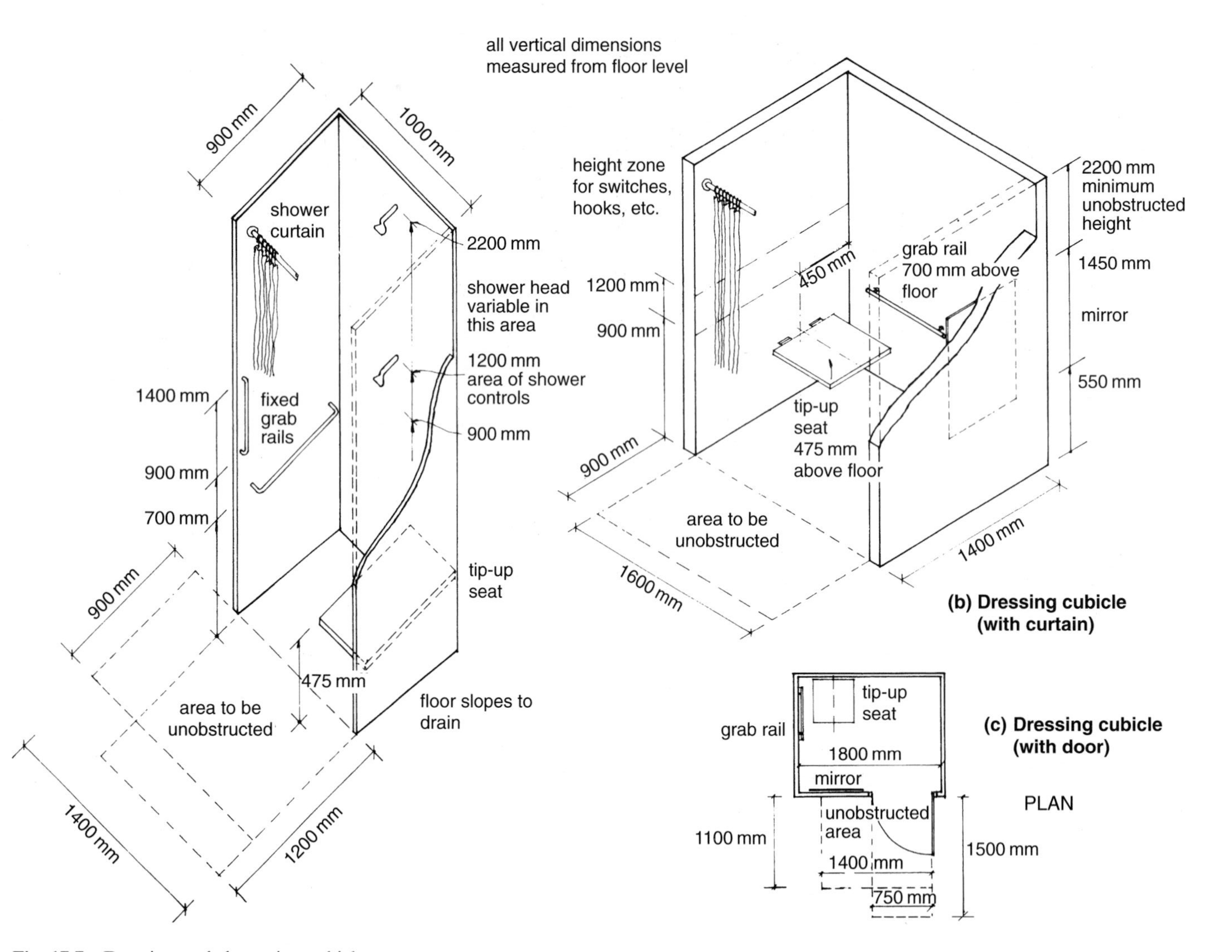

Fig. 17.7 Dressing and showering cubicles.

17.8 Sanitary conveniences

Section 4 of AD M3 sets out the requirements for the provision of sanitary conveniences suitable for both wheelchair users and ambulant disabled people.

The general principle is that where sanitary conveniences would normally be provided for able-bodied people then disabled people should also be catered for with suitable accommodation.

Some of the recommendations under the heading Staff in AD M3 would appear to be of general application and have been interpreted as such in the following discussion.

17.8.1 Design of sanitary conveniences for wheelchair users

The following general design principles should be considered:

- Unisex and/or integral conveniences may be provided under the terms of AD M3, unisex facilities being separate and self-contained whereas integral facilities are contained within the separate accommodation for men and women.
- The scale of provision will depend on the size and nature of the building and on the ease of access to the facility.
- A more flexible arrangement can be provided in large buildings by the inclusion of both unisex and integral toilet facilities.
- Travel distances should reflect the fact that some disabled people need to reach a WC quickly.
- It should not be necessary to travel more than one storey to reach a WC.
- Where more than one WC compartment is provided in a building then both left- and right-hand transfer layouts should be provided.

Individual compartments should be designed for ease of access and use and should provide:

- Adequate space for wheelchair manoeuvre and for the presence of a helper to assist with frontal, lateral, diagonal or backward transfer onto the WC.
- Hand washing and drying facilities which may be reached from the WC before transfer back to the wheelchair.
- Both unisex and integral WC compartments should be similar in layout and content.

The AD appears to show a marked preference for unisex WC accommodation and lists the following advantages over integral facilities:

- The approach to it is separated from other sanitary accommodation.
- It is more easily identified.
- There is more chance of it being available when needed.
- Assistance is permitted by helpers of either sex. (This would not be possible with integral facilities since existing custom prevents access by a helper whose sex is different from those for whom the provision is made.)

- Less space is needed overall, since duplication of facilities is necessary with integral accommodation for the same level of provision.

17.8.2 Provision of sanitary accommodation for wheelchair users

It is necessary to provide sanitary conveniences for all of a building's users whether they be visitors, customers or staff.

Disabled visitors and customers are more likely to be accompanied by a member of the opposite sex, therefore unisex facilities should be provided for them.

On the other hand, staff may be less likely to need assistance. If help is needed then it is more likely that a member of the same sex will provide it.

Therefore, for staff, the facilities may be either unisex or integral. There may be separate provision for both sexes on alternative floors if the building is provided with lifts and the sanitary conveniences are in areas with unrestricted access. In this case it should not be necessary to travel more than a cumulative horizontal distance of 40 m from a work station to the WC.

Visitors to hotels and motels will often find that guest bedrooms have en-suite sanitary accommodation. If this is the case, then this should also be the arrangement for bedrooms suitable for disabled people. Where en-suite facilities are not provided in the general sanitary arrangement then unisex facilities should be provided within easy reach. There is still, of course, the necessity to provide additional sanitary accommodation for staff and for daytime or non-resident visitors.

Figure 17.8 illustrates a typical layout for a WC compartment suitable for wheelchair users in either unisex or integral facilities. It is suitable in any of the situations referred to above.

17.8.3 Provision of sanitary accommodation for ambulant disabled people

It is permissible for certain small buildings to have stair access only (see section 17.6.3).

In these cases suitable sanitary conveniences for wheelchair users should be provided in the principal entrance storey unless this contains only vertical circulation areas or the principal entrance.

Any sanitary accommodation containing WC compartments situated on floors without lift access should contain at least one WC compartment designed for use by ambulant disabled people (i.e. people with a limited ability to walk or to support themselves). This should be in addition to any sanitary conveniences provided for wheelchair users in the principal entrance storey.

A WC compartment suitable for ambulant disabled people is illustrated in Fig. 17.9.

17.9 Audience or spectator seating

Where fixed audience or spectator seating is provided in theatres, cinemas, concert halls, sports stadia and similar buildings, then reasonable provision must be made to accommodate disabled people.

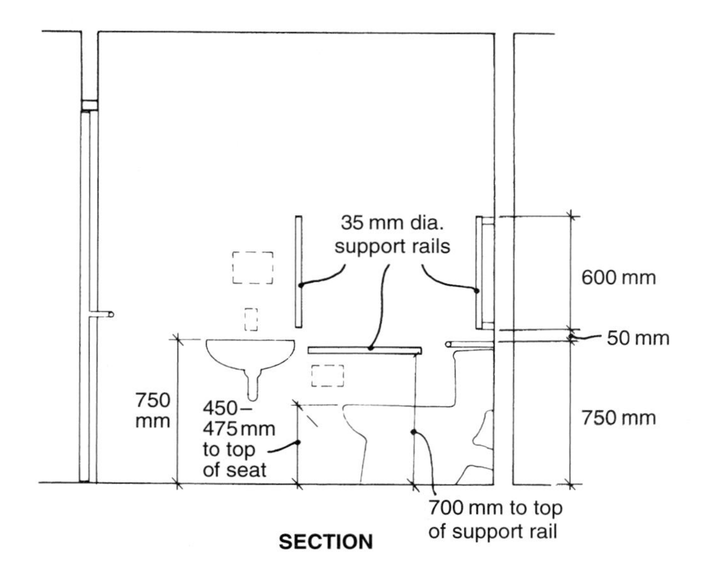

Fig. 17.8 WC for wheelchair users.

A sufficient number of wheelchair spaces should be provided which are accessible and provide a clear view of the event. The spaces may be kept clear at all times or may contain seating which can be removed easily for each occasion.

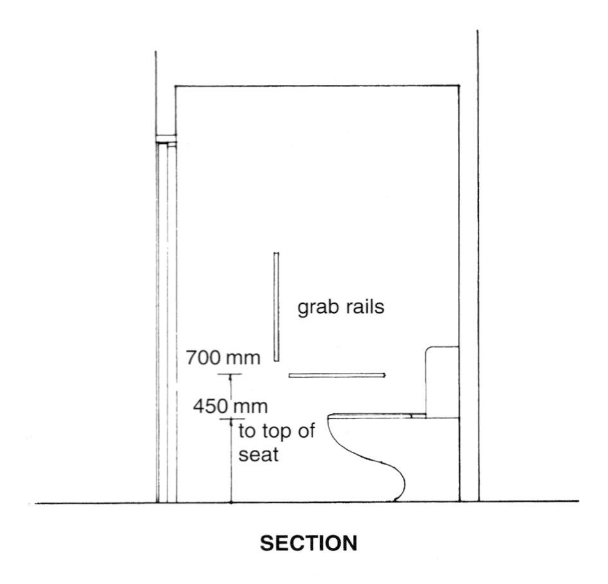

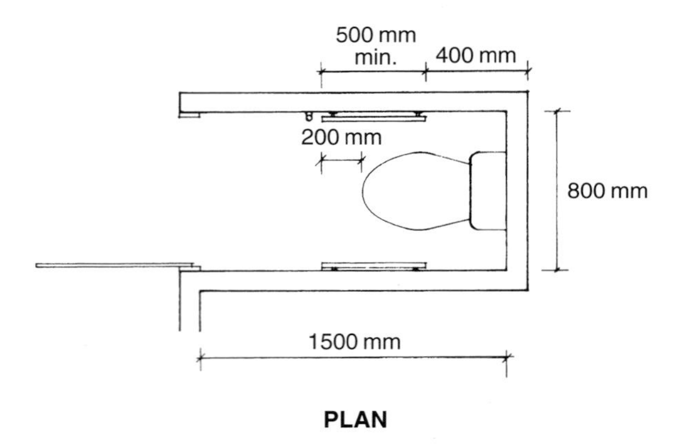

Fig. 17.9 WC for ambulant disabled people.

Wheelchair spaces should be:

- At least 900 mm wide.
- At least 1400 mm deep.
- Dispersed throughout the theatre or stadium so that disabled people have a choice of sitting next to able-bodied or disabled companions.

There should be provided at least six spaces or one for every 100 fixed audience or spectator seats available to the public, whichever figure is the greater.

In a large stadium AD M4 recommends that a smaller proportion of wheelchair spaces could be provided. Unfortunately, no further guidance is given as to what constitutes a large stadium or a smaller proportion of wheelchair spaces.

Figure 17.10 shows typical wheelchair space layouts for both theatres and sports stadia.

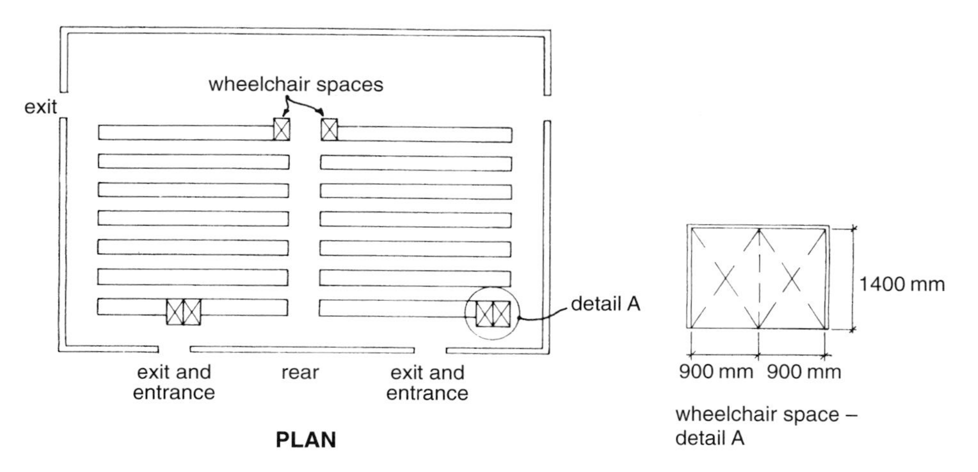

(a) Theatre – notional disposition of wheelchair spaces

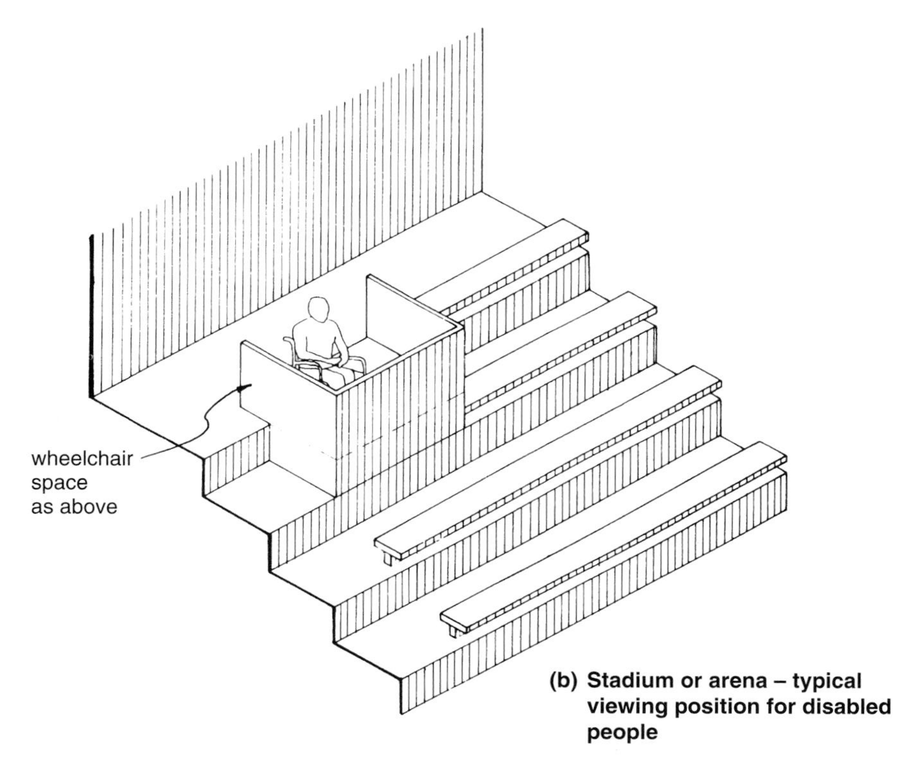

(b) Stadium or arena – typical viewing position for disabled people

Fig. 17.10 Audience or spectator seating.

Additionally, guidance on access for disabled people to sports stadia may be obtained from:

Guide to safety at sports grounds, published by the Stationery Office, 1997.
Designing for spectators with disabilities, published by the Football Stadia Advisory Council, 1993 (Available from the Sports Council).
Access for disabled people, English Sports Council Guidance Notes.

17.10 Other legislation and technical guidance

17.10.1 Access and facilities in schools and other educational buildings

Where work is to be carried out in schools or other educational establishments, the DfEE publication, Design Note 18, 1984 *Access for disabled persons to educational buildings* will satisfy the requirements of paragraphs M2, M3 and M4. The relevant paragraphs are: 2.1/2/4/6; 3.1; 4.1/2/4/6 and 5.1. However, it may also be necessary to incorporate some of the general design considerations in Design Note 18 into educational buildings, since regulation M2 also includes provisions for people with impaired hearing or sight.

17.10.2 Means of escape in case of fire

It is clear that Part M is limited to the provision of access into and around a building, and the use of that building. Therefore it does not extend to provisions for means of escape for disabled people if a fire occurs. For this, reference should be made to Approved Document B, Fire safety, and BS 5588, Part 8: *Code of practice for means of escape for disabled people* (see also Chapter 7 section 7.17.1).

17.10.3 Disability Discrimination (Employment) Regulations 1996

Under Regulation 8 of the Disability Discrimination (Employment) Regulations, an employer who installs facilities for disabled people which comply with the requirements of Part M (and continue substantially to meet those requirements in use), cannot be required to alter such facilities.

17.10.4 Workplace (Health, Safety and Welfare) Regulations 1992

It should be noted that compliance with Regulation M2, in conjunction with Part K, prevents action being taken against the occupier of a building under Regulation 17 of the Workplace (Health, Safety and Welfare) Regulations 1992 when the building is eventually in use. (Regulation 17 relates specifically to permanent stairs, ladders and ramps on routes used by people in places of work).

17.11 Access and facilities in dwellings

17.11.1 Interpretation

The following definitions of terms used in Approved Document M apply only to the provisions concerning dwellings:

DWELLINGS – In addition to dwelling-houses, flats and maisonettes, the term includes purpose-built student living accommodation (but not traditional halls of residence which only provide bedrooms and are not equipped as self-contained accommodation).

COMMON – Serving more than a single dwelling.

HABITABLE – When used to define the principal storey in a dwelling, means a room intended to be used for dwelling purposes. This includes a kitchen but not a utility room or bathroom.

MAISONETTE – A self-contained dwelling that occupies more than one storey in a building, but which is not a dwelling-house.

POINT OF ACCESS – The place where a person would alight from a vehicle when visiting a dwelling, before they approached the dwelling. This could be inside or outside the plot boundary.

PRINCIPAL ENTRANCE – The entrance that would normally be used by a visitor who was not familiar with the dwelling. In a block of flats it would be the common entrance.
PLOT GRADIENT – The slope measured between the point of access and the finished floor level of the dwelling.

STEEPLY SLOPING PLOT – A plot gradient that exceeds 1 in 15.

17.11.2 The main provisions

Reasonable provision must be made:

- For disabled people to gain access to and use the facilities in the dwelling.
- For sanitary conveniences suitable for disabled people in the entrance storey of the dwelling.

The 'entrance storey' is the storey which contains the principal entrance to the dwelling (see above for definition) and in the design of some dwellings the entrance storey may not contain any habitable rooms. If this is the case, then the option is given of placing the sanitary conveniences in either the entrance storey or a principal storey. Regulation M3 defines 'principal storey' as '*the storey nearest to the entrance storey which contains a habitable room, or where there are two such storeys equally near, either such storey*'.

Although this legal terminology may seem confusing, the intention of the regulation is simply to require sanitary conveniences suitable for disabled people to be placed in the entrance storey of the dwelling, *if this contains any habitable rooms*. Otherwise the sanitary conveniences may be placed in either the entrance storey or the nearest storey to the principal entrance which contains habitable rooms.

Therefore, in dwellings, it should be reasonably safe and convenient for disabled people to:

- Approach the principal (or other suitable alternative) entrance to the dwelling from the edge of the site or from car parking within the site.
- Gain access into the building.
- Gain access within the building.
- Use sanitary accommodation located in the nearest storey containing habitable rooms to the principal entrance.

It should be noted that the provision of access and facilities for disabled people in dwellings is to enable them to visit new dwellings and use the principal storey. The intention is that disabled occupants will be able to cope better with reducing mobility and will be able to remain living in their own homes longer than would otherwise be the case. On the other hand, the provisions are not intended to facilitate fully independent living for all disabled people.

17.12 Means of access

In general terms, a disabled person should be able to gain access into a dwelling from the point of leaving a vehicle (which is parked either within or outside the plot boundary).

In most cases it should be possible to provide a safe and convenient level or ramped approach, thereby permitting wheelchair users to gain access to the dwelling. Clearly, there will be situations on steeply sloping plots (i.e. where the plot gradient exceeds 1 in 15), where it will only be practicable to provide a stepped approach. Where this is the case the approach should be suitable for an ambulant disabled person using a walking aid.

17.12.1 The approach to the dwelling

The choice of a suitable approach to the dwelling from the point of access to the plot, will be influenced by the topography and available area of the plot, and by the distance to the dwelling from the point of access. Account may also need to be taken of local planning requirements, especially for new developments in conservation areas. Developers are advised to discuss the access requirements of Part M with their local planning authority, in conjunction with their building control supervisor (local authority or approved inspector) at an early stage in the design process, to avoid later conflicts.

It may be possible to reduce the effect of a steeply sloping plot by means of a suitable driveway. This could allow for the parking space within the plot boundary to be at a sufficiently high level to permit a level or ramped approach from the parking space to the dwelling.

The surface material of the approach to the dwelling should be firm enough to support a wheelchair and user, and smooth enough to allow satisfactory manoeuvre. Ambulant disabled people using walking aids also need to be considered. Therefore, loose surfacing materials such as gravel or shingle are unlikely to be satisfactory and the approach should be sufficiently wide (in addition to the width of the parking space) to allow safe and convenient passage.

In practical design terms the provisions can be summarised in the following paragraphs.

Provide a suitable approach:

- From a reasonably level point of access (i.e. from the vehicle parking position) to the dwelling entrance (i.e. the principal entrance or a suitable alternative entrance if it is not possible to access the principal entrance).
- With crossfalls that do not exceed 1 in 40.
- That may consist in whole or in part of a vehicle driveway.

A level approach will have:

- A gradient not exceeding 1 in 20.
- A firm and even surface.
- A minimum width of 900 mm.

A ramped approach will be needed where the overall plot gradient exceeds 1 in 20 but does not exceed 1 in 15. In this case the ramped approach should have:

- A firm and even surface.
- Minimum unobstructed flight widths of 900 mm.
- Individual flights no longer than 10 m in length with a maximum gradient of 1 in 15 (although gradients not exceeding 1 in 12 are allowed where the individual flight does not exceed 5 m in length).
- Top, bottom and if necessary, intermediate landings at least 1200 mm long, clear of any door or gate swinging across it.

A stepped approach will be needed where the overall plot gradient exceeds 1 in 15. In this case the stepped approach should have:

- Minimum unobstructed flight widths of 900 mm.
- A maximum rise of 1800 mm between landings.
- Top, bottom and if necessary, intermediate landings at least 900 mm long.
- A suitable tread profile as illustrated in Fig. 17.2.
- Uniform risers between 75 and 150 mm high and goings at least 280 mm long (measured 270 mm in from the narrow edge of the tread for tapered steps).

- Where the flight consists of three or more risers, a suitable handrail should be provided on one side. The handrail should have a profile which can be gripped, be positioned between 850 and 1000 mm above the pitch line of the flight and project at least 300 mm beyond the top and bottom nosings (see Fig. 17.11).

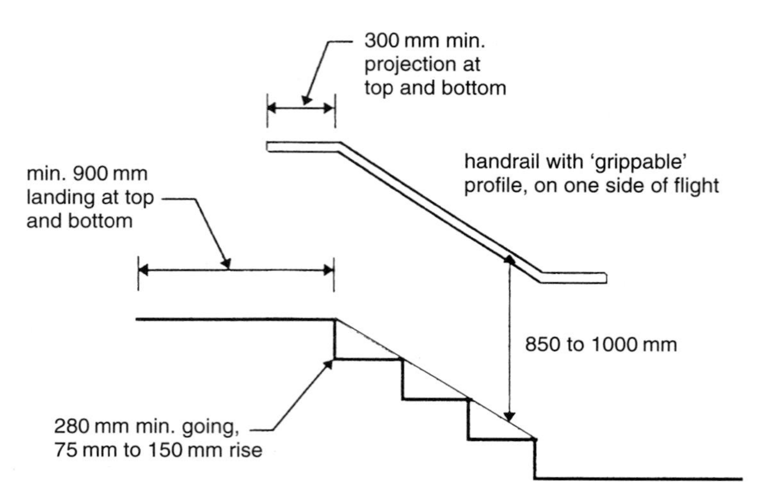

Fig. 17.11 Stepped approach to dwelling.

17.12.2 Access into the dwelling

In general, the entrance into a dwelling or a block of flats from outside should be provided with an accessible threshold irrespective of whether the approach to the entrance is level, ramped or stepped. Exceptionally, if the approach is stepped and, for practical reasons a step into the dwelling is unavoidable, it should not exceed 150 mm in height.

The design of accessible thresholds should follow the guidance contained in The Stationery Office publication *Accessible thresholds in new housing* (ISBN 0 11 702333 7).

The entrance door to an individual dwelling and/or a block of flats should be wide enough to accommodate a person using a wheelchair. This requirement can be satisfied if such a door has a minimum clear opening width of 775 mm.

17.12.3 Access within the dwelling

The requirements of regulations M2 and M3 mean that access must be facilitated to habitable rooms and to a WC (which may be in a bathroom) in the entrance storey or the principal storey of the dwelling, as appropriate.

Where it is not possible to make the principal entrance accessible and an alternative is provided instead, the route to the remainder of the entrance storey from the alternative entrance must be carefully considered, especially if the route passes through other rooms. Therefore, corridors and passageways should be wide enough to allow convenient access for a person in a wheelchair, whilst at the same time allowing for manoeuvre past local obstructions such as radiators and other fixtures.

Doors to rooms need to be wide enough to cater for both head-on approach and right angled approach from a person in a wheelchair. The rules for access within the entrance or principal storey of a dwelling are illustrated in Fig. 17.12 below.

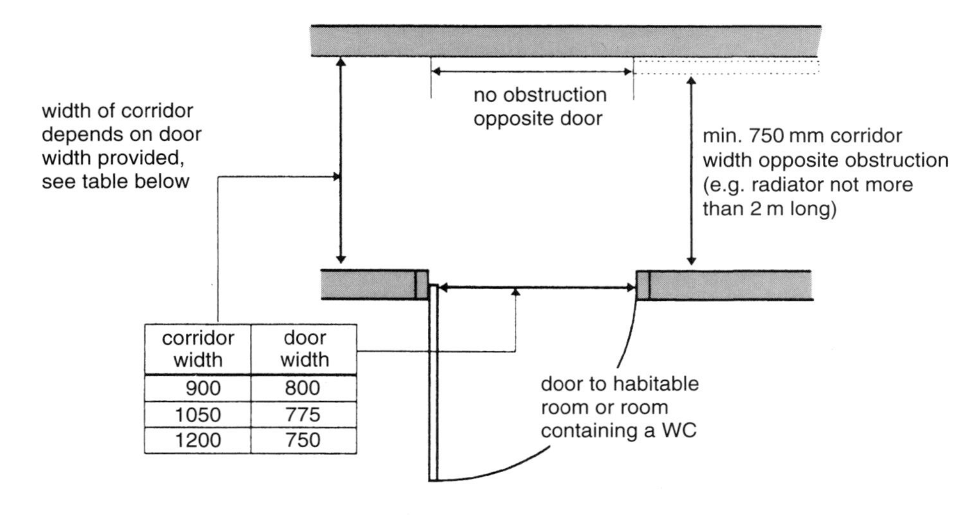

corridor width	door width
900	800
1050	775
1200	750

Right-angled approach to door

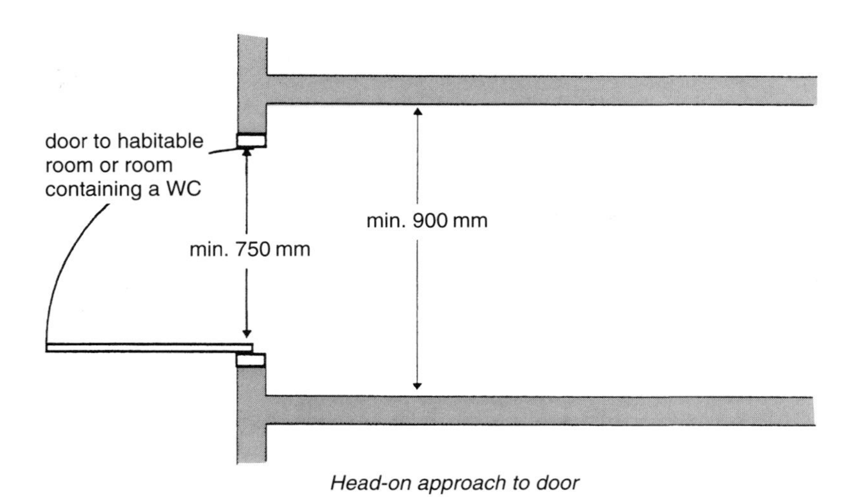

Head-on approach to door

Fig. 17.12 Corridor and door widths in dwellings.

17.12.4 Vertical circulation

Steps within the entrance storey of a dwelling should be avoided wherever possible. Sometimes, such as in the case of severely sloping plots, it may not be possible to avoid putting a change of level involving steps, in the entrance storey. In these circumstances a stair should be provided which is wide enough to be negotiated by an ambulant disabled person with assistance and with handrails on both sides. Therefore, any stair provided in the entrance storey which gives access to habitable rooms should have:

- a minimum clear width of 900 mm;
- a continuous handrail on each side, and on any intermediate landings, where the flight consists of three or more risers; and
- rise and going in accordance with the guidance for private stairs in Approved Document K (see Chapter 15).

17.12.5 Access to socket outlets and switches in dwellings

Ambulant disabled people and people who use wheelchairs are less mobile and likely to have more limited reach than able-bodied people. Therefore switches and socket outlets for such things as electrical appliances, lighting, television aerials, telephone jack points, etc., should be mounted at suitable heights so that they can be easily reached. Essentially, this means locating sockets and switches in habitable rooms between 450 mm and 1200 mm from finished floor level.

17.12.6 Common access stairs in blocks of flats

It should be possible for a disabled person to visit an occupant, on any storey in a building containing flats. The most suitable means of vertical access for a disabled person is a lift; however, AD M recognises that lifts are not always provided (and does not, at present, recommend that lifts should be installed). Therefore, where there are no passenger lifts, the common access stairs should be suitable for use by ambulant disabled people, (as well as being suitable for people with impaired sight) and be designed to have:

- step nosings which are clearly visible by the use of contrasting brightness;
- top and bottom landings which follow the guidance contained in Part K1 of Approved Document K (see Chapter 15);
- uniform risers not exceeding 170 mm in height;
- uniform goings not less than 250 mm in length (measured 270 mm in from the narrow edge of the flight for tapered steps);
- a suitable handrail on each side of flights and landings if comprising more than one riser;
- a suitable tread nosing profile (e.g. as illustrated in Fig. 17.2 but with 170 mm maximum rise and 250 mm minimum going); and
- risers which are not open.

Some of this guidance is illustrated in Fig. 17.13.

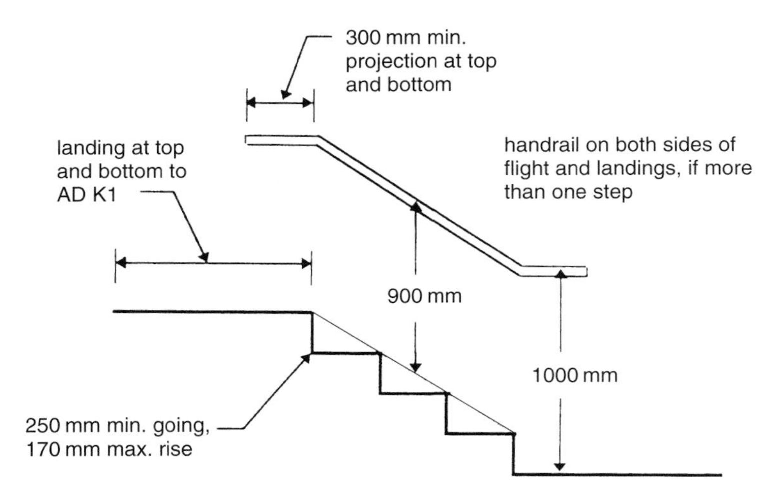

Fig. 17.13 Common access stairs in blocks of flats.

17.12.7 Passenger lifts in blocks of flats

If passenger lift access is to be provided to flats above the entrance storey in a building, it should be suitable for both unaccompanied wheelchair users and people with sensory impairments. It should also contain suitable delay systems to enable disabled people more time to enter and leave the car and lessen the risk of contact with the closing doors.

A suitable passenger lift should have:

- A minimum load capacity of 400 kg.
- An unobstructed, accessible landing space at least 1500 mm square in front of the lift doors.
- A door or doors with a clear opening width of 800 mm.
- A car with minimum dimensions of 900 mm wide by 1250 mm deep (although other dimensions may be suitable if it can be demonstrated by test evidence or experience in use, etc. that they are suitable for an unaccompanied wheelchair user).
- Landing and car controls between 900 mm and 1400 mm from landing or car floor levels and at least 400 mm from the front wall.
- Tactile floor level indication on each landing next to the lift call button to identify the storey in question, and on or adjacent to the lift buttons in the lift car to confirm the floor selected.
- Visual and audible floor indication where the lift serves more than three floors.
- A signalling system which gives a visual warning that the lift is about to stop at a floor and once stopped, a minimum of 5 seconds before the doors begin to close after being fully open.

It is necessary to ensure that the door controls can be overridden in order that lift passengers do not get caught in the closing doors. This should be a suitable

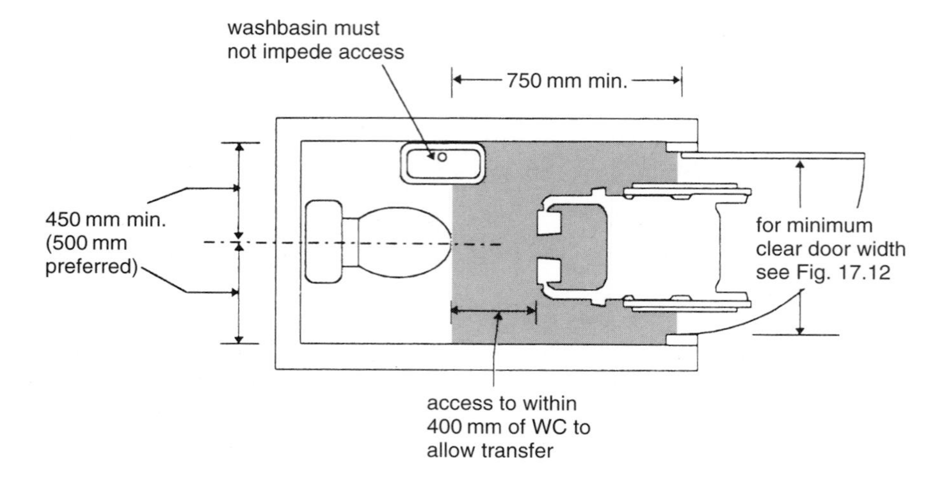

Typical WC compartment with frontal access

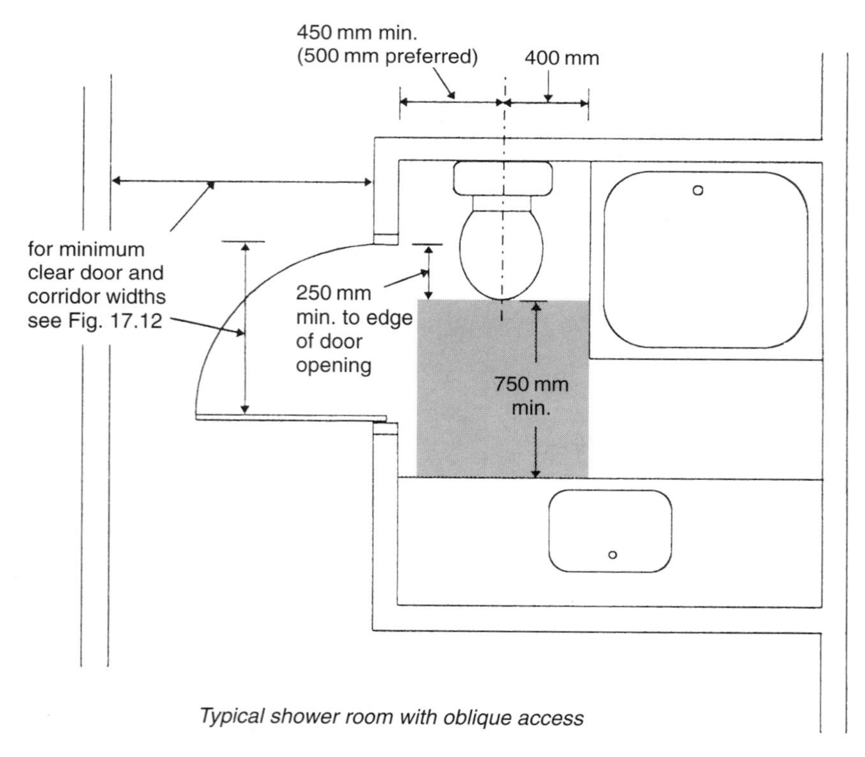

Typical shower room with oblique access

Fig. 17.14 Accessible sanitary conveniences in dwellings.

electronic system (such as a photo-eye or infrared detector), but not a door edge pressure system which might cause a disabled person to lose their balance. The door should remain fully open for at least three seconds.

17.12.8 Provision of accessible sanitary conveniences in dwellings

For dwellings which contain more than one storey, sanitary conveniences suitable for disabled people should be provided in the entrance storey, *if this contains any habitable rooms*. Otherwise the sanitary conveniences may be placed in either the entrance storey or the nearest storey to the principal entrance which contains habitable rooms. There should be no need to negotiate a stairway to reach the WC from the habitable rooms in that storey. (Obviously, for single storey dwellings and individual flats the sanitary conveniences can only be provided in the entrance storey).

Additionally, the following provisions apply to the design and location of the sanitary accommodation:

- The WC may be located in a bathroom if there is one available in the relevant storey of the dwelling.
- It is accepted in AD M that it may not always be practical for a wheelchair to be fully accommodated inside the WC compartment.
- The WC door should open outwards and should be positioned so as to allow access to the WC for a person in a wheelchair.
- The WC door should have a minimum width as shown in Fig 17.12 (and be wider if possible, so as to allow easier access and manoeuvring by wheelchair users).
- There should be sufficient space in the WC compartment to allow wheelchair users to access the WC.
- The position of the washbasin should not impede access to the WC.

Typical, suitable WC compartments are illustrated in Fig. 17.14.

17.13 Forthcoming amendments to Part M and Approved Document M

At the time of going to press, Part M and Approved Document M were nearing the end of the consultation process and it is anticipated that a new edition of Approved Document M will be published in late 2003, possibly coming into force in the summer of 2004. Amendments have become necessary, as the current guidance is outdated in respect of references to British Standards and in the context of the Disability Discrimination Act 1995 Part III, the final phase of which comes into effect in October 2004.

The proposed amendments to the requirements of Part M include:

- Omission of specific references to, and the definition of, disabled people
- Deletion of the 'Limits on application' which exempt material alterations from the requirements of Part M.

The main changes proposed to Approved Document M are concerned only with buildings other than dwellings and include:

- A major reordering and overhaul of the guidance to reflect the recommendations that BS 8300: 2001 *Design of buildings and their approaches to meet the needs of disabled people*
- Revision of Section 1 of AD M (access to buildings other than dwellings) including the need to provide stair access whenever an external ramp is provided and improved guidance on the provision of car parking
- Revision of Section 2 of AD M (access into buildings other than dwellings) including improved guidance on the design of entrances and lobbies and the provision of powered entrance doors
- Revision of Section 3 of AD M (horizontal and vertical circulation in buildings other than dwellings) including expanded guidance on the provision of lifts including platform lifts and stair lifts
- Revision of Section 4 of AD M (facilities in buildings other than dwellings) including expanded guidance on the provision of audience facilities and sleeping accommodation. New guidance on the provision of switches, outlets and controls
- Revision of Section 5 of AD M (sanitary accommodation in buildings other than dwellings) including guidance on washing facilities previously included elsewhere.

18 Glazing

18.1 Introduction

This chapter deals with safety issues associated with glazing in terms of protecting people against the risks of impact and making sure that glazed elements may be opened, closed and cleaned safely.

The 1998 edition of Approved Document N introduced new provisions governing the safe use and cleaning of glazed elements mainly in order to ensure compliance with the Workplace (Health, Safety and Welfare) Regulations 1992. Therefore, compliance with the revised regulations in Part N prevents action being taken against the occupier under the Workplace Regulations when the building is eventually in use. This involves considering aspects of design which will affect the way a building is used and applies only to workplaces.

Although dwellings are excluded from the changes, in mixed use developments the requirements for the non-domestic part of the use would apply to any shared parts of the building (such as common access staircases and corridors). With flats the situation may be less clear for although certain sections of Part N do not apply to dwellings, it is still necessary for people such as cleaners, wardens and caretakers to work in the common parts. Therefore, the requirements of the Workplace Regulations may still apply even though the Building Regulations do not.

18.2 Application

Part N of Schedule 1 to the 2000 Regulations applies to all new glazing used in the erection, extension or material alteration of a building.

It does not apply to exempt buildings (except small conservatories or porches in Class VII of Schedule 2, see section 2.5), or to replacement glazing although this may be the subject of consumer protection legislation in the future.

Requirement N1 applies to all building types whereas requirements N2, N3 and N4 apply to all building types *except* dwellings (i.e. dwelling-houses and flats).

18.3 Protection against impact

People are likely to come into contact with glazing in certain critical locations, when they are moving in or about a building. Accordingly, the glazing must:

- if broken on impact, break so that the dangers of injury are minimised (i.e. break safely); or
- resist impact without breaking; or
- be protected or shielded from impact.

Critical locations under the terms of paragraph N1 are shown in Fig. 18.1.

Two main areas may be identified where an accident may result in cutting or piercing injuries.

- Doors and door side panels between finished floor level and 1500 mm above.
- Internal and external walls and partitions between finished floor level and 800 mm above.

In doors and door side panels the main risk is in the area of door handles and push plates especially since doors are prone to stick. Also it is possible that an initial impact above waist level may result in a fall through the glass. In low level glazing away from doors the main risks are to children.

18.3.1 Possible solutions to N1

Approved Document N lists a number of solutions which can be adopted in order to minimise the risks of injury in these critical areas as follows:

(1) If breakage occurs the glazing should break safely.

The concept of safe breakage is taken from BS 6206: 1981 *Specification for impact performance requirements for flat safety glass and safety plastics for use in buildings:* clause 5.3.

A test is carried out using a leather bag filled with lead shot. This is swung pendulum-fashion to impact a sheet of safety glazing material and the results are noted.

The test material is required to remain unbroken or it may break safely as defined in one of the following ways:

(a) Cracks and fissures are allowed to develop provided it is not possible to pass a sphere of 76 mm diameter through any openings and any detached particles are of limited size; or
(b) Disintegration is allowed to occur provided the particles are of small size; or
(c) Breakage is allowed to occur provided the pieces are not sharp or pointed.

In essence a glazing material in a critical location will be satisfactory if it can be classified under the requirements of BS 6206 as Class C (i.e. it remains unbroken or breaks safely when impacted from a height of 305 mm). Additionally, if it is installed in a door or door side panel and has a pane width which exceeds 900 mm, then it should meet the requirements of Class B of BS 6206 (i.e. it should remain unbroken or break safely when impacted from heights of 305 mm and 457 mm).

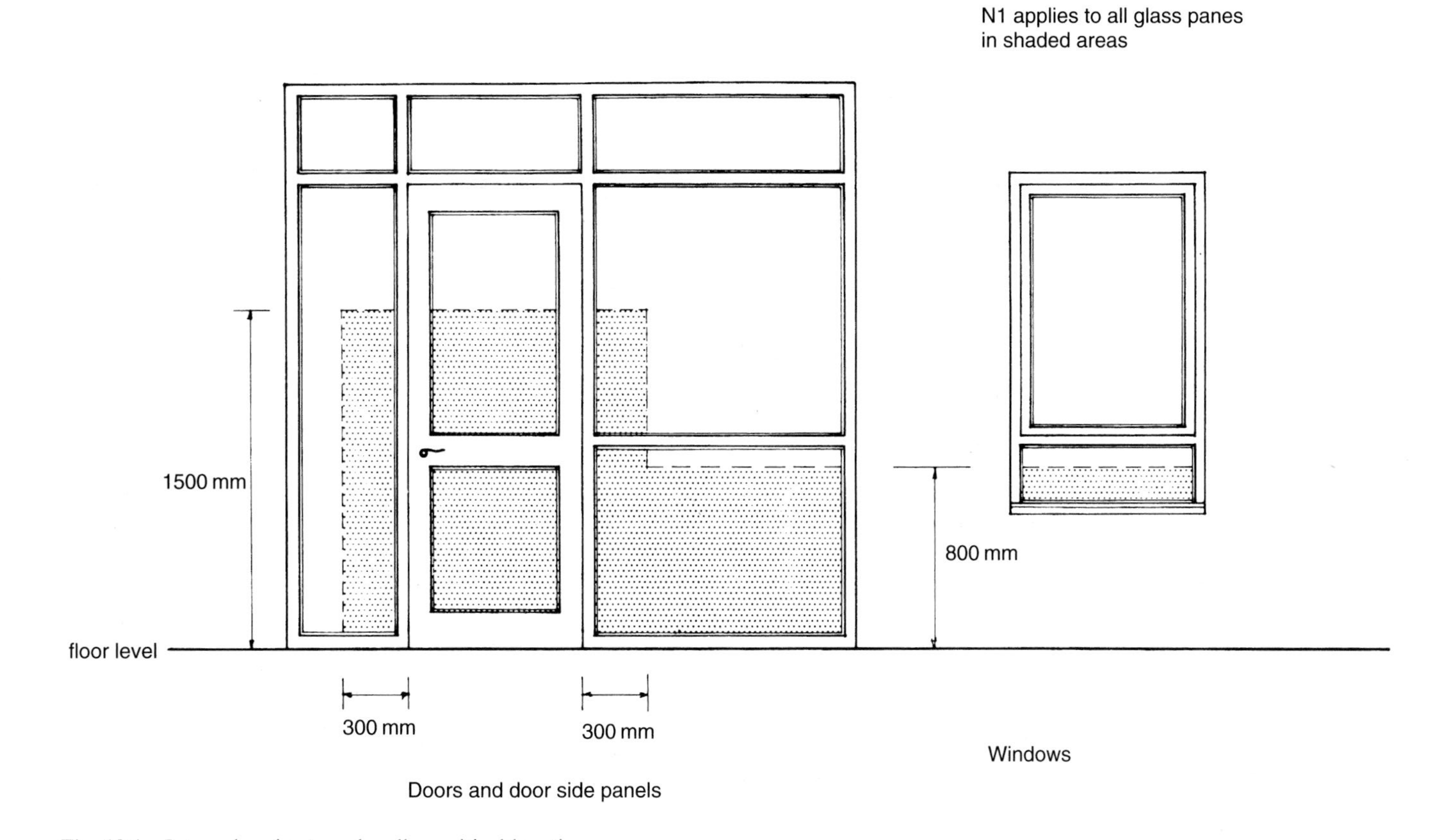

Fig. 18.1 Internal and external walls – critical locations.

(2) The glazing should be robust.

Robustness refers to the strength of the glazing material. Some materials such as glass blocks or polycarbonates are inherently strong. Annealed glass gains its strength through increased thickness and AD N1 describes the use of this material for large glazed areas forming fronts in shops, showrooms, factories, offices or public buildings. The dimensions of these glazed areas and their related glass thicknesses are shown in Table 18.1.

Table 18.1 Annealed glass – thickness/dimension limits.

Height (mm)		Length (mm)		Thickness (mm)
From	To	From	To	
0	1100	0	1100	8
1100	2250	1100	2250	10
2250	3000	2250	4500	12
3000	any	4500	any	15

Note
Annealed glass sizes and thicknesses for use in large areas to shopfronts, showrooms, offices, factories and public buildings.

(3) The glazing should be in small panes.

This relates to the use of a single pane or one of a number of panes within glazing bars, in either case having a smaller dimension not exceeding 250 mm and an area not greater than $0.5\,m^2$. Annealed glass in small panes should not be less than 6 mm in thickness although it is possible to install traditional copper or leaded lights using 4mm glass provided that fire resistance is not a factor. (See Fig. 18.2.)

(4) The glazing should be permanently protected.

Permanent protection means that the glazing should be installed behind a permanent screen which:

- prevents a sphere of 75 mm diameter touching the glazing,
- is itself robust; and
- is difficult to climb in cases where the glazing forms part of protection from falling.

Where permanent screen protection is provided then the glazing itself does not need to comply with requirement N1. (See Fig. 18.2.)

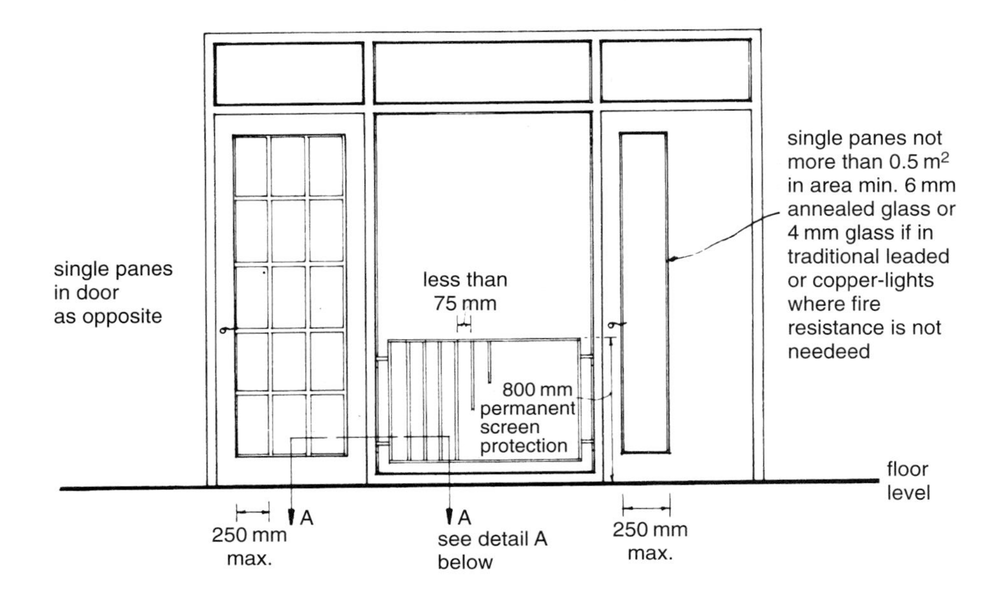

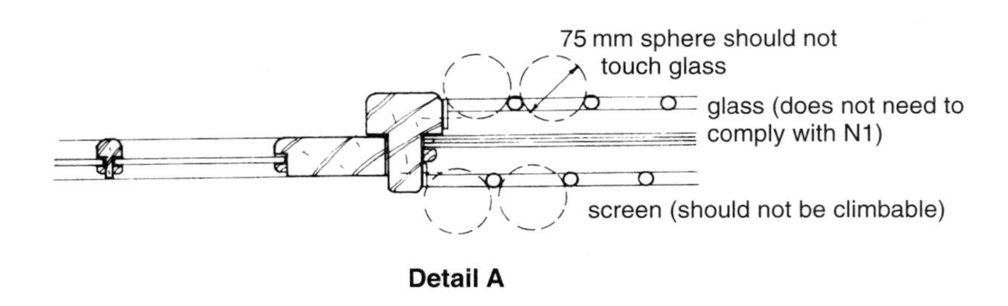

Fig. 18.2 Small panes and permanent screen protection.

18.4 Manifestation of glazing

If there is a risk that people may come into contact with large, uninterrupted areas of transparent glazing while moving in or about a building, then paragraph N2 of Schedule 1 requires that such areas must incorporate features which make the glazing apparent. As mentioned above, this requirement does not apply to dwellings.

The risk of collision and consequent injury is most serious where parts of a building or its surroundings are separated by transparent glazing and the impression is given that direct access is possible through the area without interruption. In these critical locations (i.e. internal or external walls of shops, showrooms, offices, factories, public or other non-domestic buildings) it is necessary to adopt some means of making the glazing more apparent.

This is termed 'manifestation' of the glazing in AD N2 and it may take the form of patterns, company logos, broken or solid lines, etc. marked on the glazing at appropriate heights and intervals. This is illustrated in Fig. 18.3.

It is, of course, possible to indicate the presence of glazing by other means. Such features as mullions, transoms, door frames or large push or pull handles can be effectively used and AD N2 acknowledges that where these features are incorporated in the design then permanent manifestation will not be necessary.

Some examples of these features are shown in Fig. 18.3.

18.5 Safe use of windows, skylights and ventilators

Windows, skylights and ventilators must be constructed or equipped so that they can be opened, closed or adjusted safely, if they are so positioned as to be operable by people in or about the building. This requirement does not apply to dwellings.

Compliance with Requirement N3 prevents action being taken against the occupier of a building under Regulation 15(1) of the Workplace (Health, Safety and Welfare) Regulations 1992 when the building is eventually in use. (Regulation 15(1) relates to requirements for opening, closing or adjusting windows, skylights and ventilators).

In order to meet the performance standard for safe operation of windows, skylights and ventilators controls should typically be located as follows:

- Not more than 1.9 m above the floor or other stable surface where there is unobstructed access (ignoring small recesses, such as window reveals).
- At a lower level where there is an obstruction (e.g. 1.7 m from floor level if there is an obstruction 600 mm deep (including any recess) and not more than 900 mm high).

If controls cannot be positioned at a safe distance from a permanent stable surface, then it may be necessary to install either manual or electrical remote controls.

Above ground level there may be a danger of the operator or other person falling through a window. Where this is the case suitable opening limiters should be fitted or the window should be guarded as described in Approved Document K (see Chapter 15).

18.6 Safe access for cleaning glazed surfaces

Provision must be made for glazed surfaces to be safely accessible for cleaning. This includes:

- windows and skylights; and
- translucent walls, ceilings or roofs.

Regulation N4 does not apply to dwellings, or to any of the above transparent or translucent elements if their surfaces are not intended to be cleaned. In this case, compliance with Requirement N4 prevents action being taken against the occupier of a building under Regulation 16 of the Workplace (Health, Safety and Welfare)

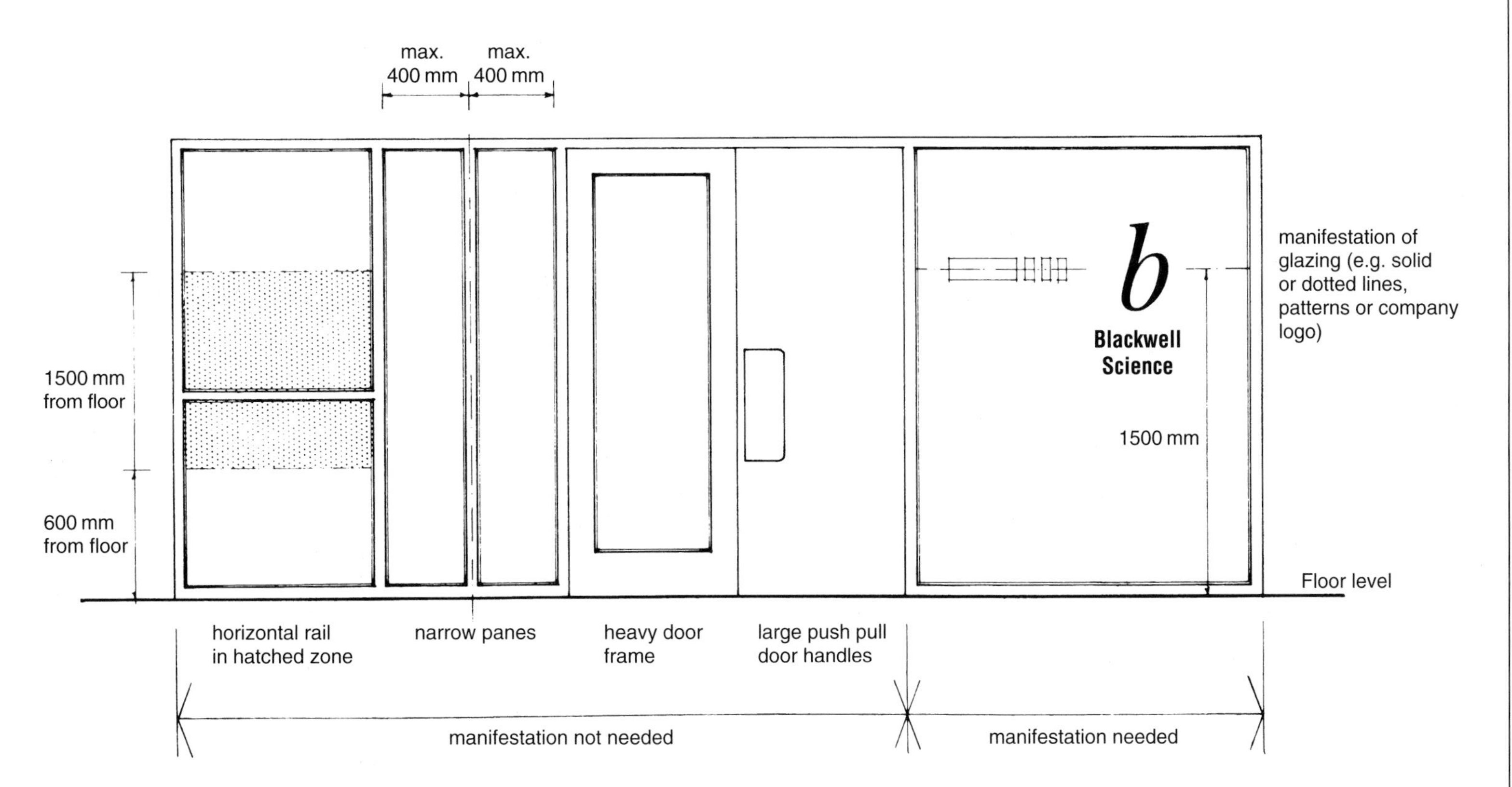

Fig. 18.3 Manifestation of large glazed areas (internal or external).

Regulations 1992 when the building is eventually in use. (Regulation 16 relates to requirements for cleaning glazed surfaces in buildings.)

In the context of Regulation N4, it will be necessary to make provision for safe means of access for cleaning *both* sides of any glazed surfaces which are positioned so that there is a danger of falling more than 2 m. Furthermore, glazed surfaces which cannot be cleaned safely by a person standing on the ground, a floor or other permanent stable surface will need to be catered for in other ways.

The following arrangements (illustrated in Figs 18.4 and 18.5) are typical examples of how it may be possible to satisfy Regulation N4.

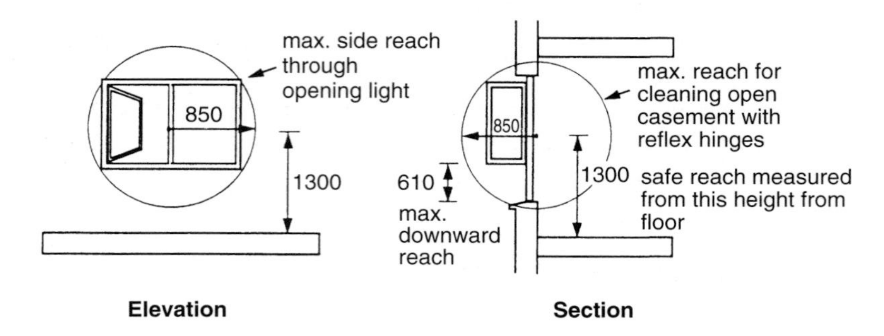

Fig. 18.4 Typical safe reaches for cleaning windows.

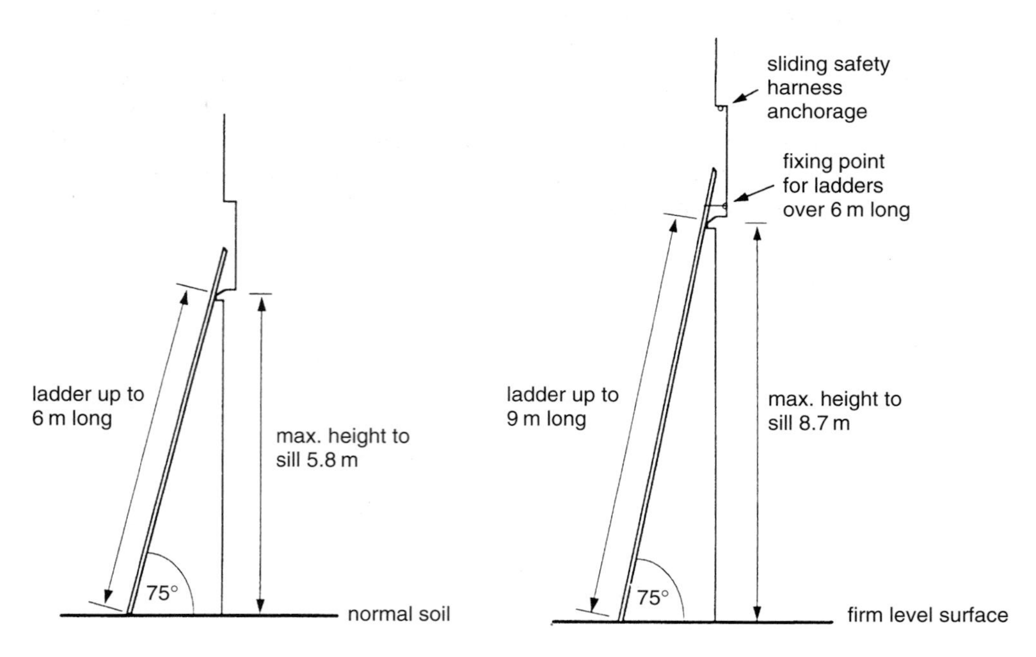

Fig. 18.5 Ladder access for cleaning windows. (**Note:** Since designers will need to relate the length of access ladders to window sill heights, the maximum sill heights from ground level are shown for both 6 m and 9 m ladders.)

- Install windows of a suitable design and size so that they can be cleaned from inside the building. Reversible type windows should be capable of being fixed in the reversed position for cleaning purposes (see Fig. 18.4). For additional information on windows see BS 8213: Part 1 *Windows, doors and rooflights*, and Approved Document K (see Chapter 15) for guidance on minimum sill heights.
- Use portable ladders up to 9 m long if there is an adequate area of firm, level ground situated in a safe place for siting the ladders. Ladders up to 6 m long may be sited on normal soil. (For ladders over 6 m long provide permanent tying or fixing points.) (See Fig. 18.5).
- Provide catwalks at least 400 mm wide with either 1100 mm high guarding or provision for anchorages for sliding safety harnesses.
- Provide access equipment, e.g. suspended cradles or travelling ladders with safety harness attachments.
- Provide adequate anchorage points for safety harnesses or abseiling hooks.
- In exceptional circumstances where other means of access cannot be used, provide suitably located space for scaffold towers.

18.7 Further references to glazing

Attention is drawn to the following Approved Documents where further information regarding glazing may be found:

- Approved Document B: *Fire safety* – guidance on fire resisting glazing and the reaction of glass to fire (see Chapter 7).
- Approved Document K: *Protection from falling, collision and impact* – guidance on glazing which forms part of protection from falling from one level to another, and which needs to provide containment as well as limiting the risk of injuries through contact. Recommendations are given concerning the heights of guarding and the means for achieving containment.

III Appendix

Appendix
Local Acts of Parliament

The following list of Local Acts of Parliament is reproduced by kind permission of Carillion Specialist Services Ltd, which is a corporate approved inspector under the Building Regulations.

The sections marked with an asterisk (*) in column 3 of the list are applicable where there are matters to be satisfied by a developer either as part of a submission under Building Regulations or which otherwise need to be addressed in parallel with that submission.

County Acts usually, but not inevitably, apply across a whole county. References to district authorities in the following list however merely illustrate those which may have been consulted, and do not necessarily constitute a complete list of districts within that county. However, where it is known that a particular section applies only in a particular district this is indicated.

The ODPM is developing proposals to repeal redundant provisions of Local Acts but this will take some time to achieve. This list was last updated on 30 September 2003.

(1) *Local Act*	(2) *Relevant sections*	(3) *Apply?*
County of Avon Act 1982	s.7 Parking places, safety requirements	*
• Bath City Council	s.35 Hot springs, excavations in certain areas of Bath	
• Bristol City Council		
Berkshire Act 1986		
• Newbury District Council	s.28 Safety of stands	
	s.32 Access for fire brigade	*
	s.36 Parking places, safety requirements	*
	s.37 Fire precautions in large storage buildings	*
	s.38 Fire precautions in high buildings	*
• Reading Borough Council		

(1) *Local Act*	(2) *Relevant sections*	(3) *Apply?*
Bournemouth Borough Council Act 1985	s.15 Access for fire brigade	*
	s.16 Parking places, safety requirements	*
	s.17 Fire precautions in certain large buildings	*
	s.18 Fire precautions in high buildings	*
	s.19 Amending s.72 Building Act 1984	*
Cheshire County Council Act 1980	s.48 Parking places, safety requirements	*
• Chester City Council	s.50 Access for fire brigade	*
• Warrington BC	s.49 Fireman switches	
	s.54 Means of escape, safety requirements	
County of Cleveland Act 1987		
• Stockton on Tees Borough Council	s.5 Access for fire fighting	*
	s.6 Parking places, safety requirements	*
	s.15 Safety of stands	
• Middlesborough Borough Council		
Clwyd Act 1985		
• Borough of Rhuddlan	s.19 Parking places, safety requirements	*
• Colwyn District Council	s.20 Access for fire brigade	*
Cornwall Act 1984		
• Caradon DC		
• North Cornwall		
Croydon Corporation Act 1960	s.93/94 Buildings of excess cubic capacity	*
	s.95 Buildings used for trade and for dwellings	
	s.79 Separate drainage systems	
Cumbria Act 1982		
• Barrow Borough Council	s.23 Parking places, safety requirements	*
• Carlisle City Council	s.25 Access for fire brigade	*
	s.28 Means of escape from certain buildings	
Derbyshire Act 1981		
• Borough of High Peak	s.16 Safety of stands	*
	s.23 Access for fire brigade	*
	s.24 MoE from certain buildings	*
	s.28 Parking places; safety requirements	*
	s.25 Fireman switches	
Dyfed Act 1987		
• South Pembrokeshire District Council	s.46 Safety of stands	
	s.47 Parking places; safety requirements	*
	s.51 Access for fire brigade	*
• Carmarthen District Council		
East Ham Corporation Act 1957		
• Newham London Borough Council	s.38 Separate drainage systems	*
	s.54 Separate access to tenements	*
	s.61 Access for fire brigade	*

(1) *Local Act*	(2) *Relevant sections*	(3) *Apply?*
East Sussex Act 1981	s.34 Fireman switches	
• Hastings Borough Council	s.35 Access for fire brigade	*
Essex Acts 1952 & 1958 (GLC areas formerly in Essex)		
Essex Act 1987		
• Uttlesford District Council	s.13 Access for fire brigade	*
Exeter Act 1987		
Greater Manchester Act 1981		
• Trafford Metropolitan Borough Council	s.58 Safety of stands	
	s.61 Parking places, safety requirements	*
	s.62 Fireman switches	
	s.63 Access for fire brigade	*
	s.64 Fire precautions in high buildings	*
	s.65 Fire precautions in large storage buildings	*
	s.66 Fire and safety precautions in public and other buildings	*
• Manchester City Council		
Hampshire Act 1983		
• Southampton City Council	s.11 Parking places, safety requirements	*
	s.12 Access for fire brigade	*
	s.13 Fire precautions in certain large buildings	*
Hastings Act 1988		
Hereford City Council Act 1985	s.17 Parking places; safety requirements	*
	s.18 Access for fire brigade	*
Humberside Act 1982		
• Gt.Grimsby BC	s.12 Parking places, safety requirements	*
	s.13 Fireman switches	
	s.14 Access for the fire brigade	*
	s.15 Means of escape in certain buildings	
	s.17 Temporary structures, byelaws	
Isle of Wight Act 1980	s.32 Access for fire brigade	*
• (Part VI)	s.31 Fireman switches	
	s.30 Parking places; safety requirements	*
Kent 1958 (GLC areas formerly in Kent)		
County of Kent Act 1981		
• Canterbury City Council	s.51 Parking places, safety requirements	*
• Rochester City Council	s.52 Fireman switches	
	s.53 Access for fire brigade	*
	s.78 Annulment of plans approvals	*
Lancashire Act 1984	s.31 Access for fire brigade	*
	s.20 Separate drainage systems	

(1) *Local Act*	(2) *Relevant sections*	(3) *Apply?*
Leicestershire Act 1985		
• Leicester City Council	s.21 Safety of stands	
	s.30 Separate drainage systems	*
	s.49 Parking places, safety requirements	*
	s.50 Access for fire brigade	*
	s.52 Fire precautions in high buildings	*
	s.53 Fire precautions in large storage buildings	
• North West Leicestershire District Council	s.54 Means of escape, safety requirements	
London Building Acts 1930–1939	s.20 Buildings of excess height or cubic capacity **N.B.:** By definition (BA 1984 s.88 & Sch.3), these Acts do not constitute Local Acts but otherwise have a similar effect (see also Chapter 5).	*
• Corporation of London		
• City of Westminster London Borough Council		
• Royal Borough of Kensington & Chelsea		
County of Merseyside Act 1980		
• Liverpool City Council	s.20 Safety of stands	
	s.48/49 Means of escape from fire	
	s.50 Parking places, safety requirements	*
	s.51 Fire and safety precautions in public and other buildings	*
	s.52 Fire precautions in high buildings	*
	s.53 Fire precautions in large storage buildings	*
	s.54 Fireman switches	
	s.55 Access for fire brigade	*
• Borough of Wirral		
Middlesex Act 1956 (GLC areas formerly in Middlesex)	s.33 Access for fire brigade	*
Mid Glamorgan County Council Act 1987		
• Merthyr Tydfil	s.9 Access for fire brigade	*
• Taff-Ely BC		
Nottinghamshire Act 1985		
• City of Nottingham		
Plymouth Act 1987		
Poole Act 1986	s.10 Parking places, safety requirements	*
	s.11 Access for fire brigade	*
	s.14 Fire precautions in certain large buildings	*
	s.15 Fire precautions in high buildings	
County of South Glamorgan Act 1976		
• Cardiff City Metropolitan District Council	s.27 Safety of stands	
	s.48/50 Underground parking places	*
	s.51 Means of escape for certain buildings	
	s.52 Fireman switches	
	s.53 Precautions against fire in high buildings	*
	s.54 Byelaws for temporary structures.	

(1) *Local Act*	(2) *Relevant sections*	(3) *Apply?*
South Yorkshire Act 1980		
• City of Sheffield Metropolitan District Council	s.53 Parking places, safety precautions	*
• Barnsley Metropolitan Borough Council	s.54 Fireman switches	
• Rotherham Metropolitan Borough Council	s.55 Access for fire brigade	*
• Doncaster Metropolitan Borough Council	s.39 Separate drainage systems	
Staffordshire Act 1983		
• Staffordshire Moorlands District Council	s.25 Parking places, safety precautions	*
	s.26 Access for fire brigade	*
	s.18 Separate drainage systems	
Surrey Act 1985		
• Guildford Borough Council	s.18 Parking places, safety requirements	*
• Spelthorne Borough Council	s.19 Fire precautions in large storage buildings	*
	s.20 Access for fire brigade	*
Tyne & Wear 1980	s.24 Access for fire brigade	*
West Glamorgan Act 1987	s.43 Parking places, safety precautions	*
• City of Swansea		
• Neath Borough Council		
West Midlands Act 1980		
• Birmingham City Council	s.39 Safety of stands	
	s.44 Parking places, safety requirements	*
	s.45 Fireman switches	
	s.46 Access for fire brigade	*
	s.49 Means of escape from certain buildings	
West Yorkshire Act 1980		
• Kirklees Metropolitan District Council	s.9 Culverting water courses	*
	s.50 Separate drainage systems	*
• Bradford City Council	s.51 Fireman switches	
Worcester City Council Act 1985		

Index